MULTIMEDIA, COMMUNICATION AND COMPUTING APPLICATION

PROCEEDINGS OF THE INTERNATIONAL CONFERENCE ON MULTIMEDIA, COMMUNICATION AND COMPUTING APPLICATION, XIAMEN, CHINA, 15–16 OCTOBER 2014

# Multimedia, Communication and Computing Application

*Editor*

Ally Leung
*Advanced Science and Industry Research Center, Hong Kong*

CRC Press
Taylor & Francis Group
Boca Raton London New York Leiden

CRC Press is an imprint of the
Taylor & Francis Group, an **informa** business

A BALKEMA BOOK

*CRC Press/Balkema is an imprint of the Taylor & Francis Group, an informa business*

© 2015 Taylor & Francis Group, London, UK

Typeset by diacriTech, Chennai, India

Published by: CRC Press/Balkema
     P.O. Box 11320, 2301 EH Leiden, The Netherlands
     e-mail: Pub.NL@taylorandfrancis.com
     www.crcpress.com – www.taylorandfrancis.com

ISBN: 978-1-138-02775-6 (Hardback)
ISBN: 978-1-315-69051-3 (eBook PDF)

# Table of contents

## 2. Multimedia processing and artificial intelligence

## 3. Information, electronic and mechanical engineering

## 4. Management engineering and computer application

*Multimedia, Communication and Computing Application – Leung (Ed.)*
*© 2015 Taylor & Francis Group, London, ISBN 978-1-138-02775-6*

# Preface

2014 International Conference on Multimedia, Communication and Computing Application (MCCA2014) was successfully held in Xiamen, China on Oct 16–17, 2014. The main purpose of this conference is to provide a common forum for experts and scholars of excellence in their domains from all over the world to present their latest and inspiring works in the area of multimedia, communication and computing application.

Multimedia, communication and computing application is an international topic. In recent years, the multimedia technology has experienced a rapid development and it is widely used in both our daily life and industrial manufacturing, such as games, education, entertainment, stocks and bonds, financial transactions, architectural design, communication and so on. Any company or institution with ambition cannot live without the latest high-tech products.

MCCA2014 has received a large number of papers and ninety-three papers were accepted after reviewing. These articles were divided into several sessions, such as multimedia processing, communication systems, artificial intelligence and application and artificial intelligence and application.

During the organization course, we have received help from many people and institutions. Firstly, we would like to show our thankfulness to the whole committee for their support and enthusiasm. Secondly, we would also like to thank the authors for their carefully writing. Lastly, the organizers of the conference and other people who have helped us would also be appreciated for their hard work.

I hope that all our participants can exchange useful information and make amazing developments in multimedia, communication and computing application after this conference.

MCCA2014 Committees

*Multimedia, Communication and Computing Application – Leung (Ed.)*
*© 2015 Taylor & Francis Group, London, ISBN 978-1-138-02775-6*

# Organizing committees

## HONOR CHAIR

E.P. Purushothaman, University of Science and Technology, India

## GENERAL CHAIR

S.K. Chen, Altair Engineering Inc., California, USA
Y.H. Chang, Chihlee Institute of Technology, Taiwan

## PROGRAM CHAIR

W.K. Jain, Indian Institute of Technology, India

## PUBLICATION CHAIR

J. Yeh, Tallinn University of Technology, Estonia

## INTERNATIONAL SCIENTIFIC COMMITTEE

S. Ravi, Pondicherry University, India
Y.L. Shang, Tongji University, China
M.S. Chen, Da-Yeh University, Taiwan
I. Saha, Jadavpur University, India
X. Lee. Hong Kong Polytechnic University, Hong Kong
Antonio J. Tallón-Ballesteros, University of Seville, Spain
J. Xu, Northeast Dianli University, China
Q.B. Zeng, Shenzhen University, China
C.X. Pan, Harbin Engineering University, China
L.P. Chen, Huazhong University of Science and Technology, China
K.S. Rajesh, Defence University College, India
M.M. Kim, Chonbuk National University, Korean
X. Ma, University of Science and Technology of China, China
L. Tian, Huaqiao University, China
M.V. Raghavendra, Adama Science & Technology University, Ethiopia
J. Ye, Hunan University of Technology, China
Q. Yang, University of Science and Technology Beijing, China
Z.Y. Jiang, University of Wollongong, Australian

# 1. Communication systems and engineering

*Multimedia, Communication and Computing Application – Leung (Ed.)*
© *2015 Taylor & Francis Group, London, ISBN 978-1-138-02775-6*

# An evolved elastic resource virtualization algorithm for OFDMA-based wireless networks

J. Wei & K. Yang
*School of Computer Science and Electronic Engineering, University of Essex, UK*

G. Zhang
*Internet of Things Research Centre, China University of Mining and Technology, Xuzhou, China*

Z. Hu
*Wireless Technology Innovation (WTI), Beijing University of Posts and Telecommunications, China*

ABSTRACT: This article investigates an evolved elastic resource virtualization algorithm (E-ERVA) for Orthogonal Frequency-Division Multiple Access (OFDMA) wireless communication systems. The objective of this algorithm is to maximize the total throughput with the constraints on individual virtual network's (VN) data rate. This is achieved by assigning each VN a set of physical resource blocks (PRBs). We obtain the performance of the proposed algorithm in a frequency-selective fading environment. The simulation compares the throughput of elastic with static resource allocation algorithms. Furthermore, E-ERVA resource utilization and delivery ratio are evaluated. The simulation results show that E-ERVA performs better in throughput, and has higher spectral efficiency when compared to the static allocation algorithm.

## 1 INTRODUCTION

An increasing number of services bring more challenges to the current wireless network, especially a shortage of spectrum resources. Hence, it makes more sense for improving such resource utilization. As a long-term solution, wireless network virtualization (WNV) is introduced to overcome this problem.

By means of WNV, spare resources from infrastructure providers (InPs) can be shared with a virtual network operator (VNO) who does not have any infrastructures. VNOs then create their own VNs to provide services to end users with leased resources. Both InPs and VNOs benefit from WNV (G. Schaffrath et al. 2009). InPs can concentrate on the maintenance of the physical equipment and save manpower. From VNOs' view, they save huge investment on the hardware and fundamental construction with resource sharing.

Q. Dai et al (2013) present a general model of the fiber–wireless (FiWi) access network virtualization that combines fiber network and wireless network. However, FiWi does not consider the differences between fiber network and wireless network.

G. Di Stasi et al. (2013) and P. Lv et al. (2012) studied the virtual network embedding in mesh network. R. Kokku et al. (2012) focused on effective wireless resources virtualization as well as admission control of WiMAX. It considers both uplink and downlink resources virtualization allocation. An efficient resource allocation algorithm for WNV using time and space division was provided by X. Zhang et al. (2012). Although higher resource utilization is achieved, Quality of service (QoS) has not been considered. A. Checco and D. Leith (2013) established a max–min fair flow rate allocation in time division multiple access (TDMA)–based mesh network.

Y. Zaki et al. (2010) provided a resource management algorithm on the long-term evolution (LTE) virtualization. This algorithm enhances the overall resource utilization and network as well as end-user performance. However, it does not provide full resource flexibility across VNOs. B. Liu and H. Tian (2013) proposed a dynamic resource allocation algorithm with the consideration of fairness algorithm for LTE system.

X. Lu et al. (2012) proposed an elastic resource virtualization algorithm (ERVA). The algorithm minimizes the throughput of VNOs and maximizes the throughput of InPs. The disadvantage of ERVA is that requests from VNOs cannot be satisfied if they exceed the service level agreement (SLA). Moreover, introducing different strategies to InPs and VNOs may complicate the algorithm. In contrast, E-ERVA allows either VNOs or

InPs to use spare resources, whereas ERVA allows only InPs to use spare resources. Thus, E-ERVA achieves higher throughput and resource utilization. In addition, E-ERVA reduces the complexity of the algorithm.

The remainder of this article is organized as follows. Section 2 illustrates and formulates the problem of resource allocation in detail. Then an elastic resource allocation algorithm is presented. Section 3 evaluates the proposed algorithm and the article concludes in Section 4.

## 2 SYSTEM MODEL AND PROPOSED ALGORITHM

This research is based on a perfect channel information assumption on both transmitters and receivers. Moreover, isolation across VNs is applied.

E-ERVA specifies VNs that are operated by InP, which is called local slice, and foreign slices represent the VNs that are operated by VNOs. In addition, E-ERVA is based on OFDMA. Hence, the resource allocation problem can be converted to allocate PRBs to both local and foreign slices to achieve maximum throughputs.

### 2.1 Problem description

E-ERVA considers a multiuser OFDMA system with $K$ slices. The $K$-th slice produces $f$ traffic flows, and each flow is assigned a set of subcarriers. Thus each flow can specify its own QoS requirements. Furthermore, the instantaneous subchannel information is known by the transmitter. Therefore, the transmitter is able to apply the allocation algorithm to assign different subcarriers to the flow according to the channel information. Table 1 indicates the major notations used in E-ERVA.

Table 1. Notations.

| Symbols | Description |
| --- | --- |
| B | Total bandwidth of the system |
| $N_0$ | Noise power spectral density |
| C | Number of system subcarriers |
| F | Number of flows |
| M | Number of foreign slices |
| $\Phi_i$ | Service level agreement of slice i |
| $r_{f,c}$ | Allocate c-th subcarrier to f-th flow |
| $r_{req}$ | Request from f-th flow in slice i |
| $p_{f,c}$ | Power allocated to c-th subcarrier in f-th flow |
| $h_{f,c}$ | Channel gain of c-th subcarrier in f-th flow |
| $BER_{f,c}$ | Bit error rate (BER) of c-th subcarrier in f-th flow |
| $I_{f,c}$ | Assignment index indicating f-th flow occupies c-th subcarrier |

One subcarrier cannot be shared by more than one flow, for each $c$ if $r_{f',c} \neq 0$, $r_{f,c} = 0$ all $f' \neq f$. In the frequency-selective fading channel, if M-ary quadrature amplitude modulation (MQAM) is used, the transmission rate of $f$-th flow that allocates the $c$-th subcarrier can be calculated.

E-ERVA aims to obtain the best assignment of $r_{f,c}$ with given transmission power and QoS constraints. To make the problem tractable, no power is needed if no data is transmitted.

### 2.2 System model

Foreign slices can be denoted as Slice 1 to M (i.e., $S_1$–$S_M$), where the local slice is represented by $S_0$ in particular. To improve the resource utilization, throughputs of both local and foreign slices should be maximized. Hence, the problem can be converted to an optimization problem that allocates subcarriers to the local and foreign slices with the QoS constraints. Mathematically, the problem can be formulated as in Equation 1.

$$\max \sum_{f=1}^{F} \sum_{c=1}^{C} I_{f,c} r_{f,c} \tag{1}$$

s.t.

$$C1: \sum_{f=1}^{F} I_{f,c} = 1$$

$$C2: \sum_{c=1}^{C} I_{f,c} \geq 1$$

$$C3: I_{f,c} = \{0,1\}$$

$$C4: \sum_{c=1}^{C} r_{f,c} \geq req_{f,S_i}$$

$$C5: \left( \sum_{i=1}^{M} \sum_{f=1}^{F} r_{req_{f,S_i}} \leq \sum_{i=1}^{M} \phi_i \right) or \left( \sum_{f=1}^{F} r_{req_{f,S_0}} \leq B - \sum_{i=1}^{M} \phi_i \right)$$

where C denotes constraints that are presented to slices. C1 describes that each subcarrier is occupied by only one flow at any time slot, while C2 illustrates that $f$-th flow can be assigned more than one subcarrier. C3 specifies the $c$-th subcarrier is occupied by $f$-th flow (value 1) or not (value 0). C4 ensures the allocated subcarriers satisfy the requests from $f$-th flow in $S_i$. Finally, C5 provides a pay as you use model to foreign slices.

The optimization problem in Equation 1 is a typical binary integer programming (BIP) problem. BIP is a special case of integer programming in which variables are required to be 0 or 1. A linear programming–based branch-and-bound method is used to solve the BIP. The E-ERVA is proposed in the following.

---

**Algorithm 1** E-ERVA

---

**Input**: M, $r_{req}$, $\Phi_i$
$\qquad$ **for** f∈ $S_i$ **do**

$$r_{f,c} = \frac{B}{C} \log_2 \left( 1 - \frac{1.5 p_{f,c} h_{f,c}^2}{\ln\left(5 BER_{f,c}\right) N_0 \dfrac{B}{C}} \right)$$

$\qquad x_i = I_{f,c}$

$\qquad q_i = r_{f,c}$

$\qquad A = r_{req_{f,s_i}}$

$\qquad Aeq = \sum_{f=1}^{F} I_{f,c}$

**end for**
**if** satisfy C5
Solve formulation (1) by branch-and-bound method
**else**

$$\sum_{i=1}^{M} r_{f,S_i} + r_{f,S_0} = B$$

**end if**

---

## 3 PERFORMANCE EVALUATION

The simulation is based on MATLAB and isolation is provided across local and foreign slices. For comparison, a static resources virtualization algorithm (SRVA) is provided. Resources no more than the SLA constraint can be allocated to foreign slices in SRVA.

The throughput is measured from the InP's side rather than individual slice. E-ERVA attempts to increase the throughputs of the whole physical network while satisfying the bandwidth requests from an individual flow of different slices.

The total bandwidth is set to 2 MHz and divided into $N = 128$ subcarriers. The bandwidth of each subcarrier is $\Delta f = 15$ kHz. Chunk-based algorithm is used in the simulation. Subcarriers are grouped into 16 chunks, and each consists of 8 subcarriers. Three slices are simulated, including one local slice and two foreign slices. In addition, each foreign slice contains two flows and the local slice contains four flows. The traffic loads follow the Poisson distribution for all flows.

Figure 1 shows the comparison of E-ERVA, ERVA, and SRVA throughput with the same signal to noise

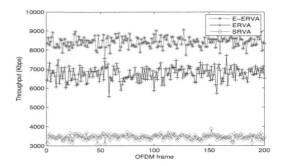

Figure 1. Throughput comparison of E-ERVA, ERVA, and SRVA.

ratio (SNR) value. Clearly, E-ERVA performs the best compared to the other two algorithms.

Figure 2 illustrates the simulation of delivery ratio under different traffic loads. The delivery ratio is defined as $Data_D/Data_R$ where $Data_D$ is the successful delivered data rate and $Data_R$ is the total amount of the request. With the traffic loads increasing, the delivery ratio decreases in both algorithms where SRVA drops dramatically.

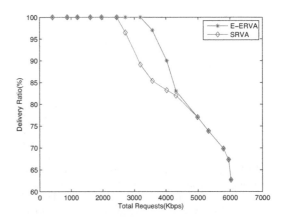

Figure 2. Delivery ratio of E-ERVA and SRVA.

Figure 3 displays the resource allocation of foreign and local slices where E represents E-ERVA and S represents SRVA. Two foreign slices (S1, S2) and one local slice (S0) are estimated. In Figure 3a, the requests from foreign slices are less than the SLA; hence, there are spare resources after satisfying foreign slices. On the other hand, the requests from local slices are over the Upp (i.e., Upp = B – SLA) while the excess request is clipped by applying SRVA. However, E-ERVA allows the local slice to use the spare resources. Thus the demand of S0 can be satisfied. Figure 3b shows the similar results. Local slice S0 shares the spare resources after satisfying local users with foreign slices by applying E-ERVA.

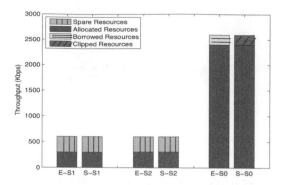

Figure 3a.    Foreign ≤ SLA, local ≥ Upp.

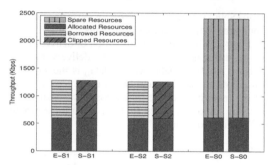

Figure 3b.    Foreign ≥ SLA, local ≤ Upp.

Figure 3.    E-ERVA's and SRVA's provisionings across slices.

Nevertheless in SRVA, requests from foreign slices cannot be satisfied due to the SLA constraint. In short, E-ERVA provides more flexibility to resources allocation. This results in higher resources utilization.

Finally, Figure 4 displays resource utilization of E-ERVA and SRVA. The resource utilization is defined as $Data_{allocated}/Data_{total}$. The resource utilization varies according to the traffic loads. When the demands increase, E-ERVA allows foreign or local slices to use the spare resources, requests can be satisfied under

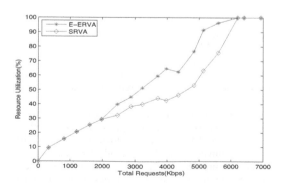

Figure 4.    Resource utilization.

E-ERVA. Nevertheless, in SRVA, foreign slices cannot occupy the spare resources. Similarly, the local slice would not allow assigning the spare resources if their requests exceed Upp. Hence, the resource utilization of E-ERVA is higher than SRVA because of the flexibility of E-ERVA.

## 4    CONCLUSION

The evaluation results show the benefits of applying E-ERVA. Compared to SRVA, E-ERVA possesses a higher throughput. This is because E-ERVA allows either foreign or local slices using the spare resources. By the means of this algorithm, the resource utilization can be improved. Moreover, the delivery ratio of E-ERVA is higher than SRVA.

E-ERVA is an evolved algorithm of ERVA. The main difference between them is that E-ERVA allows foreign and local slices to use spare resources while ERVA only allows local slice to use it. Thus, higher resources utilization can be achieved compared to ERVA. In short, E-ERVA provides more flexibility to both foreign and local slices.

## ACKNOWLEDGMENT

This work was partially funded by the EU FP7 Project EVANS (GA-2010-269323), CLIMBER (GA-2012-318939), the National Nature Science Foundation of China (No. 61471361) and the Natural Science Foundation of Jiangsu Province of China (BK2012141).

## REFERENCES

[1] Checco, A. & Leith, D. 2013. Fair virtualization of 802.11 networks.
[2] Dai, Q., Shou, G., Hu, Y. & Guo, Z. 2013. A general model for hybrid fiber-wireless (fiwi) access network virtualization. *Communications Workshops (ICC), 2013 IEEE International Conference on June 2013*: 858–862.
[3] Di Stasi, G., Avallone, S. & Canonico, R. 2013. Virtual network embedding in wireless mesh networks through reconfiguration of channels. *Wireless and Mobile Computing, Networking and Communications (WiMob), 2013 IEEE 9th International Conference on, Oct 2013*: 537–544.
[4] Kokku, R., Mahindra, R., Zhang, H. & Rangarajan, S. 2012. Nvs: A substrate for virtualizing wireless resources in cellular networks. *Networking, IEEE/ACM Transactions on, vol. 20, no. 5, Oct 2012*: 1333–1346.
[5] Liu, B. & Tian, H. 2013. A bankruptcy game-based resource allocation approach among virtual mobile operators. *Communications Letters, IEEE, vol. 17, no. 7, July 2013*: 1420–1423.

[6] Lu, X., Yang, K., Liu, Y., Zhou, D. & Liu, S. 2012. An elastic resource allocation algorithm enabling wireless network virtualization. *Wireless Communications and Mobile Computting.*

[7] Lv, P., Cai, Z., Xu, J. & Xu, M. 2012. Multicast service-oriented virtual network embedding in wireless mesh networks. *Communications Letters, IEEE, vol. 16, no. 3, March 201*: 375–377.

[8] Schaffrath, G., Werle, C., Feldmann, A., Bless, R., Greenhalgh, A., Wundsam, A., Kind, M., Maennel, O. & Mathy, L. 2009. Network virtualization architecture: proposal and initial prototype. *Proceedings of the 1st ACM workshop on Virtualized infrastructure*: 63–72.

[9] Zaki, Y., Zhao, L., Goerg, C. & Timm-Giel, A. 2010. Lte wireless virtualization and spectrum management. *Wireless and Mobile Networking Conference (WMNC), 2010 Third Joint IFIP, Oct 2010*: 1–6.

[10] Zhang, X., Li, Y., Jin, D., Su, L., Zeng, L. & Hui, P. 2012. Efficient resource allocation for wireless virtualization using time-space division. *Wireless Communications and Mobile Computing Conference (IWCMC), 2012 8th International, Aug 2012*: 59–64.

*Multimedia, Communication and Computing Application – Leung (Ed.)*
© *2015 Taylor & Francis Group, London, ISBN 978-1-138-02775-6*

# One full-diversity coding communication network transmission scheme for two users and four antennas

G.Q. Zhang, M.Q. Zou, X.H. Li & H.W. Zhang
*School of Electronic and Information Engineering, Anhui University, Hefei, China*

ABSTRACT:   To achieve high-efficiency and high-reliable data transfer between each pair of users in multi-input multi-output (MIMO) wireless communication, researchers show concerns for communication system that has become a key technology, which attracts many both at home and abroad. In this article, a coding scheme that can achieve full-rate transmission and full-diversity bit error rate (BER) performance is proposed for the network model that contains two pairs of users, and each user is fitted to four antennas. In this scheme, the orthogonal properties of Alamouti space–time code are used for alignment interference at the receiver and the joint coding technology is used to get the full-diversity coding gain performance. And the simulation results confirm the validity and correctness of the proposed coding transmission scheme.

KEYWORDS:   Multiuser MIMO system, Interference alignment, Joint coding

## 1   INTRODUCTION

As the increasing requirements of the business types and quality in the mobile communications, the architecture of wireless communication network is moving from simple single-input single-output (SISO) model to a multi-input multi-output (MIMO) model. MIMO technology can use a plurality of parallel subchannels to achieve space-division multiplexing and improve data transfer capability, and can use diversity technology to improve the reliability of data communication, which has attracted widespread attentions of researchers both at home and abroad [1]. The multiuser MIMO technology enables space-division multiplexing among multiple users, which is conducive to further increase in spectral efficiency and system capacity of the entire wireless network. But how to effectively eliminate the inter-user interference while applying full-diversity gain of multiuser MIMO system is the problem that is worth for an in-depth study currently [2][3].

Shi et al. studied an encoding and transmission method based on joint coding, which can achieve spatial diversity and suppress inter-user interference at the receiver, while being unaware of any channel state information (CSI). But it is difficult to completely eliminate the inter-user interference in testing process of this method, thus standing in the way of improving the overall BER performance to some extent [4]. Li et al. made full use of the transmitter CSI to orthogonal code, precoding transmission symbols or other

signal processing at the transmitter based on beam-forming technology, which can completely eliminate the inter-user interference at the receiver and get full spatial diversity performance. But their research only considers the simple case of two antennas for each user, and it is inapplicable under more complex multiuser MIMO environment, so the scenarios need to be extended [5]. Therefore, this article researches an effective way to realize full-diversity gain performance in a multiuser MIMO system with more antennas, which is based on joint encoding method and diversity design principles [6].

## 2   SYSTEM MODEL

For simplicity, this article focuses on the wireless communication network shown in Figure 1 with two pairs of users, and each user assembled with four antennas. The channel matrix between transmitter and receiver is represented by H, F and G, C. And each pair of antennas' channel matrix can use iid complex Gaussian variables, whose mean is 0 and variance is 1, to model. Assume that all channels are quasi-static fading channels and each transmitter knows only the CSI between the transmitters with the two receivers.

Separate the inter-user interference at receivers completely, and meet the full-rate transport requirements of 16/3[7]. Use Alamouti structure as in Equation 1 to encode modulation symbols of each transmitter.

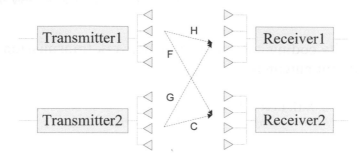

Figure 1.   Two users with four-antenna communication system model.

$$X_i = \sqrt{\frac{3P}{2}}\left(\begin{bmatrix} B_{i1}^1 & B_{i1}^3 \\ B_{i1}^5 & B_{i1}^7 \end{bmatrix} V_{i1} + \begin{bmatrix} B_{i2}^{1*} & B_{i2}^{3*} \\ B_{i2}^{5*} & B_{i2}^{7*} \end{bmatrix} V_{i2}\right) (1)$$

Where

$$B_{ij}^k = \begin{bmatrix} b_{ij}^k & b_{ij}^{k+1} \\ -\left(b_{ij}^{k+1}\right)^* & \left(b_{ij}^k\right)^* \\ 0 & 0 \end{bmatrix},$$

$$B_{ij}^{k*} = \begin{bmatrix} 0 & 0 \\ -\left(b_{ij}^{k+1}\right)^* & \left(b_{ij}^k\right)^* \\ b_{ij}^k & b_{ij}^{k+1} \end{bmatrix}$$

$b_{ij}^k$ denotes the $k$-th ($k = 1,2,\ldots,8$) symbol between transmitter $i$ and receiver $j$. $V_{ij}$ denotes the beam-forming matrices that preprocess the data between transmitter $i$ and receiver $j$, and the matrices can eliminate $B_{11}^k$ and $B_{21}^k$ at Receiver2 and eliminate $B_{12}^k$ and $B_{22}^k$ at Receiver1. It can be configured by literature methods [4], as shown in Equation 2.

$$V_{11} = \sqrt{\frac{1}{\mathrm{tr}\left(F^{-1}\left(F^{-1}\right)^*\right)}}F^{-1}, \quad V_{12} = \sqrt{\frac{1}{\mathrm{tr}\left(H^{-1}\left(H^{-1}\right)^*\right)}}H^{-1}$$

$$V_{21} = \sqrt{\frac{1}{\mathrm{tr}\left(C^{-1}\left(C^{-1}\right)^*\right)}}C^{-1}, \quad V_{22} = \sqrt{\frac{1}{\mathrm{tr}\left(G^{-1}\left(G^{-1}\right)^*\right)}}G^{-1}$$

$$(2)$$

where each beam-forming matrix should meet the requirements of the power constraint, i.e., $\mathrm{tr}\left(V_{ij}V_{ij}^*\right) = 1$.

To reach the full space diversity gain performance of each user in the multiuser MIMO systems while achieving complete separation of interference at the receivers, each of the modulation symbols at the appropriate location within the code matrix in Equation 1 needs reasonable joint coding, as is shown in Equation 3.

$$\begin{bmatrix} s_{ij}^1 \\ s_{ij}^7 \end{bmatrix} = R_1 \begin{bmatrix} b_{ij}^1 \\ b_{ij}^2 \end{bmatrix}, \quad \begin{bmatrix} s_{ij}^3 \\ s_{ij}^5 \end{bmatrix} = R_2 \begin{bmatrix} b_{ij}^3 \\ b_{ij}^4 \end{bmatrix},$$

$$\begin{bmatrix} s_{ij}^2 \\ s_{ij}^8 \end{bmatrix} = R_1 \begin{bmatrix} b_{ij}^5 \\ b_{ij}^6 \end{bmatrix}, \quad \begin{bmatrix} s_{ij}^4 \\ s_{ij}^6 \end{bmatrix} = R_2 \begin{bmatrix} b_{ij}^7 \\ b_{ij}^8 \end{bmatrix} \quad (3)$$

Where both $R_1 = \dfrac{1}{\sqrt{2}}\begin{bmatrix} 1 & e^{j(\pi/4)} \\ 1 & -e^{j(\pi/4)} \end{bmatrix}$ and $R_2 = \mu R_1$

are joint encoding matrix, $\mu = e^{j(5\pi/8)}$. It can be proved that the joint encoding can obtain the full spatial diversity gain performance at the receivers [8][9]. $s_{ij}^k$ denotes the $k$-th ($k = 1,2,\ldots,8$) symbol between transmitter $i$ and receiver $j$, which has been joint encoded. The transfer matrix of joint encoding is shown in Equation 4.

$$X_i = \sqrt{\frac{3P}{2}}\left(\begin{bmatrix} S_{i1}^1 & S_{i1}^3 \\ S_{i1}^5 & S_{i1}^7 \end{bmatrix} V_{11} + \begin{bmatrix} S_{i2}^{1*} & S_{i2}^{3*} \\ S_{i2}^{5*} & S_{i2}^{7*} \end{bmatrix} V_{i2}\right) (4)$$

where

$$S_{ij}^k = \begin{bmatrix} s_{ij}^k & s_{ij}^{k+1} \\ -\left(s_{ij}^{k+1}\right)^* & \left(s_{ij}^k\right)^* \\ 0 & 0 \end{bmatrix}, S_{ij}^{k*} = \begin{bmatrix} 0 & 0 \\ -\left(s_{ij}^{k+1}\right)^* & \left(s_{ij}^k\right)^* \\ s_{ij}^k & s_{ij}^{k+1} \end{bmatrix}$$

The transmission model of the whole multiuser MIMO wireless communication network is shown in Equation 5.

$$Y_1 = X_1H + X_2G + w_1$$

$$Y_2 = X_1F + X_2C + w_2 \quad (5)$$

where $Y_1$, $Y_2$ denote receive matrix at each receivers, respectively. $w_1$, $w_2$ denote complex additive white Gaussian noise, whose mean is 0 and variance is 1 at each receivers, respectively. Because of the structure of both receivers unanimously, the model takes Receiver1, for example, to study interference alignment and decoding methods. Substitute Equation 1 and 4 into Equation 5 and arrange, than $Y_1$ can be shown as in Equation 6.

$$Y_1 = \sqrt{\frac{3P}{2}}\left(\begin{bmatrix} S_{11}^1 & S_{11}^3 \\ S_{11}^5 & S_{11}^7 \end{bmatrix}\tilde{H} + \begin{bmatrix} S_{21}^1 & S_{21}^3 \\ S_{21}^5 & S_{21}^7 \end{bmatrix}\tilde{G}\right)$$
$$+ \begin{bmatrix} \tilde{S}^1 & \tilde{S}^3 \\ \tilde{S}^5 & \tilde{S}^7 \end{bmatrix} + w_1 \tag{6}$$

where

$$\tilde{S}^k = \begin{bmatrix} 0 & 0 \\ -f\left(s_{12}^{k+1}\right)^* - c\left(s_{22}^{k+1}\right)^* & f\left(s_{12}^k\right)^* + c\left(s_{22}^k\right)^* \\ fs_{12}^k + cs_{22}^k & fs_{12}^{k+1} + cs_{22}^{k+1} \end{bmatrix},$$

$$f = \sqrt{\frac{1}{\text{tr}\left(H^{-1}\left(H^{-1}\right)^*\right)}}, \quad c = \sqrt{\frac{1}{\text{tr}\left(G^{-1}\left(G^{-1}\right)^*\right)}},$$

$\tilde{H} = V_{11}H$, $\tilde{G} = V_{21}G$. Through the arrangement of Equation 6, the interference $S_{12}$ and $S_{22}$ are separated

as matrix $\begin{bmatrix} \tilde{S}^1 & \tilde{S}^3 \\ \tilde{S}^5 & \tilde{S}^7 \end{bmatrix}$.

To eliminate the separate matrix, use $y_{1i}$ that denotes the $i$-th row in $Y_1$, then make conjugate transformation to $y_{12}$ and $y_{15}$, and vector the transformed matrix to Equation 7.

$$\tilde{Y}_1 = \sqrt{\frac{3P}{2}}\begin{bmatrix} H_{11} & G_{11} & M_1 \\ H_{12} & G_{12} & M_2 \\ H_{13} & G_{13} & M_3 \\ H_{14} & G_{14} & M_4 \end{bmatrix}\begin{bmatrix} S_{11} \\ S_{21} \\ fS_{12} + cS_{22} \end{bmatrix} + \tilde{w}$$
$$\tag{7}$$

where $S_{ij} = \begin{bmatrix} s_{ij}^1 & s_{ij}^2 & \cdots & s_{ij}^8 \end{bmatrix}^T$, and $\tilde{h}_{ij}$ and $\tilde{g}_{ij}$ denote the $i$-th row and $j$-th column element in $\tilde{H}$ and $\tilde{G}$, respectively. As $H_{ij}$ and $G_{ij}$ has the same structure, only $H_{ij}$ is shown here.

$$H_{ij} = I_2 \otimes \begin{bmatrix} \tilde{h}_{i,j} & \tilde{h}_{i+1,j} & \tilde{h}_{i+2,j} & \tilde{h}_{i+3,j} \\ \tilde{h}_{i+1,j}^* & -\tilde{h}_{i,j}^* & \tilde{h}_{i+3,j}^* & -\tilde{h}_{i+2,j}^* \\ 0 & 0 & 0 & 0 \end{bmatrix}$$

$$M_1 = I_2 \otimes \begin{bmatrix} 0 & 0 & 0 & 0 \\ 0 & -1 & 0 & 0 \\ 1 & 0 & 0 & 0 \end{bmatrix},$$

$$M_2 = I_2 \otimes \begin{bmatrix} 0 & 0 & 0 & 0 \\ 1 & 0 & 0 & 0 \\ 0 & 1 & 0 & 0 \end{bmatrix}$$

$$M_3 = I_2 \otimes \begin{bmatrix} 0 & 0 & 0 & -1 \\ 0 & 0 & 1 & 0 \\ 0 & 0 & 0 & 0 \end{bmatrix},$$

$$M_4 = I_2 \otimes \begin{bmatrix} 0 & 0 & 0 & 0 \\ 0 & 0 & 1 & 0 \\ 0 & 0 & 0 & 1 \end{bmatrix}$$

From $M_1$ to $M_4$ structure, it is not difficult to find that the model can use $\tilde{y}_i$, which denotes the $i$-th element in $\tilde{Y}_1$, to eliminate all the interference by $\tilde{y}_2 + \tilde{y}_9$, $\tilde{y}_5 + \tilde{y}_{12}$, $\tilde{y}_8 - \tilde{y}_3$, $\tilde{y}_{11} - \tilde{y}_6$, $\tilde{y}_{14} + \tilde{y}_{21}$, $\tilde{y}_{17} + \tilde{y}_{24}$, $\tilde{y}_{20} - \tilde{y}_{15}$, and $\tilde{y}_{23} - \tilde{y}_{18}$, and to get Equation 8.

$$\hat{Y}_1 = \hat{H}_1 \cdot \begin{bmatrix} S_{11} & S_{21} \end{bmatrix}^T + \hat{w}_1 \tag{8}$$

where $\hat{H}_1 = \sqrt{\frac{3P}{2}}\begin{bmatrix} \hat{H}_{11} & \hat{G}_{11} \\ \hat{H}_{12} & \hat{G}_{12} \\ \hat{H}_{13} & \hat{G}_{13} \\ \hat{H}_{14} & \hat{G}_{14} \end{bmatrix}$, matrix $\hat{H}_{ij}$ is $H_{ij}$

eliminate its third row and sixth row, and $\hat{G}_{ij}$ is $G_{ij}$ eliminate its third row and sixth row. At this point, the interference $S_{12}$ and $S_{22}$ are completely eliminated at Receiver1.

Substitute Equation 3 into Equation 8 to get the final Equation 9 for the detection of the modulation symbol at Transmitter1.

$$\hat{Y}_1 = \bar{H}_1 \cdot I_4 \otimes \text{diag}(R_1, R_2) \cdot \begin{bmatrix} B_{11} & B_{21} \end{bmatrix}^T \tag{9}$$

where $B_{ij} = \begin{bmatrix} b_{ij}^1 & b_{ij}^2 & \cdots & b_{ij}^8 \end{bmatrix}^T$,

$$\bar{H}_1 = \begin{bmatrix} \bar{H}_1^1 & \bar{H}_1^2 \end{bmatrix}$$

$$\bar{H}_1^1 = \begin{bmatrix} H^1 & H^7 & H^3 & H^5 & H^2 & H^8 & H^4 & H^6 \end{bmatrix}$$

$$\bar{H}_1^2 = \begin{bmatrix} H^9 & H^{15} & H^{11} & H^{13} & H^{10} & H^{16} & H^{12} & H^{14} \end{bmatrix}, H^i \text{ denotes}$$

the $i$-th column in $\hat{H}_1$ can use maximum likelihood detection method to get the original modulation symbol matrix $B_{11}$ and $B_{21}$.

Although this article only studies the case of two users and four antennas, it can be further extended.

## 3  SIMULATION RESULTS

To verify the correctness and validity of the full-diversity encoded transmission scheme in this article, the two-user multi-antenna wireless communications networks can use Rayleigh fading channel model to simulate. The modulation symbols use binary phase–shift keying (BPSK) and the additive complex Gaussian noise. Use $E_b/N_0$, which is a normalized signal-to-noise ratio (SNR) measure of a system. When $E_b$ denotes the average energy that transmitter sends a bit need to use.

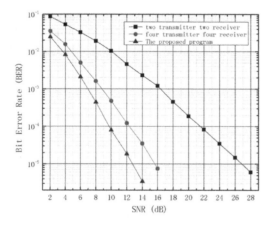

Figure 2. Comparison of several coding methods' BER performance in two-user four-antenna MIMO communication networks.

The simulation results are shown in Figure 2. In contrast to the communications network of two users with two-antenna, communication network of the two users with four-antenna is possible to obtain a lower BER and a faster rate of decline of BER because it contains greater channel capacity and uses efficient encoding schemes. If the joint coding method is not used, the detection process of the receiver will be unable to get the full spatial diversity gain, thus increasing the error rate to some extent and reducing the communication reliability of transmission.

## 4  CONCLUSION

This article, based on the joint encoding method, design principles of diversity and the study of two-user two-antenna communication network, uses Alamouti structure and beam-forming technology to extend the antenna number. It proposes an effective method to eliminate the interference and to achieve full-rate transmission, and use a full-effective way to obtain diversity gain performance in a multiuser MIMO system with more antennas. Simulation results confirm the validity and correctness of the proposed method. This method has a high excavated rate, and multiuser MIMO systems with BER performance advantages have potential applications.

ACKNOWLEDGMENTS

This article is supported by 2013 Annual Ministry of Education Doctoral Fund (20133401110003).

REFERENCES

[1] Younsun Kim, Hyoungju Ji, Juho Lee, et al. Full dimension MIMO (FD-MIMO) The next evolution of MIMO in LTE systems [J]. IEEE Wireless Communications, 2014, 21(2): 92–100.
[2] Veljko Stankovic and Martin Haardt. Generalized Design of Multi-User MIMO Precoding Matrices [J]. IEEE Transactions on Wireless Communications, 2008, 7(3): 953–961.
[3] Huang Huang and Vincent K.N. Lau. Partial Interference Alignment for K-User MIMO Interference Channels [J]. IEEE Transactions on Signal Processing, 2011, 59(10): 4900–4908.
[4] Long Shi, Wei Zhang, Xiang-Gen Xia. On Designs of Full Diversity Space-Time Block Codes for Two-User MIMO Interference Channels [J]. IEEE Transactions on Wireless Communications, 2012, 11(11): 4184–4191.
[5] Liangbin Li, Hamid Jafarkhani, Syed Ali Jafar. When Alamouti codes meet interference alignment: transmission schemes for two-user X channel [J]. IEEE International Symposium on Information Theory Proceedings, 2001(5): 2577–2581.
[6] Zhang H W, Zhang H B, Song W T, et al.. An improved group space-time block code through constellation rotation [J]. Journal of Shanghai Jiaotong University(Science), 2005, E-10(4): 349–353.
[7] S. Jafar and S. Shamai, "Degrees of freedom region for the mimo x channel," IEEE Transactions on Information Theory , vol. 54, no. 1, pp.151–170, Jan. 2008.
[8] Ma Miaoli, Giannakis G B. Full-diversity full-rate complex-field space-time coding-[J]. IEEE Transactions on Signal Processing, 2003, 51(11): 2917–2930.
[9] Damen M O, Tewfik A, Belflore J C. A construction of a space-time code based on number theory[J]. IEEE Trans Inform Theory, 2002, 48(3): 753–760.

*Multimedia, Communication and Computing Application – Leung (Ed.)*
© *2015 Taylor & Francis Group, London, ISBN 978-1-138-02775-6*

# Secure communications using different type of chaotic systems

G. Obregón-Pulido, Alan Gasca & G. Solis-Perales
*Universidad de Guadalajara, División de Electrónica y Computación, Guadalajara Jal. México*

ABSTRACT: The use of chaotic systems in security applications has been extensively studied and used due to its intrinsic advantages. Thanks to its sensitivity to initial conditions, the system has an almost unpredictable behavior. The encryption using chaotic dynamics is often based on a single system. In this paper, a system whose chaotic dynamics vary along the time is used to encrypt a message. Then to recover the message, an identifier that is able to distinguish between different chaotic behaviors is designed. Then the successful estimation of the parameters that characterize each system as well as the message is obtained.

## 1 INTRODUCTION

Chaotic systems have been widely used for various purposes, including applications in security communications using these systems as encryptors.

The increasing use of data transfers over the Internet, wireless applications among others in real time has created a technological and scientific studio where the improvement of the security is the main objective. Chaos synchronization problem was first described by Fusijaka and Yemada in 1983. The idea of using chaos as encoder has its origins in the work of Pecora and Carroll with the first application of chaos synchronization (Pecora and Carroll 1990 and Carroll and Pecora 1991) in early 1990s. Chaos theory has been exploited in a variety of fields including physical, chemical, secure communications (Kwok et al 2004, Li et al 2007) etc. Chaotic oscillators excited by sinusoidal signals increases safety in the transmission of messages (Obregón-Pulido et al 2013) and this aspect can be enhanced when considered unknown certain system parameters such as the type of chaotic system. The equations described by Lorenz, Chen and Lü attractors have in common a particular structure that, under certain transformations, their non-linear functions have the same structure as discussed below. In this work we will exploit this idea and raise a method to identify the message even when the type of chaotic system varies along the time.

## 2 STATEMENT OF PROBLEM

The chaotic encryption is carried out using three chaotic systems, Lorenz, Chen and Lu which can be described by one system and variation of some of its parameters. Then the message is injected into this system and the parameters are time varying in such manner that the three attractors are appearing along the time. After the identifier recover the message by estimating the corresponding parameters and identifying the chaotic system, the transformation based on parameter variation is as follows

$$\dot{x}_1 = x_2$$
$$\dot{x}_2 = -a_1 x_1 - a_2 x_2 + f(x_1, \eta) + m(t)$$
$$\dot{\eta} = -a_3 \eta + \alpha(x_1, x_2) \tag{1}$$

where $m(t)$ is the transmitted message. Then the equations of the three systems pass to the form (1) and we will identify each element thereof.

We take the next assumption:

Assumption 1: The message does not destroy the chaotic behavior.

### 2.1 The Lorenz Chaotic System

The Lorenz equations are the first chaotic system (Lorenz 1963) whose behavior is described by:

$$\dot{x} = a(y - x)$$
$$\dot{y} = cx - xz - y + {}^m\!/_a \tag{2}$$
$$\dot{z} = xy - bz$$

doing the change of variables $x_1 = x$, $x_2 = a(y - x)$ and $\eta = z$, the system is rewritten as:

$$\dot{x}_1 = x_2$$
$$\dot{x}_2 = -a(1-c)x_1 - (1+a)x_2 - a\eta x_1 + m$$
$$\dot{\eta} = -b\eta + x_1(x_2 / a + x_1)$$

with:

$$a_1 = a(1-c), \ a_2 = (1+a), \ a_3 = b$$

$$f(x_1, \eta) = -ax_1\eta, \quad \alpha(x_1, x_2) = x_1(x_2/a + x_1)$$

### 2.2 The Chen Chaotic System

Like the previous system, Chen's chaotic oscillator is of order 3 and its dynamics is given by the equations:

$$\dot{x} = a(y - x)$$
$$\dot{y} = (c - a)x - xz + cy + \frac{m}{a}$$
$$\dot{z} = xy - bz \tag{3}$$

taking $x_1 = x$, $x_2 = a(y - x)$ and $\eta = z$
then

$$\dot{x}_1 = x_2$$
$$\dot{x}_2 = a(2c - a)x_1 + (c - a)x_2 - a\eta x_1 + m$$
$$\dot{\eta} = -b\eta + x_1(x_2/a + x_1)$$

that is form (1) where:
$$a_1 = a(a - 2c), \; a_2 = (a - c), \; a_3 = b$$
$$f(x_1, \eta) = -ax_1\eta, \quad \alpha(x_1, x_2) = x_1(x_2/a + x_1)$$

### 2.3 The Lü Chaotic System

The system is described by:

$$\dot{x} = a(y - x)$$
$$\dot{y} = -xz + cy + \frac{m}{a}$$
$$\dot{z} = xy - bz \tag{4}$$

If we take $x_1 = x$, $x_2 = a(y - x)$ and $\eta = z$ the system takes the form:

$$\dot{x}_1 = x_2$$
$$\dot{x}_2 = acx_1 + (c - a)x_2 - a\eta x_1 + m$$
$$\dot{\eta} = -b\eta + x_1(x_2/a + x_1)$$

with the coefficients and nonlinear functions:
$$a_1 = -ac, \; a_2 = (a - c), \; a_3 = b$$
$$f(x_1, \eta) = -ax_1\eta, \quad \alpha(x_1, x_2) = x_1(x_2/a + x_1)$$

The results of these transformations can be summarized in the Table 1.

The problem is to construct a decoder that can obtain the message even when the type of chaotic system is changed.

Table 1. System parameters transformed.

|        | $a_1$     | $a_2$    | $a_3$ |
|--------|-----------|----------|-------|
| Lorenz | a(1-c)    | a+1      | b     |
| Chen   | a(a-2c)   | (a−c)    | b     |
| Lü     | −ac       | (a−c)    | b     |

## 3 RECEIVER DESIGN

Now it is possible to give the next theorem which solves the problem aforementioned. It is required that the parameters $a_1$, $a_2$ remain unknown and then we do not know the type of chaotic system which encrypts the message. Then we proceed to estimate these parameters.

Theorem 1: *Consider the outputs taken by the following equations*

$$y_1 = x_1 + x_2 \tag{5}$$
$$y_2 = x_1 + x_2 + \left( f(x_1, \eta) + m(t) \right)/g_2$$

*Then the observer given by:*

$$\dot{\hat{x}}_1 = \hat{x}_2 + (y_1 - \hat{y}_1)$$
$$\dot{\hat{x}}_2 = -\hat{a}_1\hat{x}_1 - \hat{a}_2\hat{x}_2 + g_2(y_2 - \hat{y}_1)$$
$$\dot{\hat{\eta}} = -a_3\hat{\eta} + \alpha(\hat{x}_1, \hat{x}_2) \tag{6}$$
$$\dot{\hat{a}}_1 = -\lambda_1\hat{x}_1(y_1 - \hat{y}_1)$$
$$\dot{\hat{a}}_2 = -\lambda_2\hat{x}_2(y_1 - \hat{y}_1)$$

*with the output*

$$\hat{y}_1 = \hat{x}_1 + \hat{x}_2$$

*where the positive constants* $g_2$, $\lambda_1$, $\lambda_2$ *are properly selected, is such that the signals are estimated; in other words* $e_i \to 0$, $\quad \forall i = 1, 2, \ldots, 5$ *where:*

$$e_1 = x_1 - \hat{x}_1, \quad e_2 = x_2 - \hat{x}_2, \quad e_3 = \eta - \hat{\eta}$$
$$e_4 = a_1 - \hat{a}_1, \quad e_5 = a_2 - \hat{a}_2$$

*and the recovered message has the form*

$$\hat{m} = g_2(y_2 - y_1) - f(\hat{x}_1, \hat{\eta}) \tag{7}$$

*Proof:* Consider the error system between (1) and (6):

$$\dot{e}_1 = -e_1$$
$$\dot{e}_2 = -(a_1 + g_2)e_1 + (a_2 + g_2)e_2 - \hat{x}_1 e_4 - \hat{x}_2 e_5$$
$$\dot{e}_3 = -a_3 e_3 + \alpha(x_1, x_2) - \alpha(\hat{x}_1, \hat{x}_2)$$
$$\dot{e}_4 = -\lambda_1\hat{x}_1(y_1 - \hat{y}_1)$$
$$\dot{e}_5 = -\lambda_2\hat{x}_2(y_1 - \hat{y}_1)$$

and choose the Lyapunov function

$$V(e) = c_1 e_1^2 / 2 + e_2^2 / 2 + e_4^2 / 2\lambda_1 + e_5^2 / 2\lambda_2 + e_1 e_2$$

Its derivative is given by:

$$\dot{V}(e) = -(c_1 + a_1 + g_2)e_1^2 - (a_2 + g_2)e_2^2$$

$$-(1 + a_1 + a_2 + 2g_2)e_1 e_2$$

$V(e)$ will be positive definite and $\dot{V}(e)$ negative if we choose:

$$c_1 > max\left\{1, k^2 / 4(a_2 + g_2) - (a_1 + g_2)\right\}$$

with $k = 1 + a_1 + a_2 + 2g_2$

Since $V(0) = 0$ and it's radially unbounded, it follows that $e_1 \to 0$ and $e_2 \to 0$ then by the linear independence of $\hat{x}_1$ and $\hat{x}_2$ then also $e_4 \to 0$ and $e_5 \to 0$. Finally $e_3$ also tends to zero because it is a stable linear system with an input which become zero. Then $e_i \to 0$ for all $i = 1, 2, \dots, 5$ and for the message we can write:

$$\hat{m} = g_2(y_2 - y_1) - f(\hat{x}_1, \hat{\eta}) = m(t) + f(x_1, \eta) - f(\hat{x}_1, \hat{\eta})$$

Therefore $\hat{m} \to m$, and this completes the proof.

## 4  SIMULATION

### 4.1  Function signals

We will test the identifier switching between systems Lorenz, Chen and Lü, changing them each 30 seconds of simulation. The parameters for chaotic behavior for systems (2), (3) and (4) are given in Table 2, and let the parameters of the receiver be $g_2 = 500$, $\lambda_1 = 1000$, $\lambda_2 = 2000$. A given function for m(t) = arcsin(sin(t)) [9] was used as a message to test the chaotic scrambler.

In Fig. 1 the dynamic evolution of the state $x_1$ observed which furthermore represents at the same time (at the top of the image) the obtained chaos combination of the three observed states of the simulation is shown. The difference between the state and its estimation at the beginning of the simulation and every 30 seconds when the system dynamics is modified can be seen from the figure. The figure shows that

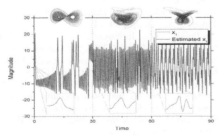

Figure 1.   Time evolution of state $x_1$ and its corresponding estimation given by the observer. At the top of the figure the corresponding time interval dynamic is shown. At the bottom the estimating is shown in detail when the dynamic is changed.

the estimation error is small ploted in the boxes one graphic above the other. In Fig. 2, the following two states are shown together with its observed state, as in the figure of $x_1$ state, the estimation is reached at the first seconds of the simulation and when switching from one attractor to another, doing the error practically imperceptible. The estimation of parameters $a_1$ and $a_2$ are shown Fig. 3, where these values vary according to the information given in Table 2. A transition is also seen when the dynamic is changed but the estimation works achieves parameter identification.

Table 2.   Parameters for chaotic behavior

|  | a | B | c |
|---|---|---|---|
| Lorenz | 10 | 3 | 30 |
| Chen | 35 | 3 | 28 |
| Lü | 36 | 3 | 20 |

Fig. 4 shows in detail the difference between the original and the recovered message. Noting that when there is a change in the dynamic, an error in the recovery message of the observer is removed successfully. Fig. 5 exhibits the estimation error where $e_m = m - \hat{m}$

### 4.2  Audio signal

Using the same scheme as in the previous simulation, we will modify the transmitted signal to have an application in audio. The audio signal sent belongs to a segment of "*Huapango de Moncayo*" by the Mexican composer José Pablo Moncayo and played by the Orquesta Simfónica Nacional de México.

Using the proposed observer (6), we estimate the transmitted signal. Fig. 6 shows the original audio and the recovered. In the zooming we only make a

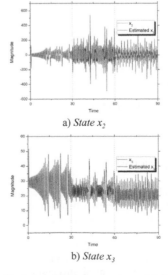

a) *State $x_2$*

b) *State $x_3$*

Figure 2.   Evolution of the states $x_2$ and $x_3$ and their observers.

15

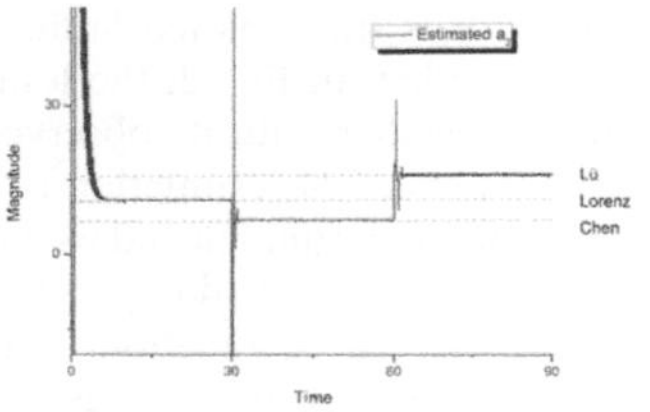


a) Estimated $a_1$

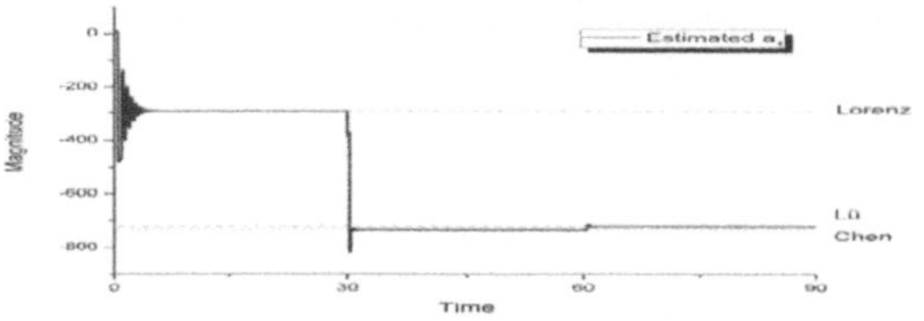


b) Estimated $a_2$

Figure 3. Estimation of parameters $a_1$ and $a_2$.

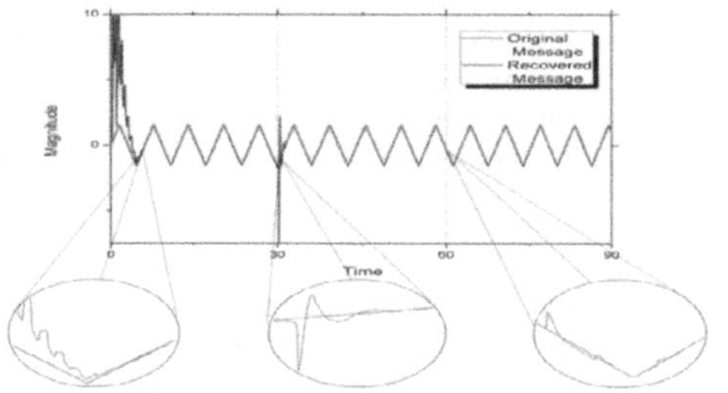


Figure 4. Recovery message using the proposed estimator.

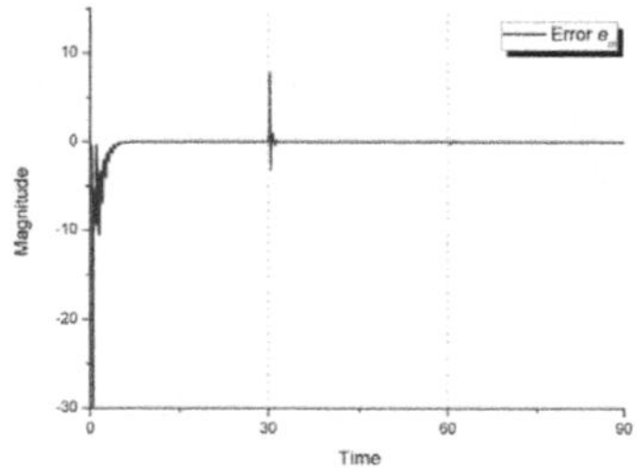


Figure 5. Error between original function signal and its estimation.

difference in the thickness of the line between the original and the recovered message for purposes of visual assessment.

## 5 CONCLUSION

It is possible to design a receiver which is able to obtain the message even when the chaotic encriptor system is changing. This is carried out by an estimator which successfully identifies the parameters $a_1$ and $a_2$ that characterize each dynamics of the type of the chaotic system (Lorenz, Chen and Lu). The encrypted message then is retrieved successfully, taking into

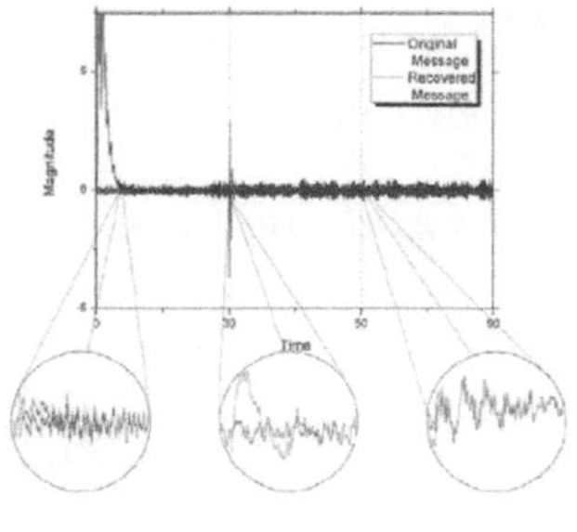


Figure 6. Audio signal and recovery message.

account that there is a short transient time to estimate the message in each change and at the beginning of the estimation. Finally, with the changes in the encriptor, the security of the transmission is improved since there is no information about the chaotic encriptor.

## ACKNOWLEDGMENT

This research has been sponsored by Universidad de Guadalajara.

## REFERENCES

Carroll T. L. and Pecora, L.M., 1991 "*Syncronizing Chaotic Circuits*", IEEE Tarns. Circ. Sys. 38, 4, 453–456.

Chiaraluce F. Cicarelli L et al, 2002, "*A new chaotic algorithm for video encryption*", IEEE Trans.Consum Electron 48, 838–843.

Cuomo K. M. and Oppenheim A. V. and Isabelle S. h., "*Spread Spectrum Modulation and Signal Masking Using Synchronized Chaotic Systems*" MIT Reserch Laboratory of Electronics Technical Report.

Fujikasa, H., Yamada, T., 1983 "*Stability theoryof synchronized motion in coupled-oscillator systems*", Progr. Theoret. Phys. 69, 332–347.

Hassan K. Khalil. 3rd ed. Prentice Hall, 2002 "*Nonlinear systems*".

Kwok, H. S., Wallace, K., Tang,S., Man, K. F., 2004, "*Online Secure Comunication System using Chaotic Map*", Internat. J. Bifurcat. Cahos, 14, 285–292.

Li. S., Alvarez G., Li Z., Halang W. A., 2007, "*Analog Chaos-based Secure Communications and Cryptoanalysis: A brief Survey*".

Lorenz, E. N. 1963, "*Deterministic nonperiodic flow*", J. Atmos. Sci. 20 pp. 130–141.

Obregón-Pulido, G. Torres-Gonzalez A. Solis-Perales G. Càrdenas Rodríguez, R. 2013, "*Estimation of sinusoidal signals using measurement of chaotic signals*", Modelling Identification and Control (ICMIC) pp. 249–252.

Obregon-Pulido G. and Cardenas-Rodriguez R., 2012, "*Relacion entre las funciones trigonometricas y las señales Diente de Sierra Triangular y Cuadrada*", Sociedad Mexicana de instrumentacion (SOMIXXVII) Culiacan Sinaloa México.

Ott E., 2002, "*Chaos in Dynamical Systems*", 2nd Ed. Cambridge University Press.

Pecora, L. M., Carroll, T. L., 1990, "*Synchronization in chaotic systems*", Phys. Rev. Lett, 64, 821–824.

# Power allocation and packet scheduling for energy harvesting communication systems

Y. Su,
*School of Software, Xidian University, China*

J. Yang & Q.H. Yang
*State Key Laboratory of ISN, School of Telecommunications Engineering, Xidian University, Xi'an, Shaanxi, China*

ABSTRACT: In this article, we study the joint-optimization power allocation and problems of multi-packet scheduling in an energy harvesting wireless communication system, in which different packets are characterized by different rewards if transmitted. With data packets and harvested energy being modeled to arrive at the source node randomly and independently, the complicated problem for maximizing the system reward is equivalently decomposed into two subproblems, and the optimal policy is acquired by two sequential steps: (1) power allocation that maximizes the system throughput is acquired within the constraint of the harvested energy and (2) the scheduling scheme of data packets based on Derman–Lieberman–Ross (DLR) theory is obtained according to power allocation and multi-packets' rewards. Simulation results demonstrate the theoretical analysis.

## 1 INTRODUCTION

Wireless nodes are capable of energy harvest; the process of collecting energy from the environment ceaselessly through various different sources makes it possible for wireless communication networks to sustain themselves with an indefinite lifetime [1]. As a result, one can predict that there shall be growing interest in wireless networks consisting of nodes of energy harvest. To utilize the energy in a best way, the systems should be optimally subjected to the constraints on the instantaneously available energy. Therefore, most works model the process of energy harvest as a set of energy packets arriving in each time-slot for analytical tractability.

An energy management model is proposed by Kansal et al. [2] to maximize the efficiency of energy usage under the assumption that the nodes can predict precisely the amount and the time of energy harvest in the future. An off-line optimization method is proposed by Ozel et al. [3] to maximize the throughput in a fading channel based on the assumption that the amount of energy harvested in the future is known in advance. A point-to-point communication with the transmitter of energy harvest is considered by Gregori and Payaro [4], wherein an algorithm that minimizes the total transmission time is proposed. The scheme of an optimal packet scheduling is studied for energy harvesting communication systems [5]. However, only one type of packet is taken into consideration, which is less attractive in information era nowadays.

In this article, we consider energy harvesting wireless communication system, where different types of packets are characterized by different rewards if transmitted. With the data packets and the harvested energy being modeled to arrive at the source node randomly and independently, we propose an algorithm to maximize the total system reward. The complicated problem is equivalently decomposed into two subproblems, and the optimal transmission policy is acquired by two sequential steps. First, we acquire the optimal power (OP) allocation that maximizes the system throughput with the harvested energy requirement based on the assumption that all types of packet are characterized by the same reward. And second, we propose the scheduling scheme of data packets based on Derman–Lieberman–Ross (DLR) theory [6], according to the obtained power allocation and multi-packets' reward.

## 2 SYSTEM MODEL

The system of an energy harvesting transmitter (TX) is shown in Fig. 1a, which has a data buffer and a battery. TX is capable of sensing a random field and generating data packets. We assume that the system is time-slotted with variable duration [7] as shown in Fig. 1b, and thus, each sensed data packet can be transmitted in a time-slot. It is obvious that the duration of a time-slot depends on the length of the data packet transmitted. Let $m \in \{1, \ldots, M\}$ be the time-slot index, namely, the transmitted data packets index, and $d(m)$ be the duration of time-slot $m$. Only one data packet is transmitted in each time-slot.

In general, the arrivals of sensed data and the harvested energy can be represented as two independently random processes. The harvested energy rate in each time-slot is assumed to be a constant and an independent identical distribution (i.i.d). Normalized to time, harvested energy is replaced by harvesting power in this article. Furthermore, constant power transmission within a time-slot is optimal as shown in the work of Yang and Ulukus [8].

We assume that the battery capacity for storing the harvested energy is infinite and the initial energy stored in the battery is zero. It is reasonable because the current battery technology makes it possible to have capacities of sufficiently large storage compared to the rate of harvested energy. Moreover, we only focus on the case that the energy consumed by the battery is only for transmission purposes [9].

Let $N(m)$ denote the data packets stored in the buffer in time-slot $m$. Each data packet is characterized by (a) the number of bits, denoted as $L_{n(m)}$, $n(m) \in \{1, \ldots, N(m)\}$ and (b) reward $r_{n(m)} = L_{n(m)}a_{n(m)}$ for completing transmission, where $a_{n(m)}$ denotes the average reward of each bit in data packet $n(m)$. The initial number of data packets is denoted as $N(0)$, which denotes the sensed data before the process of energy harvest (e.g., the node powered by solar cells can generate large number of data packets through the night, but the sensed data only can be transmitted in the day), and is a relatively large number compared to the transmitted data rate. Furthermore, the size of data buffer is assumed to be sufficiently large in this article. Let $K(m)$ denote the number of arrival data packets in time-slot $m$, each with a reward $r_{k(m)} = L_{k(m)}a_{k(m)}$, $k(m) \in \{1,\ldots,K(m)\}$, where $L_{k(m)}$ denotes the number of bits of packet $k(m)$ and $a_{k(m)}$ denotes the reward of it. If the data packet is not chosen for transmission in time-slot $m$, the reward of it will decrease by a factor $\alpha$, which can be expressed as

$$r_{n(m+1)} = \alpha r_{n(m)}. \tag{1}$$

The channel gain $h$ captures the effect of the pass loss of large scale. Given the distance from TX to RX, $h$ can be regarded as a constant [10], so the Shannon formula can be expressed as follows:

$$R_t(m) = \frac{1}{2}\log_2\left(1 + \gamma P_t(m)\right) \tag{2}$$

where $\gamma$ is a constant depending on the channel gain, the power spectral density of additive white Gaussian noise (AWGN), and the system bandwidth. $P_t(m)$ denotes the transmitting power in time-slot $m$.

Our goals is to maximize the total reward achieved by the TX until a deadline $m = M$, while maintaining the stability of the battery. This optimization problem can be formulated as

$$\max \sum_{m=1}^{M} \sum_{n(m)=1}^{N(m)} \frac{\rho_{n(m)}r_{n(m)}}{d(m)} = \sum_{m=1}^{M} \sum_{n(m)=1}^{N(m)} R_t(m)\rho_{n(m)}a_{n(m)}$$

$$s.t. \begin{cases} \sum_{m=1}^{m_0} P_t(m) \le \sum_{m=1}^{m_0} P_a(m), m_0 = 1,2,\ldots,M, \\ P_t(m) \ge 0, 1 \le m \le M, \\ \sum_{n(m)=1}^{N(m)} \rho_{n(m)} = 1, 1 \le t \le M, \\ \rho_{n(m)}(m) \in \{0,1\}, \end{cases} \tag{3}$$

where $\rho_{n(m)}$ refers to as packet scheduling indicator, which denotes whether the $n(m)$ type of packet is chosen for transmission in time-slot $m$ or not. The first constraint in Equation 3 denotes the transmit power available in each time-slot. $P_a(m)$ denotes the harvested power in time-slot $m$, which is a random variable and i.i.d with a cumulative density function (CDF) of $G_{P(z)}$.

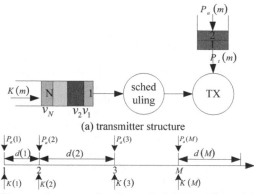

(a) transmitter structure

(b) energy harvests and packets arrivals in time-slot model

Figure 1.   System model.

The linear constraints in Equation 3 as well as the problem of mixed integer and non-convex objectives makes it difficult to find the global optimal solution, so we equivalently decompose the complicated problem into two subproblems, and then acquire optimal transmission policy by two sequential steps.

## 3   SCHEME OF OPTIMAL POWER ALLOCATION

Noting that when all data packets are characterized by the same reward, the objective function in Equation 3 is transformed into maximizing the throughput, i.e., no matter which data packet is chosen for transmission, the system reward only depends on the system throughput. Therefore, Equation 3 can be simplified as

$$\max \sum_{m=1}^{M} R_t(m)$$

$$s.t. \begin{cases} \sum_{m=1}^{m_0} P_t(m) \leq \sum_{m=1}^{m_0} P_\alpha(m), m_0 = 1,2,...,M, \\ P_t(m) \geq 0, 1 \leq m \leq M, \end{cases} \quad (4)$$

To acquire what the solution may look like, we pose the simplest version of the problem. Supposing that there is $cM$ amount of energy in the battery, where $c$ is a constant, and that the energy arrivals are zero, then Equation 4 can be rewritten as

$$\max \sum_{m=1}^{M} R_t(m)$$

$$s.t. \begin{cases} \sum_{m=1}^{M} P_t(m) \leq cM, \\ P_t(m) \geq 0, 1 \leq m \leq M, \end{cases} \quad (5)$$

By Jensen's inequality [11], the optimal strategy for Equation 5 is

$$P_t^*(m) = c, \quad (6)$$

This suggested that the optimal transmit power in each time-slot equals the means of total harvested energy. Returning to Equation 4 and noting that energy in the battery comes from the random energy arrivals, the transmit power in time-slot $m$ should tend to equal the means at time-slot $m$ according to Equation 6. So we can acquire the OP allocation expressed as

$$P_t^*(m) = \min \left( \frac{\sum_{i=1}^{m} P_a(i)}{m}, P_a(m) + \sum_{i=1}^{m} P_a(i) - \sum_{j=1}^{m-1} P_t(j) \right),$$

$$(7)$$

From the OP allocation (Equation 7), we can obtain the theorem as follows, which is very important for employing DLR theory in the next section.

*Theorem:* Let random power arrivals in each time-slot be i.i.d with CDF of $G_{P(z)}$ and $M$ be sufficiently large. The distribution function of $P_t^*(m)$ is $G_{P(z)}$, same to that of power arrivals.

The theorem is proved by verifying that the characteristic function of OP allocation is same to that of power arrivals for sufficiently large $M$. The detailed proof is omitted because of page limitation.

## 4 PACKET SCHEDULING ALGORITHM

In this section, we first introduce what the DLR theory is and how to use it. Then, the DLR theory based on task scheduling scheme for energy-harvest system is proposed.

### 4.1 DLR theory

Derman, Lieberman, and Ross introduced following optimal theories for the problems of the sequential stochastic assignment [6]. Suppose there are $N$ workers available to complete $N$ jobs. The abilities, $x_i, i = 1,...,N$, of workers to complete jobs are i.i.d. random variables with CDF of $G_W(z)$, which can be observed in advance. Associating with $j$th ($j = 1,..,N$) job is a real value $q_j$ representing its worth. When worker $i$ is assigned to the $j$th job, the reward $x_i q_j$ is obtained. Each worker is assigned to one and only one job. The goal is to assign $N$ workers the $N$ jobs so as to maximize the total reward. In particular, if $m(i)$ is defined to be the job to which $i$th worker is assigned, then the total expected reward is given by

$$\max \sum_{i=1}^{N} x_i q_{m(i)} \quad (8)$$

The following DLR theorem provides the allocation policy of the optimal resource using the available stochastic information.

*DLR Theorem:* For each $N > 0$, there exist numbers $-\infty = a_{0,N} \leq a_{1,N} \leq ... \leq a_{N,N} = \infty$ such as whenever there are $N$ assignments to make and value $q_1 \leq_2 ... \leq_N$, then the optimal choice in the first assignment by using the worker $i$ such as $x_1$ is contained in the interval $(a_{i-1,N}, a_{i,N}]$. The $a_{i,N}$ depends on $G_W(z)$ but is independent of the $q_i$ values and is calculated recursively for $N$ as follows:

$$a_{i,N} = \int_{a_{i-1,N-1}}^{a_{i,N-1}} z \, dG_x(z) + a_{i-1,N-1} G_x(a_{i-1,N-1})$$

$$+ a_{i,N-1} \left[ 1 - G_x(a_{i,N-1}) \right] \quad (9)$$

with the convention that $a_{0,N} = \infty$, $a_{N,N} = \infty$, $-\infty * 0 = 0$, and $\infty * 0 = 0$.

### 4.2 Packets scheduling algorithm

In Section 3, the OP allocations are acquired and proved to be random variables with the same CDF to that of power arrivals. Therefore, the original Equation 4 can be rewritten as

$$\max \sum_{m=1}^{M} R_m^*(m) v_{n(m)}^m, \tag{10}$$

where $v_{n(m)}^m = \rho_{n(m)} \alpha_{n(m)}$.

The set of jobs and arriving workers in the problem of stochastic assignment were considered in Subsection 4.1, which correspond to tasks in the queue and sequentially achieved data rate here, respectively. Furthermore, the performance goal given in Equation 10 is structured identically to Equation 8. Thus, the policy derived in the DLR theory can be employed here.

The energy arrivals are exponentially distributed with the parameter $\lambda$, the CDF of which can be expressed as

$$G_p(z) = 1 - e^{-\lambda z}. \tag{11}$$

Therefore, the CDF of data rate can be written as

$$F_r(y) = 1 - e^{-\frac{\lambda}{\gamma}(4^y - 1)}. \tag{12}$$

By using DLR theory, we have [11,12]

$$a_{i,N} = a_{i-1,N-1} + \frac{e^{\frac{\lambda}{\gamma}}}{\ln(4)} \left\{ E_1\left(\frac{\lambda}{\gamma} 4^{a_{i-1,N-1}}\right) - E_1\left(\frac{\lambda}{\gamma} 4^{a_{i,N-1}}\right) \right\}. \tag{13}$$

$E_1(x)$ can be expressed by convergent series as

$$E_1(x) = -\psi - \ln(x) + \sum_{k=1}^{\infty} \frac{(-1)^{k+1} x^k}{k * k!}, x > 0, \tag{14}$$

where $\psi = 0.5572$.

In summary, the DLR theory based on the scheme of task scheduling is as follows:

*Step 1*: To sort the remaining $N(m)$ types of packets in ascending order of its values in time-slot $m$.

*Step 2*: To calculate boundaries according to the DLR theory.

*Step 3*: To determine the index $n$ of the intervals that the achieved data rate corresponds to.

*Step 4*: To select task $n(m)$ for assignment from the list of tasks formed in Step 1.

### 5 SIMULATION RESULTS

In this section, the simulation results are provided to illustrate our proposed algorithm. The harvested power is assumed to be an exponential distribution. The channel factor is $\gamma$ set as 10. The initial packets stored in the data buffer $N(0)$ is set as 4, and the number of arrival data packets in each time-slot is assumed to be 1. Integer length of bits of data packet is generated randomly in the range of [1,10], and the reward of each bit in each data packet is set randomly in the range of [1,10].

In this article, we evaluate the algorithm of proposed OP to the one called Greedy [10], which uses as much energy as harvested in the present time-slot. What is more, the proposed DLR based on scheduling scheme is compared with the first in, first out (FIFO) method and the packets scheduling scheme of highest reward (HR) data.

Fig. 2 depicts the total reward achieved versus the number of data packets transmitted. We observed that the proposed algorithm outperforms other schemes, as the proposed algorithm incorporates the stochastic information of harvested energy and packet rewards into the decision-making process.

Fig. 3 demonstrates the process of each algorithm over time-slot with $M = 60$. As expected, the proposed algorithm does not accumulate the highest reward at first as it does not select the data packet with highest reward for transmission but the appropriate data packet. As the data packet is transmitted, the proposed algorithm achieves higher reward compared with other algorithms.

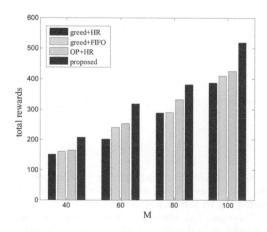

Figure 2. Reward comparison vs. different number of packets $M$.

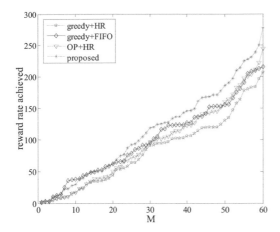

Figure 3.    Reward comparison vs. time-slot $m$, $M = 60$.

## 6    CONCLUSIONS

We studied the joint-optimization power allocation and packet scheduling problem in an energy harvesting wireless communication system. We also developed a joint-power allocation and packet-scheduling scheme by modeling the data packets and the harvested energy to arrive at the source node randomly and independently. First, the power allocation that maximizes the system throughput with the harvested energy constraint was acquired. And then, a novel DLR theory based on packet-scheduling scheme was proposed according to the obtained power allocation and multi-packets' reward. The proposed approach outperforms some other existing schemes in improving the overall system reward.

## REFERENCES

[1] S. Sudevalayam and P. Kulkarni, "Energy Harvesting Sensor Nodes: Survey and Implications," IEEE Communications Surveys and Tutorials, vol. 13, no. 3, pp. 443–461, 2011.

[2] A. Kansal, J. Hsu and S. Zahedi, "Power Management in Energy Harvesting Sensor Networks," ACM Trans. on Embeded Computer Systems, vol. 6, Sept. 2007.

[3] O. Ozel, K. Tutuncuoglu and J. Yang, "Resource Management for Fading Wireless Channels with Energy Harvesting Nodes," IEEE INFOCOM, pp. 456–460, April 2011.

[4] M. Gregori and M. Payaro, "Efficient Data Transmission for an Energy Harvesting Node with Battery Capacity Constraint," in Proc. of IEEE GLOBECOM, pp. 1–6, 2011.

[5] J. Yang and S. Ulukus, "Optimal Packet Scheduling in an Energy Harvesting Communication System," IEEE Trans. Commun., vol. 60, no. 1, pp. 220–230, Jan. 2012.

[6] C. Derman, G. Lieberman and S. Ross, "A Sequential Stochastic Assignment Problem," Management Science, vol. 18, no. 7, pp. 349–355, Mar. 1972.

[7] H. Zhang and P. Gburzynski, "A TDMA Scheme with Dynamic Frame Structure," IEEE VTC 2001, vol. 3, pp. 1407–1411, 2001.

[8] J. Yang and S. Ulukus, "Optimal Packet Scheduling in a Multiple Access Channel with Energy Harvesting Transmitters," Journal of Communications and Networks, vol. 14, no. 2, pp. 140–150, April 2012.

[9] V. Raghunathan and S. Ganeriwal, "Emerging Techniques for Long Lived Wireless Sensor Networks," IEEE Magazine Commun., vol. 44, no. 4, pp. 108–114, 2006.

[10] V. Sharma, U. Mukherji and V. Joseph, "Optimal Energy Management Policies for Energy Harvesting Sensor Nodes," IEEE Tans. Wireless Commun., vol. 9, no. 4, pp. 1326–1336, 2010.

[11] T. M. Cover and J. Thomas, Elements of Information Theory. New York: John Wiley and Sons Inc., 2006.

[12] T. Sharma, U. Mukherji, and V. Joseph, "Efficient Energy Management Policies for Networks with Energy Harvesting Sensor Nodes," IEEE Communication, control, and Computing Annual Allerton Conference, pp. 375–383, 2008.

*Multimedia, Communication and Computing Application – Leung (Ed.)*
*© 2015 Taylor & Francis Group, London, ISBN 978-1-138-02775-6*

# ID-based user off-line password authentication for mobile memory cards

N. Li, J.J. Duan & Z.L. Deng

*The School of Electronic Engineering, Beijing University of Posts and Telecommunications, China*

ABSTRACT: The authentication of the mobile memory cards is one of security issues of cards. There are many online authentication methods which are not available without network. ID-based user off-line password authentication based on ellipse weil-pairing was proposed. The scheme achieves the management of mobile memory cards and user groups, completes the user off-line authentication, and achieves the off-line update of user password. The scheme prevents parallel session attack, password guessing attack, replay attack, man-in-the-middle attack, card lost attack, and has forward, backward security, third-party security. Simulations show that the scheme is economic and practical with small calculation, small communication scale and small amount of memory.

KEYWORDS: Identity authentication, Memory card, Weil-pairing, Off-line

## 1 INSTRUCTION

The security of mobile memory devices is a big issue. People can get data in memory cards nearly arbitrarily. An easy way to ensure the information security is identity authentication before information reading. Password-based authentication scheme is a safe and convenient method, which can combine identities to unified authentication in order to strengthen the management. ID-and-password-based authentication methods depend on a remote server. This method has some problems: first, the password table is plaintext and easy to be attacked; second, the authentication is done in a remote third party and it cannot be done without network; third, the remote authentication is vulnerable to the man-in-the-middle attack. This paper proposes the ID-based of user and mobile memory cards off-line password authentication scheme. The scheme uses off-line authentication mode to achieve the management of memory cards and the user groups, completes the user off-line authentication, and achieves the off-line update of user passwords. The scheme is based on elliptic bilinear pairing authentication mode which can prevent parallel session attack, password guessing attack, replay attack, man-in-the-middle attack, card lost attack, and has forward, backward security, third-party security.

## 2 PRELIMINARIES

The authentication system of this paper is composed of remote server, users, and mobile memory cards. It is an ID-and-password-based off-line authentication that is mainly based on elliptic bilinear pairing theory.

### 2.1 Bilinear pairing

Let $G_1$ be a cyclic additive group generated by $P$ and $G_2$ be a cyclic multiplicative group. $G_1$ and $G_2$ have the same primer order q. Let : $G_1 \times G_1 \rightarrow G_2$ be a computable bilinear map, which satisfies the following properties.

1) Bilinear: $\hat{e}\,(aP, bQ) = \hat{e}\,(P,Q)^{ab}$, where $P,Q \in G1$ and a, b$\in$ Zq.

2) Non-degenerate: The map does not send all pairs in $G_1 \times G_1$ to the identity in $G_2$. That implies that P is a generator of G1, then $\hat{e}\,(P, P)$ is a generator of $G_2$.

3) Computable: There exists an efficient algorithm to compute $\hat{e}\,(P,Q)$ for any P, Q$\in G_1$.

Such a bilinear map $\hat{e}$ is called an admissible pairing, and the modified Weil or Tate pairing on elliptic curves gives a good implementation of such an admissible bilinear pairing. The group that possesses such a map $\hat{e}$ is called a bilinear group, on which the decisive Diffie–Hellman problem is easy to solve while the computational Diffie–Hellman problem is believed to be hard.

### 2.2 Related research works

There are many researches on password authentication schemes. In this section, we review some dynamic ID-based password authentication schemes, which are based on hash functions [12]. Each password authentication scheme is composed of four phases: registration phase, login phase, authentication phase, password change phase. All these schemes realize password authentication on line, but there are some problems still. We discuss them as follows.

1) Das et al's scheme[13]

It is vulnerable of smart card loss attack. It cannot avoid stolen verifier attack. It does not guarantee the forward secrecy of a system's secret key. It has low efficiency for wrong password login.

2) Liao et al's scheme[14]

It cannot avoid password guessing attack and cannot avoid stolen verifier attack. It has no forward secrecy. There is no smart card revocation. It has low efficiency for wrong password login.

3) Yoon and yoo's scheme[15]

It cannot avoid password guessing attack and has no smart card revocation.

4) Liou's scheme[16]

It is not secure for forgery attack and cannot avoid password guessing attack. It cannot avoid smart card loss attack and cannot avoid insider attack. It has no forward secrecy. It cannot fit smart card revocation. It has low Efficiency for wrong password login.

5) Wang et al's scheme and Khan et al's scheme[17,18]

The two schemes have the same flaw, which is low efficient for wrong password login.

## 3 IMPROVED SECURITY AUTHENTICATION SCHEME

This scheme applying to the system is composed of remote server, users, and mobile memory cards.

The server first initializes the authentication parameters of the cards, and binds the user to the card. Then it gives the binding card to the user, and the user gets the password at the same time. This step completes the group management of the cards and the users. The user needs to take an off-line authentication every time in order to use the card. For the first time user logs in, the legitimate user can change the password but cannot change the identity. This is conducive for the management of users and cards, and keeps the password secure. If the mobile memory card is lost, the user reports it to the server for a new card, and the old card cannot get the management service any more. Then the old card cannot be used with the time expired.

### 3.1 Scheme description

Symbols used in the scheme are as follows:

$H_1$: hash function, $H_1$: $\{0,1\}^* \rightarrow G1$;

$H_2$: hash function, $H_2$: $\{0,1\}^* \rightarrow Zq$;

$ID_i$: user identity; $PW$: user password; $ID_{card}$: the identity of mobile memory card; $\otimes$: exclusive OR; $x$: the group key of server.

This scheme is completed through four steps: user registration, user login authentication, user password update and delegation update of mobile cards.

Step 1: user registration (i.e. user initialization)

Remote server selects different keys for different groups, and selects different passwords for different users. The parameters initialization are as follows:

$$Y_1 = x \cdot H_1(ID_i) \cdot H_2(PW) \ Y_2 = x \cdot H_1(ID_{card})$$
$$RID = H_2(ID_i), J = x \otimes H_2(PW)$$
$$L = J \otimes RID = x \otimes H_2(PW) \otimes H_2(ID_i)$$

Then the server will store the parameters $(Y_1, Y_2, J, L, T)$ in the binding card with $ID_i$, in which T is the time stamp for the group of users delegated by the server. And the server will only store the group keys, user $ID_i$ and the binding card $ID_{card}$.

Step2: user login

When user logins, the user inserts the card to a PC or a card reader, then keys $ID^*$ and $PW^*$. The card will take the following calculations.

Firstly the card judges whether the time stamp is expired. If so, the authentication is denied, otherwise it takes the following two steps to authenticate:

1) User ID authentication

The card calculates:

$$J^* = L \otimes H_2(ID^*) = x \otimes H_2(PW) \otimes H_2(ID_i) \\ \otimes H_2(ID^*)$$

If $J^*$ is equal to $J$ stored in the card, it continues to the next step, otherwise, exits the authentication.

2) User password authentication

The card takes the bilinear pairing calculation:

$$\hat{e}[Y_1, H_1(ID_{card})] \\ = \hat{e}[x \cdot H_1(ID_i) \cdot H_2(PW), H_1(ID_{card})] \\ = \hat{e}[x \cdot H_1(ID_{card}), H_1(ID_i) \cdot H_2(PW)] \\ = \hat{e}[Y_2, H_1(ID^*) \cdot H_2(PW^*)]$$

If the equation holds, the user's password is correct, and the user is matched with the card. Otherwise the authentication is exited. After these two steps, the legitimacy is determined in three aspects of user identity, user password and mobile card identity. Then the card can be read normally.

Step3: user password update

Legitimate users can freely update password without a remote server. And the parameters needed to be updated at the same time are $(Y_1, J, L)$. $Y_2$ does not update. Users don't have the permission to update T. Update calculations are as follows:

$$P = H_2(PW_{new}) \quad Q = H_2^{-1}(PW)$$
$$Y_{1new} = Y_1 \cdot P \cdot Q = x \cdot H_1(ID_i) \cdot H_2(PW_{new})$$
$$J_{new} = J \otimes H_2(PW) \otimes P = x \otimes H_2(PW_{new})$$
$$L_{new} = J_{new} \otimes H_2(ID_i) = x \otimes H_2(PW_{new}) \otimes H_2(ID_i)$$

After the update, the storage parameters are $(Y_{1new}, Y_2, J_{new}, L_{new}, T)$.

Step4: delegation update of cards

When the card is lost, the legitimate user should provide an effective identification to the server. With the legitimate ID, the server will cancel the identity of the old card based on the relation of the user $ID_i$ and the card $ID_{card}$, and stop the delegation update to the old card. When the time expires, the old card cannot be used.

If the time stamp of the legitimate user's card expires, the user will report the server to extend the time. Firstly the user logins, if it is passed, the card reader will send the user's ID and the parameter $Y_2$ to the server. Then the server verifies the legitimacy of the card $ID_{card}$. At the same time if the $Y_2^* = x \cdot H_1(ID_{card})$ is equal to the received $Y_2$, the server updates the time stamp of the card. Otherwise, the server refuses to provide services.

### 3.2 Features of this scheme

Compared with the mentioned applications and security issues in section 2.2, the system has the following features:

1) The server only stores the group keys, user $ID_i$, card $ID_{card}$, and does not need to manage the password. As for users, it reduces the possibility of pw leak. As for server, it saves the memory space.

2) The group key does not store in mobile memory cards, and this overcomes the defect of password storage in the literature [13–18], makes the system more flexible and more secure. Individual user information leak will not affect the safety of the entire system. Even if the information of one group leaks, it will not affect the entire system, because of the usage of group key.

3) When the server releases the card, the group key is added to the system. This enables the group management of users and the cards.

4) In this system the relationship between the server and the user is constrained to each other. The user needs the server to initialize the cards, and the server needs the group key $x$ to achieve the management of the cards. At the same time the management of users and cards by server is completed with the help of user's passwords. The server does not control the user's passwords, but it will not complete the management of users and cards without passwords.

5) This scheme is an ID-based authentication system. The user's $ID_i$ is public. The password is changeable. But the password is corresponding to $ID_i$ that it makes user management easier.

6) There are two steps to complete the authentication in this scheme: the first step is the consistency authentication between the user identity and the password; the second step is to complete

matching authentication of the user and the mobile memory card. If the pw does not match the user's ID or the pw is wrong, the second step will not continue. Compared to the literature [13-14, 16-18], the efficiency is improved. In addition, the cost of computation is relatively small because the first step only takes the XOR operation and $H_2$ operation.

7) This scheme only involves XOR operation, hashing, elliptic multiplication and elliptic bilinear pairing operation. In section 5 of this paper, the simulation confirms that these operation times are bearable. XOR, hashing and elliptical multiplication operations on hardware have been widely implemented. Therefore, this scheme is acceptable in application.

## 4  SECURITY ANALYSIS

The security of the system has greatly improved compared to the literature [13–18].

1) Parallel session attack(SA1): In this scheme, The user only needs to provide ID and password, the rest of the authentication is done in the card. It is impossible to carry out session attack because no interaction of information is provided. Besides, what interact with server are ID, $Y_2$ and T, and useful login messages cannot be generated via this information.

2) Password guessing attack (SA2): Assuming that the attacker steals the parameters $(Y_1, Y_2, J, L, T)$ from a lost card and knows the user's ID, and then he can calculate $J = L \otimes H_2(ID) = x \otimes H_2(PW)$. Since x and PW are still unknown, and the hash function is a one-way function, the attacker is unable to guess the password.

3) Replay attack (SA3): The information interacted with server are: $ID_i$, $Y_2$ and T. There is a map of user ID and card ID in the server. So if the attacker gets a lost card, with the help of replay of intercepted information, he still cannot be accepted by the server. Because the card's ID does not match the user's ID, $Y_2^* = x \cdot H_1(ID_{card}) = Y_2$ cannot be verified. So the replay attack can be resisted.

4) Man-in-the-middle attack (SA4): The scheme is an off-line authentication. The authentication process does not need to interact with the server (except for delegation update). So the attacker cannot intercept useful information from the authentication process to implement man-in-the-middle attack.

5) Smart card lost attack (SA5): If the card is lost, the attacker can get information $(Y_1, Y_2, J, L, T)$ from the card. But the attacker cannot get the user's password from the information and cannot change the parameters of PW, J and L, because of the hardness of the elliptic curve discrete logarithm (ECDLP) and the one-way property of hash function. So the card is safe.

6) Forward and backward security (SA6): In this scheme, the user can update the password without

the server. So the attacker gets the current password which is unrelated to the previous password and following updated password.

7) Third-party security (SA7): The third party server does not store the password. Even if the server administrator gives away the user's information, only the group key X is leaked. If the card is not lost, the attacker is unable to intercept parameter $Y_1$. Otherwise, if $Y_1$ is intercepted, the attacker still cannot get the password because of the hardness of ECDLP. If the card is lost, the attacker can calculate $H_2(PW)$ from the information in the card, but he still cannot get the password because of the one-way hash function. Because of the difference of the group keys, if a group key leaks, it will not affect other group card's safety.

The security comparison with literature [13–18] is illustrated in table 1.

Y: has the ability to avoid this attack; N: does not have the ability to avoid this attack; EP: Efficient for wrong password login.

Table 1.　Comparison of security.

| Scheme | SA1 | SA2 | SA3 | SA4 | SA5 | SA6 | SA7 | EP |
|--------|-----|-----|-----|-----|-----|-----|-----|-----|
| Das[13] | Y | N | Y | N | N | N | N | N |
| Liao[14] | Y | N | Y | N | N | N | N | N |
| Yoon[15] | Y | N | Y | N | Y | N | N | Y |
| Liou[16] | Y | N | Y | N | N | N | N | N |
| Wang[17] | Y | N | Y | N | N | Y | N | N |
| Khan[18] | Y | Y | Y | Y | Y | Y | N | N |
| My scheme | Y | Y | Y | Y | Y | Y | Y | Y |

## 5　EXECUTION AND SIMULATION

This section performs this scheme in a simulation system. For the pairing-based authentication scheme, we use the weil pairing, defined over the super singular elliptic curve E/Fp : y2=x3+x with embedded degree 2, where q is a 160-bit Solinas prime q=2159+217+1 and p=2160−231−1. We implement the algorithm on the hardware platform which is an Intel Core Pentium (R) Dual-core cpu E5300 processor with 1.99GB memory, 2.6GHz cpu clock speed and a Linux 3.2.0-40-generic–pae operation system.

### 5.1　Cost analysis

Due to the resource constraints of memory cards, the authentication scheme must take efficient evaluation into consideration. In general, the efficient evaluation is usually divided into computational cost and communication cost. The computation cost of each phase is defined as the total time of various operations executed in that phase. The communication cost of authentication includes the cost of transmitting messages involved in the authentication scheme.

#### 5.1.1　Computational cost

User registration only carries out in initialization step or in re-registration step or in need of timing registration of the system. User does not need to register every time when using. So the time of registration does not affect the operation of the system and the system is not limited by the computing power of the card. But this paper still respectively gives the computation cost of the server when registration and the computation cost of the card when login.

The system parameters are selected as:

Length of group key x: 160bits, Length of ID: 13 character strings, Length of password: 13 character strings.

The running time is listed in table 2.

Table 2.　Authentication execution time (in milliseconds).

| User registration | User login | User password update | Total time |
|-------------------|------------|----------------------|------------|
| 59.146 | 59.786 | 29.573 | 148.505 |

As can be seen from the table, from the establishment of a new system to the first time to authenticate, the maximum computation time is 148.505ms. And then if no key updates, the authentication time is only 59.786ms.

#### 5.1.2　Communication cost

This scheme, as a mobile cards authentication scheme, has only one message interaction during the authentication process, which is the input of user ID and password. So the communication cost is only the cost of message transmission from user to the card, 104bits. This part of cost is so small that the system can tolerate.

Communication cost of the system is very advantageous compared with the several schemes in section2, as shown in table 3(UTS: From user to server, UTU: From server to user, UTR: From user to reader, TC: Total communication cost).

Table 3.　Comparison of communication cost (number of bits).

| Scheme | UTS | UTU | UTR | TC |
|--------|-----|-----|-----|-----|
| Das[13] | 448 | -- | 52 | 500 |
| Liao[14] | 448 | 192 | 52 | 692 |
| Yoon[15] | 448 | 192 | 52 | 692 |
| Liou[16] | 320 | 192 | 52 | 564 |
| Wang[17] | 448 | 192 | 104 | 744 |
| Khan[18] | 384 | 192 | 104 | 680 |
| My scheme | -- | -- | 104 | 104 |

## 5.2 Memory capacity

The memory capacity of the system includes server memory capacity and the card memory capacity.

The server stores the user $ID_i$, the card $ID_{card}$ and the group key X. Assuming that the $ID_{card}$ takes 80bits, $ID_i$ takes 13 character strings, that is 52bits, with 1000 users as a group. The group key takes about 10kbits, so 100,000 users need 12M memories.

One mobile card needs about 2kbits to store $Y_1, Y_2, J, L, T$ and $ID_{card}$.

Practically, the capacity of a mobile memory card can be up to several G bits, and it is normal for a server to have several hundred G of memories. So the above discussion of the hardware memory capacity is not a problem.

## 6 CONCLUSIONS

In this paper, the survey of some ID-based authentication schemes has been done. An ID-based user off-line password authentication scheme for mobile memory cards is proposed. The property and security are given. The simulation is performed, and it is confirmed from computational cost, communication cost and memory cost that the scheme is economic and practical.

## REFERENCES

[1] Yang X D, Wang C F. On-ling/Off-ling threshold proxy re-signatures. Chinese Journal of Electronics 2014 Vol.23:248–253.

[2] Lee S W, Kim H S, Yoo.KY. Improved efficient remote user authentication scheme using smart cards. IEEETransactions on Consumer Electronics2004;46(2):565–7.

[3] Ku WC, Chen SM. Weaknesses and improvements of an efficient password based remote user authentication scheme using smart cards. IEEETransactionson Consumer Electronics 2004; 50(1): 204–7.

[4] Yoon EJ, Ryu EK, Yoo KY. An improvemet of Hwang-Lee-Tang's simple remote user authentication schemes. Computers and Security2005;24:50–6.

[5] Wang XM, ZhangWF, ZhangJS, KhanMK. Cryptanalysis and improvement on two efficient remote user authentication scheme using smartcards. Computer Standards and Interfaces 2007; 29:507–12.

[6] Yang G, WongDS, WangH, DengX. Two-factor mutual authentication based on smart cards and passwords. Journal of Computer and System Sciences 2008;74(7):1160–72.

[7] Jongho Mun, Qiuyan Jin, Woongryul Jeon, Dongho Won. An improvement of secure remote user authentication scheme using smart cards.IT Convergence and Security ,2013 , Page(s): 1–4.

[8] Hsiang HC, ShihWK, Yoon-Ryu-Yoo. Weaknesses and improvements of the remote user authentication scheme using smartcards. Computer Communications 2009; 32:649–52.

[9] Homg WB, Lee CP. Security Weaknesses of Song's Advanced Smart Card Based Password Authentication Protocol.2010.9 pp.477–480.

[10] Sida Lin, Qi Xie. A secure and efficient mutual authentication protocol using hash function. 2009 International Conference on Communications and Mobile Computing.pp.545–548.

[11] Leinonen A.P. Tuikka T. Siira, E. Implementing Open Authentication for Web Services with a Secure Memory Card. Fourth International Workshop with Focus on Near Field Communication. 2012 IEEE, DOI 10.1109/NFC.2012.15.31.pp.31–35.

[12] Madhusudhan R., Mittal R.C. Dynamic ID-based remote user password authentication schemes using smart cards. Journal of network and computer applications 35(2012)1235–1248.

[13] Das ML, SaxenaA, A dynamic ID-based remote user authentication scheme. IEEE Transactions on Consumer Electronics 2004; 50(2): 629–31.

[14] Liao I,LeeCC, HwangMS. Security enhancement for a dynamic ID-based remote user authentication scheme. Proceedings of the International Conference on Next Generation Web Services Practices, NWeSP'05,Seoul,Korea, 2005,p.437–40.

[15] Yoon EJ and Yoo KY. Improving the Dynamic ID-Based Remote Mutual Authentication Scheme. Proc.OTM Workshops 2006,LNCS4277, p.499–507.

[16] Liou YP,LinJandWangSS. A New Dynamic ID-Based Remote User Authentication Scheme using SmartCards. Proceedings of 16th Information Security Conference 2006, Taiwan, p.198–205.

[17] Wang YY, KiuJY, XiaoFX, DanJ. A more efficient and secure dynamic ID-based remote user authentication scheme. Computer Communications 2009; 32(4):583–5.

[18] Khan MK, KimSK. Cryptanalysis and security enhancement of a more efficient&secure dynamic ID-based remote user authentication scheme. Computer Communications 2011; 34:305–9.

# Suppression of Es layer clutter in HFSWR using a vertically polarized antenna array

L. Zhao
*Civil Aviation Institute of Shenyang Aerospace University, Shenyang, China*

Y.Z. Liu
*Air Traffic Control Center of Northwestern Air Traffic Management Bureau of CAAC, Xian, China*

H.K. Liu
*Space Star Technology Co., Ltd, Beijing, China*

ABSTRACT: There has been Es layer clutter in the high frequency surface wave radar, and the Es layer clutter has the different arrival direction from the signal. So a vertically polarized antenna array whose mainlobe is put on the reverse direction of the signal arrival direction and whose low sidelobe is put on the direction of the signal arrival direction, can be added as auxiliary antennas in the high frequency surface wave radar system to suppress the Es layer clutter. The clutter-to-signal ratio in the outputs of the auxiliary antennas will affect the suppression of the Es layer clutter in the main antennas. The bigger of the clutter-to-signal ratio in the outputs of the auxiliary antennas, the better suppression result of the Es layer clutter in the main antennas can be achieved. Using the superdirective synthetically method, the auxiliary vertically polarized antenna array will have a lower gain on signal than the main antenna when it can maintain the same gain on Es layer clutter as the main antenna. So the clutter-to-signal ratio in the outputs of the auxiliary vertically polarized antenna array will increase, the suppression of the Es layer clutter in the main antennas which use the vertically polarized antenna array as auxiliary antenna will be more effective.

## 1 INTRODUCTION

Besides the regular D layer, E layer, F1 layer and F2 layer, ionosphere also has sporadic E layer (Es layer). Es layer is the ionized clouds with all kinds of shapes that appear in the altitude of E layer, and the peak values of the Es cloud electron density are much higher than the peak value of the E layer electron density around. The Es layer clutter (ELC) discussed in this Letter is the transmitted signal which is directly reflected back by the Es layer. ELC is a self-induced interference.

The HFSWR signal is vertically polarized. The ELC, on the other hand, could be elliptically polarized or partially polarized. In theory, it is possible to exploit this difference in polarization characteristics to suppress the ELC. Horizontally polarized antennas (HPA) can be added as auxiliary antennas in a high frequency surface wave radar (HFSWR) system that uses vertically polarized antennas (VPA). The horizontally polarized components received by the HPAs can be used to estimate the ELC component received by the VPAs (main antennas). A subtraction of this estimate from the outputs of the VPAs can then result in a suppression of the ELC in the HFSWR.

The clutter-to-signal ratio (GSR) in the outputs of the auxiliary antennas will affect the suppression of the ELC in the main antennas. The bigger of the clutter-to-signal ratio in the outputs of the auxiliary antennas, the better suppression result of the ELC in the main antennas can be achieved. In order to increase the clutter-to-signal ratio in the outputs of the auxiliary antennas, an antenna array consisted of three horizontally polarized antennas is proposed in this paper.

## 2 STRUCTURE OF THE AUXILIARY ANTENNA ARRAY

The auxiliary antenna array has the same structure with the main antenna, which is a broadside linear array composed of four vertically polarized dipoles. The auxiliary antenna arrays use the same vertically polarized dipole as the main antennas, and adopt the same placement mode of the dipoles as the main antennas. So the ELC received by the auxiliary antenna has the best correlation with the ELC received by the main antenna.

The parameters of the auxiliary antenna array are given as follow,

Type of the antenna element: dipole antenna;

Number of the vertically polarized dipoles: 4;

Spacing of the vertically polarized dipoles: 10 meter;

Length of the vertically polarized dipole: 6 meter;

Work frequency band of the antenna: 4-12MHz.

Using the superdirective synthetically method under the sidelobe restriction given by [1], the weights of every vertically polarized dipoles are calculated. The mainlobe of the auxiliary antenna array is put on the reverse direction of the signal arrival direction, and a low sidelobe is put on the direction of the signal arrival direction (for a HFSWR, it is around horizontal direction). Then the auxiliary VPAs will have a lower gain on signal than the main antenna, when it can maintain the same gain on Es layer clutter as the main antenna, so the clutter-to-signal ratio in the outputs of the auxiliary antenna array will increase.

## 3 PATTERNS OF THE AUXILIARY VERTICALLY POLARIZED ANTENNA ARRAY

Using the iterative method given by [1], the weights of every vertically polarized dipole are calculated, when mainlobe of the auxiliary antenna array is put on the reverse direction of the signal arrival direction, and a low sidelobe is put on the direction of the signal arrival direction. Using these weights, the pattern of the auxiliary vertically polarized antenna array can be achieved.

The three-dimension patterns of the auxiliary VPAs and the main VPAs are shown in figure 1 and figure 2, when the work frequency is 5MHz. From the figure 1 and figure 2, it can be seen that the auxiliary VPAs has a much lower sidelobe than that of the main VPAs on the horizontal direction which is the

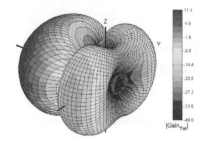

Figure 2. Three-dimension pattern of the auxiliary VPAs.

signal arrival direction in a HFSWR. And the gain of the auxiliary VPAs is only a little lower than that of the main VPAs on the vertical direction which is the ELC arrival direction in a HFSWR. It means that the clutter-to-signal ratio in the auxiliary VPAs will be much bigger than that in the main VPAs.

For further illumination, the vertical plane pattern and horizontal plane pattern of the auxiliary VPAs are compared with the patterns of the main VPAs. The horizontal plane pattern of the auxiliary VPAs is shown in figure 3. The azimuth angle $0^0$ in figure 3 is corresponding to the normal direction of the main VPAs. The solid line in figure 3 is the horizontal plane pattern of the auxiliary VPAs, and the dash line is the horizontal plane pattern of the main VPAs.

From figure 3, it can be seen that the gain on signal of the auxiliary VPAs is bout 45dB lower than that of the main VPAs.

The vertical plane patterns of the auxiliary VPAs are shown in figure 4. Figure 4a shows the vertical plane pattern for angle $0^0$, and Figure 4b shows the vertical plane pattern for angle $90^0$. In figure 4, azimuth angle $0^0$ represents the vertical direction. The solid lines in figure 4 are the vertical plane patterns of auxiliary VPAs, and the dash line are the vertical plane patterns of the main VPAs.

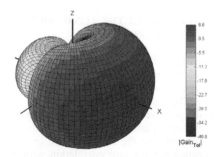

Figure 1. Three-dimension pattern of the main VPAs.

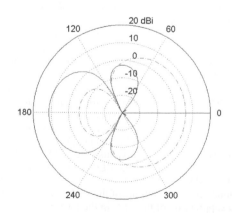

Figure 3. Horizontal plane pattern of the auxiliary VPAs.

From figure 4, it can be seen that the gain on ELC of the auxiliary VPAs is bout 5dB lower than that of the main VPAs.

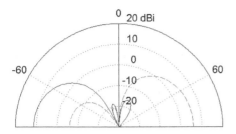

(a) Vertical plane pattern for angle $0^0$.

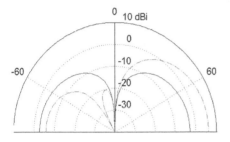

(b) Vertical plane pattern for angle 900.

Figure 4. Vertical plane patterns of the auxiliary VPAs.

## 4 ANALYZE OF THE CLUTTER- TO-SIGNAL RATIO

The clutter-to-signal ratio in the outputs of the auxiliary VPAs and main VPAs are analyzed for the same work condition. It means that the signal and ELC received by the auxiliary VPAs and main VPAs are the same. It is assumed that the magnitude of the signal is A dB and the magnitude of the ELC is B dB, the gain on signal of the auxiliary VPAs is $Gain_s$ dB, and the gain on ELC of the auxiliary VPAs is $Gain_c$ dB. Then the clutter-to-signal ratio in the outputs of the auxiliary VPAs is given by

$$CSR_{ARRAY} = B + Gain_c - A - Gain_s \qquad (1)$$

As the same, it is assumed that the gain on signal of the main VPAs is $G'ain_s$ dB, and the gain on ELC of the main VPAs is $G'ain_c$ dB. Then the clutter-to-signal ratio in the outputs of the main VPAs is given by

$$CSR_{ANTENNA} = B + G'ain_c - A - G'ain_s \qquad (2)$$

So the improvement of the clutter-to-signal ratio in the outputs of the main VPAs is given by

$$CSR_{IMPROVE} = Gain_c - Gain_s - (G'ain_c - G'ain_s) \qquad (3)$$

From the analyses above, it can be known that the gain on signal of the auxiliary VPAs ($Gain_s$) is nearly 45dB lower than that of the main VPAs ($G'ain_s$), but the gain on ELC of the auxiliary VPAs ($Gain_c$) is only 5dB lower than that of the main VPAs ($G'ain_c$). According to the formula (3), the improvement of the clutter-to-signal ratio in the outputs of the auxiliary VPAs is nearly 40 dB.

## 5 SIMULATION ANALYSIS

According to the results given in the last section, the Es layer clutter is simulated using the model given in literature 8.The same parameters are used in the simulation of Es layer clutter received by the main antenna and auxiliary antenna. Considering the different types of the main antenna and auxiliary antenna used in the actual system, a fixed phase difference, a fixed amplitude attenuation and a small random phase fluctuation are added in the simulation of the Es layer clutter received by the auxiliary antenna compared with the Es layer clutter received by the main antenna. At the same time, considering the different amplitudes of the target signals received by the main antenna and auxiliary antenna, the target signal added in the reception of the main antenna is 40 dB bigger than the target signal added in the reception of the auxiliary antenna. The spectrums of the simulation Es layer clutter in the main antenna and auxiliary antenna are shown in figure 5.

The Es layer clutter received by the main antenna was suppressed using the adaptive cancellation algorithm given in the literature 9. The cancellation results is shown in figure 6.

From figure 6, it can be seen that the Es layer clutter amplitude after cancellation has been significantly reduced, and the simulation target added appeared obviously.

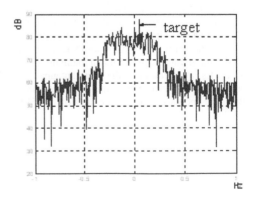

(a) Main Antenna.

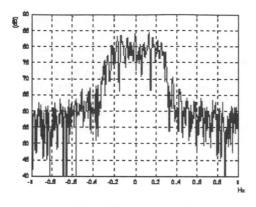

(b)Auxiliary Antenna.

Figure 5. Spectrums of the main channel and auxiliary channel.

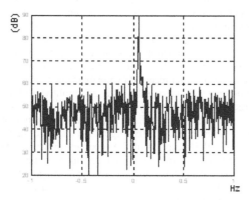

Figure 6. Spectrums after adaptive cancellation.

## 6 CONCLUSION

Using the superdirective synthetically method, the mainlobe of the auxiliary antenna array is put on the reverse direction of the signal arrival direction, and a low sidelobe is put on the direction of the signal arrival direction. Then the auxiliary vertically polarized antenna array will have a lower gain on signal than the main antenna when it can maintain the same gain on Es layer clutter as the main antenna. So the clutter-to-signal ratio in the outputs of the auxiliary VPAs will increase. The suppression of the ELC in the main antennas which uses the vertically polarized antenna array as auxiliary antenna will be more effective.

ACKNOWLEDGMENT

This work was financially supported by "the National Natural Science Foundation of China (60939002 , 61151002, U1433115)", "the Domestic civil aircraft project of Ministry of Industry and Information Technology (CXY2012SH16)" , "the Open Research Fund of The Academy of Satellite Application under grant NO. 2014_CXJJ-TX_12".

REFERENCES

[1] Yuan Liu, Weibo Deng and Rongqing Xu. A Study on the Practical Application of Superdirective Array at HF. Radar 2004.

[2] X.R.Wan, F.Cheng and H.Y.Ke. Sporadic-E Ionospheric Clutter Suppression in HF Surface-Wave Radar. IEEE International Radar Conference. 2005:742–746.

[3] X.R.Wan, H.Y.Ke and B.Y.Wen. Adaptive Cochannel Interference Suppression Based on Subarrays for HFSWR. Signal Processing Letters, IEEE. 2005, 12:162–165.

[4] H.T.Gao, X.Zheng and J.Li. Adaptive Anti- Interference Technique using Subarrays in HF Surface Wave Radar. Radar, Sonar and Navigation, IEE Proceedings. 2004, 151:100–104.

[5] A.H. Nuttall, B.A. Cray. Approximations to Directivity for Linear, Planar, and Volumetric Apertures and Arrays. IEEE Journal of Oceanic and Engineering, 2001, 26 (3): 383–398.

[6] W.L. Stutzman. Estimating directivity and gain of antennas. IEEE Antennas and Propagation Magazine. 1998, 40(4): 7–11.

[7] M.J.Lee, L.Song, S.Yoon and S.R.Park. Evaluation of directivity for planar antenna arrays. IEEE Antennas and Propagation Magazine. 2000, 42 (3): 64–67.

[8] Zhao Long. A Model for the Ionospheric Clutter in HFSWR Radar. ICIII 2008, p:179–182.

[9] Zhao Long, Zhang Ning. Adaptive Cancellation of Es Layer Interference using Auxiliary Horizontal Antenna. Journal of Systems Engineering and Electronics, 2006(6) : 313–315.

[10] Hank W.H.Leong. Adaptive Suppression of Skywave Interference in HF Surface Wave Radar Using Auxiliary Horizontal Dipole Antennas. Defence Research Establishment Ottawa, Ganada, 1999:128–132.

*Multimedia, Communication and Computing Application – Leung (Ed.)*
*© 2015 Taylor & Francis Group, London, ISBN 978-1-138-02775-6*

# The influence of angle parameters on ultraviolet communication system in the atmospheric turbulence

Y. Wang

*College of Information & Communication Engineering, Harbin Engineering University, Harbin, China*

ABSTRACT: Under the atmospheric turbulence, the relationship between the BER performance of ultraviolet communication system and angle parameters of system is studied. The results show that the BER performance is deteriorated with the increase of the transmitter apex angle and the receiver apex angle. The effect of the transmitter beam divergence on the BER performance is very weak. In addition, The BER performance can be greatly improved when the receiver field view is increased.

## 1 INTRODUCTION

Ultraviolet communication is a kind of optical wireless communication, which carries on the information transmission through ultraviolet scattering. The atmospheric ozone has a strong absorption to ultraviolet, which leads to the result that the wavelength of 200 to 280nm UV light in the near-Earth atmosphere is almost non-existent, known as the ultraviolet solar blind[1]. So the background noise by using this band of ultraviolet to communicate can be regarded as zero. By the virtue of its good security, strong anti-interference ability and non-line-of-sight communication, the ultraviolet communication becomes one of most widely applied wireless technologies. Currently, few studies have been done on how systems' angle parameters affect the performance of non-line-of-sight (NLOS) ultraviolet communication system in atmospheric turbulence. Hence, the NLOS ultraviolet communication system with different angle parameters is simulated and analyzed in this paper.

## 2 CHANNEL MODEL

The transmission of ultraviolet in atmosphere is a very complex process, including scattering, absorption and atmospheric turbulence. For the NLOS ultraviolet communication system, the transmitter (Tx) and receiver (Rx) are arranged as shown in Figure 1. $\theta_1$ and $\theta_2$ are the Tx and Rx apex angle, and $\theta_s = \theta_1 + \theta_2$ is the scattering angle. $\phi_1$ and $\phi_2$ are the transmitter beam angle and the receiver field of view. $r$ is the baseline distance between the Tx and Rx. The distances of scatter $V$ to Tx and Rx are respectively $r_1$ and $r_2$.

In order to describe the ultraviolet atmospheric channel, the following parameters can be used: Rayleigh scattering coefficient $k_s^{Ray}$, Mie scattering coefficient $k_s^{Mie}$, the absorption coefficient $k_\alpha$ and the extinction coefficient $k_e$. The total scattering coefficient is the sum of Rayleigh scattering coefficient and Mie scattering coefficient $k_s = k_s^{Ray} + k_s^{Mie}$. The extinction coefficient is the sum of the scattering

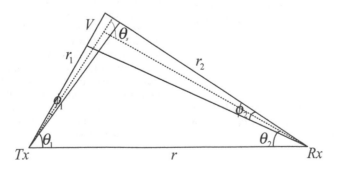

Figure 1. The NLOS ultraviolet communication link.

coefficient and the absorption coefficient $k_e = k_s + k_a$. The total path loss expression of the ultraviolet communication channel is given by[2]

$$L = \frac{96 r \sin\theta_1 \sin^2\theta_2 \left(1 - \cos\frac{\phi_1}{2}\right) \exp\left[\frac{k_e r (\sin\theta_1 + \sin\theta_2)}{\sin\theta_s}\right]}{k_s P(\mu) A_r \phi_1^2 \phi_2 \sin\theta_s \left(12\sin^2\theta_2 + \phi_2^2 \sin^2\theta_1\right)} \quad (1)$$

where the single scatter phase function is $P(\mu)$.

In the event of weak turbulence, one can accurately model the intensity of the optical field using a random variable whose probability density function is that of a log-normal random variable. Under the weak atmospheric turbulence, probability density function of the received signal optical intensity is given by[3]

$$f(I) = \frac{1}{\sqrt{2\pi}\sigma_i^2 I} \exp\left[-\frac{(\ln I - m_i)^2}{2\sigma_i^2}\right], \quad I \geq 0 \quad (2)$$

where $\sigma_i^2$ and $m_i$ is the mean and variance of $\ln I$ respectively. Although the lognormal distribution can only describe the weak turbulence model, when the receiver aperture is large enough, lognormal model can also describe the strong turbulence.

In the NLOS ultraviolet communication link with atmospheric turbulence, the optical signal will experience optical intensity scintillation caused by atmospheric turbulence. Thereby, the SNR of the receiver is defined as[4,5]

$$\langle SNR \rangle = \frac{SNR_0}{\sqrt{\left(\frac{P_{ro}}{\langle P_r \rangle}\right) + \sigma_I^2 SNR_0}} \quad (3)$$

where $P_{ro}$ is the received power without turbulence and $\langle P_r \rangle$ is the received average power in the atmospheric turbulence. $SNR_0$ is the received SNR without turbulence. Its expression is

$$SNR_0 = \sqrt{\frac{P_{r,NLOS}}{2Rhc/\lambda}} \quad (4)$$

where $R$ is the data rate, $c$ is the speed of light, $h$ is the Planck constant and $P_{r,NLOS}$ is the optical power received by the receiver.

In this paper, BPSK subcarrier intensity modulation is adopted, and the BER of the ultraviolet communication system in the atmospheric turbulence is given by[6]

$$BER_{NLOS} = \int_0^\infty Q\left(\sqrt{2}\langle SNR \rangle I\right) f(I) dI \quad (5)$$

## 3 SYSTEM SIMULATIONS

The system parameters are set as: $\lambda = 265 nm$, $K_a = 0.9$, $K_s = 0.49$, $R = 10 kbps$, $\phi_1 = 8^0$, $\phi_2 = 20^0$ and $r = 200m$. The quantum efficiency of the optical detector is $\eta = 0.8$, the detector area is $A_r = 1.77 cm^2$, the average transmitted optical power is $P_t = 1mW$, the weak turbulence is $C_n^2 = 1 \times 10^{-16} m^{-2/3}$, and the moderate turbulence is $C_n^2 = 1 \times 10^{-14} m^{-2/3}$. Under two different strength turbulence, the BER performance of ultraviolet communication system with different Tx apex angle and Rx apex angle is shown in Figure 2. It can be clearly found that the effect of turbulence strength on the system BER is obvious. When the turbulence strength changes from weak to moderate and $\theta_1 = 20^0$, the BER performance of ultraviolet communication system drops nearly three orders of magnitude. Moreover, regardless of which kind of atmospheric turbulence strength, the system BER performance will be deteriorated when the Tx apex angle or Rx apex angle is increased. For example, the BER performance of ultraviolet communication system drops nearly four orders of magnitude when the Tx apex angle changes from $20^0$ to $80^0$. Consequently, to effectively inhibit the effect of atmospheric turbulence on ultraviolet communication system, the Tx apex angle and Rx apex angle will be reduced.

Under the weak turbulence, the BER performance of ultraviolet communication system with different Tx beam divergence and Rx field view is shown in Figure 3. The BER performance of ultraviolet communication system remains nearly unchanged when the Tx beam divergence varies from $10^0$ to $60^0$. On the contrary, the increase of Rx field view can greatly improve the BER performance of ultraviolet communication system. The BER performance is improved nearly two orders of magnitude when the Rx field view changes from $10^0$ to $60^0$. As can be seen from the above analysis, the Rx field view should be increased to reduce the effect of atmospheric turbulence on the BER performance of ultraviolet communication system.

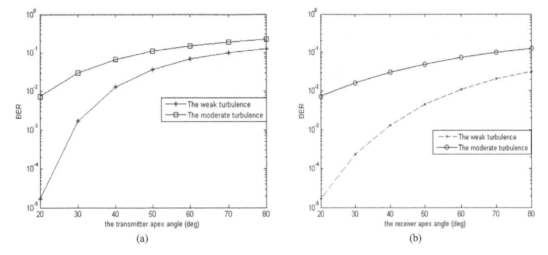

Figure 2.    The BER versus the Tx apex angle and Rx apex angle.

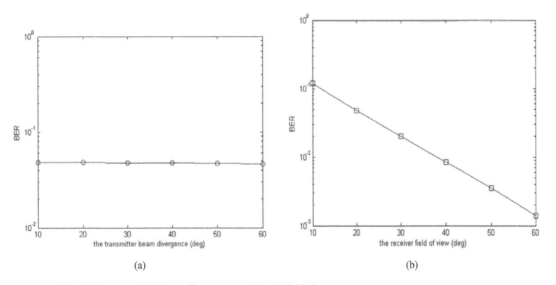

Figure 3.    The BER versus the Tx beam divergence and the Rx field view.

## 4    CONCLUSIONS

Based on the theoretical single scatting model, NLOS ultraviolet channel model is analyzed. Moreover, the atmospheric turbulence model is studied. The BER performance of the ultraviolet communication system with different angle parameters is simulated in the atmospheric turbulence. It can be seen from simulation results that the BER performance is improved with the decrease of the transmitter apex angle and the receiver apex angle. The BER performance almost never changes with the transmitter beam divergence. The increase of receiver field view can effectively suppress the influence of atmospheric turbulence on the BER performance of ultraviolet communication system.

ACKNOWLEDGMENTS

This work was financially supported by the Natural Science Foundation of China (61275082).

# REFERENCES

[1] Xu.Z & Sadler B.M.2008.Ultraviolet Communications: Potential and State-of-the-Art. *IEEE Communications Magazine* 46:67–73.

[2] Xu.Z, Ding.H.P & Sadler B.M.2008. Analytical performance study of solar blind non-line-of-sight ultraviolet short-range communication links. *OPTICS LETTERS* 33(16):1860–1862.

[3] Kiasaleh.K.2005.Performance of APD-Based, PPM free-space optical communication systems in atmospheric turbulence.*IEEE transactions on communications* 53(9) : 1455–1461.

[4] Andrews.L.C & Phillips.2005.*Laser beam propagation through random media*. SPIE Press Monograph.

[5] Ding.H.P, Chen G & Majumdar A. K.2011. Turbulence modeling for non-line-of-sight ultraviolet scattering channels. Proc. Of SPIE 8038:1–8.

[6] Gagliardi.R.M. & Karp. S.1995. *Optical Communications*, New York :John Wiley &Sons.

# A generic size neural network based on FPGA

E. Guzman-Ramírez, M.P. García & J.L. Barahona
*Electronics and Mechatronics Institute, Technological University of the Mixteca, México*

O. Pogrebnyak
*Computing Research Center - IPN, México*

ABSTRACT: In this paper we propose a generic size hardware implementation of a multi-layer perceptron artificial neural network (MLP-ANN) for general purpose applications. Our proposal includes the modeling of a set of components that can be easily interconnected to form and adapt the ANN to diverse applications. With the aim of determining the performance of our proposal, it is used to solve the XOR problem; this problem is frequently used to measure the performance of an ANN. The implementations have been developed and tested into reconfigurable hardware (FPGA). The results of these implementations have been compared with PC-based implementations.

## 1 INTRODUCTION

Respected researchers in the area of artificial neural networks (ANN) concur that an ANN has a high degree of concurrency in the different elements that comprise it. In this regard, according to Hecht-Nielsen (1990) and Freman-Skapura (1993), "an ANN is a structure of information processing, which elements are distributed, work in parallel and are interconnected in the form of directed graph".

Moreover, the structure of the field programmable gate array (FPGA) device allows exploiting the implicit concurrency in an algorithm. According to Girau (2006), compared to other technologies such as general purpose parallel computers, dedicated parallel computers and ASICs, the FPGA is the most balanced option for the implementation of ANN.

This is one of the main reasons for the increasing use of FPGAs in implementing applications that require using ANN. Below we mention some works related with described in this document.

Research by Sahin et al. (2006) shows how to efficiently use 32 bit floating-point numeric representation in FPGA-based ANN. As a part of this investigation, a VHDL library was designed for implementing ANN on FPGAs. The resultant neural networks are modular, compact, and efficient and the number of neurons, number of hidden layers and number of inputs are easily changed to adapt to applications.

In (Ortigosa et al. 2006), authors propose several hardware implementations of a multi-layer perceptron (MLP) for speech recognition applications. The hardware architectures proposed were developed in two abstraction levels: a register transfer level (RTL) and a high description level based on Handel-C.

The implementations have been developed and tested into reconfigurable hardware (FPGA) for embedded systems. The research is completed with the study of different implementation versions with diverse degrees of parallelism.

Dinu et al. (2010) present an algorithm for compact neural network hardware implementation. In order to simplify the implementation of the neuron, this proposal describes the operation of an artificial neuron, with activation step function, in terms of Boolean logic, taking advantage of the implicit properties of an FPGA. To achieve this, the mathematical model of the neuron is digitized and expressed by an optimized gate logic structure where the redundancy logic is eliminated. The modeling the neuron is performed using VHDL. This modeling is portable which allows easy reuse in various FPGA technologies.

Recently, Nedjah et al. (2012) presented the hardware architecture of an ANN with an MLP structure that exploits the inherent parallelism in the ANN. The proposed architecture is featured by allowing changing the number of inputs to the network, the number of layers that compose it and the number of neurons per layer. It is enabled by this feature of the proposed architecture that a large number of applications related to ANN that can be implemented. With the aim of optimizing the proposed architecture, the real numbers are represented using fractions of integers, thus arithmetic operations are performed with integer values and modeled easily with combinational circuits.

## 2 MLP

The multi-layer perceptron (MLP) is one of the most widely used ANNs. An MLP consists of a set of simple neurons, called perceptron, grouped in layers, with each layer fully connected to the next one and are able to adapt to different applications to respond dynamically to external stimuli. The basic concept of a single perceptron was introduced by Rosenblatt (1958). The MLP's neuron is a modification of the perceptron. The main differences between them are the activation function and the learning rule. The MLP learning rule is called back-propagation (Werbos, P. J. Beyond 1974, Rumelhart et al. 1986). Further, Hornik et al. (1989) and Funahashi (1989) showed that while single-layer networks composed of parallel perceptrons are rather limited by what kind of mappings they can represent, an MLP is capable of approximating any continuous function $f : \Re^n \rightarrow \Re^m$ to any given accuracy, provided that sufficiently hidden units are available. The system to which the ANN will be adapted is defined from a set of associations, where each association consists of an input vector and its corresponding output vector. The set of $p$ associations is defined as

$$\left\{ \left( \mathbf{x}^1, \mathbf{y}^1 \right), \left( \mathbf{x}^2, \mathbf{y}^2 \right), ..., \left( \mathbf{x}^p, \mathbf{y}^p \right) \right\} = \left\{ \left( \mathbf{x}^\mu, \mathbf{y}^\mu \right) \right. $$
$$\left. | \, \mu = 1, \, 2, \, ..., p \right\} \quad (1)$$

where, $\mathbf{x}^\mu = \left[ x_j^\mu \right]$ and $\mathbf{y}^\mu = \left[ y_i^\mu \right]_m$.

A graphic representation of MLP with a single hidden layer is shown in Figure 1.

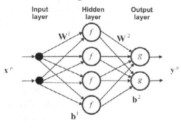

Input layer    Hidden layer    Output layer

Figure 1.    MLP with a single hidden layer.

The operation by such a feed-forward network can be defined mathematically as

$$\mathbf{y}^\mu = f \left( \mathbf{b}^2 + \mathbf{W}^2 \left( g \left( \mathbf{b}^1 + \mathbf{W}^1 \mathbf{x}^\mu \right) \right) \right) \quad (2)$$

where $\mathbf{b}^1$ and $\mathbf{b}^2$ are the bias vectors of hidden and output layers respectively; $\mathbf{W}^1$ and $\mathbf{W}^2$, are the weight matrices of hidden and output layers respectively; $f$ and $g$ are the activation functions of hidden and output layers respectively.

## 3    MODELING AND HARDWARE IMPLEMENTATION OF NEURON

An artificial neuron is a simple calculating device based on mathematical models of the biological neuron behavior. Neuron function is generated by a specific output from an input vector from the outside or other neurons.

The basic structure of a neuron, shown in Figure 2, is composed of the following elements (Widrow & Lehr 1990):

- Synaptic weight, $w_{ij}$. It refers to the strength of a connection between two neurons, and it defines the intensity of interaction between a presynaptic neuron $j$ and a postsynaptic neuron $i$.
- Propagation rule. This element computes the term known as synaptic potential $u_i$. The synaptic potential is defined as the sum of the scalar products of the incoming vector components with their respective weights.

$$u_i = \sum_j w_{ij} x_j \quad (3)$$

- Activation function, $f(u_i)$. The activation function defines the value of the neuron output, $y_i$. A sigmoid activation function is frequently used.

$$y_i = f(u_i) = \frac{1}{1 + e^{-u_i}} \quad (4)$$

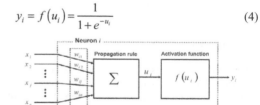

Figure 2.    Basic structure of a neuron.

Before starting the modeling, it is necessary to define the arithmetic, representation and resolution that to be used. In this respect, after an analysis of the kind of operations and information are processed, the fixed-point arithmetic and two's complement representation were chosen for the implementation of neuron. In addition, the resolution of the variables was defined as follows:

- Neuron inputs, $x_j$: 16 bits
- Neuron output, $y_i$: 16 bits
- Synaptic weights, $w_{ij}$: 16 bits
- Synaptic potential, $u_i$: 32 bits

Both neuron input and output must have the same resolution to easily manage multiple-layers processing.

Now, considering the parameters discussed, the basic structure of a neuron (Fig. 2) and the modeling of FPGA-based neuron based on methodology described in (Riesgo et al. 1999) which includes a hierarchical and modular approach was developed.

The hardware architecture of neuron is composed of three components: **synaptic weights** component, **propagation rule** component and **activation function** component (see Fig. 3); the first component stored the synaptic weights, the second component computes the synaptic potential and the third componet determines the state of the neuron (neuron output).

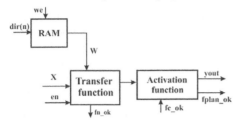

Figure 3. Block diagram of hardware architecture of neuron.

While the synaptic weights typically do not require a large amount of storage resources and the distributed RAM is an excellent option to high-performance applications that require relatively small embedded RAM, the **synaptic weights** component is implemented using this resource type.

On the other hand, the Equation 3 shows that the computation of synaptic potential consists of the summation of its inputs weighted by their associated synaptic weight. The hardware architecture used to compute this parameter is based on a typical multiply-accumulate (MAC) unit; it comprises an 8-bit multiplier and a 23-bit accumulative adder (see Fig. 4).

Figure 4. Propagation rule component, computation of synaptic potential.

Finally, the output of the **propagation rule** component is connected to the **activation function** component. The function of this component is implementing an approximation to the function expressed in Equation 4. A direct implementation of a sigmoid function is rarely used because the operations used, division and exponentiation, demand a lot of resources. Therefore, the implementation of these functions is performed using specific approximation technique. The main approximation techniques used in the hardware implementation of sigmoid functions are: approximation by combinational logic, piecewise linear approximations (PWL) and piecewise second-order approximations (Tisan et al. 2009, Tommiska 2003). In the implementation of our neuron, the PLAN (Piecewise Linear Approximation of a Nonlinear Function) technique, proposed by (Amin et al. 1997), has been used to model the approximation of sigmoid activation function. This is an efficient piecewise linear approximation technique where the multiply/add

operations of the linear interpolation are replaced by a simple gate design, which leads to a very small and fast digital approximation of the sigmoid function.

In order to obtain generic size components, the design of these components was performed by a behavioral description but keeping a low level hardware abstraction.

### 3.1 Integration of ANN

In this section, we show how to form an ANN using the neuron designed in the previous section. Figure 5 shows the integration of three neurons to form an ANN that solves the XOR problem; it is a non-linearly separable problem that is frequently used to measure the performance of an ANN.

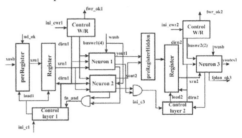

Figure 5. ANN that solves the XOR problem.

This ANN consists of a hidden layer with two neurons and an output layer with one neuron. Further including some components that complement its structure, this new components are described below.
- **Control layer** n. This component represents the control unit associated with the data path of layer "n" and it has been designed as a finite–state machine, implementing the detailed description of all necessary states, their outputs and transitions. This component manages the operation of layer "n".
- **Register**n. This component has a FIFO structure and its function is to present the j-th element of the input vector to all neurons in the layer "n".
- **preRegister**. This component is the complement of **Register** n component; it acts as the input layer of the ANN, and it has these functions: taking values of the elements forming the input vector, reorganizing them and passing them to the next component.
- **preRegisterHidden.** It fulfills the functions performed by the **preRegister** component in the remaining layers of the ANN.

### 4 RESULTS

With the aim of determining the performance of our proposal, it was used to solve the XOR problem and Iris-plant problem; both problems are frequently used to measure the performance of an ANN.

The implementations have been developed and tested into Xilinx Spartan-3E FPGA.

With aid of USB interface implementation between the proposed ANN architectures and personal computer, we can send the values of the synaptic weights which enables the ANN to adapt to the specified problem during the training process and carry out the classification process by sending a pattern to the ANN and displaying the result generated by the ANN.

Figure 6. Results obtained with ANN to perform the logical XOR function. (a) Before training process; (b) after training process; (c) classification process.

The Figure 6 shows the results generated by applying the ANN in solving the XOR problem. In Figure 6a, the virtual hyperplanes obtained from the initial values of the synaptic weights can be seen; these hyperplanes represent the functions modeled by neurons, 1 and 2, of the ANN (Fig. 5), which are defined by the following expressions

$$x_1 w_{11} + x_2 w_{12} + w_{13} = 0 \quad x_2 = -\frac{0.1}{0.5}x_1 - \frac{1}{0.5}$$
$$x_1 w_{21} + x_2 w_{22} + w_{23} = 0 \quad x_2 = -\frac{-0.7}{0.3}x_1 - \frac{1}{0.3} \quad (5)$$

where $w_{13}$ y $w_{23}$ are the bias of neurons 1 and 2 respectively.

The Figure 6b shows the results obtained after the training process. The new hyperplanes are defined by the following expressions

$$x_2 = -\frac{5.49}{-4.33}x_1 - \frac{3.29}{-4.33}$$
$$x_2 = -\frac{-5.56}{4.06}x_1 - \frac{2.25}{4.06} \quad (6)$$

Finally, Figure 6c shows the classification process where some vectors not belonging to the training set are included.

## 5 CONCLUSIONS

In this paper we propose a generic size hardware implementation of an MLP-ANN which can be easily adapted to various applications. To test the flexibility of our proposal, ANN structure was built to solve a well-known problem used frequently to measure the performance of an ANN, the XOR problem using fixed-point arithmetic 16-bit. Our proposal was compared with a PC-based similar ANN structure, modeling and implementation using C++ programming language and double-precision floating-point arithmetic, which

shows that the results of the FPGA-based ANN do not differ from the PC-based ANN results.

## REFERENCES

Amin, H., Curtis, K. M. & Hayes-Gill, B. R. 1997. Piecewise linear approximation applied to nonlinear function of a neural network. *IEE Proceedings – Circuits, Devices and Systems*, 144(6): 313–317.

Dinu A., Cirstea M. N., & Cirstea S. E. 2010. Direct Neural-Network Hardware-Implementation Algorithm. *IEEE Transactions on Industrial Electronics*, 57(5): 1845–1848.

Freeman, J. A. & Skapura, D. M. 1993. *Redes Neuronales. Algoritmos, aplicaciones y técnicas de propagación.* Addison-Wesley.

Funahashi, Ken-ichi. 1989. On the approximate realization of continuous mappings by neural networks. *Neural Networks*, 2(3): 183–192.

Girau, B. 2006. FPNA: concepts and properties (Chapter 3). *FPGA Implementations of Neural Networks*, ed. Springer-Verlag: Berlin and New York.

Hecht-Nielsen, R. 1990. *Neurocomputing.* Addison-Wesley.

Hornik, K., Stinchcombe, M. & White H. 1989. Multilayer feedforward networks are universal approximators. *Neural Networks*, 2(5): 359–366.

Nedjah, N., Martins da Silva, R. & Mourelle, L. 2012. Compact yet efficient hardware implementation of artificial neural networks with customized topology. *Expert Systems with Applications*, 39(10): 9191–9206.

Ortigosa E.M., Cañas A., Ros E., Ortigosa P.M., Mota, S. & Díaz J. 2006. Hardware description of multi-layer perceptrons with different abstraction levels, *Microprocessors and Microsystem*, 30(7): 435–444.

Riesgo, T., Torroja, Y. & de la Torre, E. 1999. Design methodologies based on hardware description languages. *IEEE Trans. on Industrial Electronics*, 4(1): 3–12.

Rosenblatt, F. 1958. The perceptron: A probabilistic model for information storage and organization in the brain. *Psychological Review*, 65: 386–408.

Rumelhart, D. E., Hinton, G. E. & Williams R. J. 1986. Learning internal representations by error propagation. *Parallel Distributed Process - ing: Explorations in the Microstructure of Cognition*, 1(8): 318–362. Cambridge, MA: MIT Press.

Sahin S., Becerikli Y. & Yazici S. 2006. Neural Network Implementation in Hardware Using FPGAs, *Proc. of the 13th International Conference on Neural Information Processing*, Springer: Hong Kong.

Tisan, A., Oniga, S., Mic, D. & Buchman, A. 2009. Digital implementation of the sigmoid function for FPGA circuits. *Acta Technica Napocensis, Electronics and Telecommunications*. 50(2): 15–20.

Tommiska, M.T. 2003. Efficient digital implementation of the sigmoid function for reprogrammable logic. *IEE Proceedings - Computers and Digital Techniques*, 150(6): 403–411.

Werbos, P. J. 1974. Beyond Regression: New Tools for Prediction and Analysis in the Behavioral Sciences. PhD Thesis, Hardvard Univ. Committee on Applied Mathematics.

Widrow, B. & Lehr, M. 1990. 30 years of adaptive neural networks: Perceptron, Madaline and Backpropagation, *Proc. Of the IEEE*, 78 (9): 1415–1442.

*Multimedia, Communication and Computing Application – Leung (Ed.)*
© *2015 Taylor & Francis Group, London, ISBN 978-1-138-02775-6*

# An improved utilization of solar node based on LEACH protocol

W.L. Bai, W.X. Jiang, W.W. Yang & Y.M. Zhang
*College of Information Engineering, North China University of Technology, Beijing, China*

ABSTRACT: LEACH is one of the hierarchical routing approaches for sensor network. In this article, we make an improvement on the LEACH protocol. This new version of LEACH protocol replaces part of the sensor nodes with solar energy nodes to prolong the lifetime of the network. In addition, the energy algorithm is improved and proposed to adapt to the new protocol. The simulation results show that this new version is more energy efficient and prolongs the lifetime of the network. Two important conclusions are obtained. One is that more solar nodes do not mean the longer lifetime and need to combine with other factors in actual network. Another is that finding the best threshold of the random number is more beneficial than pursuing a merely large number to the life extension. In practical application, these conclusions contribute to the extension of the network life cycle and reduction of cost.

KEYWORDS: Solar node, Energy, LEACH, Cluster-head (CH), Lifetime.

## 1 INTRODUCTION

WSN is a special network consisted of large number of sensor nodes with sensing, computation, and communication capabilities. The sensor nodes are equipped with a radio transceiver or other wireless communication device, a small microcontroller, and an energy source. They can work in inaccessible places, measure the environmental data, and transmit them to the base station. Combining sensor networks and cable networks with an information system has become a new field for information technology. In most WSN applications, the energy source is just a battery, but it also plays a significant role in such applications. However, the sensor nodes are usually limited by the limited energy. Therefore, saving the energy of each node or using a better energy source is an important goal, which must be considered in a protocol for WSN. In particular, the wares of wireless sending and receiving power cannot meet the needs of sensor networks, and it should be noted that the sensors transmitting information will cost more energy than performing calculations. To reduce the energy consumption of wireless sensor networks, and extend the life of the network, a large number of algorithms on Low Energy Adaptive Clustering Hierarchy (LEACH) clustering are presented. Most of the researches focus on selecting cluster stage of the cluster heads, taking energy factors into account. In essence, it improves threshold formula of LEACH algorithm, such as I_LEACH algorithm, Advanced-LEACH algorithm, and LEACH-EN-MH algorithm.

In addition, it is known that node energy consumption is mainly concentrated on the stage of the election of clustering hierarchy in LEACH clustering algorithm. So the node would conserve energy in a maximum extent possibly when an optimal algorithm is used in the process of sending packet. In this article, we propose an improved version of LEACH protocol that replaces some nodes by solar energy node to prolong the life of the node. Some energy nodes rely on solar power and the remaining nodes run on battery; solar-aware routing protocol for wireless sensor networks that preferably routes traffic via nodes is powered by solar energy.

## 2 LEACH PROTOCOL

LEACH is one of the most popular clustering algorithms for WSN. It forms clusters based on the received signal strength and uses the cluster head (CH) nodes as routers. Initially, each node in the network sends its own data, including its energy and position, to base station.

The whole process can be divided into several phases.

### 2.1 Advertisement phase

A node decides to be a CH with a probability p and broadcasts its decision. Each non-CH node determines its cluster by choosing the CH that can be reached using the least communication energy. Here, a random number $r$ between 0 and 1 is chosen by the non-CH node. The node becomes a CH node if the obtained number is less than a threshold value $T(n)$, which is calculated by the following formula.

$$T(n) = \begin{cases} \dfrac{p}{1 - p*(r \bmod \dfrac{1}{p})} & if\ n \in G \\ 0 & Otherwise \end{cases} \quad (1)$$

where $n$ = the number of nodes; $p$ = the desired percentage of CH nodes in the sensor population; $r$ = a random number between 0 and 1 that is selected by a sensor node; G = the set of nodes that were not accepted as CH in the last "$1/p$" round.

Now each nominated CH starts to advertise their own status to the rest of the nodes in the network. The non-cluster-head nodes must keep their receivers on during this phase to hear the advertisements of all the CH nodes.

### 2.2 *Cluster set-up phase*

After receiving this advertisement message the non-CH nodes decide suitable cluster for them. They will choose the CH which sent the message with the largest signal strength heard. This means the election of the CH to whom the minimum amount of transmitted energy is needed for communication. When the non-CH nodes take this decision, they will inform their respective CH by a message using CSMA MAC protocol that they want to be member of the cluster.

### 2.3 *Schedule creation phase*

After receiving all messages from the non-CH nodes, each CH includes them into their respective cluster. For each node, the CH creates TDMA schedule, which indicates that they can transmit data.

### 2.4 *Data transmission phase*

When the TDMA schedule is fixed for each node, then according to the allocated schedule each node can transmit data to its respective CH. The CHnodes must keep its receiver on to receive all the data from the nodes in the cluster. When they receive all the data from the nodes, they perform aggregation mechanism to compress the amount of data, and this data is sent to the base station. After a certain time, a new round begins with the Advertisement Phase.

## 3 IMPROVED LEACH ROUTING ALGORITHM

### 3.1 *Algorithm description*

On the basis of LEACH, the solar wireless sensor network is divided into several clusters, it is implemented in the process of selecting the head of clusters

and reconstructing the clusters cyclically, which is described by "round." Each "round" is divided into two stages: establishing the head of clusters and transmitting data. The solar sensor nodes transfer the collected data to the CH nodes and then CH nodes relay them to the base station. This stable state is keeping until its stable conditions changes, the network re-enter into the stage of establishing a new cluster and reconstructing it in another round.

Improvements of this algorithm are the establishment process of CH nodes. Taking into account the added solar sensor nodes, sensor nodes randomly generated random number formula that has changed. The improved formula is as follows:

Solar energy node must have higher likelihood of being selected as the CH node.

If the node has been elected as CH node, it should be more likely to become CH node in the next $1/p$ rounds.

The first case is easy to implement. If it is a solar energy node, it is multiplied by a larger factor to the right side of the equation. To the second case, when a node is selected as a CH node in $1/p$ rounds, the number of CH nodes should be recomputed, and denoted as cHeads. This number would keep increasing until it is equal to the number of nodes, denoted as numNodes. This is the process of selecting CH node. And at the end, the number, cHeads, should be zeroed out. Then, the next selection process would be started, as described above. According to this improved selection method, the formula of threshold value is shown as follows:

$$T_2(n) = sf(n) * \dfrac{p}{1 - \dfrac{cHeads}{numNodes}} \quad (2)$$

Where $sf(n)$ = a stratification factor. For solar energy node, it is greater than 1, otherwise it is the reciprocal of itself. In one selection process, this formula could be used for all the nodes, except for those that have been regarded as the CH node. And for the CH nodes, the threshold value is zero.

It should be noted that selection of random numbers used here is sRand and each selection random number is not the same. When random number is less than $T$, the node will be elected as CH. When a node is elected as CH, it will start to send broadcast of becoming CH to all other nodes. Then the non-CH nodes determine its cluster by choosing the CH that can be reached using the least communication energy. At the same time informing the CHthat cluster building process has been complete. At last, CH node allocates the time slot of data transmission for its cluster members in a TDMA manner.

### 3.2 *Sensor energy computation*

The energy sensor is composed of three parts: receiving, blending, and pass.

Receive energy formula as follows:

$$bits * 50.0 * NANO \tag{3}$$

Fusion energy formula is as follows:

$$5 * NANO * signals \tag{4}$$

The energy transfer formula is as follows:

$$\begin{aligned} energy = &\ bits * 10 * PICO * distance \\ &+ bits * 50.0 * NANO \end{aligned} \tag{5}$$

Where, distance is the distance between nodes and CH node.

## 4  SIMULATION AND ANALYSIS

We simulated out proposed protocol and analyzed the factors influencing the lifetime of WSN.

System simulation parameters are as follows:

Table 1.  Simulation Parameters.

| Parameter | Value |
|---|---|
| E0 | 0.5J |
| P | 0.05 |
| xMax | 1000 |
| yMax | 1000 |
| Round | 280 |
| Frame | 3 |
| Number of solar nodes | 15 |
| Number of nodes | 30 |

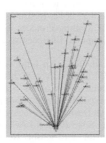

Figure 1.  Node distribution in the simulation.

Table 1 shows that there are 30 nodes and 15 solar nodes. And each node has the initial energy of 0.5 J. Frame is the number of frames that CH node sends to other nodes in the form of TDMA. We put all these nodes into a square area with length 1000 meters in this simulation environment built by OmNet++. As shown in Figure 1.

Using the data obtained from the OmNet++ simulation, the lifetime of WSN based on improved

LEACH is simulated under the following two cases by Matlab.

The first case: vary the number of solar nodes under the same random number range.

The second case: vary the random number range using one solar-powered node.

### 4.1  Vary the number of solar nodes under the same random number range

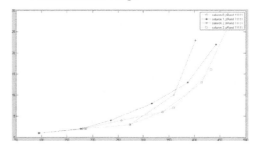

Figure 2.  Open different solar nodes.

This is the first case, we choose the number of the opening solar nodes from 0 to 3, and set the random number range 11111, where x-axis represents the running time of the node and y-axis represents the number of death nodes with the time increasing. The operation result is not very obvious, the entire network is running the longest when opening three solar nodes. The following is the case with no solar node. This shows the shortest path selection is very important. In the case of no node, if the optimal path is selected, to some extent, it will extend the survival time of the entire network. As to opening one or two solar nodes, especially in the case of two solar nodes, the network has the shortest life cycle.

Integrated on, it can be seen that the solar nodes in the network are the more the better. A best number should be determined through more experiments. And it will be very helpful to prolong the life cycle of the network and reduce the cost of production.

### 4.2  Vary the random number range using one solar-powered node

Figure 3 shows the life of the network in the same number of the solar nodes and different random number. When random number is 22, the fifteenth node was dead in 324 seconds. Similarly in case of 171, 1283, and 73664 random number, the fifteenth node was dead in 388, 407, and 436 seconds, respectively. It can be seen within the expansion of the random numbers, the protocol is more improved. On the other hand, when the random number changed from 22 to 171, the death time of the fifteenth node extended 64 seconds later. But changing from 1283 to 73664, it only gave the fifteenth node 29 seconds more of life.

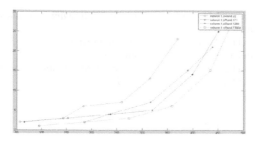

Figure 3.    Using the different random number.

This shows that the gain in life-time will decrease when the random number exceeds a certain value. In the process of building a Wireless Sensor Network, there is no need to deliberately choose too large a random number. And finding the threshold with best income is more beneficial to prolong the lifetime of the network.

5    CONCLUSION

In this work, we considered a well-known wireless sensor network routing protocol called LEACH. And we replaced part of the nodes with the solar nodes and proposed a new version of LEACH protocol with improved energy algorithm. Then the new protocol is successfully simulated and compared with the general LEACH protocol on the OmNet++ software and MATLAB. From the simulation results, we can draw the conclusion that the new protocol is more energy efficient than the LEACH protocol. It gives the network a longer lifetime. In addition, following the data of the simulation, it can be known that merely pursuing more solar nodes or a larger random number does not mean a longer lifetime. According to the specific network, an appropriate number of solar nodes and the threshold of the random number should be determined through more simulation experiments. And this research contributes to the extension of the network life cycle and reduces the cost in the process of production.

ACKNOWLEDGMENTS

This wok was supported by National Natural Science Foundation Project (NO. 61371143), Beijing Municipal Natural Foundation Project (NO. 4132026), Projects in the National Science & Technology Pillar Program (NO. 2012BAH04F00), The Digital Content Standerds and Intelligent Information Processing Platform Project for the education commission of Beijing, China (NO. PXM2013_014212_000120), The Data Acquisition and Visual Processing Platform Project for the education commission of Beijing, China (NO. PXM2014_014212_000017), And the funding from 2014 Graduate Student Innovation Platform Construction Project (NO. 14085-40).

REFERENCES

Braginsky, D. & Estrin, D. 2002, Rumor Routing Algorithm for Sensor Networks In Proc. First ACM Workshop on Sensor Networks and Applications, Atlanta: 22–31.
Heinzelman, W.R. et al. 2000, An Energy-Efficient Communication Protocol for Wireless Microsensor Networks, Journal of Networks, Vol. 6: 999–1008.
Hu, X.H. et al. 2011, Adaptive Radius Algorithm of ClusterHead Based on LEACH, Chinese Journal of Sensors and Actuators, Vol.1: 79–82.
Intanagonwiwat, C. et al. 2003, Directed Diffusion for Wireless Sensor Networking. Networking, IEEE/ACM Transactions, Vol.11: 2–16.
Lin, N. & Shi W.H. 2011, Simulation Research of Wireless Sensor Networks Based on LEACH Protocol, Computer Simulation, Vol.1: 178–181+241.
Sharma, M. & Shama, K. 2012, An energy Efficient Extended LEACH(EEE LEACH), Communicaion Systems and Network Techologies, May: 377–382.
Shah, R.C. & Rabaey, J.M. 2002, Energy Aware Routing for Low Energy Ad Hoc Sensor Netwoks, Wireless Communications and Networking Conference, Vol.1:350–355.
Voigt, T. et al. 2004, Solar-aware Clustering in Wireless Sensor Networks, Ninth IEEE Symposium on Computers and Communications, Vol.1
Wang, L. & Zhao, S.Y. 2012, Research and Improvement about LEACH Routing Protocol for Wireless Sensor Networks, Computer Engineering and Applications, Feb: 80–82.
Yektaparast, A. et al. 2012, An Improvement on LEACH Protocol (Cell-LEACH), ICACT 2012, Feb: 992–996.
Ying, J. 2008, Features of Wireless Sensor Network and Energy Optimization Strategy, Journal of Chongqing Technology and Business University, Vol.5: 537–540.
Zhang, R. & Tan, S.H. 2012, LEACH-based Clustering Routing Protocol for Sensor Networks, Computer Engineering and Design, Vol.4: 1333–1336+1346.
Zhao, F.F. & Gao, Y. 2011, Improvement and Simulation Based on LEACH Routing Protocol for Wireless Sensor Networks, Electronic Test, Vol.3: pp 47–50.
Zhong, W.P. & Gao, M.F. 2012, Simulation and Research on an Improved Algorithm of LEACH, Science Technology and Engineering, Vol.4: 786–788+803.

# Research on the novel non-uniform array structure

X.Q. Hu
*Radar Technique Staff Room of Huangpi NCO School, Wuhan Early Warning Academy, Wuhan, China*

Y.B. Liu
*Academic Branch of Huangpi NCO School, Wuhan Early Warning Academy, Wuhan, China*

X.L. Jiao
*Radar Technique Staff Room of Huangpi NCO School, Wuhan Early Warning Academy, Wuhan, China*

ABSTRACT:   To get narrow beam width and estimate coherent sources, a novel nonuniform array structure is proposed in this article. Using the whole sensors, the narrow beam can be obtained and the close signals can be separated. In the meantime, coherent signals can be estimated using part sensors. The presented nonuniform array increases the coefficient of utilization of sensors, i.e., less sensors is required. Theoretical results show that the designed novel array has no ambiguity. Two arrays are designed using this idea. Simulation experiment results demonstrate that the estimation performance of the designed array is superior to that of the uniform linear array.

## 1   GENERAL INSTRUCTIONS

There is lots of redundancy information for the uniform linear array because position difference is used to estimate the direction of arrival in spectrum estimation[1]. The equipment and cost are wasted; moreover, the computation complexity increases largely. To increase the coefficient of utilization of sensors and effective aperture of array, and to improve the resolution of array, nonuniform array was studied. Compared with the uniform linear array, nonuniform linear array has some special advantages including large array aperture and high resolution ratio. One disadvantage of the non-uniform linear array is the the ambiguity of signals[2].

Several studies were conducted in the field of array setting. The representative arrays have minimal redundancy linear, maximal continuation delay linear, and minimal interval linear. High resolution ratio is obtained using these arrays. However, these arrays are invalid in the environment of coherent signals because the great majority of estimation methods for coherent signals are used for the uniform array. To estimate coherent signals for nonuniform array, some resolutions were studied. For example, chen hui presented a novel nonuniform linear array using spacing combination method[3]. Space smoothing method is used to estimate coherent signals in the proposed array. However, the grating lobe is large. Guo yiduo presented a simple setting technique, i.e., the preceding M-1 sensors are set equably and the element spacing of the last sensor is larger, the coherent signals are estimated using the M-1 sensors, and the independent signals are obtained by the whole array. However, the improvement of the estimation performance is not obvious[4].

As we know, the representative arrays including minimal redundancy linear, maximal continuation delay linear, and minimal interval linear have narrow beam width estimating independent signals, and the nonuniform array using spacing combination method can estimate coherent sources[5–6]. Using the feature of these two types of arrays, we propose a novel nonuniform array. Both the large array aperture and coherent signals are obtained. Using the presented array, we can get narrow beam width and estimate coherent signals and independent sources.

## 2   NONUNIFORM ARRAY

The most serious problem of the nonuniform array is ambiguity. To avoid this, nonuniform array arrangement including minimal redundancy linear, maximal continuation delay linear and minimal interval linear is studied. For simplicity, we take the minimal redundancy linear.

### 2.1   *Minimal redundancy linear*

The gather of the sensors position difference is fully augmentable array in minimal redundancy linear. They can be divided into the optimal and suboptimal minimal redundancy arrays.

The optimal minimal redundancy arrays must satisfy the following conditions:

1   $\tau_{ij} \neq \tau_{kl}$ , $\forall i,j,k,l = 1,2,\cdots,M, i > j, k > l$

2   $\tau_{M1} = M_\alpha = M(M-1)/2 = 0+1+2+\cdots+M$

3   The gather of $\{\tau_{ij}, i > j\}$ is a continuous natural number when $\tau_{ij} \neq 0$

where $M$ is the number of elements. There are not same natural numbers in the gather of $\{\tau_{ij}, i > j\}$ when $\tau_{ij} \neq 0$, i.e., the redundancy is one. However, the minimal redundancy array that satisfies the above conditions does not exist. The suboptimal array takes the place of the optimal array. The suboptimal minimal redundancy array should satisfy the following conditions:

1  There exists $\tau_{ij} = \tau_{kl}$ in the gather of the sensors position difference, however, the number of the same position difference is as little as possible.
2  The position of the last sensor satisfies $M \leq n_M < M_\alpha$.
3  Except for the same number, the gather of $\{\tau_{ij}, i > j\}$ is a continuous natural number when $\tau_{ij} \neq 0$.

### 2.2  Spacing combination array

It is well known that space smoothing is a valid technique in the estimation of the coherent signals. However, the space smoothing method does not suit the nonuniform array. Therefore, the above nonuniform array is invalid when the signals are coherent. The spacing combination array that is nonuniform array is designed to estimate coherent sources. In the following, the idea of the spacing combination method is introduced.

Assuming $\Phi_i = \{d_1, d_2, \cdots, d_{q-1}\}$ denotes the $i$ th subarray of the whole array, where $q$ is the number of sensors in the subarray. $d_1, d_2, \cdots, d_{q-1}$ denotes the inter-element distance in the subarray and the first sensor is assumed as the reference element. Then the whole array can be expressed by

$$\Phi = \{\Phi_1, \Phi_2, \cdots, \Phi_{[(M-1)/(q-1)]}, d_1, d_2, \cdots, d_j\} \quad (1)$$

where $M$ denotes the number of the whole array or

$$\Phi = \{\Phi_1, \Phi_2, \cdots \Phi_i, \cdots, \Phi_2, \Phi_1\} \quad (2)$$

where $j = (N-1) \bmod (q-1)$, and the greatest common divisor in $\Phi_i$ is one. Then the greatest common divisor of $n_1, n_2, \cdots, n_N$ is one. mod denotes the modulo operation.

This array can estimate coherent signals, where the beam width is not narrow and the grating lobe is large.

## 3  A NOVEL ARRAY STRUCTURE

### 3.1  Design idea

Considering the characteristics of the above two types arrays, a novel nonuniform array is presented in this work to get the narrow beam width and the estimation of coherent signals.

Assuming it denotes the inter-element distance of the nonuniform array, such as minimal redundancy array, maximal continuation delay array and minimal interval array is $y_1 = 0.5\lambda[n_1, n_2, \cdots, n_{M1}]$, where $\lambda$ is the wavelength and $M_1$ is the number of elements. The array using the spacing combination has $M_2$ sensors and the inter-element distance is $y_2 = 0.5\lambda[m_1, m_2, \cdots, m_{M2}]$. A novel array is devised using the above two types of array except for the same elements in these elements. Assume the number of elements of the presented array is $M$ and the novel array has $y = 0.5\lambda[t_1, t_2, \cdots, t_M]$, where $t_i (1 \leq i \leq M)$ is the inter-element distance of the novel array.

### 3.2  Ambiguity analysis

For nonuniform array, ambiguity is very important. Ambiguity analysis is done here. According to document 1, the array $y_1$ is unambiguous, then

$$pt1 = \{\rho_2 \cap \rho_3 \cap \cdots \cap \rho_{M1}\} = \{0\} \quad (3)$$

where,

$$\rho_i = \{0, \pm 2k_i / n_i\} \quad k_i < n_i \quad (4)$$

$n_i (i = 2, 3, \cdots, M1)$, $k_i$ is the positive integer corresponding to $n_i$.

Assume that the parameter corresponding to array $y_2$ is $pt2$. Then

$$pt2 = \{g_2 \cap g_3 \cap \cdots \cap g_{M2}\} \quad (5)$$

where, $g_j = \{0, \pm 2l_j / m_j\} (j = 2, 3, \cdots, M2)$, $l_j$ is the positive integer corresponding to $m_j$.

Because

$$\{pt1 \cap pt2\} \subseteq pt1 = 0 \quad (6)$$

We have $0 \subseteq pt2$, and the parameter $\rho$ corresponding to the novel array $y$ has

$$\rho = \{pt1 \cap pt2\} = 0 \quad (7)$$

The above equation shows that the presented array is unambiguous.

### 3.3  The design example

Assume the wavelength is 1m. The novel array is designed by the minimal redundancy array and spacing combination array. The two types of arrays have ten elements.

1   Array one

The wavelengths of spacing combination array are one and two half wavelength, respectively. Because some elements position in this two type arrays are the same, the total number of elements in the novel array is 15. The concrete element position to half wavelength ratio is 0,1,3,4,6,7,9,10,12,13,20,27,31,35, and 36. Figure 1 shows the sensor position and direction pattern of the designed array one.

2   Array two

The wavelengths of spacing combination array are three and two half wavelength, respectively. Because some elements position in this two types of arrays are the same, the total number of elements in the novel array is 16. The concrete element position to half wavelength ratio is 0,1,3, 5,6,8,10,13,15,18,20,23,27,31,35, and 36. Figure 2 shows the sensor position and direction pattern of the designed array two.

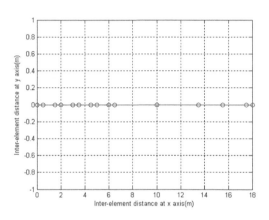

(a) Sensor position

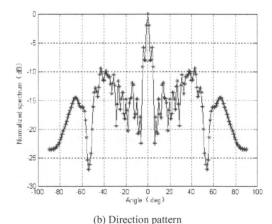

(b) Direction pattern

Figure 1.   The designed array one.

On comparing Figure 1 with Figure 2, the element number of array two is larger than that of array one and the direction pattern of array two is narrower. However, the side lobe of array two is higher.

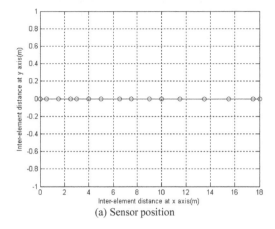

(a) Sensor position

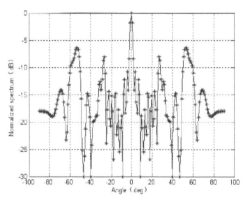

(b) Direction pattern

Figure 2.   The designed array two.

## 4   COMPUTER SIMULATIONS

### 4.1   *Simulation one*

In this section, some computer simulations are reported to illustrate the behavior of the proposed arrays. The structures of array one and array two are given above. The number of collected snapshots is 64 in this experiment. There are two coherent sources arriving at angles 20° and 30° and two independent sources arriving at angles 45° and 47°. Coherent sources are estimated by the part array using spacing combination method, and independent signals are estimated by the whole array. Figure 3 illustrates the MUSIC spectrum of the presented two arrays when SNR (signal noise ratio) is 5 dB.

Figure 3 illustrates that coherent signals can be estimated using part array and two close and independent sources can be divided exploiting the whole array. In array two, part array also can estimate independent sources. However, the estimation accuracy is worse than that of the whole array. It is clear that the estimation performance of array two is better than that of array one.

### 4.2 Simulation two

To compare the estimation performances of the presented arrays, RMSE (root mean square error) versus SNR is given in the following. The estimation performance of uniform linear whose elements number is the same as the whole array is also given for comparison. The number of collected snapshots is 64 in this experiment. There are two coherent sources arriving at angles 20° and 26° and two independent sources arriving at angles 45° and 46.5°. SNR varies from −4dB to 20 dB and the interval is 2 dB. There are 100 independent experiments.

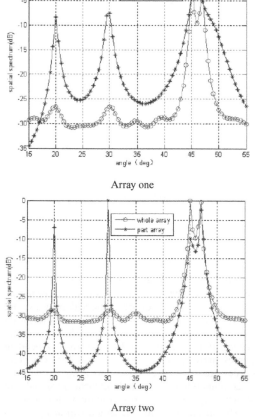

Array one

Array two

Figure 3.   MUSIC spectrum of the presented arrays.

Figure 4 shows that the estimation performance of the presented array is better than that of uniform array and the estimation performance of array two is better than that of array one. The RMSE of the proposed array is smaller than that of uniform linear and that of array two is even smaller.

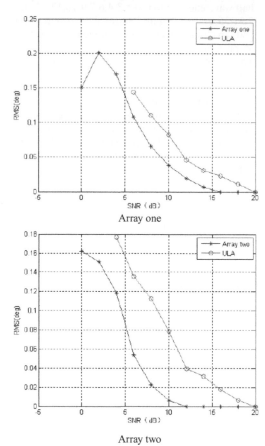

Array one

Array two

Figure 4.   Estimation performance versus SNR.

## 5   CONCLUSIONS

In this article, a novel nonuniform array is proposed using the conventional nonuniform array and spacing combination array. Exploiting the whole array narrow beam width and large valid aperture is given, and using part elements coherent signals is estimated. Large valid aperture is obtained exploiting the whole array and coherent signals can be estimated using part elements. The proposed array raises the availability of elements. Theory analysis illustrates that the designed array is unambiguous. Two arrays are given using this idea. Experimental results show that the estimation performance of presented array is better than that of uniform array. Compared to uniform array, the RMSE of the presented array is smaller. The RMSE of array two is smaller than that of array one.

Two novel setting arrays are given in this article. In practice, according to the actual need and requirement, more arrays can be obtained using this idea.

ACKNOWLEDGMENT

This work was supported by the pre-research fund program (XK2013Y004) for college.

REFERENCES

[1] Wang Yongliang, Chen Hui. Spatial spectrum estimation theory and technique[M]. Beijing: Tsinghua university press, 2004.

[2] Abramovich Y H, Gray D A. Comparison of DOA estimation performance for various types of sparse antenna geometries[C]. In Proc. EUSIPCO, Trieste, Italy, 1996, 2:915–918.

[3] Chen Hui, Wang Yongliang. Performance improvement in estimation direction-of-arrival by array geometry arrangement[J]. ACTA ELECTRONICA SINICA, 1999, 27(9):97–99.

[4] Guo Yiduo, Tong Ningning. DOA estimation algorithm for the decorrelation of non-uniform linear array[J]. Journal of projectiles, rockets, missiles and guidance, 2009, 29(1):293–296.

[5] Xu Xu, Zhongfu Ye. Adeflation approach to direction of arrival estimation for symmetric uniform linear array[J]. IEEE antennas and wireless propagation letters. 2006, 5(1):486–489.

[6] Yan Lu, Xie Mingxiang. Ambiguity and direction finding performance for non-uniformly spaced linear arrays[J]. Ship electronic engineering, 2010, 30(2):50–52,91.

*Multimedia, Communication and Computing Application – Leung (Ed.)*
© *2015 Taylor & Francis Group, London, ISBN 978-1-138-02775-6*

# An enhanced congestion avoidance mechanism for compound TCP

Y.C. Chan & Y.S. Lee
*Department of Computer Science and Information Engineering, National Changhua University of Education, Changhua, Taiwan*

ABSTRACT: Compound Transmission Control Protocol (CTCP) is widely deployed as the default transport layer protocol in the Windows operating system. It is a hybrid congestion control mechanism, in which a delay-based component is added to the standard TCP. In this article, we propose an enhanced version of CTCP named Statistic Process Control Compound TCP (SPC-CTCP). SPC-CTCP uses the information of *RTT* statistics to judge the network states and adjust its window size accordingly. The ns-2-based simulation results show that SPC-CTCP obtains more goodput as compared to that of TCP SPC and CTCP in various network environments.

## 1 INTRODUCTION

Transmission Control Protocol (TCP) is one of the core protocols of the Internet protocol suite. The behavior of TCP is tightly coupled with the overall Internet performance. To improve network efficiency, many TCP variants have been proposed. The proposed algorithms can be roughly divided into three categories, the loss-based, the delay-based, and the hybrid congestion control method.

The loss-based TCP 1 is the first development of TCP and has been widely used. Its robustness is appreciated by users. However, due to the sluggish detection of available bandwidth, it prevents the loss-based TCP form having a good performance in high-speed networks. The delay-based TCP [2,4] features a more sensitive congestion control. It adopts a more fine-grained signal, queuing delay, to avoid congestion. As compared to loss-based TCP, delay-based TCP generally performs better in homogeneous environments. Nevertheless, because of the incompatibility of the delay-based TCP with loss-based TCP, there exists no incentive for operating systems to adopt the delay-based TCP as its default transport layer protocol. The hybrid TCP [3, 6] combines the features of the loss-based and the delay-based TCPs. It becomes a new research trend for TCP congestion control.

Compound TCP (CTCP) is a variant of hybrid TCP presented by Microsoft Research [3]. It is widely deployed as the default transport layer protocol in the Windows operating system and its effectiveness in high-speed and long-distance network environments has been recognized. However, with the rapid development of wireless networking, CTCP needs further improvement.

In this article, we add a statistic process control to CTCP since the information of *RTT* statistics is reliable to judge the network states and suitable to deal with wireless random losses. Simulation results show that SPC-CTCP obtains more goodput as compared to that of TCP SPC and CTCP in different network environments.

## 2 RELATED WORK

### 2.1 Compound TCP (CTCP)

CTCP [3] is designed for high-speed and long-distance networks. It carries out the congestion control by combining the loss-based and delay-based methods. A new-state variable is introduced, namely, *dwnd* (delay window), which controls the delay-based component in CTCP. And the conventional *cwnd* (congestion window) controls the loss-based component. Then, the CTCP sending window is controlled by both *cwnd* and *dwnd*. Specifically, the CTCP sending window (*win*) is calculated as follows:

$$win = \min(cwnd + dwnd, awnd), \tag{1}$$

where *awnd* is the advertised window from the receiver. The *cwnd* is updated in the same way as in the standard TCP (Reno).

CTCP uses estimates of queueing delay as a measure of congestion. If the queueing delay is small, it assumes that no links on its path are congested, and the *dwnd* increases rapidly to improve the utilization of the network. Once queueing is experienced, the *dwnd* gradually decreases to compensate for the increase in the *cwnd*. The aim is to keep their sum approximately constant, at which the algorithm estimates are the path's bandwidth-delay products. In particular, when queueing is detected, the *dwnd* is reduced by the estimated queue size to avoid the problem of persistent congestion.

## 2.2 Principle of statistic process control

Statistic process control (SPC) was pioneered to effectively monitor and control the quality of the production process by Walter A. Shewhart at Bell Laboratories in the early 1920s. The principle of SPC [5] is to use statistic parameter estimation and hypothesis testing to monitor the whole system. It employs a time series graph (control chart) of data points with a process average, the upper and the lower control limits computation based on the data. Control limits indicate the extent of variation expected above or below an average line based on statistic probability. The control limits are defined in terms of the mean ($m$) and the standard deviation ($\Sigma$) of the distribution as follows:

$$UCL = m + K\sigma, LCL = m - K\sigma, \qquad (2)$$

where UCL and LCL are the upper and the lower control limits. $K$ is a constant specified to achieve a certain confidence level which is the percentage of measurements expected to lie between UCL and LCL. Therefore, if we know $m$ and $\Sigma$, we can predict the distribution range of the follow-up sampling and detect the unnatural symptoms on control chart.

## 2.3 TCP statistic process control (TCP SPC)

TCP SPC [4] employs $RTT$ instead of packet loss as the congestion indicator. The key innovation of TCP SPC is that more information-rich $RTT$ statistics are used instead of a single-valued smoothed $RTT$ value and SPC techniques are applied to the whole design of congestion control scheme.

Gao et al. [4] found that when network state changes, the pattern of delay variance also changes levels. Therefore, $RTT$ is suitable as congestion indicator. They used 15 consecutive $RTT$ values to calculate the mean ($m$) and the standard deviation ($\sigma$) according to the SPC standards:

$$m = \frac{\sum_{i=1}^{15} RTT_i}{15}, \qquad (3)$$

$$\sigma = \sqrt{\frac{1}{15-1}\sum_{i=1}^{15}(RTT_i - m)^2}. \qquad (4)$$

Once an $RTT$ is captured, a point is drawn on the control chart, and any change in $RTT$ values is thus recorded. To estimate the network state, they use four sets of criteria: under-load, over-load, congestion, and chart-invalidation criteria. Whenever TCP SPC captures 15 $RTT$ values, it can decide which criterion is met and adjust the window size accordingly. The simulation results show that, compared with TCP SACK, the TCP SPC throughput is improved by at least 50% in certain error-prone wireless environments.

## 3 PROPOSED SCHEME

The key idea of our proposed SPC-CTCP is adding SPC technique to the delay-based component of CTCP. Four problems need to be resolved. First, the traditional TCP acknowledgment (ACK) mechanism cannot collect the proper $RTT$ samples when packet loss event occurs because the duplicate ACK always depicts the sequence number of lost packet. Second, it is necessary to know how many samples need to be collected for network state judgment. The number of samples per network state judgment plays an important role for our mechanism. Third, the criteria for different network states should be made. The criteria in TCP SPC are not applicable for CTCP because there are two variants belonging to two different categories of TCP. Fourth, how to properly adjust the window size after the network state is to be determined. In the next section, we describe SPC-CTCP as follows.

## 3.1 TCP exact acknowledgement option

In SPC-CTCP, the $RTT$ samples are used to decide the network state. So how to collect complete and accurate $RTT$ samples is important. In a normal case, an ACK carries the next sequence number of the received packet. When packet loss event occurs, the TCP receiver always sets the sequence number of the lost packet in its ACK until the lost packet is recovered. In this period, the $RTT$ cannot be properly calculated. To overcome this issue, we add two new options in TCP.

| Kind | Length = 2 |
|------|------------|

Figure 1.   TCP EACK-permitted Option.

| | Kind | Length = 32 |
|---|------|-------------|
| Trigger Sequence Number (32) | | |

Figure 2.   TCP EACK Option.

The first one is Exact Acknowledgment (EACK)-permitted Option as shown in Figure 1. The Kind field depicting this option is EACK-permitted and its length is two bytes. When a three-way handshaking process is triggered, this option may be sent in a SYN by a TCP sender that has been extended to receive and send the EACK option once the connection is opened. When the TCP receiver has the ability to handle EACK, it will return EACK-permitted option to the sender. The negotiation is considered successful only when two ends of the connection both send EACK-permitted option in its SYN.

The second option is EACK (Exact Acknowledgement) as shown in Figure 2. The option aims to know which segment triggers the sending of the duplicate ACK. So, the Trigger Sequence Number field depicts the sequence number of the receiving packet that triggers the sending of the duplicate ACK. If an ACK is not a duplicate ACK, the EACK option is not necessary to be carried out since the *RTT* can be calculated in a normal way. When a packet loss event occurs, the TCP receives a duplicate ACK with an EACK option. Then the *RTT* can also be computed because the TCP knows the exact packet that triggers the duplicate ACK.

### 3.2 The number of RTT samples

Our new protocol adds an SPC technique to the delay-based component of CTCP. At first, SPC-CTCP needs to collect a certain number of *RTT* samples (*S*) to calculate the mean (*m*) and the standard deviation ($\Sigma$). In the SPC, the number of samples *S* is set to 15. It means when SPC get 15 *RTT*, it will calculate and check what criteria have happened to judge the network state. The problem is that if the sending window of a TCP connection is very large, e.g., 10000, the fixed value 15 may produce too many adjustments in sending window (*W*) per *RTT*. In standard TCP, the sending window is updated once per *RTT*. Thus, the meaning of the value of *S* is how long it will take to check the network state.

$$\begin{cases} S = 15, \text{ if } W \times 0.1 \leqq 15; \\ S = W \times 0.1, \text{ else}; \end{cases} \quad (5)$$

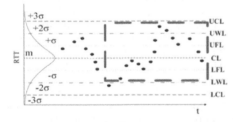

Figure 3.   The control chart of SPC-CTCP.

We think the value of *S* should be set according to the sending window size *W*. We set *S* as equation (5). When we get S samples, we take the latest 15 *RTT* in *S* to calculate the mean (*m*), standard deviation ($\Sigma$), and the control values UCL, UWL,… etc. which are used to draw the SPC-CTCP control chart as shown in Figure 3. By the control chart, SPC-CTCP can check what kinds of conditions have happened and which network criterion is met. Figure 3 shows that *S* is 22, but we only use the latest 15 *RTT* to judge the network state.

### 3.3 Network state criteria

Adopting the principles of SPC and TCP-SPC, and observing the *RTT* trends of numerous ns-2 experiments, we set three criteria for estimating the network state of CTCP connections. The three criteria are Slow-Down, Speed-Up, and Steady. The details of the three criteria are shown in Tables 1–3.

Slow-Down criterion: The value of recently collected *RTT* sample becomes significantly large or the *RTT* is in the rising trend, which means the network state is getting worse. So the sending rate should be suppressed to avoid network collapsing.

Speed-Up criterion: The value of recently collected *RTT* becomes significantly small or the *RTT* is in the decreasing trend, which means that certain network resource is released by others. So the data-sending rate could be raised to utilize residual capacities of current network.

Steady criterion: *RTT* samples present a relatively stable state, so we suggest that the current network state

Table 1.   Slow-Down criterion: It is met if one of the following conditions is satisfied.

| | |
|---|---|
| Over-UCL | 1 point is larger than UCL |
| Up-Excursion | (It will happen if all the following conditions are satisfied) |
| | 9 consecutive points keep larger than CL |
| | 10 of 11 consecutive points keep larger than CL |
| | 12 of 14 consecutive points keep larger than CL |
| Near-UWL | 4 of 5 consecutive points keep between UFL and UWL |
| Near-UCL | (It will happen if all the following conditions are satisfied) |
| | 2 of 3 consecutive points keep between UWL and UCL or larger than UCL |
| | 3 of 7 consecutive points keep between UWL and UCL or larger than UCL |
| | 4 of 10 consecutive points keep between UWL and UCL or larger than UCL |
| Point-Trend-Up | 6 consecutive points keep larger and larger |

Table 2.   Speed-Up criterion: It is met if one of the following conditions is satisfied.

| | |
|---|---|
| Over-LCL | 1 point is smaller than UCL |
| Down-Excursion | (It will happen if all the following conditions are satisfied) |
| | 9 consecutive points keep smaller than CL |
| | 10 of 11 consecutive points keep smaller than CL |
| | 12 of 14 consecutive points keep smaller than CL |
| Near-LWL | 4 of 5 consecutive points keep between LFL and LWL |
| Near-LCL | (It will happen if all the following conditions are satisfied) |
| | 2 of 3 consecutive points keep between LWL and LCL or smaller than LCL |
| | 3 of 7 consecutive points keep between LWL and LCL or smaller than LCL |
| | 4 of 10 consecutive points keep between LWL and LCL or smaller than LCL |
| Point-Trend-Down | 6 consecutive points keep smaller and smaller |

Table 3.   Steady criterion: It is met if one of the following conditions are satisfied.

| | |
|---|---|
| Stratum | 15 consecutive points keep between UFL and LFL |
| Un-Stratum | 8 points on both sides of the CL with none between UFL and LFL |
| Mutual | 14 points between UFL and LFL, above and below the CL |

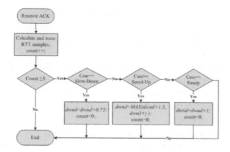

Figure 4. The flow chart of *dwnd* adjustment in SPC-CTCP.

is steady, and TCP can gently increase the sending window to probe the available bandwidth.

### 3.4 Window adjustment scheme

The SPC principles are particularly useful in a steady environment, so we judge the network state only for the adjustment of dw*nd*, which is used in the congestion avoidance phase. We do not use SPC in the slow start, fast retransmission, and fast recovery process. Like CTCP, SPC-CTCP has two windows: *cwnd* and *dwnd*. The flow chart of *dwnd* adjustment is shown in Figure 4.

When we get enough RTT samples, we can know which criterion is met. When the Slow-Down criterion is met, SPC-CTCP will decrease the *dwnd* by one quarter. When the Speed-Up criterion is met, SPC-CTCP will increase the *dwnd* by one half or $\gamma$ depending on which one is larger. The value of $\gamma$ is set at 30 as it is in CTCP. Because when the *dwnd* is small, it may be too slow by increasing one half of current *dwnd*. When the Steady criterion is met, SPC-CTCP will increase by one segment.

The *cwnd* of SPC-CTCP is the key point to keep friendly with the traditional TCP, so we just moderately modify the adjustment scheme of cwnd when it detects the packet loss event. If SPC-CTCP detects the packet loss and the network criterion which meets Speed-Up, we do not change the value of *cwnd*.

### 4 SIMULATION RESULTS

In this section, we evaluate the performance of three different TCP variants: TCP SPC, CTCP, and the proposed SPC-CTCP by ns-2. The simulation topology is shown in Figure 5. The scenario is very common now because the last hop is a wireless link. The link between $R_1$ and $R_2$ is a T3 link. The last hop between $R_2$ and the destination uses IEEE 802.11g to transmit packets. The simulation time is 200 seconds and the random loss rate of the wireless link is set between 0% and 30%. The goodput performance in the heterogeneous environment is shown in Table 4.

In Table 4, the performance of SPC-CTCP is higher than CTCP by 2.18%–17.45%, and SPC by 17.61%–102.16%. No doubt, the proposed SPC-CTCP has the best goodput performance.

Figure 5. A heterogeneous network environment.

Table 4. Goodput performance (Mbps) in heterogeneous networks.

| TCP Variant | Number of Conn. | Random Loss Rate | | | | | |
|---|---|---|---|---|---|---|---|
| | | 0% | 1% | 5% | 10% | 20% | 30% |
| SPC | 1 | 6.967 | 6.835 | 5.687 | 4.432 | 1.877 | 0.556 |
| | 2 | 7.081 | 6.53 | 6.308 | 5.06 | 1.747 | 1.45 |
| | 5 | 7.552 | 7.43 | 7.071 | 5.756 | 2.953 | 2.652 |
| CTCP | 1 | 8.539 | 8.462 | 7.03 | 5.987 | 2.486 | 0.957 |
| | 2 | 8.655 | 7.902 | 7.185 | 6.408 | 2.062 | 1.787 |
| | 5 | 8.021 | 7.902 | 7.105 | 6.188 | 3.181 | 2.954 |
| SPC-CTCP | 1 | 9.12 | 8.934 | 7.827 | 6.87 | 2.642 | 1.124 |
| | 2 | 9.148 | 8.755 | 7.88 | 7.078 | 2.278 | 1.826 |
| | 5 | 8.882 | 8.535 | 8.191 | 7.022 | 3.627 | 3.365 |

### 5 CONCLUSIONS

In this article, we propose a new variant of CTCP named Statistic Process Control Compound TCP (SPC-CTCP) for heterogeneous networks. We add the SPC technology in CTCP to improve its performance. The *cwnd* of SPC-CTCP is used to maintain the robustness as well as to keep fairness with standard TCP. The *dwnd* of SPC-CTCP is used to quickly adapt to the changing environment. We set the network criteria and *dwnd* adjustment scheme for SPC-CTCP based on SPC, TCP-SPC, and the observation of *RTT* trends in numerous ns-2 experiments. The results of ns-2-based simulation show that the SPC-CTCP has good performance in the heterogeneous networks.

### REFERENCES

[1] J. Wang, J. Wen, Y. Han, and J. Zhang, "CUBIC-FIT: A High Performance and TCP CUBIC Friendly Congestion Control Algorithm," IEEE Communications Letters, pp. 1664–1667, August 2013.
[2] W. Xiuchao, C. Mun Choon, A. L. Ananda, and C. Ganjihal, "Sync-TCP: A new approach to high speed congestion control," ICNP, pp. 181–192, October 2009.
[3] K. Tan, J. Song, and Q. Zhang, "A compound TCP approach for high-speed and long distance networks," INFOCOM, pp. 1–12, Jul. 2006.
[4] D. Gao, Y. Shu, Li Yu, and M.Y. Sanadidi, "TCP SPC: Statistic Process Control for Enhanced Transport over Wireless Links," GLOBALCOM, pp.1–5, DEC. 2008.
[5] John Wiley & Sons, "Introduction to statistical quality control.," D. C. Montgomery, 1985.
[6] X. Sun, Y. Zhang, S. Zhang, Xu Zhou, K. Niu, and J. Lin, "HHS-TCP: a novel high-speed TCP based on hybrid congestion control", ICWMMN, pp.271–275, Nov. 2013.

# High-speed radix-4 FFT processor based on FPGA

C. Li

*The Collaboration of Patent Examination for Patent Examination Center of the State Intellectual Property Office*

X.F. Li

*Department of Electromechanical Engineering, Beijing Institute of Technology, Beijing, China*

ABSTRACT: A method of implementing 256-point, high-speed, and 16-bit complex FFT is presented on the radix-4 FFT algorithm. Using a fixed geometry addressing, pipeline designing, and block floating point structure, the data is more precise with more dynamic range. The results show that the design is efficient, strongly extensive, and occupies less resource. It is a good method to meet the high-speed digital signal processing requirements.

KEYWORDS: Fixed geometry, Pipeline, Block floating point, High-speed FFT processor

## 1 INTRODUCTION

Fast Fourier transform (FFT) is an effective and fast algorithm for the discrete Fourier transform (DFT), which is the core of digital signal processing algorithms. FFT is widely used in radar, communications, image processing, signal detection, and other fields, and most of those fields call for the FFT processor with high speed and high precision of real-time processing performance. To simplify the calculation and shorten operation time to one or two magnitudes, the ideology of FFT algorithm sequentially divides the N point DFT into short sequences of DFT for calculating. Currently, the FFT algorithms can be achieved in three methods: application-specific integrated circuit (ASIC), field programmable gate array (FPOA), general-purpose processor (OPP), and digital signal processor (DSP). Considering the processing speed, bulk, and power consumption of FFT, especially in small-scale applications, FPOA processor with FFT algorithm can achieve high flexibility and cost–effectiveness.

After analyzing the radix-4 FFT algorithm, this article presents the implementation of the fixed geometry addressing by a 256-point FFT processor, which has many advantages, such as it is relatively simpler to control, highly modular, and it is expansible.

## 2 THE ANALYSIS OF FFT ALGORITHM

There is a sequence x(n), whose length is $N$, the DFT is:

$$X(\mathrm{k}) = \sum_{n=0}^{N-1} x(\mathrm{n})\, W_N^{kn},\ k = 0,1,...,N-1 \quad (1)$$

In this equation, $W_N^{kn} = e^{-j\frac{2\pi}{N}kn}$ is the rotation factor, supposing $N = 4\,\mathrm{m}$, then

$$X(\mathrm{k}) = \sum_{m=0}^{\frac{N}{4}-1} x(4\,\mathrm{m})\, W_N^{4mk} + \sum_{m=0}^{\frac{N}{4}-1} x(4\,\mathrm{m}+1)\, W_N^{(4m+1)k}$$

$$+ \sum_{m=0}^{\frac{N}{4}-1} x(4\,\mathrm{m}+2)\, W_N^{(4m+2)k} + \sum_{m=0}^{\frac{N}{4}-1} x(4\,\mathrm{m}+3)\, W_N^{(4m+3)k} \quad (2)$$

Among them, m = 0,1,…,N/4-1, supposing

$$A = \sum_{m=0}^{\frac{N}{4}-1} x(4\,\mathrm{m})\, W_N^{4mk},\ B = \sum_{m=0}^{\frac{N}{4}-1} x(4\,\mathrm{m}+1)\, W_N^{(4m+1)k}$$

$$C = \sum_{m=0}^{\frac{N}{4}-1} x(4\,\mathrm{m}+2)\, W_N^{(4m+2)k},\ D = \sum_{m=0}^{\frac{N}{4}-1} x(4\,\mathrm{m}+3)\, W_N^{(4m+3)k}$$

Then

$$\begin{cases} X(\mathrm{k}) = A + BW^P + CW^{2P} + DW^{3P} \\ X(\mathrm{k}+N/4) = A - jBW^P - CW^{2P} + jDW^{3P} \\ X(\mathrm{k}+2N/4) = A - BW^P + CW^{2P} - DW^{3P} \\ X(\mathrm{k}+3N/4) = A + j\,BW^P - CW^{2P} - jDW^{3P} \end{cases} \quad (3)$$

In this equation, k = 0,1,…,N/4-1.

The theoretical realization of the radix-4 FFT is on the basis of Equation (3).

The traditional FFT computation flow diagram uses addressing in situ, but the addressing structure of each level is different, which makes it difficult to expand and to be modular. Besides, it takes up more resources to implement it. Therefore, this work designs a fixed geometry of the addressing circuit for addressing, which has the same addressing mode in each level. The data input of radix-4 fixed geometry structure operation is in order, but the data output of that is not. Take 16 points for example, and the operation flow chart was shown in Figure 1:

The figure shows that 16-point FFT needs two operations and a full sequence operation in all, while 256-point FFT needs four operations and a full sequence operation in all.

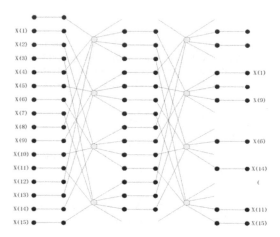

Figure 1. The 16-point FFT flow diagram with order input reverse output.

## 3 THE HARDWARE OF FFT PROCESSOR

The study designs the ping-pong rams to store data. The butterfly processing unit is implemented in parallel and pipelined structure to speed up execution rate of FFT processor. Addressing unit module is achieved by a fixed geometry structure. The radix 4 FFT processor includes butterfly processing unit, addressing

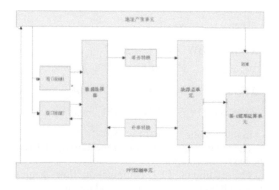

Figure 2. The implementation structure of the FFT processor.

generation unit, block floating-point unit, FFT controlling unit, storage unit, etc. The implementation chart of the FFT processor is shown in Figure2.

### 3.1  The butterfly processor

The butterfly processor is the core of the FFT processor. According to Equation (3), the parallel and pipeline structures are used to achieve the unit, and the structure is shown in Figure 3.

The four data and three twiddle factor data are simultaneously latched into the butterfly unit by the controlling unit. The data outputs are in parallel through three level pipeline.

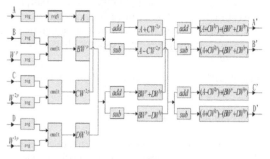

Figure 3. The implementation chart of the butterfly processor.

### 3.2  Addressing generation unit

The addressing generation unit provides the reading address of data which is input to butterfly processing unit and the writing address of data which is output to butterfly processing unit. That is the input data needed in the butterfly operation which is read separately from RAM and ROM and the output data will be written into RAM after the butterfly operation.

The fixed geometry structure is used for addressing, which is shown in Figure 1. Each level has the same reading address and the writing address for RAM addressing. Only the address of coefficient ROM is changed.

(1) RAM Reading Address: the distance between four nodes in butterfly operation is

$$r=N/4=64 \quad (4)$$

Supposing a, b, c, and d are, respectively, the address of the four nodes A, B, C, and D, then

$$\begin{cases} b=a+r \\ c=a+2r \\ d=a+3r \end{cases} \quad (5)$$

The a, which is the address of node A, always starts at 0 at the beginning of each level. The value of a will plus one in turn at the end of the butterfly operation until it completes all the butterfly operation.

(2) RAM Writing Address: the output data of the butterfly processing unit for each level is written into RAM orderly. Each level of writing address is also starting from 0 until 256 data are written into RAM.

(3) ROM reading address: the quantized rotation factors are stored in ROM. The order of the data is to be adjusted in the last level (or the fifth level). Therefore, there is no need for ROM addressing in the last level. According to Equation (3), we can find that the p-value of the coefficients participated in the butterfly operation each time decides the address. At the beginning of each level, p-value always starts from a dress 0. The p-value is the data output reversely composed of 3 bit Quaternary number plusing 1 orderly. The m ($0 < m < 5$) level, the number of butterfly groups is

$$n = 4^{m-1} \qquad (6)$$

when the quaternary number is m-1, it is assigned 0. Processing 64 times until all 256 points are completed.

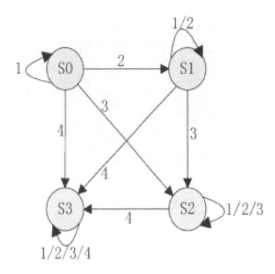

Figure 4.    State transition diagram.

### 3.3    Block floating-point unit

The control of block floating-point arithmetic is relatively simpler. Compared with floating-point algorithm, the implementation of block floating-point arithmetic is faster than that of floating-point arithmetic operations. Compared with fixed-point algorithm, the data of block floating-point arithmetic has the larger dynamic range. The block floating-point unit includes the overflow detection unit and the data selection unit, and it can be implemented by the finite state machine. The overflow detection unit detects the high four bits of the output data, and the state of the finite state machine is decided by the high four bits of the largest output data from the butterfly operation unit. The data can be divided into the following 4 conditions:

(1) No overflow: 0000 or 1111;
(2) Overflow 1: 0001 or 1110;
(3) Overflow 2: 001x or 1 l0x;
(4) Overflow 3: 01xx or 10xx.

Figure 4 is a state transition diagram of the finite state machine, where S0, S1, S2, and S3are corresponding to the four states above.

Data selection unit determines the bits of input data in the next butterfly Operation according to the state of overflow detection unit when butterfly Operation finishes each level. Take Table 1 for example.

Table 1.    The relationship between final state of overflow detection unit and the selector.

| State | S0 | S1 | S2 | S3 |
|---|---|---|---|---|
| Selection | [N-1:0] | [N:1] | [N+1:2] | [N+2:3] |

### 3.4    Control unit

Control unit coordinates all the operation units (including the butterfly processing unit, address generation unit, block floating-point unit, and other units) in the FFT processors, and has met timing requirements to complete the entire operation successfully. FFT operation involves many processing units and the finite state is used to achieve it for complex control.

In addition, because there is a large number of BLOCK RAM in FPGA, memory cells use IP cores provided by Xilinx to realize it. Thus, the design does not only save logic resources, but also has a faster read/write speed.

## 4    SIMULATION

The study is designed using the Verilog hardware to describe language. The test bench of the FFT system is based on Xilinx ISE platform, and is simulated in ModelSim SE environment. Figure 5 is the FFT processor system simulation diagram, and it shows that test signal is the halfsine wave, at 50 MHz clock frequency, and it spends 27 ms to complete the 256-point FFT. The system designs in pipeline, and 256-point data output continuously. Figures 6–8 show the results of the test signal processed in MATLAB environment. Compared with it, the results agree with theoretical values appropriately, which proves the accuracy of the processor design.

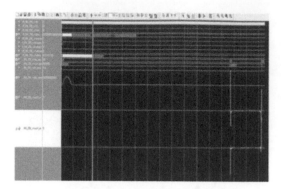

Figure 5.  256-point FFT simulation.

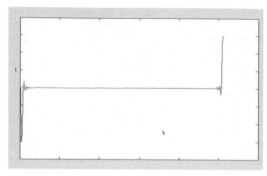

Figure 8.  FFT operation result (imaginary part).

## 5  CONCLUSION

The study analyses the feature of the radix-4 FFT algorithm, advances a new hardware implementation structure—the fixed geometry addressing structure, which can reduce consumption of resource and achieve easily larger points (such as 1024 points, 4096 points) FFT expansion. The system introduces parallel and pipeline to get a high execution rate and introduces block floating-point arithmetic to acquire high accuracy. The implemented FFT processor meets the requirements of the high-speed real-time digital signal processing

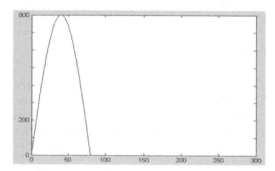

Figure 6.  Test signal.

## REFERENCES

[1] Uwe, M.B. & Liu, L., 2006, FPGA implementation of digital signal processing, Beijing, Tsinghua University Press.
[2] Uwe, M.B. & Liu, L., 2006, FPGA implementation of digital signal processing, Beijing,Tsinghua University Press.
[3] He, S.S. & Mats, T., 1998, Design and Implementation of a 1024-point Pipleline FFT Processor, IEEE Custom Integrated.
[4] Circuits Conference. LEE, H.Y, 2007, PARK In-Cheol Balanced binary-tree decomposition for area-efficient pipelined FFT processing, IEEE Transactions on Circuits and Systems.
[5] Wan, H.X & Han, Y.Q & Chen, H, 2005, A high performance FFT/IFFT for VLSI Design, Electronic Measurement and Instrument.
[6] Ding, X.L & Zurn, Z.M, 2007, 16:00 base-4FFT chip design technology research, Information Technology.

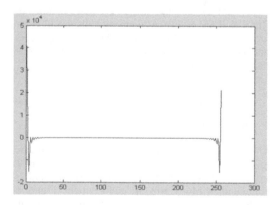

Figure 7.  FFT operation result (real part).

*Multimedia, Communication and Computing Application – Leung (Ed.)*
© *2015 Taylor & Francis Group, London, ISBN 978-1-138-02775-6*

# Markov chain-based attack on the CSMA/CA mechanism

W. Wang
*Science and Technology on Communication Information Security Control Laboratory, Jiaxing, Zhejiang, China*
*Jiangnan Electronic Communication Institute, Jiaxing Zhejiang 314001, China*

W.H. Zhao
*Nanhu College, Jiaxing University, Jiaxing, China*

ABSTRACT: The CSMA/CA (Carrier Sense Multiple Access with Collision Avoidance) mechanism is formally modeled first. Then, based on the Markov Chain theory, the analysis of the attack on CSMA/CA is introduced from two aspects which are the stochastic performance model and bandwidth share model. After discussing the typical attack methods, the performance is analyzed from the aspects of throughput, communication efficiency, and collision numbers, which validates the feasibility and efficiency of attack method based on Markov Chain theory.

## 1 INTRODUCTION

Attack to wireless communication networks is of great significance for the research on the network security. A lot of relevant researches have been made by many researchers such as Tao (2010), Wang (2010), Wang (2009), and Zhao (2008). Because of the characteristics of self-organization, multi-node, high-bandwidth, and burst communications of the future wireless networks, the application of fixed multiple access technologies may be limited, such as FDMA(Frequency Division Multiple Access) and TDMA(Time Division Multiple Access). Moreover, dynamic multiple access technology of ALOHA mechanism is not applicable to wireless networks due to its throughput. While the CSMA/CA (Carrier Sense Multiple Access with Collision Avoidance) mechanism drives more and more attention from wireless network researchers because of its better functions like timeliness and scalability, which are supportive for burst communications, prevention of hidden terminal/exposed terminal, etc. Zhang 2008 discussed the intelligent lamming attack and deception jamming in single-node and multinode collaborative methods to Ad Hoc networks based on MAC protocol and compared the interference effects by different ways. Attack is realized by periodic transmission of RTS and CTS frames by the violated node forging NAV value in the works of Noubir (2003), Raya (2004), and John (2002). Such attack results in that the nodes within one hope of the violated node wrongly update the local NAV value according to the NAV value field in RTS and CTS frames and mistake the channel as being busy, which increases the

delay of the accession to the channel. Zeng (2007) proposed an attack model on CSMA/CA, which analyzed the effectiveness and impact of the attack mode through the associated network simulation and data processing, proposed an evaluation standard for the efficiency of network attacks, and finally analyzed the improvement of the attack methods and random backoff algorithm. Cao (2008) proposed an RTS-CTS attack on IEEE 802.11 CSMA/CA, preempted the traffic channel in the attacker by modifying the contention window.

The above part discussed the possibility of attacks on CSMA mechanism from the technical aspect, but it still needs further theoretical analysis. This article focuses on research of CSMA/CA channel access mechanism and discusses the feasibility and effectiveness of the attack methods on CSMA/CA channel access mechanism through theoretical analysis and study.

## 2 FEASIBILITY ANALYSIS OF CSMA/CA ACCESS ATTACK DESCRIPTION OF THE MODEL

CSMA/CA will be a major channel access protocol of wireless networks in the future. Under this mechanism, before sending data, the node first monitors the working conditions of the channel. If the channel is busy, it will delay the data delivery to avoid a collision. The slotted random of delay is generally generated by binary exponential backoff algorithm and is freely selected by each communication node. The node selecting the minimum slotted random can access the channel to transmit data. In this protocol, it is assumed that all communication nodes randomly select slot by strictly following the protocol rules, but it lacked direct control mechanism so that the attacker can attack on network communications by making use of such flaw.

## 2.1 Operating principles of CSMA/CA

In the following analysis, it is assumed that all nodes in the wireless network implement CSMA/CA, and all unicast packets are confirmed by ACK (Acknowledgement) and it will be retransmitted in case of failing to receiving ACK. CSMA/CA is mainly related to RTS-CTS handshake and binary exponential backoff. Figure 1 shows the schematic of RTS-CTS handshake. DIFS and SIFS presented above are composed of a plurality of slots.

RTS and CTS are control frames on CSMA/CA for appointment of the next channel slot. Before sending the actual data frames, Exchanging RTS and CTS is a major channel appointment method which can solve the "hidden terminal" and "exposed terminal" problems as well as collision in case of sending a lot of data simultaneously by CSMA/CA basic access method. RTS and CTS frames contain duration field and define the data frame and ACK channel occupancy time. All workstations within the coverage of the source stations and destination stations will get medium reservation information. Figure 1 shows the schematic of RTS-CTS handshake. DIFS and SIFS mentioned above are composed of a plurality of slots.

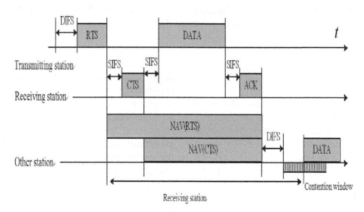

Figure 1. RTS-CTS handshake.

Binary exponential backoff mechanism: When a node sends a data frame, if it discovers a busy link, it will activate the backoff process. If the frame sent by the node does not receive ACK, it will also activate the backoff process to generate a waiting slot down counter. The node performing backoff process will determine whether each backoff slot medium is idle or not by carrier sense mechanism. In case of being idle, the counter will be -1; In case of being busy, the counter will pause and rework after media idle time reaches DIFS or EIFS. When the counter reaches 0, frames can be sent.

## 2.2 Feasibility analysis of access attack

Some symbols are described in Table 1.

The single-node bandwidth share model based on Markov Chain theory is given by introducing the CSMA/CA network operation mode and then this model is extended to fit the CSMA/CA properties under access attack.

Table 1. Symbol Description.

| Name | Description |
| --- | --- |
| $n$ | Subscript of communication node |
| $w=<w_{min}, L>$ | Backoff configuration information |
| $w=(w_1, ..., w_N)$ | Configuration set |
| $t, c, s$ | Attempt to transfer, conflict, and the probability of success within a slot |
| $T, S$ | Sum of transmission probability of all nodes and sum of the probability of success |
| $b$ | Bandwidth shared by communication nodes |

Theorem 1: Assuming ñ represents the network nodes with the following properties: ① nodes with the minimum $w_n$, min value, ② nodes with the minimum $L_n$ value of the nodes in line with ①; Assuming

$$w_{\tilde{n}} = < w_{min}, L >, \quad \tilde{w}_{min} \geq 1 + \sqrt{2 \cdot \tilde{w}_{min} \cdot 1_{\tilde{L}}} > 0 \qquad (1)$$

Wherein, $1_{(\cdot)}$ is the indicator function, and the following conditions are true:

(I) Solution to $(t_n^o(w), n=1, \ldots, N)$ is unique;

(II) Assuming $w' < w$, $w = (w_1, \ldots, w_n-1, . w, w_n+1, \ldots, w_N)$, and $w' = (w_1, \ldots, w_n-1, . w', w_n+1, \ldots, w_N)$, for bandwidth share value: $b_n(w') > b_n(w)$.

Condition (6) does not apply to the configuration information $\tilde{w}_{min} \leq 3$ and $L > 0$. However, in case of $\tilde{w}_{min} \leq 3$, the network node $n$ will occupy as large a bandwidth as possible, which will result in access attack generated by the network nodes themselves even in case of no external attack. Theorem (II) gives the possibility of access attack: to increase the transmission opportunity, the configuration information itself will enhance the rate of collisions, which makes the performance of the network seem to depend on the configuration of other nodes, for example, CSMA/CA mechanism may result in access attack.

## 3 ACCESS ATTACK METHODS FOR CSMA/CA CHANNEL ACCESS MECHANISM

By wireless networks, channel access function is generally realized by applying the network control packet/data frame. Since the control packet/frame data is not encrypted, it is easy to perform field identification and analysis of operation mechanism. In this section, a channel access disrupting attack method is proposed for CSMA/CA and the ideas for access attack to channel allocation/access/sharing mechanism are given in case of only master of control frame (RTS, CTS) field: "type" and "duration". In the course of the attack, the attacker node preempts traffic channel and sends attack data by making use of the features of CSMA/CA mechanism. This attack method can make the target node or network access points silent and even disassociate all workstations and access points, leading to network paralysis. Moreover, this method is also the necessary factor to achieve the target network non-cooperative access and target node wireless access attacks, and it also can be extended to the attack method which can serve for other network channel access mechanisms.

In the communication process, the working node will occupy the channel exclusively upon the reservation of channel through RTS and CTS until its data transmission is completed. It is assumed that the attacker has got the location and function of the control frame in "type" and "duration" field, as shown in Figure 2. In the process of the reservation of channel, the "duration" field is used to record the value of NAV, and the time of communication parties to access the wireless medium specified by the NAV. Therefore, the attacker can launch attacks to such mechanism in this way to occupy the channel, prevent workstations from data transmission, block the network access points, and paralyze the whole wireless network.

| Unknown | Type | Unknown | Duration | Unknown |
|---------|------|---------|----------|---------|

Figure 2. Channel access control frame only mastering duration field.

Way of attack by the attacker: attacks can be carried out in a variety of ways and the targets can be network access points, mobile terminal nodes, or both. An attacker needs to listen to network traffic data for NAV synchronization. In addition, this attack method can be realized by only using RTS or RTS-CTS. The main attack methods are as follows:

(a) Attack to one or several nodes: As shown in Figure 10, the attacker intercepts the RTS frame and modifies the "duration" field according to the "type" field, and then transmits an RTS frame into the target node to inform that the data is being transmitted in the current wireless network to perform the backoff process so that the target node cannot transmit data within a certain period. Attacked nodes perform the normal backoff process without any abnormal behavior. Because of the failure of receiving the RTS frame, other nodes can perform normal communications. Overall, the network communication is normal. This attack is not easy to be detected and it even can facilitate the further wireless injection attacks to the target node.

(b) Attack to access points: as shown in Figure 11, the wireless network access points and other network nodes compete for using the same common channel. Thus, the RTS frame with modified "duration" field can be resent to network access points to execute the backoff process. In the attack process, the network access point cannot send data so that it cannot respond to the requests of all nodes. Thus, all nodes are automatically disassociated with the network access point and the entire network cannot work.

Probability for the attacker to successfully preempt the channel: the minimum channel contention window is set to CW, the number of network nodes to n and the maximum retransmission time of frame to r.

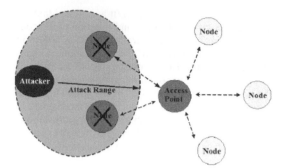

Figure 3. Attacks to nodes.

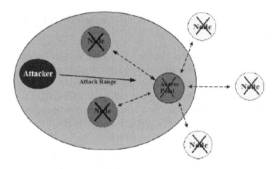

Figure 4. Attacks to the network access point.

Network nodes are under binary exponential back-off mechanism. The attacker always selects No. 0 contention window. Therefore, only the node which suffers the first collision with the attacker may suffer the second collision. After sending the first frame, the attacker will send the second frame immediately. Therefore, the NAV of the nodes suffering no collision at the first time will not be reduced by 1 for the second frame, and only the nodes suffering the first collision may select No. 0 window randomly.

In can be known from the RTS-CTS mechanism that once the attacker preempts the channel, the subsequent frames will be sent within SIFS, and other network nodes are unable to obtain the right to use the channel. Then, the probability (Prsuccess) for the attacker to preempt the channel successfully within the transmission cycle of one frame is shown below.

$$\mathrm{Pr}_{success} = 1 - \mathrm{Pr}_{fail}; \mathrm{Pr}_{fail} = \mathrm{Pr}_{fail,r};$$

$$\mathrm{Pr}_{fail,r} = \frac{1}{2^{r-2}} \mathrm{Pr}_{fail,r-1} \cdot \left(1 - \left(\frac{2^{r-2}CW - 1}{2^{r-2}CW}\right)^n\right);$$

$$\mathrm{Pr}_{fail,1} = 1 - \left(\frac{CW - 1}{CW}\right)^n$$

## 4 PERFORMANCE ANALYSIS

For the wireless network with $n = 30$, $r = 6$, CW = 32, Prfail = 8.84e-6, effect analysis diagram is given in terms of the above-mentioned attacks to CSMA/CA mechanism. Figure 5 shows the throughout in different channel load. Figure 6 shows communication efficiency in different channel load. Figure 7 shows the frequency of backoff in different channel load.

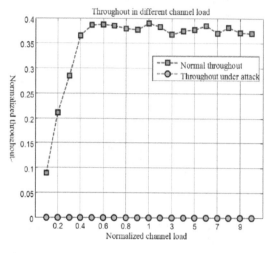

Figure 5. Throughput in different channel load.

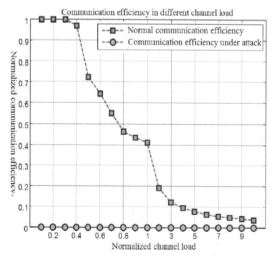

Figure 6. Communication efficiency in different channel load.

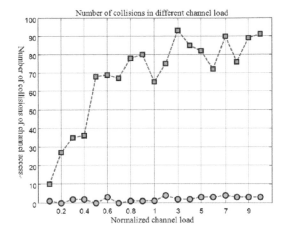

Figure 7.   Collision frequency in different channel load.

## 5   CONCLUSIONS

In this article, the CSMA/CA mechanism is formally modeled. Based on the Markov Chain theory, the analysis of attack on CSMA/CA is introduced from two aspects. They are the stochastic performance model and bandwidth share model. Finally, the specific attack methods and performance analysis are proposed and the feasibility and effectiveness of attack to CSMA/CA mechanism is verified.

## REFERENCES

Tao Y. & Liu Z.L. et al. 2010. Research on network attack situation niching model based on FNN theory. *High Technology Letters* 7:680–684.

Wang Q. & Feng Y.J. et al. 2010. Network attack model based on ontology and its application. *Computer Science* 37 (6):114–117.

Wang G. & Wang H.M. et al. 2009. Research on computer network attack modeling based on attack graph. *Journal of National University of Defense Technology* 31 (4):74–80.

Zhao F.F. & Chen X.Z. et al. 2008. Generation methods of network attack graphs based on privilege escalation. *Computer Engineering* 34 (23):158–160.

Zhang J.y. 2008. Research on ad hoc network attack based MAC protocol. *Radio Engineering* 38 (10):4–6.

Chen F. & Luo Y.X. et al. 2007. Progress of research of network attack technology. *Journal of Northwestern University: Natural Science* 37 (2):208–212.

Zhu J.W. & Han X.H. et al. 2006. Network attack plan recognition algorithm based on extended goal graph. *Chinese Journal of Computers* 29 (8):1356–1366.

Yu L. & Chen B. et al. 2006. A network attack path reconstruction program. *Journal of University of Electronic Science and Technology of China* 35 (3):392–395.

Zhang Y.G. & Li D.X. 2006. Analysis of network attack and intrusion under IPv6. *Computer Science* 33 (2):100–102.

Noubir G. & Lin G. 2003. Low power DoS attacks in data wireless LANs and counter measures. *ACM MOBIHOC* 26(12):62–69.

Raya M. & Hubaux J. E. et al. 2004. DOMINO: A system to detect greedy behavior in ieee 802.11 hotspots. *Proc. Of ACM MobiSys*, 84–97.

John R. 2002. The Sybil Attack. http://research.microsoft.com/sn/Farsite/IPTPS2002.pdf.

Zeng H.Y. & Zhang W.Z. et al. 2007. Research on the IEEE 802.11 MAC sublayer attack mode. *Network Security Technology & Application* 9:39–41.

Cao C.J. & Yang H.W. et al. 2008. RTS-CTS attack on IEEE 802.11CSMA/CA. *Communication Countermeasures* 4:32–35.

*Multimedia, Communication and Computing Application – Leung (Ed.)*
© *2015 Taylor & Francis Group, London, ISBN 978-1-138-02775-6*

# Multiple-attribute vertical handoff algorithm based on adjusting weights and fuzzy comprehensive evaluation

W.S. Sun, Y. Zhou, T. Chen & J.S. Yao
*Hangzhou Dianzi University, Zhejiang, China*

ABSTRACT: To effectively manage and utilize wireless resources of heterogeneous network, and provide a higher quality of service for the user, a vertical handoff algorithm which based on adjusting weights and fuzzy comprehensive evaluation is proposed. First, the fuzzy subordinate functions can be defined by analyzing the fuzzy comprehensive evaluation model. The normalized value and impact factors are used to construct the judgment matrix. The attribute weights are calculated by methods of AHP and entropy. Objective optimization model can be constructed by the judgment matrix and the weights. The optimum network can be selected according to the total membership grade. The feasibility and validity of the algorithm were confirmed by the simulation.

## 1 INTRODUCTION

In recent years, a rapid development of various wireless access technologies provides the user with a wide range of services across different media. At the same time, largely different characteristics like data rate for cellular networks, wireless local area network (WLAN) result in handoff asymmetry that differs from the traditional intranetwork. Cooperation of heterogeneous wireless networks can improve the capacity of networks and provide better service for users. So an appropriate handoff is especially important to choose the most suitable network access. Some previous studies on vertical handoff are based on the received signal strength (RSS), in which handoff decisions are made by only comparing the RSS with the preset threshold[1]. This handoff algorithm is relatively simple, but it only considers the influence of a single factor on the network and judgment may be inaccurate. The handoff algorithm depends on gray prediction mainly through the early prediction of the RSS to make decisions. It can achieve the purpose of reducing handover delay, but the algorithm is relatively complex[2]. Analytic Hierarchy Process (AHP) is usually to get the weight of each factors subjectively to construct the judgment matrix and select a better network access[3-4]. The theory of fuzzy has a great advantage in dealing with the uncertain problems, so many algorithms based on fuzzy logic were proposed[5-7]. Traditional method to obtain the attribute weight by AHP is always subjective, so adjusting the weight and proposing an effective algorithm is especially important. Simulation results illustrate that the proposed handoff decision algorithm can choose the optimum network through considering between the network conditions and user preference and outperform the other approaches in the number of handoffs.

## 2 ALGORITHM PROCESSES

As we all know, using only one metric for vertical handoff decision made is not efficient since existing networks overlap with each other. So more performance metric need to be taken into consideration, such as RSS, bandwidth, delay, and user preference. The case of vertical handoff algorithm involves the following four main metrics: RSS ($\alpha$), delay ($\beta$), bandwidth ($\gamma$), and user preference ($\eta$).

### 2.1 Normalized values

Different metrics have different standards, so normalization is needed to ensure that the values in different units are meaningful. The normalized function is given by (1):

$$v_i(x) = \frac{x_i - x_{min}}{x_{max} - x_{min}} \tag{1}$$

Where $x_i = \alpha, \beta, \gamma, \eta$. $v_i(\alpha)$, $v_i(\beta)$, $v_i(\gamma)$, $v_i(\eta)$ are the normalized values.

### 2.2 Fuzzy membership function

Each of the normalized parameters is assigned to three fuzzy sets by membership functions, and the fuzzy sets are: low, medium, and high. Select the suitable membership function as follows.

$$f_L(v) = \begin{cases} 1, & v < k_1 \\ \dfrac{2v - 2k_1}{k_2 - k_1}, & k_1 < v \leq \dfrac{k_1 + k_2}{2} \\ 0, & v > \dfrac{k_1 + k_2}{2} \end{cases} \tag{2}$$

$$f_M(v) = \begin{cases} 0, & v \leq k_1 \\ \dfrac{2v - 2k_1}{k_2 - k_1}, & k_1 < v \leq \dfrac{k_1 + k_2}{2} \\ \dfrac{2v - 2k_1}{k_1 - k_2}, & \dfrac{k_1 + k_2}{2} < v \leq k_2 \\ 1, & v > k_2 \end{cases} \quad (3)$$

$$f_H(v) = \begin{cases} 0, & v \leq \dfrac{k_1 + k_2}{2} \\ \dfrac{2v - k_1 - k_2}{k_2 - k_1}, & \dfrac{k_1 + k_2}{2} < v \leq k_2 \\ 1, & v > k_2 \end{cases} \quad (4)$$

## 2.3 Judgment matrix

According to the normalized parameters and the values $k_1$, and $k_2$, the membership vector can be obtained as

$$(f_L^{(v)}, \ f_M^{(v)}, \ f_H^{(v)}) \quad (5)$$

The membership degree value reflects characteristics of each metric that can be obtained through membership function and the factors which have the effect on the network selection. And influence vector are given by (6)

$$\left( J_L^{(v)}, \ J_M^{(v)}, \ J_H^{(v)} \right) = \left( \dfrac{v(x) - k_1}{k_2}, \ \dfrac{v(x) - k_1}{k_2 - k_1}, \ \dfrac{v(x)}{k_2} \right) \quad (6)$$

For instance, membership value $M$ can be computed as

$$M = \left( f_L^{(v)}, \ f_M^{(v)}, \ f_H^{(v)} \right) \bullet \left( J_L^{(v)}, \ J_M^{(v)}, \ J_H^{(v)} \right) \quad (7)$$

Judgment matrix $A$ is made up of $M$.

## 2.4 Weight

### 2.4.1 AHP

The weight relations of metrics are determined by eigenvalue method of AHP to reflect the importance and relationships of the input metrics according to the service type. The steps to obtain the weight by AHP are as follows:

Determine the metrics which have effect on the network selection; in this paper, metrics are RSS, bandwidth, delay, and user preference. Construct pairwise comparison structure and the metrics considered are at level-1.

Ensure the candidate networks at level-2. The comparison structure can be shown in figure 1.

Calculate the weights of the decision metrics.

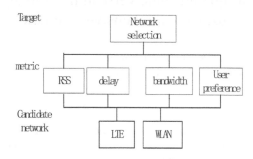

Figure 1. Pairwise comparison structure.

### 2.4.2 Entropy

The concept of entropy is a measure of the uncertain system state. Each information entropy of the decision metric can be calculated by its property. The greater the information entropy is, the smaller the weight is. The smaller the information entropy is, the greater the weight is.

$$E_{ik} = -K \sum_{i}^{M} P_{ik} \ln P_{ik} \quad (8)$$

Where $p_{ij}$ is the metric i of candidate network $k$, which contributes to the selection. $M$ is the number of the candidate networks. $E_j$ is the expectation, and

$$0 < E_j < 1, \ K = \frac{1}{\ln M}.$$

$$W_j = -\dfrac{1 - E_j}{\sum_{j=1}^{n} (1 - E_j)} \quad (9)$$

where $w_j$ is the weight, and $n$ is the number of metrics.

## 2.5 Adjusting weight

Assume that $W'_j$ is the weight obtained by AHP, and the weight vector is $W' = \left[ w'_1, w'_2, ..., w'_j \right]$, and $W'_j \in [0 \ 1]$, $\sum W'_j = 1$. $W''_j$ is the weight calculated by the method of entropy, and the weight vector is $\mathbf{W}'' = \left[ w''_1, w''_2, ..., w''_j \right]$ and $W''_j \in [0 \ 1]$, $\sum W''_j = 1$.

We can combine the two weights like this:

$$W = \lambda W' + \mu W'' \quad (10)$$

Where $\lambda$ and $\mu$ are the constants which relate to the two weights. These two values satisfy $\lambda^2 + \mu^2 = 1$ and $\lambda, \mu > 0$.

$$c_k = \sum_i a_{ij}^k w_i = \sum_i a_{ij}^k (\lambda w_i' + \mu w_i'') \qquad (11)$$

$$\text{Max}C = \sum_k c_k = \sum_k \sum_i a_{ij}^k (\lambda w_i' + \mu w_i'') \qquad (12)$$

Where $c_k$ is the performance evaluation value of each network. $a_{ij}$ is the membership value which is from the judgment matrix $A$. The highest value of $c_k$ prove that the network $k$ is the best for the user at this moment.

The adjusted weight can be computed through the Lagrange function and the partial derivative of the equation to the constants $\lambda$ and $\mu$.

$$L = \sum_k \sum_i a_{ij}^k (\lambda w_i' + \mu w_i'') + \frac{\omega}{2}(\lambda^2 + \mu^2 - 1) \qquad (13)$$

$$\begin{cases} \dfrac{\partial L}{\partial \lambda} = 0 \\[2mm] \dfrac{\partial L}{\partial \mu} = 0 \end{cases} \qquad (14)$$

The optimal value of $\lambda$ and $\mu$ can be obtained by (15) and (16):

$$\lambda = \frac{\sum_k \sum_i a_{ij}^k w_i'}{\sqrt{(\sum_k \sum_i a_{ij}^k w_i')^2 + (\sum_k \sum_i a_{ij}^k w_i'')^2}} \qquad (15)$$

$$\mu = \frac{\sum_k \sum_i a_{ij}^k w_i''}{\sqrt{(\sum_k \sum_i a_{ij}^k w_i')^2 + (\sum_k \sum_i a_{ij}^k w_i'')^2}} \qquad (16)$$

After the calculation of adjusted weight by the $\lambda$ and $\mu$, the performance value of the candidate networks can be achieved by integrating the four weight. In this way, the membership values assign to the four adjusted weight and the target network is selected.

### 2.6 Handoff decision

The vertical handoff decision can occur when a mobile node using service in one network enters to another one to obtain a high Qos. Each step can be presented as follows:

Assure the candidate networks.

Assure the metrics, obtain the membership values, and each adjusted weight.

Select the network with the largest $c_k$. If the current network is better than others, keep staying in the current one.

Otherwise, make handoff to the better one.

The flowchart of the handoff algorithm based on adjusting weights and fuzzy comprehensive evaluation is shown in Figure 2.

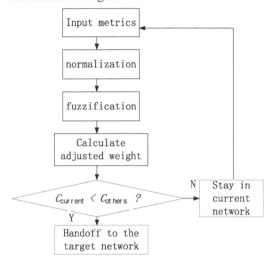

Figure 2.   Flowchart of the algorithm.

## 3   SIMULATION ANALYSIS

A heterogeneous wireless network integrating the LTE and WLAN is demonstrated in Figure 3. As we all know, LTE supports low bandwidth and relatively low data transmission rate over a large coverage area, whereas WLAN provides high bandwidth and high data transmission rate in a limited service area. Using the characteristics of the two networks to select the better one in different conditions and provide a high-quality service for the users. But if too many mobile terminals switch to the same one may have the negative effect. So many factors should be comprehensively considered in the algorithm. RSS is used to indicate the availability of a network. And the bandwidth is used to indicate the network conditions. The delay is used to reflect the performance of the handoff. Considering the user preference can satisfy different people's requirement.

According to the frequency of the LTE and WLAN, we choose the Cost231-Hata as the model to simulation. In this model, the RSS can be evaluated by(17)

$$RSS(d) = p_k - p_k(d) + C_k(\mu, \delta) \qquad (17)$$

Where $p_k$ is the transmit power, $p_k(d)$ is the path loss at distance $d$, and $C_k(\mu, \delta)$ is the Gauss random variables.

In the simulation, we consider the performance of the handoff algorithm with different numbers and velocities of mobile terminals.

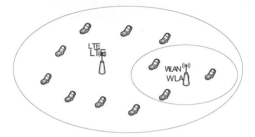

Figure 3.    Simulation model.

Table 1.    Simulation parameters.

| Parameters | Values | Parameters | Values |
|---|---|---|---|
| Transmitter power (LTE) | 43 dBm | Transmitter power (WLAN) | 20 dBm |
| Cell radius (LTE) | 1000 m | Cell radius (WLAN) | 100 m |
| Carrier frequency (LTE) | 2 GB | Carrier frequency (WLAN) | 2.4 GB |
| $k_1(\alpha)/k_2(\alpha)$ | 0.2/0.8 | $k_1(\beta)/k_2(\beta)$ | 0.3/0.7 |
| $k_1(\eta)/k_2(\eta)$ | 0.25/0.75 | $k_1(\gamma)/k_2(\gamma)$ | 0.3/0.7 |

Simulation parameter is shown in table 1. We first considered the performance under number of users ranging from 50 to 500 and the mean arrival rate equal to 10 m/s. Figure 4 shows the performance with different algorithms. Algorithms based on the RSS have the largest number of handoffs in comparison to the other two algorithms. And the algorithm based on fuzzy comprehensive evaluation (FCE), which take many metrics into consideration has the larger number of handoffs than the algorithm based on the adjusting weight and fuzzy comprehensive evaluation (AWFCE). So we can know that AWFCE can reduce the unnecessary handoffs.

Figure 5 shows the performance under different arrival rates ranging from 5 to 14 m/s with different algorithms. The simulation results show that the higher the rate, the larger the number of handoffs and AWFCE is also better than other approaches.

## 4  CONCLUSION

In this paper, we propose an algorithm which takes various input parameters into consideration and provides a generalized vertical handoff decision procedure. Simulation results confirm that this algorithm reduces unnecessary handoffs and provides a better network performance.

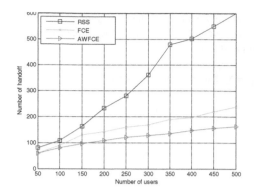

Figure 4.    Number of handoffs versus number of users.

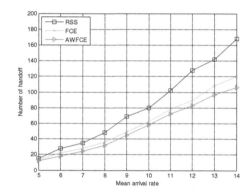

Figure 5.    Number of handoffs versus mean arrival rate.

## REFERENCES

[1] Kunarak S, Sulessathira R, Dutkiewicz E. Vertical handoff with predictive RSS and dwell time[C]. Xi'an: TENCON 2013 - 2013 IEEE Region 10 Conference. 2013:1–5.
[2] Fan Xueming, Tang Hong. Cross-layer handover optimization using grey prediction model [J]. Journal of Computer Application. 2010, 30(1):137–139.
[3] Preethi G A,Chandrasekar C. A network selection algorithm based on AHP-OW a methods[C]. Dubai: Wireless and Mobile Networking Conference (WMNC). 2013:1–4.
[4] Liu Shengmei, Pan Su. A Simple Additive Weighting Vertical Handoff Algorithm Based on SINR and AHP for Heterogeneous Wireless Networks[J]. Intelligent Computation Technology and Automation (ICICTA). 2010:347–350.
[5] Kwong Chiew Foong. Adaptive Network Fuzzy Inference System (ANFIS) Handoff Algorithm[C]. Future Computer and Communication. 2009:195–198.
[6] Qing He. A fuzzy logic based vertical handoff decision algorithm between WWAN and WLAN[J]. 2010:561–564.
[7] Aziz A, Rizvi S, Saad, N M. Fuzzy logic based vertical handover algorithm between LTE and WLAN[C]. Intelligent and Advanced Systems (ICIAS), 2010:1–4.

*Multimedia, Communication and Computing Application – Leung (Ed.)*
*© 2015 Taylor & Francis Group, London, ISBN 978-1-138-02775-6*

# Performance evaluation in PAPR for SIM-OFDM systems

X.B. Zou, Q.L. Ma & J.Q. Xie
*National Key Laboratory of Science and Technology on Communications, University of Electronic Science and Technology of China, Chengdu, Sichuan, China*

ABSTRACT: Subcarrier-Index Modulation for Orthogonal Frequency Division Multiplexing (SIM-OFDM) is a recently developed technique, where the indices of OFDM subcarriers are utilized to convey information bits. For the conventional OFDM system, the main drawbacks lie in the high Peak-to-Average Power Ratio (PAPR) at transmitter. In this paper, the performance evaluation in PAPR for SIM-OFDM system is presented and PTS method to reduce PAPR is applied. Simulation results show that SIM-OFDM system can efficiently reduce the PAPR and by using PTS method, better PAPR performance can be obtained.

## 1 INTRODUCTION

Orthogonal frequency division multiplexing (OFDM) is the key modulation technique for 4G broadband wireless communications due to its robustness against frequency-selective channel. For OFDM schemes, the wideband frequency-selective channel is divided into many narrowband frequency flat subchannel; hence, the intersymbol inference (ISI) caused by the multipath fading is efficiently removed. Furthermore, OFDM has the advantage of efficient spectral bandwidth and better performance compared with single carrier systems.

Subcarrier-index modulation for orthogonal frequency division multiplexing (SIM-OFDM) [1] is a new promising OFDM transmission technique, in which the indices of the subcarriers are used to modulate part of the information bits. In Ref. [2], the original bit mapping style of SIM-OFDM is developed to offer a better tradeoff between energy efficiency and system performance compared to OFDM. In Ref. [3], the advantage of SIM-OFDM in energy efficiency is also described. In Refs [4] and [5], the authors developed the transmit structure of SIM-OFDM by selecting more than one active subcarriers for a higher spectrum efficiency and proposed near-ML detection detectors for real implementation. Another advantage of SIM-OFDM that is shown in Refs [5]-[7] is that it is more robust to intercarrier interference (ICI) caused by the high mobile environment.

For an OFDM system, high peak-to-average power ratio (PAPR) brings on signal distortion in the nonlinear high-power amplifier, which degrades the bit error ratio (BER) performance. Hence, the problem of high PAPR in OFDM system has received remarkable attention [8]-[9].

For the sake of developing the superiority in PAPR for SIM-OFDM system, the performance evaluation in PAPR is simulated in this letter. It is shown that SIM-OFDM is able to reduce the PAPR significantly as the number of subcarriers decrease. And by using PTS method, the PAPR performance can be much reduced.

The remainders of this paper are organized as follows. Section II outlines the system model of SIM-OFDM system. Section III introduces the PAPR of SIM-OFDM and gives the PTS method to reduce PAPR. The PAPR performance of SIM-OFDM and PTS method are presented in section IV. Finally, section V concludes this paper.

The following notations are used throughout the paper. An SIM-OFDM scheme employing M-PSK/QAM is denoted by SIM-OFDM $(L, K, N) - M - PSK / QAM$, where $L$ and $K$ are the number of subcarriers and activated subcarriers in each subblock, respectively, whereas $N$ denotes the number of total subcarriers in one OFDM symbol. Furthermore, $(\cdot)^T$ represents the transpose of a vector/matrix.

## 2 SUBCARRIER INDEX MODULATION OFDM

We consider a SIM-OFDM system with $N$ subcarriers, so all the subcarriers are divided into $G$ subblocks, each containing $L=N/G$ subcarriers. Then the transmitted signal can be given as:

$$\mathbf{X} = \left[ \mathbf{Y}_0^T, \mathbf{Y}_1^T, ..., \mathbf{Y}_{G-1}^T \right]^T \tag{1}$$

where $\mathbf{Y}_g = \left[Y_{g,0}, Y_{g,1}, ..., Y_{g,L-1}\right]^T, g \in (0, G-1)$ denotes $g$-th subblock. In SIM-OFDM system, the subblock $\mathbf{Y}_g$ is the basic unit for information modulation. For each subblock, $K$ out of $L$ subcarriers are activated to transmit modulated symbols, and others transmit zeroes. The total combinations are $C_L^K$, but only $2^{\lfloor \log_2(C_L^K) \rfloor}$ combinations are permitted for modulation bits, where $C_L^K$ denotes the binomial coefficient and $\lfloor x \rfloor$ is the greatest integer smaller than $x$. Therefore, $p_1 = \lfloor \log_2(C_L^K) \rfloor$ bits are mapped into a set of subcarrier indices combination, and $p_2 = K \log_2(M)$ bits are modulated symbols transmitted by the $K$ active subcarriers, where $M$ is the modulation order of the QAM modulation scheme. After subcarrier-index modulation, the $g$-th subblock signal can be expressed as:

$$\mathbf{Y}_g = \left[0..., 0, s_1, 0, ..., 0, s_2, 0..., 0, s_k, 0, ...0\right]^T \tag{2}$$

where $s_k, k = 1, 2, ..., K$ denotes the modulary QAM constellation point. From the above analysis, the number of modulated bits for one SIM-OFDM symbol is $G(p_1 + p_2)$.

The transmitter structure of SIM-OFDM (2,1,8) is depicted in Fig. 1. As is shown in Fig. 1, there are 4 subblocks, each subblock containing two subcarriers. For each subblock, bit "0" is mapped into the first subcarrier, whereas bit "1" is mapped into the second subcarrier, and QAM symbol is transmitted by the activated subcarrier.

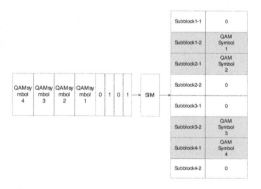

Figure 1. Transmitter structure of SIM-OFDM.

## 3 PAPR PERFORMANCE OF SIM-OFDM

### 3.1 Introduction of peak-to-average power ratio

According to Eq. (3), after IFFT processing, the based-equivalent SIM-OFDM symbol can be expressed as:

$$x_n = \frac{1}{\sqrt{N}} \sum_{k=0}^{N-1} \hat{X}_k e^{j2\pi kn/N} \qquad n = 0,1,...,N-1 \tag{4}$$

The PAPR of SIM-OFDM symbol is defined as:

$$PAPR = \frac{\max\left\{|x_n|^2\right\}}{P_x} \qquad n = 0,1,...,N-1 \tag{5}$$

where $P_x$ is the average power of transmitted symbol $\hat{X}$.

The cumulative distribution function (CDF) of PAPR is used mostly to measure the PAPR performance of a system. The complementary CDF (CCDF) denotes the probability that the PAPR of the transmitted symbol exceeds a given threshold. For OFDM system, an approximate expression is derived for CCDF of the PAPR of a multicarrier signal with Nyquist rate sampling. According to the central limit theorem, the real and imaginary parts of signal samples follow Gaussian distribution, with mean of 0 and variance of 0.5 for a multicarrier signal. Hence, the amplitude of signal follows Rayleigh distribution.

The CCDF of PAPR of a multicarrier symbol with Nyquist rate sampling can be derived as [8]:

$$P\{PAPR > z\} = 1 - P\{PAPR \leq z\}$$
$$\approx 1 - F^N(z) \qquad z \geq 0 \tag{6}$$

where $F(z)$ is the CDF of amplitude of signal and $N$ signal samples are assuming independent with each other.

### 3.2 PTS method to reduce PAPR

For an SIM-OFDM system with N subcarriers, the input data sequence is firstly partitioned into V disjoint subblocks $X^{(v)}, v = 1, 2, \cdots, V$, where all the subcarriers which are occupied by the other subblocks are set to zero, $X = \sum_{v=1}^{V} X^{(v)}$.

By multiplying a phase factor $b^{(v)} = \exp(j\theta^{(v)})$ to the $v$th subblock $X^{(v)}$, $v = 1, 2, \cdots V$, $\theta^{(v)} \in [0, 2\pi)$, the alternate frequency domain sequence can be expressed as:

$$X' = \sum_{v=1}^{V} b^{(v)} X^{(v)} \tag{7}$$

Then, one candidate time domain sequence can be given by

$$x' = IFFT\{\sum_{v=1}^{V}(b^{(v)}X^{(v)})\} = \sum_{v=1}^{V}b^{(v)}IFFT\{X^{(v)}\} \qquad (8)$$

The number of candidates is decided by the sub-blocks V and phase factor size $M$: $M^{(V-1)}$.

Finally, the candidate sequence with the lowest PAPR is selected for transmitting.

There are three different ways to divide V disjoint subblocks, such as adjacent, pseudorandom, and interleaved.

1 (a)    Adjacent division
Put a length of adjacent subcarriers into a subblock

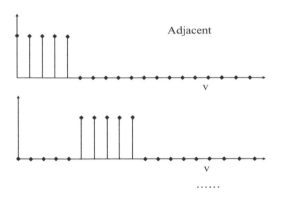

2 (b)    Pseudorandom division
Subcarriers from one subblock is chosen by pseudorandom way.

3 (c)    Interleaved division

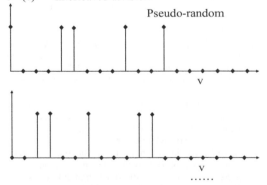

Choose subcarriers at a fixed interval into a subblock.

4    PAPR SIMULATION OF SIM-OFDM

In this section, we simulate the PAPR performance for SIM-OFDM at different configurations and the PAPR performance using different division ways of PTS.

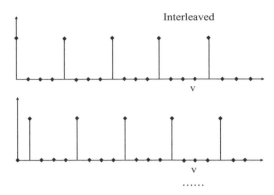

Fig. 2 shows the PAPR performance comparisons between OFDM and SIM-OFDM. It shows that SIM-OFDM can get a 0.6 dB gain comparing with OFDM. Fig. 3 shows the difference at different configurations (different L and K) with 64 subcarriers aided QPSK modulation. It is shown that SIM-OFDM system has lower PAPR compared with classical OFDM system. It is clear that the SIM-OFDM (16,1,64) is capable of achieving more than 4 dB gain at $10^{-3}$ CCDF compared with that of conventional OFDM. Furthermore, we find that the reduction gap becomes larger as the ratio $K/L$ decreases.

Fig. 4 shows the PAPR performance using the method PTS of SIM-OFDM and SIM-OFDM. Obviously, PAPR can be reduced by PTS method. Fig.5 shows PAPR performance in different division ways with 256 subcarriers aided QPSK modulation. It is shown that the pseudorandom way can get 0.2 dB performance gain at $10^{-3}$ CCDF compared with adjacent way, and adjacent division can get 0.7 dB gain than interleaved division. This is because the correlation of subcarriers with different

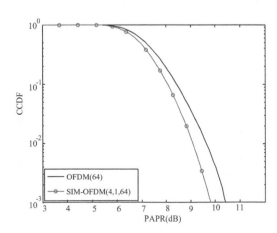

Figure 2.    PAPR performance comparison between OFDM and SIM-OFDM with 64 subcarriers.

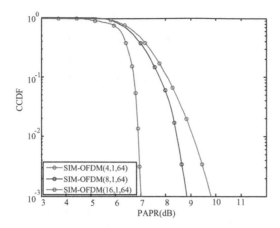

Figure 3. PAPR performance comparison between different configurations of SIM-OFDM with 64 subcarriers.

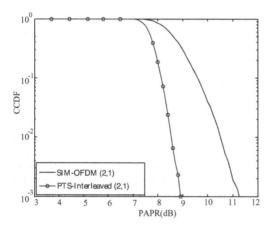

Figure 4. PAPR performance comparison between SIM-OFDM and SIM-OFDM using PTS method.

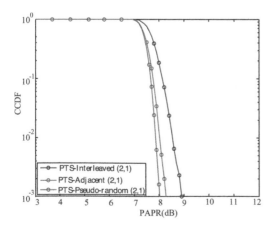

Figure 5. PAPR performance comparison between different PTS divisions with 256sucarriers.

divisions is different from each other. It is obvious that pseudorandom division has the lowest correlation character.

## 5 CONCLUSION

SIM-OFDM scheme is a novel developed OFDM technique, which can offer a better tradeoff between the energy efficiency and system performance as opposed to conventional OFDM. In this paper, the performance evaluation in PAPR and the way to reduce PAPR for SIM-OFDM system is presented. Simulation results demonstrated that SIM-OFDM system shows the PAPR performance using PTS method.

## REFERENCES

[1] R.Abualhiga and H. Haas," Subcarrier-Index Modulation OFDM,"//in Proc of the International Symposium on Personal Indoor and Mobile Radio Communications, Tokyo, Japan, 2009, pp. 13–16.

[2] Tsonev D, Sinanovic S, Haas H. Enhanced subcarrier index modulation (SIM) OFDM[C]//GLOBECOM Workshops (GC Wkshps), 2011 IEEE. IEEE, 2011: 728–732.

[3] Zhao L, Zhao H, Zheng K, et al. A high energy efficient scheme with Selecting Sub-Carriers Modulation in OFDM system[C]//Communications (ICC), 2012 IEEE International Conference on. IEEE, 2012:5711–5715.

[4] Ertugrul Basar, Umit Aygolu, Erdal Panayirci and H. Vincent Poor," Orthogonal Frequency Division Multiplexing with Index Modulation" In Global Communications Conference, Anaheim CA, 2012, PP 4741–4746.

[5] Ertugrul Basar, Umit Aygolu, Erdal Panayirci and H. Vincent Poor. Orthogonal Frequency Division Multiplexing with Index Modulation, IEEE TRANSACTIONS ON SIGNAL PROCESSING. S.H. Han and J.

[6] Y. Xiao, X. Lei, Q. Wen and S. Q. Li, "A class of low complexity PTS techniques for PAPR reduction in OFDM systems," *IEEE Signal Process. Lett.*, vol. 14, no. 10, pp. 680–683, Oct. 2007.

[7] Basar E, Aygolu U, Panaylrcl E. Orthogonal frequency division multiplexing with index modulation in the presence of high mobility[C]//Communications and Networking (BlackSeaCom), 2013 First International Black Sea Conference on. IEEE, 2013:147–151.

[8] M.Pauli, H.P. Duchenbecker. Minization of the intermodulation distortion of a nonlineary amplified OFDM signal[C]. //Wireless Communications, 1996:(4): 93–101.

[9] L.J. Cimini, N.R. Sollenberger. Peak-to-power-average power ratio by optimum combination of partial transmit sequence[J] //IEEE Communication Letters, Feb.1997:33(5)5.

*Multimedia, Communication and Computing Application – Leung (Ed.)*
© 2015 Taylor & Francis Group, London, ISBN 978-1-138-02775-6

# Fuzzy association algorithm and computer simulation of target's information from radar and AIS

Q.P. Chen, C.C. Lin, H. Lin, M. Guo, X.L. Zhang & S.S. Zhu
*Navigation Technology Institute of Jimei University, Xiamen, China*

ABSTRACT: Marine radar and AIS (Automatic Identification System) all can acquire the target's information at sea, including the ship's position, course, speed, and so on. However, they have some different characteristics. Therefore, it is very important to fuse the target's information to get the predominance. Based on the theory of multisensor information fusion, it studies on the association algorithm of the target's information from the two sensors by using multifactor fuzzy integration decision making. The mathematical models are established, including the fuzzy factor sets, membership degree, and judgment rules, and conducted the computer simulation. The results show that the algorithm can greatly improve the accuracy and reliability of the target information. It researches and discusses the relation between the association probability and the threshold.

## 1 INTRODUCTION

Marine radar is a traditional navigation equipment of ship, which can detect targets around on its own and get panoramic picture at sea. However, it has little information and low accuracy, and is hidden easily by terrain, which make it more difficult to identify a target. Especially, the ability of detection is consumedly weakened under the bad weather, like sea or rain clutter.

AIS, as a new type of navaid at sea, is more and more widely used in the ship because of its forcible requirement of IMO (International Maritime Organization) so as to improve maritime traffic safety. AIS has a lot of advantages. It can provide not only a great deal of target's information of the static state including a ship's name, ID, type but also the information of the dynamic state including a ship's position, course, speed, and so on. Moreover, it is not easily hidden by terrain, can easily identify a target, and is more accurate than radar. AIS brings on a new renovation of navigation at sea.

However, there exist disadvantages in AIS, including easy overload in communication. Further, the ships of nonforced fitted AIS, such as yacht and the floater on the sea, are detected only by radar 1.

Hence, it is worthy to fuse the targets from the two sensors to get the mutual benefits in the performance, which can improve the accuracy and reliability of the target's information.

## 2 BRIEF OF AIS AND MARINE RADAR

### 2.1 AIS

The AIS can provide information automatically and continuously for a competent authority and other ships, without involvement of ship's personnel. At the same time, it can receive and process information from other sources, including those from a competent authority or from other ships. The structure of AIS is shown in figure 1.

AIS can satisfy the following functional requirements:

- In a ship-to-ship mode for collision avoidance
- As a means for littoral States to obtain information about a ship and its cargo
- As a VTS (Vessel Traffic Service) tool, i.e., ship-to-shore traffic management

The AIS information includes four types:

Static information: MMSI (Marine Mobile Station Identification), IMO number, Call sign & name, Length and beam, Type of ship, Location of position-fixing antenna on the ship.

Dynamic information: Ship's position with accuracy indication and integrity status, Time in UTC, Course over ground, Speed over ground, Heading, Navigational status, and so on.

Voyage-related information: Ship's draught, Hazardous cargo (type), Destination, and Short safety–related information.

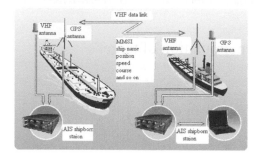

Figure 1.  AIS system composing diagram.

### 2.2  Marine Radar

Marine radar is a kind of independent navigational equipment to detect and measure the targets around by transmitting the electromagnetic waves and receiving the echo reflected from the targets around (figure 2). It can get the 6 dynamic target information as follows.

- Distance
- Bearing
- Course
- Speed
- CPA (Closest Point of Approach)
- TCPA (Time to CPA)

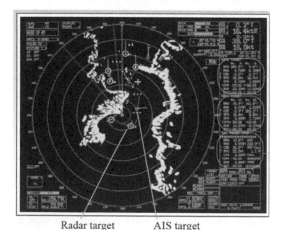

Radar target       AIS target

Figure 2.  Marine radar picture.

### 3  FUZZY ASSOCIATION ALGORITHM

The respective difference in the distance, bearing, course, and speed for the targets coming from radar and AIS are used as the fuzzy factor sets $U = \{u(1), u(2), u(3), u(4)\}$, as shown as (1).

$R(r)$ and $\theta(r)$, $R(a)$ and $\theta(a)$ are the range and bearing of the radar and the AIS, respectively. $SOG(r)$ and $COG(r)$, $SOG(a)$ and $COG(a)$ are the speed and the course over ground of the radar and the AIS,

respectively. $M$, $N$ are the target's total of radar and AIS respectively, and $L$ is an observer time. $\lambda(a)$ and $\varphi(a)$, $\lambda(o)$ and $\varphi(o)$ are the longitude and the latitude of AIS and own ship, respectively.

$$
\begin{cases}
u(1)_{ij}(k) = \left| R(a)_{ij}(k) - R(r)_{ij}(k) \right| \\
u(2)_{ij}(k) = \left| \theta(a)_{ij}(k) - \theta(r)_{ij}(k) \right| & i = 1, 2 \dots\dots M; \\
u(3)_{ij}(k) = \left| SOG(a)_{ij}(k) - SOG(r)_{ij}(k) \right| & j = 1, 2 \dots\dots N; \\
u(4)_{ij}(k) = \left| COG(a)_{ij}(k) - COG(r)_{ij}(k) \right| & k = 1, 2 \dots\dots L
\end{cases}
$$

(1)

Wherein:

$$
R(a)_{ij}(k) = \sqrt{(\lambda(a)_{ij}(k) - \lambda(o)_{ij}(k))^2 + (\phi(a)_{ij}(k) - \phi(o)_{ij}(k))^2}
$$

(2)

$$
\theta(a)_{ij}(k) = \arctan\left(\frac{(\lambda(a)_{ij}(k) - \lambda(o)_{ij})(k)}{(\varphi(a)_{ij}(k) - \varphi(o)_{ij})(k)}\right) * 180 / \pi
$$

(3)

The membership degree of the normal distribution which judges the similarity of the target information on the basic of $g$ factor is written as:

$$
\begin{aligned}
(r_{g1}(k))_{ij} &= \exp\{-\tau_g (u^2(g)_{ij}(k)/\sigma_g^2)\} \\
(r_{g2})_{ij} &= 1 - \exp\{-\tau_g (u^2(g)_{ij}(k)/\sigma_g^2)\} \\
g &= 1, 2, 3, 4 \\
i &= 1, 2, \dots\dots\dots M \\
j &= 1, 2, \dots\dots\dots N \\
k &= 1, 2, \dots\dots\dots L
\end{aligned}
$$

(4)

Where $\tau_g$ is adjustment coefficient, and $\sigma_g^2$ is the variance of the error.

So, the single-factor fuzzy judgment matrix is expressed as

$$
R_{ij}(k) = \begin{bmatrix}
(r_{11}(k))_{ij} & (r_{12}(k))_{ij} \\
(r_{21}(k))_{ij} & (r_{22}(k))_{ij} \\
(r_{31}(k))_{ij} & (r_{32}(k))_{ij} \\
(r_{41}(k))_{ij} & (r_{42}(k))_{ij}
\end{bmatrix}
$$

(5)

The fuzzy-weighed factor set A is shown below.

$$A = \begin{bmatrix} a_1 & a_2 & a_3 & a_4 \end{bmatrix} \qquad (6)$$

Here, $a_1$, $a_2$, $a_3$, and $a_4$ are the weighed coefficient in the distance, bearing, course and speed, respectively.

The integration decision making of the target information is the integrative action of the sets A and the judgment matrix $R$, which is synthesis operation [2,3,4]. The fuzzy sets $B$ is denoted as:

$$(B(k))_{ij} = A \circ (R(k))_{ij}$$

$$= \begin{bmatrix} a_1 & a_2 & a_3 & a_4 \end{bmatrix} \circ \begin{bmatrix} (r_{11}(k))_{ij} & (r_{12}(k))_{ij} \\ (r_{12}(k))_{ij} & (r_{22}(k))_{ij} \\ (r_{31}(k))_{ij} & (r_{32}(k))_{ij} \\ (r_{41}(k))_{ij} & (r_{42}(k))_{ij} \end{bmatrix} \qquad (7)$$

$$= \begin{bmatrix} (b_1(k))_{ij} & (b_2(k))_{ij} \end{bmatrix}$$

The maximum is

$$\max(b_1(k))_{ij} = \max[(b_1(k))_{i1} \ (b_1(k))_{i2}$$
$$\ldots\ldots(b_1(k))_{ij}\ldots\ldots(b_1(k))_{iN}] \qquad (8)$$
$$j = 1, 2\ldots\ldots\ldots N$$

The parameter $FT_{a1}$ is used as the threshold to judge the relevancy degree of the target information at $l$th grade. When satisfying (9), it is correlative [5,6].

$$\max(b_1(k))_{ij} \geq FT_{a1} \qquad (9)$$

## 4   COMPUTER SIMULATION

The parameters of the simulation are assumed as follows[7].

The own ship's speed is 12 kn (nautical mile/hour), and the course is 0°. The RMS ($\sigma$) of the radar error in the distance, the bearing, the speed, and the course is 30 m, 1°, 1.2 kn, and 12°, respectively.

The target course is 270°, and the speed is 11 kn. The original distance from the own ship is 4.8 nm, and the bearing is 40°. The position-fixing mode of target vessel is GPS, and the RMS ($\sigma$) of the error in the distance, the bearing, the speed, and the course is 12 m, 0.7°, 0.4 kn, and 4.8°, respectively. The parameter $\tau$g is chosen as follows.

$$\tau_1 = 0.001, \quad \tau_2 = 0.1, \quad \tau_3 = 0.1, \quad \tau_4 = 0.1$$

Figure 3 and figure 4 show the factor sets $u_1$ and $u_2$ and $u_3$ and $u_4$, respectively. Figure 5 reveals the associated ascription degree $b_1$, and figure 6 denotes the association probability $P_a$ with the different $FT_{a1}$ and $\tau_g$.

We can find out that the arithmetic is correct and effective from the simulated results. It is known that the higher the $FT_{a1}$, the lower the association probability $P_a$, and the $P_a$ can be close to 1 when $\tau_1$ is 0.01 and $FT_{a1}$ is 0.8.

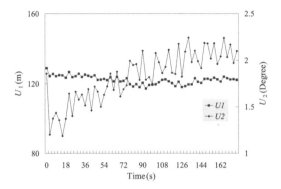

Figure 3.   U$_1$ and U$_2$ of target ship.

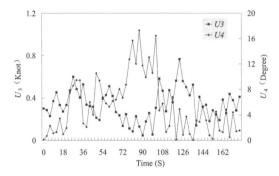

Figure 4.   U$_3$ and U$_4$ of target ship.

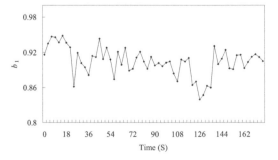

Figure 5.   Associated ascription degree $b_1$ of target ship.

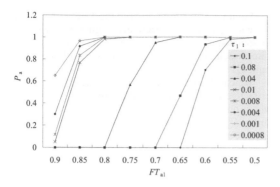

Figure 6. Association probability $P_a$ of different $FT_{a1}$ and $\tau_g$.

## 5 CONCLUSION

This paper studies and establishes the modeling of the fuzzy association of the target information from radar and AIS by the use of the multifactor fuzzy integration decision making. The computer simulation results show that the method is feasible and effective. It gets that the proper value of the $FT_{a1}$, which is about 0.8.

## 6 ACKNOWLEDGMENT

The authors thank the Main Subject Project of the Ministry of Transport of People's Republic China (2014329815090) for their support to complete this work.

REFERENCES

[1] ITU-R. Technical characteristics for a universal shipborne automatic identification system using time division multiple access in the VHF maritime mobile band. Draft *Revision of Recommendation*, ITU-R, M.1371, pp. 20–30, 2001.

[2] You H, et al. Multisensor Information Fusion with Applications. *Publishing House of Electronics Industry*: Bejing, China, pp.14–166, 2001.

[3] LIN Chang-chuan, et al, Discussion on the Method of the Fuzzy Fusion of the Object Track of Radar and AIS. *NAVIGATION OF CHINA*, 57(4), pp. 70–73, 2003.

[4] Lin Changchuan, Sun Tengda, Zhou Jianwen; Lin Hai. Development of a Display Platform of the AIS Information. *The 8th International Conference on Electronic Measurement and Instrument Proceedings*, Xi'An, China, Vol.2, pp. 2890–2893, 2007.

[5] Lin Changchuan, Lin Hai, Li Lina, Zhou Jianwen and Ou Yangping. Development of the Integrated Target Information System of the Marine Radar and AIS based on ECDIS. *The 5th IEEE WICOM 2009 Proceedings*, Beijing, China, 2009.

[6] Lin Changchuan and Dong Fang, Lin Hai, Le Lina, Zhou Jianwen and Ou Yangping. AIS Information Decoding and Fuzzy Fusion Processing with Marine Radar. *The 4th IEEE WICOM 2008 Proceedings*, Dalian, China, 2008.

[7] Ou Yangping, Lin Changchuan, Zhou Jianwen, Chen Guoquan. Study on the Target Association Arithmetic of the Marine Radar and AIS. *The 9th International Conference on Electronic Measurement & Instruments (ICEMI)*, Beijing, China, 2009.

# Data migration research between relational database and nonrelational database

D. Li & Y.F. Chen

*Management School of Xi'an University of Architecture and Technology, Xi'an, Shaanxi Province, China*

ABSTRACT: Data migration has been a database and database system, which is worth to study. It is related to the topic and we need to understand lots of problems. The topic of migration from relational database to a nonrelational database is a new field. The differences between relational databases and nonrelational database, and model designing for the subsequent relationship technology have been presented in this paper.

KEYWORDS: Data migration, Relational database, Nonrelational database, Migration model, Metadata.

## 1 INTRODUCTION

With the continuous growth and development of database theory, big data have become increasingly well known. The paper is based on large database to research relational data migration to nonrelational database. The article has three parts: The first part introduces the differences between modes. The second part gives the differences in data processing. A reliable theoretical model is proposed in the last part.

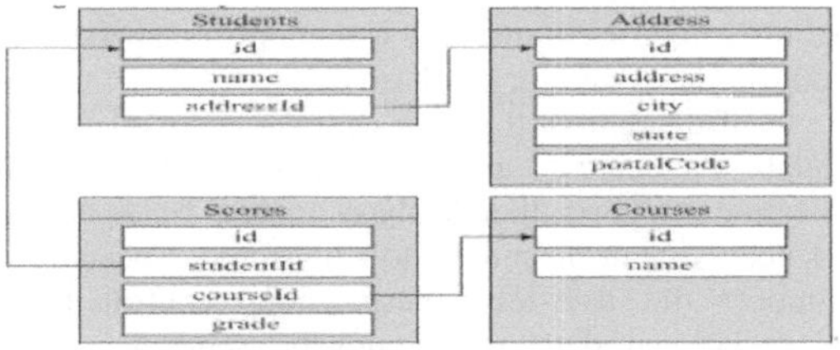

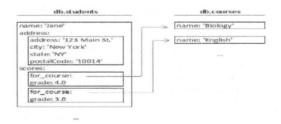

Figure 1. Relational database design.

### 1.1 Difference mode

In the data storage model, relational and nonrelational databases are very different. It is generally accepted that relational database data are divided into three layers: databases, tables, and records, known as the concept layer, logic layer, and physical layer, respectively. They strictly abide by the three paradigm constraints. Rather than a relational database, although there is also a three-tier structure: databases, collections, document object (take MongoDB as an example), it is the logical and physical layers with a clear distinction between relational databases, which led directly to the nonrelational databases not observing three paradigm rules.

Tables in a relational database require strict accordance with the three paradigms to design, corresponding to the nonrelational database collection is unrestricted, and there are great differences in the design process, which the following will illustrate.

Suppose you want to design a small database to store "Students, addresses, subjects, results" and other information, the designs of relational and nonrelational databases are shown in Figure 1 and Figure 2, respectively:

Compare the Figures 1 and 2 to satisfy the constraint paradigm. Relational database design creates

Figure 2. Nonrelational database design MongoDB map.

four tables, rather than a relational database paradigm because you do not comply with the constraint, which only designed two collections that should be address and scores and the collection incorporates a set of two students, a table, and a collection of no-one correspondence.

It is also found that the nonrelational database is stored in an object, not one of the record, which means that the field you want to expand is very convenient. In a relational database, because all the relationships among all the columns are bound to comply with the three paradigms, there will be a variety of unforeseen problems when changes happen in field extensions.

Course and the student have many relationships and to avoid data redundancy, we will propose a set course alone. This design is consistent and a relational database. Contrast to relational databases, nonrelational databases have several advantages: Firstly, the data retrieval do not need to interconnect (join) the huge overhead; Secondly, the data can be stored together to the disk more easily, and the data can be read or write more quickly. In addition, we do not need to worry about scalability issues, for the characteristic of non-relational database makes it easier to add or delete field changes.

## 1.2 *The difference between safety*

For databases, securing storage of critical information is important. It requires confidentiality, availability, and integrity. Confidentiality requires that the data can only be given to a person using the system or database roles; availability of data is that the data be stored can be used; integrity is that only one person to modify the system or database roles.

Confidentiality: the role of integrated relational database security and encryption-based communication supports the database tables, fields, rows of access control, and on top of stored procedures, support the user-level permissions control role. Nonrelational database cannot be separated from the privileges on tables, rows, or columns and be used to ensure fast access to data, for it rarely has a built-in safety mechanism. The confidentiality of data, in relational databases it may be controllde by the system level, however in the relational databases it is controlled in the application layer. Thus, confidentiality relational database is better. Availability: whether it is a relational database or a nonrelational database, availability is its basic requirements, and both of which must possess.

Integrity : To ensure the reliability of the database transaction processing, integrity, persistence data replication, and log functions, relational database has to provide basic property ACID (atomicity, consistency, isolation, durability). Rather than a relational database based on the available offer, soft state, it is eventually consistent with these three basic attributes. Though it will not be consistent after each transaction, it can ensure the final state of consistent database. This means that users may not be able to see the latest data, and they will not see the data until the last snapshot. Not all users can view the same data at the same time, and this inherent competitive condition is a real risk of database processing faces.

By comparison, it is found that relational databases and nonrelational databases are totally different. The following is the comparison between MySQL and MongoDB.

## 1.3 *Data type of problem*

Data type is the most basic thing in the database, and the database is a migration conversion of a core module as shown in Table 1.

Table 1. Correspondence between data types.

| MySQL Data Types | MongoDB Data Types |
|---|---|
| Integer or Int | 32-bit integer |
| BIGINT | 64-bit integer |
| Double(m, n) | 64-bit float |
| char(n)/varchar(n) | string |
| char | char |
| SERIAL | objectID |
| Date | date |
| Bool | bool |
| Bit | Binary |

Table 1 just described MySQL and MongoDB corresponding data types. There are many data types which support on MongoDB, such as the following special data types.

Null: null value represents the field in a relational database, but in MongoDB it can represent either a null value in this field or the field does not exist.

Regular Expressions: MongoDB in the value field can be a regular expression (need to comply with the syntax of JavaScript regular expressions), so only treated as a string processing and then converted to the field.

Code: field in MongoDB can insert JavaScript code, so this type of treatment in MySQL is used only in text or string processing.

In addition to the several special types, MongoDB as well as minimum, maximum, undefined, and embedded documents, there are an array of other special data types.

More specific data type is used only for the data for special treatment values.

## 1.4 *DDL problem*

DDL database definition language includes definitions defining the structure and method of operation, and the language used to describe the database needs to store real-world entity. Relational databases and key-value database table definitions differ as shown in Table 2.

In MongoDB, there is no need to manually create a collection, database administrators will be automatically created when inserting data collection for the first time. Therefore, the problem of relational databases is to convert between DDL nonrelational databases, and you only need to extract the relational database in each table while removing any record which is inserted into the target nonrelational databases, and you can

| Table 2. Create different database tables. | |
|---|---|
| Relational database tables (MySQL) | Nonrelational database tables (MongoDB's collection ) |
| create table dds (id int primary key, title varchar(250)); | Without displaying created |

create a relational database to achieve noncollection. Perhaps there may be a doubt that the tables in a relational database records may be incomplete, which did not fill all the fields of the table. This problem can be solved by add and delete fields because the structure of the nonrelational database is free.

## 1.5  *DML problem*

DML database is a data manipulation language to insert, delete, update, and select the operation of the data manipulation language.

Insert statement: relational database will be used to insert data statement and insert adds values (1, 'Hello'); nonrelational databases need to create an object to insert data: First create the object object = {id: 1, title: 'Hello'}; then save the object db.dds.save (object); such a data is inserted into the appropriate collection. Object is the object created, id: 1 and title: 'Hello' is to insert columns and column values, db is the database in a global variable name, dds is a collection of data, which is mentioned in the relational database form, saving data is inserted to provide a method MongoDB. Uncertain object causes the column value of the object be changed. Therefore, the migration of data to a relational database traversal out and recreating of the object is saved to the corresponding nonrelational database, which can be fully realized to solve the problem of inconsistent data.

Delete statement: delete data use a relational database statement: delete from dds, where id = 1; nonrelational database use the remove method to delete the data provided by the database: db.dds.remove ({id: 1}), similar to the delete operation in the relational database.

Query statement: It includes subquery, the query conditions, and restriction inquiries. For general inquiries, relational database use statement selected * from dds; instead of relational database queries using the find method: db.dds.find (); find () method can be used in a variety of condition parameters, so it can be done in Relations type of database query conditions, subqueries restrict the operation of a series of inquiries.

Update statements: changes are generally used to update this record. Relational database use the statement: update dds set title = 'World', where id = 1; nonrelational database use statements: db.dds.update ({id: 1}, {$ set: {title: World}});

## 2   DESIGN RELATIONAL TO NONRELATIONAL DATABASE MIGRATION MODEL

### 2.1  *Framework for data migration tool model*

Model design framework is shown in Figure 3.

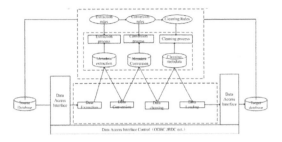

Figure 3.   Data Migration Model Framework.

Figure 3 module functions are as follows:

1  As long as the source database can belong to a relational database.
2  The target database as long as a nonrelational database can be mainly used to store data in the source database after cleaning and conversion.
3  Data interface is primarily used to access to the source database and the target database interface.
4  Data Interface Access control is mainly to access control source and target database interface. For example, ODBC or JDBC universal access.
5  The rules to develop the secondary module are in processing of metadata for setting the rule data. The main rule is set by the user of the data processing element.

### 2.2  *Processing the set of metadata and rules*

Processing the set of metadata and rules involves three steps:

1) Extracting the set of metadata and rules
The main frameworks for the extraction process are shown in Figure 4:

2) Setting rules and metadata conversion process.
The main framework of the conversion process is shown in Figure 5:

3) Washing the set of metadata and rules.
Cleaning the main framework for handling the following as shown in Figure 6:
After cleaning, all the data have the characteristics of the target metadata.

4) The source and target databases metadata design
Metadata design is shown in Figure 7:

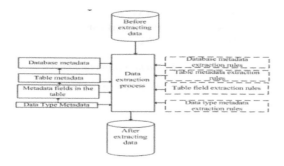

Figure 4.  Main frameworks for extraction process.

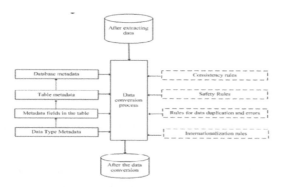

Figure 5.  Main frameworks for conversion process.

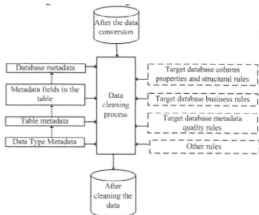

Figure 6.  Main frameworks for the cleaning process.

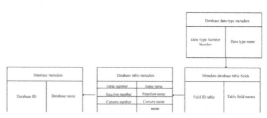

Figure 7.  Metadata Design.

## 2.3  The superiority of the model

The model compared with other database migration tool models has the following advantages:

1  High versatility. This model is based on the relationship between the types to design nonrelational databases. Therefore, it is not only applicable to relational data migration between relational databases but also applicable to nonrelational databases.

2  Good scalability. The model data are dependent on the rules set by the administrator, and the need to expand other rules is just rules set in the user module which can be added.

3  High data fault tolerance. All metadata throughout the system is running out among all separated ones, so it is extracted before the errors affecting the cleaning step, and so the data can be a good rollback to the previous state.

4  Ease of maintenance. All the operational model is therefore based on the metadata in a subsequent maintenance operation process if it is necessary to increase a portion of the metadata and it is only the metadata table needs to be maintained.

## 3  SUMMARY

Data migration between relational and nonrelational database is a very complex task and it cannot be solved by a simple research. We need to understand and master a variety of nonrelational databases so as to make a very generic data migration tool to migrate relational databases into a nonrelational databases. The process of migrating a relational database to a nonrelational database is used only when it comes to DDL and DML problems, and to make an in-depth study, we also need to study DCL, TCL, and some advanced SQL. A data migration model is presented in this paper. The use of metadata processing helps database migration process. Meanwhile, the model also have the features of a high data fault tolerance and maintainability, scalability, and versatility.

## REFERENCES

Huigang Cong, Qingdong Ren, Tianyang Li, Man Yuan. 2011. *Science Technology and Engineering Vol. 11(10):2353–2356.*

Xianghui Wang. 2011. *Technology Computer Programming Skills & Maintenance Vol.6 44–45.*

Shiyong Zheng. 2012. *Journal of hezhou university Vol.28(2) 132–135.*

Xingchun Diao, Hao Yan, Kun Ding. 2008. I *Computer Engineering Vol.34(17) :42–46.*

Long He, Jin-xian Lin. 2003. Computer Applications Vol.23(1) :31–32.

Boran Wang. 2013.D, *Journal of beijing polytechic college Vol.12(2) :26–30.*

Gang Wang, Dong Wang, Wen Li, Guangya Li. 2013. *Microcomputer Applications Vol.30(5): 1–3.*

*Multimedia, Communication and Computing Application – Leung (Ed.)*
© *2015 Taylor & Francis Group, London, ISBN 978-1-138-02775-6*

# Research on supporting technology of digital archives resource sharing and management system in the cloud computing environment

H.M. Xiao
*Jiangxi Science and Technology Normal University, China*

L. Chen, J.J. Cheng & M.M. Zhou
*Department of Information Management, Management College of Nanchang University, China*

ABSTRACT: According to the fact that digital archive resources scheduling options must follow the archive user's needs, this paper studies the scheduling mechanism of digital archive resources storage system in the cloud computing environment and proposes the dynamic scheduling mechanism and dynamic selection algorithm. Then, in view of metadata management style, metadata classification, and metadata content in digital archive resources sharing and management, this paper studies the metadata management mechanism of the distributed digital archive resources in cloud storage. Finally, in consideration of the coupling strength between cloud computing and digital archive resources, and the use ratio of resources combinations, this paper puts forward the plan how to realize "archive cloud" platform service based on Chord query algorithm.

KEYWORDS: Cloud technology, Digital resources, Archives, Service, Management system

## 1 INTRODUCTION

Cloud computing is a new internet-based computing technology, which is formed by distributed processing, parallel processing, and grid computing technology (Xiuzhen Feng et al. 2012). Cloud computing technology owns powerful parallel processing capacity and provides service to users. As for the cloud computing users, there is no need to care about the internal structure and implementation of its service. As a result, cloud service comes into being, which includes 3 patterns: software-as-a-service, platform-as-a-service, and infrastructure-as-a-service (Jeremy Leighton John, 2009).

Cloud computing is widely applied in many fields as a trend of network development and an advanced concept of processing information resources. It brings a lot of new conveniences and support to digital archive resources and saves investment cost of software and hardware during the process of establishing digital archive resources sharing and management system. At the same time, it reduces user demand for equipment maintenance knowledge. However, in order to build "archive cloud" service platform with cloud technique to allow variety of users to share digital archive resources, it requires unified format of electronic documents and digitized paper archives to remove the obstacle-"information separatism"; as a result, it lays a solid foundation for electronic documents and digitizes paper archives' archiving and

migrating in a unified standard, helps to guarantee the originality and truthfulness of electronic records, solves conflicts and disputes, reduces social transaction cost, and improves administration efficiency

From the perspective of provider and administrator of "cloud service" technology, author applies cloud computing technique to archival informatization in this digital era combined with features of cloud computing and actual situation of archival work. Then it systematically analyzes the construction of digital archive resources sharing and management system and proposes dynamic adjustment mechanism and selection algorithm on the basis of cloud computing. Owing to the commonplace between cloud computing service and digital archive resources, metadata's service mode, division of metadata and its service content in digital archive resource service system can be well studied to improve digital archives' service capability. This is of great importance to promoting the service extension, service content expansion, service mode, and innovating in coconstruction and sharing.

## 2 FEASIBILITY ANALYSIS

### 2.1 Theoretical basis analysis

From the theoretical point of view, it is reasonable to establish a digital archive resources sharing and management system based on cloud computing. And there is a relatively complete supporting system at theoretical level at present. Cloud computing theoretical system develops so fast that theoretical achievements emerge in an endless stream. All these works have a detailed introduction in cloud computing's origin, present situation, and develop tendency and provide overall theoretical support for cloud computing's application to modern network society by means of systematically analyzing and demonstrating the implementation of cloud computing system. Besides, there have been more and more theoretical achievements about digital archive resources construction and management in recent years. So there is no deny that the establishment of digital archive resources sharing and management system based on cloud computing has a solid theoretical foundation.

## 2.2 *Technical condition analysis*

From the technical point of view, cloud computing, which comes into being after distributed processing, parallel processing, and grid computing, is not a new technology. Liu Zhenpeng (2010) states that in significant measure it can be considered as a new technique scheme. At present, many domestic and foreign experts and IT organizations are studying it and have put forward some representative technological solutions. For example, Amazon has its own cloud computing web service which consists of simple storage service, elastic compute cloud, and simple queuing service. In China, plenty of experts, scholars, and institutions have proposed their own technological solutions. That is to say, from the technical point of view, it is feasible to build digital archive resources sharing and management system in cloud computing environment (i.e., "archive cloud").

A review of electronic file management in cloud computing environment, which is written by Sixin Xue and Cui Huang (2011), says that in terms of technical requirements, the fundamental functions of digital archive resources sharing and management system in cloud computing include: user identification, digital archive indexing service, human-centered store service, data exchanging service, and data scheduling service. It shows that interoperability is the core problem which must be resolved during the construction of digital archive resources sharing and management system in cloud computing environment.

## 2.3 *Practical environment analysis*

From the practical point of view, most internationally known IT enterprises have been studying and deploying cloud computing actively and have already achieved a certain degree of success. Nowadays, Over 500,000 enterprises have signed a contract to apply GAE, which owns nearly 10 million users. What's more, Chinese enterprises like Rising, Alibaba, and SDO have announced their own "cloud plan" and started undertaking to do it. So, application of cloud computing at this stage has formed a good social environment and provided some practical experience for the construction of digital archive resources sharing and management platform.

## 3 RESEARCH CONTENT

Digital archive resources management in cloud computing consists of storage resource management and computing resource management while in distributed system, and digital archive resources are divided into digital archive resources itself and metadata. Due to the absence of mechanisms and algorithms about digital archive resources and metadata dynamic management in current cloud computing framework, the author studies the definition and source of digital archive resources and metadata dynamic management as well as their coupling. So far, there have not been mechanisms and algorithms which specially direct at digital archive resources sharing and management system. On this condition, digital archive resources can be managed and reorganized in the running process of digital archive resources sharing and management system in order to improve the utilization of this system (Bing Li, 2012). Not only storage and dynamic scheduling but also metadata service of digital archive resources belongs to resources management technology in cloud storage. And that digital archive resources sharing and management cover both storage-oriented and computing-oriented system.

1 Study on scheduling service of digital archive resources sharing and management in cloud computing environment.

This study includes: theoretical significance of supporting technologies, practical significance of scheduling service, and background research; theoretical guidance of scheduling service, technical route, and solution architecture research; ontology content and method, system architecture, motivation, and mechanism.

2 Scheduling services implementation model of constructing archive resources sharing and management system.

The implementation model of digital archive resources' dynamic scheduling in "cloud service" is shown in Figure 1. It is the scheduling service technique solution of digital archive resources sharing and management system in cloud computing. Combined with the actual requirements of scheduling service, implementation model of digital archive resources dynamic scheduling service in "archive cloud" consists of resource layer, management middleware layer, and service layer.

3 Scheduling mechanism research on digital archive resources sharing and management system in cloud computing environment.

The popularity varies from data to data in cloud computing, so the value of digital archive resources sharing varies. Digital archive resource with high value means it is in high demand. When the request of digital archive resources sharing is in high-concurrency situation, average allocation method of resources sharing will certainly lead to a decline in user's resource acquisition. Thus, varying amounts of digital archive resources sharing should be allocated to different digital archive resources with varying value. This is a simple method. However, in the open source cloud computing environment based on Hadoop architecture, mechanism of this architecture lacks flexibility of dynamic adjustment in resource sharing (Zhonghua Deng, 2013). In large distributed systems, cloud computing platform is a common technology, which can improve access efficiency and fault freedom of digital archive resources sharing. Cloud computing platform not only allows digital archive resources stored in different locations but also improves efficiency of digital archive resources' reading in parallel. Above all, dynamic adjustment mechanism of digital archive resources sharing and management system keeps in accordance with system parameter in Hadoop architecture; this is one of the research emphasis.

4 Study on metadata management of digital archive resources in cloud computing.

Metadata of digital archive resources in cloud computing describes information about digital archive resources and plays an important role in distributed system. Firstly, metadata is always decoupled from digital archive resources in current cloud computing architecture and stored in memory of a single master node. However, as number of digital archive resources start growing, master node will be one bottleneck of the whole system. Hence, the analysis shows that metadata request efficiency of digital archive resources is low and resource acquisition for users will be impacted. Secondly, metadata of digital archive resources is always stored in memory of central node in cloud computing platform. Though occupying small space, metadata is limited by memory size of master node. Thus, quantity of digital archive resources supported by whole system is limited. Thirdly, single point of failure exists in single master node. If there are problems in central node, all metadata service of digital archive resource will be interrupted to figure out the above 3 problems and the author studies techniques related to distributed management of digital archive resources metadata in cloud computing environment.

5 Study on utilization rate of distributed resources in digital archive resources sharing and management system.

Utilization rate of distributed resources in digital archive resources sharing and management system needs to be further improved in cloud computing environment, because it is one of the key factors which determines service price (Zhu Xinyi & Chen Jun, 2013). As users' demands for digital archive resources are the same in cloud computing environment, resources provided by this service system is the same as resources users want to obtain. For this reason, reduplicate digital archive resources may be produced during use procedure. However, if there are a lot of reduplicate digital archive resources in the system, those resources will be wasted and the overall utilization rate is difficult to get improved (Chenhui Li, 2013). As a result, cost of cloud computing service provider will increase and competitiveness decreases. The author mainly studied digital archive resource services and their classification in cloud computing as well as coupling among digital archive resources. Digital archive resources are separated from the full high-quality digital archive resources and abstracted as a digital archive resource pool. In addition, the author studies combination and utilization of digital archive resources in detail. How to find digital archive resources with the right coupling strength is another research focus.

6 Building the "cloud" access platform and data center of digital archive resources sharing and management system.

In order to implement digital archive resources sharing and management system in cloud computing environment, the digital archive resources application and data center must be built. On the basis of existing cloud computing service platform, the "cloud service" center of digital archive resources management is established by outsourcing its work to "archive cloud" service provider.

At present, cloud computing application is still in its beginning stage and has not been widely known by most institutions and individuals. For the sake of building a digital archive resources sharing and management system platform based on cloud computing at this stage, an effective "cloud" access platform need to be established to allow varieties of departments and organizations to log on to the "archive cloud" of this system. In addition, this "cloud" access platform requires a security guaranty mechanism, which can cooperate with security system in management middleware layer to effectively verify users' identities and prevent illegal operations and hackings.

7 Implementation plan of digital archive resources sharing and management system in cloud computing environment.

For a long time, it still has difficulties in implementing digital archive resources sharing and management system because of the unified technologies and standards and the common existence of heterogeneous databases (Dayu Xu and Shanlin Yang, 2012). Therefore, the current situation of digital archive resource service should be analyzed deeply, especially technologies and standards applied to digital archive resource service. It should be found out which technologies and standards meet the requirements and standards related to digital archive resource service system further improvement. Information technology should be used regularly. Coordination between service standards of digital archive resources and other archive resources should be paid attention to. Thus, a digital archive resources sharing and management system platform within larger scope would be achieved.

8  Study on supporting technology of digital archive resources sharing and management system in cloud computing environment. This part of study includes: classification, archiving, general principles of description, requirements, elements, methods, specifications, and technical solutions; the business system of digital archive resource service; digitization, networking, and informatization requirements of digital archive resources sharing and management system.

## 4  RESEARCH APPROACH

Aiming at the above research contents, problems facing the supporting technology of digital archive resources sharing and management system can be solved from the following five aspects, so as to design the "archive cloud" service process, as shown in Figure 2. It is critical to study how to improve service capability of digital archive resources sharing and management system in cloud computing platform, how to establish the scheduling mechanism and metadata management in "archive cloud" service platform,

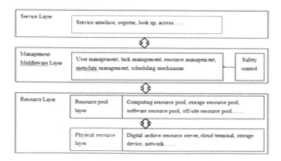

Figure 1. Implementation model of digital archive resources' dynamic scheduling service in "archive cloud."

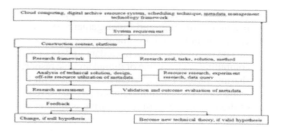

Figure 2. Supporting technology process of digital archive resources sharing and management system in cloud computing environment.

and how to improve utilization rate of digital archive resources.

## 5  SOME PROBLEMS

The purpose of this study is to build a "cloud archive" information sharing platform taking various archive resource departments around the world as "center, node, terminal." This new kind of archive resource integration model, which applies cloud computing to archive resource information sharing, is undoubtedly of great promoting value and practical significance. However, there are still some problems needed to be solved from a system point of view:

1  Reliability issues: "Cloud archive" information sharing platform should guarantee the reliability of system and various digital archive resource data. At present, as technical problems happens commonly, internet service providers usually provide foundational technologies for enterprise customers who use cloud service to run websites and store data.

2  Management issues: As the size of digital archive resource service system is very big, it is facing a great challenge to efficiently monitor, dynamically schedule and deploy the digital archive resource service. Sharing strategy and techniques is the application cores of digital archive resource service management based on cloud computing. Questions about how to improve performance of digital archive resource service system applying cloud computing, how to take "low-carbon economy" into account, how to apply strategy and algorithm of digital archive resource dynamic scheduling in cloud computing techniques, and how to meet users' needs as much as possible by means of resource sharing in premise of ensuring management quality, must be figured out in order to improve utilization and save running cost of digital archive resource service system.

## 6 SIGNIFICANCE OF THIS STUDY

In summary, from the perspective of provider and manager of cloud service technology, combined with features of cloud computing and actual situation of archive work, the author applies cloud computing technology to information system of archive work in digital era. Cloud computing technology plays an important role in construction of digital archive resource service system. And then, the dynamic scheduling mechanism and selection algorithm of digital archive resource are proposed based on cloud computing technology. Metadata's service mode, classification, and service content in digital archive resource service system are studied from the perspective of common character between cloud computing service and digital archive resource, so as to enhance the capability of digital archives. This is of great importance to service extension, business expansion, service mode, and coconstruction and sharing of digital archives. In other words, this has a great application value and practical significance to construction and service of digital archives.

From the theoretical point of view, supporting technology's framework of digital archive resource service system should be constructed and integrated with cloud computing organically. Features of "cloud storage and cloud service" bring convenience and support to digital archive resource sharing and service. Combined with key technology, management technique and method of digital archive resource sharing and service, issues like technical support theory, system architecture, storage structure, security system, calculation method, dynamic programming, scheduling service, metadata selection, adjustment mechanism and coupling relationship between digital archive resources are researched based on "cloud platform." And then metadata service's content and method, as well as cloud storage system's technical solution of digital archive resource service are proposed. This is of very great theoretical significance to enriching and supplementing research object and practicing content of archival science discipline system in cloud era.

From a practical point of view, implementation of digital archive resource service system's supporting technology should be planned and designed systematically. "Archive cloud" platform of digital archive resource dynamic scheduling service is built with the actual requirements of this system based on cloud computing. This allows digital archive resource to reflect the whole picture of society and undertake responsibility of forming memory of a country and a nation. At the same time, archives department can provide better service. It is in line with the concept and goal of a service-oriented government. This is not only of great importance to promoting service extension, business expansion, and service mode of digital archive resource as well as innovation of coconstruction and sharing service system but also of great practical significance to realization of citizens' rights and construction of a harmonious society.

## ACKNOWLEDGMENTS

Science and Technology in Jiangxi province department of education project (GJJ13082): "Digital archive resources management and service system integration research in cloud computing environment"*

Social science fund project in Jiangxi province (13TQ10): "Research on public digital cultural heritage protection technology of cloud computing"*

## REFERENCES

[1] Xiuzhen Feng & Peng Hao. 2012. Study on cloud service mode of information resource in cloud computing environment. *Computer Science* (10):110–115.
[2] Jeremy Leighton John. 2009. The future of saving our past. *Nature. London*:775–778.
[3] Jim Ericson. New Ideas for Old Information. Software/service providers are a natural fit for a growing archival headache. For many organizations. the cloud looks like the best answer.
[4] Qingxuan Zhu & Yunyun Sang & Yun Fang. 2011. Study on archive information resource sharing mode based on cloud computing. *Lantai World* (7):8–9.
[5] Jie Wen. 2011. Study on construction of digital archives based on cloud computing. *Archives & Construction*(1):46–49.
[6] Zhenpeng Liu & ning Zhang & Zhaoling Bian. 2010. Discussing the application of cloud computing technology to archives. *Lantai World* (16):33–34.
[7] Xuemei Yin. 2009. From cloud computing to personal digital archives. *Archives of Shanxi* (2):23–24.
[8] Sixin Xue & Cui Huang. 2011. A review of electronic file management in cloud computing environment. *Beijing Archives* (9):65–67.
[9] Gengda Jin & Jiaun He. 2005. Metadata issue and its countermeasure in application of archive information resource integration management and integration service mode. *Archives Science Bulletin*(5):54–58.
[10] Lingling Zheng. 2010. Study on urban construction archive data's integration management application – taking Dongguan digital urban construction archive management platform for example. *Archives Science Bulletin*(5):91–94.
[11] Bing Li. 2012.Study on key techniques of dynamic resource management in cloud computing environment. Beijing University of Post and Telecommunications.

[12] Zhonghua Deng. 2013. Cloud system and service mode of information resource – information resource cloud and knowledge service. Information Studies: *Theory & Application* (1):6–9.

[13] Jiejing Cheng et al. 2010. First discussion about digital archives service and management in cloud computing environment . *Archives Science Study* (6):71–76.

[14] Xinyi Zhu & Jun Chen. 2013. The shallows of cloud service platform architecture of digital archives. *Lantai World* (2):17–18.

[15] Li Chenhui. et al. 2013. A resource management model based on heterogeneous cloud computing platform. Information Studies: *Theory & Application* (1):105–107.

[16] Dayu Xu & Shanlin Yang & He Luo.2012. Management method of multisource information resource in cloud computing environment.Computer Integrated Manufacturing Systems (9):2028–2039.

*Multimedia, Communication and Computing Application – Leung (Ed.)*
© 2015 Taylor & Francis Group, London, ISBN 978-1-138-02775-6

# A scheduling strategy of cloud resource with low consumption based on GA

W.H. Hu, Y.L. Zhang & Q. Zheng
*Hangzhou Dianzi University, Zhejiang, China*

ABSTRACT: The low utilization of cloud resources and the high consumption of cloud data center are the main problems of cloud computing. With the low consumption as the goal the studying is the deployment between virtual machines and physical machines in the cloud data center. Given the advantage in the global convergence and adaptive of GA, we use it to assess each scheduling scheme of cloud resource. The low consumption is achieved in these two respects: Minimizing the number of opened physical machines and loading these physical machines as possible; minimizing the migration costs if necessary. Besides, I combine the idea of load balancing. The results show that each physical machine concentrates on the resource utilization in the threshold infinitely but does not exceed it with relatively short convergence time. Besides, it improves the utilization of the physical machines and reflects the low consumption.

## 1 INTRODUCTION

Cloud computing has been widely used in academia and industry. Green energy and scalability, scheduling mechanism, copy policy of the cloud data centers are all hot researches. Because of the expanding of these centers' scales, the providers of these resources get less benefit. The resource utilization of multi-dimensional becomes lower while the consumption becomes more and more. Solving these problems can make cloud computing provide greater benefits for us. Resource scheduling is a key technology of cloud computing, involving two aspects[1]: between tasks and virtual machines, virtual machines and physical machines. The former has been studied extensively, the latter relatively little. The selection and use of appropriate relationship between the virtual machines and physical hosts can not only improve the utilization of resources, but also reduce consumption. This article is centered on the low consumption of date centers, researching viable scheduling policy of the cloud resource.

## 2 PROBLEM DESCRIPTION

GA simulates Darwinian's evolution of natural selection, or "survival of the fittest". It has good adaptability and global optimality, powerful search capabilities, parallel computing, and less information, being a good solution to many practical problems which is widely used in cloud computing research. Through it we can well get a variety of potential mappings between virtual machines and physical machines which meet the consumption requirement.

This paper studies the low-consumption schedule of cloud resources from the following three perspectives:

1 By the dynamic migration and consolidation techniques of the virtual machines, it dynamically mergers the virtual machines on physical machines with low utilization to the physical machines with higher utilization based on the current resource needs, ensuring the physical machines not overloaded. Then the idle physical machines can be changed into power-saving mode or even off.

2 If there are no reasonable allocation strategies of the cloud resource, it will be prone to load imbalance, emerging the failures of some physical machines. Therefore, to coordinate the loads of all the physical machines by a suitable mechanism of load balancing can improve the utilization of cloud resource. Besides, it can reduce the consumption of the cloud data center and improve the performance of the entire system.

3 Load balancing is usually achieved by the migration of virtual machines, so the consumption of migration must be considered. Frequent migration of virtual machines may cause some bottlenecks of communication and other issues, so it should try its best to make migration less. And this is also an aspect of reducing consumption. The migration of virtual machines is a complex issue. This paper considers minimizing the number of the migrating.

## 3 STRATEGIC ANALYSIS

### 3.1 GA

GA (Genetic algorithm)[2–3] changes the initial population into the next generation through operating the genes such as natural selection, crossover and mutation. Each population is a chromosome: Selection operator selects a part of the genes on the chromosome to produce offspring. Usually, the genes with higher degree of adaptation are easier to produce offspring and be retained: Crossover operation exchanges two parts genes of the chromosome to produce a recombinant one; mutation operation changes some genes according to certain laws. The specific description of these concepts is as follows.

#### 3.1.1 Encoding and decoding

The encoding and decoding of the mapping between virtual machines and physical machines is the primary problem to be solved of GA. They affect the calculation method of the crossover operator, mutation operator and other genetic operators, largely determining the evolutionary efficiency. The appropriate method of encoding and decoding can quickly get a stable global optimal solution. Currently, people have proposed many different encoding methods [4] which are mainly divided into three categories: Binary, floating-point, symbol. I use the tree encoding method in the symbol method, and decoding is a reverse process based on the corresponding encoding method.

The root node indicates controller of the data center defined by the system. The second layer has N nodes representing physical hosts with the M leaf nodes representing virtual machines. And all these make up a chromosome:

$$C = \{Control(H_1(VM_1,VM_2,VM_3), H_2(VM_4,VM_5,VM_6),$$
$$H_3(VM_7,VM_8,VM_9)\} \tag{1}$$

#### 3.1.2 Fitness function

The fitness function [5–6] is the basis of survival of the fittest. It assesses the extent to adapt to the environment of each individual in the population based on the requirements and obtains the corresponding values. The individual with higher value has better adaptability and is easier to be left. At present, there are linear acceleration and nonlinear acceleration. In order to simplify the content of the study, I use the most original fitness function and don't utilize the acceleration method.

Traditional GA mainly has two evaluation criteria for packing problems: Minimizing the number of boxes; try to make all the boxes fill under the premise of a minimum number of boxes. This article uses the second evaluation criteria combining load balancing. Actually, there are kinds of load. Here, we consider CPU, memory, hard disk, and bandwidth:

$$Load(i) = \omega 1 \times Load(i)_{CPU} + \omega 2 \times Load(i)_{Mem}$$
$$+ \omega 3 \times Load(i)_{Disk} + \omega 4 \times Load(i)_{BW} \tag{2}$$

Wherein, $\omega 1$, $\omega 2$, $\omega 3$, $\omega 4$ are weighting factors, and $\omega 1 + \omega 2 + \omega 3 + \omega 4 = 1$.

The load function of each physical machine can be expressed as follows:

$$f1 = \frac{\sum_{j=1}^{m_i} Load(i)_j}{Load(i)} \tag{3}$$

In this formula, Load $(i)_j$ represents the loads of the j-th virtual machine on the i-th physical machine, $m_i$ refers to the total number of virtual machines on the physical machine, and Load (i) represents the maximum load of the physical machine. We can see that the larger is molecule (physical machine tends to be full) and the value of the load function, the greater is the value of the fitness function.

The fitness function is defined as follows:

$$F(S,T) = A \times f1 + B \times f2 \tag{4}$$

In this formula, A, B are weighting factors, and A+B=1. $f2$ is the function of load balancing, and it will be described in the following.

#### 3.1.3 The basic operators of GA

1 Selection operator
This operator well leads the search to the solution space meeting the constraints. And the size of the selection intensity affects the rationality of the convergence speed. An appropriate selection strategy ensures the search refrain from precocity and slowly converge speed. This article uses a commonly used algorithm, and roulette wheel selection algorithm, and is based on the fitness proportional. Wherein, the calculation based on fitness proportional is defined as follows:

$$P(S_i) = \frac{F(S_i,T)}{\sum_{i=1}^{D} F(S_i,T)} \tag{5}$$

In this formula, $F(S_i, T)$ represents the i-th individual's fitness, D represents the size of the population.

Dividing the disk into D parts based on the fitness proportional, and the i-th segment angle is $2\pi P(S_i)$. Firstly, generat number k randomly ( k is between 0 and 1), and rotate the disk until it stops. When it stops at the i-th sector, it satisfies the following formula:

$$P(S_1)+P(S_2)+\cdots+P(S_{i-1})<k\le P(S_1)+P(S_2)+\cdots+P(S_i) \quad (6)$$

Then choose the i-th individual.

From the foregoing, the greater is the value of the individual's fitness, the bigger is the area of the segment. Then its probability to be selected is bigger.

2  Crossover operation

Two parent individuals obtained by the selection operation can get a new individual according to some laws by replacing a part of the structure. Then we can change the structure of spanning tree, improving the search capabilities of the GA.

The available resources of the present physical machines can be expressed as:

$$Load(i)' = Load(i)-\sum_{j=1}^{m_i} Load(i)_j \quad (7)$$

In this formula, $m_i$ represents the total number of virtual machines of the i-th physical machines.

The crossover strategy is shown in Figure 1, and GEN represents the current generation, $GEN_{max}$ represents the maximum generation:

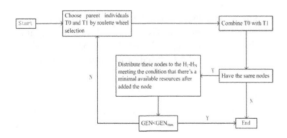

Figure 1.   Crossover process.

3  Mutation operation

This operation can increase the diversity of the population, expand the scope of the search, avoid premature phenomenon, and improve the performance of GA. Firstly, select an individual to perform the mutation operation according to some probability. Then randomly select two target physical machines of the

individual, and two virtual machines to be exchanged partly on the two target physical machine by some laws (For example, these two virtual machines are at the same location on each physical machine which is used by this paper). This article's probability formula is defined as follows:

$$P_m = \exp\left(-\alpha \times t \times \sqrt{M}/D\right) \quad (8)$$

In this formula, $\alpha \in R^+$ is a constant, t is the generation. D is the size of the population. M is the number of the virtual machines.

3.1.4  *Termination condition of GA*

The algorithm won't exit until the individual meets the given requirements or it exceeds the given maximum generation. Then choose the most appropriate solution among these schemes according to certain criteria. Requirements given in this article are load balancing, the least costly of migration. To simplify the calculation, migration cost is defined as the number that virtual machines need to be migrated.

3.2  *Load balancing*

Load balancing[8-10] has been used in many areas, ensuring a rational allocation of resources and avoiding imbalance problems. It can achieve the sharing and maximal utilization of resources. Load balancing is widely used to assign tasks to virtual machines which ensure the equal distribution; load balancing can also be used to get the allocation between virtual machines and physical machine. There are many similarities between the two programs, and the former's algorithms can be reasonably used to the latter which is studied by this paper. In recent years, there have been many intelligent heuristic algorithms which can be well combined with the balancing algorithms, such as Annealing algorithm, GA, Ant algorithm. In this paper, load balancing is combined with GA.

The present load of the i-th physical machine:

$$Load = \sum_{j=1}^{m_i} Load(i)_j \quad (9)$$

When VMr is added to or removed from the i-th physical machine, the load of this physical machine changes as follows:

$$\hat{Load}(i)=\begin{cases} Load(i)+Load(i)_r \\ Load(i)-Load(i)_r \end{cases} \quad (10)$$

This paper uses the standard deviation of the load changes to measure the load balancing:

$$\sigma_i = \sqrt{\frac{1}{N} \sum_{i=1}^{N} \left( Load(i) - \overline{Load} \right)} \qquad (11)$$

In this formula,

$$\overline{Load} = \frac{1}{N} \sum_{i=1}^{N} Load(i) \qquad (12)$$

In order to achieve load balancing, all $\sigma_i \leq \sigma_0$, and $i \in [1, N]$. Here it is 0.5. Then defining the $f2$ mentioned above as follows:

$$f2 = \frac{1}{C + D \cdot f_T} \qquad (13)$$

$$f_T = \begin{cases} 1 \ , & \sigma_i \leq \sigma_0 \\ \lambda \ , & \sigma_i > \sigma_0 \end{cases} \qquad (14)$$

In this formula, $\lambda$ is a constant which is bigger than 1.

## 4 ALGORITHM PROCESSES

1 The stage of the initialized allocation of virtual machines

If we turn on all the physical machines at first, it will cause a large consumption. Therefore we define the laws as follows: put the virtual machines into the physical machines one by one, and choose the opened physical machine which is closer to its full load if this virtual machine is added. If all the opened physical machines will overload, turn on a new physical for this virtual machine. Repeat this process until all the virtual machines are allocated. To simplify the research, this paper assumes that all properties (CPU, memory, etc.) are the same, giving the number of virtual machines M and the number of physical machines N.

2 The stage that all virtual machines have been allocated

1 Calculate the total number of physical machines that should be opened:

$$N' = \left\lceil \frac{\sum_{j=1}^{M} Load(i)_j}{\rho} + \tau \right\rceil \qquad (15)$$

In this formula, the numerator represents the total load of all virtual machines, and $\rho$ represents the threshold of a physical machine's load when $\tau$ is a tuning constant.

2 When there are $N'$ physical machines in the initialized allocation, it comes to an end if it satisfies load balancing. Otherwise, it runs this algorithm presented above based on the previous mapping relationship, getting the collection $\{S^*\}$. The scheme from the collection which has the least cost of migration is the best distribution. When $\Delta N = N - N' > 0$, here, I migrate all the virtual machines from the physical machine which has a minimum of virtual machines, then shut down the physical machine.

3 Sequentially remove the virtual machines above to the physical machine which has a maximal Load(i) if added current virtual machine, get the collection $\{S'^*\}$. The scheme from the collection which has the least cost of migration is the best distribution. Then end the algorithm.

3 Supplementary stage

The second stage is an ideal case of a steady state. In practice, it is a complicated situation as there may be many virtual machines added continuously or at different time with different intensity. If there're some virtual machines added at one point, perform the two stages above in turn.

## 5 EXPERIMENT RESULTS

1 Analyze the scheduling algorithm through the mapping between virtual machines and physical machines

Supposing there have been 12 virtual machines assigned to work on five physical machines, and, D=50, $P_c$=0.6, $\alpha$ =1.5, $\rho$ =65. In this paper, the algorithm is achieved by Matlab, the mapping before and after the algorithm are formula (16) and (17):

$$Control\{H_1\{VM_1, VM_2, VM_3\}, H_2\{VM_4, VM_5, VM_6\},$$
$$H_3\{VM_7, VM_8\}, H_4\{VM_9, VM_{10}, VM_{11}\}, H_5\{VM_{12}\}\} \ (16)$$

$$Control\{H_1\{VM_1, VM_9, VM_{11}\}, H_2\{VM_2, VM_4, VM_5, VM_7\},$$
$$H_3\{VM_3, VM_8\}, H_4\{VM_6, VM_{10}, VM_{12}\}\} \qquad (17)$$

The respective values are as follows:

We can see that after this scheduling algorithm, the load of each physical machine approaches the threshold of 65 with the change of the system load less than the certain value which is calculated to be 0.25. Therefore, this scheduling algorithm has global convergence with relatively short time.

Table 1. The load of each virtual machine.

| Virtual machines | Load(%) | Virtual machines | Load(%) |
|---|---|---|---|
| VM$_1$ | 25.1 | VM$_7$ | 14.6 |
| VM$_2$ | 18.4 | VM$_8$ | 41.3 |
| VM$_3$ | 22.3 | VM$_9$ | 17.5 |
| VM$_4$ | 16.7 | VM$_{10}$ | 25.8 |
| VM$_5$ | 13.4 | VM$_{11}$ | 20.3 |
| VM$_6$ | 21.2 | VM$_{12}$ | 16.2 |

## 2 Analyze the scheduling algorithm from the perspective of execution time

Assuming there're 100 virtual machines (each with load in the range of 5% to 30%) $P = 80$. I randomly select four physical machines to be observed:

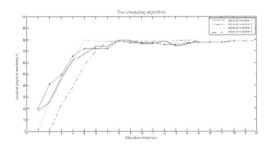

Figure 1. The scheduling algorithm.

From the diagram, we can see that the time before 7min is the stage of the initialized allocation of virtual machines, and the load of each physical machine is close to 80% but less than it. The time after 7min is the stage to migrate virtual machines. Around 20min, the load of each physical machine has been stable with good load balancing, each physical machine being closer to its full load. After calculation, there're totally 22 physical machines opened with the average load to be 78.6%.

## 6 CONCLUSION

This paper studies a scheduling strategy of cloud resource with low consumption based on GA combined with load balancing, making physical machines running tend to be full and minimizing the number of the physical machines. Though we have gotten good results, there're some shortages. Lack of results compared with other existing algorithms, we haven't quantified consumption of physical machines' open and close to compare the migration consumption from which we can get a better solution. If these problems can be improved to be more practical, it is an important direction for future research. If these shortages can be improved, it will be more practical which will be an important direction for future research.

## REFERENCE

[1] Zhao Chunyan, 2009. Research and implementation of job scheduling algorithm in cloud environment [D]. Beijing: Beijing Jiaotong University:48–53.
[2] Fonseca C M, Fleming P J, 1995. An overview of evolutionary algorithms in multiobjective optimization[J]. Evolutionary computation: 3(1):1–16.
[3] Beasley D, Martin R R, Bull D R, 1993. An overview of genetic algorithms:Part 1. Fundamentals[J]. University computing, 15:58.
[4] Sun Xin, 2012. Research for mechanism of efficient resource scheduling of cloud data centers[D]. Beijing: Beijing University of Post and Telecommunications:109–114.
[5] Wang Shun, 2013. Research for scheduling algorithm of load balancing based on the minimum migration consumption of cloud resources[D]. Xi'an: Electronic Technology University :52–58.
[6] Zhang Xiaoqing, 2013.Research for optimization method of resource supply in cloud environment[D]. Wuhan:Wu Han University of Technology :121–135.
[7] Liu Zhanghui, Wang Xiaoli, 2012. A algorithm of load balancing for a group of virtual machines of cloud computing with GA[J].Journal of Fuzhou university (natural science ), 40(4):453–458.
[8] Braun T D, Siegel H J, Beck N, etal, 2001.A comparison of eleven static heuristics for mapping a class of independent tasks onto heterogeneous distributed computing systems[J].Journal of Parallel and Distributed computing, 61(6):810–837.
[9] Shi Yangbin, 2011.A algorithm of load balancing based on virtual machines' dynamic migration in cloud environment[D].Shanghai: Fudan Univ :31–42.
[10] Zhang Z?, X Zhang.A load balancing mechanism based on ant colony and complex network theory in open cloud computing federation, in 2010 2nd International Conference on Industrial Mechatronics and Automation (ICIMA). 2010:240–243.

# Item-based collaborative filtering recommendation algorithm based on MapReduce

W.H. Hu, F. Yang & Z.W. Feng
*Department of Computer Science and Technology, Hangzhou Dianzi University, Hangzhou, Zhejiang Province, China*

ABSTRACT: With the rise of big data, the E-commerce website wants to extract information left by users to provide them products and services in more intelligent ways. However, the sharp increment of the number of users and commodities makes the utility matrix extremely sparse. Besides, the traditional collaborative filtering algorithm did not consider the real-time attribute of the demand for goods and shopping preference by users. This led to a serious decline in the quality of recommendation system 1. So this paper will present an item-based collaborative filtering recommendation algorithm based on MapReduce, and it calculated the similarity based on frequent sequential patterns of click streams, introduced time factor and made multi-pattern fusion from multiple dimensions of different weights, and combined machine-learning algorithm to improve the timeliness and accuracy of the recommendation system.

## 1 INTRODUCTION

Generally, collaborative filtering recommends items by calculating the similarity between the user and items. It recommends items which are favored by the user who is similar to the target user. The collaborative filtering algorithm can find the user set in which the user's behavior of interest are close to the current user's, and recommends the favorite items to the current user on the basis of the user set. A collaborative filtering technology developed so far shows many inadequacies due to its cold start, and the need to hit a specific label for commodities, on one hand, with the increasing amount of mining data, the user-item assessment matrix will be extremely sparse, resulting in a poor synergistic effect, on the other hand, the traditional collaborative filtering algorithm does not consider the real-time attribute of the demand for goods and shopping preference by the users.

Item-based collaborative filtering, ItemCF calculates the similarity between items by analyzing user's behavior records [2]. The algorithm is considered that there's a great similarity between commodity A and B because most of the users who have clicked A also clicked B. Besides users' recent behavior, can reflect their interests better, than before. Therefore, when predicting the users's current interest, we should increase the weight of the recent behavior, and give priority to those commodities that are similar to the users' recent interest [3].

## 2 COMMODITY DIMENSIONS AND MODEL FUSION

In order to describe the commodity accurately, we should synthetically calculate and rank from multiple dimensions. The recommended commodity will have maximum target value after fusing all dimensions. As shown in Figure 1, we can calculate and describe the similarity from different perspectives such as text similarity, users' vote, price of commodities, the role of its own property, users' clicks and buying behavior, etc.

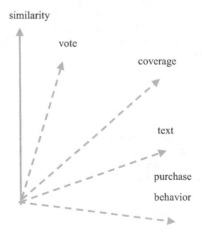

Figure 1. Description from different dimensions.

The actions of users browsing and shopping goods in e-commerce sites can be broadly grouped into three steps: (1) View, click on the link to enter commodity's detail page from the website home page, (2) Add to cart, add the commodity be concerned to cart, and (3) Order and buy goods. From the dimension of the user's interest, all the goods in web site can be divided into clicked item, added to shopping cart items, purchased items, and other items. In this process, the product can be divided into different weights:{no click item}<{clicked item}<{added to shopping cart items}<{purchased items}[4, 5]. User's data has real-time characteristic, and the long-term feature, the data includes frequent sequential patterns of click streams,V2V (view to view), B2B (buy to buy), users' interest preference, and purchasing power, etc. Product data includes the product of the title, category, brand, price, etc.

## 2.1 Frequent sequential patterns of click streams

Assuming frequent click-stream sequential patterns I={I1, I2, …In} shows an association between items, user $u$ generated interest on item $j$ after clicking on an item $i$ in I can be expressed as the difference credibility of $j$ in I, and the number of item sets including item $j$. As shown in the Formula 1.

$$r_{uj} = \frac{I \cup \{j\}}{I} - \frac{j}{N} \tag{1}$$

## 2.2 V2V and B2B

During the process of users' browsing commodities, a user's interest in a commodity is different under different conditions, specifically including V2V (View to View), B2B (Buy to Buy),V2B (View to Buy). These patterns will produce different recommended weights.

## 2.3 Users' interest preferences

User's long-term interest preference is the characterization of user's long-term shopping habits and purchasing power, it indicates that whether the user is interested in the current item or not. Including brand preference, category preference, age, sex and other characteristics of the user.

## 2.4 Purchasing power

The purchasing power generated from users' long-term online shopping process will be considered as a priority at the same time. Recommending luxury to an actual use based on buyer is unrealistic. Therefore,

the purchasing power of users should be considered as a one-dimensional factor.

## 2.5 Description of commodities and keywords

The calculation of the similarity of product title mainly focused on the similarity of the text. The calculation method is shown as Formula 2. It transformed commodity description into keywords, and mapped into the vector space calculating the cosine similarity. Wherein $v_u$ and $v_i$ are the keywords eigen vectors of users and recommended item.

$$sim(u,i) = \frac{v_u \cdot v_i}{\|v_u\| \cdot \|v_i\|} \tag{2}$$

# 3 ITEM-BASED COLLABORATIVE FILTERING RECOMMENDATION ALGORITHM BASED ON MAPREDUCE

## 3.1 Item-based CF recommendation algorithm

Item-based CF Recommendation algorithm mainly can be divided into two steps:

1. Similarity calculation between items.
2. Generate a recommendation list for the users based on their historical behavior and the similarity matrix of the items.

Calculating the similarity between the items [6] can be expressed as:

$$w_{ij} = \frac{|N(i) \cap N(j)|}{|N(i)|} \tag{3}$$

Denominator $|N(i)|$ is the number of users who like item $i$, molecular $|N(i) \cap N(j)|$ is the number of users like item $i$ and $j$ at the same time. Therefore, the formula above can be interpreted as the proportion that the number of users like items $j$ accounted for that of $i$. However, when item $j$ is very hot, so $w_{ij}$ will be large, and close to 1. Therefore, the formula will cause any item which will have great similarity with the hot one. To avoid recommended hot items, so with the following improvements:

$$w_{ij} = \frac{|N(i) \cap N(j)|}{|N(i)|^{1-\alpha} |N(j)|^{\alpha}} \tag{4}$$

Formula 4 restricted on the weight of item $j$, reducing the similar problems between many items and hot one. However, distinguishing similarity between items just from the number of people still exists a large deviation. The credibility of a user $u$ interested in an item $j$ can be defined as Formula 5. Generally speaking, the less commodity a user click on a website, the higher the credibility each click is.

$$w_u = \frac{I \cup \{j\}}{I} \tag{5}$$

Obviously, frequent sequential patterns have higher relevant credibility. Generally speaking, the shorter step length a user on a website, that is, the fewer number of items that the user clicks, the greater interest of the user for the current item is. After improvement, the way of similarity calculation between two items is as shown in Formula 6.

$$w_{ij} = \frac{\sum\limits_{u \in U(I_i) \cap U(I_j)} (w_u^2 \times f(|t_{ui} - t_{uj}|))}{\sqrt{\sum\limits_{u \in U(I_i)} w_u^2} \sqrt{\sum\limits_{u \in U(I_j)} w_u^2}} \tag{6}$$

In Formula 6, $w_u$ indicates the credibility of the current user $u$ click on an item, time factor $f(t_{ui}\text{-}t_{uj})$ indicates the impact of different time between user $u$ click on item $i$ and $j$. factor $f$ introduces a time variable. The implication is that the longer different time between user's action on item $i$ and $j$, the lower correlation between items is. Usually, it is easy to find a lot of time decay function. We use the decay function shown in Formula 7 in this article. $\alpha$ is a time decay parameter, if a user's interests changed quickly in a system, its value should be greater, otherwise , it will be smaller.

$$f(|t_{ui} - t_{uj}|) = \frac{1}{1 + \alpha |t_{ui} - t_{uj}|} \tag{7}$$

### 3.2 Machine learning algorithm of mode fusion

In order to improve the shortcomings the single CF algorithm brings, we fuse different dimensions of recommended effect [7]. Mode fusion problem is a typical regression problem essentially. So all regression algorithm can be used to model fusion. This paper uses logistic regression algorithm as a machine-learning fusion algorithm. Each algorithm of different dimension is one-dimensional feature of machine learning, and the recommended values of mode fusion Top-N is the sort of recommended target values to be calculated ultimately.

Logistic regression model is a commonly used machine-learning method. It is used to estimate the possibility of the occurrence of an event. For example, we estimate the possibility of the possibility of a user

purchasing a commodity using known information, etc. The formal definition of logistic regression model is that vector $\bar{X} = (x_1, x_2, ..., x_n)$ represents a vector having N independent variables, $P(Y=1|X)$ represents the probability of occurrence of the event Y under the condition of the variable X, $g(x)=w_0+ w_1x_1+ ... + w_nx_n$ is the weighted linear of N independent variables of vector $\bar{X}$. The logistic regression model can be expressed as shown in Formula 8.

$$P(Y = 1 | X) = \frac{1}{1 + e^{-g(x)}} \tag{8}$$

The machine-learning process of logistic regression model is the process of fitting the learning parameters through the training data. The prediction process is generating a predicted value through the learned predictive function according to the characteristics of test data. The size of the predicted value represents the relative size of the probability of occurrence of an event. For Top-N recommendation problem, sort the predicted value for each training sample in descending order, and top N value is the recommended result.

As we can see in Table 1, the goal of the main commodity Item is to recommend an optimal element from the optional set {Item$_1$}, {Itam$_2$}, {Item$_3$}......, {Item$_n$} by means of Mode Fusion Algorithm (MFA), which would fuse attributes in different dimensionality, including V2V, B2B, customer long-term preference and purchasing power. Finally, the MFA results in a set of target value. Then rank these target value in descending order, and Ton-N is the final recommended result.

Table 1. Mode Fusion of Click Flow Continual Sequence.

| Main commodity set | Optional | V2V | B2B | Purchasing power | | Target |
|---|---|---|---|---|---|---|
| Item | Item$_1$ | 0.021 | 0.51 | 0.47 | ...... | 0.53 |
| | Item$_2$ | ...... | 0.49 | ...... | ...... | 0.48 |
| | Item$_3$ | ...... | 0.33 | ...... | ...... | 0.45 |
| | ...... | ...... | ...... | ...... | ...... | ...... |
| | Item$_n$ | ...... | 0.58 | ...... | ...... | 0.31 |

### 3.3 Implementation of item collaborative filter recommend algorithm, by Map Reduce

This part will provide the compute transplant under a mass of data of the algorithm which mentioned above. The first Map Reduce program aims to implement the compute of different weight in each dimensionality. The second Map Reduce program is used to implement the compute of Mode Fusion and rank of Top-N. The input of traditional synergy filter algorithm is 2 matrix, one User-Item matrix, which is used to record the user preference, and another Item Matrix which recording the characteristic information of recommended item. The dot product between an arbitrarily row vector in the User-Matrix and an arbitrarily column vector in the Item-Matrix is exactly the degree of interest of the user in the User-Matrix to the recommended item in the Item-Matrix. In this paper, the User-Item Preference Matrix is switched to Item-Item co-occurrence Matrix $C$, in which item is dimensionality. $C[i][j]$ records the number of users who like both item $i$ and item $j$. In addition, Co-occurrence Matrix $C$ in this article records not only the number of users who like both item $i$ and item $j$, but also the time difference between different user-item $i$ and $j$. The implementation of item CF, based on the similarity of items, is as follows:

1) *Map function*

Input User-Item Preference Matrix $U$ records the reliability of user click. Co-occurrence Matrix $C$ records the number of users who like this item and the different time of item click between them. In the Map, task matrix $U$ and $C$ will be switched to Key-Value Pair (KVP) separately, and the U-KVP take $(i, k)$ as the key, $(U, j, u_{ij})$ as the value, the C-KVP takes $(i, k)$ as key, $(C, j, c_{ij})$ as the value. The time attenuation parameter **a** is also input in this task. $w_u^2$ in Formula 6 requires compute, this step can be completed in Map Function, so the output of Map Function is $((i, k)$, $(U, j, w_u^2))$ and $((i, k), (U, j, f(|t_{ui} - t_{uj}|)))$.

2) Reduce function

Values with the same $(i, k)$ key are input in the Reduce task by means of matrix multiplication. In other words, the function will calculate the dot product of item $i$ and item $j$ when there exists a user who like both item $i$ and item $j$, which is the similarity between $i$ and $j$. The output of Reduce Function is the similarity between a given target item $I$ and other optional item, which is expressed in the formulation of $(I_i, W_{ij})$. Implementation of Mode Fusion in logic regression model, MapReduce algorithm is as follows,

3) Map Function

First KVP $(I_i, W_{ij})$ from last MapReduce task will be received in this function, KVPs with the same $I_i$ will be sent to a same Map task, and value $W_{ij}$ is the computation to the weight of item $j$. Then the logic regression model Formula 8 is called. Finally, the recommend comprehensive similarity $W_{target}$ to item $j$ after fusion of weight in different dimensionality. The output of Map function is $(I_i, (I_j, W_{target}))$.

4) Reduce Function

Rank the input $(I_i, (I_j, W_{target}))$ in descending order of $W_{target}$, then output Top-N.

Reduce function finally output the Top-N recommendation result items $(I_i, I_j)$ to given item $I_i$.

## 4 EXPERIMENT AND ANALYSIS OF RESULT

This experiment employed Hadoop-1.01 to build a small-scale Hadoop cluster with 8 nodes as the experimental platform. The configuration of each nodes in this cluster is as follows, Intel(R) core(TM) i3 CPU, 4G internal storage. Each nodes is equipped with Redhat 6.1 operating system. The test data in this article is the vertical e-commerce web log data provided by click flow data set of KDD cup. The data set is divided into two parts, training set and test set. The training set is used to provide recommended commodity item set, and the test set is used to assess the algorithm index. This article assesses the fusion result of experimental data and recommendation accuracy by Mean Average Precision (MAP) and recalling rate. MAP@n represents the Mean Average Precision top-N recommended results of all users. The formula definition is as Formula 9 and 10. $P(k)$ which is the accuracy of top-k recommended results, while $m$ is the total number of recommended results actually accepted by users. AP@n is the Mean Average Precision of top-n recommended results of an individual.

Recalling rate represents the ratio of associated commodity quantity in the $k$ commodities recommended to user $u$ to all commodity object quantity associated with user $u$. As the Formula 11 shows, recall @n is the recalling rate of top-n recommended results.

$$AP@n = \sum_{k=1}^{n} P(k) / m \qquad (9)$$

$$MAP@n = \sum_{i=1}^{N} AP@n_i / N \qquad (10)$$

$$recall@n = \frac{1}{U} \sum \frac{|R_u \cap T_u|}{T_u} \qquad (11)$$

This experiment employed six data characteristics of user, frequent sequences, V2V, B2B, brand _preference, category _preference and purchasing _power. Logic regression model is applied to fuse above six characteristics to train. Predictions are given in Table 2.

Table 2. The effect of user data fusion.

| Feature | MAP@5 |
| --- | --- |
| V2V | 0.3381 |
| V2V+B2B | 0.3505 |
| V2V+B2B+ frequent_sequences | 0.4032 |
| V2V+B2B+frequent_sequences+ category_ prefer +brand_prefer+purchasing_power | 0.5105 |

As shown in Table 2, mode fusion experiments conducted in accordance with itemized added, firstly, it tested the characteristic effect of V2V dimension, this is one of the most important recommended strategies of algorithms ItemCF, the similarity of single V2V is 0.3381, then the B2B joined, the B2B feature can usually indicates the user's interest, however, due to the limited test data of B2B in this paper, the rise of MAP @ 5 is not very significant. We can see a obvious improvement of the experimental value, after the frequent sequences dimension was joined. Finally, according to the results after all six dimensions fusion, the final value of MAP @ 5 reached 0.5105. The effect of the improvement is more obvious than traditional single CF. Overall this indicates that these six characteristics enhance the degree of optimization of the recommended item of the experimental commodity.

We calculate the recall rate of the consolidated recommendation algorithm. The experimental results are described in Table 3, as can be seen from the experimental results, from the comparison of recall rate in three intervals, the consolidated multidimensional recommendation algorithm which fuse multi-dimensional characteristics, has a higher degree of accuracy than single-mode CF algorithm. After mode fusion recall@50 relative increased 30.2%, recall@100 relative increased 15.7%, and recall@200 relative increased 10.8%.

In the field of machine learning, we generally use Area Under the ROC Curve (AUC), as a performance evaluation to measure the superiority of the algorithm [8]. ROC space use false positive rate (FPR) as the horizontal axis, and true positive rate (TPR) as the vertical axis, to form a two-dimensional space. Generally, the larger value of a program of AUC, more excellent the performance a program will be. As shown in Figure 2, the consolidated multidimensional recommendation algorithm which fuse multi-dimensional characteristics, has a significantly improved AUC values than single-mode CF algorithm. Therefore, the recommended performance is better.

Table 3. Recall rate of the consolidated recommendation algorithm.

| No. | Feature | recall@50 | recall@100 | recall@200 |
|-----|---------|-----------|------------|------------|
| 1 | V2V | 15.91% | 22.45% | 34.31% |
| 2 | 1+B2B | 16.54% | 23.16% | 40.06% |
| 3 | 2+ frequent_ sequences | 17.23% | 24.09% | 41.60% |
| 4 | 3+ category_ prefer | 18.02% | 24.96% | 42.23% |
| 5 | 4 +brand_ prefer | 18.94% | 25.32% | 43.45% |

## 5 CONCLUSION

This paper presents an item-based collaborative filtering recommendation algorithm based on MapReduce by using logistic regression model. It fuses items from a number of different dimensions, and gives the algorithm transplant of MapReduce in the case of massive data. On one hand, it avoided the problem of utility matrix sparsity caused by sharp increment of the number of users and commodities, on the other hand, it introduced time factor when calculating the similarity between items. Because users usually have greater interest in recent commodity they clicked

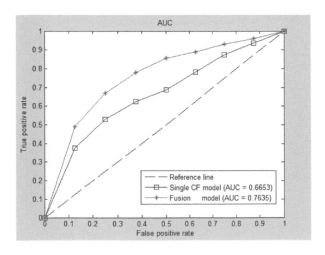

Figure 2. Comparison of two algorithms.

on, so it combines the characteristics of the users' long-term preference. During the process of fusion mode, the program avoids the single characteristic of traditional ItemCF recommendation algorithm. The experimental results show that the algorithm improves the accuracy of the recommendation, and enhances the recommendation effect.

# REFERENCES

[1] ZHANG Liang,BAI Lin-sen,and ZHOU Tao Crossing Recommendation Based on Multi-B2C Behavior.Web Sciences Center, University of Electronic Science and Technology of China,2013, 42(1):154–162.

[2] Sarwar B, Karypis G, Konstan J, et al. Item-based collaborative filtering recommendation algorithms[C]// Proceedings of the 10th international conference on World Wide Web. ACM, 2001:285–295.

[3] XIANG Liang. Practice of Recommendation system [M]. BeiJing: People's Posts and Telecommunications Press, 2012:147–160.

[4] XIONG Xin, WANG Wei-ping, and YE Yue-xiang. Personal recommendation algorithm based on concept hierarchy[J]. Computer Applications, 25(5), 2005:1006–1010.

[5] LEEWP, LIU CH, LU CC. Intelligent agent-based systems for personalized recommendations in Internet commerce[A]. Expert Systems with Applications 22[C], 2002. 275–284.

[6] CHO YH, KIM JK, KIM SH. A personalized recommender system based on Web usage mining and decision tree induction[J]. Expert Systems with Applications, 2002, 23(3): 329–342.

[7] DAHLEN B J, KONSTAN J A, HERLOCKER J L, et al.Jump starting movielens: user benefits of starting a collaborative filtering system with "dead data", TR98-017[R]. Minnesota, USA: University of Minnesota.

[8] http://zh.wikipedia.org/wiki/ROC

*Multimedia, Communication and Computing Application – Leung (Ed.)*
*© 2015 Taylor & Francis Group, London, ISBN 978-1-138-02775-6*

# A nutch-based method of real-time theme searching for massive data

Q. Wu, Y. Yu, L. Wu & Y. Yao
*Hangzhou Dianzi University, Hangzhou, Zhejiang Province, China*

ABSTRACT:  In the face of vast amounts of data, how to accurately and effectively find useful information becomes a research hotspot. The development of vertical search engines in real time is to solve the needs of Internet users' searching for large-scale time-sensitive data. Web sorting algorithm in search engine is to put the related and authoritative pages in the front of massive pages' search results and to help users to locate the information rapidly. It is the key to search engine. Sorting algorithm in the nutch search engine is a basic ranking synthetical model. In order to meet the needs of professional users, the theme correlation factor and web authoritative factor are added into the nutch web pages scoring formula. Experiments show that the improved algorithm can improve the accuracy of search results, which can play a role in practical applications.

## 1 INTRODUCTION

How to efficiently, quickly, and accurately find the required information are current problems, which need to be considered about most of the search engines. Search engine, after a user submits the search term, it will return large result sets, and then find the time you spend with the needed information. Based on the random walk model, PageRank algorithm [1] has been very successful in solving the scheduling problem page, but with the internet geometric data growth, the classic PageRank algorithm [2] has been unable to meet the needs of users. According to the algorithm, many scholars put forward their own improvements. Topic-Sensitive PageRank algorithm [3] and MP-PageRank algorithm [4] by changing the way of the average weight, to some extent, fixed a topic drift problem. Nutch [5] as an open source search engine provides a good choice for researchers. Nutch sorting algorithm [6–7] is an online analysis algorithm for the design of a general search engine like the PageRank. The sorting algorithm has obvious defects, such as weight equally linked pages; emphasis on the history pages; ignoring the relevance of web content. These factors will affect the quality of web page sorting. This paper provides some improvements: to distinguish the on-site and off-site link weights; to add the time factor, and to accelerate old page weight decline rate; based on the vector space model similarity computing page content.

## 2 MATHEMATICALFORMULATIONS

Web search results sorting algorithm is the most key technology of search engine, there are a variety of factors which influence the outcome of sorting, among which the most important arc the user's query and topic relevance of web content and web links to authoritative.

### 2.1 Vector space model

The vector space model represents each document into a vector form [8]. If there are topics, thesaurus $K(k_1,k_2,k_3,...k_n)$ and pages $K(w_{i1},w_{i2},w_{i3},...w_{in})$, $n$ is the number of keywords in the theme thesaurus $K(k_1,k_2,k_3,...k_n)$, $W_{ij}$ is the right value of the theme keyword $K_j$ on the page $P_i$, which reflects the ability to express the text message. $W_{ij}$ using TF-IDF algorithm, the formula is:

$$W_{ij} = TF \times IDF = (1 + \ln(TF)) \times \ln \frac{N}{n_{k_j}} \qquad (1)$$

From Formula 1, we can know that if the keyword $K_j$ appears in a document, the higher the frequency, the lower the frequency of occurrence in other documents, the larger amount of information contained high keyword, the greater the weight.

### 2.2 Page rank algorithm

PageRank algorithm can be explained by the user in the network topology of the random walk model. Assuming that the unified probability d for each user in the current page, and select the hyperlink to access to the next link, when users browse a web page and do not continue clicking the hyperlink, with probability

1-d start a new migratory behavior at any time, the user P is the probability for the page:

$$PR = (1 - d) + d \sum_{i=1}^{n} \frac{PR_{(T_i)}}{C_{(T_i)}} \qquad (2)$$

$\dfrac{PR_{(T_i)}}{C_{(T_i)}}$ shows PR value of the webpage $T_i$ split $C_{(T_i)}$ copies of the distributed web P. in other words, vote for web p. d Average of 0.85.

### 2.2.1 Real-time data fetching strategy

Obviously, in some specific areas, the site will be more frequently changed. In addition, even in the same area, the hot object is often much quicker than other objects in the field of change. In today's information, timely access to useful information has become more and more important. So if we lose real-time search engine search results, data for the user would not only make no sense, but also sometimes even bring the negative impact.

In a real-time search engine, we need to grab the data on the Web page regularly to ensure local data maintaining high freshness. EPA : The extended PageRank algorithm

The improved page-ranking algorithm based on PageRank algorithm called EPA.

The EPA can be divided into the following three basic steps.

Concrete steps which we can get from the following block diagram.

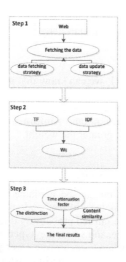

Figure 1. Three basic steps of EPA.

Step 1: Set the initial URL, Nutch crawler to crawl strategy based on the initial URL and get real-time data to the network crawling pages. Then download the data to extract information in order to carry out the theme parsing filter.

Step 2: Information will be filtered out of the data represented by the vector model to extract features on the topic. Determine the subject library, to prepare for the topic similarity judgment.

Step 3: PageRank algorithm reuse topic relevance factors and improved theme for massive data search.

A, B, C constitute a comprehensive ranking model which, considers topic relevance, real-time, and authoritative. Firstly, through the module A crawl down the data on the web using real-time strategy. Then, the data are passed to the module B, and the weights of this data processing. Finally to the module C, using PageRank algorithm can improve the ranking results of the final.

### 2.3 The distinction between stations and inbound links

The classic PageRank algorithm evenly distributed PR value of web P is assigned to a web page link which pointing to a web P. In other words, the PR value of the page P which related to the penetration and PR value of page P. So the network topology determines the importance of the node. However, an authority of the pages and the uneven of linked pages, are not very objectively reflecting the importance of the linked content. Moreover, some web designers will add site navigation and advertising pages link to cheat, in order to improve the site's search rankings. So classic PageRank evenly page weight is not reasonable. In this paper, to improve the type 2 is as follows:

$$PR_{(p)} = (1 - d) + d[(1 - a) \sum_{j \in V_2} \frac{PR_j}{C_j} + a \sum_{i \in V_1} \frac{PR_i}{C_i}] \qquad (3)$$

a $(0 < a < 1)$ is the proportion of factors outside the control station and outbound links. During normal inbound links than outbound links to better reflect the importance of objectively linked content, a take 0.75. For all of it into a chain of pages P, V1 represents the web page which differs from P in the same set of Web sites. $c_i$, $c_j$ denotes the number of voting page link $i$ and $j$ all linked pages. V2 represents all the pages into a chain of P which has the same site collection of pages. By controlling inbound and the proportion of inbound links to distribution value. Avoiding the classical algorithm equally weights considered cheating by the designer to improve the site address right weight, and to improve search quality.

### 2.4 Time attenuation factor

The classic PageRank algorithm weight distribution was carried out on the web page through the link relations. The longer the page exists on the Internet the more page links. Even if the content of the new web page is more important, links to other pages also need to be processed. In view of the problem focusing on PageRank's history page, the time factor is introduced to review the old web to do a weight decay process. In Formula 3, based on the improvement the following formula:

$$PR_{(p)}=(1-d)+d[(1-a)\sum_{j\in V_2}\frac{PR_j\times 2^{\frac{\lambda}{t}}}{C_j}+a\sum_{i\in V_1}\frac{PR_i\times 2^{\frac{\lambda}{t}}}{C_i}] \quad (4)$$

In the formula, $t$ is a time of existence of the page, $\lambda$ is a constant. The formula by adding the time decay factor $\lambda$ can inhibit the right to exist for a long time on the Internet site of the re-growth rate, making the new station bury in the old station.

### 2.5 Content similarity calculation

PageRank algorithm is the use of the network structure of reverse link information to sort of a web page, it has nothing to do with the theme of sorting algorithm. However, users generally choose the topic of the current page link to a larger page jump when browses the web. For classic PageRank topic drift problem exists, it should be added to the theme correlation factor in the algorithm to improve the rights of the heavy topic page. Suppose you have page $P_1(W_{11}, W_{12}, W_{13}, ... W_{1n})$ and topic vector $P_2(W_{21}, W_{22}, W_{23}, ... W_{2n})$, the web pages $P_1$ using the theme of the similarity between feature vector cosine of the Angle formula:

$$Sim(P)=Sim(P_1,P_2)=\cos(P_1,P_2)=\frac{\sum_{j=1}^{n}w_{1j}\times w_{2j}}{\sqrt{\sum_{j=1}^{n}w_{1j}^2}\sqrt{\sum_{j=1}^{n}w_{2j}^2}}) \quad (5)$$

In the formula, $W_{ij}$ is calculated according to the type 1 corresponding weights.

Finally, a comprehensive website ranking model which has the theme topic of the vector and the page Sim(P) and improved PR value assigned different weights, the page similarity score function by the equations 4 and 5 we have:

$$S = r_1 \times Sim(P) + r_2 \times PR_{(p)} \quad (6)$$

$r_1$, $r_2$ respect relevance weight of the theme and PR value of the weight. $r_1 + r_2 = 1$. Combining the topic

relevance and link analysis, can not only reflect the authority of link, but also can reflect the correlation of website theme, improve the quality of the sorting result, and enhance the user's search experience.

## 3 THE EXPERIMENT AND ANALYSIS

With more widespread exposure of food safety information, people are more eager to learn timely information on food safety, so the food information sites are likely to become people's favorite. Therefore, in this paper, food safety information on the Internet for data collection is conducted.

Nutch search engine is a set of search systems, including reptiles, page analysis, Lucence indexing, sorting, and so on.

Experiments based on Nutch web crawler technology, food safety information on the Internet for data acquisition and text information extraction, and then based on Lucene [9] [10] for full-text search tool kit for local data segmentation, inverted index, index retrieval and improved correlation degree sorting process to achieve a final design for food safety information with vertical search engines.

The effect of search engine query results is usually measured by a precision filter irrelevant documents to find the ability to focus on examination of the experimental sort algorithm improved precision, so that the first N query results and queries related to the number of pages and N ratios. In this study, some of the food safety website (Chinese food safety nets, throw out the window, etc.) home address as the entry address Nutch crawler collection, a total of 11275 crawling pages, these pages as an experimental document set [11]. After sorting algorithm for improved Nutch web through the search page, enter the query words, before and after Nutch improve the precision and effectiveness of comparative experiments sort. Select the four key areas to search for food safety in the experiment. For every ten records for precision analysis and comparison, the result is shown in Figure 2.

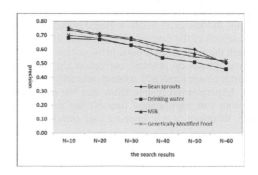

Figure 2. Algorithm improved the keywords precision comparison chart.

The four former algorithms to improve keyword search, and calculate the average, with the average of the results in Figure 2, the comparison results shown in Figure 3.

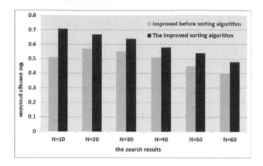

Figure 3. algorithm improvement before and after improving the average precision comparison chart.

As can be seen from Figure 2, the results of each keyword is not with the same precision, and first ten records of precision are basically on top of 0.6, indicating that the improved sorting algorithm can distinguish between pages, the relevant high degree of pages at the front of the search results.

By modifying the sorting algorithm, the improved precision, Nutch is significantly higher than the previous improvements, especially the forefront of the search results, the previous improved algorithm is superior to the improved, which can be known from Figure 3. It shows that the algorithm optimizes the sorting quality, more in line with the requirements of users with realistic feasibility.

## 4 CONCLUSION AND FUTURE WORK

It is very high that the vertical search engine based on Nutch theme similarity requirements for web page sorting algorithms. The classic PageRank algorithm average PR value out of the distribution chain, the chain does not consider the authority of the page, and ignores the theme and content of the query, easily leads to topic drift problem. Based on the analysis and research of the original algorithm, the distinction between the right to stand outside the station link with heavy emphasis on problem correction algorithm in the history pages and add content similarity calculations. Experimental results show that the improved algorithm can significantly improve the user's precision, therefore, it will improve user search results.

The next step is to consider adding user habits factor in Nutch score in order to further improve the sorting effect.

ACKNOWLEDGMENTS

This work was supported by the third level of 2011 Zhejiang Province 151 Talent Project and National Natural Science Foundation of China under Grant No.61100043.

REFERENCES

[1] Xing W,Ghorbani A.Weighted pagerank algorithm[C]. Fredericton:Proceedings of the 2rd Annual Conference on Communication Networks and Services Research, 2004:305–314.
[2] Pasquinelli M. Google's pagerank algorithm: a diagram of cognitive capitalism and the rentier of the common intellect[J]. Deep Search, 2009(3):152–162.
[3] Haveliwala. Topic-Sensitive PageRank: A Context-Sensitive Ranking Algorithm for Web Search[J]. IEEE Transactions on knowledge and data engineering, 2003, 15(4):784–796.
[4] Richardson M,Domingos P. The intelligent surfer: probabilistic combination of link and content informaionin in PageRank[J]. Advances in Neural Information Processing Systems, 2002, 14(3):1 441–1 448.
[5] Zhang Wen Long, Liu Yi Wei, Sun Jie. The vertical search engine based on Nutch study [J]. Journal of nankai university (natural science edition), 2012, 45(2):37–44.
[6] Tao Lin, Zhan Chao, Qiang Bao Hua. Based on the Hadoop Nutch research and implementation of web page sorting algorithm [J]. Guilin University of Electronic Technology, 2013, 33(2):139–143.
[7] Pan Tao, Liang Zheng You. The Nutch page sorting effect in improving methods[J]. Computer engineering. 2010, 26(13): 42–44.
[8] Eric J Glover, Kostas Tsioutsiouliklis, Steve Lawrence, etal. Using web structure for classifying and describing web pages[C]. New York:Proceedings of the 11th international conference on World Wide Web, 2002: 562–569.
[9] Li Yong Chun, Ding Hua Fu. The research and application of full text retrieval of Lucene [J]. The computer technology and development, 2010, 20, 2:12–15.
[10] Pasquinelli M. Google's pagerank algorithm: a diagram of cognitive capitalism and the rentier of the common intellect[J]. Deep Search, 2009(3):152–162.
[11] Wan J,Pan S.Performance evaluation of compressed inverted index in lucene[C]. IEEE: Research Challenges in Computer Science, 2009:178–181.

*Multimedia, Communication and Computing Application – Leung (Ed.)*
*© 2015 Taylor & Francis Group, London, ISBN 978-1-138-02775-6*

# Research and improvement of multiple pattern matching algorithm in Linux intrusion detection

F.Y. Wang & H.T. Chen
*Tianjin Key Laboratory of Intelligence Computing and Novel Software Technology, Tianjin University of Technology, Tianjin, China*

ABSTRACT:   With the development of Internet, the Linux operating system is widely used, and at the same time, the means of network intrusion attack is also constantly changed. Intrusion detection system can find the intrusion behavior through comparing the packets captured with the known network intrusion feature library. AC algorithm has the advantage of simultaneously doing multiple pattern match searching; however, there will be many unnecessary comparison, which will affect the efficiency of the intrusion detection system. To improve the matching efficiency, according to the automaton building of finite state of AC algorithm, this article uses the bouncing ideas of BM algorithm to optimize the displacement of text string and then an improved algorithm combining the bidirectional AC algorithm and BM algorithm is proposed. The experimental results show that the matching time and the number of matches are effectively reduced.

KEYWORDS:   AC algorithm, BM algorithm, Multiple pattern matching algorithm, Intrusion detection

## 1   INTRODUCTION

With the rapid development of network information technology, the dependence on network and information technology strengthened, and all kinds of potential safety problems also increased. We need to take effective measures to improve the performance of network security protection, and the intrusion detection system is based on the collection and analysis of several key points in the computer network or computer system, then we need to find whether there is a violation of security policy and the signs of being attacked in the network or system. The performance of IDS is closely related to efficient pattern matching algorithm.

By recording the work time of each part of intrusion detection system, M. Fisk et al. concluded that the string matching time accounted for 31% of the total execution time of Snort intrusion detection system, which showed that the improvement of efficiency of pattern matching algorithm can improve the processing speed of intrusion detection system. The early version of the Snort system adopted BM algorithm, and the algorithm can optimize the digits of text string needing to move, which greatly reduced the number of time of movement of the text string, and then the improved BM algorithm arise. However, with the improvement of network bandwidth, the rule base of intrusion detection system is also becoming larger, and simple pattern matching algorithm cannot satisfy the development of intrusion system. C. Jason Coit et al. proposed algorithm combined with the bidirectional AC algorithm and BM algorithm. They built multiple

pattern string to a tree, and then formed the finite state automaton, and there can be more pattern string matching operation through one search, and at the same time, it can reduce unnecessary matching, which greatly reduced the number of time of matching.

However, the testing efficiency of AC_BM algorithm is not ideal with respect to the problem that the length of text strings is too big, so this article proposed the DAC_BM algorithm, which run the finite state automaton of two pattern strings concurrently, and improve the time performance of the algorithm.

## 2   RELATED RESEARCH

### 2.1   *AC Algorithm*

The algorithm is a classical multiple pattern matching algorithm, which is proposed by Alfred V.Aho and Margaret J. Corasick in 1974. For a given text with length and pattern set $P\{p_1, p_2, \ldots p_m\}$, through the pretreatment on pattern set, the influence of the size of pattern set on the speed of matching algorithm is eliminated, within the time complexity $O(n)$, all target pattern in the text is found, which has nothing to do with the size of pattern set.

In the preprocessing stage, the classical AC algorithm consists of three parts, namely goto table, fail table, and output table. Goto table is the state automaton, which is composed of all patterns in the pattern set P, for a given set $P\{p_1, p_2, \ldots p_m\}$, the step to construct goto table is as follows: for each pattern $p_i$ [1…j] (1<= i <m+1) in P, sequentially input the letter

it contained to automaton, initial state is D[0], if the current state of automaton is D[p], there is no available movement for the current letter $p_i[k](1<=k<=j)$ in $p_i$, then the total number of states of state machines is smax + 1, and make $p_i[k]$ the current state, then let the shift position $D[p][p_i[k]]=smax$, if there is available shift $D[p][p_i[k]]=q$, then turn to state D[q], at the same time, taking out the next letter $p_i[k+1]$ of pattern string, then repeat the above judgment process.

Fail table is that when it is in a state of the state machine D[p], the input character c makes D[p][c] = 0, the state machine needed to shift.

The output table is the matched pattern string in the matching process when achieving a state of the state transition automaton, take pattern set p = {he, she, his, hers} as an example, its goto table, fail table and ouput table are shown in Figure 1:

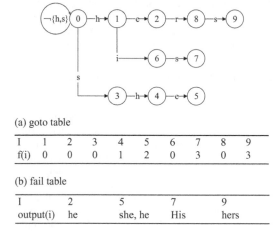

(a) goto table

| I | 1 | 2 | 3 | 4 | 5 | 6 | 7 | 8 | 9 |
|---|---|---|---|---|---|---|---|---|---|
| f(i) | 0 | 0 | 0 | 1 | 2 | 0 | 3 | 0 | 3 |

(b) fail table

| I | 2 | 5 | 7 | 9 |
|---|---|---|---|---|
| output(i) | he | she, he | His | hers |

(c) output table

Figure 1. The corresponding forward finite state automaton of pattern set P.

In the searching stage, the algorithm uses the three tables to scan text, and find out the matched pattern string in the pattern set P.

### 2.2 BM algorithm

BM algorithm is a single pattern matching algorithm first proposed by Robert S. Boyer and J. Strother Moore in 1977, the idea of the algorithm is that: First, do pretreatment on pattern P and align the left of pattern P at the left of pattern T, then do comparison from right to left, if the match fails, the BM algorithm uses bad character rules and good suffix rules to calculate the distance that pattern P shifts to the right, the maximum between them is taken as the distance of the shift to the right, then continue to move pattern P to the right, start a new round of matches until the match is successful.

Hypothesis: When mismatch of pattern P occurs, the matched part is u, there is mismatch between a in

text string T and b in pattern string P, which is shown in Figure 2.

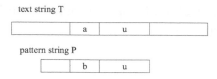

Figure 2. Mismatch between text string and pattern string.

(a) Bad character rules

Assuming the pattern string is P[1...m], the length of matched part of u is $i$, then we need only to analyze the situation of mismatch of text string a in pattern string P.

(1)If the mismatch character a in the text string exists in pattern string P, and the position of a in pattern string P is j, if a occurred repeatedly in pattern string P, then the character a close to b in pattern string P is chosen, and the shift distance of pattern string P to the right is (m-j-i), which makes text string T align with the rightmost a of pattern string P, as shown in Figure 3.

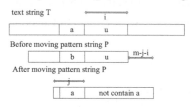

Figure 3. The first case of bad character rules.

(2) If there is no adaptation of character a in pattern string P, then the distance the pattern string P move to the right is (m-i), until that the first character of pattern string P is aligned with the next character of character a in text string T, as shown in Figure 4.

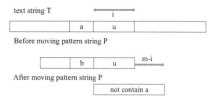

Figure 4. The second case of bad character rules.

Above all, assuming that the distance P moves to the right is Jump(x), m is the length of pattern string P, Last(x) is the rightmost position of character x in pattern string P.

$Jump(x) =$

$$\begin{cases} m - j, x \text{ not in pattern string } p \\ \quad which\ don't\ match; \\ m - Last(x), x \text{ in the right end of pattern string } p \\ \quad which\ don't\ match; \end{cases} \quad (1)$$

(b) Good suffix rules

(1) If the unmatched part of pattern string P has the same field u' with the matched field u, and if there are multiple child fields, then we choose one close to b, and u in the text string is aligned with u' in the pattern string P, as shown in Figure 5.

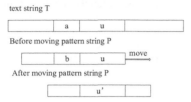

Figure 5. The first case of good suffix rules.

(2) If u' in (1) do not exist in pattern string P, but there exists a child string u' that starts from the first position of pattern string P and is the same to the substring in the rightmost of u, then we align u" with the longest substring, and continue matching. As shown in Figure 6.

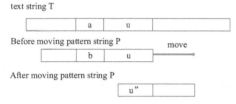

Figure 6. The second case of good suffix rules.

In conclusion, assuming that the length of pattern string P is $m$, the value of function move(x) is the character position of the first character of u' or u" in P, Shift(x) represents the moving distance to the right of pattern string P.

$$shift(x) = m + 1 - move(x) \qquad (2)$$

In the BM algorithm, the bigger one of Jump(x) and Shift(x) is used as the leap distance.

## 3  ALGORITHM

This article proposes a new algorithm combining a bidirectional AC algorithm and BM algorithm. Bidirectional AC algorithm is based on the original AC unilateral matching algorithm, and a reversed finite state automaton is added, which is refined as DAC algorithm. The algorithm (DAC_BM algorithm) used in this article combined the jump way of bad character rules and good suffix rules of BM algorithm, and matching from the positive and negative directions at the same time to speed up the matching.

In DAC algorithm, the pattern character string is first loaded, then forward finite state automaton and reverse finite state automaton is established, respectively, according to the pattern string. Then we need to make sure the center of string needed to match and

the match position of positive and negative finite state automaton. The starting point to match of the positive and negative state automaton is related to the center point of string to be matched and the length of the longest pattern string. The length of the longest string in the pattern string is the size of repeat area of bidirectional matching, then this area is divided averagely and set on both sides of the center point. Then this area is analyzed to judge whether it matches with the longest pattern string; if it mismatches, then forward finite state automaton starts matching from the left side of the region to the right, and the reverse state automaton starts matching from the right side of this region to the left. If there is a successfully matched string, the matched pattern string is output through the output function, the aforementioned string to match is the network packets, and various rules in the intrusion detection system is the pattern string. DAC_BM algorithm is based on the DAC algorithm and combined with BM algorithm, which increases the move distance of pattern string and reduces the matching time.

Each pattern string in the pattern string set is arranged reversely, taking pattern set p = {he, she, his, hers} as an example, the reverse goto table, fail table, and output table are shown in Figure 7:

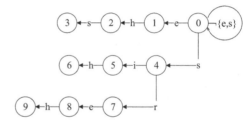

(a) reverse goto table

| i | 1 | 2 | 3 | 4 | 5 | 6 | 7 | 8 | 9 |
|---|---|---|---|---|---|---|---|---|---|
| f(i) | 0 | 0 | 4 | 0 | 0 | 0 | 0 | 1 | 2 |

(b) reverse fail table

| i | 2 | 3 | 6 | 9 |
|---|---|---|---|---|
| output(i) | he | she | his | hers |

(c) reverse output table

Figure 7. The corresponding reverse finite state automaton pattern set of P.

## 4  EXPERIMENTAL ANALYSIS

To test the performance of DAC_BM algorithm, compared with basic AC algorithm, the snort intrusion detection system 2.9.2.0 is used, and the experiment was conducted under Ubuntu 12.04, CPU Intel® Core™ i3 CPU 550, 3.3 GHz, and memory 1.0 GB.

The test data used in this article is the intrusion detection test data set of 1999DARPA, which included a large number of normal network traffic and all kinds of attacks, and this article selected inside.tcpdump data set in the first week from Monday to Friday and then inside.tcpdump data set on Friday in the third week to test the performance of the algorithm.

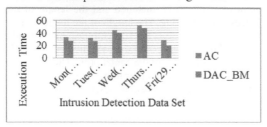

Figure 8. The comparison of time of algoritm between different intrusion detection datasets.

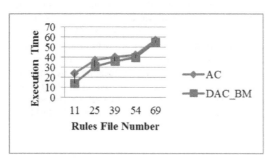

Figure 9. The comparison of time of algorithm with different rule file.

The comparison of test results of time of algorithm between different intrusion detection dataset is as follows:

Table 1. The test results of time performance of algorithm. (time/s).

| Test dataset (inside.tcpdump) | AC | DAC_BM |
|---|---|---|
| Monday (325MB) | 33.35 | 27.35 |
| Tuesday (325MB) | 32.55 | 26.68 |
| Wednesday (367MB) | 44.67 | 39.46 |
| Thursday (527MB) | 52.10 | 48.14 |
| Friday (294MB) | 28.49 | 20.20 |

Table 2. The matching time of detecting all packets with different rules file (time/s).

| The number of rule file | 11 | 25 | 39 | 54 | 69 |
|---|---|---|---|---|---|
| AC | 24.17 | 37.32 | 40.48 | 42.67 | 57.19 |
| DAC_BM | 14.29 | 30.73 | 36.33 | 39.88 | 55.09 |

The comparison of time of algorithm with different rules file is as follows:

The above experiments show that the performance of DAC_BM algorithm proposed in this article is promoted by 20.9% averagely compared with the basic AC algorithm under the condition that the number of rule files is the same, when the number of rule file is different, the increase amplitude of performance of DAC_BM will be reduced with the increase of the number of rule file, but the matching time to detect all packets will be still less than the basic AC algorithm.

## 5 CONCLUSION

In this article, several algorithms of intrusion detection system are simply introduced and analyzed. DAC_BM algorithm used the advantages of the multimodal search of AC algorithm and the optimization of displacement of BM algorithm, and the reverse automaton is added to match reversely, which greatly reduced the detection time of intrusion detection system under high-speed network environment. The experiment shows that the algorithm is more effective to reduce the number of matching, the matching time is shortened and the detection efficiency of the Snort intrusion detection system is improved, which has broad application prospects for the protection of network environment. To further optimize the algorithm, fuzzy clustering algorithm and neural network algorithm can be combined with this algorithm, which can be better applied in the intrusion detection system.

REFERENCES

Boyer R.S &Moore J.S. 1997.A fast string searching algorithm. *Communications of the ACM* 20(10):762–772.
Hou Zheng-feng & Zhang Xiao-le.2011. Research and improvement of AC-BM algorithm. *Chinese Journal of Scientific Instrument* 3(2):216–221.
Lutz, M.J et al. 2002. Flexible Pattern Matching in Strings. *Cambridge University Press* 35(9):81.
Sun Wen-jing & QianHua. 2013. Improved BM Algorithm andIts Application in Network Intrusion Detection. *Computer Science* 40(12):174–176.
Wang Zheng-cai et al. 2013. Multi-objective AC-BM Algorithm Based on Automata Union Operation. Computer Science40(6):119–123.
Yang Chao. 2011. Two-Way AC Algorithm and its Application to Intrusion Detection System. *Computer Systems & Applications* 20(3):222–225.
YANG Wen-jun & WEI Zhan-guo. 2009. Efficient Pattern Matching Algorithm for Intrusion Detection Systems. *Journal of Chinese Computer Systems* 30(11):2189–2194.

# Research and improvement of Apriori algorithm based on the matrix decomposition affairs

Q. Wu & Y. Yao

*Hangzhou Dianzi University, Hangzhou, Zhejiang Province, China*

ABSTRACT: Apriori Algorithm is a typical algorithm frequently applied in the field of data mining. It can generate a large number of candidate itemsets through scanning the database several times which means that the traditional Apriori algorithm has its shortcomings, especially the low efficiency. This article proposes an improved Apriori algorithm called FApriori. The FApriori adopts the method of compression matrix to store the object's information so that it can calculate the support count faster by scanning the database . The experimental results show that the larger the data set, the more efficient the FApriori algorithm in data mining.

## 1 INTRODUCTION

Because of the accumulation of data, and the pressing needs for information and knowledge of the market competition, data mining has become a very hot research topic. The explosion of data indicates many important information. A simple query and statistics have been unable to meet the needs of doing business. We need to research the data mining further.

Association rules of mining finds the interesting relation in a large amount of itemsets [1]. From a large number of transaction records, it can be found that there is an interesting correlation between diapers and beer. Seemingly, they are irrelevant goods, but there is a great increase in sales of these two goods. That is because after having bought some goods, the customers will buy some other goods.

The most classical algorithm for mining relevant rules is Apriori algorithm. Apriori algorithm is able to do layer-by-layer search via iteration method, that is, to scan the database and produce frequent itemsets from a large number of candidate items. Then acquires frequent itemset k+1 from itemset k by connecting. But this algorithm method has two defects: (1) it produces a large number of candidate itemsets and (2) it requires to scan the database many times. And when the database is particularly big, the mining efficiency is extremely low[2].

This article proposes a method of mining frequent itemset rules for relevant rules based on decomposition Transaction Matrix (FApriori algorithm). This algorithm can compress the relevant information in the database and store it in memory in the form of decomposition Transaction Matrix and then mine the frequent itemsets from the decomposition Transaction Matrix. This algorithm proposes a method for connecting frequent itemsets without any comparison, which saves the operation time of the algorithm. This article proposesa new method for calculating the number of candidate itemsets support, which solves the problems about repeatedly scanning database in Apriori algorithm.

## 2 RELEVANT CONCEPTS

Set I={i1,i2,…, im} is a collection of projects, transactional database D={t1,t2,…,tn} consists of a series of transactions with unique TID, each transaction ti(i=1,2,…,n)is corresponding to a subset of the I. To effectively explain the improved algorithm, we need to use the following definition and properties:

1) Definition 1: set I1⊆I, support of itemset I1 in dataset includes I1 transactions percentage in D.

$$Support(I_1) = \frac{\|\{t \in D \mid I_1 \subseteq t\}\|}{\|D\|} \qquad (1)$$

2) Definition 2: A definition is given by satisfying a certain confidence level like I1⇒I2 of association rules in the I and D. The so-called confidence refers to the number of transaction contains I1 and I2 and contains I1 the ratio of the number of transactions[10].

$$Confidence\ (I_1 \Rightarrow I_2) = \frac{Support\ (I_1 \cup I_2)}{Support\ (I_1)}$$

Where $I_1, I_2 \subseteq I$,    $I_1 \cap I_2 = \varnothing$ \qquad (2)

3) Definition 3: Set transaction database D matrix is expressed as the affairs:

$$
M = \begin{pmatrix} D_1 \\ D_2 \\ \vdots \\ D_m \end{pmatrix} = \begin{matrix} I_1 & I_2 & \cdots & I_n \\ \begin{pmatrix} d_{11} & d_{12} & \cdots & d_{1n} \\ d_{21} & d_{22} & \cdots & d_{2n} \\ \vdots & \vdots & & \vdots \\ d_{m1} & d_{m2} & \cdots & d_{mn} \end{pmatrix} & \begin{matrix} T_1 \\ T_2 \\ \vdots \\ T_m \end{matrix} \end{matrix}
$$

4) Definition 4: Column (row) vector is defined as matrix column (row) a collection of all items, such as column vector of the j-th column: $D_j = (d_{1j} d_{2j} \cdots d_{mj})$, row vector of the i-th row: $D_i = (d_{i1} d_{i2} \cdots d_{im})$.

Nature 1: If a transaction T, the number of the item less than 2, then we can delete the corresponding row vector in the firm in the matrix.

Nature 2: If a column vector of Boolean matrix is less than the minimum support count, the column can be deleted.

## 3 APRIORI ALGORITHM

Apriori algorithm is a classic algorithm of mining frequent itemsets. In Apriori algorithm, the frequent itemset 1 (L1) is obtained by traversing the database. If L1 is not empty, connected by an L1 itself, it can generate candidate itemsets C2, scan the database, select all count in the C2 meeting minimum support itemset of frequent itemsets 2 (L2). After the above steps, then connect, prune, and scan database search iteration step by step, till all the frequent itemsets are dug up.

1) Connection. Two frequent (item set k - 1 connection is a prerequisite for k - 2) items are the same, before the first k - 1 item, L1[1] = L2[1] ∩ … ∩ L1[k-2] = L2[k-2] ∩ L1[k-1] < L2[k-1]. To meet the conditions of frequent itemsets, Lk k - 1-1 with its own connection generates candidate itemsets k Ck.

2) Pruning. Ck is a superset of Lk. All the frequent itemsets are included in the Ck, k. But Ck may not contain frequent member. Each candidate can be established by the method of scanning the database directly on the number of support so as to determine the Lk. But Ck can be very big, the number of scanning database directly will have a lot of I/O overhead[3]. So we can prune Ck, and delete a subset of frequent candidate itemsets, to compress the size of the candidate itemsets.

## 4 FAPRIORI ALGORITHM

### 4.1 The algorithmic thought

First, scan transactional database and create a BooleMatrix, which each row stores a transaction information. Support counting of each item in a matrix is calculated by counting. The result L1 has been got by ranking a matrix in descending order according to the support size. Its value is {I1,I2,…,In}. The columns, the minimum support counting are less than setting value of minimum support in a matrix which has deleted. The remaining items can be drawn into frequent itemsets 1.

The transactional matrix was scanned reversely from the last column to the first. The entire matrix was decomposed into several sub matrices, which were respectively mined for frequent itemsets with the FApriori algorithm..

### 4.2 Algorithm description

Scanning database establishes Boolean matrix D, and each row in D stores a compressed transaction information. The m transaction and the n project in the database may establish a matrix D[m][n]. To encode each transaction T, the way of encoding is to delete the item not meeting the minimun support count in each transaction , and to record the remaining items to get T' according to the order of L1. We code a length of n row vector composed of 0 and 1. If records T Ij is contained in item, its corresponding position dj(1≤j≤n) or less value is 1, otherwise 0. According to the ascending order of frequent itemsets, 1 M' can be broken down into b matrix. Record for each matrix decomposition: $M_1, M_2 \cdots M_b$.

Scanning the matrix M' b column, the column $D_b = (d_{1b} d_{2b} \cdots d_{ab})$, judgment $d_{ib} (1 \le i \le a)$, if $d_b = 1$, the extraction $(d_{i1} d_{i2} \cdots d_{ib})$, combined into submatrices, is represented as

$$
Mb = \begin{matrix} I_1 & I_2 & \cdots & I_b \\ \begin{pmatrix} d_{11} & d_{12} & \cdots & 1 \\ d_{21} & d_{22} & \cdots & 2 \\ \vdots & \vdots & & \vdots \\ d_{a1} & d_{a2} & \cdots & 1 \end{pmatrix} & \begin{matrix} T_1 \\ T_2 \\ \vdots \\ T_a \end{matrix} \end{matrix}
$$

Through a case, the author introduces a process of the algorithm. Transaction data are shown in Table I.

Table 1. Transaction Data.

| TID | The list of item ID |
| --- | --- |
| TI | I1, I2, I6 |
| T2 | I2, I3 |
| T3 | I1, I2, I4 |
| T4 | I1, I3, I5, I6 |
| T5 | I2, I5, I6 |
| T6 | I1, I2, I5, I6 |
| T | I5, I6 |

110

Scanning database, becoming the matrix:

$$
\begin{array}{cccccc}
I_1 & I_2 & I_3 & I_4 & I_5 & I_6
\end{array}
$$

$$
M[7][6] =
\begin{array}{|cccccc|c}
1 & 1 & 0 & 0 & 0 & 1 & T_1 \\
0 & 1 & 1 & 0 & 0 & 0 & T_2 \\
1 & 1 & 0 & 1 & 0 & 0 & T_3 \\
1 & 0 & 1 & 0 & 1 & 1 & T_4 \\
0 & 1 & 0 & 0 & 1 & 1 & T_5 \\
1 & 1 & 0 & 0 & 1 & 1 & T_6 \\
0 & 0 & 0 & 0 & 1 & 1 & T_7
\end{array}
$$

We get the column vector according to the matrix: {I1: 4, I2: 5, I3: 2, I4: 1, I5: 4, I6: 5}, the value of row vector: {T1:3, T2:2, T3:3, T4:4, T5:3, T6:4, T7:2}. Frequent 1 – can get the matrix M' from descending order:

$$
\begin{array}{cccccc}
I_2 & I_6 & I_1 & I_5 & I_3 & I_4 \\
\text{Sup } 5 & 5 & 4 & 4 & 2 & 1
\end{array}
$$

$$
M'[7][6] =
\begin{array}{|cccccc|c}
1 & 1 & 1 & 1 & 0 & 0 & 4 \\
0 & 1 & 0 & 1 & 1 & 0 & 4 \\
1 & 1 & 1 & 0 & 0 & 0 & 3 \\
1 & 0 & 1 & 0 & 0 & 1 & 3 \\
1 & 1 & 0 & 1 & 0 & 0 & 3 \\
1 & 0 & 0 & 0 & 1 & 0 & 2 \\
0 & 1 & 0 & 1 & 0 & 0 & 2
\end{array}
$$

The sum of deleting is equal or less than 2 in some colum and row vectors, as follows M:

$$
\begin{array}{cccc}
I_2 & I_6 & I_1 & I_5 \\
\text{Sup } 4 & 4 & 3 & 3
\end{array}
$$

$$
M''[5][4] =
\begin{array}{|cccc|}
1 & 1 & 1 & 1 \\
0 & 1 & 0 & 1 \\
1 & 1 & 1 & 0 \\
1 & 0 & 1 & 0 \\
1 & 1 & 0 & 1
\end{array}
$$

At last, by matrix decomposition affairs, we can find all the frequent itemsets.

Table 2. The process of matrix decomposition affairs

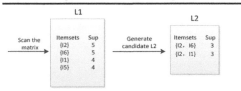

Scanning the database only once, we are able to get all the frequent itemsets which means that the work efficiency of the algorithm have been improved greatly.

### 4.3 Improvement in connection

The itemsets in Apriori algorithm are sorted by the dictionary, and in FApriori algorithm, each element of frequent itemset by support counting is ranged from the largest to the smallest. The frequent item K set is generated by the connection of a frequent K-1 itemset L1 and other frequent k-1 itemset L2 when the first item's support counting of L2 is higher than that of L1 and the remaining K-2 items are the same, that is, $L2[1] > L1[1] \cap L2[2] = L1[2] \cap \ldots \cap L2[k] = L1[k]$. The L2 first item is just put into L1 during the L1 and L2's connection. All the frequent K items, generated from the L1 and other frequent K-1 items' connection are sorted together. So, for frequent K itemsets, the first item which is different and the rest which is same are stored together. So the next connection can be done by taking out two itemsets in a given set [7].

### 4.4 Improvement in support counting's calculation

Apriori algorithm was used to calculate the support counting: each candidate itemset in the Ck was scanned according to the database to get the size of support counting, so as to verify whether it has joined the Lk or not.

FApriori does not need a database retrieval, but it can get the support counting by calculating each candidate itemset. Each frequent itemset has a set of rows, indicating the rows where frequent itemsets appears in the decomposition affairs matrix. If candidate itemsets were not pruned to be generated from two frequent itemsets, the row sets of candidate itemsets can be obtained by 'AND operation' in the two row sets of frequent itemsets. When calculating the support counting, the candidate itemsets' support counting can be obtained by adding the corresponding support counting into each row just after the retrieval for the corresponding decomposition affairs matrix according to the row set of candidate itemsets.

With the help of the improved calculation, each support counting can be obtained in table only by an "AND operation" of the row set of frequent itemsets without retrieval in the database, and then support counting of each candidate itemsets can be obtained by adding the support counting into each row. This requires a simple operation and there is no need to scan the database any more. In other words, the improved algorithm can greatly save the I/O cost.

## 5 RESULTS AND ANALYSIS

To verify the improvement of Apriori algorithm, the Apriori algorithm and the improved FApriori algorithm were tested. The two algorithms are realized in using the JAVA programming language on the 2.6 GHz CPU PC with 2 GB memory. The data for mining is the dense datasets Mushroom recording a total of 9416 records and 80 projects in the UCI machine learning database. The length of the average transaction is 32. The time, which is generated from the frequent itemsets under the different minimum support degree, can be seen in Table 3. The consumed time of the two algorithms tends to be closer to the support degree. The advantages of the FApriori algorithm were not reflected for the generated frequent itemsets, and is less when support degree is larger. Furthermore, scanning matrix also consumed a certain amount of time.

## 6 CONCLUSIONS

Our algorithm improves the association rules of mining in Apriori algorithm and uses decomposition affairs matrix to store the transaction database. The algorithm takes the advantage of together-stored characteristics for frequent items without comparison and judgment of whether they can be connected or not during the connection operation, which leads to direct connection and time-saving comparison. Using record row set to calculate the support counting can reduce the time of scanning and save the IO cost. The experiment shows that the algorithm's efficiency is obviously higher than that of Apriori algorithm under the lower supporting degree.

Table 3.    The result of experiment

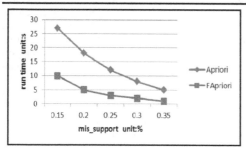

## 7    ACKNOWLEDGMENTS

This work was supported by the third level of 2011 Zhejiang Province 151 Talent Project and National Natural Science Foundation of China under Grant No.61100043.

## REFERENCES

[1] G.X. Cui, L. Li, K.K. Wang. "Research and Improvement of Apriori Algorithm in Association Rules Mining,"J. Computer application, vol.11, pp:2952–2955, 2010.

[2] S.Fu, D. Song. "Research and Improvement of Apriori Algorithm on Matrix.," J. Microelectronics and computer, vol.29(5), pp:156–160, 2012.

[3] C.G. Qian, Y.H. Dong. "Improvement of FP-growth Algorithm on the Matrix Decomposition Affairs," J. Journal of hubei vocational and technical college, vol.15(1), 2012.

[4] W.J.Yuan, J.Yan."Data Mining Association Rules Apriori algorithm".J.Computer Systems & Applications, vol.22(4).pp.121–124, 2013.

[5] Y.Z.Wang, X.L. Le, X.Q. Chen. "Network Big Data: Status and Prospects". J. Journal of Computers, vol.36(6). pp. 1125–1138, 2013.

[6] H.Huang, X.D. Yi, S.S. Li, etc. "Massive high-performance computers for data processing platform and evaluation".J.Computer Research and Development,vol.1. pp. 357–361, 2012.

[7] T.Xu, X.Dong. "Mining frequent patterns with multiple minimum supports using basic Apriori," C.//Natural Computation (ICNC), 2013 Ninth International Conference on. IEEE, pp: 957–961, 2013.

[8] G.J.Mao. "the Theory and Algorithm on data mining," M. Beijing: Qinghua University Press, 2007.

[9] H.V. Nguyen, E. Muller, K. Bohm ."4S: Scalable subspace search scheme overcoming traditional Apriori processing," C.//Big Data, 2013 IEEE International Conference on. IEEE, pp: 359–367, 2013.

[10] J.L.Chen, K.l.Xiao. "BISC: A Bitmap Itemset Support Counting Approach for Efficient Frequent Itemset Mining," J. ACM Transactions on Knowledge Discovery from Data (TKDD), vol.4(3),2010.

*Multimedia, Communication and Computing Application – Leung (Ed.)*
© *2015 Taylor & Francis Group, London, ISBN 978-1-138-02775-6*

# Analysis of SDN security applications

Y.F. Yang, Y. Chi, D. Li & W. Zhang
*The College of Computer, Nanjing University of Posts and Telecommunications, Nanjing, Jiangsu, China*

ABSTRACT: With the growing maturation of standards and technologies of SDN, the benefits of SDN that obtains from the application fields will attract more and more attention. While the idea that control and forwarding have been separated is simple, SDN has the features of being programmable and centralized control, global view and rapid deployment, and it is to solve the problems that traditional networks cannot deal with and make the network more secure and reliable. In this paper, four SDN security applications include congestion control, traffic restrictions, load balancing and security policy conversion are mainly analyzed. At last, we provide the prospects of the problems that need to be solved by SDN security application model in the future.

KEYWORDS: SDN, Security applications, Centralized control, Global view.

## 1 INTRODUCTION

In 2006, Martin Casado et al. in Stanford University proposed the management structure for the enterprise network called SANE [1], based on 4D [2] architecture, which defined a protective layer that can manage all the connections between the link layer and IP layer. A logic central server controlled the protective layer. SANE was mainly applied in enterprise networks, focused on security control, and not yet implemented complex-routing decisions. Since it had not been tested massively, the actual deployment was still relatively difficult. In 2007, Ethane [3] extended on the basis of SANE, aiming to use a centralized controller, so that the network administrators could easily define network-flow-based security control strategies. Then these security policies were applied to a variety of network equipment to achieve the security control of communications in the entire network. Inspired by SANE and Ethane, Martin Casado and his mentor, Professor Nick McKeown, found that if they could generalize the design of Ethane and separate the two modules of the traditional network device: data forwarding and route control, then use the standardized interfaces to manage and configure a variety of network devices through a centralized controller, it would provide more possibilities for the design, management and use of network resources and make it easier to promote the innovation and development of the network. In the year of 2008, they proposed the concept of OpenFlow [4] on ACM SIGCOMM conference, and further abstracted the core theory of SDN (Software Defined Network) [5] in the following year. SDN treated all the network devices as the managed resources, abstracted the details of

underlying network devices, and also provided a unified management view and programming interfaces for the upper application. In 2009, SDN was nominated for Technology Review's annual top ten frontier technologies and started to gain the recognition and support of academia and industry. After several years of research and practice, ONF (Open Network Foundation) [6], a non-profit organization established by the Internet companies and telecommunications operators in 2011, gave the framework of SDN. From the bottom up, the framework was divided into infrastructure layer, control layer and application layer. Many network operators, equipment suppliers and technology providers set up a network function virtualization industry alliance to introduce the concept of SDN to the telecommunications industry.

With the continuous development of SDN, SDN-related security problems are gradually revealed, and any kind of network or application is inseparable with security. SDN has no exception. If the security is not considered in the initial construction of a network, the network will spend a lot of resources addressing the security problems at a later stage. There are four SDN security problems: (1) Congestion control problems, once the network is attacked, the TCP congestion control decisions cannot solve network congestion quickly and accurately. So we should think about whether we could use the idea of centralized control of SDN and take advantage of the global network view to solve the problem of network congestion control. (2) Traffic restrictions problems, when the network detects abnormal traffic, the core problem is how to locate and isolate malicious nodes quickly, and use the function of traffic restrictions to protect the interests of users and enterprises.(3) Load- balancing problems, when the server is attacked and cannot deal with many network packet requests, a crash will occur. How to prevent in advance to ensure network security is a key research question. (4) Security policy conversion problems, we need to consider the possibility of establishing an security policy conversion mechanism to convert the global security policies into simple security rules that the underlying switches can handle. It can simplify the programmers' work and the security policies can be arranged quickly and efficiently. In this paper, the existing SDN security applications will be analyzed to solve these problems, and we wish that the paper will enlighten some inspirations for the research of SDN security problems.

## 2 SDN FRAMEWORK

The more mature framework of SDN is the three-tier architecture proposed by ONF, which consists of the infrastructure layer, control layer, and application layer, as shown in Figure 1. Network infrastructure acts as the role of forwarding plane in original switching/routing design, which is also called OpenFlow switch. Control layer makes a resource abstraction for the underlying network infrastructure, provides a global network abstract view for the upper application, and gets rid of the problem that hardware network equipment and network control functions are bundled together. The application layer programs for the network abstraction is provided by the control layer through the open interfaces to manipulate a variety of traffic models and the network traffic generated by applications. It makes the traffic generated by applications perceive the network to achieve network intelligence. The control software on the control layer interacts with the switching/routing and other network devices on the infrastructure layer through the control data plane interface and communicates with a variety of applications on the application layer through an open API.

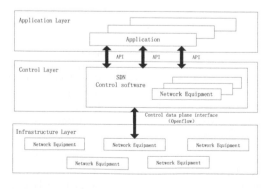

Figure 1. SDN framework.

Although SDN research is generally based on OpenFlow protocol, SDN is not the same as OpenFlow. SDN is a framework and an idea that can be achieved by a variety of ways, with OpenFlow being one of them. Our research should focus on the nature of SDN, namely we should think about how to achieve security applications when the network is centrally controlled.

## 3 SDN SECURITY APPLICATIONS

The problem we care about is how to take advantage of the centralized control of SDN to implement the associated security applications after understanding the framework and implementation of SDN. In the next step, we will analyze the corresponding security applications for SDN in four aspects: congestion control, traffic restrictions, load balancing and security policy conversion.

### 3.1 Congestion control application

Most schemes focus on the functions of routing and forwarding, but ignoring the advantages provided by the congestion control of SDN. It is a key point that how we can solve the congestion problem quickly when the network is attacked. Given the centralized control of SDN, we should consider whether we could take the advantage of a global network view on the controller to make the congestion control decisions faster and more accurate. Therefore, we can establish a system that can observe the static and dynamic changes of computer networks in real time, simplify the adjustment of different TCP parameters and promote to change the protocol itself dynamically. We would be able to obsserve the congestion conditions of the network in time by programming the congestion control procedures on the SDN controller. When a certain point is congested, the SDN controller will automatically change the TCP variants or adjust the TCP parameters of the point to solve the congestion problems quickly and ensure the fluency of network operation.

Monia Ghobadi et al. in the University of Toronto proposed OpenTCP [7] as a system for dynamic adaptation of TCP that could easily switch the different TCP variants in the automatic (or semi-supervised) manner or adjust the TCP parameters according to the network conditions. For example, OpenTCP could use different TCP variations in the data center. What kind of variants we should use was decided ahead of time by the congestion control strategy defined by the network operator. Since OpenTCP could solve the network congestion problems quickly, when attackers attacked a device or multiple devices on the network, and it led to the network congestion, OpenTCP could play a key role.

Figure 2 shows a schematic diagram about how OpenTCP works in the case of the network being attacked to congestion. Oracle, switches, and CCA (Congestion Control Agent) are three components of OpenTCP, where Oracle is a SDN controller application at the core of OpenTCP and collects information about the underlying network and traffic. When the network is under attack to congestion, Oracle can make changes in accordance with TCP protocol according to the congestion control strategies defined by network operators. Finally, it distributes the update message (CUE) to terminal hosts. Switches are compatible with SDN and can not only forward data but also collect traffic statistics. The CCA is a kernel module installed at each terminal host. It receives update messages from Oracle through the switches, and is responsible for modifying the TCP protocol stack on each host.

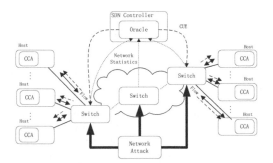

Figure 2.    Congestion control application.

OpenTCP can take the advantage of the global network view to solve the problem of network congestion and reflect the features that congestion control procedures are easy to program, the switching of TCP variants and the adjusting of TCP parameters are achieved automatically. OpenTCP will play a significant role in promoting the research of SDN security problems, as long as knowing in which line or which location congestion occurs, it can solve the congestion problem by the feedback information quickly, ensure the network operation unobstructed and make the network more secure and reliable.

## 3.2  *Traffic restriction application*

The purpose of traffic restrictions is to limit the traffic of malicious nodes to ensure the operation of other normal traffic. The SDN controller is able to adopt a global view to understand the changes of network topology and network traffic, which

provides a convenient platform for detecting the position of malicious nodes. We can consider establishing a centralized control framework to observe the operation of the network traffic, timely detect the position of abnormal traffic and isolate the malicious node to avoid affecting the operation of other normal traffic.

Lalith Suresh, Julius Schulz-Zander and Ruben Merz jointly proposed an SDN framework called Odin [8]. It was built on an abstract LVAP (Light Virtual Access Point) that greatly simplified client management. The design goal of Odin was to enable programmers to ignore changes of the authorizer, authenticator, and client state machine. The core component of Odin was the abstract LVAP that could virtualize association state and separate them from the physical access point. Multiple clients connected to a single physical access point were considered as a set of isolated clients connected to different ports of a switch logically. The abstract LVAP provided a straightforward programming model for Odin.

As shown in Figure 3, Odin is composed of the Odin Master, multiple Odin agents and a set of applications, where Odin Master is an application that works on an OpenFlow controller, passes the OpenFlow protocol to switches and access points and uses a custom protocol to make every Odin agent run on access points. Odin has a global view of dynamic changes of flows in the network, clients connected to the network and the network infrastructure. Therefore, when the abnormal traffic is detected, Odin will isolate the malicious nodes and limit the flow to ensure network's security.

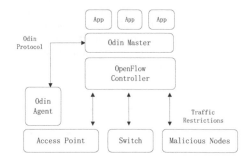

Figure 3.    Traffic restriction application.

We can limit the traffic of malicious nodes by using Odin without affecting the operation of other normal traffic. Thanks to the advantages of the centralized control of SDN, we can easily detect the position of abnormal traffic and take appropriately isolated measures to restrict malicious traffic to ensure the safe operation of normal network traffic.

115

### 3.3 Load-balancing application

Load balancing is an optimization based on the network link performance and user policy, which provides a cheap, effective, and transparent way to expand the bandwidth of network devices and servers, increases throughput, enhances the capability of network data processing and improves the flexibility and availability of the network. The traditional solutions to load balancing generally need to use a gateway or router to monitor and count server workloads in the entrance of server clusters and dynamically allocate user requests to the server where load is relatively light. Since all network devices can use the centralized control and management through OpenFlow, the simultaneous application server load can be timely transmitted to the OpenFlow Controller, thus making OpenFlow very suitable for the work of load balancing.

Nikhil Handigol et al. in Stanford University designed Aster*x [9] to solve the problem of load balancing. Aster*x used Host Manager and Net Manager to monitor the workloads of servers and network respectively, and then Host Manager and Net Manager sent the feedback to Flow Manager. So Flow Manager could redefine the OpenFlow rules on network devices according to the real-time load information, and the user requests (network packets) were adjusted and distributed on the basis of the capacity of servers.

When the server cannot handle the large number of network packet requests which will cause a collapse, it is a key problem that how to prevent the collapse in advance and ensure network security. As shown in Figure 4, the hacker attempts to cause the collapse of a server by sending many malicious request packets to the server. Aster*x has a global view of the changes of server loads and reasonably allocates network packets to multiple servers for processing without causing any serious server load. Precisely because of the advantage of centralized control, we can understand the conditions of server loads and take measures to maintain network security timely.

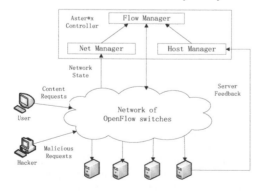

Figure 4.    Load balancing application.

Aster*x can observe the status of server loads in a global view and allocate the loads to multiple servers in the same region in time. However, if multiple servers within a region cannot deal with the requests of network packets in the region, whether we can send network packets to the servers in another region to reduce the loads? It requires the coordination and cooperation between the SDN controllers, which is a problem we should think about deeply in the future.

### 3.4 Security policy conversion application

Above three applications use the specific controller to achieve the appropriate security application, but if the controller has no specific function, how to deploy security policies of the upper application to the underlying switches quickly and efficiently? We can install a mechanical device on the controller to convert a high-level security policy into multiple low-level security rules. The future SDN platform should support the automatic conversion of security policies across multiple switches. It would simplify programming by allowing programs being written on an abstract switch to run over a more complex network topology and simplify analysis by converting a security policy spreading over multiple switches into a series of single security rules. A sound and complete set of axioms need to be established in order to achieve the conversion between security policies and security rules and to maintain the implication of the forwarding security policy while rewriting of security rules across multiple switches. These axioms are the basis to create and analyze algorithms for optimizing the rewriting of security rules.

Nanxi Kang et al. in Princeton University proposed the idea that SDN controller could convert policies [10] automatically. The target of their work was to establish a mechanical device to achieve policy conversion. By using this mechanism, the policies could be distributed to a set of switches, and the rules distributed to multiple switches could be combined into one. The goals of policy/rule conversion include: (1) policies could be decomposed, and policies, which were too large to fit on any one switch could be spread over multiple switches. (2) policies were easy to program, and it was easier for the programmers to write policies for a switch. (3) policies could be transplanted, and the policies of a network topology could be converted to run on another network topology. (4) a policy could be analyzed, and the policy distributed on multiple switches could be converted into a series of simple rules, making it easier for programmers to analyze and observe whether the policy met invariants or not. (5) policies could be automatically converted,

116

and the controller converted policies automatically and the programmers did not need to pay attention to the details.

Security policy conversion application is built on the idea that the SDN controller converts the policies automatically, as shown in Figure 5. The global security policy is converted into multiple security rules by a mechanical device and the controller distributes the rules to multiple switches. In order to achieve security policy conversion correctly and effectively, we can plan to develop a common security policy rewriting framework. The framework should satisfy two properties: (1) rewriting framework is reliable and any combination of rewriting axioms can guarantee to generate a list of rules which are semantically equivalent. (2) rewriting framework is complete, any policy, which is semantically equivalent can be generated by the application axiom.

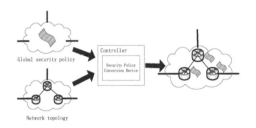

Figure 5.  Security policy conversion application.

The controller in the security policy rewriting framework can be installed a mechanical device to convert the security policy. The global security policy can be divided into multiple local security rules which not only reduces the burden of the programmers, but also makes the switch execute the security policy of upper application quickly. Although the security policy rewriting framework is still in the theoretical stage, the idea of security policy conversion would have a profound effect on solving the security problem of SDN.

## 4  CONCLUSION

SDN is an emerging network architecture that belongs to the research scope of network technology of the next generation, but it has a big difference with the research direction of other technologies of the next generation network: SDN can not only inherit the existing network technology, but also develop independently without depending on the existing network technology; SDN not only obeys the current new trend of applications, but also meets the idea that the control and forwarding

are separated. In addition, SDN aims to abstract and simplify the control plane of the existing complex network, make the control plane develop independently and innovatively and make the network application-oriented programmable. Though using SDN in various types of network is becoming more and more popular, the research on SDN security application is still in its infancy. In this paper, four aspects, including congestion control, traffic restrictions, load balancing and security policy conversion will be used to analyze the corresponding SDN security applications. It is believed that the paper will have some enlightening for the researchers engaged in the security of SDN. While there are still many unsolved problems in the field of SDN, such as the security problem of northbound interface, the backup problem of control information, the conflict problem of security rules of application, the interaction problem between eastbound interface and westbound interface, the security problem of controllers, the portability problem of SDN, the consistency problem of control logic and so on, it is hoped to find a way to solve these problems in the next study and have a more in-depth understanding of SDN.

## ACKNOWLEDGMENT

The research is supported by National Natural Science Foundation of China (Grant No. 61272422).

## REFERENCES

Casado M, Garfinkel T, Akella A, Freedman MJ, Boneh D, Mckeown N, Shenker S, 2006. SANE: A protection architecture for enterprisenetworks. In: Proc. of the 15th Conf. on USENIX Security Symp. Vancouver: USENIX Association. 137−151.

Greenberg A, Hjalmtysson G, Maltz DA, Myers A, Rexford J, Xie G, Yan H, Zhan J, Zhang H, 2005. A clean slate 4D approach to network control and management. ACM SIGCOMM Computer Communication Review,35(5):41−54.

Casado M, Freedman MJ, Pettit J, Luo J, Mckeown N, Shenker S, 2007. Ethane: Taking control of the enterprise. In: Proc. of the SIGCOMM 2007. Kyoto: ACM Press. 1−12.

Mckeown N, Anderson T, Balakrishnan H, Parulkar G, Peterson L, Rexford J, Shenker S, Turner J, 2008. OpenFlow: Enabling innovationin campus networks. ACM SIGCOMM Computer Communication Review,38(2):69−74.

McKeown N, 2009. Software-defined networking[J]. INFOCOM keynote talk.

Open Networking Foundation, Software-defined networking: the new norm for networks,white paper, Apr.2012.

Ghobadi M, Yeganeh S H, Ganjali Y, 2012. Rethinking end-to-end congestion control in software-defined networks[C]//Proceedings of the 11th ACM Workshop on Hot Topics in Networks. ACM: 61–66.

Suresh L, Schulz-Zander J, Merz R, et al, 2012. Demo: programming enterprise WLANs with odin[J]. ACM SIGCOMM Computer Communication Review, 42(4): 279–280.

Handigol N, Seetharaman S, Flajslik M, et al, 2010. Aster* x: Load-balancing as a network primitive[C]//9th GENI Engineering Conference (Plenary).

Kang N, Reich J, Rexford J, et al, 2012. Policy transformation in software defined networks[C]//Proceedings of the ACM SIGCOMM 2012 conference on Applications, technologies, architectures, and protocols for computer communication. ACM:309–310.

*Multimedia, Communication and Computing Application – Leung (Ed.)*
*© 2015 Taylor & Francis Group, London, ISBN 978-1-138-02775-6*

# The study and design of dynamic deploy in honeynet

K.H. Zhang
*Linyi University, Shandong, China*

ABSTRACT: Aiming at the disadvantage of static deployment and maintenance in honeynet, we propose an idea of dynamic deployment and relevant algorithms. The dynamic honeynet can monitor itself and real-time network environment, collects clews, and automatically determines what type of honeypots and topology should deploy and select. When security situation was changed, it can adjust the structure of honeynet timely. A prototype is designed based on the Honeyd which is a famous open source software.

KEYWORDS: Honeynet, Dynamic deploy, Grey forecast, Biology imitation.

## 1 INTRODUCTION

Honeynet system is the resources of the active defense in network security, to attract hackers, the worm scanning, attack and penetration invasion as the main design purpose. However, the honeynet system cannot improve the security of the network performance directly, the main function of the two. First of all, by attracting invasion to learn and understand the information of the attacker, it can analyze attacker's new attack strategies and invasion tools. For the worm virus, it can capture the new varieties and extract detection feature codes. Thus honeynet capture and analyze the attack behavior, which can find security vulnerabilities and potential risks in the network, which is of great help to network defense work. It is convenient for network security management in possible attacks before repairing security vulnerabilities, avoiding the happenings of attack. The second important function is to distract the attacker's attention and consume attack resources. Honeynet system can be completely used as an enhancement scheme in traditional network security system, through a dedicated Honeynet in a controlled environment, giving the attacker completely real system interaction, so that the attacker has run into a lot of Honeynet consuming the energy of the invaders and reducing their judgment, thus it protects indirectly the security of the network from the side.

In this paper, Section 2 introduces related concept and model of the network security situation prediction; in Section 3, honeynet dynamic deployment system structure are introduced, the author puts forward the strategy of the dynamic deployment, bionic deployment algorithm and realization technology; Section 4 presents two sets of experiment scheme, verified the feasibility and effectiveness of the model; Section 5 is a summary of this article.

## 2 SECURITY SITUATION GRAY PREDICTION ALGORITHM

Network security situation assessment refers to reflecting the system security elements through access to a computer network environment, integrated analysis of various factors, and on the basis of the analysis to predict the trend of future [8]. YongWei et al has provided the physical network to predict the change of the security situation using the data, in this reference to the prediction model. But considering the physical network and honeynet's data, it achieved a good effect.

The grey system theory is founded by professor Julong Deng in 1982. Grey system research theory and methods of some small samples and the uncertain information, it can look for rules from a limited, discrete data, and then set up grey model to forecast [9]. In the application of the most common GM (1, 1) single variable mathematical model of the first-order grey differential equation form, this model can weaken the randomness of the original random sequence. The degree of change in the host, IP activity, and honeynet activity predicts the current network security posture, while GM (1, 1) of the single variable for security situation prediction results. The method to construct the prediction model is to make a list of raw data discrete repeatedly accumulative to get regular accumulation generation sequence, and the accumulative sequence modeling. The modeling steps are as follows:

### 2.1 A cumulative processing the original data sequence

$$X_{(t)}^{(0)} = \{X_{(1)}^{(0)}, X_{(2)}^{(0)}, ..., X_{(n)}^{(0)}\}$$

$$X_{(t)}^{(1)} = \sum_{i=1}^{t} X_{(i)}^{(0)} (t = 1, 2, ..., n).$$

New sequence is $X_{(t)}^{(1)} = \{X_{(1)}^{(1)}, X_{(2)}^{(1)}, ..., X_{(n)}^{(1)}\}$

### 2.2 Single variable one-order grey differential equation

$$\frac{dx^{(1)}(t)}{dt} + \alpha x^{(1)}(t) = \mu$$

With $\alpha$ is coefficient for development, reflects the development situation of the prediction model, the greater the absolute value the faster speed of the development; $\mu$ is grey action, reflects the stimulus of external to the system. These two parameters is undetermined parameter of prediction model, it can use the least squares method to calculate.

$$(\alpha, \mu)^T = (B^T B)^{-1} B^T y$$

$$B = \begin{bmatrix} -0.5\left[x_{(1)}^{(1)}\right] + x_{(2)}^{(1)} ......1 \\ ...... \\ -0.5\left[x_{(n-1)}^{(1)}\right] + x_{(n)}^{(1)} ......1 \end{bmatrix}$$

$$y = \left[X_{(2)}^{(0)}, X_{(3)}^{(0)}, ..., X_{(n)}^{(0)}\right]$$

(c) Calculate to get the predicted accumulative sequence value: $\hat{X}_{(t+1)}^{(1)} = \left[X_{(1)}^{(1)} - \frac{\mu}{\alpha}\right]e^{-\alpha t} + \frac{\mu}{\alpha}$

Finally, the result of the restore is the predicted value.

$$\hat{X}_{(t+1)}^{(0)} = \hat{X}_{(t+1)}^{(1)} - \hat{X}_{(t)}^{(1)}$$

Prediction module based on timing collect value of three indicators of the current network system, forms the original forecast data column. $t$ is monitoring data

collected for different time points, predictive value are obtained by GM (1, 1) model treatment.

## 3 DYNAMIC HONEYNET DDH

### 3.1 Dynamic deployment strategy

Honeynet required to reconfigure honeynet node types, quantity and location when invasion occurs in the dynamic deployment process. Each honeynet node can be thought of as independent individuals, invasion attack occurred as a food source, therefore the entire physical and virtual honeynet is thought of as a closed-loop biological environment. So, the honeynet dynamic deployment problem is similar to the automatic migration process of animals according to the changes of the food source. Animals with change of the food source in the nature in the following four actions: birth, competition, migration, and death. Converting the judgment of the conditions is on the basis of the distribution and quantity of the food source. In this simulation, biological characteristics is used to formulate honeynet dynamic deployment strategy and algorithm.

### 3.2 Dynamic deployment algorithm

#### 3.2.1 Initialization algorithm

HFA express honeynet agency collection, HFAi express the $i$th honeynet agent;

Random probability P1 and random vector R (R1,R2,...,Rn);

For i = 1 to length (HFA)

HFAi with probability P1 selected node type TYPEi from MDB;

HFAi with (TYPEi, Ri) as parameters generated node in the area i;

End For

End of the algorithm.

#### 3.2.2 Competitive algorithm

Variable NewHFType expresses the most active node type; Variable NewIPlocation expresses the most active node position, the value has four scope; Hi expresses honeynet nodes, Hi.Type expresses the type of node, Hi.life_values the life value of node; Alpha, beta represents growth and failure step length of life value.

Assuming that the current moment is Ti, establishing a honeynet active queue QueActiveHoney{HS} according to the Master for virtual node vector <ActiveNodeVec>.

For i =1 to length(QueActiveHoney)
 IF NewHFType < Max(Type) Then
 NewHFType = Max(Type).
 Building queue QueActiveIP according to the physical network IP activity vector;
 For i =1 to length(QueActiveIP)
 IF NewIPlocation <Max(IPlocation) Then
 NewIPlocation = Max(IPlocation)
 5) For i =1 to length(HFA)
 IF (Hi∈ HFAi) and (Hi.Iplocation∈ NewIPlocation) and (Hi.Type = NewHFType)
 Hi.life_value += α
 Else Hi.life_value -= β
 Update QueActiveHoney and QueActiveIP
 End of the algorithm

### 3.2.3 *The migration algorithm*

$\delta1, \delta2$ is migration conditions respectively, $\varepsilon$ is the size of the migrated nodes life value;

1) Set the initial threshold $\delta1$, $\delta2$, $\varepsilon$;
2) For i =1 to length(HFA)
   a) While HFAj ∈ Neighbour(HFAi)
      IF (HFAj.Hi.life_value <$\delta1$) And (HFAi.Hi.life_value >$\delta2$)
      Migrate HFAi to HFAj;
      HFAi.Hi.life_value =$\varepsilon$;
   b) IF (HFAj.Hi.life_value = HFAi.Hi.life_value)
      Swap(HFAi ,HFAj);
      HFAi.Hi.life_value =$\varepsilon$;
      HFAj.Hi.life_value =$\varepsilon$;
      End For
3) End of the algorithm.

### 3.2.4 *Die algorithm*
1) Die set threshold is $\Phi$, death rate is $\Psi$,
2) For i =1 to length(HFA)
   For j =1 to length(HFAi)
   IF Hij .life_value < $\Phi$
   Destroy(Hij); count++;
3) IF count >$\Psi$
   Call the algorithm 3.3,3;
4) End of the algorithm

### 3.2.5 *Described of dynamic deployment algorithm*
1) Algorithm 3.3.1 create node of each agency jurisdiction in the honeynet
2) Calling algorithm 3.3.2 through the competitive learning within a certain time period;
3) When GPM informs MASTER start-up dynamic deployment orders, we call algorithm 3.3.3 and algorithm3.3.4;

4) Jump to step 2 to continue;
5) End of the algorithm.

## 4 DYNAMIC DEPLOYMENT EXPERIMENT

Honeynet configured for information is as follows: because there are four positions, so the position A, B, D set a HFA; Set three HFA in the position C. Set a CNA for the four location of physical network respectively. First of all, putting forward a set of static deployment scheme based on the experience of experts, static deployment scheme as shown in Table 1.

Table 1. Static deployment information table.

| Location and mold | A | B | C | D |
| --- | --- | --- | --- | --- |
| high | 0 | 3 | 12 | 6 |
| medium | 1 | 2 | 3 | 1 |
| low | 5 | 6 | 12 | 3 |
| sum | 6 | 11 | 27 | 10 |

Dynamic deployment solution: select honeynet node type random probability P1 is 0.35, set agent under the jurisdiction of the each honeynet the number of virtual node random vector R (10, 15, 20, 20, 20, 15); Life growth and virtual node failure step into α= 0.9, β= 0.7; Virtual node migration conditions $\delta1$= 2.2, $\delta2$ = 5.3, after the migration, life value reset epsilon $\varepsilon$ is 0.5; Die threshold value $\Phi$ is 0.4, death rate $\Psi$ is 4.

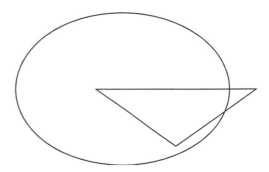

Figure 1. Dynamic deployment node map.

Abscissa represents the sample points in Figure 1, the vertically distributed in four different positions for virtual node number. Located in different position on the number of virtual nodes by different color histogram, it says most of them are located on the Intranet C position. What can be seen from the trend of five sampling number of virtual nodes early in the four position gradually increase, this is due to the increase of the number of nodes, the decrease of migration and death number. After middle and later, virtual node number is stable gradually, this shows that the new node of the creation, migration, and the old node of the demise reach a state of equilibrium.

## 5 CONCLUSION

Honeynet dynamic deployment scheme is proposed in this paper, which has good application value for the active network defense system. The deployment of the honeynet node runs a daemon in the physical computer memory and the dynamic deployment scheme does not need security experts. Hence, the hardware and software management of honeynet costs are low. The study is not only for the current honeynet researchers with good inspiration, but also for the real application prospect. In the next step, we will provide more strict and precise definition of decoy model and give more detailed division of honeynet node type to improve the decoy effect of the dynamic deployment.

## REFERENCES

[1] Ling chen, Hao Huang. 2005. "The design of the network trap type intrusion prevention model"[J]. Computer Application, 25(9)2074–2077.
[2] Hua Kuang, Xiang Li,. 2005. "The design of the network under Win32 honeypot with VC + + realize"[J], Computer Application, 25(12):150–151.
[3] PROVOS N. 2004. A virtual honeypot framework[A]. Proceedings of 13th USENIX Security Symposiu-m[C]. San Diego, CA, USA:1–14.
[4] Lianying Zhou, Deng Cao, NianYI Yuan. 2005."Analysis and research of virtual honeypot system fra-mework Honeyd"[J]. Computer Engineering and Applications, 27(1):137–140.
[5] Jianwei Zhuge, Xinhui Han. 2007 "HoneyBow: A honeypot technology based on high interactive aut-omatically capture of malicious code"[J]. Journal on Communications, 28(12):8–13.
[6] Jie Wang, Huaping Hu, Yong Tang. 2006. "The design and implementation of distributedvirtual trap network"[J]. Computational Engineering and Science, 28(2):33–35.
[7] Libo Ma, Haixin Duan, Xing Li.2005. "Honeypot deployment analysis"[J. Journal of dalian university of technology, 45(10)suppl:150–155.

*Multimedia, Communication and Computing Application – Leung (Ed.)*
© *2015 Taylor & Francis Group, London, ISBN 978-1-138-02775-6*

# Two layers trusted identities in cyberspace

B. Huang
*College of Telecommunications and Information Engineering, Nanjing University of Posts and Telecommunications, Nanjing, China*

ABSTRACT: Every communication entity in physical space has an image in cyberspace which is called silicon creature here. To enhance cyber security, privacy protection, and network confidence, it is necessary to construct trusted identity for the silicon creature in cyberspace. Based on the principle that the human consists of body and soul, a silicon creature is thought to be made up of body and soul. Then, the identity for the silicon creature is split into two parts: Body ID and Soul ID. On the basis of Body ID and Soul ID, the research methods on identifying the communication entity in cyberspace are proposed.

## 1 INTRODUCTION

"On the Internet, nobody knows you are a dog" is a popular saying used to describe the anonymity of the Internet. It is still used today when we talk about the issues about online identity. Human, dog or something else in physical space enters the cyberspace through network terminal equipment and become a number of the "Silicon Creatures" in cyberspace. The silicon creatures are engaged in various "activities" in cyberspace like human in physical space. It is well known that every creature in physical space has a unique identity, namely DNA. Therefore, we hope to find and construct the only identity for the silicon creatures in cyberspace like what we have done in physical space. Here, we give the silicon creature's identity a name – Network DNA or Net-DNA just as the creatures' biological DNA in physical space. A biological DNA can be used to uniquely identify a creature, and it is hoped that a Net-DNA can be used to uniquely identify a silicon creature too. It is very important for the network security, privacy protection, and network confidence. At present, American has explicitly put forward the "The National Strategy for Trusted Identities in Cyberspace" plan, which enhanced the identity problem in cyberspace to the national strategic level (Schmidt, 2011).

It is Dr. Liu Nanjie who firstly put forward the network DNA clearly in some document. In 2003, funded by HUAWEI company, Dr. Liu Nanjie presented the concept of "communication fingerprint" and patented it (Liu and Wang 2005). Here, communication fingerprint is a collection of the features and characteristics when the communication entity is left in the process of the network activities.

In 2005, Tadayoshi Kohno proposed a method to identify the remote physical device by measuring

equipment clock offset characteristics (Kohno & Broido 2005). And it is a fingerprint method in fact. In 2008, Sergey Bratus researched the terminal equipment behavior in wireless networks according to the protocol stack ability to process the exception messages by an active induced method (Bratus & Cornelius 2008). In 2011, Nam Tuan Nguyen used non parametric Bayesian estimation methods to provide network security through researching equipment fingerprint (Nam & Zheng, 2011.). There are some other papers (Sasikanth, & Amit, 2012; Xiao, & Yan, 2013) which have also been written on this object.

These researches were focused on the identification problem of communication equipment, but did not involve the person or thing which used the communication equipment.

This paper presents a new method to identify the silicon creature and form the silicon creature's Net-DNA. Based on the truth, a human in physical space has body and soul, and it is reasonable to think that the silicon creature in cyberspace is also made up of soul and body. Then, the Net-DNA for the silicon creature is split into two parts: Body ID and Soul ID. This makes it easy to get and use the silicon creature's Net-DNA.

## 2 BODY ID AND SOUL ID FOR A SILICON CREATURE

A person in physical space possesses body and soul and DNA can be extracted from the body. Therefore, we think that a silicon creature can also be made up of body and soul and is similar to net-DNA. Here, "body" means the communication terminal equipment by which people or things go into the cyberspace, and "soul" means network behaviors and habits. It is different from the physical space. The silicon creature's soul can be carried on different body (through different network terminal into the network). So the silicon creature's net-DNA will be made up of two parts as is shown in Figure 1.

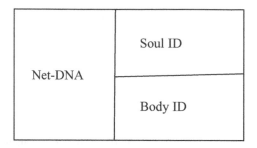

Figure 1. Two layers Net-DNA structure.

Table 1. Soul ID contents.

| Soul ID | Social attribute | Social circles |
| | | Interpersonal relationship |
| | | Routing attributes |
| | Personal attribute | Frequently visited sites and contents |
| | | Shopping habits |
| | | Favorite music and videos |
| | | Network games |
| | | Chat habits |
| | | Commonly used software |
| | | Password setting habits |

## 2.1 The body ID layer

Communication entity goes into the cyberspace by using different communication terminal equipment. And the communication equipment has its own hardware and software. We extract the core components from the hardware and software, make up a sequence number and make sure that different equipment has only one and different sequence number. Here, we give this sequence number a name -Body ID.

For example, a person goes into the vehicle network through a car, and the car is communication equipment now. As you know every car has its unique VIN number. We can extract the car VIN number and take it as the Body ID.

## 2.2 The soul ID layer

A person or something else goes into the cyberspace through communication terminal equipment, roams in the cyberspace and shows some unique "cyberspace properties". We extract some of the cyberspace properties which are not changed or almost not changed, and format a sequence number. Here, this sequence number is called Soul ID.

The cyberspace properties include: social circles, interpersonal relationships, routing attributes, shopping habits, hobbies, interest on music and video, network games, chat habits, commonly used passwords, commonly used software, frequently visited sites and contents, and so on.

The social circles, interpersonal relationship, and routing attribute for silicon creatures in cyberspace are classified as "social attribute"; Shopping habits, favorite music and videos, online games, chat habits, password setting habits, commonly used software, frequently visited sites and contents are summed up as "personal property", as shown in Table 1.

The extraction of Soul ID is a very complex process and it involves big data, data mining, statistic inference, and some other various aspects of theory and technique (Cabena & Pablo 1997; Liu 2007; Ian & Eibe 2011).

## 3 THE IMPLEMENT TO IDENTIFY A SILICON CREATURE

The silicon creature's Net-DNA is split into two parts: Body ID and Soul ID. If we get one silicon creature's Body ID and Soul ID, we can identify who this silicon creature is and know whether it can be trusted or not. But in practical application, it is difficult to extract the Body ID and Soul ID. And sometimes it is not necessary to require both Body ID and Soul ID to uniquely identify a silicon creature. For that reason, we should apply this concept in accordance with local condition.

### 3.1 Only using body ID

Sometimes, we only want to identify the communication terminal equipment, and think that the user is always using this equipment. Take the car identity in the vehicle network as an example. If we take the traffic as the object of the research, it is enough to get the car identity. That is to say it is enough to use Body ID to identify the car and driver in the cyberspace. And when it is difficult to extract the Soul ID, we have to use the Body ID to identify a silicon creature.

### 3.2 Only using soul ID

When it is not feasible to extract the Body ID, we only research the Soul ID. There are some deviations if we use only Soul ID to identify a communication entity in physical space. We should remember this point.

### 3.3 *Using both body ID and soul ID*

It is perfect to use both Body ID and Soul ID to identify a silicon creature. Soul ID can identify the user who is using the communication equipment and Body ID can identify the communication equipment. The user and communication equipment form a communication entity in physical space, and the Body ID and Soul ID form a silicon creature in cyberspace. They are mirror images of each other. It should be noted that Soul ID is the essence, and it can be carried on different Body ID.

## 4 CONCLUSIONS

In order to solve the problem of network security, privacy protection and network confidence, we must solve the problem of trusted identity in cyberspace at first. Every communication entity has an image in cyberspace which is called a silicon creature here. We think that a silicon creature is made up of "body" and "soul", and "body" means the communication terminal equipment by which people or things goes into the cyberspace, and "soul" means network behaviors and habits in cyberspace. Then, it is reasonable to think that the silicon creature can be identified by its soul and body. Give its soul and body an identity respectively and call them Body ID and Soul ID. Therefore, a silicon creature can be determined by its Soul ID and Body ID. But the detail extraction method for the Body ID and Soul ID is not given here and this is also author's further work.

ACKNOWLEDGMENT

This work was supported by the Scientific Research Foundation of Nanjing University of Posts and Telecommunications (Grant No. NY211036).

REFERENCES

Schmidt, H. 2011. National Strategy for Trusted Identities in Cyberspace. http://www.whitehouse.gov/b log/2010/06/25/national-strategy-trusted-identities-cyberspace.

Liu, N.J. et al. 2005. Communication fingerprint system and its acquisition, management method. *Chinese patent number:200510135987.9.*

Kohno, T. et al .2005. Remote physical device fingerprinting. *IEEE Transactions on Dependable and Secure Computing* 2(2):93–108.

Bratus, S. et al. 2008. Active behavioral Fingerprinting of Wireless Devices. *In proceedings of the 1st ACM Conference on Wireless Network Security*, March 31-April 02, 2008, Alexandria, VA, USA.

Nam T.N. & Zheng, G.B. 2011. Device Fingerprinting to Enhance Wireless Security using Nonparametric Bayesian. *IEEE International Conference on Communications*, 10-15 April 2011, Shanghai, China.

Sasikanth, A. & Amit, B. 2012. Privacy in mobile technology for personal healthcare. *ACM Computing Surveys* 3:1–54.

Xiao, L. & Yan, Q. 2013. Proximity-Based Security Techniques for Mobile Users in Wireless Networks. *IEEE Transactions on Information Forensics and Security* 8(1):2089–2100.

Cabena, P. & Pablo h.j. 1997. *Discovering Data Mining: From Concept to Implementation*. New Jersey: Prentice Hall.

Liu, B. 2007. *Web Data Mining: Exploring Hyperlinks, Contents and Usage Data*. New York: Springer.

Ian, H.W. & Eibe F. 2011. *Data Mining: Practical Machine Learning Tools and Techniques*. Amsterdam : Elsevier.

# Parallel signal processing software

X. Li
*Shenyang Institute of Automation, Chinese Academy of Sciences, Shenyang, Liaoning, China*
*University of Chinese Academy of Sciences, Beijing, China*

J.S. Du, J.T. Hu & X. Bi
*Shenyang Institute of Automation, Chinese Academy of Sciences, Shenyang, Liaoning, China*

ABSTRACT: At present, parallel signal processing tasks is applied more and more widely. Traditional way of development of parallel signal processing tasks is custom development. This development way needs manual programming, the development cycle is long, and the platform portability is poor. The parallel signal processing software discussed in this paper will make revolutionary change in the way of development of parallel signal processing tasks. The goal of parallel signal processing tasks is to achieve fully graphical and modularized development under an integrated development software environment. The software can automatically generate codes. Besides, the software provides the function of predicting and evaluating the performance. Furthermore, the software has generalization of hardware platform.

KEYWORDS: Parallel signal processing, Performance prediction, Graphical, Generalization

## 1 INTRODUCTION

Traditional way of development of parallel signal processing tasks is custom development. This development way is restricted by the experience of developers. The development cycle is long and the efficiency is low. Besides, the signal processing system has poor platform portability, with the high maintenance and upgrade cost.

The aim of the software in this paper is to solve these problems in the process of the development of parallel signal processing system. The name of this software is parallel signal processing laboratory, referred to as PspLab. PspLab software is an integrated development environment of parallel signal processing tasks to achieve the graphical, modularized, and cross-platform development. PspLab sets up an extensible signal processing functional module library and an extensible hardware module library. Users can drag and drop modules in module library to build signal processing flowchart, hardware topology chart, and the mapping relationship between the two. Based on these, PspLab can automatically generate codes and runs. PspLab can monitor the real-time running state of the code. Besides, PspLab can predict and evaluate the overall performance of the signal processing system, so as to guide the developer to design the system. PspLab can be applied to any kind of hardware platform. Therefore, PspLab really achieves generalization (Axel & Mario, 1998), expansibility, and high efficiency of development.

## 2 DESIGN AND REALIZATION OF THE SOFTWARE

The developing procedure used in PspLab is shown as below. Users drag and drop functional modules in functional module library to build signal processing flowchart, drag, and drop hardware modules in hardware module library to build hardware topology chart. Then users assign each functional module in the flowchart to the hardware module in the hardware topology chart. This assignment is called mapping. PspLab can predict and evaluate the overall performance of the signal processing system after mapping. PspLab can also automatically search the optimal mapping solution. Based on the signal processing flowchart, the hardware topology chart, and the mapping solution, PspLab can automatically generate the high-level language code, then load and run. While the code is running, PspLab can read the values of variables in the code in real time and draw the statistical figure.

PspLab includes the following main functions: management of module library, modeling of signal processing flowchart, modeling of hardware topology chart, mapping, automatic generation of codes, and prediction and evaluation of performance.

### 2.1 Management of module library

In the PspLab, there are two kinds of module libraries: functional module library and hardware module library. Functional module library is used to build the signal processing flowchart, and the hardware module library is used to build the hardware topology chart.

PspLab achieves an extensible functional module library. The functional module library provides usual functional modules. Developers of signal processing system can also add self-defining modules to the library. This is important to ensure the expansibility of PspLab. When adding a new functional module, users only need to fill in the properties of the functional module according to the guidance of creating new modules. Every functional module includes the following properties: module name, module category, information of input ports, information of output ports, and information of parameters. PspLab stores information of the functional module library in the form of XML (Eddy, 1999) to facilitate extension. Besides, PspLab provides functions about the functional module library as follows: editing modules, removing modules, displaying modules according to the categories and fuzzy query.

The hardware module library is also extensible, and is also stored in the form of XML. It is similar to the functional module library. So it is not going to introduce in detail.

## 2.2 Modeling of signal processing flowchart

Building signal processing flowchart is the core operation of PspLab. Users can build signal processing flowchart by graphical and modularized method to complete the design of signal processing system. A simple example of signal processing flowchart is shown in Figure 1.

### 2.2.1 Operation of building signal processing flowchart

First, drag and drop modules. Users drag and drop the functional modules to the signal processing flowchart from the functional module library.

Second, connect lines. Connect lines between input ports (red dots) and output ports (blue dots) of

modules, and then the relationship between two modules is set up.

Third, edit properties. Functional modules have some properties, so do the lines between modules. These properties are used to describe information in signal processing flowchart. PspLab provides the function to edit the properties. Type of property value in PspLab is very flexible. The type can be int, float, double, and array.

### 2.2.2 Hierarchical package and decomposition of functional modules

PspLab achieves the function of hierarchical package and decomposition. Hierarchical package refers to combine a number of separate modules in a signal processing flowchart into a single functional module, and the function does not change. Hierarchical package also supports nested package of multiple layers. Decomposition refers to decompose a packaged module into separate modules. When users develop a very large signal processing system, the number of signal processing function modules in flowchart can be a lot. At this time, packaging the adjacent modules into a single module can simplify the flowchart. The packaged module can be dragged and dropped to other signal processing flowcharts.

### 2.2.3 RML language

After building one signal processing flowchart, how to save the information of the flowchart? How to get the information from a flowchart? On the basis of XML language, PspLab defines a language that describes specially the signal processing flowchart. We call it RML (Radar Markup Language) language. The reason of using XML language is that it is open and extensible. In addition, XML language is independent from hardware platform. A simple example of RML hierarchical structure is shown in Figure 2.

Figure 2.   RML hierarchical structure.

Meanings of main nodes in RML are shown in Table 1.

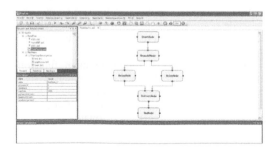

Figure 1.   A simple example of signal processing flowchart.

Table 1. Meanings of main nodes in RML.

| Node | Meaning |
|---|---|
| rml | the root node, showing the file type |
| system | the signal processing system |
| application | the signal processing flowchart |
| actor | the function module, including some attributes as follows. "name" attribute: the module name; "class" attribute: the class name of the module; "inport" attribute: the amount of the input ports of the module; "outport" attribute: the amount of the output ports of the module; "x" attribute: the horizontal ordinate in the flowchart; "y" attribute: the vertical ordinate in the flowchart |
| editable | the editable attributes of the module |
| params | the parameter list |
| param | the parameter of functional module, including some attributes, such as the parameter name, the parameter value, and the parameter type |
| port | the input or output port. Each "port" node includes some attributes, such as the port name, the data transfer direction, and the port type |
| connect | the link between two function modules |
| src_port | the source port |
| dest_port | the destination port |
| src_actor | the source function module |
| dest_actor | the destination function module |
| Subsystem | the hierarchical package module. Each "Subsystem" node includes some "actor" nodes, "connect" nodes, and "Subsystem" nodes |

## 2.3 Modeling of hardware topology structure

The modeling of hardware topology structure achieves the generality on hardware platforms of the PspLab. Users can drag and drop the modules in the hardware module library to build a hardware topology chart in a fully graphical and modular way. Similar to the functional module library, the hardware module library is also extensible. Therefore, PspLab is applicable to any kind of hardware platform.

Similar to modeling of signal processing flowchart, in order to save and analyze the hardware topology chart, on the basis of XML language, PspLab defines a language that describes specially the hardware topology chart. We call it HTL (Hardware Topology Language) language. HTL describes the information of the hardware topology chart, including processor type, processor amount, link structure, and communication link between processors.

The basic idea and the grammar of HTL are similar to RML. And the operation of building the hardware

topology structure chart is also similar to building the signal processing flowchart. So they are not introduced in detail in this paper.

## 2.4 Mapping function

Mapping function is allocating the function modules in the signal processing flowchart to the hardware modules in the hardware topology chart. That means mapping decides which signal processing programs run on which hardware processors. After the mapping solution is determined, PspLab can predict and evaluate the performance of the signal processing system.

PspLab provides two mapping modes: the graphics mode and the forms mode. Mapping solution is also saved in the form of XML.

## 2.5 Performance prediction and evaluation

After building the signal processing flowchart, the hardware topology chart, and the mapping solution, PspLab can predict and evaluate the overall performance of the signal processing system. Performance prediction and evaluation is crucial to the development of signal processing system, but is the difficulty in the software design.

For a long time, at the beginning of the system design, it is very difficult for developers to propose accurate demand of the software and hardware according to system index. The aim of performance prediction and evaluation is to guide developers to grasp demand of software and hardware on the macro level at the beginning of system design. It can greatly improve the speed and success rate of system development.

### 2.5.1 Performance prediction

PspLab combines the analytical method with the simulation method (Vikram & Mary, 2004) to achieve performance prediction. The nature of performance prediction is to simulate actual running state of the signal processing system. Actual running state is reflected through the state change of each processor along the time axis. In order to simulate the state change, PspLab carries out the time sequences simulation. PspLab builds three classes: Actor, Core, and Link. Actor expresses the functional module in the signal processing flow, Core expresses the virtual calculation unit, and Link expresses the link structure of the hardware platform. Actor contains the port information and running time information of the functional module. Actor and Core simulate calculation behavior of the system. Core is built to solve the calculation problem of Actor. When Actor needs to calculate, it calls Core to achieve. Core returns the calculate result to Actor. Then Core clears itself and waits to be called next time. Link contains the information of links between functional modules. Link simulates the communication behavior of the system. It is called when a functional module communicates with another functional module.

PspLab gets the information from the signal processing flowchart, the hardware topology chart, and the mapping solution, and then fills the information into objects of class Actor, Core, and Link. Actor has three states: idle, ready, and running. Core has two states: ready and running. Link has two states: ready and running. Each class has different state shift mechanisms. The object of each class can make a request and reply to the outside request. State shift may happen during this process. State shift cases of class Actor, Core, and Link are shown in Figure 3, Figure 4, and Figure 5, respectively. The label near each arrow expresses request message before state shift or response message after state shift.

After initializing objects of class Actor, Core, and Link, PspLab activates the first functional module and launches the time sequences simulation. The simulation process goes forward along the time axis. The time axis is advance sign set to ensure the simulation to go ahead successfully. Its response message

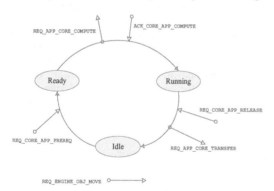

Figure 3. State shift of functional module (Actor).

Figure 4. State shift of calculation unit (Core).

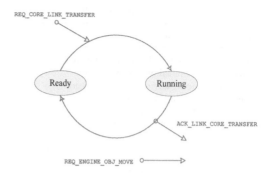

Figure 5. State shift of link structure (Link).

is REQ_ENGINE_OBJ_MOV, which is shown in Figure 3. Moving step of the time axis is the time from now to next event occurrence of any object. When arrives at this time point, some objects which meet the conditions are activated, and other objects which do not meet the conditions keep the original state. The activated objects make the state sequence changed; by this process, the simulation can achieve successfully. The time point mentioned above should be recorded. By these time points, PspLab can get a wealth of information, such as running time and communication time of each functional module. Furthermore, according to these time points, the calculation of PspLab can arrive at performance evaluation indexes.

### 2.5.2 Performance evaluation

Performance evaluation adopts the following evaluation indexes.

Through putting rate of the system: the maximum rate at which the system can receive and deal with the input data correctly (the minimum time interval of data input: $1/T_{IN,min}$).

Processing delay: the time interval between the time when the input data is received by the first level functional module and the time when the same data is sent by the last level functional module.

Utilization efficiency of processors: it is expressed by Equation 1 shown below.

$$\eta = \frac{\sum_{j=1}^{N} T_{j,exe}}{NT} \tag{1}$$

where $\eta$ = utilization efficiency of processors; $N$ = amount of processors; $T$ = total running time of the system; and $T_{j,exe}$ = running time of processor j.

Load balance degree of processors: utilization efficiency of a single processor is expressed by Equation 2 below.

$$\eta_j = \frac{T_{j,exe}}{T} \qquad (2)$$

Load balance index of processors is expressed by Equation 3 below.

$$K = 1 - \sqrt{\frac{\sum_{j=1}^{N}(\eta_j - \eta)^2}{N}} \qquad (3)$$

where K = load balance index of processors.

Overhead ratio between processing and communication: $\lambda = T_{exe}/T_{com}$. $T_{exe}$ is processing time of processors. $T_{com}$ is communication time of processors.

System Scale: the total hardware resources needed to accomplish the whole signal processing task, including the amount of processors, memorizers, and processing boards.

Cost: price of current system scale.

### 2.6 *Automatic code generation*

The automatic code generation replaces the traditional manual programming way. PspLab analyzes the signal processing flowchart, the hardware topology chart, and the mapping solution, and get the useful information from these three files. Then according to the code template, PspLab generates the C++ code, which can be compiled and run on the relevant hardware platform.

The software will be used on different hardware platforms, and the code generated on each platform must be different. Therefore, we must design a reasonable mechanism of generating code, so as to make the function of automatic code generation expand easily to different hardware platforms.

PspLab uses the abstract factory pattern in the design of the automatic code generation. The abstract factory pattern provides an interface which creates a series of related or dependent objects without specific class (Liang et al. 2006). The abstract factory pattern promotes the object-oriented programming to the interface-oriented programming (Wang & Wang, 2005), which fits "OCP (Open Closed Principal)" (Alan & James, 2004).

We design an abstract class used to generate code. Each subclass of this abstract class is relevant to the code template on a kind of hardware platform. Creation of special object is postponed to the subclass. The upper layer program of generating code only deals with the abstract class. So when the hardware platform is changed, we just need to add a subclass which represents the code template on the new hardware platform. We do not need to modify the upper layer program and the analysis program.

## 3 CONCLUSIONS

PspLab software in this paper has been used on several different kinds of hardware platforms, such as multi-DSP platform and Power PC platform. Test results shown that as an integrated development environment of parallel signal processing system, PspLab achieves graphical, modularized, and cross-platform development way successfully. PspLab replaces the traditional manual programming way, improves the development efficiency significantly. Furthermore, performance prediction and evaluation of PspLab can help developers to grasp demand of software and hardware on the macro level at the beginning of system design, which greatly improves the success rate of system development.

## REFERENCES

Alan, S. & James, R.T. 2004. *Design Patterns Explained.* Beijing: Tsinghua University Press.
Axel, H. & Mario, D.C. 1998. Performance and dependability evaluation of scalable massively parallel computer systems with conjoint simulation. *ACM Transactions on Modeling and Computer Simulation* 8(4):333–373.
Eddy, S.E. 1999. *XML Command Explained.* Beijing: Publishing House of Electronics Industry.
Liang, W., Zheng, F., Du, Y. & Dang, L. 2006. The application of abstract factory pattern in .net multi-layer distributed programs. *Computer Era* 3:27–29.
Vikram, S.A. & Mary, K.V. 2004. Parallel program performance prediction using deterministic task graph analysis. *ACM Transactions on Computer Systems* 22(1):94–136.
Wang, Y. & Wang, Y. 2005. *Object-oriented Guide to Practice.* Beijing: Publishing House of Electronics Industry.

*Multimedia, Communication and Computing Application – Leung (Ed.)*
© 2015 Taylor & Francis Group, London, ISBN 978-1-138-02775-6

# BIC-based speaker segmentation using multi-feature combination

J.L. Chen & C.S. Cai
*Communication University of China, Beijing, China*

ABSTRACT: Since MFCC and F0 both contain speaker-turn information, it is expected that combining them in some form can be helpful in detecting speaker-change points. In this paper, we utilize the BIC-based speaker segment method, and combine MFCC with F0 in form of $\Delta BIC_M + \Delta BIC_F$ and $\Delta BIC_M \times \Delta BIC_F$ investigate their effect on the performance of system. The experiment results show that two types of combinations can obtain better performance of the BIC-based speaker segmentation system than those without it.

## 1 INTRODUCTION

The task of speaker segmentation is to mark where speaker changes occur in a speech signal automatically. The speech signals need to be parameterized prior to segmentation. Speech parameterization is the process of extracting a set of features from the speech waveform, and the features must represent the speech signal reasonably and have adequate distinguishing capacity between sounds.

There are various speech features in speaker segmentation, such as Linear Prediction Coding (LPC), Mel-Frequency Cepstral Coefficients (MFCC), Linear Predictive Cepstral Coefficients (LPCC), Perceptual Linear Prediction (PLP), and Neural Predictive Coding (NPC). Nevertheless, MFCC is the most popular feature because it is based on the known theory of the human ear's critical bandwidth frequency.

MFCC feature contains both speaker information and content information. For MFCC-based speaker segmentation, speaker information is the positive information that used to distinguish speeches from different people, but the speech content information is negative information, which degrades the performance of the speaker segmentation system. Pitch period is one of the most important features in speaker identification and verification system, which carries unique speaker information and little content information. So we have reason to expect that the combination of pitch period and MFCC can improve the performance of speaker segmentation. This paper gives brief introduction to the effect of combining MFCC with pitch period on the performance of speaker segmentation. Besides, it also explains how BIC-based method is used to detect speaker turns.

The structure of this paper is as follows. In section 2, we give a brief introduction to MFCC and pitch period extraction. The third part is about the BIC-based segmentation and feature combination.

In section 4, it describes experiments and the results. Finally, a conclusion is made in section 5.

## 2 FEATURE EXTRACTION

### 2.1 MFCC extraction

MFCC is one of the most popular parameterized representatives of the speech signals in various speech processing systems, because it is based on the human ear's critical bandwidth frequency. Psychophysical studies have shown that human perception of the sound frequency for speech signals does follows 'Mel' scales instead of a linear scale. On 'Mel' scale, for each tone with an actual frequency, f, measured in Hz, a subjective pitch is obtained by following equation:

$$f_{mel} = 2595 \log_{10}(1 + \frac{f}{700}) \qquad (1)$$

MFCCs are a set of DCT (Discrete Cosine Transform) separate parameters, which are computed through a transformation of the logarithm- mically compressed filter-output energies, driving through a perceptually spaced triangular filter bank that processes the Discrete Fourier Transformed (DFT) speech signal.

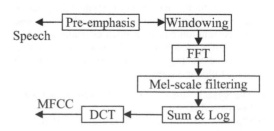

Figure 1. Block diagram for extracting MFCC.

The procedure for extracting MFCC parameters is illustrated in Figure 1.

1) Pre-emphasis is performed

$$s_p(n) = s(n) - \alpha \cdot s(n-1) \qquad (2)$$

where s(n) denotes the speech signal; sp(n) denotes the pre-emphasized signal; $\alpha$ is a constant with value between 0.9 and 1.

2) The pre-emphasized signal is windowed by a Hamming window

$$s_w(n) = s_p(n) \cdot w(n) \qquad (3)$$

where w(n) denotes a Hamming window.

3) Fast Fourier transform (FFT) is performed to obtain the spectrum

$$S_w(k) = \sum_{n=0}^{N-1} s_w(n) \cdot \exp(-\frac{2\pi kn}{N}) \qquad (4)$$

4) Performing Mel-filtering and log operations to get sub-bands' log-energies

$$e_j = \log\left( \sum_{k=Kl_j}^{Kh_j} M(j,k) \cdot |S_w(k)|^2 \right) \qquad (5)$$

where $Kh_j$ and $Kl_j$ are the frequency indexes corresponding to the high and low cutoff frequency of the j-th triangle filter respectively, and M(j, k) is the amplitude of the k-th discrete frequency index of the j-th triangle filter.

5) Performing DCT to obtain MFCC

$$C_i = \sum_{j=1}^{J} D(i,j) \cdot e_j \qquad (6)$$

where D(i, j) is the Z and j-th column elements of the DCT transform matrix.

## 2.2 Pitch detection

The task of pitch detection is to estimate the pitch or fundamental frequency (F0) of a speech. we choose an autocorrelation based on pitch estimator to detect the pitch of a speech. Figure 2 shows the block diagram for extracting pitch period (or F0), which includes following steps:

1) The speech signal is being filtered by a 900Hz low-pass filter.
2) Segments of 30msec length are selected at 10msec intervals.

3) Check whether a segment is silent . If the segment is classified assilent segment, then no further action is taken.
4) Clip the speech signal and COMPUTING the autocorrelation function.
5) Locate the largest peak of the autocorrelation function and compare the peak value with a fixed threshold. If the peak value falls below the threshold, the segment is classified as unvoiced, and if it is above the threshold, the pitch period is defined as the location of the highest peak.

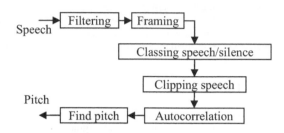

Figure 2.  Block diagram for extracting pitch period.

## 3  SPEAKER SEGMENTATION

### 3.1  BIC-based segmentation

BIC is a maximum Bayesian model selection criterion penalized by the model comp- lexity. Given a data set $\mathbf{Z} = \{z_1, z_2, \ldots, z_N\}$, and a candidate-model set $\mathbf{MC} = \{M_1, M_2, \ldots, M_K\}$, the purpose of model selection is to choose the model that best fits the distribution of $M_i$ from $\mathbf{MC}$. When using BIC for model selection, the BIC value of $M_i$ for $\mathbf{Z}$ is

$$BIC(M_i, \mathbf{Z}) = \log p(\mathbf{Z} | \hat{\theta}_i) - \frac{1}{2}\lambda P_i \log N \qquad (7)$$

Where $\lambda$ is a penalty factor, $\hat{\theta}_i$ is the maximum-estimate of the parameter set of $M_i$, and $P_i$ is the number of parameter of $M_i$. The model with the largest BIC value will be under selection.

For speaker segmentation, two different models are adopted, and $\Delta$BIC serves as the distance measurement between two speech segments. Given two audio segments represented by feature vectors, $\mathbf{X} = \{x_1, x_2, \ldots, x_{Nx}\}$ and $\mathbf{Y} = \{y_1, y_2, \ldots, y_{Ny}\}$, and $\mathbf{Z} = \mathbf{X} \cup \mathbf{Y} = \{z_1, z_2, \ldots, z_{Nz}\}$, where $Nz = Nx + Ny$, the following two hypotheses are evaluated

$$H_0: \quad z_1, z_2, \ldots, z_{N_z}, \sim N(\mu_Z, \Sigma_Z),$$

$$H_1: \quad x_1, x_2, \ldots, x_{N_x}, \sim N(\mu_X, \Sigma_X)$$

$$\text{and } y_1, y_2, \ldots, y_{N_r}, \sim N(\mu_Y, \Sigma_Y),$$

Under $H_0$, X and Y are derived from the same multivariate Gaussian distribution, that is, the two speech segments come from the same speaker. Under $H_1$, X and Y are modeled by distinct multivariate Gaussian distribution, that is, the two speech segments come from two different speakers.

$$\Delta BIC_{(X,Y)} = BIC(H_1, \mathbf{Z}) - BIC(H_0, \mathbf{Z})$$

$$= \log p(\mathbf{X} \mid \hat{\mu}_X, \hat{\Sigma}_X) + \log p(\mathbf{Y} \mid \hat{\mu}_Y, \hat{\Sigma}_Y)$$

$$- \log p(\mathbf{Z} \mid \hat{\mu}_Z, \hat{\Sigma}_Z) - \frac{1}{2}\lambda\left(d + \frac{d(d+1)}{2}\right) \log N_Z \quad (8)$$

$$= \frac{N_Z}{2} \log \left|\hat{\Sigma}_Z\right| - \frac{N_X}{2} \log \left|\hat{\Sigma}_X\right| - \frac{N_Y}{2} \log \left|\hat{\Sigma}_Y\right|$$

$$- \frac{1}{2}\lambda\left(d + \frac{d(d+1)}{2}\right) \log N_Z$$

where $\hat{\mu}_Z$, $\hat{\mu}_X$, and $\hat{\mu}_Y$ are the sample mean vectors of Z, X, and Y respectively. $\hat{\Sigma}_z, \hat{\Sigma}_x$, and $\hat{\Sigma}_Y$ are sample covariance matrices of **Z**, **X**, and **Y** respectively; and $d$ is the dimension of the feature vector. The larger the value of $\Delta BIC$ is, the less similar the two segments will be. When $\Delta BIC$ is greater than a specified threshold, usually 0, the hypotheses $H_1$ holds, that is, the two speech segments come from two different speakers; While $\Delta BIC$ is less than a specified threshold, usually 0, the hypotheses $H_0$ holds, that is, the two speech segments come from the same speaker.

## 3.2 Feature combination

Because both MFCC and pitch period contain speaker information, it's expected that the comb- ination of these two features is good for speaker segmenting. How to combine different features is not a trivial problem since they have diverse origins. Some researchers combined the MFCC and pitch period features at the likelihood stage. The MFCC coefficients are treated as the $S_m$ stream and the pitch-period feature is treated as the $S_p$ stream. Each source of information is formed in a statistical model, and the combined likelihoods are combined using

$$\log p(s_m, s_p \mid \theta) = w_m \log p(s_m \mid \theta_m) + w_p \log p(s_p \mid \theta_p) \quad (9)$$

Where $w_m$ an $w_p$ are weights, and $w_m + w_p = 1$.

In this paper, we first determine $\Delta BIC_M$ using $S_m$ stream which comprised by MFCC coefficients, and determine $\Delta BIC_F$ using $S_p$ stream which comprised by pitch periods. Then $\Delta BIC_M, \Delta BIC_F,$

$\Delta BIC_M + \Delta BIC_F$ and $\Delta BIC_M \times \Delta BIC_F$ are used to locate the speaker change points respectively.

## 4 EXPERIMENTS AND RESULTS

### 4.1 Experiment data and setups

In general, there are three primary decisive factors which have been used for speaker segmenting and clustering research and development: broadcast news audio, recorded meetings, and telephone conversations. Each type of data from them has different characteristics in the quality of the recordings, the amount and type of non-speech sources, the number of speakers, the durations and sequence of speaker turns. In our experiments, broadcast news are chosen as experimental data.

General broadcast news usually contains backgr- ound music, commercials and speeches of news-. To concentrate our research on speaker segmentation, we banishedthe background music, commercials, and environmental noise from the broadcast-news data. About 30 minutes of broadcast news contain- ing 200 speaker turns were evaluated. The duration of each speaker turn varies from 2s to15s. We used 12 dimensional MFCCs plus pitch periods as fea- ture vectors, and all vectors were computed within frames of 32ms. The length of an analysis window is usually 5s.

### 4.2 Performance measures

In speaker segmentation, there are two types of errors (i.e. False alarms and Missed detections). False alarms occur when a speaker turn is detected although it does not exist, while Missed detections occur when the process does not detect the existing speaker turn. Both of the false alarm rate (FAR) and the missed detection rate (MDR) are used to evalu- ate the performance of speaker turn detection system. FAR and MDR are defined as

$$FAR = \frac{FA}{GT + FA}$$

$$MDR = \frac{MD}{GT} \quad (10)$$

where FA denotes the number of false alarms, MD denotes the number of miss detections, and GT denotes the number of the ground-truth speaker turns. The other two widely used performance methods are the precision (PRC) and recall (RCL), and they are defined as

$$PRC = \frac{CFC}{CFC + FA}$$

$$RCL = \frac{CFC}{CFC + MD} \qquad (11)$$

where CFC denotes the number of correctly in detecting change points. In this paper, we use FAR and MDR to evaluate the system performance.

### 4.3 *Experiment results*

The results of the BIC-based speaker segmentation usually change according to the penalty factor $\lambda$ in the calculation of BIC. Typically, both of FAR and MDR change in opposite directions with varying $\lambda$, so we use FAR-MDR curve to investigate the improvement of the performance obtained by combining MFCCs with pitch periods. FAR-MDR curves is shown in Figure 3.

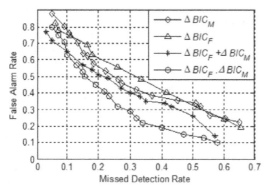

Figure 3. MDR-FAR curves obtained by using MFCCs, Pitch Periods and their combination.

From the chart above, we can see that the BIC-based speaker segmentation method with only MFCCs as features has better performance comparing with the segmentation method with only pitch periods or fundamental frequencies (F0), and both combining MFCC with F0 features by $\Delta BIC_M + \Delta BIC_F$ and $\Delta BIC_M \times \Delta BIC_F$ can get better performance of the BIC-based speaker segmentation method comparing with those regarding only MFCCs or F0s as features. The reason is that both $\Delta BIC_M$ and $\Delta BIC_F$ contain some speaker-turn information, and their waves have some similarities, and their location distribution

of local-maximum peaks typically corresponding to speaker-change points also have some common Therefore, as we expected, combining $\Delta BIC_M$ with $\Delta BIC_F$ can improve the performance of the speaker segmentation system.

## 5 CONCLUSION

In this paper, we investigate the effect on the improvement of the BIC-based speaker segmentation system performance with MFCCs and F0s. Since both MFCC and F0 contain speaker-turn information, it is expected that combining them in some forms can be helpful in detecting speaker-change points. In our experiments, we use $\Delta BIC_M + \Delta BIC_F$ and $\Delta BIC_M \times \Delta BIC_F$ to combine MFCC with F0, and evaluate their effect on the performance of the system. The experiment result shows that the combination of MFCCs with F0s by $\Delta BIC_M + \Delta BIC_F$ and $\Delta BIC_M \times \Delta BIC_F$ can obtain better performance of BIC-based speaker segmentation system comparing with those regarding only MFCCs or F0s as features.

## REFERENCES

Lawrence R. Rabiner & Ronald W. Schafer. 2011. Theory and Applications of Digital Speech Processing.

MargaritaKotti, Emmanouil Benetos, and Constantine Kotro-poulos. 2008. Computationally Efficient and Robust BIC-Based Speaker Segmentation. IEEE TRANSACTIONS ON AUDIO, SPEECH, AND LANGUAGE PROCESSING. VOL .16, NO.5.

M. Pardo, Roberto Barra-Chicote. 2012. Speaker Diarization Features: The UPM Contribution to the RT09 Evaluation. IEEE TRANSACTIONS ON AUDIO, SPEECH, AND LANGUAGE PROCESSING. VOL.20, NO.2.

Sue E. Tranter, 2006. An Overview of Automatic Speaker Dia- rization Systems. IEEE TRANSACTIONS ON AUDIO, SPEECH, AND LANGUAGE PROCESSING. VOL.14, NO.5.

Shih-Sian Chen, Hsin-Min Wang, and Hsin-Chia Fu. 2010. BIC-Based Speaker Segmentation Using Divide-and- Conquer Strateges With Application to Speaker Diariza- tion. IEEE TRANSACTIONS ON AUDIO,SPEECH, AND LANGUAGE PROCESSING. VOL.18, NO.1.

Wei-Qiang Zhang, Dengzhou Yang, Jia Liu, Xiuguo Bao, 2010.

Perturbation Analysis of Mel-Frequency Cepstrum Coefficients. ICALIP2010:715–718.

*Session 2. Multimedia processing and artificial intelligence*

*Multimedia, Communication and Computing Application – Leung (Ed.)*
© 2015 Taylor & Francis Group, London, ISBN 978-1-138-02775-6

# A novel trust management based on cloud model for wireless sensor networks

C. Yan

*International School, Beijing University of Posts and Telecommunications, Beijing, China*

ABSTRACT:   Trust management is a complementary security method, which can prevent wireless sensor networks from internal attacks. Many papers focus on the trust management without uncertainties of trust being fully considered. Therefore, this paper puts forward a novel trust management based on cloud model, which has advantage in dealing with uncertainties and is used to calculate the direct trust value. During the update process, a decay time factor is adopted to weigh the current trust information and the history information. Additionally, in order to avoid dishonest recommendations, we use dynamic aggregation which adopts weights of recommended trust values to calculate the ultimate indirect trust value. The results of simulation show that our algorithm can deal well with uncertainties of trust and effectively defend against bad-mouthing attack.

## 1   INTRODUCTION

Wireless sensor networks (WSNs) are composed of many sensor nodes that have capability to sense, process information, and collaborate with each other (Lopez, J. et al. 1989). As emerging technologies, WSNs have a variety of applications in the fields of battle field surveillance, emergency response, accident detection, and so on (Huang, Y. M. 2009, Lai, C. F. 2011). However, WSNs are often randomly deployed in unattended or even hostile territories, thus they are sensitive to the physical capture. What is more, the open communication channel makes it an ideal medium for adversaries to launch internal attacks (Han, G., Jiang. et al. 2014). Therefore, it is necessary to solve the security problems of WSNs.

The outside attacks can be defended by cryptographic measures. However, when a node is compromised, the corresponding cryptographic keys are also compromised, and the adversary can become legitimate network member (Kumar, G., Titus, I., & Thekkekara, S. I. 2012). Therefore, a complementary security means is indispensable.

As an additional means of security in WSNs, trust management has gained global recognition (Khalid, O. et al. 2013). The core algorithm of trust management is trust evaluation. By evaluating a node's trust value based on its previous behavior, it is possible to calculate how much this node can be trusted to perform a particular task (Lopez, J. et al. 1989). There are a variety of literatures on trust evaluation. Zhan et al. presented a resilient trust model which integrated past history and recent risk to accurately identify the current trust level (Zhan, G. et al. 2011). Based on fuzzy logic systems, Zarci et al. proposed a novel congestion control scheme, which enabled the nodes to investigate the behavior of their neighbors (Zarei M. et al. 2009). However, many existing literatures ignored the fact that trust is a concept with many uncertainties and failed to accurately grasp the uncertainty of trust.

To address these problems, we propose a novel trust management based on cloud model for wireless sensor networks. The main contributions of this paper are as follows. The direct trust value is calculated by cloud model, which has advantage in dealing with uncertainties. The update of direct trust value adopts a decay time factor to weigh the current trust information and the history information. Indirect trust value from common neighbor nodes shared by both subject and object node is obtained through conditional transitivity. To prevent false recommendations, the ultimate indirect trust value is obtained by dynamic aggregation which adopts weights of recommended trust values.

## 2   THE STANDARD CLOUD

The standard cloud is a special cloud model which consists of many clouds. Each cloud represents a corresponding trust level and has a certain concept. For ease of comprehensive evaluation, the concept is mapped to [0, 1] interval and the evaluation level is reflected by dividing the interval. Supposing the number of trust levels is $n$ with $n-1$ subintervals, and each subinterval can be expressed by [ $f_{min}^i, f_{max}^i$ ], where $f_{min}^i$ and $f_{max}^i$ represent the upper limit and the lower limit of this subinterval, respectively. The standard cloud can be described by Algorithm 1.

**ALGORITHM I:**   THE STANDARD CLOUD

**Input:** $n$-$1$ subintervals

**Step 1:** $Ex_i = \left( f_{\min}^i + f_{\max}^i \right) / 2$;

**Step 2:** If $i$=1, $En_1 = \left( Ex_2 - Ex_1 \right) / 3$;

    else $En_i = \left( Ex_i - Ex_{i-1} \right) / 3$;

**Step 3:** $He_i = \delta$

**Output:** $TC_i \left( Ex_i, En_i, He_i \right)$, $i$=1,2,3,…,n.

where $\delta$ is a constant, which is obtained by computing. So as to make the evaluation easier, we set $\delta = 0.05$.

# 3   TRUST EVALUATION MODEL BASED ON CLOUD

Trust is the confidence that node i has on node j about how node j will perform as expected (P. Trakadas. et al. 2009). In this paper, it is assumed that node i evaluate the trust value of node j by both direct trust value and indirect trust value.

## 3.1   The direct trust evaluation

Supposing there are m attribute evaluations of the target node, the reverse cloud generator is used to combine these evaluations. Detailed process is shown as Algorithm 2.

**ALGORITHM II:**   THE STANDARD CLOUD

**Input:**   trust vector $V_g$ ( $1 \leq g \leq m$ ) and trust discrete scale W.

**Step 1:** $N_g = \sum_{k=1}^{L} v_k$ , where $N_g$ is the number of interactions, $1 \leq g \leq m$ .

**Step 2:** $\overline{X_g} = \frac{1}{N_g} \sum_{k=1}^{L} \left( v_k * w_k \right)$ , where $\overline{X_g}$ is the mean value.

**Step 3:** $S_g^2 = \frac{1}{N_g - 1} \sum_{k=1}^{L} \left[ v_k * \left( w_k - \overline{X_k} \right)^2 \right]$

where $S_g^2$ is the variance.

**Step 4:** $Ex_g = \overline{X_g}$

**Step 5:** $En_g = \sqrt{\frac{\pi}{2}} * \frac{1}{N_g} \sum_{k=1}^{L} \left[ v_k * \left| w_k - \overline{X_k} \right| \right]$

**Step 6:** $He_g = \sqrt{\left| S_g^2 - En_g^2 \right|}$

**Output:** $TC_g \left( Ex_g, En_g, He_g \right)$. Regarding each evaluation as a cloud drop, we put it into the reverse cloud generator and obtain the characterization parameters: $Ex_g$, $En_g$, $He_g$.

Combining m attribute evaluations, we can obtain the direct trust value of node $i$ on node $j$, which is denoted as $DTC_{ij}(Exd_{ij}, End_{ij}, Hed_{ij})$. It can be computed by Algorithm 3.

**ALGORITHM III:**   THE DIRECT TRUST VALUE OF NODE J ON NODE I

**Input:** $TC_g \left( Ex_g, En_g, He_g \right)$, $1 \leq g \leq m$ ;

**Step 1:** $Ex_{dij} = \frac{1}{m} \sum_{g=1}^{m} Ex_g$

**Step 2:** $En_{dij} = \frac{1}{m} \sqrt{\sum_{g=1}^{m} \left( Ex_g \right)^2}$

**Step 3:** $He_{dij} = \frac{1}{m} \sum_{g=1}^{m} He_g$

**Output:** $DTC_{ij} \left( Ex_{dij}, En_{dij}, He_{dij} \right)$

## 3.2   The update of direct trust value

As time goes on, the trust value will decay. And each interaction between nodes affects the trust value. Hence, it is necessary to update the direct trust value based on the history information and current information. We adopt β as the time factor, and the update process can be described by Algorithm 4.

From the equation, we can see that as β increases, the history information affects less on $DTC_{ij}(Ex_{dij}, En_{dij}, He_{dij})$ while the current information affects more, which is in accord with the fact that trust vale will decay as time goes on.

**ALGORITHM IV:**   THE UPDATE OF DIRECT TRUST VALUE

**Input:** $DTC_{ij}^{''} \left( Ex_{dij}^{''}, En_{dij}^{''}, He_{dij}^{''} \right)$ is the history direct trust, $DTC_{ij}^{'} \left( Ex_{dij}^{'}, En_{dij}^{'}, He_{dij}^{'} \right)$ is the current direct trust, and β is the time factor.

**Step 1:** Compute the current direct trust $DTC_{ij}^{'} \left( Ex_{dij}^{'}, En_{dij}^{'}, He_{dij}^{'} \right)$ according to Algorithm 2 and Algorithm 3.

**Step 2:** Compute the direct trust $DTC_{ij} \left( Ex_{dij}, En_{dij}, He_{dij} \right)$ after updating

$Ex_{dij} = \left( 1 - \beta \right) Ex_{dij}^{''} + \beta Ex_{dij}^{'}$

$En_{dij} = En_{dij}^{''} + \frac{Ex_{dij} - Ex_{dij}^{''}}{Ex_{dij}^{'} - Ex_{dij}^{''}} \left( En_{dij}^{'} - En_{dij}^{''} \right)$

$He_{dij} = He_{dij}^{''} + \frac{Ex_{dij} - Ex_{dij}^{''}}{Ex_{dij}^{'} - Ex_{dij}^{''}} \left( He_{dij}^{'} - He_{dij}^{''} \right)$

**Output:** the direct trust $DTC_{ij} \left( Ex_{dij}, En_{dij}, He_{dij} \right)$ after updating

### 3.3 Indirect trust value

Supposing the indirect trust value of node $i$ on node $j$ can be obtained through $s$ recommended nodes, and the number of recommendation nodes $s$ is decided by nodes' distribution and transmission radius avoiding trust recycle recursion and reducing the network communication payload, the indirect trust values are confined to direct trust value of the common neighbors shared by both node $i$ and node $j$.

Using $RTC_{ij}^k$ to denote the indirect trust value of node $i$ on node $j$ through the recommended node $k$, it can be calculated by Algorithm 5.

---

**ALGORITHM V:** THE INDIRECT TRUST VALUE OF NODE I ON NODE J THROUGH THE RECOMMENDED NODE K

**Input:** $DTC_{dik}\left(Ex_{dik}, En_{dik}, He_{dik}\right)$ is the direct trust value of node $i$ on node $k$, and $DTC_{dkj}\left(Ex_{dkj}, En_{dkj}, He_{dkj}\right)$ is the direct trust value of node $k$ on node $j$

**Step 1:** $Ex_{ij}^k = Ex_{dik} * Ex_{dkj}$ ;

**Step 2:** $En_{ij}^k = \min\left(\sqrt{En_{dik}^2 + En_{dkj}^2}, 1\right)$;

**Step 3:** $He_{ij}^k = \min\left(\sqrt{He_{dik}^2 + He_{dkj}^2}, 1\right)$

**Output:** $RTC_{ij}^k\left(Ex_{ij}^k, En_{ij}^k, He_{ij}^k\right)$

---

There are $s$ recommended nodes, and node $i$ would obtain $RTC_{ij}^k$ $(k=1,2,...,s)$. Then, node $i$ would aggregate those recommendations. Due to the existence of malicious nodes that may offer dishonest recommendation, we introduce Algorithm 6 to adjust weights of recommended trust values.

---

**ALGORITHM VI:** THE WEIGHTS OF RECOMMENDED TRUST VALUES

**Input:** $DTC_{ij}\left(Ex_{dij}, En_{dij}, He_{dij}\right)$ and $RTC_{ij}^k\left(Ex_{ij}^k, En_{ij}^k, He_{ij}^k\right)$ ;

**Step1:** $u_k = 1 - \dfrac{\sqrt{\left(Ex_{ij}^k - Ex_{dij}\right)^2 + \left(En_{ij}^k - En_{dij}\right)^2 + \left(He_{ij}^k - He_{dij}\right)^2}}{\max\left\{\left|RTC_{ij}^k\right|, \left|DTC_{ij}\right|\right\}}$

**Step 2:** $w_k = \dfrac{u_k}{\sum_{k=1}^{s} u_k}$

**Output:** the weights of recommended trust values $w_k$ $(k = 1, 2, ..., s)$

---

Finally, we can get the indirect trust value $RTC_{ij}(Ex_{rij}, En_{rij}, He_{rij})$ by Algorithm 7.

---

**ALGORITHM VII:** THE DYNAMIC AGGREGATION

**Input:** $RTC_{ij}^k\left(Ex_{ij}^k, En_{ij}^k, He_{ij}^k\right)$, and $w_k$

**Step 1:** $Ex_{rij} = \sum_{k=1}^{s}\left(w_k * Ex_{ij}^k\right)$;

**Step 2:** $En_{rij} = \min\left(\sqrt{\sum_{k=1}^{s}\left(w_k * En_{ij}^k\right)^2}, 1\right)$

**Step 3:** $He_{rij} = \min\left(\sqrt{\sum_{k=1}^{s}\left(w_k * He_{ij}^k\right)^2}, 1\right)$

**Output:** indirect trust value $RTC_{ij}\left(Ex_{rij}, En_{rij}, He_{rij}\right)$

---

### 3.4 Integrate trust value

The integrated trust value of node $i$ on node $j$ based on the direct trust value and indirect trust value can be computed by Algorithm 8.

---

**ALGORITHM VIII:** THE INTEGRATED TRUST VALUE

**Input:** $RTC_{ij}\left(Ex_{rij}, En_{rij}, He_{rij}\right)$ and $DTC_{ij}\left(Ex_{dij}, En_{dij}, He_{dij}\right)$

**Step 1:**: $Ex_{ij} = w_d * Ex_{dij} + w_r * Ex_{rij}$

**Step 2:** $En_{ij} = \min\left(\sqrt{\left(w_d * En_{dij}\right)^2 + \left(w_r * En_{rij}\right)^2}, 1\right)$

**Step 3:** $He_{ij} = \min\left(\sqrt{\left(w_d * He_{dij}\right)^2 + \left(w_r * He_{rij}\right)^2}, 1\right)$

where $w_d + w_r = 1$

**Output:** $TC_{ij}\left(Ex_{ij}, En_{ij}, He_{ij}\right)$

---

## 4 SIMULATION

We use Matlab platform to analyze our method. In the simulation, three attribute evaluations are used to compute the direct trust value, and they are history reputation, network speed, and transaction time. Here, we set five trust levels: level I denotes "totally untrust," level II denotes "untrust," level III denotes "a little trust," level IV denotes "trust," and level V denotes "totally trust." The trust levels of those attribute evaluations are shown in Table 1.

Table 1.   The statistical results.

| Trust levels | Attribute Evaluations | | |
| --- | --- | --- | --- |
| | History reputation | Network speed | Transaction time |
| I | Bad | Very slow | Long |
| II | A little bad | Slow | 20 days |
| III | General | General | 15 days |
| IV | Good | A little fast | 10 days |
| V | Very good | Fast | 5 days |

### 4.1   *Evaluation of trust value*

The interval [0,1] is divided into [0,0.4], [0.4,0.6], [0.6,0.8], and [0.8,1]. According to Algorithm 1, we can obtain the standard cloud which is shown in Table 2.

Table 2.   The numerical characteristics of standard cloud.

| Trust levels | Ex | En | He |
| --- | --- | --- | --- |
| I | $Ex_1=0$ | 0.0667 | 0.05 |
| II | $Ex_2=0.2$ | 0.0667 | 0.05 |
| III | $Ex_3=0.5$ | 0.1 | 0.05 |
| IV | $Ex_4=0.7$ | 0.0667 | 0.05 |
| V | $Ex_5=1$ | 0.1 | 0.05 |

Let node i and node j interact with each other 1000 times. The statistical results are shown in Table 3.

Table 3.   The statistical results.

| Trust levels | I | II | III | IV | V |
| --- | --- | --- | --- | --- | --- |
| History reputation | 13 | 35 | 124 | 810 | 18 |
| Network speed | 9 | 87 | 213 | 302 | 389 |
| Transaction time | 6 | 47 | 52 | 843 | 52 |

Then, the statistic of the attribute evaluations can be calculated by Algorithm 2. Results can be referred to Table 4.

Table 4.   The statistic of the attribute evaluations.

| | Ex | En | He |
| --- | --- | --- | --- |
| I | 0.6540 | 0.1090 | 0.0879 |
| II | 0.7243 | 0.2688 | 0.0557 |
| III | 0.6775 | 0.0896 | 0.1138 |

According to Algorithm 3, the direct trust value of node i on node j is $DTC_{ij}(0.6853, 0.1012, 0.0858)$.

### 4.2   *Evaluation of the weight scheme*

Fig. 1 shows the change of recommended nodes' weight with different proportion of malicious nodes. Here, in order to clearly see the function of our weight scheme which is described as Algorithm 6, we suppose node *1* gives the dishonest recommendation when the proportion of malicious nodes is 10%, node *1* and node *2* do when the proportion is 20%, and so on. When the proportion of malicious nodes is 10%, node *1* becomes malicious and its weight decreases. As the proportion increase to 30%, the same results happen to node *3*. Node *7* is a normal node and its weight increases with the increase of the proportion of malicious nodes. Hence, once a node provides dishonest recommendation, its weight will falls while the normal node's weight increases. The reason is that we distribute weights calculated by Algorithm 6, which can mitigate the effect of dishonest recommendation on the final indirect trust value. Hence, our method can defend against bad-mouthing attack effectively.

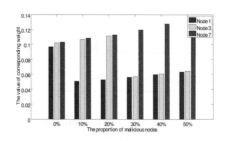

Figure 1.   The change of recommended nodes' weight with different proportion of malicious nodes.

## 5   CONCLUSION

In this paper, we propose a novel trust management based on cloud model, which can deal with uncertainty of trust well. A decay time factor is adopted to update the direct trust value, which enhances the algorithm's accuracy and dynamism. The ultimate indirect trust value is obtained by dynamic aggregation which adopts weights of recommended trust values in order to avoid false recommendations. The Matlab platform is used to test the performance of our method, and simulation results show that the proposed algorithm can deal with uncertainties of trust well and effectively defend against bad-mouthing attack

# REFERENCES

Lopez, J., Roman, R., Agudo, I., & Fernandez-Gago, C. 2010. Trust management systems for wireless sensor networks: Best practices. Computer Communications 33(9):1086–1093.

Huang, Y. M., Hsieh, M. Y., Chao, H. C., Hung, S. H., & Park, J. H. 2009. Pervasive, secure access to a hierarchical sensor-based healthcare monitoring architecture in wireless heterogeneous networks. IEEE Journal on Selected Areas in Communications 27(4):400–411.

Lai, C. F., Chang, S. Y., Chao, H. C., & Huang, Y. M. 2011. Detection of cognitive injured body region using multiple triaxial accelerometers for elderly falling. Sensors Journal 11(3):763–770.

Han, G., Jiang, J., Shu, L., Niu, J., & Chao, H. C. 2014. Management and applications of trust in Wireless Sensor Networks: A survey. Journal of Computer and System Sciences 80(3):602–617.

Kumar, G., Titus, I., & Thekkekara, S. I. 2012. A comprehensive overview on application of trust and reputation in wireless sensor network. Procedia Engineering 38:2903–2912.

Khalid, O., Khan, S. U., Madani, S. A., Hayat, K., Khan, M. I., Minallah, N., & Chen, D. 2013. Comparative study of trust and reputation systems for wireless sensor networks. Security and Communication Networks 6(6):669–688.

Zhan, G., Shi, W., & Deng, J. 2011. SensorTrust: A resilient trust model for wireless sensing systems. Pervasive and Mobile Computing 7(4):509–522.

Zarei. M, Rahmani. A.M, Sasan. A & Teshnehlab. M. 2009. Fuzzy based trust estimation for congestion control in wireless sensor networks. International Conference on Intelligent Networking and Collaborative Systems, pp. 233–236.

Yang. M, Wang. L.N & Lei. Y.D. 2009. A novel cloud-based subjective trust model. International Conference on Multimedia Information Networking and Security, pp. 187–190.

He, R., Niu, J., & Zhang, G. 2005. CBTM: A trust model with uncertainty quantification and reasoning for pervasive computing. Parallel and Distributed Processing and Applications, pp. 541–552.

P. Trakadas, S. Maniatis, P. Karkazis, T. Zahariadis, H.C. Leligou & S. Voliotis. 2009. A novel flexible trust management system for heterogeneous wireless sensor networks. International Symposium on Autonomous Decentralized Systems, pp. 1–6.

*Multimedia, Communication and Computing Application – Leung (Ed.)*
*© 2015 Taylor & Francis Group, London, ISBN 978-1-138-02775-6*

# Object tracking by fusing multi-feature based on infrared particle filter

Y.W. Chong, Z.W. Wang & Y.Y. Wang
*LIESMARS, Wuhan University, Wuhan, China*

R. Chen
*Wuhan Technical College of Communications, Wuhan, China*

Y. Li
*Institute of Technology, Vocational School, Zhejiang Normal University, China*

ABSTRACT: Under the condition of infrared target tracking, it is difficult to predict the existence of background interference, since there is a contradiction among the real-time effectiveness of the algorithms. This paper presents a particle filter tracking algorithm which is an adaptive fusion of color features and edge features. By using the natural framework of the particle filter, the aim of the condition of infrared is to select color features and edge features that can represent the target information to best to construct the multi-feature model of the target. By means of dynamic space model, its target is to improve the particle filter tracking algorithm, predict the motion state of the particles and to overcome the effects of environment mutations on tracking stability. Experimental results show that the proposed algorithm can overcome the interference of all kinds of background clutter and noise which could ensure the robustness and accuracy of tracking.

KEYWORDS: Infrared, Multi-feature fusion, Particle filter, Dynamic spatial model, Target tracking.

## 1 INTRODUCTION

Infrared motion tracking has been a hot research topic in the field of computer vision. It is also a core technology in infra-red precision-guided and infra-red warning algorithm. Jianfu Li[1] proposed a pedestrian tracking in infrared image sequences using wavelet entropy features even if it does not work in complex backgrounds. Nummiaro et al.[2] used color feature to track targets which has a great robustness when objects were shortly blocked and deformed in shape. Vignon et al. [3] had put forward a tracking algorithm based on Hausdorff distance. The algorithm's purpose is to calculate the target's shape information for target tracking. But when the target is similar to the other object or background, it will have a larger calculation and the tracking failed easily. Therefore, tracking, only based on a single feature, has bad robustness in complex background. Liu [4] proposed a fusion-tracking algorithm based on color and edge features. When using the Kalman filter algorithm for the first time to predict the next target trajectory, the trajectory of objects is almost nonlinear. Chen [5] had put forward that tracking algorithm based on multiple cues. A particle filter with a certain weight coefficient could use color features and edge characteristics of the target to build the likelihood function which acted as a basis for target tracking. The key of this algorithm is to give each feature appropriate weight, although they often failed tracking under complex background. Komeili [6] Proposed an algorithm combing variety of observations and space scattering particles by using weighted color, edge and texture features. However, a lot of features will lead to an extremely complex operation which are not suitable for real-time applications [7].

In this paper, we built multi-feature model with the help of a particle filter natural framework and achieved the online update feature weights.

## 2 FEATURE FUSE ADAPTIVELY

In order to improve tracking algorithm for particle filter algorithm, the method is using dynamic space model [7] to predict the particle state and to enhance the stability of the track.

### 2.1 *Feature fusion*

Bhattacharyya coefficient[8] defines the distance between the target and the target candidate. That is, the greater its values, the higher the similarity between the two distributions will be. The likelihood function of status is positively related to the distance between a candidate target and the reference target in the current state, That is to say, the more similarities the candidate target and the reference target have, the greater chance the likelihood probabilities [9] according to B's coefficient of calculating the similarity between the candidate model($H_i$) and target model($H_0$) of $x_k^i$:

$$\hat{p}(z_k|x_{k-1}^i) = \frac{1}{\sqrt{2\pi}\sigma} \exp(\frac{1-\rho[H_i,H_0]}{2\sigma^2}) \quad (1)$$

When the $\hat{p}(z_k|x_{k-1}^i)$ reach to the maximum, the target model and candidate mode have the most similarities;

Besides, according to the color feature's similarity of goals and candidate, we can get the likelihood function of the color characteristics:

$$\hat{p}_{color}(z_k|x_{k-1}^i) = \frac{1}{\sqrt{2\pi}\sigma_{color}} \exp(\frac{1-\rho[H_{color\ i},H_0]}{2\sigma_{color}^2}) \quad (2)$$

likelihood function of FDF feature is:

$$\hat{p}_{FDF}(z_k|x_{k-1}^i) = \frac{1}{\sqrt{2\pi}\sigma_{FDF}} \exp(\frac{1-\rho[H_{FDF\ i},H_0]}{2\sigma_{FDF}^2}) \quad (3)$$

Where $\hat{p}_{color}(z_k|x_{k-1}^i)$ represents the likelihood function of color, $\hat{p}_{FDF}(z_k|x_{k-1}^i)$ represents the likelihood function of FDF features, $H_{color,i}$ represents the candidate model of each particle's color characteristics, $H_{PDF,i}$ represents the candidate model of each particle's FDF features. We can get the likelihood function of the target's color characteristics and the FDF features by the formula (2) and (3).

Because of the multi-objective optimization problem, the weighting method is a common method [10] for a maximization problem which has targets $k$. It can be expressed like this:

$$y = \max \sum_{i=1}^{k} w_i(x)f_i(x)\ s.t.\ 0 \le w_i(x) \le 1$$

$$\sum_{i=1}^{k} w_i(x) = 1 \quad (4)$$

Where $x \in R^m$ is the decision vector, $y \in R^m$ is the target vector, $f_i(x), i = 1,2,\dots,k$ is the objective function, $w_i(x)$ is the system constraints. According to this theory, setting the joint likelihood function of the target as follows:

$$\hat{p}(z_k|x_{k-1}^i) = \alpha_i \hat{p}_{color}(z_k|x_{k-1}^i) + \beta_i \hat{p}_{FDF}(z_k|x_{k-1}^i)$$

$$s.t.\ 0 \le \alpha_i \le 1$$

$$0 \le \beta_i \le 1$$

$$\alpha_i + \beta_i = 1 \quad (5)$$

Where $i$ is particles, $\alpha_i$ is the weights of each particle's color characterized in that target joint likelihood

function, $\beta_i$ is the weights of each particle's FDF features. The larger the feature weight is, the more the tracking results would rely on. By adjusting the weights of each feature can realize the goal's reliable tracking.

### 2.2 Feature weight update

Assume the differences between the color characteristics and the FDF's likelihood function of all the particles are settled as $k$, It would seem to be like this:

$$r_k^i = \hat{p}_{color}(z_k|x_{k-1}^i) - \hat{p}_{FDF}(z_k|x_{k-1}^i) \quad (6)$$

By the formula (5) and (6) we can get:

$$r_k^i = \frac{1}{\sqrt{2\pi}\sigma_{color}} \exp(-\frac{1-\rho[H_{color\ i},H_0]}{2\sigma_{color}^2})$$
$$- \frac{1}{\sqrt{2\pi}\sigma_{FDF}} \exp(-\frac{1-\rho[H_{FDF\ i},H_0]}{2\sigma_{FDF}^2}) \quad (7)$$

If $\sigma_{color}^2 \ge \sigma_{FDF}^2$, $\sigma_{color}^2$ can be expressed as:
$\sigma_{color}^2 = \sigma_{FDF}^2 + d^2$; Otherwise: $\sigma_{FDF}^2 = \sigma_{color}^2 + d^2$

According to the Second-order Maclaurin formula, when $d$ is small

$$r_k^i = \begin{cases} \frac{1}{2}(\frac{d}{\sigma_{color}})^2 [1-(\frac{\rho[H_{color\ i},H_0]-\rho[H_{FDF\ i},H_0]}{\sigma_{color}})^2] \\ \times \hat{p}_{color}(z_k|x_{k-1}^i) \quad while \quad \sigma_{color}^2 \ge \sigma_{FDF}^2 \\ \frac{1}{2}(\frac{d}{\sigma_{FDF}})^2 [1-(\frac{\rho[H_{FDF\ i},H_0]-\rho[H_{color\ i},H_0]}{\sigma_{FDF}})^2] \\ \times \hat{p}_{FDF}(z_k|x_{k-1}^i) \quad otherwhise \end{cases} \quad (8)$$

For the color likelihood function and the FDF likelihood function, $d$ is small enough, and

$$d^2 \ll \sigma_{FDF}^2\ s.t\ d^2 \ll \sigma_{color}^2\ \text{Then}\ \frac{1}{2}(\frac{d}{\sigma_{color}})^2\ \text{and}$$

$\frac{1}{2}(\frac{d}{\sigma_{FDF}})^2 \sim \varepsilon$ ($\varepsilon$ is infinitesimal) is small enough.

$r_k^i \triangleq \rho[H_{color,i},H_0] - \rho[H_{FDF,i},H_0]$, indicates that the change of the particles' weight depends on the change of the differences between the characteristics of the model, so the differences between each particle characteristic likelihood function reflect the changes in the environment. Weight changes normalized:

$$\tilde{r}_k^i = \frac{r_k^i}{\sum_i |r_k^i|} \quad (9)$$

Through $r_k^i$ to adjust the weight of each component of each particle combined characteristics.

To update color feature weight: $\alpha_i = \alpha_i + \tilde{r}_k^i$    (10)

To update FDF feature weight: $\beta_i = \beta_i - \tilde{r}_k^i$    (11)

Then to achieve the updating right joint likelihood function's component. According to environmental changes, making the joint likelihood function has more representational to the current target.

### 2.3 Particle weight update

The change of the color weights $\alpha_i$ and FDF weight $\beta_i$ causes the change of formula (5). For the particle filter, the individual particles which is likelihood function indicates the weight of the particle. That is the possibility of the target in the current location. The changes of the target function cause the changes in the weight of the particles. Update the weight of each particle:

$$\omega_k (i-1) \times P(z_k | x_{k-1}^i) \qquad (12)$$

Where $\omega_k (i-1)$ represents the weight of particle $i$ at the time of $k-1$, $\omega_k (i)$ is the weight of particle $i$ at the present time, $\hat{p}(z_k | x_{k-1}^i)$ is the observation likelihood function of particle $i$ fat the present time, i.e., the above joint likelihood function. Particle right re-normalization.

Then, complete the weights update of the particles. The joint probability likelihood function $\hat{p}(z_k | x_{k-1}^i)$ changes with the environment, the weight of the particles also was updated which make the movement of the particle filter target tracking more stable.

### 2.4 Particle forecast

**A, B** are constant which can be obtained by learning the image sequence. According to the dynamic space model and the history motion, the state of particles whose aim is to predict the current state of the particle act like this:

$$x_k^i = \boldsymbol{A}x_{k-1}^i + \boldsymbol{B}w_{k-1} \qquad (13)$$

Thus ensured the particle filter tracking algorithm without losing track target. It should be set at once when the environment mutate, for it helps to enhance the stability of the track.

## 3 EXPERIMENTAL RESULTS AND ANALYSIS

In order to confirm the robustness of our method, we conducted a test in the six video sequences. The test

video comes from OSU heat walking database [11], Terravic infrared motion database [12], Audio-Visual Vehicle (AVV) database which contains a variety of anomalies.

### 3.5 Multiple sequence tracking results

Each row from top to bottom as an image sequence S1–S6, where S1–S4 are obtained from OSU heat walk database, S5 is obtained from the database infrared motion Terravic, S6 is obtained from AVV data set, the uniform size of the image is 240 * 320, and the interesting objects are all pedestrians. All sequence parameters are the same in the tracking system.

### 3.6 Performance comparison

To validate the advance between the proposed algorithm and the current algorithm except visual comparison of the effect track, proposed displacement error rate (DER) could be taken into consideration. The compared quantitative indicators track the performance characteristics of the fusion algorithm and that is the difference between the actual position and the calculated position. The rate holds the actual position, which is defined as follows:

$$DER = \frac{|rc - xc|}{|rc|} \qquad (0 \leq DER \leq 1) \qquad (14)$$

where $xc$ is the calculated position which coordinates of the target. $rc$ is the actual position coordination. The smaller DER is, the better the track is. The proposed algorithm could be compared with the [3,6–10] (denoted 1st–6th Method). The DER benchmark of the seven algorithms at database of Terravic infrared motion image sequence S5 is shown as below:

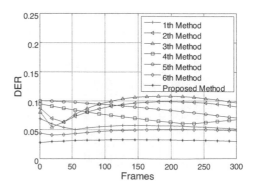

Figure 1. DER between proposed algorithm and other algorithms.

Seen from Figure 2, the advanced feature fusion algorithm's DER values fluctuate between 0 and 0.12. Method 1 is a fusion of color features and the FDF features Kalman filter tracking algorithm; Method 2 represents a fusion of color with fixed weights FDF features and particle filter tracking algorithm; Method 3 fuzzy methods are to achieve feature fusion; Method 4 equals to each feature component of the expectation and variance to build a joint likelihood function; Method 5 is based on the multi-objective optimization evolutionary algorithm feature fusion; Method 6 is PSO-based integration strategy.

According to DER definition, DER fluctuated mostly in fuzzy algorithm. The proposed likelihood function based on the characteristics of adaptive multi-feature fusion particle filter tracking algorithm. DER values small and smooth movements. This is because the proposed dynamic particle tracking model is based on environmental changes adapting the weights of each feature to the complex environment of the moving target tracking. However, FDF features just need to calculate the gradient of the four directions. The calculation is smaller than the need to calculate nine directions' HOG. Therefore, in the tracking performance with high accuracy, it takes less time to calculate the edge feature.

To further evaluate the performance of the current advanced and proposed algorithm, we measure each fusion algorithm by calculating the proportion of the correct tracking of each sequence [10] :

$$p = 1 - \frac{F_{error}}{F_{total}} \quad (15)$$

Where $F_{error}$ represents the wrong tracks in a sequence, $F_{total}$ means the total number of the sequence. Test sequences S1-S6 by [3,6,7,8,9,10]

(denoted 1st–6th Method) characteristic fusion algorithm and the proposed method, the test results were shown as below:

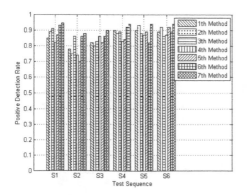

Figure 2. Positive detection rate of each tracking algorithm.

Taking into account the particle filter, it is a probabilistic algorithm, so there will be differences in the results of each test. Each test runs 12 times, taking the average results. The track errors of the proposed features are less than other tracking methods, so the tracking accuracy rate is higher. In most of the tracking sequences, Methods 4 and 5 have a larger probability of adverse outcomes. Method 6 is close to this proposed method, but this method has the smallest tracking error.

## 4 CONCLUSION

According to the characteristics of the infrared image, select a weak correlation of the color characteristics from the strong complementary and edge features is needed. With the natural framework of particle filter, its aim is to build the color characteristics between likelihood function and FDF edge features likelihood function. And then propose particle filter dynamic model according to the environmental change to set different weights to the two features. When the target color and the background are similar, color characteristics will not be sensitive for the target identification, then use the edge features to supplement.

ACKNOWLEDGMENTS

This paper was supported by the National Natural Science Foundation of China (41271398), the Fundamental Research Funds for the Central Universities (204201kf0242, 204201kf0263), Natural Science Foundation of Hubei Province of China (2012FFB04204).

REFERENCES

[1] Li Jianfu, Wang Yang. Pedestrian Tracking in Infrared Image Sequences using Wavelet Entropy Features[J]. *IEEE Second Asia-Pacific Conference on Computational Intelligence and Industrial Applications*, 2009:288–2690.
[2] Nummiaro, Koller, Van. An adaptive color-based particle filter[J].*Image and Vision Computing*, 2012, 21(1):99–110.
[3] Vignon, Lovell, Brian.Rober General Purpose Real-time Object Tracking Using Hausdorff Transforms[J]. *In Proceedings of IPMU*, 2012:1–6.
[4] Liu Jin. Research and Implementation Based on multi-feature fusion particle filter algorithm[J]. *Computer Measurement & Control*, 2013, 11(5):1307–1309.
[5] Chen Shanjing.Study of particle filtering algorithm based on multi-feature fusion[J]. *Computer Engineering*, 2011, 37(7):179–180.
[6] Komeili M, Armanfard N, Kabir E. A fuzzy approach for multi-feature pedestrian tracking with

particle filter[J]. *IEEE International Symposium on Telecommunications*, 2008, 2(8):570–576.

[7] Yao Haitao, Zhu Fuxi,Chen Haiqiang. Face Tracking Based on Adaptive PSO Particle Filter[J].*Geomatics and Information Science of Wuhan University*, 2012, 37(4):493–497.

[8] Wang Shupeng, Ji Hongbing. Based target tracking method color space model[J]. *Xi'an Electronic Science and Technology Report*, 2009, 34(4):77–78.

[9] Jia Yonghong, Li Deren.An Approach of Classification Based on Pixel Level and Decision Level Fusion of Multi-source Images in Remote Sensing[J]. *Geomatics and Information Science of Wuhan University*, 2001, 26(5), 431–433.

[10] Wang Huan,Wang Jiangtao, Ren Mingwu, etc. A Robust multi-feature Fusion Target Tracking Algorithm[J]. *Chinese Journal of Image and Graphics*, 2011, 14(3):489–498.

[11] Mostaghim, Teich, Quadtrees. A Data structure for restorin-g Pare to sets in Multi-object Evolutionary Algorithms with Elitism[A]. *Springer Verlag*, 2005:812–104 .

[12] OSU Thermal Pedestrian Database: www.vcipl .okstate.edu.

*Multimedia, Communication and Computing Application – Leung (Ed.)*
© *2015 Taylor & Francis Group, London, ISBN 978-1-138-02775-6*

# The performance characteristics of color language on the digital photography using warmed color works for case analysis

S.H. Wu & J.H. Chen
*Department of Digital Media Design, College of Creative Design, Asia University, Taichung City, Taiwan*

M.Y. Liu
*Department of Industrial Design, Chaoyang University of Technology, Taichung City, Taiwan*

ABSTRACT: In our daily lives, color plays a vital role and it can be easily seen everywhere. The color can stimulate the visual sensory of humans and induce their physical cognition. Besides, each color has its own symbol and image. Color difference will result in dissimilar psychological reaction. Color is also an important factor of the manifestation of digital photography. Meanwhile, it plays an important role in the style of photography as well as in the situational expression. In this study, the purpose is to discuss the effects of warmed color image on the works of digital photography using the method of literature review and case analysis of digital photography under the observation of the perspective of chromatics. In conclusion, it is suggested that creators should make it the priority to handle the factors of color image based on the emotion image of picture color when they are engaged in the creation of digital photography.

## 1 INTRODUCTION

Color is one of the most important constituent elements in photography. Except for displaying the appearance of things which is photographed, it can also transmit the particular situation and make a good impression of the picture. Basically, the forming of color is a kind of simple physical phenomenon. It is the image of vision observation when the light is projected on human's brain through reflection or transmission in the nature. Color itself does not process any emotion and perceptual factors such as cold and warm, light and heavy. However, many people often combine the color with the mental feeling and further produce the emotion factors. They also impart colors with the meaning of culture and spirit value through the ocular characteristic of color. For example, red is a kind of joyous and festival color throughout China. With the differences of era, environment, region, race, and belief, the simple relation of color will further have deeper thought and more complicated meaning (Chang & Chang 2012).

In 1838, Daguerre, a French physicist, invented the photographic technique (also named Daguerreotype), and now this technique has a lot of different forms (Memes 2010). Especially, the recording methods of photographing image have been upgraded, changing from traditional photographic film to digital photo storage devices such as compact flash (CF) and secure digital (SD) memory card in recent years (Fujiwara et al. 1988, Chandra et al. 2000). Digital

photography with digital imaging devices has converted the optical image into digital record message and also changed the presentation style of the image. Therefore, the digital camera possesses the function of designing. Digital photography has created a lot of colorful images and affected the interpretation of the visual sense. To achieve the original target of design, the creators and/or photographers themselves are able to operate the items including the selection of photographic subject, shooting angles capture, screen constitutes configuration, and the color arrangement of light. Changing the previous visual experience will make the image creative to have broader display techniques.

## 2 LITERATURE REVIEW

Human vision mainly relies on light-sensitive cells in the retina of the eye. When the sensitive cells are stimulated by different wavelength light, the signals of perception from sensors are sent to the brain and enable us to see the colors. In other words, color is a function of human visual system and is not an intrinsic property. There are six psychological primary colors including red, yellow, green, blue, white, and black in the field of color psychology (Hall 2014, Kaiser 1997). Especially, black color is an absence of light stimulation to the human visual cells according to the conception of physic. However, a large number of people consider that no light stimulation only lacks a feeling in psychology and black color is actually a kind of sensation (Oh & Lai 1983, Cheng & Lin 2007).

When the light approaches the visual receptor in the brain, the generation procedure of color awareness is approximately divided into three stages. The first is physical stage. It is a nature and quantity of the light. The second is physiological stage. As producing color signal of visual cells are responsive to the light, and it is then transmitted to the brain. The third, psychological stage is based on the idea that the mental awareness is changed after the visual cells accepting the light as shown in Figure 1. Briefly, color perception is a result of the complex combination of different responses of these three procedures, which have been described above.

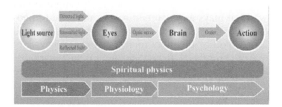

Figure 1. Diagram of color perception process.

In addition, there exist a lot of differences in human brains to the response of color perception. Owing to the stimuli of various colors, it will also bring about different consequences and impacts to people. There are roughly two aspects including 1) the visibility of color, such as visual acuity, legibility, and attention; and 2) determination of color perception, such as color weight, color temperature, the distance sense of color, and the judgments of positivity and negativity. The production of color psychological effects is mostly due to the different factors including gender, age, life, and culture. Moreover, this difference is mainly derived from the different reactions mentally. The combination of numerous conceptions often affects our color perception such as habits, prejudices, and images. If the color is the one or more concepts, that combined with each of the color or those of various concepts lead out from color can be referred to as color association. The content of color association is also divided into two types. One is associated with the specific objects. For examples, the color of yellow, red, and blue are associated to banana, flame, and sky, respectively. The other is associated with the abstract ideas or emotions. For examples, the color of white, red, and purple are, respectively, associated to purity, passion, and mystery. If this kind of abstraction association became a common experience and/or reaction, then the color will be fixed in a specific expression and gradually build its own conceptual

meaning. This is called the color symbolism (Chen & Yang 2008).

In terms of imagination, colors are able to be divided into two groups containing warm and cold color. In color wheel system (Fig. 2), if the brightest yellow and the darkest purple are linked to be an axial line, the colors such as red, orange, and yellow on the upper right side are referred to as warm colors because they remind us of things like the sun or fire. Then the colors on the lower left side are cool colors because they remind us of things like water or grass.

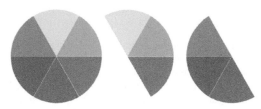

Figure 2. Color wheel, central: warm colors, right side: cool colors.

## 3 CREATIVE WORKS FOR CASE ANALYSIS

This study was divided into two parts containing literature investigation and photography creation. In the beginning, the literature and theory of color were made as an analysis and summarization. Then the technique of digital photography was applied on the practical creation works which the themes were based on the symbolization and image of color. The authors would like to expect that the color language was presented in the form of creation works and hope to effectively convey the important information of color which constructed the specific style and content of digital photography through the research processes and experimental results of this study.

When we see a kind of color, except for response the feeling in physic aspect, our body will also reply immediately to a feeling, which is difficult to describe in words. Then we name it the impression and this is the color image. Due to the limitation of paper pages, the following analysis and description are priorities with the warmed color photographic creations.

### 3.1 Red color image

It is well known that the color red has the longest wavelength in the spectrum; its wavelength ranges from 610 nm to 700 nm. When it penetrates into the air, it formed a minimum angle of refraction and the linear distance of radiation is farther. The image on the retina of eyes is in the deepest position. This phenomenon will give us a visual sense of approaching expansion. Therefore, it is known as advance color.

The color coordinates of red, similar color, and color association are shown in Table 1.

Red color easily attracts human's attention and it is widely used in various media. In addition to having a better photopic effect, it even has been used to convey the impression and spiritual meanings which usually represent the dynamic, active, enthusiastic, warm, and forward meanings.

Table 1.   Red color coordinates.

| Red | |
|---|---|
| HSB* | (0°, 100%, 100%) |
| RGB | (255, 0, 0) |
| LAB | (54, 81, 70) |
| CMYK | (0 %, 96%, 95%, 0 %) |
| Web Color** | #FF0000 |
| Concrete association | Fire, blood, apple, cherry |
| Abstract association | Passion, positive, joyous, dangerous, angry |
| Similar color | |

*Color Data Source : Photoshop CS6 color picker, finished by author.
**Web Color Data Source : http://en.wikipedia.org /wiki/X11_color_names

In our life, red color is often used as a symbol of excitement and joy such as luxurious red carpet and beautiful beauty. However, fire and blood are both red, in which case the color is usually considered as a symbolic color containing dangerous, disaster, explosion, and terrorist meaning (Fig. 3).

Figure 3.   The rays of reddening sunlight occupied more than half of the picture composition, and this symbolized the red of fire as well as blood, and it had a disaster and horrible feeling. (Photographed by M. Y. Liu).

### 3.2   Orange color image

Orange is a mixture color of red and yellow, which is a pigment secondary color. In the spectrum, orange is between red and yellow, and the range of wavelengths is between 590 nm and 620 nm. The penetration efficiency of orange is next to the red color in air and its feeling of color is warmer than red. The bright orange color can give us a feeling of solemn and noble. So it belongs to a psychological color. The color coordinates of orange, similar color, and color association are shown in Table 2.

In nature, pomelo, rice, and illumination all have a rich orange color. The color of orange gives us a very mild feeling and it also possesses some color perception like lively, prosperous, happy, bright, warm, and moving feeling. Therefore, women usually like to make this color as a decorative color.

Moreover, the orange is also available as the restaurant's layout color. In the restaurant, it is utilized to increase appetite because these colors are warm colors and they may have the function of attracting attention, suggesting aroma and increasing appetite.

Table 2.   Orange color coordinates.

| Orange | |
|---|---|
| HSB* | (39°, 100%, 100%) |
| RGB | (255, 165, 0) |
| LAB | (76, 28, 79) |
| CMYK | (0%, 46%, 91%, 0 %) |
| Web Color** | #FFA500 |
| Concrete association | Tangerine, orange, flame, leaf in autumn |
| Abstract association | Happiness, lively, warm, ripe, distinct |
| Similar color | |

*Color Data Source : Photoshop CS6 color picker, finished by author.
**Web Color Data Source : http://en.wikipedia.org /wiki/X11_color_names

In addition, the orange is also available as a celebration color. It can also be used as a rich and honorable color such as many decorations in the palace. Orange color has a higher photopic vision. The use of this color in industrial safety is a warning color such as locomotives, climbing clothing, backpacks, and life jackets. However, orange also has a bright and dazzling property; it sometimes gives us a negative and vulgar imagery. Furthermore, orange color is also likely to cause people's visual fatigue, as shown in Figure 4. In this photographic work, the creator effectively took advantage of the feature of orange perception, and he also further matched the sharp sense of the triangle to magnify the visual pressure of this work.

### 3.3   Yellow color image

The wavelength range of yellow is approximately from 570 to 585 nm. Yellow light is produced by the mixture of red and green light. The complementary color of yellow is blue, but most designers still replace it with purple. Due to the wavelength difference of yellow, the characteristic of thickness and weakness are not liable to be distinguished. The color coordinates of yellow, similar color, and color association are shown in Table 3.

The objects of yellow irradiated with yellow light will cause the phenomenon of eclipse. Meanwhile,

Figure 4. Orange color is likely to cause vision fatigue. The creator uses the feature and also adds the sharp sense of the triangle, thus the visual pressure of this work is more obvious. (Photographed by M. Y. Liu).

Figure 5. In the composition of this work, the effect of light and shadow is used to form a feeling of split and reflection picture. Yellow color strengthens the wise and noble feeling. (Photographed by M. Y. Liu).

this color can symbolize a kind of bitter, morbid, and perverse meaning and it cannot afford to dilute with white color. When contrasting with the dark color, the visual effect can be enhanced. The yellow has a strong light sense, and it can give us the impression like bright, brilliant, brisk, and clean.

Yellow has the highest brightness, and it gives us a bright and brilliant feeling, which is the symbol of wisdom, power and pride. In Chinese history, lots of emperors and religious leaders are conventionally dressed up in bright yellow apparels. Places like palaces and temples are often strengthened with yellow, which gives us a feeling of nobility, wisdom, honor, dignity, and kindness (Fig. 5).

Table 3. Yellow color coordinates.

| Yellow | | |
| --- | --- | --- |
| HSB* | (60°, 100%, 100%) | |
| RGB | (255, 255, 0) | |
| LAB | (98, -16, 93) | |
| CMYK | (10%, 0%, 83%, 0%) | |
| Web Color** | #FFFF00 | |
| Concrete association | Banana, egg yolk, the sun | |
| Abstract association | Brightness, hoping, frivolous, clear, noble | |
| Similar color | | |

*Color Data Source : Photoshop CS6 color picker, finished by author.
**Web Color Data Source : http://en.wikipedia.org/wiki/X11_color_names

## 4 CONCLUSIONS

With the advance of technology, the function of photography will become more diverse. The photographers can even enhance the image quality of digital photography by using computer software. Dr. Mitchell, a professor of MIT Media Lab., had even said that "if the digital technology escapes from the spirit of human, it will lose its meaning" (Mitchell 1992). Color is not just a simple color tone but also further contains many distinguishing characteristics. Therefore, by virtue of the association of color image, a piece of digital photography work can pass on the message which its creator wants to express.

When the creators are using the digital photographic techniques, which may be through the visual design and composition of the picture and/or through the manifestation of color and line, they appropriately guide the viewer's visual into the psychological level to enjoy their works. At this moment, the creative process and the work will lead to a better resonance with the viewers.

## REFERENCES

Chandra, S., Ellis, C.S. & Vahdat, A. 2000. Managing the storage and battery resources in an image capture device (digital camera) using dynamic transcoding. Proceeding WOWMOM '00 Proceedings of the 3rd ACM international workshop on Wireless mobile multimedia. P. 73-82. Retrieved from http://cseweb.ucsd.edu/~vahdat/papers/wowmom00.pdf.

Chang, Y. & Chang, C. 2012. On the psychological language of color. *Youth Literator* 17:125. (In Chinese).

Chen, J. H. & Yang, D. M. 2008. *Visual Design*. 208 pp. Chuan Hwa Press. Taiwan. ISBN:978-9572168844 (In Chinese).

Cheng, K. Y & Lin, P. S. 2007. *Color Plans*. 174 pp. Art Wind Hall Press. Taiwan. ISBN:979-9578494427 (In Chinese).

Fujiwara, T., Nishihara, S. & Hirose, S. 1988. Color flow-visualization photography and digital image processing techniques. *JSME international journal* 31(1):39–46.

Hall, R.H. 2014. Color perception. Retrieved from http://web.mst.edu/~rhall/web_design/color_perception.html.

Kaiser, P.K. (1997). *The Joy of Visual Perception*: A Web Book. Retrieved from http://www.yorku.ca/eye/noframes.htm.

Memes, J.S. 2010. *History and practice of photogenic drawing on the true principles of the Daguerrotype*. 104 pp. (Original author L. J. M. Daguerre in 1839, trannslated by J. S. Memes) Nabu Press. (Feb. 2010) ISBN: 978-1144073846.

Mitchell, W.J. 1992. *The reconfigured eye: visual truth in the post-photographic era*. Cambridge MA: MIT Press. ISBN 978-0262631600.

Oh, S.M. & Lai, L. Y. 1983. *Practice chromatology*. 126 pp. Lion Art Press. Taipei. ISBN:957-9420289. (In Chinese)

*Multimedia, Communication and Computing Application – Leung (Ed.)*
© 2015 Taylor & Francis Group, London, ISBN 978-1-138-02775-6

# High speed network camera based on linear CCD

C. Li
*The Collaboration of Patent Examination for Patent Examination Center of the State Intellectual Property Office*

X.F. Li
*Department of Electromechanical Engineering, Beijing Institute of Technology, Beijing, China*

ABSTRACT: A structure of the linear CCD camera is offered firstly including the design method of main function blocks. Secondly, the linear CCD driving circuit is recommended, and so does the network interface. The performance of the camera is proved in a series of tests. A prototype instrument is now employed to inspect the chassis in a safety check-up station, and the flexibility and expandability of the system is demonstrated successfully.

KEYWORDS: Linear CCD camera, Image, Acquisition, ARM microprocessor

## 1 INTRODUCTION

Linear charge coupled device (CCD) is equipped with image scanning, non-contact measurement, displacement detection and the bar code reading, etc. so that it has small size, lightweight, low power consumption, etc. And as it has the features of excellent precision, resolution, sensitivity and dynamic behaviour. So the price is very high.

An image acquisition system adopting the linear CCD sensor is presented in this paper. Through the network interface, the image can be transmitted to the computer connected in the LAN. The system is employed to inspect the chassis in a safety check-up station to acquire the chassis photos for the checking of secreting objects or persons.

The core part of the camera are the ARM microprocessor and the linear CCD modular which are cooperating with the CCD driving circuit, signal amplifying circuit, analog- to- digital convertor, data buffer chip, and the time sequence generating circuit of CPLD.

This camera system adopts the ARM microprocessor and CPLD as the core part of the camera design, including linear array CCD chip, CCD-drive circuit, amplifier circuit, AID-converter circuit, data- cache chip, timing-control circuit based on CPLD and the data-transfer circuit based on ARM microprocessor. The schematic is shown as Fig. 1.

## 2 SYSTEM OVERALL DESIGN

The working process of the hardware system is described as follows: Firstly, according to the timing characteristics of linear array CCD, drive timing

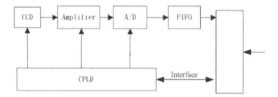

Figure 1. Schematic of hardware circuit.

needed is provided by the CPLD. Secondly, three analog CCD output signals are amplified into the high-speed AID conversion chip to be digitized. Because of the high frequency of the ARM and the relatively slow speed of AID conversion, the first in first out memory between them (FIFO) is added into them in order to solve the matching speed problem. When there are data in the FIFO, EF pin is set high for ARM processor interrupt and ARM will read out the data through the way of DMA. When a frame data of the linear array CCD is finished, RST pin of FIFO will be set low by CPLD for its reset. Finally, the processing and transferance of the data is completed in ARM. CPLD is sued for AID converter chip initialization, and to provide timing of work to match the clock to the AID, FIFO chip and CCD. The system can complete the collecting work without ARM processor systems, which improves the system's working efficiency.

## 3 DRIVER CIRCUIT DESIGN OF LINEAR ARRAY CCD CHIP

CCO is an image sensor in conversion type, while the MOS type semiconductor device uses charge as a signal. The basic structure of MOS capacitors is closely arranged, which can store charge by exciting incident light in the CCD photosensitive element and can directionally transmit charges into the form of charge pockets and complete the conversion of optical signals into electrical signals by the drive of the appreciate clock. Charge-coupled device is different from the ordinary MOS type semiconductor devices, which needs a complex drive pulse to work properly.

Circuit design of CCD drive timing is one of the keys for CCD data acquisition circuit design. There are many ways to generate the drive timing, of which the following methods are commonly used: EPROM, direct digital circuits driving method, SCM method, and special IC. As the CPLD has high integration, so it has a greater advantage in speed and timing control. CPLD logic can be reprogrammed according to the needs after circuit design is completed. So choosing CPLD to design CCD driving circuit is an efficient, flexible method.

This paper uses Toshiba CCD linear array chip TCD2252D as the image sensor. TDC2252D is a colour image sensor with high sensitivity and low-dark current, which has 2,700 valid pixel with the smallest size of 8, and the size between the lines is 64. TDC2252 adopts built-in sample hold circuit and clamp circuit which now has been widely used in close measurement, non-contact non-destructive testing, remote sensing and industrial inspection and other fields.

TCD2252D requires a total of six road-driving pulses, respectively, being the two-phase clock pulses FI and F2, reset pulse RS, integration time pulse SH, sample hold pulse SP, clamp pulse CP. The six-way drive pulse timing relationship is shown in Fig. 2. FI and F2 is a square wave pulse with the duty cycle of 1: 1 and frequency of 0.5MHz. The duty cycle of RS, SP and CP pulse are 1:7, and the frequency is square wave pulse of I MHz (typical values here). When the SH pulse is high, FI must remain high, then all FI electrode would form the potential well and SH's high would make potential well under the FI electrode communicate with storage potential well of MOS capacitor .

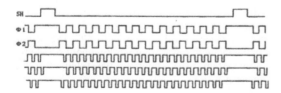

Figure 2. TCD2252D driving timing.

The signals pocketed in MOS capacitor are transferred into the F I electrode potential well of transfer

gate shift register up and down through the transfer gate. When the SH turns from the high into the low, low barrier formed by the SH will isolate the potential well of the MOS capacitors store gate and the potential well of Fl electrode. Then the storage gate potential will turn into the light integral state, while the shift register will make the signal charge pockets be transferred into the Fl electrode transfer bit-by-bit under the drive of the two-phase clock pulse of Fl and F2, and normally outputs through the output circuit by the output OS pin. The timing logic is designed with the simulation results shown in Fig. 3.

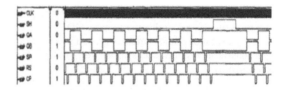

Figure 3. TCD2252D driving timing simulation.

## 4 NETWORK TRANSFER DESIGN

### 4.1 *Hardware design*

Embedded processor AT9 1RM9200 is manufactured by Atmel. The 10/100 Mbit/s adaptive Ethernet network interface DM9200 is the core part of the hardware design of network transfer with the function of acquisition and network transference of images and data.

### 4.2 *Software design*

The operating system IS requires the system having real-time, can be cut, and having a network of support functions for multi-task scheduling and running and so on. Therefore, the system uses the embedded Linux operating system, which has a strong network support functions, including a variety of Internet network protocols besides TCPIIP protocol.

Network transfer of this system is based on TCPIIP protocol, using the BSD Socket interface for web application development of Linux operating system. The socket has a similar function which can open the file called Socket 0, which returns from an integer Socket descriptor. Then the connection establishment, data transfer and other operations are also implemented by the Socket.

The design uses streaming Socket, thus providing reliable, connection oriented communication flow, and uses the TCP protocol to ensure data accuracy and orderliness and uses a simple socket server/client. Communication flow is shown in Fig. 4.

Figure 4.   Communication flow.

1  Open Socket descriptors and establish binding:  In order to implement input/output, the Socket must be established first. Socket program can call the function, which returns a handle similar into file descriptor. The socket function prototype is:
int socket(int domain, int type, int protocol);
2  Socket Configuration: After a socket descriptor returns through the socket call, the socket must be configured. The socket client stores information of the local and remote in socket data structure. Client of no connection socket and connection-oriented socket server can configure the local information by calling the bind function. After binding function associates socket to a port, the port service request can be listened.
3  Connection establishment: Connect function starts and directly connects to a remote host. The connection-oriented client socket needed to use this socket is connected with the remote host. Connectionless protocol never establishes a direct connection, and it only listens to the port in the agreement and the customer requests passively. Listen function makes socket in a passive listening mode, and it can establish an input data for the socket queue, and stores the arrival of service requests in this queue until the program begin to deal with them.

Accept 0 function makes server to receive client requests. After the establishment of input queue, the server calls accept function and then sleeps and waits for the client connection requests.

4  Data transfer: The two functions of Send ( )and recv ( )are used for connection-oriented socket to

transfer data. Function prototype of send ( ) is: int send(int sockfd, const void *msg, int len);
Function prototype of recvOis:
int recv(int sockfd,void *buf,int len);
5  Ending transfer:When all the data operation is ended, the close 0 function will be called to release the socket, thereby to suspend the operation of any data on the socket.

The shutdown 0 function can be called to close the socket. This function only can stop data transferring in one direction, and another direction of data transfer can be continued.

## 5   APPLICATION OF THE SYSTEM IN VEHICLE CHASSIS SCANNING SYSTEM

After the images and data acquisition system hardware and software design are completed, server-side receiver of the PC-side and image display interface are programed, and then the image data will be sent to the server via LAN and  images underneath the vehicle will eventually be displayed on the PC side. The actual image of vehicle is shown in Fig. 5.

Figure 5.   The actual image of the vehicle.

The system's image resolution is 2700 points per frame, and each pixel is 2 bytes, and the network transfer speed is100Mbps; Through field tests, when the speed is less than 30kmh, the system can clearly and completely display the images underneath the vehicle.

## 6   CONCLUSION

The paper introduces the linear array CCD camera, the overall structure of the network, and analyses some of the features and implementation methods and focuses on the description of a linear array CCD driving circuit design process. CPLD-based linear array CCD drive circuit greatly simplifies the application of linear array CCD circuit. Through the modification process, the sampling rate of the CCD chip and the exposure time can be changed to enhance system flexibility and increased system stability. What's more, the appropriate modifications of drive circuit can be applied to other kind of signal of CCD chips.

The ARM-embedded system can be easily transplanted into the embedded operating system, easily to be added with TCP/IP and other network communication protocols. The entire camera system using modular design concept, is able to get the different resolutions of images by hardware and software changes only. Currently, the system has been successfully integrated in the vehicle chassis scan system. In addition, through the appropriate changes, it can be applied into industrial inspection, traffic monitoring and other occasions too.

REFERENCES

[1] Wang, Q.Y., 2000, CCD application technology, Tianjin University Press.
[2] Liu S. & Zhao K.S. & Long Z.C. 2007, Optoelectronics Laser, 18(11), 1296–1298.
[3] Hu, J.H& Xia, Z.F & Shi, C., 2007, Yangtze University Institute of Research, 4(3), 77–81.
[4] Shi, C.& Li, Z.H& Xia, Z.F.,2006,Wuhan Institute of Technology, 28(6), 27–29.
[5] Wei, X.J & Zheng, X.F. & Ding, T.F., 2008, Microelectronics and Computer, 25(8), 232–235.
[6] Peng, G.S., 2007, Electric Power Automation Equipment, 27(1), 87–89.

# Evaluation of impact of various writing tools on individual electro-encephalogram (EEG) characteristics of designer

S.H. Wu & J.H. Chen
*Department of Digital Media Design, College of Creative Design, Asia University, Taichung City, Taiwan*

H.F. Lee
*Department of Industrial Design, Chaoyang University of Technology, Taichung City, Taiwan*

K.S. Yao
*Center for General Education, National Taitung Jr. College, Taitung City, Taiwan*

ABSTRACT: With the advance of technology in recording electroencephalogram (EEG) signals, a physiological indicator of brain activity, researchers attempt to convert it directly into an objective index for evaluation of creativity. To assess the various differences of designer personal EEG, it was observed by using writing brush, marker pen, ball-point pen and pencil for practice activity. EEG was recorded at FP1, T5, T6 and Fpz points according to the international 10 20 system. These results showed that the change of subject brainwaves such as alpha and theta waves was significant different ($P<0.05$). Especially, the effect of beta wave on the right brain was the highest after using a writing brush, and it was even up to 94%. In conclusion, these evidences reveal that the different writing tools have an impact on changing the brainwave of designers'. Particularly, the writing brush and pencil play an important potential to affect the creator's brainwaves.

## 1 INTRODUCTION

As early as in 1875, Richard Carton, a physician in Liverpool, first described the electroencephalogram (EEG) activities of animals' in the world. Subsequently, the measurable electrical signals recorded from the exposed cerebral hemispheres of various animals such as rabbit, monkey and dog have attracted much attention (Swartz 1998). Hans Berger, a physiologist and psychiatrist, succeeded in recording the first human EEG in 1924 (Haas 2003). Until 1938, electroencephalography had been widely recognized by researchers in this field; it has also been applied in practical use in diagnosis in the United States, England and France (Wiedemann 1994). From that moment, understanding the activity of human brain has entered a new stage. Brainwaves are approximately divided into four patterns according to the frequencies, which are in the range of 0.5–3.5 (Delta), 3.5–7 (Theta), 7–13 (Alpha) and 13–30 (Beta) cycles per second (Hz). Nowadays, EEG is a new kind of tool for measuring neurological system signals. Although EEG is more complex, most recently, it has been widely used as a physiological measurement tool for evaluating the changing of mental processes by psychologist and physiologist due to the property of recording brainwave signals in direct (Gevins 1997, Herrmann 1996).

Yepsen (1987) pointed out that the method of biological feedback (also named as EEG research) enabled us to convert thinking. In other words, the frequency signals of brainwave are further processed and encoded by using a brainwave collector, which enable the subjects to feel their own brainwave messages and learn how to trigger certain physiological mechanism by employing a conscious mental process. Squire and Kandel (1999) had published a paper: *Memory: from Mind to Molecules*, which further reported that the brain research on human was the last piece of wild land, and it would be playing an important era for studying the relationship between the human brain and mental in the 21st century.

With the continuous advance of techniques in recording EEG signals, a physiological indicator of brain activity, many researchers attempt to convert the quantitative data of EEG directly into an objective index for assessment of creativity promotion. For example, Chao & Lee (2006) had said, in the pattern of more exerting creativity, there might be a hidden trend, but the mixture proportion of various types of brainwaves include β, α, θ and δ waves should depend on the differences of personal reasons or activities. And it should not be considered that there is a simplistic norm to directly assess the creative capability. Therefore, the research relationship between EEG and design creativity should seem to focus on the personalized and in-depth directions in future. EEG can help designers to understand the relationships between the operation of their intrinsic creative abilities and brain waves by using a collector instrument of human brainwave. It also enables the creators to realize the process of self-understanding faster and deeper. The authors had reported previously that various factors such as visual animation and illustrated creation are able to influence the designer EEG (Wu 2007, Wu & Lee 2007, Wu et al. 2009, Wu et al. 2011). Therefore, the main purpose of this study is to further evaluate the influences of various writing tools such as Chinese writing brush, marker pen, ball-point pen and pencil on the individual brainwave of creators'.

## 2 EXPERIMENT INSTRUMENT AND PROCEDURES

### 2.1 *Experiment instrument*

The equipment of EEG 2000 type (purchased from Mindquest Inc. Taiwan) includes hardware and software provided by the Lab. of Creative and Energy established in Chaoyang University of Technology in Taiwan (Fig. 1). The operation of brainwave collector was conducted in a quiet indoor environment surrounded by interference-free of strong electromagnetic. To rapidly obtain the quantitative data, the electrode positions of EEG (EEG 2000 type) were set up at FP1, T5, T6 and Fpz points according to the user's manual and the international 10-20 system (Oostenveld & Praamstra 2001), as shown in Figure 2.

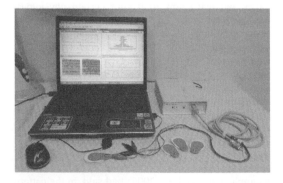

Figure 1. EEG 2000 type machine, a brainwave collector purchased from Mindquest Technology Co. Ltd. in Taiwan. URL http: //www.mindquest.com.tw.

### 2.2 *Procedures*

The experimental procedures in recording designer EEG were previously described by Chao & Lee (2006)

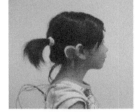

Figure 2. Setup the demo of EEG electrodes, which positions were located at FP1, T5, T6 and Fpz points respectively, according to international 10–20 system (As cited in Wu, 2007).

and Wu (2007). Briefly, the operator had to first keep the feelings of calm, cozy and relax during the tested period. For comparison, keeping an eye on white paper was carried out for 3 minutes. Subsequently, the changes of EEG were observed under this condition by using different writing tools such as Chinese writing brush, marker pen, ball-point pen and pencil for writing practice. After the experiments, all the raw data were further calculated using the method of analysis of variance (ANOVA).

## 3 RESULTS AND DISCUSSION

Neuropsychological researches have demonstrated that the alpha wave of human brainwaves' is in the range of 8 to 12 Hz frequency, which represents an adult existed in the main rhythmic activity under the quiet and arousal condition (Kiloh et al. 1981). Those whose hearts are kept in a quiet condition, $\alpha$ wave of the brain has a gradually promotion trend. However, it rarely appears in that of anxious people, replaced by a more high frequency of $\beta$ wave (Zou 2000).

Based on the phenomenon, the authors would like to further understand the possible relationship between the writing tools and personal EEG of the designer. Various pens such as Chinese writing brush, marker pen, ball-point pen and pencil were tested in this study. These results showed that the relative intensity of $\beta$ wave of the right brain was apparently increased under the condition of various writing tools. And the ranges of promotion percentage were up to 56~94%. Especially, the use of writing brush was the highest and even achieved 94% (Fig. 3).

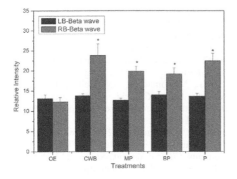

Figure 3. Effect of different writing tools such as Chinese writing brush (CWB), marker pen (MP), ball-point pen (BP) and pencil (P) on the designer brainwaves. The changes of β waves are observed. Data are expressed as the mean relative intensity of brainwave power and compared with the control group (Open eyes, OE; $n$=5). One asterisk (*) represents 0.05 level or statistical significance.

The power of $\alpha$ wave has a slight upward trend at 2~7% in the right brain. On the contrary, it was obviously shorten from 8% to 13% in the left brain (Fig. 4). In other words, various writing tools such as writing brush, marker pen, ball-point and pencil utilized in this study were able to slightly change the capacity of $\alpha$ wave of the human brain. These experimental results seemed to be similar to the report previously described by Chao (2007), which pointed out that β and $\alpha$ waves would be affected as the subjects were engaged in activities like turning the head or creating a design. Moreover, lots of research also revealed that the intensities of beta and alpha waves will be suppressed when the loadings of human memory activities are increasing (Gevins et al. 1997, Sterman et al. 1994).

The power of $\theta$ and $\delta$ waves in the right brain also had clearly demotion phenomenon, which were from 4% to 28% (Figs 5, 6). Among those conditions, the use of writing brush and pencil for writing practice enabled the $\theta$ waves to reduce 18% and 16%, whereas the $\delta$ waves were 27% and 28% respectively. Moreover, the power of $\theta$ wave in the left brain was gradually promoted at the range of 8% to 13%. Gundel & Willson (1992) also reported that the intensity of $\theta$ waves depends on the difficulty of the work. Therefore, the upward phenomenon of brainwaves' may be related to the loadings of mental tasks or attention activities (Gundel & Willson 1992).

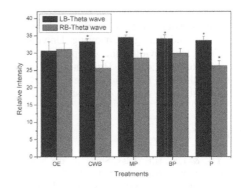

Figure 5. Effect of different writing tools such as Chinese writing brush (CWB), marker pen (MP), ball-point pen (BP) and pencil (P) on the designer brainwaves. The changes of $\theta$ waves are observed. Data are expressed as the mean relative intensity of brainwave power and compared with the control group (Open eyes, OE; $n$=5). One asterisk (*) represents 0.05 level or statistical significance.

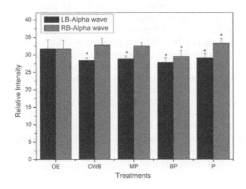

Figure 4. Effect of various writing tools such as Chinese writing brush (CWB), marker pen (MP), ball-point pen (BP) and pencil (P) on the designer brainwaves. The changes of $\alpha$ waves are observed. Data are expressed as the mean relative intensity of brainwave power and compared with the control group (Open eyes, OE; $n$=5). One asterisk (*) represents 0.05 level or statistical significance.

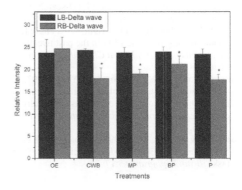

Figure 6. Influence of various writing tools such as Chinese writing brush (CWB), marker pen (MP), ball-point pen (BP) and pencil (P) on the designer brainwaves. The changes of $\delta$ waves are observed. Data are expressed as the mean relative intensity of brainwave power and compared with the control group (Open eyes, OE; $n$=5). One asterisk (*) represents 0.05 level or statistical significance.

# 4 CONCLUSIONS

In conclusion, these evidences demonstrate that the change of personal EEG of designer may be related to the various writing tools used in the creation work. Especially, the writing brush and pencil play an important potential to affect the alpha and theta waves in creators' in this research. Therefore, it is very important for those people engaged in the creation/design to choose a suitable writing instrument for their own work for enhancing the creation efficiency in future.

# ACKNOWLEDGMENTS

The authors would like to deeply thank the Lab. of Creative and Energy, Chaoyang University of Technology for providing the equipment of brainwave, Mr. Ren-Yu Lee, Ms. Hui-Chu Chen and Ms. Tiffany for their aid in recording the EEG. The help was essential for the success of this study.

# REFERENCES

Chao, Y.S. & Lee, H.F. 2006. Study on the brain wave measures in design education – In the case of students enrolled at 2004 academic year in department of visual communication design at Chaoyang University of Technology. *Conference of technology and teaching held at Ming-Chi University of Technology, Taipei, 11 November 2006.* (In Chinese)

Chao, Y.S. 2007. A Preliminary Pilot Study on the brain wave measures in design education –In the case of ten students in department of visual communication design at Chaoyang University of Technology (Unpublished master's thesis). Retrieved from http://ethesys.lib.cyut.edu.tw/ETD-db/ETD-search/view_etd?URN=etd-0827107-114203. (In Chinese)

Gevins, A. 1997. Hans Berger was right: What I have learned about thinking from the EEG in the past twenty years. *Electroencephalography and Clinical Neurophysiology 103*(1):5–6.

Gevins, A., Smith, M.E., McEvoy, L. & Yu, D. 1997. High-resolution EEG mapping of cortical activation related to working memory: Effects of task difficulty, type of processing, and practice. *Cerebral Cortex* 7(4): 374–385.

Gundel, A. & Wilson, G.F. 1992. Topographical changes in the ongoing EEG related to the difficulty of mental tasks. *Brain Topography* 5:17–25.

Herrmann, N. 1996. *The whole brain business book.* McGraw Hill Professional Press, New York pp.334. ISBN:978-0070284623.

Haas, L.F. 2003. Hans Berger (1873–1941), Richard Caton (1842–1926) and electroencephalography. *Journal of Neurology, Neurosurgery & Psychiatry* 74 (1):9.

Kiloh, L.G., McComas, A. J., Osselton, J.W. & Upton, A.R.M. 1981. *Clinical electro encephalography.* Butterworth-Heinemann Press, London pp.292. ISBN: 978-0407 001602.

Oostenveld, R. & Praamstra, P. 2001. The five percent electrode system for high-resolution EEG and ERP measurements. *Clinical Neurophysiology* 112:713–719.

Squire L.R. & Kandel E.R. 1999. *Memory: from Mind to Molecules.* Scientific American Library, New York (Henry Holt and Company, Repress in 2003) pp.254.

Sterman, M.B., Kaiser, D.A., Mann, C.A., Suyenobu, B.Y., Beyma, D.C. & Francis, J. R. 1994. Application of quantitative EEG analysis to workload assessment in an advanced aircraft simulator. *Proceedings of the human factors and ergonomics society* 1:118–121.

Swartz, B. E. 1998. The advantages of digital over analog recording techniques. *Electroencephalography and Clinical Neurophysiology* 106 (2):113–117.

Wiedemann, H.R. 1994. Hans Berger. *European Journal of Pediatrics 153*(10):705.

Wu, S.H. & Lee, H.F. 2007. Application of electroencephalogram (EEG) measurement on the creativity design education: The effect of illustrated creations on personal EEG characteristics of designer. *Proceedings of international conference of art design and science,* pp. 238–245. *Taipei, 23 July 2007.* (In Chinese)

Wu, S.H. 2007. Effects of writing practice with a brush dampened ink of four print primary colors on the EEG activities of designer (Unpublished master's thesis). Retrieved from http://ethesys.lib.cyut.edu.tw/ETD-db/ETD-search-c/view_etd?URN=etd-0802107-104359. (In Chinese)

Wu, S.H., Lee, H.F. & Ma, K.Y. 2009. Application of electroencephalogram (EEG) measurement on the creativity-design education: The effect of color-elements of design on personal EEG characteristics of designer. *MingDao Journal of General Education* 6:253–266. (In Chinese)

Wu, S.H. & Lee, H.F. 2011. Application of electroencephalogram (EEG) measurement on the creativity design education: The effect of basic geometric vision-elements on personal EEG characteristics of designer. *Journal of National Taitung College* 1:1–9. (In Chinese)

Yepsen, R.B. 1987 *How to Boost Your Brain Power: Achieving Peak Intelligence,* Memory *and Creativity.* Rodale Press. Emmaus, Pa. (ISBN: 978-0878576531)

Zou, M.D. 2000. An exploration of bio-feedback technique of EEG-alpha rhythm. *Beijing Biomedical Engineering* 19(1):38–41. (In Chinese)

*Multimedia, Communication and Computing Application – Leung (Ed.)*
© *2015 Taylor & Francis Group, London, ISBN 978-1-138-02775-6*

# Application of Radio Frequency Identification technology in information system for warehouse management

S.Y. Zhu, J.Y. Liu, L.H. Zhao & X.Y. Wang
*Logistics Department, College of Automation, Beijing Union University, China*

ABSTRACT: In order to meet the increasing requirements of reducing manual work intensity and improve the work efficiency in warehouse management of manufacture enterprises, advanced information technologies focus on the warehouse storage management flow. We have applied Radio Frequency Identification (RFID) technology in the warehouse management and developed a Management Information System system (MIS) which integrates with the data acquisition system based on RFID and Office Automation (OA) system for storage and inventory operation. The information on RFID is designed for the management purpose. Interaction between the data acquisition system based on RFID and the OA system is conducted with relational databases. The MIS is designed with Microsoft Visual Studio 2010 and Lotus Domino Designer and has been tested. Results have shown that the MIS can collect and manage the stored product information automatically and work efficiently by incorporating the field operation and the senior management, and that it can provide an effective solution for RFID application and system integration in the warehouse management.

## 1 INTRODUCTION

With the rapid development of modern economy and society in China, new requirements are put forward for the manufacture enterprises to improve work effectiveness and efficiency in warehouse management. In the past, traditional enterprises always use manual approach and bar code to manage the product. However, the low efficiency with high error rates often happens in the manual approach. The visibility, limited information and the vulnerable characteristics have restrained the application of bar code, especially in warehouse management. By contrast, RFID can provide much better services for management, such as sending information automatically and no need to check products one by one for efficiency. Moreover, the RIFD tag can carry much more information than the barcode, such as the expiry date, product number, storage position, production information and so on. If a bar code carries so much information, larger size may result in failure to read.

The RFID technology has been applied in many fields, such as agriculture, industry, daily life and military affairs. Poon et al (2011) have studied the application of RFID in small batch replenishment, and concluded that the proposed system using RFID technology and GA mechanism has provided effective and efficient replenishment routes for the production floor and warehouse. Moreover, compared with the traditional manual approach, RFID/GA approach is clearly superior. Lim et al (2013) have pointed out that studies showing linkage between RFID benefits can strengthen the case for RFID adoption. RFID solutions are treated as a closed loop system and its benefits are less exploited across the supply chain, in spite of recommendations in the earlier works. Chow et al (2006) have studied RFID applied in the management system for warehouse operations and pointed out that automatic data identification RFID technology can help to provide a high-level of logistics service satisfaction in resource planning and execution. Poon et al (2009) have studied management system using RFID to manage order-picking operations and realized that RFID can be used in different types of warehouse and give out references to help enterprises to select the most appropriate RFID equipment. Dong et al (2009) have studied the major technology of warehouse management system based on RFID and using RFID for the overall architectural design, and they have designed the hardware which has the following modules: the main control module, electronic tag reading and writing module, memory and operating clock module, LCD display module. Yang et al (2011) have studied and designed the in-out and inventory process management by using RFID, and analyzed the C/S and B/S architectures of information system.

In China, RFID is not used efficiently and the advantages compared with bar code are not shown sufficiently during warehouse management. Many enterprises only use the RFID tag as the product ID for management. Also, there exist fusion, integration problems between the OA system and data acquisition systems. To work out an effective solution, we have designed the management information system

(MIS) by using RFID and combining with OA for the warehouse storage management.

This paper is organized as follows: first, we introduce the system design; then the developed MIS is presented and tested; finally the results are discussed.

## 2 WAREHOUSE MANAGEMENT INFORMATION SYSTEM DESIGN WITH RFID

### 2.1 *Warehouse management information system configuration based on RFID*

The MIS configuration based on RFID technology is shown in Figure 1. With RFID technology, the product information can be collected automatically and quickly. The information system is generally divided into two parts: the automatic data acquisition system (hardware) and PC software system (software).

The automatic data acquisition system is composed of RFID reader and tags with working frequency of

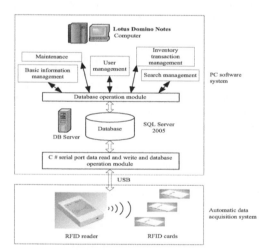

Figure 1. Information system configuration based on RFID.

13.56MHz based on ISO/IEC 14443 A. When the tag enters the working range of the reader, it obtains energy from the reader and sends its information to the RFID reader, which then is sent to the PC through the USB interface.

The PC software system consists of two parts. One is a software module developed with C # which collects the tag information and stores it in the database established on SQL Server 2005. The other one is the automated system developed in Lotus Domino Notes with which the warehouse management process is realized based on RFID. It includes the Inventory transaction management, Database operation, User management, Maintenance, Basic information

management and Search management modules. The Database operation module is used to read RFID tag information and manage generated information during warehouse management with the automatic system through SQL Server 2005.

### 2.2 *RFID tag information design*

After production, the RFID tags are written with information about product code, product number and expiration date, as is shown in Figure 2. This information is used during the warehouse management process for collecting stored product information automatically and efficiently. The tag is fixed on the dispatch unit.

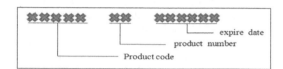

Figure 2. Information in the RFID tags.

### 2.3 *Inventory control*

Inventory control is essential to manage manufacturing and storage. To design the information management for warehouse management, we have to determine the inventory demand of safety stock which provides a reference for managers in response to the market and reduce the waste of funds and materials. To get the safe stock, demand forecasting should be made first. Methods, to forecast the demand, can be classified as: time classification and quantitative/qualitative classification. For a manufacturing enterprise, we have decided to choose the way of short-term forecasting by adopting the method of combining with qualitative analysis and quantitative analysis.

The safety stock can be calculated as

$$SS = z\sqrt{\sigma_D^2 \overline{L} + \overline{D}^2 \sigma_L^2} \qquad (1)$$

where, $\sigma_D$= the standard deviation of demand in lead time; $\sigma_L$ = the standard deviation of lead time; $\sigma_L$ = average daily demand during lead time; $\overline{L}$ = the average time of lead time; $z$ = safety factor.

Safety stock has three types: the demand is changeable and the lead time is fixed; Lead time changes, and the demand is fixed; both lead time and demand change randomly. Enterprises can choose or alter strategies depending on different situations. Here we suppose the lead time is fixed and the demand is changeable. The safety stock is calculated with the formula (1) and if the inventory is lower or higher than the safety stock, there will be an alarm through the information system. In the process of inventory control, we manage the inbound and outbound process with the principle of "first-in first-out".

## 2.4 *Analysis of warehouse management workflow*

The warehouse management workflow with RFID is shown in Figure 3. The MIS is designed accordingly. In inbound process, RFID reader reads the tag information automatically and then a storage plan is made and submits to the audit management group who determine whether to store the products according to the factors such as customer's order and safety stock. If the product is not allowed to store, the inbound process is stopped and the products will be waiting for further instructions: customer emergency orders; customer need product, but inventory is insufficient and cannot allocate location effectively. Except for the above situations, products will be stored in a designated position. The inventory database is updated in time to ensure the real-time change, accuracy and validity of data. In outbound process, considering the customers' demands, warehouse group will submit a dispatch plan. The management group decides whether to dispatch it or not depending on the inventory situation. In this process, outbound products are executed by following the principle of "first in - first out". After review from management group, the warehouse group will generate an outbound order timely. In this process the RFID technology is used to check if the outbound products are correct. If there is no mistake, the outbound will be carried out and the database is updated.

## 2.5 *Relational database design with SQL server 2005*

In this paper, there are four main entities: product, department, warehouse and user. The E-R design is shown in Figure 4. In inventory management, the four entities are indispensable and will be fully used. They support the normal establishment and management of warehouse.

The two-dimensional relational tables are designed as follows, according to the E-R diagram:

1  Product information table (product id, product name, on-hand quantity, outbound quantity, product specifications, the product price, amount, positions, expire date, storage time, the delivery time, handlers)
2  The staff table (user id, name, gender, department number, login password)
3  Department table (department id, department name)
4  The warehouse information table (warehouse id, name, size)
5  Inventory list (inventory id, product number, product name, the actual inventory, original data quantity, expire date)

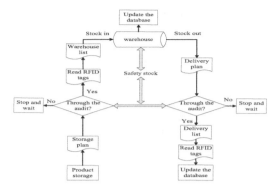

Figure 3.   Operation flow.

6  Inbound schedule table (inbound plan id, product number, product name, inbound quantity, price, sum, expire date, state)
7  Inbound order (warehousing list id, product id, product name, quantity, price, sum, handlers, positions, expire date, storage time, state)
8  Outbound schedule table (outbound plan id, product id, product name, outbound quantity, price, sum, positions, expire date, state)
9  Outbound order (outbound order id, product id, product name, outbound quantity, price, sum, handlers, position, expire date, outbound date, state)

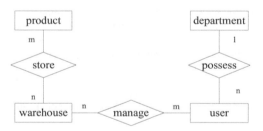

Figure 4.   E-R Design.

## 3   SYSTEM TEST AND DISCUSSION

Based on the above system design, the MIS based on RFID is developed and tested. Data acquisition is based on RFID.

The data acquisition system can read and write the RFID tags automatically. The hardware is shown in Figure 5 which consists of RFID reader, tags, and the USB cable and interface.

### 3.1 *C# data collection and storage software*

The data collection and storage software developed with C# in Microsoft Visual Studio 2010 is shown in Figure 6, which can collect and download data through the serial communication port-USB interface, and it can control the read and write on tags. It also communicates with the SQL server database.

Figure 5. Hardware of the data acquisition system.

Figure 6. Data collection and storage software interface.

### 3.2 The automatic system software

The automatic system software is developed in Lotus Domino Notes. Different access purviews are designed for different roles. For an administrator, who has the highest purview, the designed interface is shown in Figure 7.

On the left of interface, it includes two modules: form and view action. Managers can manage user, department, warehouse, product information and

Figure 7. The administrator interface.

Figure 8. User management interface.

so on. Also, managers can search product information by using the inventory search function. The users with other roles only have limited access to the system management function. Taking the user management function, for example, its interface is shown in Figure 8 on which the manager can add, update or delete a user.

We have tested the above MIS based on RFID. The results show that this system can work effectively.

## 4 CONCLUSIONS

This paper designs an MIS for warehouse management based on RFID technology. In China, the RFID is usually used as an ID tag instead of an information carrier for different management purposes. We include the product code, product number and expire time in the RFID tags for warehouse management. Also, there exist fusion and integration problems between OA system and data acquisition system. That is why we use the Lotus Domino Notes to design the information system software. System test shows that the system can work effectively. The proposed information system based on RFID provides one referred solution for modern warehouse management of manufacture enterprises.

ACKNOWLEDGMENT

This work is partially supported by "New Start" Academic Research Projects of Beijing Union University (ZK10201401).

REFERENCES

[1] Dong L. & Sun, L. 2009. Research on the major technology of warehouse management system based on RFID. Microcomputer Information 25(2): 202–203, 209.
[2] Chow, H.K.H., Choy, K.L., Lee, W.B. & Lau, K.C. 2006. Design of a RFID case-based resource management system for warehouse operations. Expert Systems with Applications 30:561–576.
[3] Lim, M.K., Witold, B. & Stephen, C.H.L. 2013. RFID in the warehouse: A literature analysis (1995–2010) of its applications, benefits, challenges and future trends. Int. J. Production Economics 145:409–430.
[4] Poon, T.C., Choy, K.L., Chow, H.K.H., Lau, H.C.W., Chan, F.T.S. & Ho, K.C. 2009. A RFID case-based logistics resource management system for managing order-picking operations in warehouses. Expert Systems with Applications 36:8277–8301.
[5] Poon, T.C., Choy, K.L., Chan, F.T.S., Ho, G.T.S., Gunasekaran, A., Lau, H.C.W. & Chow, H.K.H. 2011. A real-time warehouse operations planning system for small batch replenishment problems in production environment. Expert Systems with Applications 38:8524–8537.
[6] Yang, Y. & Hu, K. 2011. Analysis and design of Logistics & Storage management system basedon RFID. Modern Electronics Technique 34(22):200–201, 210.

*Multimedia, Communication and Computing Application – Leung (Ed.)*
© 2015 Taylor & Francis Group, London, ISBN 978-1-138-02775-6

# A remote user authentication method based on the voiceprint characteristics

H.W. Wang & J.J. Li
*China Tobacco Zhejiang Industial Co., Ltd., Hangzhou, Zhejiang, China*

ABSTRACT: Speaker recognition is one of the biometric identification technologies. Personality characteristics extracted from the speaker's voice can be used to distinguish different speakers. Therefore, the speaker recognition can be widely applied in the field of information security. However, since the voice signals are often affected by the remote environment, mismatch and other potential factors, the recognition accuracy is not ideal. The speaker recognition technology is rarely applied to remote authentication systems in real environment now. This paper proposes a remote user authentication method based on human-machine characteristics and the GMM - UBM system. This system used Mel-Frequency Cepstral Coefficients (MFCCs) and pitch frequency characteristics to enhance the robustness of speech recognition system. At the same time, the machine features and speech features are banded in the recognition process to improve the recognition accuracy. The method has achieved a good experiment result in a real environment.

## 1 INTRODUCTION

Nowadays, the congenital safety defect of traditional identification methods, such as keys, passwords and certificates becomes more highlighted. And a single short password is vulnerable to be attacked and cracked. Voiceprint recognition for smart phone users has obvious advantages [1-2]. Users won't remember passwords and operate easily. But, the disadvantage is the limited accuracy, which is less than 90%. In this paper, the combination of GMM-UBM system and machine features has higher recognition accuracy. Using logarithmic likelihood ratio can reduce the influence of the threshold. We will use the log-likelihood ratio as our scoring criteria to estimate performance of the system. At the same time, the user s' biological features and terminal machine characteristics are bounded. The recognition threshold will be affected by the machine features which are delivered in SSL channel. In other words, if the machine features were certified and passed in the earlier step and the result will react to the threshold. Threshold can decrease slightly and the higher false probability can be allowed to improve the accuracy.

## 2 SIGNAL PRE-PROCESSING AND MIXTURES OF GMM-UBM

Speaker recognition, which is also known as Voiceprint recognition, is an important technology in the field of information security [3-5]. Unique personality characteristics extracted from the speaker's voices can be the theoretical basis to distinguish the different speakers quite well. In the whole process of speaker recognition, the most important thing is the selection and confirmation of the speaker characteristic parameters. We will adopt the GMM - UBM system, and the confirmation of model's mixed number is the focus of this article.

### 2.1 *Signal pre-processing*

The acquisition of voice signals is mostly under the noisy environments, so speech samples must be pre-filtered. In this paper, we use wiener filter for signal pre-processing. After filtering, the non-silent segments of the voice signal need to be an endpoint detected before the extraction of pitch frequency. Voice endpoint detection means to confirm the location of the starting point and end point of non-silent segment. The results are shown in Fig. 1.

The role of the pre-processing is to extract from the speech signal features that convey speaker-dependent information. In addition, techniques to

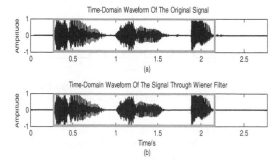

Figure 1. The endpoint detection technology can determine the boundaries of the non-silent segments.

minimize confounding effects from these features, such as linear filtering or noise, may be employed in the pre-processing. In our paper, a dual-threshold method was used.

## 2.2 The mixtures of GMM-UBM

In addition, the mixtures of Gaussian components are very important parameters. Theoretically, the higher number of Gaussian components can fit the Gaussian, the better the distribution function is. Actually, the optimal mixed value is related to the scale of experimental data. In a word, the selection of mixtures is cautious and will be adjusted. The mixtures will show through DET curves in later experiments.

## 3 PITCH DETECTION AND MFCCS FOR GMM-UBM SYSTEM

In this paper, we will extract the pitch frequency and MFCCs from voice signals as personality characteristics. The pitch period is one of the most important parameters of the speech signals, and be regarded as an important feature that reflects the character of the excitation source. It will be used to distinguish the gender roughly but effectively.

### 3.1 Overview of pitch detection

We often use next three physical characteristics to describe speech signals: loudness, pitch and timbre. The pitch detection results are shown in Fig. 2.

The vibration period of the vocal cords is defined as the pitch period, whose reciprocal is called pitch frequency. Since the shape of the human's vocal tract is easy to change, and the possible scope of pitch may

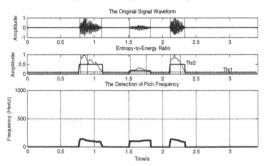

Figure 2. The pitch detection method can extract the pitch frequency from the speeches. The mean of pitch values can distinguish the gender roughly.

be broad. In addition, the pitch frequency is easily influenced by human emotion and intonation. All of these uncontrollable factors make the estimation of pitch hard. In this paper, our algorithm joined with pitch frequency is beneficial to improve our recognition rate in noisy environment.

### 3.2 Mel-frequency cepstral coefficients

The voice signals after wiener filter must be the pre-emphasis before the extraction of MFCCs. The pure voice signal will go through a high-pass filter, and the high frequency part of the signal will be raised. We can choose a simple FIR filter to achieve the process. The software implementation is as follows:

$$H(z) = 1 - a \cdot z^{-1} \quad (1)$$

Where $a$ is a constant close to 1, here $a$ =0.9375.

MFCCs are proposed based on the auditory mechanism, and can veritably reflect the auditory features.

MFCCs can simulate the perceptive parameters that are based on the Mel Frequency. The MFCCs is proposed first by Davis and Mermelstein in 1980s. Before it, the LPCS and LPCCs were regarded as the main method and widely used in the study. The Mel frequency and the actual frequency are linked in the following formula.

$$m = 2595 \log_{10}\left(1 + \frac{f}{700}\right) \text{ Or } m = 1127 \log_e\left(1 + \frac{f}{700}\right) \quad (2)$$

The above formula will convert Hertz to MEL.

### 3.3 GMM-UBM verification system

Gaussian Mixture Models are the basic components of the GMM - UBM system. A GMM is used in speaker recognition applications as a generic probabilistic model for multivariate densities capable of representing any densities, which is suitable for unconstrained and text-independent applications. The use of GMMs for text-independent speaker identification was introduced early in [6-7]. A further extension of the GMM-UBM system to it was described and evaluated on some speech corpora in [8]. A very important step in the implementation of the above likelihood ratio is the selection of the actual likelihood function $p(X|\lambda)$. The choice of this function is mainly dependent on the features that are used as well as specifics of the application. For example, in text-dependent applications, where there is strong prior knowledge of the spoken text, additional temporal knowledge can be incorporated by using hidden Markov models (HMMs) as the basis for the likelihood function [9]. For a $D$-dimensional feature vector, the mixture density used for the likelihood function is defined as:

$$p(x|\lambda) = \sum_{i=1}^{M} \omega_i p_i(x) \quad (3)$$

168

The density is a weighted linear combination of $M$ unimodal Gaussian densities, $p_i(x)$, each parameterized by a mean $D \times 1$ vector, $\mu_i$, and a $D \times D$ covariance matrix, $\Sigma_i$.

The mixture weight $\omega_i$, which satisfies $\sum_{i=1}^{M} \omega_i = 1$

Collectively, the parameters of the density model are denoted as $\lambda = \{\omega_i, \mu_i, \Sigma_i\}$, and where $i = 1, \ldots, M$. While the general model form supports full covariance matrices, we use only diagonal covariance matrices. The system is referred to the Gaussian Mixture Model-Universal Background Model (GMM-UBM) the speaker detection system.

Often, we used $p(X|\lambda)$ instead of $P(\lambda|X)$ to describe the output probability of GMM. A log-likelihood ratio method had been used by Higgins to calculate the model score in 1991. The log-likelihood ratio is computed as follows:

$$L = \frac{p(X|\lambda)}{p(\lambda|X)} \qquad (4)$$

Where $L$ is the score of GMM-UBM model, and will be used to compare with the threshold value. And is the parameter of tester's GMM-UBM model. And is the parameter of target tester's GMM-UBM model. Often, the logarithm of this statistic is used to give the log-likelihood ratio:

$$\Lambda(X) = \log(p(X|\lambda)) - \log(p(X|\bar{\lambda})) \qquad (5)$$

### 3.4  GET and the confirmation of the threshold

A single, speaker independence, threshold is swept over the two sets of scores and the probability of miss and probability of false alarm are computed for each threshold. The error probabilities are then plotted as detection error tradeoff ($DET$) curves to show system performance[10]. We concern two probabilities for $DET$: False Alarm Probability ($FA$) and Miss Probability ($FR$).These two kinds of error probabilities are not only related to the characteristic parameters and system modeling, but also are related to the threshold value. FA and FR curve have an intersection, where is called the Equal Error Rate (EER). $DET$ is more intuitive to find an intersection. Machine characteristics were verified before speaker verification, once the verification of machine characteristics was passed, the threshold can adjust slightly to the direction where the FR is lower. In other words, the threshold can be lower than the recognition rate when the machine characteristics are legal and correct. The new threshold can be:

$$\Lambda(X) = \log(p(X|\lambda)) - \log(p(X|\bar{\lambda})) + \eta \qquad (6)$$

Here $\eta$ is the correction coefficient.

The mixture of speaker recognition model and machine characteristics has a good recognition rate in a real environment. The machine characteristics which must be unique can be the hard disk serial number, the memory serial number, Media Access Control Address ($MAC$), etc.

## 4  EXPERIMENTAL RESULTS AND ANALYSIS

### 4.1  Experiment of pitch frequency

In this section, we will present experiments and results using the adaptive GMM-UBM system. Experiments are conducted on private corpora. The first voice corpora are recorded through a mobile phone. There are 50 testers in total. Each person reads 10 sentences in Chinese. And each sentence lasts for about 5 seconds. We do the record work in the laboratories, classrooms, and the playground where are noisy. We use this voice database to extract the pitch frequency in average.

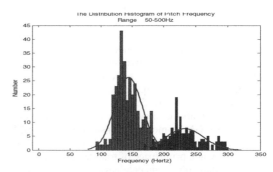

Figure 3.  The distribution histogram of pitch frequency.

The experimental result shows that the men's pitch frequency is generally lower than the women's. So the pitch frequency can be used to distinguish the gender roughly.

### 4.2  Experiment about the effect of mixtures

The experiments presented here show the general effects on the performance of various components and parameters of the GMM-UBM system. The second corpus adopted in experiments is also recorded through a mobile phone. There are 36 volunteers in total. Each person reads 10 sentences in Chinese. And each sentence lasts for about 10~15 seconds or so. The original signals' sampling frequency is 44100 Hz, which is changed to 16 kHz in our experiments. All the voice signals are pre-processed and framed. Then, we extracted 20-order MFCCs. We usually discard the first parameter of MFCCs which repents the $DC$

component. In our paper, we used GMM-UBM system as our speaker recognition model. The selection of the mixtures of GMM or UBM that we had discussed above is a hard work. So we assume the UBM and GMM have the same mixture coefficient firstly. Here is *DET* curve for UBM-GMM system with the same mixtures. The results are shown in Fig. 4.

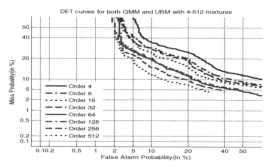

Figure 4. DET curves for systems using both GMM and UBMs with 4–512 mixtures.

Actually, the mixtures of UBM are restrained by the scale of training date. So we need to further confirm in order to find optimal mixtures both of UBM and GMM model. Let the mixtures of UBM is 64, and change the GMM's mixtures in order to find an optimal value. We always hope that the UBM has large mixtures of his adequate experimental data. The results are shown in Fig. 5.

The curve above shows the ERRs of different mixtures of GMM models. Considering the UBM represents all voice print space, while GMM just represents all voice print space of training data. And the training data can't adequately cover the tester's print space. The best mixtures will be used in our system.

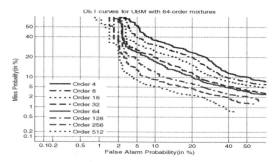

Figure 5. DET curves for systems using GMM with 16–2048 mixtures.

## 5 CONCLUSIONS

This paper firstly introduced the pitch frequency and the extraction of the MFCCs as the speakers' personality characteristics. We have described the major elements of the GMM-UBM system used for high-accuracy speaker recognition. The GMM-UBM system is operated around the optimal likelihood ratio test for detection, using simple but effective Gaussian mixture models for likelihood functions, a universal background model for representing the competing alternative speakers, and a form of Bayesian adaptation to derive hypothesized speaker models. At the receiving end, the whole voice signals are pre-progressed. Then we extracted the pitch frequency and MFCCs as the speaker personality characteristics to train GMM-UBM model. The machine characteristics are verified before the signal processing, and the result will react to the threshold. The probability of miss will decrease by adjusting the threshold, and ignore the influence of the probability of false alarm deliberately. The verification technology of machine characteristics has been very perfect. While the GMM-UBM system has proven to be very effective for speaker recognition tasks, there are several open areas where future research can improve or build on from the current approach.

## REFERENCES

[1] J. Portel et al. 2013, Speaker Verification using Secure Binary Embeddings, in Proc. Eusipco, Marrakech, Morocco.
[2] M. Pathak & B. Raj 2012, Privacy-Preserving Speaker Verification as Password Matching, in Proc. ICASSP, Kyoto, Japan.
[3] Selva N.S. & Selva K. R. 2013, Language and Text-independent Speaker Identification System Using GMM [J] WSEAS Trans. on Signal Processing.
[4] Reynolds D.A. & Roserc 1995. Robust Text-independent Speaker Identification Using Gaussian Mixture Speaker Models [J] IEEE Transactions on Speech and Audio Processing.
[5] Rose, R.C. & Reynolds, D.A. 1990, Text-independent speaker identification using automatic acoustic segmentation, in Proceedings of the International Conference on Acoustics, Speech, and Signal Processing.
[6] Reynolds, D. A. 1994, Experimental evaluation of features for robust speaker identification, IEEE Trans. Speech Audio Process.
[7] Reynolds, D.A. 1995, Speaker identification and verification using Gaussian mixture speaker models, Speech Commun.
[8] Reynolds, D.A., Automatic speaker recognition using Gaussian mixture speaker models, Lincoln Lab. J. 8, 1996, 173–192.
[9] G.E. Kopec 1986., Formant tracking using hidden Markov models and vector quantization, IEEE Trans. Acoust., Speech, Signal Processing.
[10] Martin, A. et al. 1997, The DET curve in assessment of detection task performance, in Proceedings of the European Conference on Speech Communication and Technology.

*Multimedia, Communication and Computing Application – Leung (Ed.)*
*© 2015 Taylor & Francis Group, London, ISBN 978-1-138-02775-6*

# The design and implementation of topic detection based on domain ontology

Q.H. Hu & Y. Luo

*Telecommunications Engineering with Management, Beijing University of Posts and Telecommunications,*
*Beijing, China*

ABSTRACT: The imprecise of keywords is considered to be a common problem with traditional method. There is a new approach combined with domain ontology. This new approach can improve the accuracy of tough match. The category in certain degree of the semantic information is achieved through domain ontology. The ontology enriched literal information between terms so that documents share the same literal meaning. In clustering process, "equal relation" and "hyponymy relation" are two clustering standards set by ontology to build a words mapping. After the mapping existing, an evaluation between traditional method and domain ontology was implemented based on K-means clustering algorithm. At the final step, the experimental results showed that on otology method had better performance against the traditional method.

KEYWORDS: Topic detection, Domain ontology, Evaluation

## 1 INTRODUCTION

Owning to the development of the Internet and the explosion of media information, it looks to be quite difficult for people to read all news and stories produced every day. A suitable choice is intended on some preferable topics. Therefore, figuring out potential topics automatically from a lot of documents becomes a very interesting and meaningful task.

Numerous academic communities have noticed this fundamental problem and defined it as Topic Detection. A traditional and popular architecture is set to divide this task into two major processes. The first step is to cluster all target documents into different groups taking into account their similarities, which are named "clustering process." The second step is to produce potential topics on the summary of each group, which is called "summarizing process." Figure 1 shows this architecture.

In the two processes above, the clustering process acts as a major and critical role. In fact, how to summarize a topic is a very subjective behavior that different people with different understanding abilities will have different approaches. For example, some readers may be helpful at generalization, so they may represent a topic as a set of keywords. But some other people may prefer to represent a topic with several vigorous sentences or even a few paragraphs. Therefore, it is hard to evaluate the summarizing process.

Although a topic's representation is diverse, it has only one theme and one major idea, which means that

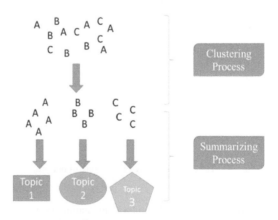

Figure 1. Two processes to detect topics—clustering process and summarizing process.

all of its related documents should share the same internal semantic meaning. Thus, a summary label (such as topic 1, 2 ... n) can be assigned to each topic and it becomes easier for readers only to judge which topic a document should belong to. Namely, it is possible to evaluate the clustering process. Thus, the improvement of the clustering process will be primarily concentrated on in the following paper.

Traditional clustering methods are on the basis of textual feature vectors consisting of a set of keywords and corresponding weights. These keywords are directly extracted from documents. Therefore, any two keywords will be judged as distinct as long as they are not equal to each other literally. So words such as "Bing Gan" (the abbreviation of "hepatitis C") and "Bingxing Ganyan" (hepatitis C) will be regarded as different ones. Another problem arises when the topic is very general, for example, to "epidemic diseases," documents on "SARS" and "H1N1" should have been clustered together. However, "SARS" and "H1N1" will be regarded as irrelevant words due to their inequality on the literal level. All of above problems are noticed as the loss of semantic information. Obviously, it should be avoided during clustering.

Next, the ontology method is introduced to improve the traditional clustering process. Ontology is a term coming from philosophy, but in common words it can be regarded as a set of concepts and there are various relations between them. Generally, ontologies are indicated for a particular domain. With the help of the domain-based ontology, we are able to enrich the semantic information so that the results of the clustering process will be improved.

The research and experiment are carried out within a narrowed domain—"Public health emergencies" and domain-based ontology is created. The comprehensive architecture of our project can be represented as Figure 2.

Figure 2.   The comprehensive architecture of our project.

To achieve the goal, the implementation process is subdivided into six major stages: 1) Data Collection, 2) Text Representation, 3) Documents Clustering, 4) Topic Summarization, 5) Ontology Introduction, and 6) Comparison. The project concentrates on the processing of Chinese texts, so all of the research data were coming from the Internet in Chinese. After this experiment, the final results display obvious advantages comparing with traditional methods.

## 2   BACKGROUND AND RELATED WORK

### 2.1   *Background*

Document Clustering is a process to cluster a lot of documents into groups or clusters so that documents within a group exhibit strong intracohesion

while documents from different groups are mutually exclusive. Since the characteristic of each group is not predefined before clustering, Document Clustering was regarded as an unsupervised learning process, which means documents are clustered into groups without prior knowledge. There are two typical types of clustering algorithms: flat clustering and hierarchical clustering. Flat clustering supposes data samples (i.e., documents) were on the same flat plane (a super plane if the vector dimension is larger than three) and the clustering algorithm directly partitions them into different groups or clusters without constructing hierarchical relationship between documents. That just says all documents belong to the same level. K-means clustering [3] is a typical and important flat clustering algorithm. It aims to iteratively partition each document into a cluster with the nearest distance (largest similarity) to the cluster center and eventually minimize an objective function. Unlike flat clustering, hierarchical clustering outputs clustered groups along with hierarchical relations such as a tree's structure. The number of final clusters does not have to be predefined before clustering and its hierarchical structure can provide richer information than flat clustering. However, the quality of the feature vector will directly influence final clustering results. So the loss of semantic information between keywords is as well the problem. In addition, HAC is more time consuming and complicated than K-means clustering. Also, the total number of clusters is well known for us from the testing data in this project. Therefore, we eventually chose the simple and efficient K-means clustering to demonstrate research.

Figure 3.   An example ontology related to the domain of "camera."

Ontology is a term coming from philosophy, but in computer science, its definition has been adapted. Now most researchers agree with this statement: An ontology is defined as a formal and explicit specification of a shared conceptualization [4][5]. In most time, an ontology is determined under a specific domain. A typical ontology is constructed with a limited number of concepts (conceptual terms) and relations between them. Figure 3 provides an example of ontology in the domain of "camera."

*http://www.xfront.com/owl/ontologies/camera.owl*

Obviously, an ontology is a kind of consensus towards a given domain and it provides the semantic interoperability so that associated concepts can have semantic relations or direct mappings between each other. These enrichments of semantic information can be used in many fields such as data mining, information retrieval, information extraction, coreference resolution, and so on. An example applicable in information retrieval is: If a query, says "PC," is failed and no related documents are retrieved, the searching engine can try some other queries such as "computer," which shares the semantic meaning with "PC" and recommend these results to users.

## 2.2 Related work

J. Allan's introduction to topic detection and tracking [7] gave a systematic introduction of topic detection. Qiu's research [8] on topic detection and Zhang's survey [1] on technologies of topic detection introduced the traditional clustering strategy to detect topics. Then Gong [9] implemented a system on automatic detection of hot topics for practical data on the Internet. In Gong's research, a hot topic detection scheme has been built which is on the basis of outdated document clustering. However, this method depends on brutal matches between keywords and fails to consider the loss of semantic information when computing similarities between feature vectors.

Some researchers have noticed this kind of problem and tried to introduce ontology method to improve them. Dou and his students made much effort on the introduction of ontology into fields like information retrieval, data mining, and document processing. [10, 11] Similarly, Yang [12] brought in this ontology method to enhance document clustering process and got very positive experimental results.

For the further study, the ontology method is proposed to improve the clustering process in order to finally improve topic detection results. Unlike Yang's work, a domain-based ontology will be introduced whose advantages of its many aspects will be taken as well.

## 3 DESIGN AND IMPLEMENTATION

### 3.1 Design of our experiment system

A comprehensive functional block diagram of our experimental system is designed as Figure 4. The experiment is narrowed into a specific domain—"Public health emergencies." Figure 5 shows parts of our domain ontology.

Two plans are constructed to utilize the ontology. Through the observation of testing documents and ontology, it is noticed that there are two types of

Figure 4. The functional block diagram of our experiment system.

functional semantic mappings between words. The first type is "equal relation." In Figure 5, those words in blue circles have equal relation with some concept words in the ontology, which is a good illustration of "equal relation." Another type is "hyponymy relation." In Figure 5, words "Jiaxing Liugan (H1N1)" and "Feidianxingxing Feiyan (SARS)" have the same father concept "epidemic disease," which is an example of "hyponymy relation." The advantages of these two types of relations will be taken to ensure that improvements. But the strategy should accord with the clustering requirements, namely the **clustering granularity**.

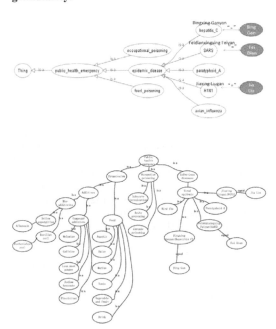

Figure 5. Part of an ontology in the domain of "Public health emergencies."

Clustering granularity determines the generality of a cluster, i.e., the generality of the corresponding topic. This concept was suggested because there existed some issues when preparing the testing data. The testing documents were gathered from the real Internet and labeled by different people. They were invited to partition testing documents into topics manually according to their own knowledge.

As a result, the labeling results are very different. If the total number of topics are not preset, some people may produce more topics than others because they partition several documents into a more specific topic, as shown in Figure 6. Hence, a concept is proposed called clustering granularity to describe the expected generality of a cluster or topic. The finer the clustering granularity is, the more specific topics will be identified. Built on above analysis, the enhanced ontology method is planned to be tested on different clustering granularities.

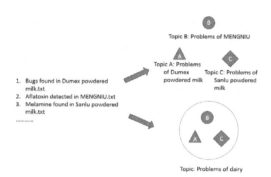

Figure 6. Different people will detect different results with different clustering granularity.

Now, it moves back to the discussion of the use of ontology. According to different clustering granularity, the different combinations of "equal relation" and "hyponymy relation" will be used. For a finer clustering granularity, only "equal relation" will be enough. For a coarser clustering granularity, "hyponymy relation" can be further included so that documents sharing the same conceptual contents will be clustered together to obtain better results.

As discussed in the background section, the representation of a topic is subjective. Therefore, a simple and clear way is utilized to represent topics— to select top $n$ keywords from a cluster and all of its document titles as the final topic.

### 3.2 *Implementation*

During the experiment, the operating system is Windows 7 and the programming language is Java under the IDE environment of MyEclipse 8.0.

Since many Chinese news needed handling, the documents were mainly collected from popular Chinese news portals. A preliminary list of topics was created at the beginning and several volunteers were ordered to collect related news and stories for each topic by querying relevant keywords in famous searching engines.

After this process, there were totally 1035 documents. Then, another group of volunteers was asked to cluster these documents by hand and detect topics based on their own knowledge. But two sets of rules were set before this manually detection. One is to detect topics that are more specific and the other one is to figure out topics that are more general. After the integration and the elimination of duplicated documents of these two sets of human labelling results, two sets of testing data were prepared finally.

The first set of testing data was in accordance with more specific topics, which contained 45 topics and 1012 documents in total. Topics numbered from 0 to 44 were called *45-topic* testing data. The second set of testing data was in accordance with more general topics, which contained 28 topics and 1012 documents in total. Topics numbered from 0 to 27 were named *28-topic* testing data.

## 4 EXPERIMENTS AND EVALUATION

To evaluate both traditional and ontology method clustering, four evaluation values are prepared: Precision, Recall, F-score, and Purity. Precision, Recall, and F-score values are common.

Purity is a value which reveals the purity of clustering results. Each result cluster is assigned to the class that is most frequent in this cluster, and then the purity is the number of correctly assigned documents dividing the total number of documents, which is,

$$purity(\Omega, \mathbb{C}) = \frac{1}{N} \sum_{k} \max_{j} |\omega_k \cap c_j|$$

where $\Omega = \{\omega_1, \omega_2, ..., \omega_K\}$ is the set of resulting clusters and $\mathbb{C} = \{c_1, c_2, ..., c_J\}$ is the set of real classes (namely clusters of testing data detected by human beings). Each $\omega_k \omega_k$ or $c_k c_k$ refers to a group of documents.

Traditional K-means algorithm is implemented by randomly selecting initial seeds which may cause issues for fair evaluations. Thus, we also implemented an optimized K-means clustering centers selection algorithm based on the method introduced in Ref. [13] during our experiments.

Table 1 shows the statistics of the final results. These results were visualized and a set of bar charts was produced as Figure 7 and Figure 8.

From the bar chart for 45-topic testing data, it shows that:

1 For the average result from random selection of seeds, F-score of ontology method is 2.4% larger than traditional method and Purity is 1.4% larger.

Table 1. Final results of traditional and ontology method for 45-topic and 28-topic testing data based on random and optimized seeds initialization.

| Type | Traditional Method Clustering | | | | Ontology Method Clustering | | | |
|---|---|---|---|---|---|---|---|---|
| | Precision | Recall | F-score | Purity | Precision | Recall | F-score | Purity |
| Average_45 | 0.562 | 0.779 | 0.652 | 0.758 | 0.587 | 0.800 | 0.676 | 0.772 |
| Optimized_45 | 0.680 | 0.812 | 0.740 | 0.832 | 0.740 | 0.820 | 0.778 | 0.854 |
| Average_28 | 0.467 | 0.401 | 0.431 | 0.687 | 0.594 | 0.502 | 0.543 | 0.745 |
| Optimized_28 | 0.517 | 0.383 | 0.440 | 0.708 | 0.612 | 0.488 | 0.543 | 0.753 |

**Average_45: the average result based on random selection of seeds for 45-topic testing data.**
**Optimized_45: the result based on optimized (adapted) selection of seeds for 45-topic testing data.**
**Average_28: the average result based on random selection of seeds for 28-topic testing data.**
**Optimized_28: the result based on optimized (adapted) selection of seeds for 28-topic testing data.**

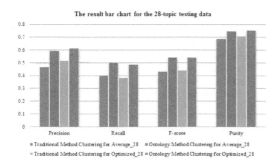

Figure 7. The result bar chart for the 45-topic testing data.

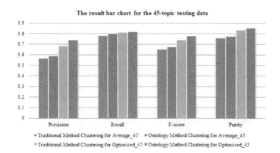

Figure 8. The result bar chart for the 28-topic testing data.

2 For the result from optimized selection of seeds, F-score of ontology method is 3.8% larger than traditional method and Purity is 2.2% larger.

From the bar chart for 28-topic testing data, it shows that:

1 For the average result from random selection of seeds, F-score of ontology method is 11.2% larger than traditional method and Purity is 5.8% larger.

2 For the result from optimized selection of seeds, F-score of ontology method is 10.3% larger than traditional method and Purity is 4.5% larger.

In a word, the ontology method indeed displays advantages on clustering process on both testing data sets. According to our representation of a topic in this paper, better clustering results also lead to better topic detection results, meaning that our ontology method defeated traditional method on final topic detection.

Additionally, three more natures of the ontology method can be summarized:

1 On the same testing data, both of the average result from seeds which are randomly selected and the result from seeds which are selected via the optimized algorithm have better results when using ontology method than the traditional method. That is to say: 1) the domain-based ontology method works better in terms of statistics, even though the initial condition is highly random; 2) when the selection of seeds of K-means algorithm has been optimized so that traditional method can perform stably, the ontology method is still able to make improvements.

2 For both the coarser clustering granularity (28-topic) and the finer clustering granularity (45-topic), appropriate combination of ontology knowledge works better than the traditional method.

3 For the finer clustering granularity, the ontology method improves 1–3% on F-score and Purity than the traditional method. While in the coarser clustering granularity, ontology method improves 4–11% on F-score and Purity. That just says the ontology method works even better when the clustering granularity is coarser.

For nature 1 and 2, the reason is obvious—the utilization of domain-based ontology enriches semantic information so documents can be clustered more appropriately during the clustering process.

For nature 3, the ontology works better when the clustering granularity is coarser. It is valid and believed because a coarser clustering granularity makes the final cluster more general so that our ontology method can bring in a higher level of concepts and more relations. This will lead to better results.

## 5 CONCLUSIONS AND FUTURE WORK

In this paper, we pointed out that traditional clustering methods calculate similarities between documents by thorough feature vectors. But this approach depends too much on the hard match between keywords, meaning that two documents could be recognized as similar ones only if they share common keywords in their feature vectors literally. Obviously, it is lack of semantic information.

Then we proposed our ontology method to overcome the loss of semantic information. The domain ontology owns rich semantic and conceptual information that can be utilized to guide and improve various data mining tasks. For clustering process, our ontology method brought in two kinds of relations from the domain ontology—"equal relation" and "hyponymy relation"—to construct word mapping. By doing this, similar documents could be successfully clustered together if their keywords share the same semantic meaning.

We implemented both the traditional method and our ontology method on the basis of K-means clustering algorithm and tested them on two sets of testing data. In both situations, our ontology method produced better results than the traditional method. Our ontology method enriched semantic information and successfully improved results for topic detection in almost every aspect than the traditional method.

However, for this project, there are still some other experiments worth doing. For example, we can also implement the hierarchical clustering algorithm and test our ontology method and the traditional method on it. Also, many more relations and knowledge of the ontology can be utilized in the future. Moreover, since the ontology method is suggested to solve the problem of the loss of semantic information, it can be introduced to many other similar tasks, such as documents classification and information extraction.

## REFERENCES

[1] Zhang Xiaoyan, Wang Ting (2009). *Research of Technologies on Topic Detection and Tracking.* Journal of Frontiers of Computer Science and Technology, 347–357.

[2] G. Salton, A. Wong, C. S. Yang (1975). *A vector space model for automatic indexing,* Communications of the ACM, v.18 n.11, p.613–620.

[3] Christopher D. Manning, Prabhakar Raghavan & Hinrich Schutze (2008). *Introduction to Information Retrieval.* Cambridge University Press. (978-0-521-86571-5).

[4] T.R. Gruber (1993), *A translation approach to portable ontology specifications,* Knowledge Acquisition 5(2) 199–220.

[5] R. Studer, V.R. Benjamins and D. Fensel (1998), *Knowledge engineering: principles and methods,* Data Knowledge Engineering 25(1) 161–197.

[6] Grigoris Antoniou, Frank van Harmenlen (2004), *A Semantic Web Primer,* MIT Press. (ISBN 0-262-01210-3).

[7] J. Allan (2002). *Introduction to Topic Detection and Tracking.* Kluwer Academic Publishers, page 1–16.

[8] Qiu Likun, Tao Ran, Long Zhiyi, et al (2007). *Internet-oriented Study on Topic Detection.* NetSec 2007.

[9] Gong Haijun (2008). Research on technologies of hot topic detection on the Internet. Degree thesis, Huazhong Normal University.

[10] Daya C. Wimalasuriya and Dejing Dou (2010). *Ontology-based information extraction: An introduction and a survey of current approaches.* Journal of Information Science, page 306–323.

[11] Daya C. Wimalasuriya and Dejing Dou (2009). Using Multiple Ontologies in Information Extraction. Conference on Information and Knowledge Management, CIKM 2009.

[12] Yang Cailian (2008). Researches on document clustering based on ontology. Degree thesis, Liaoning Normal University.

[13] Han Lin-bo, Wang Qiang, Jiang Zheng-feng, et al (2010). *Improved k-means initial clustering center selection algorithm.* Computer Engineering and Applications, page 150–152.

[14] Ren Jiabiao, Yin Shaohong (2010). *An Effective Method for Initial Centrepoints of K-means Clustering.* JISUANJI YU XIANDAI. (1006-2475(2010)07-0084-03).

[15] Jiawei Han, Micheline Kamber (2000). *Data Mining: Concepts and Techniques [M].* Morgan Kaufmann.

[16] Liu Kaipin (2009). The design and implementation of classification module for telecommunication documents on the basis of domain ontology. Degree thesis, Beijing University of Posts and Telecoms.

[17] Michael W.Berry and Malu Castellanos (2007). Survey of Text Mining: Clustering, Classification, and Retrieval.

[18] Diarmuid O. Seaghdha (2011). *An assignment for semantic analysis.* 2011 Summer school on machine learning by HIT and MSRA.

*Multimedia, Communication and Computing Application – Leung (Ed.)*
© *2015 Taylor & Francis Group, London, ISBN 978-1-138-02775-6*

# ACE algorithm in the application of video forensics

S. Jia, C.H. Feng, Z.Q. Xu, Y.Y. Xu & T. Wang
*LIESMARS, Wuhan University, Wuhan, China*

ABSTRACT: Digital video forensics is an effective way to test the authenticity of the video content. In some specific detection algorithms of frame deletion, frames with sudden lighting changes will interfere with the detection of tampered video sequences and increase the false alarm rate. In this paper, the color difference between the frame with sudden lighting changes and its reference frame is analyzed, and the Automatic Color Equalization (ACE) algorithm is used to remove the falsely detected frames caused by lighting changes and to reduce the false alarm rate. Moreover, an improved algorithm is proposed to speed up the computation based on Local Linear Looking Up Table (LLLUT) method and multithreaded parallel programming. Experiment results show that with the help of the proposed ACE algorithm in detection of frame deletion, the false alarm rate caused by sudden lighting changes is effectively reduced, achieving good results in video forensics.

## 1 GENERAL INSTRUCTIONS

Nowadays, with the popularity of digital cameras and editing softwares of visual media, digital images and videos can be easily acquired and processed. Meanwhile, some mature and simple-operated editing softwares make it easy and popular to tamper with the digital image and video. The tampered image or video may mislead the public, even lead to serious consequences. For example, the disguised photos or surveillance videos presented as evidence in the court may forge a false alibi to affect the judgment of the case t; if the press releases tampered images or videos, it will cause false or inflammatory public opinion. Therefore, how to identify the authenticity and integrity of the visual media has become a key technical problem to be solved.

Early on, scientists used digital watermarking technology to identify the authenticity of visual media[1]. Digital watermarking technology needs to embed specific validation data in visual media during their production as the basis for the late detection. However, because most acquisition equipments do not have the function of embedding such kind of data, thus they limit the application of digital watermarking technology. In this case, the technology of digital visual media forensics has emerged. After over ten years of development, it has become an interdisciplinary field with diverse theoretical foundation, diverse implementation methods and divergent thinking. Through analyzing the intrinsic features in visual media, it can recover the history of getting and post-processing the visual media to determine the credibility of them. However, the forensic process is affected by many factors, such as lighting changes and sudden zooming

when videos are shot, and these factors have become the interference for existing image and video forensics. Thus, researchers have been working on more efficient methods of video forensics.

While we are developing the video forensics technology, media forgers begin to work on methods to hide the fingerprints and fool the forensic techniques, namely the anti-forensics. Compared with the researches on video forensics, studies on anti-forensics are fewer. However, a research on it is of great significance. On the one hand, the means of tampering to conceal or erase the traces of tampering can be learned. On the other hand, more efficient, practical video forensics technology can be developed on the basis of anti-forensics, which is also called as anti-anti-forensics, thereby intensifying the fight against such crimes to ensure the security of visual media.

The methods of video tampering include deleting objects and adding objects. Frame deletion belongs to the former and is a typical video manipulation act. This paper makes a further research based on an existing frame deletion detection algorithm proposed by[2], which has a high true positive rate (TPR) and a relatively high false alarm rate (FAR). The false detection originates from different factors, such as relocated I-frames, sudden lighting changes, frames with sudden zooming, which can be called as interference frames. Among them, sudden lighting changes are one of the main causes of false detection. Therefore, in this paper, we first analyze the color difference between the frame with sudden lighting changes and its reference frame, which will be different from that of frame deletion point (DP). According to the difference, the automatic color equalization (ACE)[3] algorithm is used to distinguish the interference frames caused by natural lighting changes. On the other hand, if the features of DPs in detection algorithm are learned by forgers, then man-made lighting changes may be added in the video sequences to hide the DPs. The similar method can be applied to anti-anti-forensics and the interference frames caused by anti-forensics can also be distinguished, thereby reducing the FAR. Secondly, regarding the feature of a large number of frames in video sequences, an improved algorithm is proposed based on local linear looking up table (LLLUT)[4] method and multithreaded parallel programming to speed up the computation. In the end, the proposed method is tested by experiments.

The rest of the paper is organized as follows. In Section 2, the principle and implement of the proposed algorithm to reduce FAR in frame deletion detection is presented. The speed optimization of ACE algorithm is provided in Section 3, and the experimental results and discussions are given in Section 4.

## 2 PRINCIPLE AND IMPLEMENT OF THE PROPOSED ALGORITHM

### 2.1 *False detection in the frame deletion forensics*

One effective detection algorithm of frame deletion is proposed in paper [2]. It locates the exact frame deletion point (DP) based on the temporal difference of DP and its reference frame. It analyzes each frame according to the characteristics of the video frames. For tampered videos, frame deletion will affect the continuity and differences between two adjacent frames. Intuitively, since several frames are cut, the temporal difference of DP and its reference frame become larger, therefore the most straightforward character of DP is a larger motion residual.

However, in the tampered sequences, there are still no lack of interference frames which have similar statistical characters with DP, such as frames with rapid motion content, the relocated I-frames, frames with sudden lighting changes and frames with sudden zooming. After extensive observation, it has been found that the differences of mean motion residual of adjacent quick motion frames are relatively continuous within a short period of time. Whereas for DP, the temporal motion could be several times larger than its adjacent frame, thus the mean motion residual of DP shows an obvious increase. And the fluctuation character of the distribution of DP and relocated I-frame are essentially different, which can be used for distinguishing the two of them. Moreover, sudden lighting changes and sudden zooming in videos can also have large motion residual and cause interference frames in detection. Since the sudden lighting changes are one of the main reasons of the false detection. Therefore,

in this paper, we aim at using the ACE processing to distinguish this interference.

### 2.2 *Principle of ACE to reduce the FAR of frame deletion detection*

The difference between DP and frame with sudden lighting changes is first analyzed in tampered video sequences. Denote one DP by $I_d$ and extract its previous frame and denote it by $I_d - 1$; similarly, denote one frame with sudden lighting changes by $I_l$ and its previous frame by $I_l - 1$. The specific analysis is as follows.

1. After extensive observation, it has been found that there are usually no sudden lighting changes in DPs with limited number of deleted frames, therefore the paper assumes that all DPs do not have sudden lighting changes. Consequently, the $I_d$ and $I_d - 1$ are basically similar in lighting conditions (brightness, contrast, et al).

2. The difference between frame with sudden lighting changes and its previous one is relatively evident in lighting conditions, which means that the $I_l$ and $I_l - 1$ are obviously different in lighting conditions.

Thereinto, lighting conditions can be changed by ACE algorithm. It mimics some characteristics of the Human Visual System (HVS)[5], whose efficiency is due to its smart adaptive mechanisms, in particular color constancy and lightness constancy. It occupies a unique place in color processing algorithms. Use ACE algorithm to process the current frame $I$ and its previous frame $I - 1$ respectively, contrast each frame's gray-level histogram before and after the equalization and denote their difference values by $\Delta_I$, $\Delta_{I-1}$. Therefore, the differences in lighting conditions of two adjacent frames can be quantitatively reflected by the parameter $\delta_I$, which is defined as equation

$$\delta = \left| \Delta_I - \Delta_{I-1} \right| \tag{1}$$

Thus, we conclude that
1. the frame deletion point $I_d$ satisfies

$$\delta_d = \left| \Delta_{I_d} - \Delta_{I_d-1} \right| \approx 0 \tag{2}$$

2. the frame with sudden lighting changes $I_l$ satisfies

$$\delta_l = \left| \Delta_{I_l} - \Delta_{I_l-1} \right| > \beta \tag{3}$$

Where the threshold value $\beta$ is bigger than zero; it is determined by the specific content of different videos, and the larger the value is, the greater the difference between two images in lighting conditions will be. Namely, the frame $I_l$ is different from its previous frame $I_l - 1$, which is caused by sudden lighting changes in $I_l$. Therefore we can exclude $I_l$ from the suspicious DPs. That is, the interference frames caused by lighting changes can be distinguished and accordingly the FAR can be reduced.

## 2.3 Implement of ACE to reduce the FAR of frame deletion detection

Apply ACE algorithm to the detection of frame deletion of tampered video sequences and the specific steps are as follows.

3.1 Detection of frame deletion: use the known algorithm to detect the tampered video sequences with frame deletion. Compare the detection results with the features of the tampered video sequences, and calculate the FAR $e_1$ as

$$e = \frac{n_e}{n} \qquad (4)$$

Where $n_e$ is number of error frames, $n$ is the total number of video frames.

2. ACE processing: denote each suspicious DP in detection results by $I$ frame in turn, and extract its previous frame and denote it by $I-1$. Use ACE algorithm to process each $I$ and $I-1$ frames, and calculate their difference values $\Delta_I, \Delta_I$ of gray-level histograms before and after the equalization. Record the corresponding value of $\delta$ in a table.

3. Threshold setting: set the threshold value $\beta$ according to the record results (which is always between 0.03 and 0.04, observed from a large number of experimental data), regard the frames whose $\delta$ is greater than $\beta$ as the frames with sudden lighting changes and remove them from the detection results. Calculate the FAR $e_2$ after processing.

The FAR reduces from $n_1$ to $e_2$, and we can use $r$ to express the reduction rate, which will reflect the reduction effect on the FAR of the proposed algorithm.

## 2.4 Application of ACE in anti-anti-forensics

Anti-forensics manipulation means extra tampering manipulation in order to conceal or erase the traces of the previous tampering. In tampering videos with frame deletion, the forger may learn the features of DPs in detection algorithm and tries to hide them by adding man-made interference frames with similar features, such as partial or full adjustments in lighting conditions by reducing some frames' contrast, changing the brightness, et al. Use the above method in the detection results, after anti-anti-forensics, we can also distinguish the interference frames caused by

man-made lighting changes effectively and improve the detection rate of tampered videos.

## 3 SPEED OPTIMIZATION OF ACE ALGORITHM

The traditional automatic color equalization (ACE) algorithm has prominent color correction effect, capable of handling multi-lighting conditions, such as underexposure, overexposure, low contrast conditions, color cast. However, because it processes images by point to point calculation, its computational complexity is high and will increase rapidly as the image size increases. Therefore, it is not suitable for most application scenarios. But in video forensics, with a large number of frames in video sequences, the image processing needs to be accomplished "instantaneously" to achieve high efficiency to improve the FAR in detection. To meet the above requirements, this paper optimizes ACE algorithm from two perspectives: improving the algorithm based on LLLUT method and using parallel processing.

## 3.1 Algorithm level optimization

ACE algorithm is mainly divided into two stages. The first stage accounts for a chromatic spatial adaptation (responsible of color constancy) and merges with the gray world and white patch approaches and performs a sort of lateral inhibition mechanism, weighted by pixel distance. The result is a local-global filtering. The second stage, with dynamic tone reproduction scaling, configures the output range to implement an accurate tone mapping. It maximizes the image dynamic, normalizing the white at a global level only. No user's supervision, no statistics and no data preparation are required by the algorithm. In this paper, ACE algorithm optimization is first based on its acceleration algorithm, LLLUT method. The main idea of this method is to apply the color equalization algorithm to a small sub-sampled version of the input image (by introducing a sub-sampling factor $d$) and to use a modified look-up table technique to maintain the local filtering effect of the color equalization algorithm. The method increases the speed of color-filtering algorithms, reducing the number of pixels involved in the computation by sub-sampling the input image. But its processing efficiency is still too low to process video sequences rapidly. Therefore, this paper simplifies LLLUT algorithm appropriately to achieve further optimization of ACE algorithm.

The LLL method uses the filter $f$ on a sub-sampled version ($I^s$) of the original image ($I$) to limit the number of pixels involved in the computation but with some improvements[6]. The idea of LLLUT is to consider the pixel of position $(x, y)$ and then to generate LUT functions depending on the value of $I$ in $(x, y)$ and on the values of the pixels in suitable subsets $\Omega$ of $I^s$ and $O^s$. The LLL technique can be symbolically described by the following equation

$$O_{LLL}(x, y) = LLL_{(x,y)}(I_{(x,y)}, I_\Omega^S, O_\Omega^S) \qquad (5)$$

The scheme of LLL method used in ACE algorithm is shown in Figure 1 and its main steps are shown in [6].

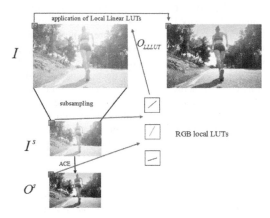

Figure 1. LLLUT scheme.

The original LLLUT algorithm computes the final output value by weighting three contributors $O_k$ $(O_{\Omega_{centre}}, O_{\Omega_{\Delta x}}, O_{\Omega_{\Delta y}})$ and using the Euclidean distance $d_k$ between the centers of $\Omega$ and $(x, y)$ to improve the processing effect. However, in this paper, ACE algorithm is mainly used to help distinguish the interference frames with sudden lighting changes, therefore, to simplify the algorithm and gain much faster processing speed, we only defines $\Omega_{center}$, the first subset that will be used to perform a linear interpolation. The experiment presented later shows that the simplified algorithm obtains significantly faster speed and better effect.

### 3.2 Parallel processing of ACE

Parallel computing is designed to use multiple computing resources to solve the problem of computing simultaneously. Since ACE algorithm computes pixels of R, G, B channels separately, we can take full advantage of the common multi-core processors, apply multithreaded parallel programming techniques to ACE optimization algorithm, and further improve the processing speed without affecting the color correction effect. This paper parallelizes the serial

program based on the shared memory programming model OpenMP [7] (Open Multi Processing). The specific steps are as follows.

1. Optimize the serial ACE program. Improve the serial program to get the best performance when it computes with single-core.

2. Parallelize the program. Use the guidance command sections to allocate the independent calculations of R, G, B channels of ACE algorithm for three threads to calculate at the same time.

3. Set parallel compilation environment. Add the header file "omp.h" in the beginning of the program; modify the program attributes to support OpenMP; set the properties of correlated variables as private or shared according to the needs of the thread.

4. Compile the parallel program. Compile and generate the executable files.

5. Verify the results. Record and compare the results (processing effect and computation time) of the serial ACE program and parallel processing results to verify that parallel processing will improve the calculating speed without affecting the image processing effect.

6. Test the program. When the parallel program is running, open the Task Manager of the computer, observe the usage and record of CPU to learn the effect of parallelization.

7. Improve the parallelization scheme based on the performance of the program.

## 4 EXPERIMENTAL RESULTS AND DISCUSSIONS

This section evaluates the performance of the proposed ACE algorithm to reduce the FAR caused by natural or man-made sudden lighting changes in detection of frame deletion. First, we test the efficiency of the proposed ACE optimization algorithm. Then, the accuracy and feasibility of locating frames with sudden lighting changes are verified by the proposed algorithm. The performance of the proposed algorithm to reduce the FAR caused both by natural and anti-forensics are drawn in the last.

The computing environment is under windows systems, Intel i3 processor (dual-core four threads); and the programming language is C/C++ in Microsoft Visual Studio 2005.

### 4.1 Effects of ACE optimization algorithm

Test the optimization algorithm with one low-light library image (300*400), and compare the results of ACE processing before and after optimization, which is shown in Table 1 and Figure 2. The experimental data of traditional ACE algorithm, simplified

LLLUT algorithm and the proposed algorithm (simplified LLLUT with parallel computing) with the sub-sampling factor ($d$) equal to 4 are recorded. And the data include the mean, standard deviation and entropy of the gray scale image, and the calculation time.

Table 1. Quality parameters of library image.

|  | original image | traditional ACE | simplified LLLUT | proposed algorithm |
|---|---|---|---|---|
| mean | 60.9856 | 127.5091 | 125.5870 | 125.5821 |
| standard deviation | 45.3014 | 56.4399 | 58.7103 | 58.7126 |
| entropy | 3.5239 | 3.8692 | 3.8872 | 3.8876 |
| calculation time(s) | / | 340.860 | 1.213 | 0.678 |

(a) original image    (b) traditional ACE    (c) proposed algorithm

Figure 2.  Comparison of processing results.

Table 1 shows the different effects of the three methods to process the same image. By comparing the data of the first column and those of the other three columns, it has been found that the ACE algorithm has effectively corrected the image color with higher mean, higher standard deviation and entropy values. That is, the image has higher quality in brightness, contrast and information quantum. Besides, the optimization algorithms have little difference with the traditional ACE in processing effects, which means that the calculation error produced by simplifying the algorithm is little and parallel computing can not affect the color equalization effect. It is also evident that the proposed algorithm has effectively improved the computing speed. The visual effects of traditional ACE and the proposed optimization algorithm can be seen from Figure 2, and both of them have corrected the underexposed original image effectively.

In summary, the experimental results in Table 1 and Figure 2 show that the optimization algorithm achieves good processing results assessed whether by subjective observation or objective quality criteria.

Speed-up ratio is the time-consuming ratio of the same tasks respectively running on a single processor system and a parallel one, which is used to measure the performance and effects of parallelization. Table 2 shows the speed-up ratio of images with different sizes in the parallel computing with the sub-sampling factor $d$ equal to 4.

Table 2 shows that images with different sizes have gotten different speed-up ratio, but all of them are less than 3. The time consumption results from the system overhead, such as creation and switch of OpenMP threads. Moreover, when each section is executed at the same time, the final calculation time depends on the slowest one.

Therefore, after applying the above parallelized ACE optimization algorithm to video sequences, such as a video sample with 200 frames (size of 300*400), it shows that it takes about 19 hours to process all the frames with traditional ACE algorithm, but only 2 minutes is taken in the proposed algorithm. It is thus obvious that the computing speed has been significantly improved.

### 4.2  Detection of frames with sudden lighting changes

Before the experiments of detecting DPs and interference frames in video forensics, we first verify the accuracy and feasibility of ACE processing to locate the frames with sudden lighting changes. That is, we experiment to testify that the ACE processing can distinguish the lighting changes frames among the suspicious DPs but not the DP frames themselves.

Select crew.yuv as the video sample, whose total number of frames is 285; DPs are the 15th and 25th frame; frame number set with sudden flashlight is F={2,46,73,75,123,130,136,143,159,165,183,206, 212,218,261,265}

Calculate the difference value $\Delta$ of all frames in the video sample before and after ACE processing by the above proposed method, and set the threshold value $\beta$ as 0.04 to get the frame number set L, which records the frames with relatively strong lighting changes.

L={2,46,123,130,136,143,165,183,206,218, 261,265}

By comparing re the numbers in set F and L, we can find that $L \subseteq F$, which means that all the frames located by the proposed method are the true frame points with sudden lighting changes. Therefore,

Table 2.  Speed-up ratio of the proposed algorithm.

|  | 200*300 | 300*400 | 600*700 | 800*800 | 1600*1600 |
|---|---|---|---|---|---|
| speed-up ratio | 1.731 | 1.774 | 1.983 | 2.020 | 2.041 |

we verify the accuracy and feasibility of detecting frames with sudden lighting changes by the proposed ACE algorithm and the differences of gray-level histograms.

### 4.3 Effect of ACE algorithm on frame deletion detection

Select crew.yuv as the tampered video sample, whose DPs are the 15th and 25th frame. Do the experiment as follows.

1. Detection of frame deletion. Get the results detected by the known detection algorithm, which are shown in Table 3. Compare the frames with the features of the video sample, and calculate FAR $e_1$ as 3.86%.

2. ACE processing. Extract the related frames and use ACE algorithm to process them and calculate their difference values of gray-level histograms before and after the equalization. Record the corresponding value of $\delta$ in Table 4.

Threshold setting: set the threshold value $\beta$ as 0.04 according to the recorded values, regard the frames whose $\delta$ is greater than $\beta$ as frames with sudden lighting changes and remove them from the detection results. Therefore, only the 15th, 25th, 94th frames have few obvious changes in lighting conditions.

The detection results after removing are shown in Table 5.

Compare the detection results after processing with the features of the video sample, as is shown in Table 6, and the calculated FAR $e_2$ is equal to 0.35%. We can see that the reduction rate $r$ of FAR is 90.93%, reducing from 3.86% to 0.35% in this video sample sequences. It is clear that the proposed algorithm effectively reduced FAR.

Experiment on more 20 tampered video samples (each with 200 frames) with interference frames caused by sudden natural lighting changes. After the proposed ACE optimization algorithm processing with $\beta$ equal to 0.03, the impact on FAR is shown in Table 7.

As can be seen from Table 7, the experiment has significantly reduced FAR by distinguishing the interference frames caused by sudden lighting changes. The frames have not been distinguished, because there is sudden zooming or the lighting changes is too weak to reach the threshold value. Therefore, it can be concluded that the proposed method can distinguish all the interference frames caused by relatively strong lighting changes in detection of frame deletion, thus reducing the FAR effectively.

### 4.4 Application in anti-anti-forensics

Select anti.yuv as the video sample with frame deletion, whose total number of frames is 100; DPs are the 15th and 25th frame; frame numbers with anti-forensics processing by artificially enhancing the image brightness to conceal DPs are 10, 39, 75. Experiments of anti-anti-forensics are as follows.

### 4.5 Detection of frame deletion

Get the results detected by the known detection algorithm, which are shown in Figure 8.

Table 3. Results of DP detection.

| Suspicious DPs | 15 | 25 | 46 | 94 | 123 | 130 | 136 |
|---|---|---|---|---|---|---|---|
| | 143 | 165 | 183 | 206 | 261 | 265 | |

Table 4. $\delta$ values.

| $\delta_{15}$ | $\delta_{25}$ | $\delta_{46}$ | $\delta_{94}$ | $\delta_{123}$ |
|---|---|---|---|---|
| 0.015045 | 0.020088 | 0.057041 | 0.026280 | 0.081915 |
| $\delta_{130}$ | $\delta_{136}$ | $\delta_{143}$ | $\delta_{165}$ | $\delta_{183}$ |
| 0.110238 | 0.089100 | 0.098300 | 0.096787 | 0.047058 |
| $\delta_{206}$ | $\delta_{261}$ | $\delta_{265}$ | | |
| 0.191584 | 0.076090 | 0.068230 | | |

Table 5. Detection results after processing.

| Suspicious DPs | 15 | 25 | 46 |
|---|---|---|---|

Table 6. The comparison results.

| True DPs | 15 | 25 |
|---|---|---|
| False DPs | 94 | |

Table 7. Effect on FAR of more videos.

| | 2 video samples | 4 video samples | 14 video samples |
|---|---|---|---|
| FAR before processing (%) | 1.5 | 1 | 0.5 |
| FAR after processing (%) | 0.5 | 0.5 | 0 |

Table 8. Results of DP detection.

| Suspicious DPs | 10 | 15 | 25 | 39 | 46 | 75 |
|---|---|---|---|---|---|---|

### 4.6 ACE processing

Record the related values in Table 9.

Table 9. $\delta$ values.

| $\delta_{10}$ | $\delta_{15}$ | $\delta_{25}$ | $\delta_{39}$ | $\delta_{46}$ | $\delta_{75}$ |
|---|---|---|---|---|---|
| 0.117372 | 0.015045 | 0.020088 | 0.060869 | 0.057041 | 0.142337 |

3.Threshold setting: set the threshold value as 0.04 and remove the frames whose value is greater from the detection results. The results after removing are shown in Table 10.

Table 10. Detection results after processing.

| Suspicious DPs | 15 | 25 |
|---|---|---|

By comparing the detection results with the features of the video sample, it has been found that the rest of frames are all the true deletion frames, and the FAR is reduced to 0. Therefore, it concludes that the proposed method can effectively distinguish the interference frames not only by natural lighting changes, but also by man-made lighting changes of anti-forensics.

## 5 CONCLUSIONS

In this paper, we first analyzes the principle that the automatic color equalization can reduce the FAR in a particular forensic algorithm of video frame deletion, and then optimize the ACE algorithm based on LLLUT method and multithreaded parallel programming to speed up the calculation in video sequences. At last we test the performance of the proposed optimization algorithm to distinguish interference frames caused by lighting changes. Experiment results have shown that the proposed algorithm is quite effective for distinguishing the frames with sudden lighting changes formed both naturally and man-madely. 70% of the tested video samples' FARs caused by lighting changes have been reduced to 0. However, there still exist some shortcomings in the proposed algorithm and its application in video forensics, for example, if the true frame deletion points have sudden lighting changes, the method will remove them from detection results as interference frames, thus reducing the TPR. Therefore, in the future, we intend to extract new features to distinguish the frame deletion points from frames by sudden lighting changes.

ACKNOWLEDGMENTS

This work is supported by the Major State Basic Research Development Program (No. 2011CB302306, 2011CB302204) and the National Natural Science Foundation of China (No. 41371402).

REFERENCES

[1] FENG Chun-hui, XU Zheng-quan, ZHENG Xing-hui, et al. Digital visual media forensics[J]. Journal on Communications, 2014, 35(4):155–165.
[2] Feng C, Xu Z, Zhang W, et al. Automatic location of frame deletion point for digital video forensics[C]// Proceedings of the 2nd ACM workshop on Information hiding and multimedia security. ACM, 2014: 171–179.
[3] Gatta C, Rizzi A, Marini D. Ace: An automatic color equalization algorithm[C]//Conference on Colour in Graphics, Imaging, and Vision. Society for Imaging Science and Technology, 2002, 2002(1): 316–320.
[4] Rizzi A, Gatta C. A local linear lut method for increasing the speed of generic image filtering algorithms[R]. Technical report, University of Milano, Milano, Italy, 2002.

[5] Rizzi A, Gatta C, Marini D. Color correction between gray world and white patch[C]//Electronic Imaging 2002. International Society for Optics and Photonics, 2002: 367–375.

[6] Gatta C, Rizzi A, Marini D. Local linear LUT method for spatial colour-correction algorithm speed-up[C]// Vision, Image and Signal Processing, IEE Proceedings-. IET, 2006, 153(3): 357–363.

[7] Parallel programming in OpenMP[M]. Morgan Kaufmann, 2001.

*Multimedia, Communication and Computing Application – Leung (Ed.)*
© *2015 Taylor & Francis Group, London, ISBN 978-1-138-02775-6*

# Modified fuzzy clustering with partial supervision algorithm in classification and recognition of pulmonary nodules

Q.P. Li, H. Liu & Z.Y. Su
*Department of Computer Science and Technology, Shandong University of Finance and Economics, Jinan, China.*
*Digital Media Technology Key Lab of Shandong Province, Jinan, China*

ABSTRACT: Accurate classification and recognition of pulmonary nodules are important and key processes of lung cancer Computer-Aided Diagnosis (CAD) system. In this paper, to improve the accuracy, we propose a modified partial supervision fuzzy clustering algorithm based on the annotation information offered by doctors in LIDC database. First, all pulmonary nodules are segmented from the CT images. Second, according to the lesion characteristics of pulmonary nodules, we extract a set of mainly shape-based feature vectors. Finally, we calculate the reference membership by exploiting the classification information of labeled samples in the process of clustering, and use the reference membership to guide the clustering process of the testing samples, for helping the testing samples to cluster more accurately. Experimental results show that the proposed method can outperform the traditional algorithm in classification and recognition.

## 1 INTRODUCTION

In 2009, the National Cancer Institute estimates that 22.7% of all cancer deaths in China were due to lung cancer. This indicated that lung cancer was the leading cause of cancer deaths in China (NIE, shengdong et al., 2009). The early diagnosis of lung cancer is one of the key issues in addressing to reduce the mortality of lung cancer (Jemal A et al., 2008). Therefore, research on the lung cancer computer-aided diagnosis techniques has been increasingly important. To promote the development of lung cancer CAD technology (K DOI. 2005; Ayman El-Baz et al., 2013), the National Cancer Institute (NCI) of the United States established a lung CT image database—Lung Imaging Database Consortium (LIDC), which contained 1012 chest cases and 1356 nodules at the time of this study (Han fangfang et al., 2013). Each nodule case contains the painting boundary coordinates and vision features from four radiologists. And the malignancy ratings are set by the visual decision of the four radiologists from one to five. The bigger value represents more suspicion of malignancy (Wiemker Rafael et al., 2009).

The main working process of the CAD are segmentation of pulmonary nodules, feature extraction and classification and recognition. The key technique includes image processing and machine learning. First, because the image segmentation technology has been mature enough (Dehmeshki J et al., 2008), we

can acquire satisfying result using the existing image segmentation technology (Jiang X et al., 2007). Second, through analyzing and interpreting the lesion characteristics of pulmonary nodules in LIDC database, we extract a set of mainly shape-based feature vectors which can represent the feature of the pulmonary nodules completely. Finally, because the main basis of identifying the CAD system's availability is the accuracy of classification, the focus of our research is to classify the pulmonary nodules. In this paper, we divide the extracted nodules into three categories, they are malignant nodule, benign nodule and false-positive nodule. Since the classification and recognition of pulmonary nodules are relatively professional works, we can't get satisfactory results only by means of extracting a set of scalar feature vectors to describe pulmonary nodules, and then conducting classification and recognition. The traditional fuzzy clustering with partial supervision algorithm introduces the marker information, and uses classification information of the labeled sample to guide the clustering process (Witold Pedrycz et al., 1997), but the algorithm cannot achieve satisfactory accuracy. To improve the accuracy of classification, we improve the traditional algorithm. First, we calculate and select the labeled samples whose distance between testing samples are less than a certain threshold. Second, we exploit their classification information to calculate the testing samples' reference membership. Finally, the reference membership is used to guide the clustering process of the testing samples, it can help the testing samples to cluster more accurately. Experimental results show that the proposed method can achieve higher accuracy of classification and recognition.

## 2 PREPROCESSING

### 2.1 Segmentation of pulmonary nodules

Segmentation of pulmonary nodules is the first key link of lung cancer CAD. The precision of segmentation directly affects the succeeding work. By using the existing lung CT image segmentation algorithm, we can segment out pulmonary nodules with high accuracy. This paper quotes the FRFCM segmentation method (Zheng Fuhua et al., 2013), and uses this method to segment out pulmonary nodules from the lung CT image in LIDC database. Figure 1 shows the segmentation result.

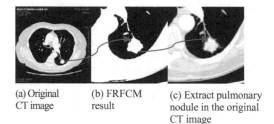

(a) Original CT image    (b) FRFCM result    (c) Extract pulmonary nodule in the original CT image

Figure 1. Example of extracting pulmonary nodule.

### 2.2 Feature extraction of pulmonary nodules

The feature extraction of pulmonary nodules is related to the accuracy of classification and recognition. The diameters of the pulmonary nodules are between 3 and 30 mm in general. The LIDC database provides a description of nine pathologic characters of pulmonary nodules, including subtlety, sphericity, internal structure, calcification, margin, lobulation, spiculation, texture and malignancy (Samuel G. Armato et al., 2011 ). As is shown in the Figure 2, among the nine pathologic characters, the lobulation and spiculation can best characterize the malignancy of malignant nodule, and the more obvious the lobulation and spiculation are , the higher malignant degree of the pulmonary nodules is. Moreover, the lobulation and spiculation are the manifestation of the shape feature. In this paper, through experimental verification and study, we extract a set of mainly shape-based feature vectors of the pulmonary nodules, including gray variance, gray histogram entropy, circularity, radial distance mean and variance, boundary roughness, compactness, shape invariant moment H0,H2,H3,H4, eleven in total.

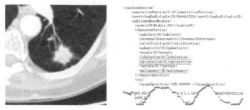

(a) Pulmonary nodule (b) XML document in LIDC database

Figure 2. Example of pulmonary and its pathologic characters.

## 3 MODIFIED FUZZY CLUSTERING WITH PARTIAL SUPERVISION

### 3.1 The traditional fuzzy clustering with partial supervision

Witold Pedrycz et al. proposed a fuzzy clustering with partial supervision algorithm. The algorithm divides $N$ samples into $C$ fuzzy groups, searches the clustering center of each group and updates the clustering centers through updating fuzzy membership of the samples related to each clustering. The key idea of the clustering algorithms with partial supervision is to take advantage of the available classification information and apply it actively as part of the optimization procedures (Witold Pedrycz et al., 1997).

$$J = \sum_{i=1}^{c}\sum_{k=1}^{N}u_{ik}^{p}d_{ik}^{2} + \alpha\sum_{i=1}^{c}\sum_{k=1}^{N}(u_{ik} - f_{ik}b_{k})^{p}d_{ik}^{2} \quad (1)$$

Where $c$ is the amount of clusters, $N$ is the amount of samples, $u_{ik}$ is the fuzzy membership of the $k$ th sample with respect to the $i$ th cluster center, $d_{ik}$ denotes Euclidean distance between the $k$ th sample and the $i$ th cluster center, $\alpha(\alpha >= 0)$ denotes a scaling factor whose role is to maintain a balance between the supervised and unsupervised component within the optimization mechanism, here $\alpha$ takes the value of the rate $N / M$ where $M$ denotes the number of labeled samples, $b_{k} = b[k]$ $k = 1, 2, ..., N$ is a two-valued (Boolean) indicator vector with 0-1 entries to distinguish between labeled and unlabeled samples, the value of the labeled samples are 1 and 0, otherwise, $F = [f_{ik}]$ $i = 1, 2, ..., c, k = 1, 2, ..., N$ is the membership matrix of the labeled samples. The component of supervised learning encapsulated in the form of $b$ and $F$ contributes additively to the objective function.

Using the standard technique of Lagrange multipliers to minimize the objective function under the constraint of $\sum_{i}^{c} u_{ik} = 1, k = 1, 2, ..., N$, we obtain the expression of the fuzzy membership and the clustering center:

$$u_{ik} = \frac{1}{1+\alpha} \left[ \frac{1+\alpha\left(1-b_k\sum_{i=1}^{c} f_{ik}\right)}{\sum_{j=1}^{c} \dfrac{d_{ik}^2}{d_{jk}^2}} + \alpha f_{ik} b_k \right] \quad (2)$$

$$v_i = \frac{\sum_{k=1}^{n} u_{ik}^2 x_k}{\sum_{k=1}^{n} u_{ik}^2} \quad (3)$$

$$\|U'-U\| = \sum_{i=1}^{c}\sum_{k=1}^{N}(u_{ik}' - u_{ik})^2 < \varepsilon \quad (4)$$

The algorithm updates the fuzzy membership matrix and the cluster centers iteratively according to the expression (2), (3). The termination criterion is as the expression (5), where $\varepsilon$ is a small number that can be set by the user.

### 3.2 Modified fuzzy clustering with partial supervision

The traditional fuzzy clustering with partial supervision algorithm introduces the marker information, and uses the labeled sample's classification information to guide the labeled samples cluster faster, and then guide the clustering process, but the result is not satisfactory. To make full use of the marker information, we propose a modification to the traditional FCM algorithm. We calculate and select the labeled samples whose distance between testing samples are less than a certain threshold, and calculate the reference membership according to the expression (6). Finally, we use the reference membership to guide the clustering process of the test samples. The modified objective function is given by

$$J = \sum_{i=1}^{c}\sum_{k=1}^{N} u_{ik}^p d_{ik}^2 + \alpha \sum_{i=1}^{c}\sum_{k=1}^{N}(u_{ik} - f_{ik}b_k)^p d_{ik}^2$$
$$+ \beta \sum_{i=1}^{c}\sum_{k=1}^{N}(u_{ik} - \overline{b}_k f_{ref\,ik})^p d_{ik}^2 \quad (5)$$

$$f_{ref\,ik} = \frac{1}{N_k} \sum_{\substack{x_j \in N_k \\ d_{jk} \le \varepsilon_2}} f_{ij} \quad (6)$$

Where $\beta(\beta \ge 0)$ stands for a parameter adjusting the effect from the supervised to the unsupervised component within the optimization mechanism, here

$\beta$ takes $\alpha/2$, $\overline{b_k}$ is a two-valued (Boolean) indicator vector likes $b_k$, but the opposite values where the labeled samples are 0 and 1 otherwise. The expression of the $f_{ref\,ik}$ is shown in (6), we calculate and select the labeled samples whose distance between testing samples are less than the threshold $\varepsilon_2$, and calculate the average of the labeled samples' membership, that is $f_{ref\,ik}$ is the reference membership of the testing sample and $k$ is to the clustering center $i$.

Using the standard technique of Lagrange multipliers to minimize the objective function, we obtain the expression of the fuzzy membership $u_{ik}$ and the clustering center $v_i$ in the same way of the traditional one:

$$u_{ik} = \frac{1}{1+\alpha+\beta} \left[ \frac{1 + (1-b_k\sum_{m=1}^{c} f_{mk}) + (1-\overline{b}_k\sum_{m=1}^{c} f_{ref\,ik})}{\sum_{m=1}^{c} \dfrac{d_{ik}^2}{d_{mk}^2}} + \alpha b_k f_{ik} + \beta \overline{b}_k f_{ref\,ik} \right] \quad (7)$$

$$v_i = \frac{\sum_{k=1}^{N}\left(u_{ik}^2 + \alpha(u_{ik} - b_k f_{ik})^2 + \beta(u_{ik} - \overline{b}_k f_{ref\,ik})^2\right)x_k}{\sum_{k=1}^{N}\left(u_{ik}^2 + \alpha(u_{ik} - b_k f_{ik})^2 + \beta(u_{ik} - \overline{b}_k f_{ref\,ik})^2\right)} \quad (8)$$

The algorithm updates the fuzzy membership matrix and the cluster centers iteratively according to the expression (7), (8). The termination criterion is the maximum iteration number and as is shown in the expression (5), where $\varepsilon$ is a small number that can be set by the user.

## 4 EXPERIMENTAL RESULTS

This paper uses experimental CT images provided by the Lung Image Database Consortium (LIDC). We select 188 CT scans including 147 malignant nodules, 149 benign nodules and 156 false-positive nodules. For the whole datasets, the nodules were labeled by five levels for malignancy with number 1 to 5. It means more suspicious of malignancy with bigger values. Therefore, the malignant nodules are corresponding to the 4 or 5 level of malignancy which are annotated by three or four radiologists in LIDC database. Similarly, we set the nodules with malignant level 1,2 and 3 as benign nodules. The false-positive nodules are segmented out, but there is no annotation information in the LIDC database. 47 malignant nodules, 42 benign nodules and 55 false-positive nodules are chosen as the labeled nodules respectively, and the other nodules are taken as testing samples. We will use two evaluation methods of accuracy and average accuracy to present the performance of the traditional algorithm and ours.

Figures 3, 4 and 5 demonstrate the clustering results of partial malignant nodules, benign nodules and false-positive nodules. The results in Fig.3, Fig.4 and Fig.5 show that most of the nodules can be correctly classified in our method, but there are also some nodules which are incorrectly classified to other classification because of less obvious characteristic, for example, the false-positive nodules 3q38, 3q85 are incorrectly classified to the malignant classification in Figure 3.We can see in the figure that the nodule 3q38 has spiculation in the edge and the nodule 3q85 has lobulation like malignant nodule.

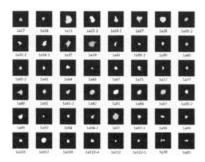

Figure 3.  Clustering results of partial malignant nodules.

Figure 4.  Clustering results of partial benign nodules.

Figure 5.  Clustering results of partial false-positive nodules.

In order to further demonstrate the performance of the algorithm proposed by this paper, Table 1 shows the performance of two algorithms. We can see in the table that the accuracy of each classification and the average accuracy are all higher than the traditional algorithm.

## 5  CONCLUSION

In this paper, a set of mainly shape-based feature vectors which can represent the feature of the pulmonary nodules completely are extracted, through analyzing and interpreting the lesion characteristics of pulmonary nodules in LIDC database. To improve the accuracy of classification ,We present a modified fuzzy clustering with partial supervision algorithm, calculating and selecting the labeled samples whose distance between testing samples are less than a certain threshold, and using their classification information to calculate the testing samples' reference membership. The reference membership can guide the testing samples to cluster more accurately. Experimental results of our proposed algorithm and the traditional algorithm illustrate that our proposed algorithm outperforms the traditional algorithm. The method can

Table 1.  Performance compare of our algorithm and the traditional algorithm.

| Algorithm | Traditional algorithm | | | Proposed algorithm | | |
|---|---|---|---|---|---|---|
| | Cor* | Err* | Acc*** | Cor* | Err* | Acc*** |
| Malignant nodule | 62 | 15 | 80.5% | 63 | 6 | **91.3%** |
| Benign nodule | 77 | 20 | 79.3% | 90 | 21 | **81.1%** |
| False-positive nodule | 87 | 47 | 65% | 86 | 42 | **67.2%** |
| Average accuracy | | | 73.4% | | | **77.6%** |

*Cor is the abbreviation of correct classify number.
**Err is the abbreviation of error classify number.
***Acc is the abbreviation of accuracy.

provide real aided diagnosis for early detection and screening lung cancer to some extent.

# 6 ACKNOWLEDGEMENTS

This research is supported by the National Natural Science Foundation of China (NSFC)(Nos. 61272245, 61103117 and 61173174), Shandong Province Young and Middle-Aged Scientists Research Awards Fund (No. BS2011DX025), Jinan Science and Technology Development Plan (No. 201401216).

# REFERENCE

Ayman, El-Baz. & Garth M.Beache.et al. 2013. Computer-Aided Diagnosis Systems for Lung Cancer: Challenges and Methodologies. International Journal of Biomedical Imaging.

Dehmeshki, J & Amin, H. 2008. Segmentation of pulmonary nodules in thoracic CT scans: a region growing approach [J]. IEEE Transaction on Medical Imaging 27(4):467–80.

DOI, K. 2005. Current Status and Future Potential of Computer-Aided Diagnosis in Medical Imaging. The British Journal of Radiology 78(S3-S19).

Han, Fangfang. & Zhang, Guopeng. & Wang, Huafeng. et al. 2013. A Texture Feature Analysis for Diagnosis of Pulmonary Nodules Using LIDC-IDRI Database. IEEE.

NIE, shengdong. & SUN, xiwen. & CHEN, zhaoxue. 2009. Progress in Computer-Aided Detection for Pulmonary Nodule Using CT Image. Chinese Journal of Medical Physics. 26(2):1075–1079.

Jemal, A. & Siegel, R. et al. 2008. Cancer statistics a Cancer Journal for Clinicians. 58(March–April):71–96.

J, Yang. & V, Honavar. 1998. Feature subset selection using a genetic algorithm. IEEE Intelligent Systems. 13(2):44–49.

Jiang, X. & S, Nie. 2008. Segmentation of Pulmonary nodule in CT image based on level set method. IEEE/ICME International Conference on Complex Medical Engineering, Beijing, 2698–270.

Samuel G. Armato. & Geoffrey, McLennan. et al.2011.The Lung Image Database Consortium (LIDC) and Image Database Resource Initiative (IDRI): A Completed Reference Database of Lung Nodules on CT Scans. Medical Physics.38(2):915–931.

Wiemker, Rafael. & Bergtholdt, Martin. & Dharaiya, Ekta. et al. 2009. Agreement of CAD features with expert observer ratings for characterization of pulmonary nodules in CT using the LIDC-IDRI database, Medical Imaging 2009: Computer-Aided Diagnosis, 72600H.

Witold, Pedrycz, Senior Member, IEEE. & James Wletzk. 1997. Fuzzy Clustering with Partial Supervision. IEEE Transaction on System, Man, and Cybernetics, PartB: Cybernetics, 27(5):787–795.

Zheng, Fuhua. & Zhang, Caiming. & Zhang, xiaofeng. & Liu, Yi. 2013. A fast anti-noise fuzzy C-Means algorithm for image segmentation, ICIP, 2728–2732.

# Feature selection for phoneme recognition using a cooperative game theory based framework

J. Ujwala Rekha
*Department of Computer Science and Engineering, JNTUH College of Engineering Hyderabad, India*

K. Shahu Chatrapati
*Department of Computer Science and Engineering, JNTUH College of Engineering Manthani, India*

A. Vinaya Babu
*Department of Computer Science and Engineering, JNTUH College of Engineering Hyderabad, India*

ABSTRACT: Human speech results from the air being forced from the lungs, through the vocal chords and along the vocal tract through the mouth and nose. The resultant sound and the mixture of frequencies which are produced by these different sources of sounds determine the unique sound i.e., a phoneme. The problem is to determine individually, the mixture of frequencies that characterize each phoneme. Mel Frequency Cepstral Coefficients called MFCCs are the most common features used in Automatic Speech Recognition (ASR) systems which represent the frequency-intensity pattern of speech. Consequently, coefficients of MFCCs which are relevant for identifying each phoneme are to be determined individually. This endeavor can be mapped to feature selection, one of the fundamental problems in machine learning, pattern recognition, and statistics. This paper proposes a cooperative game theory based framework to find individually the mixture/subset of coefficients relevant for identifying each phoneme.

## 1 INTRODUCTION

Human speech is produced by vibrations of the air exhaled from the lungs, which is then modulated and shaped by the vocal cords. The output of the vocal cords is further shaped by resonances of the vocal tract as the air is pushed out through the mouth and nose. The sounds produced by the combination of different sources, modulated by the shape and size of these sources determine unique sound - i.e., a phoneme, resulting in a shaped spectrum with energy peaks (poles) and zeros depending on the properties of these sources. MFCCs are derived from the mel-scale filter bank output using the discrete cosine transform. The low-order MFCCs describe the slowly varying spectral envelope while the higher-order ones describe the fast variations of the spectrum. Depending on the sound produced some MFCCs have higher values (peaks) while other MFCCs have lower values (zeros). Therefore, the coefficients of MFCCs that are relevant for identifying each phoneme are different.

Conventional ASR systems look at all of the MFCCs in each frame, for training and classification (Juang et al. 1993, Young, S. et al. 2006). Due to the "curse of dimensionality", some studies advocate the reduced feature set that characterizes phonetically different sounds while preserving the discriminability amongst them (Paliwal, K.K. 1992, X.Wang et al. 2003). Dimensionality reduction can be performed by either feature transformation and/or feature selection. While feature transformation generates a reduced set of new features, feature selection chooses a subset of the original features.

The most popular feature transformation techniques for dimensionality reduction used in current ASR systems are Principal Components Analysis (PCA) and Linear Discriminant Analysis (LDA) (Paliwal, K.K. 1992, Wang et al. 2003, Zahorian et al. 2011). PCA is accomplished by projecting data from a higher dimensional space to a lower dimensional space, such that the error incurred in reconstructing the data represented in the higher dimensional space to reduced dimensionality is minimized. On the other hand, the objective of LDA is to perform dimensionality reduction so that the ratio of "between class variance" to "within class variance" is minimized. In the research by Singh-Miller et al. (2007) Neighborhood Components Analysis (NCA) method for acoustic modeling is proposed. NCA learns a projection of acoustic vectors that optimizes a criterion which is closely related to the classification accuracy of a nearest-neighbor classifier. A variant of LDA called Heteroscedastic Discriminant

Analysis is proposed in (Kumar et al. 1998, Zhang et al. 2005); it is a generalization of LDA to handle heteroscedasticity where "within class variance" is not same for all classes.

There are published works on feature selection, coupled with multi-stream ASR systems. Multi-stream ASR systems concatenate multiple feature streams for training and classification (Abdulla et al. 2003, Christensen et al. 2000, Ellis & Daniel P.W. 2000, Ellis et al. 2000). The characteristic of such systems is to choose an ensemble/subset of multiple feature streams for concatenation that satisfies the objective criterion. Generally, the objective criterion for ASR systems is maximizing accuracy. In (Ellis & Daniel P.W. 2000, Ellis et al. 2000) four feature streams-Perceptual Linear Predictive (PLP) Cepstral coefficients, their deltas, modulation-filtered spectrogram features which are split into two banks covering 0–8 Hz modulation frequencies and 8–16 Hz are considered. The conditional mutual information is used in Ellis et al. (2000) to determine the combination of feature streams which are jointly advantageous. In (Abdulla et al. 2003, Christensen et al. 2000) each feature stream corresponds to a different frequency band. Hill climbing approach for feature selection at the level of individual features rather than feature streams is investigated in Gelbart et al. (2009). It considers two feature streams-MFCCs and PLP coefficients, and selects some MFCCs and some PLP coefficients discarding other coefficients.

In this paper, we propose a cooperative game theory based framework for feature selection to individually determine coefficients of MFCCs which are relevant for identifying each phoneme. The essential idea behind our endeavor is that, not all coefficients of MFCCs are relevant for modeling a phoneme, and that each phoneme is characterized by a different set of coefficients. However, it is not feasible to run experiments combinatorially on all subsets of the feature-coefficients to select the best performing subset. Therefore, a feature ordering method is desired to retain coefficients that provide the greatest contribution to recognition performance while discarding coefficients that contribute the least. At each step, our feature selection algorithm either adds a coefficient with the greatest contribution to the subset of previous step or discards a coefficient with least contribution from the subset.

## 2 GAME THEORY

Game theory is a branch of applied mathematics used in diverse disciplines such as social sciences, biology, engineering, computer science and particularly economics for strategic decision-making.

Specifically, it is the study of mathematical models of conflict and cooperation among intellectual, rational decision-makers (Roger, B. Myerson 1991).

A cooperative game is a game where groups of players form coalitions, and a payoff which is a real-valued function is associated with each coalition which denotes the profit achieved by the coalition in the game; hence the game is a competition among coalitions of players, rather than among individual players. Mathematically, a cooperative game is defined by a pair $(N,v)$, where $N=\{1,2,...,n\}$ is the finite set of players and $v(S)$ is a characteristic function defined as $v:2^N\{R\}$. The characteristic function $v(S)$, describes *collective payoff* a group of players can achieve by forming a coalition $S \subseteq N$. A *payoff profile* is a distribution of collective payoff to each of the players. An efficient payoff profile is one in which the sum of each player's payoff is $v(N)$. In game theory, *Shapley* value is a solution concept for cooperative games (Lloyd S. Shapley 1953) that distributes collective payoff among the players in a fair way. Let the marginal contribution of a player $i$ in coalition $S \cup \{i\}$ be calculated as $v(S \cup \{i\})-v(S)$, then the Shapley value that distributes collective payoff is defined as follows

$$\Phi_i(v) = \sum_{S \subseteq N \setminus \{i\}} \frac{|S|!\,(|N|-|S|-1)!}{|N|!} \\ v(S \cup \{i\})-v(S) \qquad (1)$$

where the sum ranges over all $S \subseteq N$ not containing player $i$.

## 3 FEATURE SELECTION WITH COOPERATIVE GAME THEORY

Feature selection chooses a subset of features that are most relevant to modeling and/or classification while discarding features that are redundant and irrelevant. By translating feature selection into cooperative game theory framework, MFCCs are mapped to N players, and the payoff represented as $v(S)$ is the measure of the performance of the classifier $f(S)$ generated using the set of coefficients S.

Let S be a set of coefficients and $S' \subseteq S$; P a set of phonemes; then $acc_p(S')$ is the accuracy of the classifier $f(S')$ in recognizing the phoneme $p \in P$; $conf_{p,p'}(S')$ is the number of times p is confused with p' by the classifier $f(S')$ divided by the total number of instances of p where $p' \neq p$. The more the difference between $acc_p(S')$ and $conf_{p,p'}(S')$, the better the classifier $f(S')$ distinguishes p from p'. The measure that determines how best a classifier $f(S')$ distinguishes p from p' can be calculated as $v_{p,p'}(S')=acc_p(S')-conf_{p,p'}(S')$ where p'≠p. Thus, the set of coefficients S' that best distinguishes p from p' is the maximizer of

$$\max_{S'} \; v_{p,p'}(S') \qquad (2)$$

Therefore, a classifier $f(S')$ that increases $acc_p(S')$ while simultaneously decreasing $conf_{p,p'}(S')$ for $\forall p' \in P \wedge \neq p'$ is the best classifier for phoneme p and S' is called the winning coalition for the phoneme p. The measure that determines the performance of a classifier $f(S')$ in recognizing a phoneme p while distinguishing p from all p'≠p can be calculated as

$$v_p(S') = acc_p(S') - \sum_{\forall p' \in P \wedge p' \neq p} conf_{p,p'}(S') \qquad (3)$$

and the winning coalition S' is the set of coefficients that is the maximizer of

$$\max_{S'} \; v_p(S') \qquad (4)$$

It is impractical to run experiments combinatorially on all subsets of the feature coefficients to select the subset S' that maximizes either (2) or (4). Therefore, a feature ordering algorithm based on Shapley value is used to retain coefficients which provide the greatest contribution to recognition performance and discard coefficients that contribute the least. The marginal contribution of a coefficient $i$ in the coalition S' in recognizing p while distinguishing p from all p'≠p can be determined as

$$cont_p(\{i\}, S') = v_p(S' \cup \{i\}) - v_p(S') \qquad (5)$$

Shapley value $\Phi_p(i)$ which determines the actual contribution of a coefficient $i$ in the coalition S' for recognizing a phoneme p can now be defined as follows:

$$\Phi_p(i) = \sum_{S' \subseteq S \setminus \{i\}} \frac{|S'|! \, (|S| - |S'| - 1)!}{|S|!} v_p$$
$$(S' \cup \{i\}) - v_p(S') \qquad (6)$$

Computation of Shapley value requires summing over all possible subsets of coefficients that are impractical. Therefore, our feature selection algorithm (FSA) given in Algorithm 1 begins with S'=S and at each step, generates classifiers $f(S' \cup \{i\})$ (for all $i \in$ S-S') and classifiers $f(S'-\{i\})$ (for all $i \in$ S'). That is, it generates only |S| subsets instead of $2^{|S|}$ subsets

in a single step. Then at each step, it computes the contribution of each coefficient $i$ in the coalition S' from the new Shapley value determined according to (7) and either adds a coefficient with the greatest contribution to the coalition S' or discards a coefficient with the least contribution from the coalition S'.

$$\Phi'_p(i) = \begin{cases} \dfrac{|S'-1|}{|S'|} \sum_i v_p(s') - v_p(s' - \{i\}) \text{for all } i \in S' \\ \dfrac{1}{|S-S'|} \sum_i v_p(S' \cup \{i\}) - v_p(S') \\ \qquad \text{for all } i \in S - S' \end{cases} \qquad (7)$$

---

**Algorithm 1**   Feature-Selection-Algorithm (FSA)

---

**Input**
  i. A sampling dataset T
  i. S={$c_1,...,c_n$}

**Output**
  i. Selected feature subset S' that best recognizes phoneme p∈ P while distinguishing p from all p'≠p
1. Initialize S'=S and R'=Φ
2. Generate classifier $f(S')$ from dataset T
3. Repeat step 4 until there is no change in S'
4. For each p'∈ P∧p'≠p
    4.1. For each $c_i \in$ S'
        4.1.1. Generate classifier $f(S'-\{c_i\})$
        4.1.2. Compute contribution of $c_i$ in coalition S' using formula (7)
    4.2. choose min$c_i$ whose contribution is the least
    4.3. For each $c_j \in$ R'
        4.3.1. Generate classifier $f(S' \cup \{c_j\})$
        4.3.2. Compute contribution of $c_j$ in coalition S'∪{$c_j$} using formula (7)
    4.4. choose max$c_j$ whose contribution is the greatest
    4.5. if $v_{p,p'}(S' \cup \{max c_j\}) \geq v_{p,p'}(S'-min c_i)$
        then S'= S'∪{max$c_j$} and R'=R'-{max$c_j$} else S'={min$c_i$} and R'=R'∪{min$c_i$}

---

## 4   EXPERIMENTS AND RESULTS

### 4.1  Speech Database

In order to evaluate the performance of the FSA, TIMIT acoustic-phonetic database is used. This database is a phonetically transcribed corpus and contains a total of 6300 sentences, 10 sentences spoken by 630 speakers of 8 major dialect regions of the United States (Lemel, L. et al. 1986). TIMIT corpus is subdivided into two portions for training and testing containing 4620 and 1680 sentences respectively. It is partitioned into Training Set (TRS) and Test Set (TES) in such a way that the speakers of two sets are different (speaker independent). In our experiments, a Minimal Training Set (MTS), which is a subset of the original

Table 1.    Recognizer built using reduced feature set vs. recognizer built using all of the MFCCs.

| Phonemes | Features Relevant | Number of Features Relevant | TIMIT TrainingSet Full Feature Set | Minimal Training Set Reduced Feature Set | TIMIT Training Set Reduced Feature Set | TIMIT Test Set Reduced Feature Set |
|---|---|---|---|---|---|---|
| ax | 1,2,3,4,6,7,8,9,10,11 | 10 | 40.4 | 47.79 | 46.2 | 43.9 |
| m | 0,1,2,3,4,5,7,8,9,10,11 | 11 | 43.6 | 51.7 | *42.1 | *38.1 |
| b | 0,1,2,3,4,7,8,9,10,11 | 10 | 70.9 | 74.81 | 71.5 | 72.2 |
| iy | 0,1,2,3,4,5,6,7,8,10,11 | 11 | 59.2 | 64.09 | 59.9 | 60.3 |
| v | 0,2,3,4,5,6,7,8,10,11 | 10 | 61.5 | 67.45 | 61.5 | 64.1 |
| ey | 0,1,2,4,5,6,7,10,11 | 9 | 58.3 | 63.04 | 59.2 | *57.5 |
| t | 0,2,4,10,11 | 5 | 37.6 | 44.4 | 38.9 | *36.7 |
| ae | 1,2,3,4,5,6,7,8,9,10 | 10 | 59.2 | 63.32 | *58.2 | *58.7 |
| d | 0,1,2,5,6,7,8,9,10 | 9 | 14.8 | 17.62 | 16.4 | 15.8 |
| n | 2,10 | 2 | 23.3 | 40.19 | 34.5 | 35 |
| ay | 1,2,3,4,5,6,7,10 | 8 | 53.4 | 55.74 | 53.5 | *52.5 |
| z | 0,1,2,4,5,6,7,8,10 | 9 | 61.8 | 70 | 63.6 | 62.4 |
| ih | 1,2,3,4,5,6,7,8,9,10,11 | 11 | 30.9 | 34.75 | *29.9 | *29.9 |
| l | 0,1,2,3,4,5,6,7,10 | 9 | 31.3 | 39.83 | 32.7 | 31.4 |
| aa | 0,1,2,3,9,11 | 6 | 40 | 51.87 | 43.5 | 40.2 |
| sh | 1,2,5,6,7,8,10,11 | 8 | 76.4 | 79.87 | 79.7 | *75.8 |
| axr | 0,1,2,3,5,7,8,11 | 8 | 55.4 | 61.28 | 57.9 | 60.6 |
| jh | 0,2,4,6,7,8,9,10,11 | 9 | 48.2 | 62.77 | 48.3 | 52.7 |
| aw | 0,1,4,5,6,7,8,10,11 | 9 | 54.9 | 64.52 | *53.6 | *48.8 |
| ah | 0,1,3,4,5,6,7,8,11 | 9 | 38 | 39.27 | 41.4 | *33.1 |
| p | 0,2,3,4,6,7,8,10 | 8 | 77.9 | 75.69 | *72.4 | *69 |
| s | 0,1,2,5,6,7,8,9,10,11 | 10 | 79.6 | 80.18 | 81.5 | 79.6 |
| uw | 0,1,2,4,5,6,8,9,10,11 | 10 | 44.7 | 54.83 | 45.6 | *42.4 |
| ao | 1,2,3,5,6,9,10 | 7 | 46.5 | 58.78 | 51.7 | 51.7 |
| er | 0,1,3,4,5,6,7,8,10 | 9 | 35.5 | 47.84 | *33.7 | *28.2 |
| k | 0,1,3,4,7,9,11 | 7 | 47 | 47.28 | 47.2 | *43.2 |
| eh | 1,2,4,5,6,7,8,9 | 8 | 40.8 | 45.75 | 43.7 | 43.9 |
| ng | 0,1,4,5,6,8,9,11 | 8 | 64.7 | 74.07 | 67 | 64.8 |
| y | 0,1,2,3,4,5,6,7,8,10 | 10 | 75.9 | 78.93 | 76.5 | *72.7 |
| ch | 0,1,2,3,4,6,8 | 7 | 62 | 73.61 | *57 | *55.8 |
| w | 0,1,2,3,4,5,8,11 | 8 | 77.9 | 83.81 | 78.2 | 80.8 |
| hh | 0,1,5,6,7,9,11 | 7 | 52.9 | 59.71 | 58.8 | 55.8 |
| uh | 1,2,3,4,5,6,8,11 | 8 | 31.1 | 53.57 | 32.1 | 31.9 |
| f | 0,1,2,3,4,5,6,7,8,9,10 | 11 | 86.1 | 90.4 | 86.2 | 87.4 |
| ow | 0,3,4,5,6,7,9 | 7 | 43 | 50.85 | *41.5 | *34.2 |
| g | 0,3,4,5,6,8,11 | 7 | 30.6 | 43.73 | 37.9 | 36.3 |
| dh | 0,1,2,3,4,5,7,8,9,10 | 10 | 30.1 | 37.75 | 32.1 | 35.3 |
| zh | 0,1,2,3,4,5,6,7,8,9,10,11 | 12 | 72.4 | 78.57 | 72.4 | *60.8 |
| oy | 0,1,2,4,5,10,11 | 7 | 61.3 | 60 | *56.2 | *53.5 |

(*Instances where accuracy, using FSA is less than Baseline System)

TIMIT training set is obtained by randomly selecting 600 sentences. Also, in our experiments the original 61 TIMIT phonemes are mapped to a smaller set of 39 phonemes as in Lee, K. F. & Hon, H. W. (1989).

### 4.2    Baseline System

Each sample of the speech signal is represented by a 12-dimensional feature vector containing 12 MFCCs. An HMM-based recognizer is developed that models each phoneme by a context-independent five-state (including entry and exit states) left-to-right HMMs. In the baseline system, all the original TIMIT TRaining Sets (TRS) are used for training HMMs. Moreover, feature vector containing 12 MFCCs is used in the process of training HMMs. The accuracy of this system in recognizing each of the phonemes is listed in column 4 of Table 1.

### 4.3  Experiments using an FSA

Feature Selection Algorithm (FSA) is run for each of the 39 phonemes using an HMM based recognizer to individually determine the mixture/subset of coefficients relevant to identify each phoneme. Each phoneme is modeled by a context-independent five-state (including entry and exit states) left-to-right HMM. The recognizer/classifier wrapped in our FSA will be built from the Minimal Training Set (MTS) described in section 4.1 to individually predict the subset of coefficients relevant to identify each phoneme. Thus, a classifier/recognizer each phoneme uses for identifying it (obtained using FSA) is built from MTS. The accuracy of each of classifiers'/recognizers' in recognizing the phoneme for which it was built is listed in column 5 of Table 1.

Later, the subset of coefficients predicted by the FSA on MTS is used for training HMMs on the complete TRaining Set (TRS) and tested on the TIMIT TEst Set (TES). The idea behind not running the FSA on the complete training set (TRS) is because it is time-consuming. Column 6 and 7 in Table 1 list the accuracy of the classifiers'/recognizers' built from TRS and TES respectively in recognizing the phoneme for which it was built using the reduced feature set (obtained using FSA).

### 4.4  Discussion of Results

- In most of the cases, recognizer obtained using the reduced feature set performed better than the recognizer built from all the MFCCs
- An interesting case is obtained for the phoneme *n*. Two features were sufficient to recognize phoneme *n* accurately. Moreover, the percentage increase in the accuracy of classifying phoneme *n* using the reduced feature set is more than 50%.
- Whenever there is a decrease in accuracy using the reduced feature set, the percentage of decrease in accuracy is less than 15%.
- Reduced feature set obtained on the Minimal Training Set (MTS), gave better results even on complete TIMIT TRaining Set (TRS) and unseen TIMIT TEst Set (TES).
- The average number of relevant features is 8.8.

## 5  CONCLUSION

Subsets of coefficients relevant for identifying individual phonemes were obtained using cooperative game theory based framework. In most cases, recognizer using the reduced feature set outperformed the recognizer built from all the coefficients. However, it will be more appealing if one can justify the relevance of features on some physical grounds so that it can be used with other databases and for other recognition tasks. Moreover, feature selection of features in a single stream is investigated which can easily be extended to select features from multiple streams.

## REFERENCES

[1] Abdulla, Waleed H., and Nikola Kasabov. 2003. Reduced feature-set based parallel CHMM speech recognition systems. *Information Sciences* 156(1): 21–38.

[2] Christensen, Heidi, Borge Lindberg, and Ove Andersen. 2000. Employing heterogeneous information in a multi-stream framework. *In: Proc. Int. Conf. on Acoustics, Speech, and Signal Processing*, vol. 3, pp. 1571–1574.

[3] Ellis, Daniel PW. 2000. Stream combination before and/or after the acoustic model. *In: Proc. Int. Conf. on Acoustics, Speech and Signal Processing*, vol. 3, pp. 1635–1638.

[4] Ellis, Daniel PW, and Jeff A. Bilmes. 2000. Using mutual information to design feature combinations. *In: Proc. 6th International Conference on Spoken Language Processing*, pp. 1–4 .

[5] Gelbart, David, Nelson Morgan, and Alexey Tsymbal. 2009. Hill-climbing feature selection for multi-stream ASR. *In: INTERSPEECH*, pp. 2967–2970.

[6] Juang, Biing-Hwang, and Lawrence Rabiner. 1993. Fundamentals of speech recognition. *Signal Processing Series*. Englewood Cliffs, NJ : Prentice Hall.

[7] Kumar, Nagendra, and Andreas G. Andreou. 1998. Heteroscedastic discriminant analysis and reduced rank HMMs for improved speech recognition. *Speech communication* 26(4) : 283–297.

[8] Lee, K. F. and Hon, H. W. 1989. Speaker-Independent Phoneme Recognition Using Hidden Markov Models. *IEEE Transactions on Acoustics, Speech, and Signal Processing*, 37(12), pp. 1641–1648.

[9] Lemel, L., R. Kassel, and S. Seneff. 1986. Speech database development: Design and analysis. *In: Proc. DARPA Speech Recognition Workshop*, Report no. SAIC-86/1546.

[10] Lloyd S. Shapley. 1953. A Value for n-person Games. In Contributions to the Theory of Games, volume II, by H.W. Kuhn and A.W. Tucker, editors. *Annals of Mathematical Studies* v. 28, pp. 307–317. Princeton University Press.

[11] Paliwal, K. K. 1992. Dimensionality reduction of the enhanced feature set for the HMM-based speech recognizer, *Digital Signal Processing* 2(3): 157–173.

[12] Roger, B. Myerson. 1991. Game theory: Analysis of conflict.

[13] Singh-Miller, Natasha, Michael Collins, and Timothy J. Hazen. 2007. Dimensionality reduction for speech recognition using neighborhood components analysis. *In: INTERSPEECH*, pp. 1158–1161.

[14] Wang, Xuechuan, and Kuldip K. Paliwal. 2003. Feature extraction and dimensionality reduction algorithms and their applications in vowel recognition. *Pattern recognition* 36(10) : 2429–2439.

*Multimedia, Communication and Computing Application – Leung (Ed.)*
© *2015 Taylor & Francis Group, London, ISBN 978-1-138-02775-6*

# A solution to the reconstruction of a realistic knee model based on CT images

J.F. Li, J.X. Chen, R.F. Li, H.W. He & J.Q. Mo
*Guangdong Provincial Key Laboratory of Computer Integrated Manufacturing System, School of Electromechanical Engineering, Guangdong University of Technology, Guangzhou, China*

ABSTRACT: Virtual knee arthroscopic surgery requests the high quality of the virtual three-dimensional model of knee. The high-quality virtual three-dimensional model can contribute to the immersion of the system, and it is necessary to meet the demand of real-time interaction. This paper provides an analysis of the CT images of knee and puts up a method to reconstruct a three-dimensional knee model that is based on the patient's CT images. By handling the voxel existing in the three-dimensional data field that is composed of CT sequence images and converting geometric data to graphic data, we can get the model of knee through CT three-dimensional reconstruction. Finally, an illustrative example is verified.

KEY WORDS: Reconstruction of three-dimensional model, CT images, Arthroscopic surgery, Volume rendering

## 1 INTRODUCTION

An eligible knee model is an important part of the composition of the virtual training system for knee arthroscopy, and a realistic knee model allows the operator to get a better sense of immersion. In addition, the knee model corresponding to geometric topology directly determines the surgical cutting, deformation, and other real-time interactive operations [1].

At present, two methods are used to build a three-dimensional model of the knee. The first method is based on a three-dimensional modeling software (such as 3DS Max), which by virtue of its experiences builds a three-dimensional model of the knee. As a tool, it features a small amount of data, but the data are not precise enough and the model is not realistic. The second method is based on medical tomographic images (such as CT and MRI), which gets a three-dimensional model of the knee through algorithm processing of medical images. This method can get a realistic and accurate model of the knee, as well as getting a large amount of data. Depending on different rendering algorithms, the 3D reconstruction of medical tomographic images is divided into surface rendering and volume rendering. Surface rendering focuses on information on the surface of the organ tissue, ignoring internal details, with good fidelity and a small amount of data. Volume rendering retains the internal details of the overall tissue and maps the organs and tissues completely, but the model contains a large amount of data [2].

This paper presents a three-dimensional model of the knee using the volume rendering method based on light projection method, which is based on CT images used in the knee provided by the Guangzhou Orthopedics and Traumatology Hospital. Each knee CT image resolution is $512 \times 512$, and the total is 230.

## 2 CT IMAGE PREPROCESSING

As the knee CT image generation and transmission in the procession are often subject to various interfering factors, the quality of image will be decreased. CT images are preprocessed before being used to finish reconstruction, so as to improve the efficiency of the three-dimensional reconstruction and get a more accurate and realistic three-dimensional model of the knee.

### 2.1 Image cutting

The original CT image of the knee includes complete information on the two legs of the knee, so here is a professional graphic image processing tool for CT image batch crop, so that the CT images leave only the image information side of the knee. Knee CT images before cutting and after cropping are shown in Fig. 1. After cutting, it reduces the amount of data of CT images, which can accelerate the speed of reconstruction.

 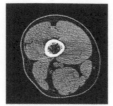

a. Before CT image cropping     b. After CT image cropping

Figure 1.   CT image cropping.

## 2.2  *Image enhancement*

Image enhancement depends on the application's needs, projecting artificial suppression or elimination of certain information. In general, it is the enhancement of interesting information and reduction or elimination of unwanted information (noise and interference).

Median filtering is used for CT image denoising and it makes the image clearer. A two-dimensional median filter can be used in Equation 1. The method discussed in this paper used a $3 \times 3$ sliding window for each $3 \times 3$ pixel gray values of the pixels within the field of any sort and took the middle of the gray values as the gray values of all pixels [3].

$$g(m,n) = Median\{f(m-k, n-l), (k,l) \in W\} \qquad (1)$$

$W$ denotes the vertical plane window (usually $2 \times 2$ or $3 \times 3$), and $f(m,n)$ and $g(m,n)$ denote the original and processed images, respectively.

## 3  THREE-DIMENSIONAL MODEL

After CT image optimization and processing, a three-dimensional model can be reconstructed. Because each knee CT image data value represents a group of two-dimensional matrices, the 230 knee CT images are sequentially superposed together, constituting a three-dimensional volume data field. Assuming three-dimensional volume data of the voxels can absorb and reflect light of small particles, we simulate the light passing through the translucent material. After generating the light, calculating intensity, and synthesizing and displaying optical image, it can finally achieve a three-dimensional model of knee reconstruction.

## 3.1  *Light generation*

Ray casting belongs to volume rendering algorithms. Volume rendering methods with the highest image quality help to store details, especially for blurring of features, which give voxel model featuring a high degree of correlation between organs and tissues. It is assumed that the three-dimensional volume data are evenly distributed on the grid or rules of grid points. Fig. 2 is a schematic diagram of a three-dimensional data field. The three-dimensional data field is generated by superposition of the multilayer knee CT image. Each data field represents a voxel grid, which is the smallest volume data cell in the three-dimensional data field [4].

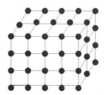

Figure 2.  Rule structure data field.

Knee CT image reconstruction is based on the requirements of the three-dimensional model, and the first step is to simulate the light generated. We assume that each point on the screen issues a ray, and rays pass through a three-dimensional volume data field. A voxel is treated as particles that have physical properties, and we should study the relationship between the three-dimensional volume data field and the voxel when light passes through them.

## 3.2  *Intensity calculation*

The data of a 3D volume data field are expressed in different tissues of the knee. The muscle, fat, bone, and other organs have different meanings in the data. To display the internal structure by opacity and to get the transparent image, the transparency value should be given to the type of a (a is equal to 0 means image is completely transparent, and a is equal to 1 means image is completely opaque).

In this example, according to the outcomes of several experiments different sets of gray values are shown in Table 1. Different gray values of fat, muscle, and bone with the best three-dimensional display are shown.

Table 1.  Knee opacity settings.

|  | Fat | Muscle | Bone |
| --- | --- | --- | --- |
| Gray value | 0-69 | 70-219 | 220-255 |
| Opacity | 0.3 | 0.5 | 1.0 |

After setting the opacity, each voxel is assigned a different color value (R, G, B). $P_i$ will be the percentage of the first $i$ substances, and $C_i = (a_i R_i, a_i B_i, a_i)$ is the color value of the first $l$ substances; then, the color value of the voxel is shown in Equation 2:

$$C = \sum_{i=1}^{n} P_i C_i \qquad (2)$$

The light's test direction passes through the three-dimensional volume data field and intersects with the voxel. Selecting $K$ equidistant sampling points on the ray, opacity and color values from a sample point of the last eight voxels do trilinear interpolation, and then the opacity and color values of the sampling point are solved. Fig. 3 shows the radiation emitted from the screen through a three-dimensional volume data field, performing schematic trilinear interpolation.

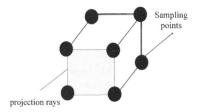
Sampling points

projection rays

Figure 3. Trilinear interpolation values.

Trilinear interpolation comprises opacity and color interpolation, assuming the value of the volume data with eight voxel vertices $P_i$ (i = 0, 1, 2, ... ,7) is $V_i$ (i = 0, 1, 2, ... ,7). The interpolation is based on the values of the eight vertices of the estimated value of $V_p$ that can represent the voxels estimating a point $P(x, y, z)$ of the internal volume data $V(x, y, z)$. Because the trilinear interpolation assumes that the data changes in two points of the lines have little gaps, three-dimensional volume data can be set with an arbitrary voxel vertex at the origin, which are the coordinate values. The obtained voxel values $P(x, y, z)$ of the trilinear interpolation can be estimated by the trilinear function shown in Equation 3:

$$V_p = (1-x)(1-y)(1-z)V_0 + (1-x)y(1-z)V_1 +$$
$$(1-x)(1-y)zV_2 + (1-x)yzV_3 + \qquad (3)$$
$$x(1-y)(1-z)V_4 + xy(1-z)V_5 +$$
$$x(1-y)zV_6 + xyzV_7$$

In this paper, vtkPiecewiseFunction () is used to set opacity, and AddPoint () method of this class calls for specific settings. Color values are set completely by vtkColorTransferFunction () and use the member function AddRGBPoint () for specific settings.

### 3.3 Synthesis and display of the optical image

After intensity calculations, the resampling process is finished by all sampling points on all rays emitted from the screen. Then, compound the optical image along the ray direction. There are generally two types

of compounding, front to back and back to front. This paper uses the method of front to back to get synthesis of opacity and color value and obtain the color value of the pixel [5].

$C_{now}$ and $a_{now}$ represent the color and opacity of the first $i$ voxel value. $C_{in}$ and $a_{in}$ represent the color and the opacity of the entered first $i$ voxel. $C_{out}$ and $a_{out}$, respectively, are the result of color values and opacity values after the first voxel, and the equations are shown as follows(4 and 5)

$$C_{out} = C_{in}(1-a_{now}) + C_{now}a_{now} \qquad (4)$$
$$a_{out} = a_{in} + a_{now}(1-a_{in}) \qquad (5)$$

When the calculated color values and opacity values for each pixel are on the screen, and then after format conversion, the geometric data format of the medical image will be converted into three-dimensional graphical entities. Finally, we get a three-dimensional model of the knee.

Computer configuration: Memory 4G, processor Intel (R) Core (TM) i5-3470, and graphics NVIDIA Quadro600.

Tool: Visual Studio 2010, VTK.

The result of rendering the three-dimensional model of knee is shown in Fig. 4.

Figure 4. The result of rendering three-dimensional model of the knee.

After contrasting the outcomes of the repeated experiments, it costs 5 seconds to finish reconstruction with the preprocessing images, which is 5 times faster than no preprocessing. This shows that pretreatment processing is very important for the reconstruction of 3D model of the knee with preprocessing, which can reduce the cost of the system.

## 4 TRIANGULAR FACES CUT

It will need a large amount of data to finish reconstruction of the three-dimensional model of knee, and the model contains a large number of vertices and triangular faces. Therefore, it is necessary to cut a triangular faces model and reduce the amount of data output from the knee model. The method using triangular faces cut is vtkDecimatePro(). After an 80% reduction in triangular facets, knee model data are greatly reduced and the model of the knee vision does not have any distortion.

Ultimately, the three-dimensional model of the knee has a simple interactive feature, such as translation, rotation, scaling, etc. This model can be used as vtkWriter () to store the three-dimensional model of reconstruction and import knee arthroscopy training system.

## 5 CONCLUSION

After analyzing CT images of the knee and the structured three-dimensional volume data field built by knee CT images, giving its physical attributes voxels, we make the optical model. Then, we assume that the light is emitted from each pixel on the screen. According to the simulation principle of light energy to penetrate opaque substances, we can efficiently draw three-dimensional model of the knee joint. Experimental results show that the three-dimensional model of the knee reconstructed by the algorithm is realistic and accurate and able to satisfy the requirements of the virtual knee arthroscopy training system.

ACKNOWLEDGMENTS

This work was financially supported by the National Natural Science Foundation of China (61300106 and 51275094) and Science and Technology Program of Guangzhou, China (No. 2010J-D00341).

REFERENCES

[1] Xu, F. 2010. Research on Key Techniques of Virtual Surgery Training System. Beijing: Beijing Institute of Technology.
[2] Kumar, T.S., Rakesh, P.B. 2011. 3D reconstruction of facial structures from 2D images for cosmetic surgery. Recent Trends in Information Technology (ICRTIT), 2011 International Conference on IEEE, 2011: 743–748.
[3] Liu, G.H., Guo, W.M. Improved median filtering denoising algorithm analysis. Computer Engineering and Applications, 46(10): 187–189.
[4] Frieder, G., Gordon, D., Reynold R.A. 1985. Back-to-front display of voxel-based objects, IEEE Computer Graphics and Applications, 5(1): 52–60.
[5] Hong, T., Pan, Z.F., Lin, L.B. 2011. Application and implementation of the three-dimensional reconstruction of VTK medical images. Computer Systems and Applications, 20(4): 127–130.

*Multimedia, Communication and Computing Application – Leung (Ed.)*
© *2015 Taylor & Francis Group, London, ISBN 978-1-138-02775-6*

# Clustering method of pedestrian body based on nonparametric estimation

W. Jin, B. Li, H.B. Qu & J.Q. Wang
*Beijing Institute of New Technology Applications, Beijing Academy of Science and Technology, Beijing, China*

ABSTRACT: To reduce the noise in pedestrian body detection, a kernel function–based nonparametric clustering algorithm is proposed that can classify feature points and remove noise in color space. First, the body in the video is detected using Haar-like operator and integral image, and then the noise is removed. By computing the gradient of feature points in color space and moving the points to local extremum of feature space by using iteration, feature points are classified and the noise is removed. The experiment results show that the method proposed in this paper is superior to *K*-mean and mixture Gaussian model with respect to detection accuracy and convergence time.

KEY WORDS: Nonparametric estimation, Kernel function, Local extremum, Body detection, Cluster analysis

## 1 INTRODUCTION

Pedestrian detection is effective and fast. However, problems of detection precision such as some background and noise are detected during the pedestrian detection. The source of the noise and background is very broad, such as image noise caused by motion blur and sunshine, the noise of the objects as pedestrians in the background, and so forth. The distribution of the noise and background information is unknown. Therefore, while the detected noise is removed, the classification of the detected pedestrians becomes a crucial step in the detection process.

There are many types of the study about tracking pedestrians, such as mobile sensor and fixed sensor in terms of acquisition methods, two dimensions and three dimensions in terms of data-processing methods, and having models and no models in terms of cluster. Theoretically, Bobick [1] and Ross Cutler and Larry Davis [2] detected and identified human bodies through video data; Haritaoglu et al. [3], Isard [4], and Ramanan [5] modeled the human body as different shapes to track human body movements. Ioffe and Forsyth [6] and Ramanan and Forsyth [7] proposed to detect human body in single images through probability methods. Owing to the variety of the background and noise, the detection value is distributed with multimode in the feature space. For the given video images, because it is impossible to determine the number of the people and to get the distributing information of detected background and noise, the fitting with the traditional classic parameterized method of estimating probability density is unable to

be conducted. More importantly, when the automatic pedestrian detection system is built, the automatic tracking of human movements should be realized with the decrease of artificial intervention or even without artificial intervention. Therefore, without the foreknowledge of the detection value and noise distribution, it is necessary to construct a distribution model of pedestrians, background, and noise in the multidimensional space and to apply this model to the cluster analysis of the detection results. Traditional algorithms such as *K*-mean and mixture Gaussian model with multimode classification in multidimensional feature space need the foreknowledge of *K*-mean and mixture Gaussian model, parameter *k*— the concrete number of people in the video cannot be predicted, therefore, these algorithms cannot be applied to the cluster of detected pedestrians.

In this paper, a mean shift–based nonparametric density estimation cluster method is proposed to realize cluster of any distribution in the feature space. It is a clustering method of nonparametric multimodel, by which: (1) the gradient of probability density function is estimated using kernel function, (2) the local extremum point (namely the mode) of probability distribution in the feature space is searched iteratively, (3) the original feature point is moved to the local maximum point in the neighborhood, and (4) the cluster of data points is realized in the feature space. The experiment results of this method show that if convergence is faster, then cluster is more accurate than that in *K*-mean and mixture Gaussian model, and that nonparametric cluster of pedestrians is achieved so that detection accuracy is improved.

## 2 MEAN SHIFT–BASED CLUSTERING METHOD

In the given $d$-dimensional space $R^d$ there are $n$ sample points $x_i$, $i = 1, \ldots, n$. The basic form of mean-shift vector in point $x$ is defined as

$$M_h(x) = \frac{1}{k} \sum_{x_i \in S_k} (x_i - x) \tag{1}$$

where $S_h$ is a high-dimensional ball area with the radium of $h$, and it satisfies the collection of $y$ points with the following relations:

$$S_h(x) \equiv \{ y : (y - x)^T (y - x) \leq h^2 \tag{2}$$

where $k$ indicates that there are $k$ points in $n$ sample points $x_i$ which fall into area $S_h$.

From (1), it can be seen that $(x_i - x)$ is the offset vectors of the sample point $x_i$ relative to the point $x$, and that mean-shift vector $M_h(x)$ is the sum and then the average of the offset vectors relative to point $x$ of $k$ sample points which fall into area $S_h$.

In the given $d$-dimensional space $R^d$, there are $n$ sample points $x_i$, $i = 1, \ldots, n$. $K_h(x)$ is the kernel function in this space. The probability density estimation function [19] of point $x$ is

$$\hat{f}(x) = \frac{1}{n} \sum_{i=1}^{n} K_H(x - x_i). \tag{3}$$

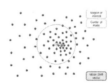

Figure 1. Schematic diagram of mean-shift vector.

Figure 1 shows the schematic diagram of mean-shift vector. In this figure, the area set by the big circle is $S_h$, small dots are sample points $x_i \in S_h$ which fall into area $S_h$, the point set by the yellow circle is the benchmark $x$ of mean shift, and the arrow shows the offset vector of the sample point relative to the reference point $x$. Because the nonzero probability density gradient points to the direction of the maximum increase of probability density, the average offset vector will point to the area with the most sample distribution, namely, the gradient direction of probability density function. Therefore, the sample points in the area $S_h$ fall more into the direction along the probability density gradient, and the corresponding mean-shift vector $M_h(x)$ points to the direction of the probability density gradient.

Two types of kernel functions, unit uniform kernel function and Unit Gaussian kernel function (Figure 2), are often used in the mean-shift algorithm.

Unit uniform kernel function is

$$F(x) = \begin{cases} 1 \text{ if } \|x\| < 1 \\ 0 \text{ if } \|x\| \geq 1 \end{cases}. \tag{4}$$

Unit Gaussian kernel function is

$$N(x) = e^{-\|x\|^2}. \tag{5}$$

Figure 2. Schematic diagram of kernel function.

In practice, radially symmetric kernel function is more suitable, therefore unit Gaussian kernel function is adopted in this paper [20].

Usually, diagonal matrix $H = diag[h_1^2 L h_d^2]$ or unit matrix $H = h^2 I$ is used by bandwidth parameter. With the consideration of $K(x) = k(\|x\|^2)$, in the given $d$-dimensional space $R^d$ there are $n$ sample points $x_i$, $i = 1, \ldots, n$, the probability density estimation function can be transformed into [21]

$$\hat{f}(x) = \frac{1}{nh^d} \sum_{i=1}^{n} k\left( \left\| \frac{x - x_i}{h} \right\|^2 \right) \tag{6}$$

where $h$ is the bandwidth parameter, $k(x)$ is the kernel function used for estimating the density function and it satisfies the condition that the estimation of gradient $\nabla f(x)$ of $\int k(x) dx = 1$ probability density function $f(x)$ is

$$\hat{\nabla} f(x) = \frac{2}{nh^{d+2}} \sum_{i=1}^{n} (x - x_i) k'\left( \left\| \frac{x - x_i}{h} \right\|^2 \right) \tag{7}$$

where the new function $g(x)$ is used to define $k'(x)$, and the new kernel function $G(x)$ is obtained.

Formula (7) is rewritten into

$$\hat{\nabla} f_{h,k}(x) = \frac{2}{nh^{d+2}} \left[ \sum_{i=1}^{n} g\left( \left\| \frac{x - x_i}{h} \right\|^2 \right) \right]$$

$$\left[ \frac{\sum_{i=1}^{n} x_i g\left( \left\| \frac{x - x_i}{h} \right\|^2 \right)}{\sum_{i=1}^{n} g\left( \left\| \frac{x - x_i}{h} \right\|^2 \right)} - x \right] \tag{8}$$

where $\sum_{i=1}^{n} g(\left\|\frac{x-x_i}{h}\right\|^2)$ is a positive number,

$\frac{2}{nh^{d+2}} \sum_{i=1}^{n} g(\left\|\frac{x-x_i}{h}\right\|^2)$ is the probability density

$\hat{f}_{h,G}(x)$ calculated on point $x$ by using the kernel

function $G(x)$, and $\dfrac{\sum_{i=1}^{n} x_i g\left(\left\|\frac{x-x_i}{h}\right\|^2\right)}{\sum_{i=1}^{n} g\left(\left\|\frac{x-x_i}{h}\right\|^2\right)} - x$ is the

mean-shift vector.

In (8), the part in the second bracket on the right is the mean-shift vector, and the part in the first bracket is the estimation of probability function f(x) labelled as $\hat{f}_G(x)$ with G(x) as the kernel function. $\hat{f}(x)$ defined by (8) is relabeled as $\hat{f}_K(x)$, so (8) can be rewritten as

$$\hat{\nabla}f(x) = \nabla\hat{f}_K(x) = \frac{2}{h^2}\hat{f}_G(x)M_h(x) \qquad (9)$$

$$M_h(x) = \frac{1}{2}h^2\frac{\nabla\hat{f}_K(x)}{\hat{f}_G(x)}. \qquad (10)$$

It can be shown from (10) that mean-shift vector $M_h(x)$ calculated on point $x$ by using the kernel function $G$ is proportional to the gradient of probability function $\hat{f}_K(x)$ estimated using the kernel function $K$.

If $x$ in (8) is put outside the sum, the following formula can be obtained:

$$M_h(x) = \frac{\sum_{i=1}^{n} x_i g\left(\left\|\frac{x-x_i}{h}\right\|^2\right)}{\sum_{i=1}^{n} g\left(\left\|\frac{x-x_i}{h}\right\|^2\right)} - x. \qquad (11)$$

If the first item on the right in (11) is labeled as $m_h(x)$, the following formula can be obtained:

$$m_h(x) = \frac{\sum_{i=1}^{n} G(\frac{x_i-x}{h})x_i}{\sum_{i=1}^{n} G(\frac{x_i-x}{h})}. \qquad (12)$$

## 3 CLUSTERING PROCESS

The classification of feature points in the feature space can be seen as a clustering process where the gradient of each point $x$ in the feature space is calculated, and this point is moved iteratively to its closest local extremum point by using the gradient direction calculated at this point. Using the mean-shift vector,

the feature point can be moved to the local extremum point in its neighborhood in the multidimensional space, and the feature points which are moved to the same local extremum point can be regarded as the same cluster [11].

A starting point $x$ is set, the kernel function is G(x), $\varepsilon$ is permitted, and mean-shift algorithm is conducted circularly with the following three steps until the termination condition is met:

1. Calculating $m_h(x)$
2. $x$ is assigned to $m_h(x)$
3. If $\left\|m_h(x) - x\right\| < \varepsilon$, the circulation ends; if not, step (1) is performed.

In the actual calculation, the dimension number of the color space needs to be reduced to reduce the calculation amount of the system. The order of magnitude of RGB component is reduced to 8, and a 512 dimensional color space with 8×8×8 is obtained. In this 512 dimensional color space, the number of pixels of each color is calculated, and the feature vectors corresponding to the detected image are acquired. After the image is quantified to the feature points in the 512 dimensional color space, the clustering algorithm of each feature point is conducted to acquire image cluster.

## 4 EXPERIMENT RESULTS AND ANALYSIS

### 4.1 Comparison of the clustering algorithms

To obtain the performance of the proposed clustering method regarding the realization of data cluster, the mean-shift algorithm was compared with the traditional K-means algorithm with the assumption that the foreknowledge necessary for cluster had been acquired. The data sets which were needed in the comparison included Iris data sets in the Italian UCI database, Wine data sets, and the data sets including pedestrians, noise, and background. The cluster of the data sets was conducted using mean-shift algorithm, K-means algorithm, and mixture Gaussian model and ran on frequency of 3.1 GHz INTEL dual-core CPU. The experimental results are shown in Table 1.

From Table 1, it can be seen that the maximum accuracy of K-means algorithm is still lower than the accuracy of mean-shift algorithm, so there are great differences among the clusters of K-means algorithm and the result of K-means algorithm is not stable. By contrast, the cluster accuracy of mean-shift algorithm is higher than that of K-means algorithm using the same data set. Because both the mean-shift algorithm and K-means algorithm are iterative process, the convergence time has a direct influence on the execution speed of the two algorithms. From Table 2, it can be seen that the execution time of the two algorithms is

Table 1. Comparison of clustering accuracy and convergence time of three algorithms.

| Data sets | Accuracy of k-means algorithm(%) | | Accuracy of mixture Gaussian model (%) | | Accuracy of the algorithmin this paper (%) | |
|---|---|---|---|---|---|---|
| | maximum | minimum | maximum | minimum | Maximum | minimum |
| Iris | 89.21 | 55.34 | 88.45 | 79.22 | 93.82 | 85.43 |
| Wine | 70.52 | 56.83 | 85.22 | 75.46 | 88.56 | 80.55 |
| Pedestrians | 65.84 | 52.93 | 76.34 | 66.83 | 83.25 | 78.45 |

Table 2. Convergence time of three methods.

| Data sets | $K$-Means algorithm (s) | The algorithm in this paper (s) |
|---|---|---|
| Iris data sets | 0.05 | 0.03 |
| Wine data sets | 1.1 | 0.05 |
| Pedestrians | 3.6 | 0.08 |

close only by using the Iris data sets, so the execution time of $K$-means algorithm is much more than that of mean-shift algorithm using other data sets, especially the data sets with high-dimensional pedestrians and noise. The reason is that each point in the data sets needs to be searched and the distance among them needs to be calculated when the $K$-means algorithm is conducted. When there are a large number of the vector dimensions in the data sets, more convergence time of the algorithm is needed.

### 4.2 Relationship between the choice of bandwidth and convergence time

The mean shift–based clustering time depends on the convergence time of feature points in the feature space, and the convergence time is influenced by the bandwidth parameter of the kernel function. Table 3 shows the numerical relationship between bandwidth parameter and convergence time.

In Table 3, the pedestrians and noise are detected using pedestrian detectors, and then these detected values are mapped to 512 dimensional feature space. Then mean shift cluster algorithm is conducted on 500 samples, and the convergence time is acquired. Table 3 shows that the choice of bandwidth parameter of the kernel function is related to convergence time, and that when $h$ is 0.26, the convergence time is the least.

Table 3. Relationship between bandwidth and convergence time of cluster algorithm

| Number of sample (piece) | Bandwidth parameter (h) | Convergence time (s) |
|---|---|---|
| 500 | 0.12 | 0.0523 |
| 500 | 0.2 | 0.0322 |
| 500 | 0.26 | 0.0273 |

## 5 CONCLUSION

The cluster of pedestrians is an important preprocessing stage after pedestrian detection. The detected data include noise, background, and pedestrian body, therefore, sorting should be conducted and noise should be removed so that the data can be used for the subsequent study of human body. In this paper, mean shift–based nonparametric probability density estimation method is proposed. In this method, with the kernel function, the sample points in the feature space are used to estimate the arbitrary probability distribution in the space. This nonparametric probability distribution function can be adopted for calculation to acquire the core of mean shift—the mean-shift vector of sample points in the space. With this vector, the local extremum of spatial probability distribution can be searched iteratively at a high speed, which is getting the modal point of the spatial distribution. Meanwhile, the iterative search is one process to move the feature points to local extremum, and to cluster the points to different types according to their similarities measured by spatial distance. The experiment results show that this algorithm is effective and fast, and its running on the serial images including body can detect pedestrians and separate the noise. With the appropriate kernel function bandwidth $h$, the iterative times of the algorithm can be reduced to a minimum, and the convergence time can be cut significantly so that the quick cluster of detected values can be achieved.

ACKNOWLEDGMENT

This work was supported in part by Young Core Plan of the Beijing Academy of Science and Technology.

REFERENCES

[1] Bobick, J. Davis. The recognition of Human Movement Using Temporal Templates. PAMI. 2001. 23(3): 257–267.
[2] Ross Cutler, Larry Davis. Robust periodic motion and motion symmetry detection. IEEE Computer Society Conference on Computer Vision and Pattern Recognition. 2000. 2:2615.

[3] Haritaoglu I, Harwood. D, Davis. L. VV: real-time surveillance of people and their activities. IEEE Trans Pattern Analysis and Machine Intelligence, 2000, 22(8): 809–830.

[4] M. Isard, A. Blake. Condensation – Conditional density propagation for visual tracking. IJCV. 1998. 29(1):5–28.

[5] Deva Kannan Ramanan. Tracking People and Recognizing their Activities. University of California, BERKELEY. PH. D Dissertation. 2005.

[6] S. IOFFE, D. A. FORSYTH. Probabilistic Methods for Finding People. International Journal of Computer Vision. 2001, 43(1): 45–68.

[7] D. Ramanan and D. A. Forsyth. Finding and tracking people from the bottom up. CVPR, 2003.

[8] Sullivan, S. Carlsson. Recognizing and Tracking Human Action. Proceedings of the 7th European Conference on Computer Vision. 2002: 629–644.

[9] http: //www. scs. Leeds. ac. uk/imv.

[10] Haritaoglu I, Harwood. D, Davis. L. VV: real-time surveillance of people and their activities. IEEE Trans Pattern Analysis and Machine Intelligence, 2000, 22(8): 809–830.

[11] A. D. Wilson, A. F. Bobick, and J. Cassell. Recovering the temporal structure of natural gesture. In Proceedings of the Second International Workshop on Automatic Face and Gesture Recognition. 1996.

[12] N. Oliver, A. Garg, E. Horvitz. Layered representations for learning and inferring office activity from multiple sensory channels. CVIU. 2004, 96(2):163–180.

*Multimedia, Communication and Computing Application – Leung (Ed.)*
*© 2015 Taylor & Francis Group, London, ISBN 978-1-138-02775-6*

# An image registration algorithm based on Fourier-Mellin transform and feature point matching

X.R. Wang & Y. Ding
*Key Laboratory of Dynamics and Control of Flight Vehicle, Ministry of Education, Beijing Institute of Technology, Beijing, China*

ABSTRACT: To achieve multi-sensor image registration, an image registration algorithm based on Fourier-Mellin transform and feature point matching is proposed in this paper. The initial registration parameters are calculated using Fourier-Mellin transform. Then the feature points extracted from the images are mapped according to the initial parameters. A point matching and optimizing process is accomplished which gives some one-to-one matched point pairs. The registration parameters are finally calculated using the least square method. The experiment with the visible and near-infrared images shows that the presented algorithm can realize multi-sensor image registration through one circulation, it effectively avoids the feedback process. This algorithm is brief and the computing time and the registration result can meet the application requirements.

KEYWORDS: Image registration, Fourier-Mellin transform, Harris corner detection, Feature point matching

## 1 INSTRUCTIONS

Because of the differences among the viewing angles, acquisition time and many other imaging factors, the images about the same target from multi-band sensors usually contain redundant information (Zhang 2008,Ni & Liu 2004). In order to utilize more information in these images, a precise registration processing is necessary to make sure the feature information extracted from the images can be used and therefore the registration result can meet various application requirements. In the registration, according to certain similarity criteria, confirmation of the conversion parameters is important to accomplish the best matching in pixel level between two images (Wang et al. 2009). Different from the image registration methods based on gray information, transform domain, control points and elastic model (Yuan et al. 2009, Yi et al. 2008), an image registration algorithm based on Fourier - Mellin transform and feature point matching is presented in this paper.

Firstly, the frequency distributions of two images are analyzed using Fourier-Mellin transform. The initial registration parameters are calculated based on the phase correlation theory (Li et al. 2006). Then, feature points extracted from the images with certain rules are mapped and matched according to the initial parameters (Hu et al. 2013). The optimum registration parameters are finally calculated using those one-to-one matched point pairs.

## 2 PRINCIPLES OF OBTAINING THE REGISTRATION PARAMETERS BASED ON FOURIER-MELLIN TRANSFORM

For two images that only have translation, the translation parameters are recorded as $(\Delta x, \Delta y)$. $f_1(x, y)$ denotes the original image and $f_2(x, y)$ denotes the translated image. The relation between $f_2(x, y)$ and $f_1(x, y)$ can be described by the Equation 1:

$$f_2(x, y) = f_1(x - \Delta x, y - \Delta y) \tag{1}$$

Take the Fourier transform of $f_1$ and $f_2$. The transformation in frequency domain are denoted as $F_1(\mu, v)$ and $F_2(\mu, v)$. According to the Fourier transform displacement theory (Jiao 2010), $F_1$ and $F_2$ should meet the Equation 2.

$$F_2(\mu, v) = F_1(\mu, v) e^{-j(\mu \Delta x + v \Delta y)} \tag{2}$$

The cross power spectrum is

$$\frac{F_1(\mu, v)\overline{F_2}(\mu, v)}{\left| F_1(\mu, v) F_2(\mu, v) \right|} = e^{j(\mu \Delta x + v \Delta y)} \tag{3}$$

$\overline{F_2}(\mu, v)$ represents the complex conjugate of $F_2(\mu, v)$. It is showed in Equation 3 that the phase offset of $f_1(x, y)$ and $f_2(x, y)$ equals to the phase of the cross power spectrum (Gan & Hua 2006). Take the inverse Fourier transform with Equation 3, there will be a pulse in $(\Delta x, \Delta y)$. Therefore, the translation parameters ($\Delta x$, $\Delta y$) are calculated by searching the peak in the inverse Fourier transform of the cross power spectrum.

For two images that have translation, rotation and one-scale scaling (the scaling parameters are the same in two directions of one image) at the same time, the translation parameters are recorded as ($\Delta x$, $\Delta y$), the rotation parameter is recorded as $\theta_0$ and the scaling parameter is recorded as $a$. The relation between $f_2(x, y)$ and $f_1(x, y)$ can be described by the Equation 4:

$$f_2(x, y) = f_1 \left[ a \left( x \cos\theta_0 + y \sin\theta_0 \right) - \Delta \right] x,$$
$$a \left( -x \sin\theta_0 + y \cos\theta_0 \right) - \Delta y \tag{4}$$

The amplitudes of $F_1(\mu, v)$ and $F_2(\mu, v)$ should meet the Equation 5:

$$\left| F_2(\mu, v) \right| = a^{-2} \left| \begin{array}{c} F_1 \left[ a^{-1} \left( \mu \cos\theta_0 + v \sin\theta_0 \right), \right] \\ a^{-1} \left( -\mu \sin\theta_0 + v \cos\theta_0 \right) \end{array} \right| \tag{5}$$

Given: $\mu = \rho \cos\theta$ and $v = \rho \sin\theta$. Where $\rho$ represents the polar vector of the point ($\mu, v$) in polar coordinate system, and $\theta$ represents the corresponding argument of it. In this way, variables of $F_1$ can be transformed to the polar coordinate system as showed below.

$$\begin{cases} a^{-1} \left( \mu \cos\theta_0 + v \sin\theta_0 \right) = \\ \dfrac{\rho}{a} \left( \cos\theta \cos\theta_0 + \sin\theta \sin\theta_0 \right) = \dfrac{\rho}{a} \cos\left( \theta - \theta_0 \right) \\ a^{-1} \left( -\mu \sin\theta_0 + v \cos\theta_0 \right) = \\ \dfrac{\rho}{a} \left( \sin\theta \cos\theta_0 - \cos\theta \sin\theta_0 \right) = \dfrac{\rho}{a} \sin\left( \theta - \theta_0 \right) \end{cases} \tag{6}$$

$s$ and $r$ denote the amplitudes under polar coordinate. They can be described by rewriting the Equation 5 with Equation 6.

$$s(\rho \cos\theta, \rho \sin\theta) = a^{-2} r \left[ \dfrac{\rho}{a} \cos\left( \theta - \theta_0 \right), \dfrac{\rho}{a} \sin\left( \theta - \theta_0 \right) \right] \tag{7}$$

The equation above can be simplified into: $s(\theta, \rho) = a^{-2} r[(\theta - \theta_0), (\rho/a)]$. Given $\rho = e\lambda$ and $a = e^b$, the amplitudes under log-polar coordinate can be described by the Equation 8.

$$s_1(\theta, \lambda) = e^{-2b} r_1 \left( \theta - \theta_0, \lambda - b \right) \tag{8}$$

It is obvious that the Equation 8 has the same form as the Equation 1. Therefore, the rotation parameter and the scaling parameter ($\theta_0, a$) can be calculated in the similar way like the translation parameters described above. They are the peak ($\theta_0, b$) in the inverse Fourier transform of the cross power spectrum calculated from the Fourier transform of the Equation 8. Notice that $a = e^b$.

## 3 FEATURE POINT MATCHING ALGORITHM BASED ON FOURIER-MELLIN TRANSFORM

Images at different wave bands reflect variant feature information since the imaging mechanisms are diverse (Liu et al. 2008). In view of the unique features in visible images and near-infrared images, a feature point matching algorithm based on Fourier-Mellin transform is discussed to realize the precise registration of them.

### 3.1 Selection of the origin and Cartesian coordinate system conversion

For both visible and near-infrared images, the lower left corner is selected to be the origin $O$, and the Cartesian system is established as showed in Figure 1.(a). $x$ represents the width of the image and $y$ represents the height. ($j_0, i_0$) represents any point in the image and ($x_0, y_0$) represents the centre point of the image.

Cartesian system conversion is discussed here because it is very important in the rotation process. Firstly, it converts the origin to the centre point O' to establish the new Cartesian system which is showed in Figure 1.(b). Then, the new coordinate system is rotated at a degree $\theta_0$ as showed in Figure 1.(c). Finally, the image is converted to the Cartesian system showed in Figure 1.(d). The origin is the new lower left point O" and the centre point is ($x''_0, y''_0$).

Equation 9 describes the pixels conversion relation of each point. ($j_0, i_0$) represents the point before conversion and ($j_3, i_3$) represents the corresponding point after coordinate transformation.

$$\begin{bmatrix} i_3 \\ j_3 \\ 1 \end{bmatrix} = \begin{bmatrix} \cos\theta_0 & \sin\theta_0 & -y_0 \cos\theta_0 - x_0 \sin\theta_0 + y''_0 \\ -\sin\theta_0 & \cos\theta_0 & y_0 \sin\theta_0 - x_0 \cos\theta_0 + x''_0 \\ 0 & 0 & 1 \end{bmatrix} \begin{bmatrix} i_0 \\ j_0 \\ 1 \end{bmatrix} \tag{9}$$

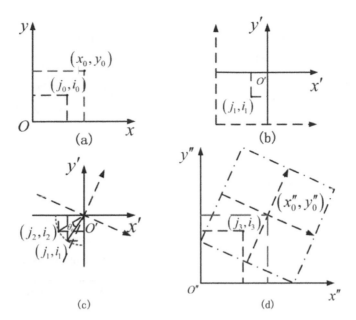

Figure 1.   Cartesian coordinate system conversion diagram.

### 3.2   Coordinate transformation between Cartesian system and polar coordinate system

$\rho$ denotes the polar vector in polar coordinate system of any point $(j_l, i_l)$ in Cartesian system, and $\theta$ denotes the corresponding amplitude of it. Equation 10 describes the conversion between these two coordinates.

$$\begin{cases} \rho = \left[ \sqrt{i_1^2 + j_1^2} \right] \\ \theta = \left[ \arctan(i_1 / j_1) \right] \end{cases} \tag{10}$$

Where $[\cdot]$ represents the truncating operation of integer.

Since the transformation from Cartesian system to polar system is not a symmetrical process, many blank pixels without mapping will appear in the image converted if the Equation 10 is used directly.

The algorithm discussed in this section uses an inverse mapping method to accomplish the conversion. The image width and height are set to represent $\rho$ and $\theta$ respectively. $\rho_{max}$ and $\theta_{max}$ are given in the Equation 11.

$$\begin{cases} \rho_{max} = \min(h/2, w/2) \\ \theta_{max} = 360 \end{cases} \tag{11}$$

$h$ represents the original image height and $w$ represents the width in Equation 11. In this way, the inverse mapping conversion is described by

$$\begin{cases} i_1 = [\rho \sin\theta] \\ j_1 = [\rho \cos\theta] \end{cases} \tag{12}$$

Figure 2 shows the conversion from Cartesian system to polar coordinate system.

### 3.3   Coordinate transformation between Cartesian system and log-polar coordinate system

$\varepsilon$ denotes the log-polar vector in log-polar coordinate system of any point $(j_l, i_l)i$ n Cartesian system, and $\theta$ denotes the corresponding amplitude of it. Equation 13 describes the conversion between these two coordinates.

$$\begin{cases} \varepsilon = \left[ \ln\left( \sqrt{i_1^2 + j_1^2} \right) \right] \\ \theta = \left[ \arctan(i_1 / j_1) \right] \end{cases} \tag{13}$$

$\varepsilon$ is magnified $k$ times in the process of the conversion. The modified conversion method is described by the Equation 14.

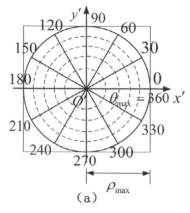

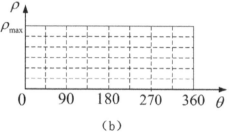

(b)

Figure 2.   Coordinate system conversion diagram.

$$\begin{cases} \varepsilon = \left[ k \cdot \ln\left( \sqrt{i_1^2 + j_1^2} \right) \right] = [k \cdot \ln \rho] \\ \theta = \left[ \arctan\left( i_1 / j_1 \right) \right] \end{cases} \quad (14)$$

Therefore, $\varepsilon'(\rho)$ is

$$\varepsilon'(\rho) = \frac{k}{\rho} \quad (15)$$

There are three regions under discussion:

1  The distance in Cartesian system is stretched when converted to log-polar vector as $\varepsilon'(\rho) > 1$. The stretched region marked in Figure 3.(b) is corresponding to the centre region in Figure 3.(a);

2  There is a one-to-one correspondence between one horizontal line in log-polar coordinate system and one circle which has the center at the origin and a radius of $k$ in Cartesian system as $\varepsilon'(\rho) = 1$. The one-to-one relation is showed in Figure 3 using the bold lines;

3  When $\varepsilon'(\rho) < 1$, on the contrary, there is a constringent matchup between two coordinate systems. The constringent region marked in Figure 3.(b) is corresponding to the surrounding region in Figure 3.(a).

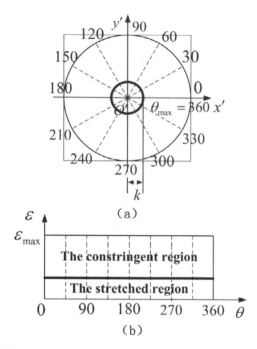

(a)

(b)

Figure 3.   Coordinate system conversion diagram.

### 3.4 *The calculating process of rotation parameter*

The visible image is the registered image and the near-infrared image is the reference as showed in Figure 4.

(a) Visible image.

(b) Near-infrared image.

Figure 4.   The original images.

210

(a) Visible image.

(b) Near-infrared image.

Figure 5. The Fourier spectrum images.

The calculating process has four steps as follows:

1 Take the FFT (fast Fourier transform) of Figure 4.(a) and 4.(b). The Fourier spectrum images are transformed following the rules discussed in section 3.1. Figure 5 is the results that centered at the origins;

2 Transform the images in Figure 5 to the polar coordinate system following the rules discussed in section 3.2. The interpolation method used here is the nearest interpolation method;

3 In order to unify the image data, the spectrum images under polar coordinate system are extended from the lower left point. The blank area is filled up with pixels 128 as showed in Figure 6;

4 Take the FFT of Figure 6 and the cross power spectrum is calculated using the results. Then it takes the inverse Fourier transform with the cross power spectrum and searches the peak of it. The pixel value in width of the peak is the rotation parameter $\theta_0$.

### 3.5 The calculating process of scaling parameter

Figure 4.(a) is rotated by an angle $\theta_0$ calculated in section 3.4 to be the new registered image. The near-infrared image is still the reference. New inputs are showed in Figure 7.

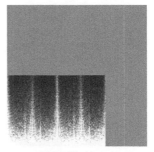

(a) Visible image.

(b) Near-infrared image.

Figure 6. The expanded images in polar coordinate.

(a) Rotated visible image.

(b) Near-infrared image.

Figure 7. The new images to be registered.

Take the FFT of Figure 7.(a) and 7.(b). The Fourier spectrum images are showed in Figure 8.

(a) Rotated visible image.

(b) Near-infrared image.

Figure 8. The Fourier spectrum images.

(a) Rotated visible image.

(b) Near-infrared image.

Figure 9. The expanded images in log-polar coordinate.

The Fourier spectrum images are transformed following the rules discussed in section 3.3. There will be many irrelevant pixels in the stretched region under log-polar coordinate system. In view of the errors caused by these pixels, only the one-to-one region and the constringent region are retained in the extended images. The interested regions are showed in Figure 9.

The following calculating process is the same as described in section 3.4. The pixel value in height of the peak is the scaling parameter $b$ under log-polar coordinate system where $a=e^b$.

### 3.6 The calculating process of translation parameters

Figure 7.(a) is scaled by the scaling parameter $a$ to be the new registered image. The input images for this process are showed in Figure 10.

Take the FFT of Figure 10 and the cross power spectrum is calculated using the results. Take the inverse Fourier transform of the cross power spectrum and the pixel values of the peak are the translation parameters ($\Delta x$, $\Delta y$).

### 3.7 The extraction of image feature points

For both visible image and near-infrared image, Harris operator is selected to extract the feature points. Harris operator uses a small window to

(a) Scaled and rotated visible image.

(b) Near-infrared image.

Figure 10. The input images to be registered.

observe the image and defines the feature point as the points that the gray value changes evidently when the small window moving through it (Li et al. 2007). It is a simple costing method and has considerable stability at the same time. The precision and the number of feature points extracted are revisable by modifying the threshold.

Specifically there are five steps:

1  Directional derivative is calculated using the Prewitt operator as the difference operator. The horizontal and vertical directional derivatives are saved in the arraies $Ix$ and $Iy$;
2  The matrix $M$ is calculated by

$$M = \begin{bmatrix} I_x^2 & I_x I_y \\ I_x I_y & I_y^2 \end{bmatrix} \tag{16}$$

$I^2x = Ix \times Ix$ and $I^2y = Iy \times Iy$ in Equation 16.
3  In order to suppress the isolated points and salient points, the matrix $M$ is preprocessed by Gaussian kernel smooth filter. $M'$ denotes the resul;
4  $cim(x,y)$ is defined by Equation 17.

$$cim(x, y) = \frac{I_x^2 \times I_y^2 - \left(I_x I_y\right)^2}{I_x^2 + I_y^2 + 0.000001} \tag{17}$$

The matrix $cim$ is calculated using the elements in matrix $M'$. In particular, infinitesimal 0.000001 is added to the denominator in the Equation 17 to avoid the error caused by divisor 0;

5  A circulation is set in the neighborhood of every pixel to find the local maximum points. Then the local maximum point is marked to be the feature point if $cim(x,y)$ of it is greater than the given threshold.

## 3.8 The feature point matching method

Figure 11 shows the process of matching.

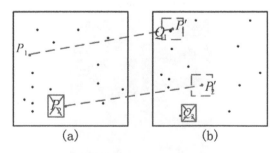

(a)                              (b)

Figure 11.   The schematic of feature points matching.

The feature point matching method can be expressed as follows:

1  The feature point $P$ in the registered image showed in Figure 11.(a) is mapped to the reference image showed in Figure 11.(b) according to the initial conversion parameters. $P'$ denotes the mapped point in Figure 11.(b);
2  The matching search is done in the 7×7 neighborhood of $P'$. If there exists a feature point $Q$ in the reference image in the 7×7 neighborhood, the point $Q$ is marked as the matching candidate of $P$. Such as $P_1$ and $Q_1$ in Figure 11;
3  If there is no referenced feature point in the 7×7 neighborhood, the point $P$ is deleted from the registered image as the point $P_2$ showed in Figure 11.(a);
4  When the whole matching search is over, all feature points in the reference image are checked one by one to delete those without mark.

The search neighborhood determined by the initial parameters is rather small so that the matching search can result in a series of one-to-one matched points after one search. There is no need for a traditional optimizing procedure.

## 3.9 The calculating process of optimum registration parameters

The least square method is used to calculate the optimum registration parameters.

Suppose there are $N$ pairs matched feature points. $(x_{Pi}, y_{Pi})$ $(i=1,2,3,...,N)$ denotes the feature points in the registered image and $(x_{Qi}, y_{Qi})$ $(i=1,2,3,...,N)$ denotes the corresponding feature points in the reference image. The conversion between them is described by the Equation 18.

$$\begin{bmatrix} y_{Qi} \\ x_{Qi} \\ 1 \end{bmatrix} = \begin{bmatrix} a\cos\theta_0 & a\sin\theta_0 & a\left(-y_0\cos\theta_0 - x_0\sin\theta_0 + y_0''\right) - \Delta y \\ -a\sin\theta_0 & a\cos\theta_0 & a\left(y_0\sin\theta_0 - x_0\cos\theta_0 + x_0''\right) - \Delta x \\ 0 & 0 & 1 \end{bmatrix} \begin{bmatrix} y_{Pi} \\ x_{Pi} \\ 1 \end{bmatrix} \tag{18}$$

Given:

$$
\left\{
\begin{array}{l}
B = \begin{bmatrix} y_{Q1} & y_{Q2} & \cdots & y_{QN} \\ x_{Q1} & x_{Q1} & \cdots & x_{QN} \\ 1 & 1 & \cdots & 1 \end{bmatrix} \quad and \quad A = \begin{bmatrix} y_{P1} & y_{P2} & \cdots & y_{PN} \\ x_{P1} & x_{P1} & \cdots & x_{PN} \\ 1 & 1 & \cdots & 1 \end{bmatrix} \\[20pt]
W = \begin{bmatrix} a\cos\theta_0 & a\sin\theta_0 & a\left(-y_0\cos\theta_0 - x_0\sin\theta_0 + y_0''\right) - \Delta y \\ -a\sin\theta_0 & a\cos\theta_0 & a\left(\ y_0\sin\theta_0 - x_0\cos\theta_0 + x_0''\right) - \Delta x \\ 0 & 0 & 1 \end{bmatrix}
\end{array}
\right.
$$

$$(19)$$

The Equation 18 can be simplified into $B=WA$. It is proved that the least square solution $W$ can be described as Equation 20 when matrix $A$ has a rank greater than 6.

$$W = BA^T \left(AA^T\right)^{-1} \tag{20}$$

The optimum registration parameters are calculated from the matrix $W$.

## 4 EXPERIMENTAL RESULTS AND ANALYSIS

In order to verify the algorithm based on Fourier-Mellin transform and feature point matching presented above, we collected several images at visible and near-infrared bands to be the inputs of registration algorithm using the CMOS industrial camera and optical filters.

The camera model is UI-3240CP-NIR-GL and the resolution is 1280×1024. We perform the algorithm using Visual C++ 6.0 in computer with CPU of Inter Core I5, CPU frequency of 2.60 GHz and memory size of 4.0GB.

### 4.1 The preprocessing of the registered image and the reference image

The original visible image and the near-infrared image collected in the experiment are showed in Figure 12.

It is observed that the main scene is similar in two images but several items are not the same because of the tiny movements in the viewing angle, time and distance. Also the gray level and the contrast ratio are different because of the imaging mechanisms.

The original images collected are adjusted to 640×512 so that the amount of calculation can be reduced in the process. Two squares with a size of 512×512 are captured from the centre of the images to verify the algorithm. Figure 13 is the original registered image and the reference image.

### 4.2 The initial registration parameters

Table 1 is the results of the initial parameters calculated in the experiment.

(a) Visible image.

(b) Near-infrared image.

Figure 12. The original images collected (1280×1024).

(a) Visible image.

(b) Near-infrared image.

Figure 13. The original images captured (512×512).

Table 1. The results of the initial registration parameters.

| $\theta_0$ | $a$ | $(\Delta x, \Delta y)$ |
|---|---|---|
| 3° | 0.98758 | (-60,21) |

### 4.3 The results of feature point extraction

The threshold $T$ is set to 1500 in the Harris method. The numbers of feature points in Figure 13.(a) and (b) are 490 and 143 in the experiment.

### 4.4 The result of feature point matching

The matching process with the 7×7 neighborhood results in 49 pairs one-to-one matched point pairs showed in Figure 15.

### 4.5 The optimum registration parameters

15 pairs out of 49 pairs are chosen randomly as a group. Using the least square method, the optimum parameters are calculated by taking the average of three groups. Table 2 is the results of optimum registration parameters. The image registration result is showed in Figure 16.

(a) Visible image.

(b) Near-infrared image.

Figure 15. The matched feature points.

Table 2. The results of the optimum registration parameters.

| $\theta_0$ | $a$ | $(\Delta x, \Delta y)$ |
|---|---|---|
| 2.9° | 0.99 | (-59,20) |

(a) Visible image.

(b) Near-infrared image.

Figure 14. The feature points images.

Figure 16. Image registration result.

## 5 CONCLUSIONS

This paper presents an image registration algorithm based on Fourier-Mellin transform and feature point matching. The initial registration parameters are calculated based on the Fourier-Mellin transform. Then the feature points extracted by Harris operator are mapped and matched according to the initial parameters. The optimum registration parameters are calculated by the least square method using the best matched pairs. The experiment with visible and near-infrared images collected by industrial camera shows that the presented algorithm can efficiently realize multi-band image registration through one circulation, and avoid the feedback optimization. The process is brief. The computing time and the registration result can meet the application requirements.

## REFERENCE

[1] Gan Houji, Hua Wenshen. Post processing technique for correlation results based on Fourier-polar trandformation[J]. Journal of Applied Optics, 2006, 29(3): 466–472.

[2] Hu Yongli, Wang Liang, Liu Rong, etc. A Coarse-to-Fine Registration Method for Satellite Infrared Image and Visual Image[J]. Spectroscopy and Spectral Analysis, 2013, 33(11): 2968–2972.

[3] Jiao Jichao. A Fast Image Registration Algorithm Based on Fourier-Mellin Transform for Space Image[J]. ACTA ARMAMENTARII, 2010, 31(12): 1551–1556.

[4] Li Bo, Yang Dan, Zhang Xiaohong. Noval Image Registration Based on Harris Multi-Scale Corner Detection Algorithm[J]. Computer Engineering and Applications, 2007, 42(35): 37–40.

[5] Li Xiaoming, Zhao Xunpo, Zheng Lian. An Image Registration Technique Based on Fourier-Mellin Transform and Its Extended Applications [J]. Chinese Journal of Computers, 2006, 29(3): 466–472.

[6] Liu Xiaojun, Yang Jie, Sun Jianwei, etc. Image registration approach based on SIFT[J]. Infrared and Laser Engineering, 2008, 37(1): 156–160.

[7] Ni Guoqiang, Liu Qiong. Analysis and prospect of multi-source image registration techniques[J]. Opto Electronic Engineering, 2004, 31(9): 1–6.

[8] Wang Kunpeng, Xu Yidan, Yu Qifeng. Classfication and State of IR/Visible Image Registration Methods[J]. Infrared Technology, 2009 (5): 270–274.

[9] Yi Z, Zhiguo C, Yang X. Multi-spectral remote image registration based on SIFT[J]. Electronics Letters, 2008, 44(2): 107–108.

[10] Yuan Jinsha, Zhao Zhenpin, Gao Qiang, etc. Review and prospect on infrared/visible image registration[J]. Laser & Infrared, 2009, 39(7): 693–699.

[11] Zhang Honglin. Proficient Visual C ++ typical digital image processing algorithm and implementation [M]. Beijing: Post & Telecom Presss.2008.7,430–433.

*Multimedia, Communication and Computing Application – Leung (Ed.)*
© *2015 Taylor & Francis Group, London, ISBN 978-1-138-02775-6*

# An improved edge-detection algorithm according to wavelet transform

Y.Z. Yang, H. Xu & Y. Ding
*Key Laboratory of Dynamics and Control of Flight Vehicle, Ministry of Education, Beijing Institute of Technology, Beijing, China*

ABRSTRACT: Considering the problems in the traditional algorithm of edge detection, such as fuzzy, discontinuous, and so on, according to the multiscale characteristics of the edge and noise, and the basic principle of wavelet transform, this paper presents a method to get the local maxima of module by interpolation, and then associates with the image edge by setting dual-threshold. The experimental result with Lenna image shows that the proposed algorithm extracts the image details effectively, and the continuity of the image edge is satisfying.

KEYWORDS: Wavelet transform, Edge detection, Dual-threshold

## 1 INTRODUCTION

The edge detection is an important basis for the image analysis. In an image, the edges can be expressed as some kinds of points that have pixel value mutations and that carry important information of the image. Unfortunately, these mutation points are corresponded to not only the edge of the image but also the noise points, which are the high-frequency part of the frequency domain. It is difficult to remove the noises and retains the image edge details at the same time. For example, some algorithms introduce noises to the image while extracting the edge. Other algorithms, although removes the noises effectively, it causes the lack of edge details at the same time. Therefore, an improved edge-detection algorithm according to wavelet transform is proposed in this paper. As a multiscale edge-detection method, wavelet transform can distinguish the signals and the noises by the difference of multiscale propagation characteristics, and it obtains high precision in the edge extraction, while effectively suppressing the noise.

This paper uses wavelet transform to get the information of horizontal and vertical directions, introduces directional derivative which reflects direction feature, and puts forward a dual-threshold edge-detection algorithm according to the local maxima of module of wavelet.

## 2 BASIC PRINCIPLE OF EDGE-DETECTION ALGORITHM ACCORDING TO WAVELET MODULUS MAXIMA

Considering the input image $f$, $(x, y)$ is the coordinates of the current point in the image, $f(x, y)$ stands for the pixel value and the value range is [0, 255], 0 means black and 255 means white. Take a two-dimensional smooth function $\theta(x, y)$ which is computed by

$$\begin{cases} \int_{-\infty}^{+\infty} \int_{-\infty}^{+\infty} \theta(x, y) dx dy = 1 \\ \lim_{|x|+|y| \to +\infty} \theta(x, y) = 0 \end{cases} \tag{1}$$

Under a certain scales, the new image smoothed by the smoothing function is expressed as

$$F(x, y) = (f * \theta_s)(x, y) \tag{2}$$

where $\theta_s(x, y) = (1/s) \theta(x/s, y/s)$.

Define two-dimensional wavelet by $\theta(x, y)$

$$\begin{cases} \psi^1(x, y) = \dfrac{\partial \theta}{\partial x}(x, y) \\ \psi^2(x, y) = \dfrac{\partial \theta}{\partial y}(x, y) \end{cases} \tag{3}$$

Record

$$\begin{cases} \psi_s^1(x, y) = \dfrac{1}{s} \psi^1\left(\dfrac{x}{s}, \dfrac{y}{s}\right) \\ \psi_s^2(x, y) = \dfrac{1}{s} \psi^2\left(\dfrac{x}{s}, \dfrac{y}{s}\right) \end{cases} \tag{4}$$

A two-dimensional wavelet transform along the horizontal and vertical directions is, respectively, conducted with the input image, the DWT coefficients $WT^1f(s, x, y)$ and $WT^2f(s, x, y)$ reflect gradient of horizontal and vertical directions. The synthesis of gradient modulus value is proportional to the modulus of the gradient vector $\nabla(f*\theta_s)(x, y)$ and the proportionality coefficient is $s$ (Hong et al. 2014, Jin 2010).

Record the modulus value of the composite gradient as $Mf(s, x, y)$, thus it is calculated by

$$Mf(s,x,y) = \sqrt{|WT^1 f(s,x,y)|^2 + |WT^2 f(s,x,y)|^2} . \quad (5)$$

The phase angle $Af(s, x, y)$ is calculated by

$$Af(s,x,y) = \arctan\left(\frac{WT^2 f(s,x,y)}{WT^1 f(s,x,y)}\right). \quad (6)$$

Therefore, the local module maxima of the composite gradient are proportional to the gradient vector, which is probably the edge point. Linking these points along a particular direction leads to the result of edge detection under the scale (Zhao et al. 2009).

## 3 AN IMPROVED ALGORITHM OF EDGE DETECTION ACCORDING TO WAVELET TRANSFORM

### 3.1 Selection of wavelet basis and scale

A different wavelet basis leads to different results of image decomposition. The wavelet basis used in this paper is the Cubic B-Spline wavelet.

Wavelet function with arbitrary order of regularity can be constructed using B-Spline function. Besides, B-Spline function is a smooth function which is regular and compactly supported. Comparing with other spline function, B-Spline function has the smallest support base in the same scale space. Wavelet structured by B-Spline function has a regularity in the m-2 order expect the first and second orders (Li & Liang 2008). The filter coefficients of the Cubic B-Spline wavelet in this paper are shown in the following table, in which G stands for high-pass filter and H stands for low-pass filter.

Table 1. Filter coefficient table.

| n | H | G |
|---|---|---|
| −2 | 0.125 | 0 |
| −1 | 0.375 | −1 |
| 0 | 0.375 | 0 |
| 1 | 0.125 | 1 |
| 2 | 0 | 0 |

Wavelet and scaling functions are shown in Figure 1(a), low-pass filter coefficients in Figure 1(b), and high-pass filter coefficients in Figure 1(c).

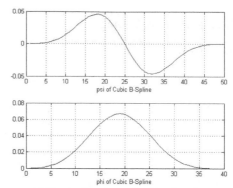

(a) Wavelet function and scaling function.

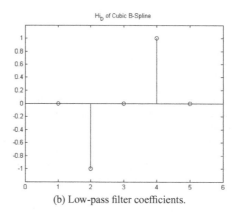

(b) Low-pass filter coefficients.

(c) High-pass filter coefficients.

Figure 1. Schematic diagram of the Cubic B-Spline wavelet.

Decompose Lenna image with the Cubic B-Spline Wavelet, the result in scale $2^1$ is shown in Figure 2(a), the result obtained using 5–3 wavelet is shown in Figure 2(b).

(a) Cubic B-Spline wavelet.

(b) 5–3 wavelet.

Figure 2. Results of image decomposition in scale $2^1$.

From the images, we can get that there are more details in horizontal and vertical directions in Figure 2(a) than in Figure 2(b). Thus the Cubic B-Spline wavelet has a better effect in edge detection.

The noises will be suppressed when detecting image edge in a large scale, but some edge details will be suppressed at the same time, which will cause blurring and skewing of the edge. In a small scale, the edge is clear, but the noises cannot be removed effectively and this will have great influence on the result of edge detection. Therefore, an appropriate scale is

very important in edge detection. The results proved by repeated experiments show that decomposing the image in scale $2^2$ is the best.

### 3.2 Determination of the local maxima of module

The gradient matrix of horizontal direction $gx$ and direction $gy$ can be calculated by decomposing an image with the Cubic B-Spline Wavelet in scale $2^2$. Then the gradient modulus value of each point can be calculated by

$$g(x,y) = \sqrt{gx^2 + gy^2}. \qquad (7)$$

Then we can get the gradient modulus matrix $G[x, y]$ (Hu et al. 2006, Gao et al. 2006).

The coordinate of a point is $(x_0, y_0)$ and the positions of its eight neighborhoods are shown in Figure 3.

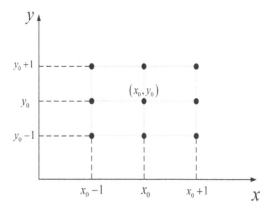

Figure 3. Position of the eight neighborhoods.

Set the point $(x_0, y_0)$ as the origin and establish a rectangular coordinate, as shown in Figure 4. $\theta$ represents the gradient angular of the point and

$$\theta = \arctan \frac{gy[x_0, y_0]}{gx[x_0, y_0]} \qquad (8)$$

The two-dimensional space is divided into four regions according to $\theta$, considering the horizontal gradient $gx[x_0, y_0]$ and the vertical gradient $gy[x_0, y_0]$, we can obtain the value of interpolation coefficient $a$, the gradients of the point ($gradient_1$ and $gradient_2$) can be calculated using interpolation method.

Region I, $\theta \in (0, \pi/4)$ and $\theta \in (0, \pi/4)$, in this region $gx[x_0, y_0] > gy[x_0, y_0]$, and the interpolation coefficient $a = (gy[x_0, y_0])/(gx[x_0, y_0])$:

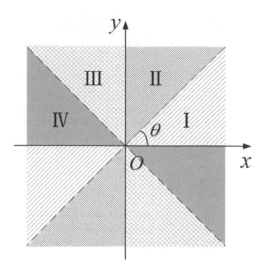

Figure 4.   Regional diagram of the gradient angular.

$$\begin{cases} \text{gradient}_1 = a * g(x_0+1, y_0+1) + (1-a) * g(x_0+1, y_0) \\ \text{gradient}_2 = a * g(x_0-1, y_0-1) + (1-a) * g(x_0-1, y_0) \end{cases} \quad (9)$$

Region II, $\theta \in (\pi/4, \pi/2)$ and $\theta \in (5\pi/4, 3\pi/2)$, in this region $gy[x_0, y_0] > gx[x_0, y_0]$, and the interpolation coefficient $a = (gx[x_0, y_0])/(gy[x_0, y_0])$:

$$\begin{cases} \text{gradient}_1 = a * g(x_0+1, y_0+1) + (1-a) * g(x_0, y_0+1) \\ \text{gradient}_2 = a * g(x_0-1, y_0-1) + (1-a) * g(x_0, y_0-1) \end{cases} \quad (10)$$

Region III, $\theta \in (\pi/2, 3\pi/4)$ and $\theta \in (3\pi/2, 7\pi/4)$, in this region $gy[x_0, y_0] > gx[x_0, y_0]$, and the interpolation coefficient $a = (gx[x_0, y_0])/(gy[x_0, y_0])$:

$$\begin{cases} \text{gradient}_1 = a * g(x_0-1, y_0+1) + (1-a) * g(x_0, y_0+1) \\ \text{gradient}_2 = a * g(x_0+1, y_0-1) + (1-a) * g(x_0, y_0-1) \end{cases} \quad (11)$$

Region IV, $\theta \in (3\pi/4, \pi)$ and $\theta \in (7\pi/4, 2\pi)$, in this region $gx[x_0, y_0] > gy[x_0, y_0]$, and the interpolation coefficient $a = (gy[x_0, y_0])/(gx[x_0, y_0])$:

$$\begin{cases} \text{gradient}_1 = a * g(x_0-1, y_0+1) + (1-a) * g(x_0-1, y_0) \\ \text{gradient}_2 = a * g(x_0+1, y_0-1) + (1-a) * g(x_0+1, y_0) \end{cases} \quad (12)$$

If $\text{gradient}_1$ $\text{gradient}_2$ and the gradient modulus value $G[x_0, y_0]$ satisfy

$$\begin{cases} G[x_0, y_0] > \text{gradient}_1 \\ G[x_0, y_0] > \text{gradient}_2 \end{cases} \quad (13)$$

Gradient of the point $(x_0, y_0)$ is the local maxima of module, that is, it may be an edge point, then set its pixel value to 128. Traverse all the pixels in the image, and we can obtain all the possible edge points.

### 3.3 Edge points selecting and chaining method based on dual-threshold

To establish a gradient histogram according to the gradient modulus value $G[x_0, y_0]$, set $T_h$ as the high threshold, the number of the points which the gradient modulus is smaller than the high threshold can be calculated by $n_h = T_h \times n$, where $n$ is the total point number of the image. The gradient modulus $G_h$ corresponded with $n_h$ can be obtained through the gradient histogram. With the low threshold as $T_l$, we can get $n_l$ and $G_l$ by the same way.

Traverse all the possible edge points (points whose pixel value are 128), treat point $(x_0, y_0)$ as the starting point of an edge chain if $G[x_0, y_0] \theta G_h$, and changes its pixel value from 128 to 255. Its eight neighborhoods are searched for the possible edge points in a certain order, their gradient modulus are judged whether bigger than the low threshold $G_l$, the first point that meets the condition is added to the edge chain, and its pixel value is changed to 255. The current point is set to be a new starting point, and then the aforementioned steps are repeated until there is no possible edge point in its eight neighborhoods, thus we obtain a complete edge chain. The next edge chain is searched for until all the edge chains are found.

If not 255, changing pixel value of the points to 0 after all the edge points are marked, all the points whose pixel value remain 255 are the edge points.

## 4   EXPERIMENTAL RESULTS AND ANALYSIS

To verify the effect of the algorithm in this paper, we perform the simulation by using Visual C++6.0 in computer with CPU of Intel Core I5, CPU frequency of 2.6 GHz and memory size of 4.0 GB. The size of the Lenna image is $256 \times 256$.

The standard Lenna image is shown in Figure 5(a). Figure 5(b) shows the simulation result of edge detection by setting the threshold $T_h = 0.857$, $T_l = 0.4$. Figure 5(c) is the simulation result by setting the threshold $T_h = 0.86$, $T_l = 0.4$. Running time of the algorithm is 46 ms.

Figure 5 shows that the edge-detection algorithm in this paper gives good results. Lots of edge details are retained and most of the noises are removed. The edges are thin, clear, and linked well.

To test the effect of algorithm proposed in this paper, several traditional algorithms of extracting edge of image are given to serve as a contrast.

(a) Standard Lenna image.

(a) Roberts operator.

(b) Thresholds are 0.857, 0.4.

(b) Sobel operator.

(c) Thresholds are 0.86, 0.4.

Figure 5.   Results obtained by the edge-detection algorithm.

(c) Traditional wavelet transform.

Figure 6.   Results of traditional algorithms in edge detection.

Figure 6(a) shows the result obtained using Roberts operator, Figure 6(b) shows the result obtained using Sobel operator, and Figure 6(c) shows the result obtained using traditional wavelet transform method.

Figure 6 shows that edges are thick and details are obscure when using traditional algorithms in edge detection and there are lots of noises. Extracting edges with traditional wavelet in scale $2^2$ has good

effect, but there are some isolated points, and the continuity of the edges is not good enough.

5   CONCLUSION

The improved algorithm of edge detection according to wavelet transform in this paper can effectively remove the noises and retain the image details at the same time. Results obtained by standard Lenna image shows that selecting and chaining edge by setting dual-threshold ensures the edges precise and good linking, and noises are suppressed. It is proved that the algorithm is fast and reliable.

## REFERENCES

[1] Gao Guorong, Liu Ran, Yi Xuming. A Revised Image Edge Detection Method Based on Wavelet[J]. J. Wuhan Univ. (Nat. Sci. Ed.), 2006, 51(5): 615–619.

[2] Hong Xueting, Shi Xiaowei, Qian Yixian. Optical Correlation Eecognition Based on Edge Detection of Wavelet Transform[J]. Laser and Optoelectronics Progress 2014, 51(4): 112–117.

[3] Hu Xuejuan, Ruan Shuangchen, Liu Chengxiang. An approach of near infrared image feature extraction based on wavelet transform[J]. 2006.

[4] Jin Jing. Application of Wavelet Transform in Edge Detection[D]. Zhejiang University. 2010.

[5] Li Qing, Liang Yan. Image Edge Detector Based on Cubic B-spline Wavelet[J]. JOURNAL OF WUT (INFORMATION & MANAGEMENT ENGINEERING), 2008, 29(12): 1–3.

[6] Zhao Zhigang, Yang Yingping, Jiang Aixiang, et al. Image edge detection based on directional derivative and cubic B-spline wavelet[J]. Computer Engineering and Applications. 2009, 45(23): 176–178.

*Multimedia, Communication and Computing Application – Leung (Ed.)*
© *2015 Taylor & Francis Group, London, ISBN 978-1-138-02775-6*

# A mutation-based fault injection method for C program

H. Song, Y.W. Wang, G.N. Qian & *Y.Z. Gong*
*State Key Laboratory of Networking and Switching Technology, Beijing University of Posts and Telecommunications, Beijing, China*

Y.W. Wang
*State Key Laboratory of Computer Architecture, Institute of Computing Technology, Chinese Academy of Sciences*

ABSTRACT:    Fault injection plays a critical role in the verification of software's reliability. This paper presents a fault injection method by mutation. Moreover, a strategy of using semantic-based mutators is proposed to improve the efficiency of mutation testing. We implemented the method in FIT, an automatic fault injection tool. Experiments prove that the method is effective and the tool can produce lots of mutants which show that mutation testing is satisfying.

KEYWORDS:    Software test, Fault injection, Mutation test

## 1    INTRODUCTION

Fault Injection plays an important role in software testing. How the software works after faults occur is a key criterion to measure its reliability (Sullivan & Chillarege 1991). Therefore, researchers put forward the concept of mutation testing, a fault-based testing technique, to ensure the software quality (Chen & Gu 2012). This paper presents an automatic mutation-based fault injection method for C language program. With this method, a fault injection tool (FIT) is designed and implemented. Meanwhile, in order to reduce the cost of mutation testing, this paper gives some semantic-based mutators for C program.

The paper is organized as follows: Section 2 introduces the design of the FIT and the process of the automatic fault injection. Section 3 introduces how the semantic-based mutators work. Section 4 introduces the experiment and the results. Section 5 concludes this paper tells people what to do next.

## 2    DESIGN OF FIT

The fault injection method was adopted in the development of FIT, which introduces faults into a C program. FIT gets the information of C program with Static Analysis (WEN & WANG 2005). It can confirm the Injection Point (IP) in the way of tree-traversal, to find specific syntax structures which could be an IP. When an IP's location is confirmed, the FIT will produce a mutant by string-replacement. A mutant is a piece of code with faults.

### 2.1    Process of fault injection

Here is the process turning the original code into a mutant:

1) With Static Analysis (SA), an Abstract Syntax Tree (AST) is built, whose nodes contain syntax and sematic information of the original code.

2) Select a model of mutators from the mutator library as the standard mutator. A transformation rule that generates a mutant from the original program is known as a mutator.

3) Visit the AST and build a one to many mapping relationship between the standard mutator and syntax tree nodes, that is, a mutator corresponding to different types of syntax tree nodes

4) Pick a tree node and combine it and the standard mutator as an IP. The IP contains both information of the original code and fault code. This is the key step of the process because it is the time when the mutation will happen.

5) A piece of fault code is built by string replacement based on the IP.

The automation of the fault injection would be achieved by five steps above-mentioned. The method of picking the tree nodes decides what kind of order the mutants come out, sequentially or randomly. If we choose more than one mutators, it will generate a lot of mutants provided for testing.

### 2.2    FIT architecture

According to the process of Fault Injection designed, the author developed the fault injection tool (FIT) which meets the requirement of mutation testing.

The Fig. 1 describes the architecture of the FIT. The following modules are contained by FIT:

1) Static Analyzer: responsible for SA of the original code. It includes three processes: Lexical Analysis, Syntactic Analysis and Semantic Analysis. And there will be a completed AST in the end.

2) Mutator Library: modeling mutators and storing them. It also provides the APIs to visit the models.

3) AST Visitor: visit the AST in depth-first-search. The visitor collects specific tree nodes and puts them in a set. Each mutator is corresponded to a set, which means there is a one-to-many relationship between the mutator and the tree nodes.

4) Semantic Checker: responsible for checking the semantic of the original code's context. It can narrow the range of the mutation by excluding the fault causing semantic violation.

5) IP Generator: select a tree node from the mutator-based set. The tree node selected contains the code's information while the standard mutator contains mutation information. Merge them as an IP and mutating.

6) File Generator: generate the files with the mutants.

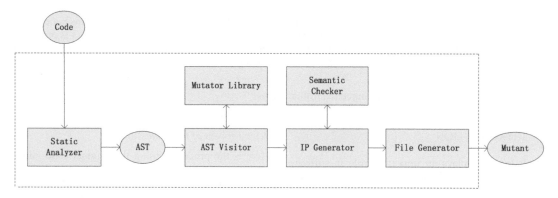

Figure 1.    Architecture FIT.

## 2.3  *How to get an IP*

How to get an IP is the critical problem of the fault injection because an IP which is confirmed makes sure of the fault injection's success. In the way of traverse of the AST, we can get an IP accurately and rapidly.

The following algorithm describes how to solve the problem:

INPUT: the root of AST, the selected mutator

OUTPUT: the set of tree nodes passed the semantic check

```
1 for mutator in mutator_lib
2     visit(root, mutator, node_set)
3 return node_set
4 visit(node, mutator, node_set)
5 if node.type is corresponded to mutator
6     then semantic_check(node)
7         if node.mutation_candidates ≠ Ø
8             then node_set[mutator].add(node)
9 for child in children(node)
10     visit(node, mutator, node_set)
```

In Line 5, the mutator is one-to-many to the type of node, so it will find the type in a mutator-based map. In Line 6, it will do the semantic check to the left nodes. There is a list called mutation candidates in the mutator. The semantic check will narrow the candidates by excluding the fault causing semantic violation after integrating the information of original code stored in the node. In Line 7, if the candidates list isn't empty, which means the node is available. Then put it into the set.

For example, the node of additive expression is a=1.2+1.3 and the mutator is the replacement of arithmetic operators, the initial candidates list is {-, *, /, %}. After the semantic check, it will be {-, *, /} because the float number cannot do the remainder operation.

## 3   THE SEMANTIC-BASED MUTATORS

### 3.1  *Background*

A transformation rule that generates a mutant from the original program is known as a mutator. In 1989, Agrawal proposed a comprehensive set of mutators for the ANSI C programming language. There were 77 mutators defined in this set, which was designed to follow the C language specification. These operators are classified into variable mutation, operator mutation, constant mutation, and statement mutation. But when these mutators are used in the automatic mutation testing, the mutants failed to compile are generated because the mutants may break the semantic. In 2013, Richard Baker tested an onboard software system in mutation testing. There are 3,149 mutants built and among them 594 failed to compile, about 18%. These failed mutants were generated automatically even though they could not be used in testing. It wasted time and increased cost. That is why the semantic-based mutators were put forward.

### 3.2 *Mutators*

FIT has six kinds of mutators based on semantic as is shown in Table 1.There are ARER, AROR, ASOR, CORP, STER, VARP.

Table 1. Semantic-based mutators.

| Mutator | Description |
|---------|-------------|
| ARER | array element replacement |
| AROR | arithmetic operator replacement |
| ASOR | assignment operator replacement |
| CORP | constant replacement |
| STER | structure element replacement |
| VARP | variable replacement |

The followings introduce each of mutators and how they work:

1) ARER -- If the index is constant, replace it with the array length, resulting in an array of cross-border failures.

2) AROR -- The arithmetic operators are "+", "-", "*", "/", "%", "++", "--". The first five operators are binary operators and the left are unary operators. The mutator allows one operator to be replaced by another which belongs to the same type. In the C program, the arithmetic operators are used in basic type and pointer type. The module operator can only be used between the integer data, which means if "%" in the candidate list and at least one of the operands is float, "%" will be removed from the list. There are three cases when the pointer type is involved:

(1) pointer – pointer
(2) integer + pointer
(3) pointer +/– integer

If the operator is used like case (1) and (2), no mutation should occur. In case (3), it only allows the operators to replace h the other one.

3) ASOR — The mutation rules of assignment operators are the same as that of arithmetic operators. When assignment replacement happens, we check out the rules and the corresponding arithmetic operator obeys. What needs to be noticed is the replacement between "= =" and "=". We must make sure that the left operand is non-constant otherwise it would fail to compile. For example,

*if ( NULL == ptr )*

......

In the conditional expression, "==" cannot be replaced by "=", otherwise an error would be thrown.

4) CORP -- The constant in C language includes Integer Constant, Float Constant, String Constant, and Symbolic Constant and so on. When we do divide or module operation, 0 cannot be right operated. If the constant is used in switch-case structure as the condition, the value of the mutated constant cannot be equal to other constants.

5) STER -- The member variance of a structure must be replaced by another one from the same structure with the same type. If we cannot find such a variance, STER stops. For example,

*struct TreeNode {*
 *int val;*
 *struct TreeNode *left;*
 *struct TreeNode *right;*
*};*

*struct root;*

According to STER, *root.left* can be replaced by *root.right*, not *root.val*.

6) VARP -- VARP allows a variance replaced by ones which are declared in the same scale or upper ones. The types of them need to be the same too. To simulate real programming environment, the author use the edit distance (Ristad & Yianilos 1998) to choose the mutant. The smaller the edit distance between two variances is, the higher the priority of the candidate variance is. We will choose the candidate with higher priority to mutate.

## 4 EXPERIMENT

We designed an experiment to verify the effectiveness of the FIT. Four C programs were picked for injecting faults. With the logic of them were unknown in advance, we injected ten faults into the programs both artificially and automatically. It is required that the types of mutators are as many as possible and the locations of IP are random. We got the time cost of both, and calculated the success of compilation rate (SCR) and the fault covered block coverage (FC) for FIT.

The Table 2 shows the result of the experiment. Compared with the traditional injection method, FIT's time cost can be ignored. The mutants built by FIT passed the compilation so that all of them can be used in mutation testing.

Table 2. Experiment result.

| Programs | | | Results | | |
|---|---|---|---|---|---|
| Name | Block | Artificially | Automatically | | |
| | | Time | Time | SCR | FC |
| days.c | 22 | 8min | <1s | 100% | 45.5% |
| bonus.c | 12 | 7min | <1s | 100% | 100% |
| division.c | 7 | 5min | <1s | 100% | 85.7% |
| equation.c | 9 | 7min | <1s | 100% | 88.9% |

## 5  CONCLUSION

This paper offers a fault injection method based on mutation. It is proved that this method can implement automatic fault injection in C program. Compared with the traditional method, this way can obviously improve the efficiency. At the meantime, this paper provides a way to generate mutators based on semantic check, and make sure mutants generated from this approach can satisfy semantic rules for compiler. The next step is to apply this method in mutation testing to investigate how it works for the future work.

## ACKNOWLEDGMENT

This work was supported by the National Grand Fundamental Research 863 Program of China (No. 2012AA011201), the National Natural Science Foundation of China (No. 61202080), the Major Program of the National Natural Science Foundation of China (No. 91318301), and the Open Funding of State Key Laboratory of Computer Architecture (No. CARCH201201).

## REFERENCES

Sullivan, M., & Chillarege, R. (1991, June). Software Defects and their Impact on System Availability: A Study of Field Failures in Operating Systems. In FTCS (pp. 2–9).

Chen, X., & Gu, Q. (2012). Mutation Testing: Principal, Optimization and Application. Jisuanji Kexue yu Tansuo, 6(12), 1057–1075.

WEN, C. C., & WANG, Z. S. (2005). Static analysis research of software automatic test [J]. Computer Engineering and Design, 4, 046.

Agrawal, H., Demillo, R., Hathaway, R., Hsu, W., Hsu, W., Krauser, E., ... & Spafford, E. (1989). Design of mutant operators for the C programming language. Technical Report SERC-TR-41-P, Software Engineering Research Center, Department of Computer Science, Purdue University, Indiana.

Baker, R., & Habli, I. (2013). An empirical evaluation of mutation testing for improving the test quality of safety-critical software. Software Engineering, IEEE Transactions on, 39(6), 787–805.

Ristad, E. S., & Yianilos, P. N. (1998). Learning string-edit distance. Pattern Analysis and Machine Intelligence, IEEE Transactions on, 20(5), 522–532.

Multimedia, Communication and Computing Application – Leung (Ed.)
© 2015 Taylor & Francis Group, London, ISBN 978-1-138-02775-6

# Comparison of different soft decision bit metrics for non-square 32-ary QAM constellation

C. Li, Y.P. Cheng, Y.M. Zhang & Y.Z. Huang

*College of Communication Engineering, PLA University of Science and Technology, Nanjing, China*

ABSTRACT: This paper discusses different calculation methods of soft bit metric for MIL-STD-110B non-square 32-ary QAM constellation. Specifically, we propose an improved simplified algorithm to compute soft bit metrics for MIL-STD-110B 32-ary QAM constellation. Moreover, we investigate the performance and computation complexity of different soft bit metric calculation methods. The findings in this paper show that MIL-STD-110B 32-ary QAM constellation has the lower peak-average power ratio and achieves similar bit error rate (BER) performance as that of standard non-square 32-ary QAM constellation, which suggests MIL-STD-110B 32-ary QAM constellation is more appropriate to the practical systems.

## 1 INTRODUCTION

To improve transmission reliability and increase bandwidth efficiency in data transmission, it is an efficient way to combine advanced channel encoding with high order modulations in modern wireless communications. However, the potential power of advanced channel encoding or coded modulation schemes cannot be fully exploited due to the employment of hard decision based demodulation. Motivated by this, the coded modulation scheme based on soft decision has received much attention in recent years [1].

Due to the outstanding error correcting performance, low density parity check (LDPC) codes introduced by Gallager has become a key enabling technology for future wireless standards such as WiMax [2, 3]. The key idea of LDPC codes is based on the theory of random encoding introduced by Shannon, and soft-input-soft-output (SISO) decoder with belief propagation (BP) iterative decoding algorithm [4]. Hence, high order modulations with LDPC codes are considered in this paper. We investigate the two following, fundamental and interesting questions: 1) how to calculate the soft bit information for BP iterative decoding algorithm and 2) how to design the simplified soft bit metric methods for high order modulations.

Multilevel quadrature amplitude modulation (M-ary QAM) has become an attractive modulation technique for high-data-rate communication and high spectrum efficiency. According to the difference of modulation level, M-ary QAM can be classified into two categories, i.e. , square M-ary QAM and non-square M-ary QAM. To improve the reliability of decision, M-ary QAM demodulation using soft decision has been broadly investigated for M-ary QAM. However, most works focus on the investigation of simplified soft bit metrics for square M-ary QAM to reduce the cost of implementation [4-5]. In [4-5], various approximated bit metrics using the distance between the received signals and the hard decision line are proposed for 16-ary QAM and 64-ary QAM constellations, respectively. In [6], the authors proposed a recursive soft demodulation algorithm taking advantage of the relation between different bits in the modulated symbol. In [7, 8], the authors investigated simplified methods to compute the soft bit information for non-square M-ary QAM constellation.

Resorting to the result in [9], MIL-STD-110B 32-ary QAM constellation has lower peak-average power ratio (PAPR) than that of standard M-ary QAM constellation. However, to the best of authors' knowledge, simplified soft bit metric calculations for such constellations have not being-well investigated. Based on this, simplified soft bit metric calculation methods are designed for MIL-STD-110B 32-ary QAM constellation in this paper.

## 2 SYSTEM MODEL

Let us consider a system model shown in Figure 1. Prior to transmission, a sequence of binary information $b_i \in \{0,1\}$ is encoded by LDPC, and the coded sequence is $c = [c_1, c_2, \cdots, c_N]$. Then, the coded sequence is directly mapped to M-ary modulation symbol $s_n \in S$, where S is the modulation constellation set with cardinality $M = 2^K$. Without loss of generality, we consider MIL-STD-110B 32-ary QAM [10], and we have $M = 32$ and $K = 5$. The example of MIL-STD-110B 32-ary QAM symbol mapping is shown in Figure 2.

Let $r_n$ denote the baseband received signal corresponding to the transmitted symbol $s_n$, and we have

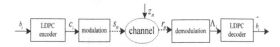

Figure 1. System Model.

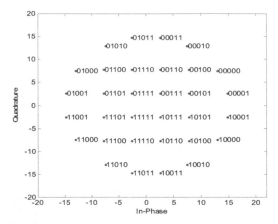

Figure 2. Bits to symbol mapping for MIL-STD-110B 32-ary QAM constellation.

$$r_n = r_{In} + jr_{Qn} = \sqrt{E_S}\, s_n + z_n, \tag{1}$$

where $E_s$ represents the received signal symbol energy, and $z_n \sim \mathcal{N}(0, N_0)$ is complex white Gaussian noise.

## 3 SOFT BIT LOG-LIKELIHOOD RATIO CALCULATION

In this section, we analyze different calculation algorithms of soft bit log-likelihood ratio (LLR), which can be used in the demodulator for the calculation of a posteriori LLR.

### 3.1 *Exact LLR calculation*

In this subsection, we first review the calculation of exact posteriori LLR, based on which the simplified algorithms are investigated.

Let $\Lambda(b_{ni}|r_n)$ indicate the LLR of $i$ the bit $b_{ni}$ in $r_n$, which is given by

$$\Lambda(b_{ni}|r_n) = \ln\left(\frac{\Pr(b_{ni} = 0 \mid r_n)}{\Pr(b_{ni} = 1 \mid r_n)}\right). \tag{2}$$

Define $S_i^0 = \{s_n : b_{ni} = 0\}$ and $S_i^1 = \{s_n : b_{ni} = 1\}$ as the subsets of symbol index corresponding to $b_i = 0$ and $b_i = 1$, respectively. Then, we have

$$\Pr(b_{ni} = 0|r_n) = \sum_{s_n \in S_i^0} \Pr(s_n|r_n) \tag{3}$$

$$\Pr(b_{ni} = 1|r_n) = \sum_{s_n \in S_i^1} \Pr(s_n|r_n) \tag{4}$$

By invoking the concepts of probability theory and assuming the equiprobable of the transmitted symbols, we have

$$\Lambda(b_{ni}|r_n) = \ln\left(\frac{\Pr(r_n \mid b_{ni} = 0)}{\Pr(r_n \mid b_{ni} = 1)}\right) = \ln\left(\frac{\sum_{s_n \in S_i^0} p(r_n \mid s_n)}{\sum_{s_n \in S_i^1} p(r_n \mid s_n)}\right) \tag{5}$$

where

$$p(r_n|s_n) = \frac{1}{\pi N_0} \exp\left(-\frac{(r_n - s_n)^2}{N_0}\right) \tag{6}$$

Finally, substituting (6) into (5), the LLR of $i$ th bit $b_{ni}$ can be expressed as

$$\Lambda(b_{ni}|r_n) = \ln\left(\frac{\sum_{s_n \in S_i^0} \exp\left(-\left(|r_n - s_n|^2\right)/N_0\right)}{\sum_{s_n \in S_i^1} \exp\left(-\left(|r_n - s_n|^2\right)/N_0\right)}\right) \tag{7}$$

It is noted from (7) that the computational complexity of LLR grows exponentially with the modulation constellation size $M$, which makes the soft demodulator intractable. Hence, the sub-optimum soft decision demodulators with the reduced complexity are highly desirable in practical systems.

### 3.2 *Max-log LLR calculation*

As is shown in Eq. (7), $\Lambda(b_{ni}|r_n)$ contains the summation of the exponent function. According to [5], the largest exponents in the above summations will be dominant since the exponential is a rapidly growing function of its argument. Hence, by taking

$$\ln\left(\sum_i e^{x_i}\right) \approx \max(x_i),$$

an approximation of $\Lambda(b_{ni}|r_n)$ is given by

$$\Lambda(b_{ni}|r_n) = \frac{1}{\sigma^2}\left\{\min_{s_n \in S_i^1}\left(|r_n - s_n|^2\right) - \min_{s_n \in S_i^0}\left(|r_n - s_n|^2\right)\right\} \tag{8}$$

Although the approximated expression (8) has reduced the computing complexity, it still needs to search all the points in the constellation.

### 3.3 Simplified LLR calculation

Due to the lower PARP, MIL-STD-110B 32-ary QAM constellation has been broadly adopted in high frequency communications to improve power efficiency. Hence, we study the simplified methods for MIL-STD-110B 32-ary QAM constellation in this subsection.

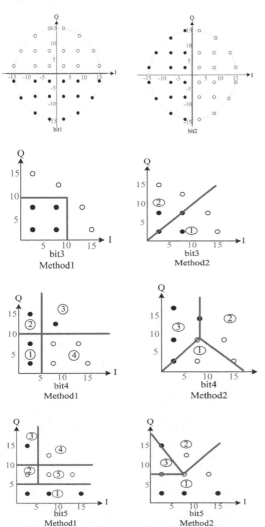

Figure 3. Partitions of MIL-STD-110B 32-ary QAM constellation.

Before involving into the detail analysis, we first give the partition $(s_i^0, s_i^1)$ for the $i$ th bit $b_i$ in Figure 3, where the hollow circle represents $b_i = 0$ and the solid circle denotes $b_i = 1$.

As is shown in Figure 3, the distance between the received symbol and the hard decision boundary as the LLR in the Method1, which is similar to the method used in the square QAM. Therefore, the

simplified LLR algorithm for MIL-STD-110B 32-ary QAM is given by
Method 2:

$$\Lambda_1 = r_Q \tag{9}$$

$$\Lambda_2 = r_I \tag{10}$$

$$\begin{cases} \Lambda_3 = |r_I| - 10 \\ \Lambda_3 = |r_Q| - 10 \end{cases} \tag{11}$$

$$\begin{cases} \Lambda_4 = |r_I| - 5 \\ \Lambda_4 = |r_I| - |r_Q| \\ \Lambda_4 = 10 - |r_Q| \end{cases} \tag{12}$$

$$\begin{cases} \Lambda_4 = |r_Q| - 5 \\ \Lambda_4 = |r_I| - 5 \\ \Lambda_4 = 10 - |r_Q| \end{cases} \tag{13}$$

In the equations (11), (12) and (13), each sub-equation is corresponding to the areas divided in Figure 3, Method2, respectively.

### 3.4 Complexity analysis

We have analyzed the computational complexity for the two kinds of simplified LLR methods, the exact LLR algorithm, and Max-log LLR algorithm in this subsection. To make comparison fair, the division and subtraction operations are regarded as the multiplication and addition operations, respectively. The complexity comparison can be seen clearly in Table 1. It suggests that the two simplified algorithms have much lower complexities compared with that of exact and Max-log schemes.

Table 1. Computational complexity analysis.

| Algorithm | Exponent | Multiplication | Additive | Comparison |
|---|---|---|---|---|
| Exact | 37 | 96 | 246 | -- |
| Max-Log | -- | 69 | 101 | 150 |
| Method1 | -- | 1 | 5 | 7 |
| Method2 | -- | -- | 3 | 3.25 |

## 4 SIMULATION RESULTS

In this section, we compare the performance of different soft bit metric calculation methods for standard non-square 32-ary QAM and MIL-STD-110B non-square 32-ary QAM in AWGN channels, respectively. Without any loss of generality, a rate-1/2 LDPC with code length 360 is used, and BP iterative decoding algorithm is adopted in the following simulation.

Figure 4 plots BER performance of four different soft bit metric algorithms for the two different 32-ary QAM scheme in AWGN channels, respectively. It can be seen clearly in Figure 4 that the two different simplified calculation methods have almost the same performance as that of exact LLR scheme and Max-log LLR scheme. However, the computational complexities of Method 1 and Method 2 are much lower than that of exact LLR scheme and Max-log LLR scheme as analyzed in the above section. On the other hand, the proposed simplified methods for MIL-STD-110B non-square 32-ary QAM, i. e., Method 1 and Method2, have about 0. 5dB and 0. 3 dB performance losses for that of exact LLR algorithm. These results suggest that the proposed simplified algorithms are valid for MIL-STD-110B 32-ary QAM. Moreover, we can see that MIL-STD-110B 32-ary QAM and standard 32-ary QAM achieve similar BER performance in coded modulation systems.

To exploit the advantages of MIL-STD-110B 32-ary QAM, Figure 5 plots the complementary cumulative distribution function (CCDF) versus

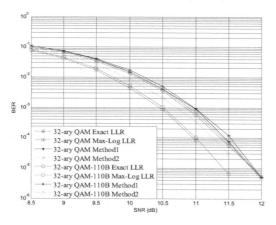

Figure 4. BER comparison for standard non-square 32-ary QAM and MIL-STD-110B non-square 32-ary QAM.

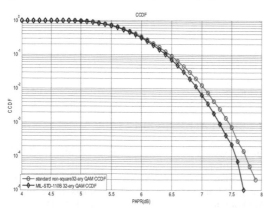

Figure 5. PAPR comparison performance between two different 32-ary QAM.

PAPR performance for the two 32-ary QAM constellations. In this simulation, the square-root-raised-cosine filter with the roll-off factor 0.25 is adopted in the transmitter and receiver, and the oversample rate is 8. As can be clearly seen in the figure, MIL-STD-110B 32-ary QAM has lower PAPR compared with that of square 32-ary QAM, which indicates that MIL-STD-110B 32-ary QAM has more power efficiency.

## 5 CONCLUSION

In this paper, we investigate different soft bit metric calculation methods for the two different 32-ary QAM constellations. Based on the conventional soft bit metric calculation method for standard non-square 32-ary QAM, we extend the methods to the MIL-STD-110B 32-ary QAM constellation. Simulation results show that simplified LLR calculation methods with negligible performance degradation have much lower complexity than that of exact LLR method and Max-log LLR method. In addition, MIL-STD-110B 32-ary QAM has lower peak-to-average power ratio than that of standard non-square 32-ary QAM, which suggests that MIL-STD-110B 32-ary QAM constellation has more power efficiency and is more appropriate to wireless communications in some practical scenario.

## REFERENCES

[1] D. Liang, X. N. Soon, and H. Lajos, "Soft Decision star QAM aided BICM-ID", *IEEE Signal Processing. Lett.*, vol. 18, no. 3, pp. 169–172, Mar. 2011.

[2] Q. Li, G. Li, W. Lee, M.-I. Lee, D. Mazzarese, B. Clerckx, and Z. Li, "MIMO techniques in WiMAX and LTE: A future overview," IEEE Commun. Mag. , vol. 48, no. 5, pp. 86–92, May 2010.

[3] E. Soleimani-Nasab, M. Matthaiou, and M. Ardebilipour, "Multi-relay MIMO systems with OSTBC over Nakagami-m fading channels", IEEE Trans. Veh. Technol., vol. 62, no. 8, pp. 3721–3736, Oct. 2013.

[4] D. J. C. MacKay and R. M. Red, "Near Shannon limit performance of low-density parity-check codes", Electron. Lett. , vol. 32, pp.1645–1646, 1996.

[5] F. Tosato and P. Bisaglia, "Simplified soft-output demapper for binary interleaved COFDM with amplification to HIPERLAN/2", *IEEE International Conference on Communications*, vol. 2, pp. 664–668, 2002.

[6] J. Zhao, J. Zhang, "Bit confidence-based soft decision metric generation for QAM", Journal of Electronics & Information Technology, vol. 31, pp. 985–988, 2009.

[7] L. Li, D. Divsalar, and S. Dolinar, " Iterative demodulation, demapping and decoding of coded non-square QAM", IEEE Transactions on Communications, vol. 53, no. 1, pp. 16–19, Jan. 2005.

[8] L. Wang and D. Xu, "Simplification of bit metric calculation for 32QAM", Journal of Applied Sciences-Electronics and Information Engineering, vol. 28, no. 4, pp. 331–336, Jul. 2010.

[9] J. G. Proakis, *Digital Communications*, 4th ed. McGraw-Hill, 2001.

[10] MIL-STD-188-110B. 27 April, 2000.

*Multimedia, Communication and Computing Application – Leung (Ed.)*
© 2015 Taylor & Francis Group, London, ISBN 978-1-138-02775-6

# The measurement method of human body bust based on Kinect depth image

Y.W. Chong, Y.Y. Wang, S.M. Pan & Z.W. Wang
*Wuhan University, LIESMARS, Wuhan, China*

H.Q. Dai
*Chinese Academy of Science, Institute of Semiconductors, Beijing, China*

Y. Li
*Institute of Technology, Vocational School, Zhejiang Normal University, Jinhua, China*

ABSTRACT: The noncontact measuring method has gradually taken the place of the traditional manual measuring method because of its high speed, accuracy, high efficiency, and convenience. The scanners of the three-dimensional body scanning technology are too expensive and inconvenient to be moved. Because of the visible light data acquisition method, the technology of body measurement based on RGB image is vulnerable to environmental factors such as light and obstructions and it is not accurate and environmentally adaptive. To solve these problems, this paper proposes a new method of human body measurement technology to calculate human skeleton information. First, the method involves gathering human depth image by Kinect sensor infrared coding technology. Second, it involves analyzing and identifying depth images, obtaining joints of the human body skeleton, and fitting sizes of human body busts on the basis of human skeleton measurement. Experiments proved that this method is able to accurately measure and estimate bust size, get greater efficiency, and has environmental suitability and wider applicability. In a word, this method can be applied to obtain the body's surface size quickly and efficiently.

## 1 INTRODUCTION

Kinect is a new kind of three-dimensional camera for motion sensing. It is able to obtain accurate depth and location information without influences of illumination and light factors. With the rapid development of technology, automation, individualization, and body measurement are becoming more and more important in the apparel industry [1].

There are two kinds of apparel anthropometric methods, and they are contact measurement and non-contact measurement methods [2]. Contact measurement method involves original manual measurement including Martin measurement method and gypsum finalize skill [3,4]. Noncontact measurement method includes color photographic measuring method and three-dimensional body scanning measurement method. Three-dimensional body scanning technology needs expensive equipment. So, it cannot be widely used in body measuring. Photographic measurement is susceptible to the influence of illumination and obstacles and is unable to achieve real-time measurement. The body measurement method based on Kinect depth image is able to detect the human body and obtain the size of human body parts from depth images without the influences of illumination and color factors [5].

## 2 KINECT EQUIPMENT AND DATA ACQUISITION

### 2.1 *Structure of Kinect equipment*

Kinect is a kind of a portable motion sensing input device. Kinect consists of 6 major components, including infrared projector, RGB camera, infrared camera, multiarray microphone, motor adjusting camera angle, and logical circuit [6]. The image resolution of the depth stream is $640 \times 480$ pixels [7]. The depth data obtained by the Kinect are composed of 2 bytes [8]. The view field of depth sensor seems like a pyramid with an elevation angle of 57° and a maximum detectable range of 4000 mm [9].

### 2.2 *Measuring chest depth using Kinect*

Human body can be recognized when the Kinect device detects a human [10]. We can get the human body skeleton, which consists of 20 joint nodes, and obtain the length between different joint nodes. First, we should measure the chest depth of human body by Kinect. The chest of a woman's body in profile is obvious and outward. So, we can find out the maximum thickness of a woman's body in profile. The maximum of the thickness is defined as chest depth. The position of the maximum thickness of a woman's body in profile is the woman's chest. The chest is located between the left shoulder and the spine joint. So, we traverse the thickness of body side between the shoulders' joint and the spine joint by Kinect and find the maximum of the body's thickness. First, we should determine the height of the shoulder joint and the spine joint. The height of the shoulder joint is defined as *Shoulder_Height*. *Shoulder_Height* is the sum of length of the bone from joint Foot Left, Ankle Left, Knee Left, Hip Left, Hip Center, Spine to Shoulder Center, as shown in formula (1).

$$Shoulder\_Height = Length(HipLeft,$$
$$KneeLeft, AnkleLeft, FootLeft) +$$
$$Length(ShoulderCenter, Spine, HipCenter) \quad (1)$$

The height of the shoulder joint is defined as *Spine_Height*. It is the sum of length from joint Foot Left, Ankle Left, Knee Left, Hip Left, Hip Center to Spine, as shown in formula (2). The maximum width is named as chest depth. As shown in figure 1, chest depth is the length between point M and point N. At the same time, record the corresponding height of point N as Height_MN. Height_MN is the height of chest.

$$Spine\_Height = Length(HipLeft, KneeLeft,$$
$$AnkleLeft) + Length(Spine, HipCenter) \quad (2)$$

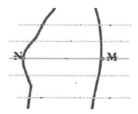

Figure 1.   Determine the chest depth.

### 2.3   *Measuring chest width using Kinect*

We have got the height of the chest, Height_MN, according to formula 2. So, we can determine the location of woman's chest and get the chest width

data. In figure 2, the chest width of a woman's body is the length between point $M_1$ and point $N_1$.

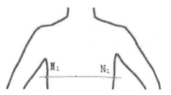

Figure 2.   Determine the chest width.

### 2.4   *Measuring shoulder width using Kinect*

According to human body physiology, the bust size of human body is associated with chest depth, chest width, and shoulder width. We find the Right Shoulder joint and Left Shoulder joint based on the skeleton of Kinect device. Shoulder width is the distance between Right Shoulder joint and Left Shoulder joint, as shown in formula (3).

$$Shoulder\_Width = Length\ (ShoulderRight,$$
$$ShoulderLeft) \quad (3)$$

## 3   FITTING CHEST REGRESSION FUNCTION BASED ON KINECT MEASURING DATA

### 3.1   *Verifying the reliability of Kinect measuring data*

In order to verify the accuracy and reliability, we design an experiment making a comparison between Kinect measuring data and traditional measuring data. We randomize the selection of 140 women between the ages of 18 and 40 years and measure their chest depths, chest widths, and shoulder widths in two different ways. We compare the measuring results of the two different measuring methods by mean variation and relative average deviation. Mean variation is denoted by $\bar{d}$. The formula of mean variation is shown as formula (4). Formula (5) shows the formula of relative average deviation.

$$\bar{d} = \frac{\sum_{i=0}^{n} |x_i - \bar{x}|}{n} \quad (4)$$

$$s = \frac{\sum_{i=0}^{n} |x_i - \bar{x}| / n}{\bar{x}} \times 100\% \quad (5)$$

Table 1 shows the comparison between Kinect measuring data and traditional manual measuring data. The mean variations of chest depth, chest width, and shoulder width are all less than 0.5 cm. The relative average deviations of chest depth data, chest width data, and shoulder width data are still less than 0.5%. The experiment results show that there is no significant difference between Kinect measuring data and traditional manual measuring data. The human body size data measured by Kinect are reliable and can be used for regression and curve fitting of bust.

Table 1. Difference between Kinect measuring data and traditional manual measuring data.

| Kinect measuring data | Mean variation | Relative average deviation |
|---|---|---|
| Chest depth | 0.23 cm | 0.25% |
| Chest width | 0.34 cm | 0.36% |
| Shoulder width | 0.37 cm | 0.21% |

### 3.2 Regression and curve fitting based on Kinect

Human physiology and anthropometry of apparel industry reveal that there is a positive correlation between women' bust and three sets of measuring data [2], which include chest depth, chest width, and shoulder width. A multiple linear regression model is established to fitting bust through multiple linear regression function of bust. Fitting result is formula (6).

$$Bust = 0.8677 + 1.9561 \times chest\_depth + 0.8408 \times chest\_width + 0.3962 \times shoulder\_width \qquad (6)$$

The evaluation parameter of fitting result is correlation coefficient. The correlation coefficient is regarded as $r^2$. The result of correlation coefficient is $r^2 = 0.98216$. The evaluation parameter proves that regression and curve fitting give good results.

## 4 THE EXAMINATION OF REGRESSION FUNCTION OF BUST

### 4.1 Under circumstances of natural light and body not being occluded

In order to verify the accuracy of regression function (6) under the circumstances of natural light and body not being occluded, we randomly select 20 women between the ages 18 and 40 years. Kinect is used to measure their chest depth, chest width, and shoulder width. Plug the three sets of data into the linear regression function (6) and get the fitting data of bust.

Table 2 shows Kinect fitting bust, manual measuring bust, and their difference of ten experimenters. Experiment shows that the average error of fitting bust of 20 experimenters is 1.24 cm, and the average relative error is 1.46%. The experiment results show that there is no significant difference between Kinect measuring fitting bust and traditional manual measuring bust. The established multiple linear regression function of bust is very accurate.

Table 2. Kinect fitting error of bust under normal circumstances.

| Manual measuring bust/cm | Fitting error of bust/cm | Kinect fitting bust/cm |
|---|---|---|
| 79.22 | 78.46 | −0.76 |
| 77.21 | 77.64 | 0.43 |
| 75.60 | 75.31 | −0.29 |
| 81.62 | 80.52 | −1.10 |
| 89.85 | 91.06 | 1.21 |
| 78.30 | 78.16 | −0.14 |
| 83.14 | 82.85 | −0.29 |
| 79.52 | 80.67 | 1.15 |
| 76.43 | 77.59 | 1.16 |
| 77.94 | 77.26 | −0.68 |

### 4.2 Under circumstances without light and body not being occluded

Similarly, we randomly select the same 20 women under the circumstances of darkness and body not being occluded. Then, we measure their chest depth, chest width, and shoulder width using the Kinect and plug the three sets of data into (6) and get the fitting data of bust. Table 3 shows Kinect fitting bust, manual measuring bust, and their difference of ten experimenters. Experimental data of Kinect show that the average error of fitting bust of 20 experimenters is 1.52 cm, and the average relative error is 1.79%. The experiment results show that there is no significant difference between Kinect measuring fitting bust and traditional manual measuring bust in the dark environment.

Table 3. The Kinect fitting error of bust in the environment without light.

| Manual measuring bust/cm | Fitting error of bust/cm | Kinect fitting bust/cm |
|---|---|---|
| 79.22 | 81.42 | −1.07 |
| 77.21 | 77.95 | −0.39 |
| 75.60 | 74.86 | 0.53 |
| 81.62 | 80.31 | −1.16 |

(Continued)

Table 3.   (*continued*)

| Manual measuring bust/cm | Fitting error of bust/cm | Kinect fitting bust/cm |
|---|---|---|
| 89.85 | 88.96 | 1.01 |
| 78.30 | 78.79 | 0.65 |
| 83.14 | 83.63 | 0.48 |
| 79.52 | 78.84 | 0.82 |
| 76.43 | 77.68 | 1.41 |
| 77.94 | 77.26 | 1.16 |

### 4.3   *Under circumstances of natural light and body being partly occluded*

Under circumstances of natural light and part of the body being occluded, we carry out the experiment in the same way. Table 4 shows the Kinect fitting bust, manual measuring bust, and their differences of ten experimenters. Experimental data show that the average error of fitting bust of 20 experimenters is 1.63 cm, and the average relative error is 1.92%. The experiment results show that there is no significant difference between Kinect fitting bust and traditional measuring bust. The multiple linear regression function of bust is very accurate.

Table 4.   The Kinect fitting error of bust in the environment of body being occluded.

| Manual measuring bust/cm | Fitting error of bust/cm | Kinect fitting bust/cm |
|---|---|---|
| 79.22 | 77.96 | −1.26 |
| 77.21 | 77.85 | 0.64 |
| 75.60 | 76.32 | 0.72 |
| 81.62 | 80.35 | −1.27 |
| 89.85 | 89.00 | −0.85 |
| 78.30 | 79.20 | 0.90 |
| 83.14 | 82.34 | −0.80 |
| 79.52 | 78.32 | −1.20 |
| 76.43 | 77.92 | 1.49 |
| 77.94 | 76.56 | −0.70 |

### 4.4   *Environment adaptability test*

Figure 3 shows the differences of Kinect fitting bust between natural light environment and dark environment when the human body is not being blocked. Experiment reveals that the differences of Kinect fitting bust between normal environment and dark environment are both less than 1.5 cm. The small differences of experiment show that Kinect measuring and fitting method of bust girth is not vulnerably affected by light.

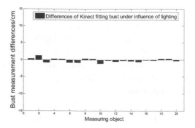

Figure 3.   Differences of light effect experiment.

Figure 4 shows the differences of Kinect fitting bust girth between circumstance of human body being blocked and circumstance of human body not being blocked in the environment of natural light. The experiment reveals that the differences of two different blocking conditions of fitting bust are all less than 1.5 cm. As the experiment results suggest, the measurement method of the bust based on Kinect is not vulnerably affected by light and occlusion of the body. The method has the characteristics of good environmental adaptability and stability.

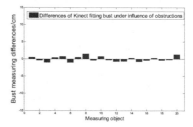

Figure 4.   Differences of blocking effect experiment.

## 5   CONCLUSIONS

In the paper, we gather human depth image and human skeleton from Kinect. We can obtain sizes of chest depth, chest width, and shoulder width on the basis of human skeleton and joints measurement. Finally, we get the sizes of human bodies' bust by the means of regression and curve fitting. Experiment results suggest that the measurement method of bust based on Kinect has high accuracy of data measurement and bust fitting result.

The advantages of obtaining human body size information based on Kinect are quickness and effectiveness, and it can also obviate the interference of light, color, or other obstacles. The measurement method of bust based on Kinect depth image can be applied to some more complex environments and we can obtain ideal bust measurement results. In this paper, we only measure the size of the bust by Kinect. In the next step, we will measure other sizes related to the apparel industry, including waistline, hipline, thigh circumference, and so on. The noncontact measurement technology of human body based on Kinect will adapt to the pursuit of personalized and rapid consumption trend.

ACKNOWLEDGMENTS

This paper was supported by the National Natural Science Foundation of China (41271398), Fundation Research Funds for the Central Universities (204201kf0242 and 204201kf0263), Shanghai Aerospace Science and Technology Innovation Fund Projects (SAST201425), and Natural Science Foundation of Hubei Province of China (2012FFB04204).

REFERENCES

[1] Zhao Jingmiao. The Research of Non-contact Body Measurement System-Digital Image Processing and Anthropometric information obtaining [D]. Tianjin: Tianjin University of Technology, 2004.
[2] Xu Feng, Zhang Hao, Zheng Rong. Prediction for bust girth using regression analysis from photographic and anthropometric data [J]. Journal of textile, 2006, 27(8):49–52.
[3] Peng Ronghua, Zhong Yuexian, Zhang Wuming. The study of the system for non-contact body 3D measurement [J]. Measurement Technique. 2004, (2):36–38.
[4] Li Jun, Liu Suyan, Zhang Yun. Application of three-dimensional body measurement system [J]. Journal of Donghua University (Eng. Ed.). 2005, 22(4):134–138.
[5] Fan Jianying, Yu Shuchun, Wang Yang, Image Segmentation Based on Edge Fusion of Normal Component [J]. Computer Engineering. 2010, 36(17):221–222, 225.
[6] Erica Naone. Microsoft Kinect: How the Device Can Respond to Your Voice and Gestures [Z]. Pioneering with Science and Technology. 2011, 4(04):8–83.
[7] Yang Dongfang, Wang Shicheng, Liu Huaping, Scene Modeling and Autonomous Navigation for Robots Based on Kinect System [J]. Robot, 2012, 34(5):581–589.
[8] Garcia J, Valencia E S, Zalevsky Z, et al. Range mapping using speckle decorrelation: The United US7433024B2[p]. 2008-10-07.
[9] Microsoft Kinect for Windows SDK Programming Guide [M]. USA: Microsoft Research, 2011.
[10] Li Hongbo, Ding Jianlin, Ran Yongguang, Human Identification Base on Kinect Depth Image [J]. Digital Communication, 2012, (4):21–26.

*Multimedia, Communication and Computing Application – Leung (Ed.)*
© *2015 Taylor & Francis Group, London, ISBN 978-1-138-02775-6*

# Image denoising method of edge-preserving based sparse representations

S.K. Liu, Y.F. Chen, K. Zhao & L.B. Zhao
*Department of Computer Science and Technology, Beijing Institute of Technology, Beijing, China*
*China Aerospace Science & Industry Corp, Beijing, China*

ABSTRACT: Utilizing sparse and redundant representations framework for image denoising is currently a relatively new denoising method. This paper attempts to make an overview of sparse model denoising based on the understanding and analysis of recent literatures. In this paper, the effective representation of characteristics and edge information of noisy images is studied. To make the edge retaining of images effective, a method of edge-preserving denoising is proposed. The comparative experimental results are given to demonstrate that the proposed algorithm has better efficiency in denoising and enhances the capacity of retaining edge and gets a better visual effect of denoising image.

KEYWORDS: Image denoising, Sparse representation, Edge-preserving denoising, Retaining edge

## 1 INTRODUCTION

Images are often infected by noise in the process of acquisition and transmission in the digital form, so image denoising has become an important part of the image processing application. Image denoising problem can be generally formulated by

$$I_{noisy} = I + v \tag{1}$$

Where $I_{noisy}$ is a noisy image, $v$ is zero-mean white and homogeneous Gaussian noise, and $I$ is the original image.

Extensive studies have been conducted in removing noise from the original image, which is inverse problem. Sparse representation was first suggested in the statistical literature [1]. This method is similar to Fourier analysis and wavelet analysis, in which the image is represented as a linear combination of basic functions. What is different is that the expansion coefficients of the image in this basic function are zero and the number of nonzero coefficients is small. The basic functions are called atoms, and the collections of atoms is called dictionary. In order to achieve sparse representation, the dictionary that is used to express the image is usually over complete.

Sparse decomposition can concentrate the energy of a signal on a few coefficients, which is not only beneficial to signal compression and coding but also suitable for removing noise from images. Because the existence of the image signal is self similar, ideal and nonnoisy signal can be represented sparsely. And due to the absence of structure, the noise cannot be sparse representation. If the noisy image is decomposed sparsely, the sparse component corresponds to the desired signal and the residual between the noise images and the sparse component corresponds to the noise. Utilizing this feature, the sparse component can be extracted from the noisy image by sparse decomposition and the original image will be reconstructed by the sparse component, which can achieve the purpose of removing noise.

At present, researchers pay close attention to the dictionary and sparse reconstruction algorithm in the research of image denoising method based on sparse representation. There are many ways for the selection of dictionary: the orthogonal basis, such as the Fourier basis; wavelet basis; and dictionary based on learning, which is obtained by training samples. The typical sparse reconstruction algorithm includes the greedy algorithm based on the minimization of the norm, such as matching pursuit; orthogonal matching pursuit[2]; linear programming algorithm based on the minimization of the norm, such as basis pursuit method[3]; gradient projection method[4]; and statistical optimization reconstruction algorithms, such as Bayes reconstruction.[5]

The sparse denoising method is used in reconstruction error minimization as fidelity[6], which can make the reconstruction error minimum, but it has disadvantages in edge information recovery. In order to effectively denoise and maximize the recovery of geometric features of the image, the paper sets different edge-preserving operators according to the different regions, which are divided into edge details and smooth regions based on the gradient information, so that the image edge in the new denoising algorithm can be recovered better.

## 2 THE IMAGE DENOISING MODEL OF SPARSE EDGE PRESERVING

### 2.1 The image model of the sparse representation

According to the theory of calculation of harmonics, the $u \in R^{N*1}$ signal can be expressed as a linear combination of atoms; if the atoms are $d_1, d_2, d_3, \ldots d_n$, then

$$u = d_1 k_1 + d_2 k_2 + d_3 k_3 + .. + d_n k_n \qquad (1)$$

The theory of image sparse representation[7] is that there is a proper dictionary $D = [d_1, d_2, d_3, \ldots d_n]$ that makes $u$ the sparse representation. And the sparse matrix $K = [k_1, k_2, k_3, k_4 \ldots k_n]^T$ will have zero elements rarely: $\|k_n\|_0 << N$. The noise model for the image is that

$$y = x + v \qquad (2)$$

where $y$ is the noisy image and $x$ is the original image. Based on the theory of sparse representation, we should make variance between the sparse representation and the samples reaching the minimum, and the objective function is constructed as follows:

$$\hat{a} = argmin \| a \|_0 \ s.t. x = Da \qquad (3)$$

where $\|a_i\|_0$ is the 0-norm, which stands for the count of all the nonzero entries in $a$. According to the description of the model of sparse representation and the noisy image, the model of image denoising of sparse representation can be obtained:

$$y = x + v = Da + v \qquad (4)$$

where $y$ is the noisy image, $x$ is the original image, $D$ is the over complete dictionary, $a$ is the sparse matrix, and $v$ is the noise. For a given $D$, $a$ can be obtained by the optimization problem as follows:

$$\hat{a} = argmin \| a \|_0 \ s.t. \| Da - y \|_2^2 \le M \qquad (5)$$

So, we get $x = Da$. It is a typical NP hard problem to obtain the unique solution of the formula for a redundant dictionary. The problem can be transformed into the 1-norm of $a$ in the case that $a$ is sufficiently sparse. Then (5) can be converted into

$$\hat{a} = argmin \| a \|_1 \ s.t. \| Da - y \|_2^2 \le M \qquad (6)$$

The above is a convex optimization problem. The optimization problem can be expressed by Lagrange multiplier method so that the aforementioned optimization task can be changed to

$$\hat{a} = argmin_a \| Da - y \|_2^2 + \mu \| a \|_1 \qquad (7)$$

in which the constraint becomes a penalty and $\mu$ is a Lagrange multiplier that is used to balance the fidelity term and the sparse prior term.

### 2.2 The image denoising model of sparse edge preserving

From the theory of sparse representation, the useful information of the image has a certain structure and the structure and atomic structure are consistent. But the noise does not have this feature and the useful information and noise are effectively separated by sparse decomposition, which could achieve the purpose of denoising. For the low-SNR image, the hard threshold is set as the termination condition of sparse representation. When the SNR is very low, the noise has larger variance compared with the image signal. If better denoising effects are our goal, we have to set the lower threshold, but the denoising algorithm needs more time and ignores the matching of local geometric features because of too much noise. In order to maintain the edge information of the image much better, this paper proposes the image denoising model of edge preserving based on sparse representation.

$$\hat{a} = argmin_a \| \lambda Da - y \|_2^2 + \mu \| a \|_1 \qquad (8)$$

$$l = \frac{1}{1 + f_n} \qquad (9)$$

$$f_n = normal(gradient(d_n^T a)) \qquad (10)$$

Where $D = [d_1, d_2, d_3, \ldots d_n]$; gradient $(d_n^T a)$ represents the gradient image pixel value; $f_n$ is the normalization of the pixel gradient, which limits $f_n$ in the range of $[0,1]$ so that the data processing is more convenient and the convergence speed becomes faster; and $\lambda$ is the edge-preserving operator. When $f_n$ is in the range of $[0,1]$, $f_n$ increases and the gradient of the point is larger, so the change of gray is larger and the value of the gradient direction is greater.

Note the aforementioned formulas (9) and (10). It is suggested that the edge-preserving operator depends entirely on the gradient value; thus, the solution of the gradient completely determines the effect of image restoration. In order to get the gradient information effectively from the noisy image block, as the traditional gradient operators that deal with gradient information such as Sobel operator, Roberts operator, and Prewitt operator cannot operate effectively in the noise environment, the LOG operator, which is robust to noise and has a discrete point is used to calculate the gradient in this paper. We proposed the low gradient threshold $P$, and when the gradient is lower $P$ is set to 0. The value of $P$ is Mean$(f_n)$ + RMS$(f_n)$, where Mean$(f_n)$ is the average of $f_n$ and RMS is the variance of $f_n$. So, when $f_n$ is lower than Mean$(f_n)$ +RMS$(f_n)$ $f_n$ is zero.

## 2.3 The image denoising process of edge preserving based on sparse representation

In order to recover well the edge information of denoising based on the sparse representation, this paper proposes the edge-preserving operator based on gradient information. The denoising algorithm of sparse representation includes two important parts, dictionary learning and sparse decomposition. In view of the new model, the K-SVD algorithm is used to generate the dictionary and the OMP (orthogonal matching pursuit) algorithm, which has the characteristic of simple calculation and is fast when used for image restoration. The basic idea is to get the combinations of atom from the dictionary, which is the most match of the residual signal by iteration and it will be terminated when the iteration times exceed a certain value.

In summary, the denoising method of sparse edge preserving presented in this paper is as follows:

1 Initialization:
Train sample. First, a two-dimensional image signal was transformed into a one-dimensional signal and then the one-dimensional signal was normalized; initial training samples, $Y = [y_1, y_2, ..., y_n] \in R^{N*1}$. The training samples were expressed by column vector, and the initial dictionary is the DCT dictionary.
2 Dictionary learning:
The dictionary $D$ is learned by the K-SVD algorithm based on the initial DCT dictionary [8].
3 Sparse decomposition:
The variables are assumed such that $r^0 = y$, $D = [d1,d2,d3, ... dn]$, $\lambda = \frac{1}{1} + normal(gradient(y))$
a. Find out the maximum index $I$ of the inner product between the residual $R$ and column $D_i$ of dictionary $D$.
$I_t = argmax_{i=1...N} |< r_{t-1}, d_i >|$;
b. Update the index
set $\Lambda_t = \Lambda_{t-1} U\{I_t\}$, recording the atomic set of reconstruction that is in the dictionary.

$D_t = [D_{t-1}, d_{I_t}]$;

c. We can get $a_t$ by least squares

$$\hat{a}_t = argmin \| \lambda Da - y \|_2^2$$

d. Update residuals $r_t = y - \lambda D_t a_t$, $t=t+1$;
e. To determine whether $t$ meets $t > K$ (the sparse coefficient). If the condition is satisfied, the iteration should be stopped; if it is not ,then we should stop $a$;
4 The dictionary and the sparse coefficient that can be obtained by step 3 are multiplied by the result image. In order to get a better edge-preserving image, we can get the edge matrix through edge detection in the first iteration (the part of the edge is 1 and the part of the nonedge is 0); then, we can only increase the effective part through the residual illustrations (image of the new algorithm minus the image of the OMP algorithm). Then, add the residual illustrations to the image of the OMP algorithm.

## 3 RESULTS

In order to verify the better denoising effect and stronger ability to retain the characteristics, Barbara, Lena, and Peppers were added to the standard difference sigma = 20, 40, 60, 80 of the Gauss noise and we use the OMP algorithm and the new algorithm to denoise. In order to ensure the accuracy of experimental results, the experiment is performed in the same environment (noise addition and dictionary training). Comparing the PSNR (Peak Signal-to-Noise Ratio) (see Table 1) and visual effects, we can see the effect of the proposed algorithm: the larger the PSNR, the better the denoising effect.

Table 1. Summary of the denoising PSNR results in [dB].

| Image | Standard deviation | PSNR/dB | |
|---|---|---|---|
| | | OMP | The proposed algorithm |
| Barbara | 20 | 30.80 | 30.82 |
| | 40 | 26.85 | 26.88 |
| | 60 | 23.70 | 23.73 |
| | 80 | 21.74 | 21.79 |
| Lena | 20 | 31.08 | 31.10 |
| | 40 | 27.24 | 27.28 |
| | 60 | 25.02 | 25.05 |
| | 80 | 22.99 | 23.12 |
| Peppers | 20 | 30.04 | 30.06 |
| | 40 | 26.71 | 26.74 |
| | 60 | 24.46 | 24.50 |
| | 80 | 22.34 | 22.39 |

239

The image of Peppers and Cameraman are added to the Gauss noise with a standard deviation of 40 and

Figures 1 and 2 are the results of the OMP algorithm and the proposed algorithm, respectively.

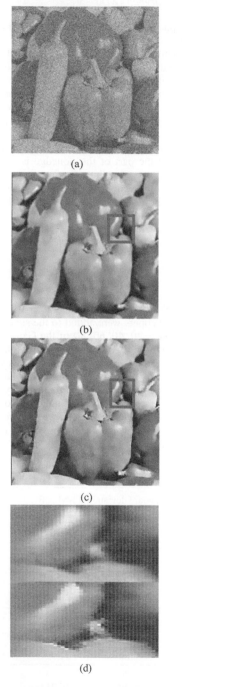

(a)

(b)

(c)

(d)

Figure 1. Image deblurring performance comparison for Peppers image. (a) The noisy image; (b) the OMP algorithm (PSNR = 26.71 dB); (c) the proposed algorithm (PSNR = 26.74 dB); (d) close-up views (top shows the details of (b) and bottom shows the details of (c)).

(a)

(b)

(c)

(d)

Figure 2. Image deblurring performance comparison for Cameraman image. (a) The noisy image; (b) the OMP algorithm (PSNR = 25.35 dB); (c) the proposed algorithm (PSNR = 25.40 dB); (d) close-up views (top shows the details of (b) and bottom shows the details of (c)).

In Figures 3 and 4, we can see the respective residual images between the results of the two algorithms and the original image. The original image is added to the Gauss noise with a standard deviation of 40, and the dictionary is obtained by the K-SVD algorithm.

## 4 CONCLUSIONS

This paper proposed a new image denoising method of sparse edge preserving, and the experiments proved the denoising and edge-preserving effect. From the horizontal results, we can see that the PSNR of the new algorithm is higher than that of the OMP algorithm in the same condition of noise, and with the increase of noise the effect of new algorithm gets better, which shows that the new algorithm has great improvement on the OMP algorithm. It is clearly seen that the edge feature of image in the proposed algorithm remains a good improvement from the residual image.

However, the robustness of the edge-preserving operator, which is proposed in the new sparse edge-preserving model, to noise still needs to be further strengthened.

It will exert a great influence on the results of the extraction of gradient information in a large amount of noise information. The excellent extraction algorithm of gradient information can greatly enhance the PSNR of the reconstructed image. In this paper, we use the traditional K-SVD algorithm to obtain the atom without other optimization such as clustering. Therefore, further research could be conducted in view of the aforementioned questions.

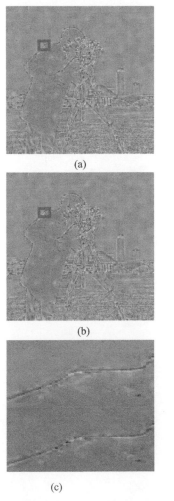

Figure 3. The comparison of residual images in terms of deblurring performance for Cameraman image. (a) The noisy image; (b) the OMP algorithm; (c) close-up views (top shows the details of (a), and bottom shows the details of (b)).

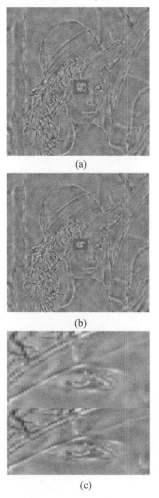

Figure 4. The comparison of residual images in terms of deblurring performance for Lena image. (a) The noisy image; (b) the OMP algorithm; (c) close-up views (top shows the details of (a), and bottom shows the details of (b)).

241

REFERENCE

[1] Aharon, M., Elad, M., Bruckstein, A. 2006. K-SVD: An Algorithm for Designing Over Complete Dictionaries for Sparse Representation. *IEEE Transactions on Signal Processing 54(11): 12–16.*

[2] Patyic, Rezaiifarr, Krishnaprasad PS. 1993. Orthogonal matching pursuit: recursive function approximation with applications to wavelet decomposition. *IEEE Press1: 40–44.*

[3] Chen SS, Donoho DL, Saundem M A. 2001. Atomic decomposition by basis pursuit. *SIAM Review, 43(1): 129–159.*

[4] M Figueiredo, R Nowak and S Wrigllt. 2007. Gradient projection for sparse reconstruction. *IEEE Journal of Selected Topics in Signal Processing, 12, l(4): 586–597.*

[5] H Zayyani, M Babaie-Zadeh and C Jutten. 2009. Bayesian pursuit algorithm for sparse representation. *Proceedings of the ICASSP, 1549–1552.*

[6] Bartuschat D, Borsdorf A, K Sstler H. 2009. A parallel K-SVD implementation for CT image denoising. *Technical report, Department of Computer Science 10th (System Simulation), Friedrich-Alexander-University of Erlangen-Nuremberg.*

[7] Michael E, Michal A. 2006. Image denoising via sparse and redundant representations over learned dictionaries [J]. *IEEE Transactions on Image Processing, 15(12): 3736–3745.*

[8] Yang J, Abdesselam B, Son I P. 2009. A new approach to sparse image representation using MMV and K-SVD. *Springer: Advanced Concepts for Intelligent Vision Systems: 200–209.*

*Multimedia, Communication and Computing Application – Leung (Ed.)*
© *2015 Taylor & Francis Group, London, ISBN 978-1-138-02775-6*

# Intelligent human–computer interaction in document retrieval system

Y.E. Lv
*Linyi University, Shandong, China*

ABSTRACT: Traditional information retrieval systems use click buttons, drop-down boxes, check boxes, and text input as the main ways to interact, but they cannot provide good service for less professional users. In this paper, we have researched intelligent human–computer interaction of documents retrieval systems and designed an intelligent man–machine interface based on natural language, which can progressively understand users through natural language dialogue approach and help users find the most relevant digital document to its retrieval target. This paper discusses the natural language processing of intelligent human–computer interaction systems approaches. For natural language processing, we adopt network analysis algorithms based on syntactic transfer.

KEYWORDS: Human–computer interaction, Document retrieval, Human–machine interface, Natural language processing

## 1 INTRODUCTION

With the advance of science and technology, people's understanding of the natural world is getting deeper and deeper. People's dependence on information resources is also increasing. In this information society, there is message exchange not only between people but also between man and machine. In other words, each application system is inseparable from the exchange of information between man and machine, and the information retrieval system is no exception. Moreover, almost all members of the society need a variety of means such as newspapers, radio, television, Internet, conferences, and other interpersonal communication to get information. User demand for information has gradually expanded from a single academic research into all areas of social life.

Similarly, information retrieval systems are also facing the challenge that provides service for more nonprofessional users. Traditional information retrieval systems take click buttons, drop-down boxes, check boxes, and text input as the main ways to interact, which are characterized by the following:

1 The input way is complex, and ordinary users may find it difficult to learn to master;
2 Key word selection is needed, and the user must know the field key word
3 The content of the input is demanding. Invalid input often fails to return results;
4 The lack of a feedback mechanism is not conducive to improving the search strategy.

In response to these shortcomings of traditional human–computer interaction, we use an intelligent man–machine interface based on natural language processing. Natural language is the most friendly and concise human–computer interaction. With the dialogue-based approach, the user's request can be speculated, which directs the user to the target document in the document retrieval system.

## 2 DESIGN OF HUMAN–COMPUTER INTERACTION

HCI (human–computer interaction) is about the design, evaluation, and implementation of interactive computer systems and the major phenomena surrounding these areas of scientific research. As an important part of a computer system, HMI (human–machine interface) mainly refers to the means of communication between humans and computer systems, including human–machine two-way exchange of information. Interaction mainly involves how to determine man–machine communication and focuses on the design process and design methods of human–computer interaction, and human–computer interaction technology mainly focuses on the study of how to implement this means of communication using hardware and software.

In this document retrieval system, human–computer interaction is still in the form of a man–machine interface, which depends on characteristics of electronic documents. But unlike the traditional interactive human–machine interface, the document retrieval system designs an intelligent man–machine interface of natural language processing for nonprofessional users and the retrieval process is more like the natural language dialogue process of a user and an administrator. Natural language processing is a core part of the man–machine interface.

The main task of the man–machine interface is to extract the query request from the user's answer. Then, the extracted key words match the concept of the ontology. Because the form of the problem of interactive systems is simple, such as "Which is your interested domain?" and "Please choose a more specific domain?," the user's answer is very simple and the grammar of the answer sentence covers only a small part of the English grammar subset, so it can be described by the following rewritten rules of context-free grammar:

S→NP* VP*
NP→PRON
NP→ART|NUM (ADJ*) N
VP→VP+ NP*
VP→V* NP*

Among them, VP is a verb phrase; it can be a verb in Garbin language, the verb plus its complement, plus a double object verb form. NP is a noun phrase, as subject or object; it can be a pronoun or noun phrase. ART represents articles, NUM represents numerals, and ADJ represents adjectives. "*" Is the Kleene star, indicating that the grammatical category can appear more than 0 or 0. "+" Is Kleene plus, indicating that the grammatical category can appear once or more than once. Table 1 shows the common answers and their syntax descriptions:

Table 1. Common syntax descriptions.

| Syntax Description | Sample Sentence |
| --- | --- |
| S(NP) | Papers on artificial intelligence. |
| S(VP(VNP)) | I need some documents on automatics. |
| S(VP(VP(VNP))) | I want to find something about database. |

From Table 1, we can see that a lot of ingredients of the user's answers are not relevant to the query request, such as the subject part and the predicate part. So, the main natural language processing module extracts the user to answer the object or object complement part by analyzing the syntax of the sentence. Here, the parsing is through a recursive transition network.

## 3 NATURAL LANGUAGE PROCESSING ALGORITHMS

Common applications of natural language processing contain machine translation and an automated

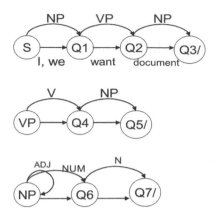

Figure 1. Parsing transfer network.

answering system. The former tries to translate a natural language to another with the computer, whereas the latter is a high-performance Internet-based software system. After special treatment for certain areas of knowledge, users can ask questions in the natural language form through the browser and the system can automatically give the answer and the appropriate recommendations.

Different from the automatic answering system based on natural language processing, the documents retrieval systems based on natural language processing ask questions by the computer and answer questions by the people. The focus of an interactive system is to extract query objective information from a user's answer.

The parsing algorithm based on the transfer recurrent network is as follows:

Input: ontology $T_1, T_2, …, T_n$ of sentence string S; here, set m ≦ n;

Output: the user answers questions;

Step 1 word segmentation of input string

1.1 add terminator $ at the end of the string. Read characters of a string in turn until spaces character; denote the string as $w_1$, i = 1, S = S-$w_1$;

1.2 read into characters of string S. On encountering spaces, go to step 1.3; on encounter terminator, go to step 1.4;

1.3 denote the characters (without spaces) as $w_i$, i = i + 1, S = S-$w_i$, go to step 1.2;

1.4 denote the characters (without terminator) as $w_n$, n = i + 1, get user answer sentence of string S' = ($w_1, w_2, … , w_n$).

Step 2 parse string S and access object components.

2.1 read $w_1$; if $w_1$ is a pronoun, indicating that the sentence has an object, go to step 2.2; otherwise, indicating that the sentence does not have a subject, go to step 2.4;

2.2 read $w_3$; if it is "to," indicating that the syntax is subject + predicate + predicate complement, S" = $w_5$,

$w_6$, $w_n$, go to step 2.3; otherwise, indicating that the syntax of answer is subject + predicate + object, S'' = $w_3$, $w_4$, $w_n$, go to step 2.4;

2.3 With stopword vocabulary in the database, remove S'' of all the articles, numerals, prepositions (e.g., on and about), and high-frequency words in front of these prepositions; the rest of the phrase contains a collection of query key words extracted;

2.4 read the string; if it is a verb, then let S'' = $w_2$, $w_3$, $w_n$, go to step 2.3.

Step 3 generate the corresponding questions according to query.

3.1 calculate the similarity of the resulting phrases from step 2 and the ontology concepts; sort the similarity descending and obtain similarity sequence $t_1$, $t_2$, ... , $t_m$, let $t = t_1$;

3.2 if $t_i$ is already the leaf node of the ontology tree, then return the documents under $t_i$ to Set, and go to step 3.3; otherwise, go to step 3.4;

3.3 if there is user feedback that Set contains no relevant documents, then find the node $t_i + 1$ whose similarity is slightly below the node $t_i$, $t_i + 1 = t$, and go to step 3.2;

3.4 if $t_i$ is not a leaf node of the ontology tree, automatically generates a choice questionwhich allows users to select from a related areas, receives sentence answer, and go to step 1.

When the document retrieval system application environment is a remote multiuser-distributed environment, steps 1 and 2 can be performed locally on the client side. After parsing is complete, the system gets the search request, and then sends it to the server, the server-side retrieval key words in the ontology library build the appropriate questions and send them to the client. The client asks questions and the user answers, and the client continues processing the user's answer sentence. At this time, if the system still uses B/S structure, the server will generate a lot of natural language processing, thus seriously affecting the efficiency of the system. With the use of C/S structure, users simply install the client on the local program that provides the retrieval interactive interface and built-in syntax analysis module, which can greatly reduce the workload on the server side to improve the efficiency of the system.

ACKNOWLEDGMENTS

We acknowledge the support of the natural science foundation of Shandong Province, China (No. ZR2014FL017).

REFERENCES

[1] Guarino N, Masolo C, Vetere G. OntoSeek. 1999. Content 2 Based Access to the Web. IEEE Intelligent Systems, 14(3):70–80.
[2] Shun S B, Motta E, Domingue J. ScholOnto. 2000. An Ontology-Based Digital Library Server for Research Documents and Discourse. Intl J Digital Libraries, 3(3):237–248.
[3] Tom Gruber. 1995. Toward principles for the design of ontologies used for knowledge sharing.
[4] Cheng Hian Goh. 1997. Representing and Reasoning about Semantic Conflicts in Heterogeneous Information Sources [M1, PhD, MIT].
[5] Heiner Stuckenschmi dt, Holger Wache. 2000. Context modelling and transformation for semantic interoperability [R]. In: Knowledge Representation Meets Databases (KRDB 2000).
[6] Bruce RS. 1997. Information Retrieval in Digital Libraries: Bringing Search to the Net. Science (1).
[7] Guarino N, Masolo C, Vetere G. OntoSeek. 1999. Content-Based Access to the Web. IEEE Intelligent Systems, 14(3):70–80.
[8] Shun S B, Motta E, Domingue J. ScholOnto. 2000. An Ontology-Based Digital Library Server for Research Documents and Discourse. Intl J Digital Libraries, 3(3):237–248.
[9] Jensen K, Binot J L. 1987. Disambiguating prepositional phrase attachment by using online dictionary definitions [J]. Computational Linguistics, 13(3):3–4.
[10] Honglan Jin, Kam Fai. 2002. A Chinese Dictionary Construction Algorithm for Information Retrieval, Proceedings of the ACM Transactions on Asian Language Information Processing (TALIP) 1(4): 281–296.

# Document retrieval based on multi-ontology

Y.E. Lv
*Linyi University, Shandong, China*

ABSTRACT:   Ontology has good concept hierarchies and logical reasoning support, and thus it is very suitable for special document retrieval such as academic and professional literatures. Ontology facilitates people from a variety of angles and ways to retrieve a document. So, the introduction of ontology in digital document information retrieval is important and also has theoretical and practical significance. This paper explains how to build multi-ontology based on semantic information and how to organize the body of literatures. Here, we build two semantics ontologies, one of which describes documents classification and the other describes publication time, and propose the document retrieval algorithms and the similarity calculation algorithm based on ontology.

KEYWORDS:   Multibody, Document retrieval, Document organization, Similarity calculation

## 1 INTRODUCTION

Conventional key word–based information retrieval technology cannot meet the needs of both semantics and knowledge. So, finding new ways of information retrieval has become a research hot spot. Ontology has good concept hierarchies and logical reasoning support, and thus in information retrieval, especially in knowledge-based retrieval, ontology has been widely used.

The design ideas of ontology-based information retrieval basics can be summarized as follows:

(1) With the help of experts in the field, establish ontology.

(2) Collect data sources, and with reference to the established Ontology store the data collected in the metadatabase (relational databases, knowledge bases, etc.) with the prescribed format.

(3) The query converter converts queries into a prescribed format in accordance with Ontology, and match the qualifying Metaphase data collection with the help of Ontology.

(4) Customize search results and return to the users.

This paper, with reference to the aforementioned design ideas, achieves document retrieval based on two ontologies that are able to describe document classification and publication time.

## 2 ORGANIZATION OF MULTI-ONTOLOGY

In information retrieval system, ontology is mainly used to describe the semantic information of retrieval objects. For documents, semantic information is available to describe the document classification, publication time, document subject, publication journals, etc.

Based on the ways in which users retrieve, we construct two ontologies: one is the classification ontology, which adopts the taxonomy "Chinese Library classification." The other is publication time ontology, which is built according to the objects' published time.

## 3 DESIGN OF THE DOCUMENT RETRIEVAL SYSTEM

### 3.1 *Components of the document retrieval system*

Document retrieval system mainly consists of the following components (shown in Figure 1):

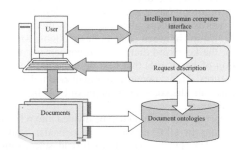

Figure 1.   Components of the DRS.

(1) Intelligent human–computer interaction retrieval interface: The main role of this interface is to carry out man–machine natural language dialogues. The intelligent human–computer interaction retrieval

interface consists of dictionaries and natural language processing mechanism and ontology-based reasoning mechanism. It can be used as a simple conversation in English or Chinese exchanges, and it can carry heuristic questions in the form of natural language statements and accept the users' natural language answers.

(2) Dictionary and knowledge base: Dictionary and knowledge base consists of the thesaurus, rule base, and corpus. The system analyzes input sentences and extracts search requests according to the users' answers by using common syntactic rules, common corpus, and relevant context.

(3) Domain ontology library: Domain ontology library consists of more than one domain ontology, as well as various concepts of the ontology of a collection of related key words. Domain ontology describes the relationship between domain-specific concepts. For this discussion, without loss of generality, only the following two ontologies are considered: one is document classification ontology, and the other is publication time ontology.

(4) Document database: The system collected documents published in ACM, IEEE, and other journals and conference proceedings related to artificial intelligence, knowledge management, information retrieval, databases, and other aspects of more than 600 papers between 1995 and 2014.

### 3.2 Document retrieval system

In the initial stage, the retrieval system issues paper document database literature with multi-ontology. This process is to allocate literature to each leaf node of the ontology. First, each document extracts metadata, including the title of the article, author, abstract, key words, and access paths and then generates the document metadata description. Each ontology has a concept, which describes the key words concept dictionary and hierarchical relationships between concept descriptions. Among them, the key word dictionary of concepts is a collection of key words that best describe the field of literatures. Here, documents organizational methods are introduced, as an example of CLC literature organization.

A certain document D will be divided into n fields according to CLC classification methods, with a set $D = \{D_1, D_2, \ldots, D_n\}$. In each field, select some key words that best represent the domain knowledge; $D_i$ is the set of key words of the i-th field. For document A, select some of the most representative key words of the content of the document, represented by the set $S(A)$. Order vector $= \{x_1, x_2, \ldots, x_n\}$; each component represents the semantic relevance of document A and

the i-th field, which is calculated by the following formula:

$$A(X_i) = \frac{|S(A) \cap D_i|}{|D_i|} \tag{1}$$

Select the highest weight, $\max\{A(X_i)\}$, where the value of i represents the field of the document. Then, the document is linked to the corresponding leaf node.

In the retrieval phase, the system asks questions and receives answers via the intelligent human–computer interaction retrieval interface. Initially, the user is at the root of the ontology concept tree R; R is a larger discipline in the CLC system, which corresponds to the TP (automation technology, computer technology) in this paper. We extract the noun ingredients from the users' answers with natural language processing and find the child node R of $T_i$ with vocabularies. Here, $T_i$ corresponds to a subcategory of the discipline R. If $T_i$ is not a leaf node of the ontology concept tree, then continue to ask the users' needs to retrieve a specific subclass of $T$, $T_{i1}, T_{i2}, \ldots T_{in}$ in which subcategories, such as $T_i$, are already leaf nodes; then, the documents under the node will be assigned to return to the users.

When a user's search request involves two or more semantic information, through different issues, extract the corresponding semantic information and find the collection $set_{t1}, set_{t2}, \ldots, set_{tm}$ by mapping semantic information and each leaf node; then, take the intersection $SET = set_{t1}, set_{t2}, \ldots, set_{tm}$. The intersection SET is a collection of documents required by the users.

## 4 CONCEPT RELAXED BASED ON MULTI-ONTOLOGY

From the concept map of ontology, the concept relaxed is to relax the users' search request to its father node, which is to select the appropriate results from the node's brother nodes.

Algorithms for optimal relaxed path consist of three components: the establishment of the relaxed path, empowerment of the relaxed path to ensure that E(Q) minimum, and achievement of the optimal relaxed solution. The algorithm is described as follows:

Input: the users' search requests $Q = (q_1, q_2, q_m)$, ontology of $T_1, T_2, \ldots T_n$, Here, $m \leqq n$;

Output: optimal relaxed solution SET;

Step 1 Establish the relaxed path based on search request1.1 find m ontologies related to m semantic information from ontology of $T_1, T_2, \ldots T_n$, and renumber them as $T_1, T_2, \ldots T_m$;

1.2 find $T = (t_1, t_2, \ldots, t_m)$, which satisfies $Q = (q_1, q_2, \ldots q_m)$. If $t_i \in T$ is a leaf node, make seti assigned to documents under $t_i$; otherwise, $set_i$ is assigned to the documents under all leaf nodes of $t_i$;

1.3 Order $SET = set_{t1} \cap set_{t2} \cap \ldots \cap set_{tm}$. If $SET \neq \Phi$, return SET; otherwise, establish a path $p_i = (t_i, f(t_i). f(f(t_i))...r_i)$ $(i = 1 \ldots n)$, where $f(t_i)$ is the father node of $t_i$ and $r_i$ is the root node of path $p_i$. Go to step 2;

Step 2 empowerment of relaxed path

2.1 For each ontology tree, calculate tree height hi; make maxh = max$(hi)$ $(i = 1 \ldots n)$;

2.2 for each edge $e(t_i, f(t_i))$ of the relaxed route, the layers of $t_i$ is given by $hi$; then, $N(e) = n^{\max h - hi}$; go to step 3;

Step 3 seek the best solution to ensure that E(Q) is minimum; it must be ensured that each added edge weight is minimum

3.1 if there is $t_i \neq r_i, t_i \in I$, order $i = 0$, go to step 3.2; otherwise (all concepts are at the root), prompt the users "The system has not documents."

3.2 for $t_i \in T$, if there is an edge e of weights that satisfies $e(t_i, f(t_i)) = n^i$, go to 3.3; otherwise, go to 3.4;

3.3 make $set_{ti} = set_{f(ti)}$, $t_i = f(t_i)$. If $SET = set_{t1} \cap set_{t2} \cap \ldots \cap set_{tm}$ is not empty, return SET to the user and the algorithm ends; otherwise, go to 3.2;

3.4 order $i = i + 1$. If $i \leq \max h - 1$, go to 3.2; otherwise, prompt the users "system has not documents" and the algorithm ends.

Example: the retrieval system based on multi-ontology is composed of three subontologies X, Y, and Z. The weighted relaxed path of retrieval request $Q = (X3, Y2, Z5)$ is shown in Figure 2, where n = 3 and maxh = 3. Shadowed boxes and thick lines show the relaxed path. Assuming the result set is not empty when relaxed to the concept (X1, Y1, Z1), the final E(Q) = 3 + 9 + 9 + 1 + 3 + 9 = 34 and the intersection of the documents set of X1, Y1, Z1 is the optimal solution to relax.

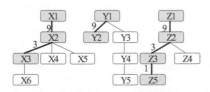

Figure 2. Multiconcept map of weighted relaxed path.

Another case is that when the finesse of various subontologies is very different, the numbering method must be different. Figure 3 shows the two numbering methods of an ontology: the left figure uses the same numbering methods, and the right figure uses different numbering methods which numbers the detailed classification ontology a smaller numbers (1,2,4,8) and numbers the coarser classification ontology a larger numbers (3,12). In order to improve the recall and precision of the retrieval system without causing flooding result, priority should be given to the finer classification ontology to concept relax.

| X1 | | Y1 | | X'1 | | Y'1 |
|---|---|---|---|---|---|---|
| 8 | | 8 | | 12 | | 8 |
| X2 | | Y2 | | X'2 | | Y'2 |
| 4 | | 4 | | 3 | | 4 |
| X3 | | Y3 | | X'3 | | Y'3 |
| | | 2 | | | | 2 |
| | | Y4 | | | | Y'4 |
| | | 1 | | | | 1 |
| | | Y5 | | | | Y'5 |

Figure 3. Weights relaxed path notation of different precision.

## 5 SIMILARITY CALCULATION BASED ON MULTI-ONTOLOGY

Most retrieval systems are based on Boolean logic mode or classification mode, such as the vector space and probability retrieval mode. In such a mode, a document or query is described as containing a variety of weight collections of words. The weights indicate the importance of a word in a document or query. If the word weight is 1 or 0, this is a Boolean model; if the word weight lies between 0 and 1, it belongs to a classification model. In the Boolean model, retrieval paradigm is based on key word matching. Its effectiveness will depend on the users' accuracy to construct the query term and users' familiarity to this retrieval service. Untrained users select important terms and concepts and then consider not only certain rules of logical combination but also changes of synonyms, antonyms, roots, and wildcard symbols such as the approximate character. In rating mode, retrieval paradigm is based on the assessment of the query and the document similarity, which is based on the weighting function.

Because the ontology can be viewed as a classification tree T, depth (A), which is the number of edges from root to A, indicates the depth of node A; depth (root) = 0. LCA (A, B) is the nearest common ancestor node of two leaf nodes A and B; then, the structural similarity of A and B can be calculated using the following formula:

$$similarity_{structure}(A, B) = \left( \frac{2 \times depth(LCA(A, B))}{depth(A) + depth(B)} \right)$$

Because the publication time ontology also can be viewed as a tree, the similarity of the time ontology can also be calculated by the aforementioned formula. But because publication time has many features of natural numbers, we can take advantage of it; it will be more convenient and accurate to calculate the similarity of two documents on publication time.

$$Similarity_{publishTime}(A, B) =$$

$$\frac{c}{a * diff(A, B)^2 + b * diff(A, B) + c}$$

If the two documents' publication time values are represented by time A (t) and B (t), then $diff(A, B) = A(t) - B(t)$ represents the publication time difference between the two documents. According to different archive scales and retrieval particle sizes, this difference can be taken yearly, monthly, or daily at different levels. According to the general search habits, the recent publication time of documents is usually close to the retrieval request. For example, if the user needs to retrieve documents in 2012, then 2011 and 2013 documents are closer to the user's requirements compared to 2010 and 2014 documents. So, similarity calculations with polynomials are more reasonable. The parameters a, b, and c can be taken as different values according to the needs. For example, if the user thinks that the later publicated documents are more valuable than the earlier ones, with the respect to the 2012 documents, the 2013 documents are more valuable than the 2011 documents. So, the value of b should be taken positively. Then, there will be similarity publishment time(2012,2013) > similarity publishment time(2012,2011), so the documents of 2013 are more likely to be submitted as the results to users.

The method of calculating the similarity between the concepts and the bodies of the documents can also be used to calculate the semantic similarity between two documents; the formula is as follows:

$$Similarity_{Semantc}(A(X), B(X)) = \frac{A(X) \circ B(X)}{\| A(X) \| \times \| B(X) \|}$$

Here, o represents the dot product between vectors. Set a threshold $\alpha$ $(0 \le \alpha \le 1)$; when $Similarity_{Semantic}(A(X), B(X)) \ge \alpha$, it can be considered that document A and document B are similar

from the perspective of key words, because the two documents have a great association with a particular field. So, these two documents can be considered as having similarity in semantics.

In conclusion, for the document retrieval system based on multi-ontology, the similarity calculation between two documents, taking into account their respective classification, the publication time and similarity of key word set can be summarized as follows:

$$Similarity(A, B) = \alpha * Similarity_{Structure}(A, B)$$
$$+ \beta * Similarity_{PublishTime}(A, B)$$
$$+ \gamma * Similarity_{Semantic}(A, B)$$

ACKNOWLEDGMENT

We acknowledge the support of the natural science foundation of Shandong Province, China (No. ZR2014FL017).

REFERENCES

[1] Guarino N. Semantic Matching. 1997. Formal Ontological Distinctions for Information Organization, Extraction, and Integration. In: Pazienza M T, eds. Information Extraction: A Multidisciplinary Approach to an Emerging Information Technology, Springer Verlag, 139–170.

[2] Gruber T R. 1995. Towards Principles for the Design of Ontologies Used for Knowledge Sharing. International Journal of Human-Computer Studies, 43:907–928

[3] Guarino N, Masolo C, Vetere G. OntoSeek. 1999. Content-Based Access to the Web. IEEE Intelligent Systems, 14(3):70–80.

[4] Shun S B, Motta E, Domingue J. ScholOnto. 2000. An Ontology-Based Digital Library Server for Research Documents and Discourse. Intl J Digital Libraries, 3(3):237–248.

[5] Tom Gruber. 1995. Toward principles for the design of ontologies used for knowledge sharing.

[6] Cheng Hian Goh. 1997. Representing and Reasoning about Semantic Conflicts in Heterogeneous Information Sources [M1, PhD, MIT].

[7] Heiner Stuckenschmi dt, Holger Wache. 2000. Context modelling and transformation for semantic interoperability[R]. In: Knowledge Representation Meets Databases (KRDB 2000).

*Multimedia, Communication and Computing Application – Leung (Ed.)*
© *2015 Taylor & Francis Group, London, ISBN 978-1-138-02775-6*

# Human action recognition based on fusion of slow features

P.F. Wang, G.Q. Xiao & X.Q. Tang
*College of Computer and Information Science, Southwest University, Chongqing, China*

ABSTRACT: Slow Feature Analysis (SFA), which originated from the principle of visual nerve, extracts invariant and slowly varying features from video information. In this paper, a fusion method based on SFA is proposed to recognize human actions. The main contributions of this paper are as follows: 1) Multifeatures of the supervised SFA, including Histogram of Gradient (HoG), Histogram of optical Flow (HoF), and Histogram of SIFT (HoSIFT), are investigated to capture local information; 2) the low-level feature fusion is utilized to concatenate the aforementioned supervised SFA-based feature vectors; 3) and the methods of recognition accuracies are compared with and without fusion. To evaluate our approach, a multiclass support vector machine (SVM) has been used for recognizing human actions. Experimental results on the KTH and Weizmann data sets show, in particular, that the fusion method based on SFA boosts the performance of action recognition significantly.

KEYWORDS: Human action recognition, Supervised SFA, Low-level feature fusion

## 1 INTRODUCTION

As an important part of computer visuals, human action recognition shows an extensive prospect for application in human–computer interaction, video archive indexing, and video surveillance. The key issue of human action recognition is feature extraction. Aiming at feature extraction, many works of different models have been reported.

In this paper, a fusion method based on slow feature analysis (SFA) is proposed to recognize human actions. SFA originates from the principle of visual nerve so that high-level responses of retinal receptors in a video tend to vary slowly for a long time. According to the slowness principle, Wiskott and Sejnowski [1] proposed a concept of SFA and used mathematical functions to prove unsupervised SFA. In the task of human action recognition, SFA learning strategies, including supervised SFA, discriminative SFA, and spatial discriminative SFA, have been developed to obtain accepted results [2]. However, discriminative SFA and spatial discriminative SFA require a large quantity of computation. In addition, many works only handle simple features such as the pixel gray. Therefore, we propose a fusion method based on SFA that employs low-level feature fusion to concatenate multifeature vectors of the supervised SFA. The framework of the proposed algorithm is shown in Fig. 1. The steps are as follows:

First, we extract multiple feature descriptors, including histogram of gradient (HoG), histogram of optical flow (HoF), and histogram of SIFT (HoSIFT).

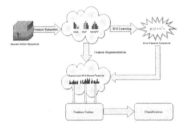

Figure 1. Flowchart of the fusion method based on SFA.

HoSIFT, the proposed descriptor derived from SIFT, is described in part A of section 3.

After extracting features, supervised SFA is applied to learn functions of slow feature for each action category. Then, the supervised SFA-based features are computed with all slow feature functions.

For feature fusion, we utilize low-level feature fusion, so the fusion feature vector is formed by concatenating the supervised SFA-based feature vectors of HoG, HoF, and HoSIFT.

Finally, a multiclass Support Vector Machine (SVM) is used to recognize actions. To validate and evaluate our proposed method, we have conducted a set of experiments on the KTH and Weizmann human action data sets.

The remainder of this paper is organized as follows: Section 2 reviews related works on human action recognition. Section 3 introduces our proposed algorithm. The experimental results and the analysis of recognition performance are reported in Section 4. Section 5 concludes this paper.

## 2 RELATED WORK

It is previously mentioned that feature extraction is the key issue of human action recognition. In terms of the features that are used to describe human actions, existing methods can be mainly divided into three categories: global feature based methods, local feature based methods, and biological feature based methods.

According to some holistic properties of moving objects, global feature based methods extract holistic features, such as object shape [8], silhouette [9], trajectories of key joints [10], and motion energy information (MEI) [11]. However, the global feature based methods require both accurate foreground segmentation and perfect object tracking. Therefore, these approaches are sensitive to background noises and tracking errors.

Recently, many local features have been applied to overcome the effects of both background noises and tracking errors. Local feature based methods usually include some processing steps, such as interest point detection, local descriptor representation, coding quantization, and temporal representation. For local features, many effective methods have been proposed. Cho and Byun [12] proposed an overlapping multifeature descriptor of HoG and HoF, which catches local and temporal information, using a Hankel matrix representation. In addition, they used a random forest classifier to cope with noise of descriptor, which is produced from no-activity frames of action sequences. SIFT [13], similar to the static image feature HoG or HoF, is used to characterize the local motion of the detected interest points. Scovanner et al. [14] proposed a 3D-SIFT descriptor so that the spatiotemporal information can be encoded by histogram. Babu et al. [15] utilized optical flow–based complex-valued features extracted from action sequences for human action recognition. They proposed the fast learning of a fully complex-valued neural (FLFCN) classifier to boost recognition performance.

Furthermore, biological feature based methods are motivated by researches on biological vision. Jhuang et al. [3] applied a biologically inspired model to extract concentrated local motion regions in the visual cortex [4]. Wiskott and Sejnowski [1] proposed unsupervised SFA to extract the invariant and slowly varying features from video information. Mathematical function was utilized to prove the fact that the solution of SFA is equal to the generalized eigenvalue problem. Berkes and Wiskott [5] employed SFA to investigate many properties in complex cells of primary visual cortex, including motion direction selectivity and edge orientation selectivity. Franzius et al. [6] proposed a hierarchical model to learn invariant object representation based on slow feature analysis.

Recently, Thériault et al. [7] have dealt with classification of video sequences that are composed of dynamic natural scenes, based on SFA. Zhang Zhang and Dacheng Tao [2] proposed another three kinds of SFA learning strategies that originated from unsupervised SFA strategy [1], including supervised SFA, discriminative SFA, and spatial discriminative SFA, to learn slow feature functions from a multitude of training cuboids obtained by random sampling in object boundaries. Afterward, they used accumulated squared derivative (ASD) to represent action sequences. Although improved performance has been listed in several data sets, discriminative SFA and spatial discriminative SFA require a large amount of calculation.

## 3 METHODOLOGY

### 3.1 *Feature extraction*

Recently, some researchers have tried to develop SFA algorithms to improve recognition performance [1], [2], [6]. Instead, we consider multifeatures of the supervised SFA, such as HoG, HoF, and HoSIFT, capturing more local information for better recognition performance. We first extract features, including HoG, HoF, and SIFT. As the resolution ratios of frame for KTH data set [16] and Weizmann data set [8] are different, the numbers of patches in each frame for them are different. The size of a patch is set as $24 \times 20$, and the resolution ratios for KTH and Weizmann data sets are $120 \times 160$ and $144 \times 180$, respectively. The patch of HoG is built by discretizing the gradients into eighteen bins. The patch of HoF, similar to that of HoG is constructed by discretizing the optical flow into eighteen bins.

Owing to the different numbers of SIFT keypoints extracted from each frame, we propose a descriptor that is defined as the histogram of SIFT (HoSIFT) based on the SIFT descriptor (a vector of 128 dimensionality). First, we use the k-means algorithm to cluster the SIFT points in each frame with a fixed number of clusters (set as 15). According to the cluster centers, each SIFT point in the frame is assigned to a cluster center; thus, a set of histograms is constructed for each frame. The histograms of each frame are defined as the HoSIFT descriptor.

### 3.2 *Slow feature function learning*

After feature extraction, slow feature functions will be learned from the extracted HoG, HoF, and HoSIFT, using the supervised SFA strategy. In mathematical terms, the task of learning slow feature functions can be stated as follows:

Given an $N$-dimension feature vector $x(t) = [x_1(t),...x_N(t)]^T, t \in (t_0, t_1)$, such as HoG, HoF, or HoSIFT, a set of slow feature functions $g(x) = [g_1(x),...,g_1(x)]^1$ are found by SFA, so the $M$-dimension output vector $y(t) = [y_1(t),..., y_M(t)]$ with $y_m(t) = g_m(x(t)), m \in \{1,...,I\}$ varies as slowly as possible. For each $m \in \{1,...,I\}$,

$$\Delta(y_m) = \langle \dot{y}_m^2 \rangle_t \text{ is minimal} \qquad (1)$$

Under the constraints

$$\langle y_m \rangle_t = 0 \text{ zero mean} \qquad (2)$$

$$\langle y_m^2 \rangle_t = 1 \text{ unit variance} \qquad (3)$$

$$\forall m' < m, \langle y_{m'} y_m \rangle = 0 \qquad (4)$$

where $\dot{y}$ denotes the operation of computing the first-order derivative of $y$ and $\langle y \rangle_t$ is the time averaging of $y$. Equation (1) is the objective function that minimizes temporal variation of the output vector, where the temporal variation is measured by the average of the squared first-order derivative.

Wiskott and Sejnowski [1] proved that the solution of slow feature function such as $g_m(x) = w_m^T x$ is equivalent to the generalized eigenvalue problem. Therefore, the solution of the generalized eigenvalue problem is an important processing step for learning slow feature functions. The generalized eigenvalue equation is defined as follows:

$$AW = BW\Lambda \qquad (5)$$

where

$$A = \langle \dot{x}\dot{x}^t \rangle_t \qquad (6)$$

$$B = \langle xx^T \rangle_t \qquad (7)$$

Here, $A$ indicates the expectation of the covariance matrix of the temporal first-order derivative of the feature vector, $B$ indicates the expectation of the covariance matrix of the feature vector, $\Lambda$ is a diagonal matrix of $N$ generalized eigenvalues $\lambda_i(1 \le i \le N)$, and $W$ is the generalized eigenvector matrix that consists of $N$ corresponding eigenvectors $w_i(1 \le i \le N)$. Small eigenvalues in the diagonal matrix $\Lambda$ reflect slow variations of output vectors.

According to the order of eigenvalues, we select the $I$ eigenvectors $w_1, w_2..., w_I(I \le N)$ that correspond to the eigenvalues $\lambda_1 \le \lambda_2 \le ... \le \lambda_1$. The $I$ eigenvectors $w_1, w_2,..., w_I$ are the weights of slow feature functions $g_1(x), g_2(x),...g_I(x)$, respectively. The slow feature functions are denoted as $F = \{g_1(x),...g_I(x)\}$.

Supervised SFA is a kind of learning strategy rooted in unsupervised SFA. We utilize the supervised SFA to learn slow feature functions for each action category. Then, the supervised SFA-based features are computed with all slow feature functions organized by action categories. Given a video, the values of elements in the supervised SFA-based feature vector corresponding to the action category that the action in the video belongs to should be smaller than those of the other action categories. Therefore, the supervised SFA-based features of different action categories are significantly different. The supervised SFA-based features of HoG are shown in Fig. 2. The equation of supervised SFA-based feature is described in detail below.

### 3.3 Supervised SFA-based feature representation and fusion

After learning all slow feature functions of HoG, HoF, and HoSIFT, the supervised SFA-based feature is computed by

$$v_i^z = \sum_{s=1}^{K} [C^z(s+1) \otimes F_i - C^z(s) \otimes F_i]^2, i \in \{1,...,I\} \qquad (8)$$

where $C^z(s)$ denotes the $z$th type of feature in the $s$th frame, which is HoG, HoF, or HoSIFT; $\otimes$ is the mapping operation from $x$ to $y_m$ in the slow feature function $y_m = w_m^T x$; $F_i$ is the slow feature function of the $i$th action class; and $v_i^z$ is the accumulated squared derivative. The supervised SFA-based feature vector is denoted as $V^k = [v_1^k,...v_I^k]$.

We use the low-level feature fusion that concatenates the supervised SFA-based feature vectors of HoG, HoF, and HoSIFT to form a fusion feature vector, i.e., the fusion feature vector $V$ is

$$V = [V^G, V^F, V^S] \qquad (9)$$

where $V^G, V^F, V^S$ are the supervised SFA-based feature vectors of HoG, HoF, and HoSIFT, respectively.

In addition, it is necessary to normalize the supervised SFA-based feature vectors with the L1-norm. Finally, we utilize a multiclass Support Vector Machine (SVM) for action recognition.

Figure 2. Actions of bend, jack, and jump on the Weizmann data set and their respective supervised SFA-based features of HoG.

## 4 EXPERIMENTAL RESULTS

In this section, we report the experimental results on the Weizmann data set and KTH data set to evaluate the performance of the proposed algorithm.

The KTH data set includes six categories of single personal actions (waving, clapping, boxing, running, jogging, and walking) performed by twenty-five people in four scenarios of outdoors (scenario1), scale-varied outdoors (scenario2), outdoors with different clothes (scenario3), and lighting-varied indoors (scenario4). In total, the KTH data set has 598 action sequences. The Weizmann data set of 90 action sequences contains ten categories of single-person actions including bending, jacking, jumping, pjumping, running, siding, skipping, walking, waving 1 (one-hand wave), and waving 2 (two-hands wave). The dimensionality of the supervised SFA-based feature is equal to the total number of the learned slow functions. For each action category, the numbers of slow feature functions of HoG, HoF, and HoSIFT are 50, 50, and 15, respectively. Thus, for the KTH data set the dimensionality of the supervised SFA-based feature of HoG or HoF is $50 \times 6 = 300$, and the dimensionality of the supervised SFA-based feature of HoSIFT is $15 \times 6 = 90$. So, the dimensionality of the fusion feature is $300 + 300 + 90 = 690$. Similarly, for the Weizmann data set the dimensionalities of the supervised SFA-based features of HoG, HoF, and HoSIFT are 500, 500, and 150, respectively. Accordingly, the dimensionality of the fusion feature is $500 + 500 + 150 = 1150$.

Testing strategies of classifier can be splitted into Leave-One-Out (LOO) and Random-Split-Data (RSD). Our experimental results are based on the RSD testing strategy. The KTH data set contains 598 action sequences, which are divided into six categories. We randomly selected 60 action sequences of each category to train the classifier, and the remaining action sequences of each category are for test. Table 1 compares the average recognition accuracies of our proposed method with state-of-the-art results.

Table 1. Comparison between the proposed method and previous methods on the KTH data set.

| Methods | Test. Str. | Accuracy |
|---|---|---|
| Dollar [17] | LOO | 81.17% |
| Niebles [18] | LOO | 83.33% |
| Schuldt. [16] | RSD | 71.72% |
| JHuang [3] | RSD | 91.7% |
| Laptev [19] | RSD | 91.8% |
| Zhang [2] | RSD | 88.83% |
| Supervised SFA of HoG | RSD | 91.1% |
| Supervised SFA of HoF | RSD | 89.45% |
| Supervised SFA of HoSIFT | RSD | 91.57% |
| Fusion of supervised SFA | RSD | 94.3% |

The supervised SFA of HoG, HoF, or HoSIFT achieves a higher recognition accuracy than that of Zhang [2], which adopts the same supervised SFA strategy. From Table 1 it can be observed that the recognition result of the feature fusion is significantly improved compared with the results of the state-of-the-art methods. Fig. 3 shows the confusion matrix of recognition on the KTH data set by fusion of supervised SFA. In the confusion matrix, distinguishing between jogging and running is difficult owing to the two similar types of actions.

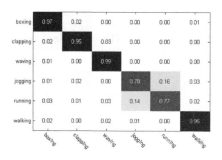

Figure 3. Confusion matrix of recognition on the KTH data set by fusion of supervised SFA.

For the Weizmann data set, we split the data set into a learning set with 5 randomly selected action sequences of each category and a test set with the remaining 4 action sequences of each category. Table 2 shows the average recognition accuracies of some existing methods and our proposed approach.

Similarly, the method of feature fusion achieves the highest recognition accuracy among all methods of Table 2. Fig. 4 shows the confusion matrix of recognition on the Weizmann data set by fusion of supervised SFA.

Table 2. Recognition accuracies of existing methods and our proposed approach for the Weizmann data set.

| Methods | Test. Str. | Accuracy |
|---|---|---|
| Niebles [18] | LOO | 72.8% |
| Liu [20] | LOO | 90.4% |
| Klaser [21] | LOO | 90.5% |
| Ali [10] | LOO | 92.6% |
| Zhang [2] | RSD | 86.40% |
| Supervised SFA of HoG | RSD | 91.17% |
| Supervised SFA of HoF | RSD | 86.5% |
| Supervised SFA of HoSIFT | RSD | 92.67% |
| Fusion of supervised SFA | RSD | 95.83% |

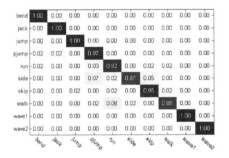

Figure 4. Confusion matrix of recognition on the Weizmann data set by fusion of supervised SFA.

## 5 CONCLUSION

In this paper, we propose a feature fusion method based on SFA for human action recognition. We investigated multifeatures of the supervised SFA, including HoG, HoF, and HoSIFT, to capture local information. Further, low-level feature fusion has been used to concatenate the aforementioned feature vectors. The experimental results demonstrate that the proposed method outperforms the state-of-the-art methods. In the future, we will exploit a more robust fusion method based on various types of supervised SFA features to recognize challenging real-world actions.

## ACKNOWLEDGMENT

In this paper, research was sponsored by Key Technologies Research and Development Program of China (Grant No. 2013BAD15B06).

## REFERENCE

[1] L. Wiskott & T. Sejnowski, 2002. Slow Feature Analysis: Unsupervised Learning of Invariances. Neural Computation, vol. 14, no. 4: 715–770.

[2] Zhang Zhang & Dacheng Tao, 2012. Slow Feature Analysis for Human Action Recognition. IEEE Trans. Pattern Analysis and Machine Intelligence, vol. 34, no. 3: 436–450.

[3] H. Jhuang & T. Serre & L. Worf & T. Poggio, 2007. A Biologically Inspired System for Action Recognition. Proc. IEEE Int'l Conf. Computer Vision, INSPEC 9849024: 1–8.

[4] M. Giese & T. Poggio, 2003. Neural Mechanisms for the Recognition of Biological Movements and Action. Nature Rev. Neuroscience, vol. 4: 179–192.

[5] P. Berkes & L. Wiskott, 2005. Slow Feature Analysis Yields a Rich Repertoire of Complex Cell Properties. J. Vision, vol. 5, no. 6: 79–602.

[6] M. Franzius & N. Wilbert & L. Wiskott, 2008. Invariant Object Recognition with Slow Feature Analysis. Proc. 18th Int'l Conf. Artificial Neural Networks: 961–970.

[7] Christian Thériault & Nicolas Thome & Matthieu Cord & Patrick Pérez, 2014. Perceptual Principles for Video Classification with Slow Feature Analysis. IEEE Journal of Selected Topics in Signal Processing, vol. 8, no. 3: 428–437.

[8] L. Gorelick & M. Blank & E. Shechtman & M. Irani & R. Basri, 2007. Actions as Space-Time Shapes. IEEE Trans. Pattern Analysis and Machine Intelligence, vol. 29, no. 12: 2247–2253.

[9] D. Weinland & E. Boyer, 2008. Action Recognition Using Exemplar-Based Embedding. Proc. IEEE Int'l Conf. Computer Vision and Pattern Recognition, INSPEC 10140035: 1–7.

[10] S. Ali & A. Basharat & M. Shah, 2007. Chaotic Invariants for Human Action Recognition. Proc. IEEE Int'l Conf. Computer Vision, INSPEC 9849077: 1–8.

[11] A. Boblck & J. Davis, 2001. The Recognition of Human Movement Using Temporal Templates. IEEE Trans. Pattern Analysis and Machine Intelligence, vol. 23, no. 3: 257–267.

[12] S.Y. Cho & H.R. Byun, 2011. Human Activity Recognition Using Overlapping Multi-feature Descriptor, Electronics Letters, vol. 47, no. 23: 1275–1277.

[13] D.G. Lowe, 2004. Distinctive Image Features form Scale-invariant Key-points. Int. J. Comput. Vis., vol. 60: 91–110.

[14] P. Scovanner & S. Ali & M. Shah, 2007. A 3-Dimensional SIFT Descriptor and Its Application to Action Recognition. Proc. ACM Int'l Conf. Multimedia: 357–360.

[15] R. Venkatesh Babu & S. Suresh & R. Savitha, 2012. Human Action Recognition Using a Fast Learning Fully Complex-valued Classifier. Neurocomputing 89 (2012): 202–212.

[16] C. Schuldt & I. Laptev & B. Caputo, 2004. Recognizing Human Actions: A Local SVM Approach. Proc. IEEE Int'l Conf. Pattern Recognition, vol. 3: 32–36.

[17] P. Dollar & V. Rabaud & G. Cottrell & S. Belongie, 2005. Behavior Recognition via Sparse Spatiotemporal Features. Proc. IEEE Int'l workshop Visual

Surveillance and Performance Evaluation of Tracking and Surveillance 2005: 65–72.

[18] J.C. Nibles & H. Wang & L. Fei-Fei, 2007. Unsupervised Learning of Human Action Categories Using Spatial-Temporal Words. Int'l. Computer Vision, vol. 79, no. 12: 2247–2253.

[19] I. Laptev & M. Marszalek & C. Schmid & B. Rozednfeld, 2008. Learning Realistic Human Actions from Movies. Proc. IEEE Int'l Conf. Computer Vision and Pattern Recognition, INSPEC 10140060: 1–8.

[20] J. Liu & S. Ali & M. Shah, 2008. Recognizing Human Actions Using Multiple Features. Proc. IEEE Int'l Conf. Computer Vision and Pattern Recognition, INSPEC 10139832: 1–8.

[21] A. Klaser & M. Marszalek & C. Schmid, 2008. A Spatiotemporal Descriptor Based on 3D-Gradients. Proc. British Machine Vision Conf.: 1–10.

*Multimedia, Communication and Computing Application – Leung (Ed.)*
© *2015 Taylor & Francis Group, London, ISBN 978-1-138-02775-6*

# Study on improved ID3 algorithm and its application in GOLF decision tree

Q.X. Meng, X.D. Liu & L. Liu
*Chinese Highway Engineering Consulting Group Co., Ltd, Beijing, China*
*China Trans Geomatics Co.Ltd, Beijing, China*

ABSTRACT: Machine learning has been one of hot points and a difficult research field in data mining. We have chosen the classic GOLF data set and Car Evaluation data set in University of California Irvine (UCI) machine learning database to study the decision tree. To overcome the problems of low efficiency, poor scalability, and integration with Iterative Dichotomizer 3 (ID3) algorithm, we put forward an improved ID3 algorithm based on querying database, and introduce the integration of the algorithm and database to avoid the time consumption of inside and outside storage exchanging into the training data, to improve the algorithm efficiency. Experiments show that, time consumption of the improved ID3 decision tree algorithm is less than that of the original ID3 algorithm in decision tree generation. With similar attribute number and attribute value of samples, when the amount of training data increases, the improved ID3 algorithm is more obvious and scalable.

KEYWORDS: Data mining, Machine learning, Iterative Dichotomizer 3 (ID3) algorithm, Decision tree.

## 1 GENERAL INSTRUCTIONS

With the increasing application of information technology, the amount of data stored in the database increases rapidly. Extraction of the effective value and knowledge from the huge database and a large number of complex information to improve the utilization of information has initiated a new research on data mining technology [1]. Currently, the most common method used in data mining is the decision tree algorithm, and the Iterative Dichotomizer 3 (ID3) algorithm is the most widely used algorithm in decision tree generation.

Quinlan[2] proposed the famous ID3 algorithm, which has four major problems. First, the calculation of information gain depends on attributes with high value, but which are usually not optimal. Second, the ID3 is a nonincremental algorithm. Third, ID3 algorithm is a univariate decision tree, so many complex concepts are difficult to explain and the interaction of the attributes is not robust enough, which would easily result in that the subtree of decision tree is reduplicate and some properties are tested many times in one route of the tree. At last, the noisy immunity is less strong, and it is more difficult to control the positive and negative training routes. Quinlan[3] proposed the C4.5 algorithm that made the classification rules easier to understand and more accurate. Rather choosing attributes with high value of the ID3 algorithm, Wang and Jiang[4] put forward a method to process continuous attributes by discretization of the continuous attributes value first. With two original ID3

algorithms,[5] Chai and Wang[6] revised the algorithm to improve the efficiency and accuracy of the classification. Yu[7] proposed an improved ID3 algorithm (AAID3) on the basis of the attributes value problem. Wang and Hu[8] discussed the basic idea and realization method of the ID3 algorithm according to the problem on choosing multiattributes and improved the algorithm by introducing weight conception. In addition, Duan et al.[9] proposed a fast corner detection algorithm on the basis of fuzzy ID3 decision tree. In Liu and Li[10] introduced the significance of attributes by rough sets theory to improve the ID3 algorithm, and proposed the IF-ID3 algorithm by using Maclaurin formula to simplify the ID3 algorithm.

## 2 ID3 DECISION TREES ALGORITHM

Quinlan[2] put forward the famous ID3 algorithm by information theory, and ID3 is summarized and simplified in 1983 and 1986, which had become a typical decision tree algorithm.

### 2.1 *Principle of ID3 algorithm*

The core idea of the ID3 algorithm is to introduce mutual information concept in information theory. It is used to select the node attributes as the testing conditions in decision-making, to measure the judgment of attributes classification. However, Quinlan[3] identified the mutual information as *information gaining* in the ID3 algorithm.

In the ID3 algorithm, according to the definition of information theory, the training sample set is D, which is corresponding to the source input state set, the target attribute value of D set is C, which is corresponding to output state set. Let $C = \{C_1, C_2, ..., C_n\}$, where $n$ is the number of possible values of C. Assume that frequency of occurrence of the samples with property of $C_i$ for the training set D is $p_i (i = 1, 2, ..., n)$, corresponding with probability of message occurrence, the information entropy of training set D is defined as

$$\text{Entropy}(D) = -\sum_{i=1}^{n} p_i \log_2 p_i. \qquad (1)$$

Quinlan information gain is understood as the difference of impure degree (entropy) of the sample sets between before and after division (corresponding with uncertainty mutual information). Assuming that we use attribute A to divide the set D, then the information gaining could be defined as follows:

$$\text{Gain}(D, A) = \text{Entrophy}(D) - \text{Entrophy}(D|A) \qquad (2)$$

where the Entropy (D|A) is the entropy of D divided by attribute A, corresponding with conditional entropy in information theory. Similarly, it is defined as follows: the assumption that the attribution A has different $m$ values, the sample set D is divided into $m$ subset, $D = \{D_1, D_2, ...D_m\}$, the sample number of each subset is $|D_i| (i = 1, 2, ..., m)$, and $|D|$ denotes the sample number in set D. Thus,

$$\text{Entrophy}(D|A) = \sum_{i=1}^{m} \frac{|D_i|}{|D|} \text{Entrophy}(D_i). \qquad (3)$$

In the attributes test condition, if the information is greater, the classification capability of attribution is more powerful. By calculating the value of each attribute information gain in D, the attribute with the largest gain could be chosen as classification attributes.

ID3 is one of the most classical algorithms, but it is not perfect, as there are still some shortcomings of the algorithm.[13][14] Owing to memory limitation and file storage mode, it is difficult to carry out the seamless integration between algorithm and database application, so the classical ID3 algorithm is lacking in scalability, integration, and efficiency.

### 2.2 Improved ID3 algorithm

According to the aforementioned analysis, this paper proposes the database query technology to improve the algorithm. Using the JDBC method, we combine JAVA language with ORACLE DBMS. Owing to the advantage of using ORACLE database management

system, we could directly query the data in ORACLE database with high efficiency, and connect the query result to the JAVA application program. Then, we could process and analyze the results in the procedures, and get each piece of information in the condition attributes to establish the ID3 decision tree by recursive according to the ID3 algorithm theory. In this process, continuous database querying is used to obtain the classification rules. In the realization process of the algorithm, in addition to control the algorithm step in Java runtime environment, the rest could be operated in ORACLE database management system. With the help of the existing data query, the process does not need to read and converse the data in text form, which would greatly improve the algorithm efficiency and make the algorithm easier to implement.

The basic idea: before the decision tree is built, first the number of statics sample should be set, the number of attributes and each attribute name, and the number and the name of classification of the whole classification data set, return the result from database query, and gradually establish a classification decision tree by recursive. In the process of building decision tree, we should pay attention to whether it could meet the termination condition. If the conditions are satisfied, the procedures exit. The improved algorithm flow chart is shown in Figure 1.

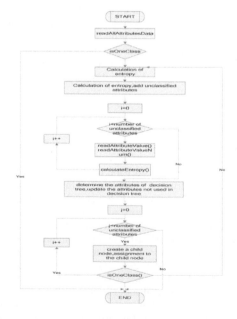

Figure 1. Improved algorithm flow chart.

# 3 THE EXPERIMENT AND ANALYSIS

## 3.1 Data preparation

The GOLF decision is proposed by Quinlan, the inventor of the ID3 algorithm. In 1985, he introduced the ID3 algorithm in detail in the paper of "Decision tree induction"by the case of whether playing golf.

The data preparation process is data collection, including cleaning, conversion, and reduction. In this paper, we directly use the existing mature data, whose data environment is similar. Therefore, the data preparation is relatively simple. At the same time, the data set the small number, so we no longer need to carry out the work of data reduction. Accordingly, the data is directly imported into in the GOLF table of the database, as shown in Table 1.

Table 1.    Converted data.

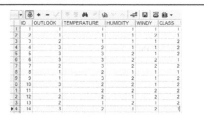

### 3.1.1  Data preprocessing

We generated the decision tree by the improved ID3 algorithm with the data after conversing in the GOLF table in Oracle database.

1) Calculation of entropy

2) Calculation of the information gain

According to the calculation, the maximum value is Outlook information gain, which is on behalf of the most classification Contribution. Outlook is selected as the decision attribute to classify the sample set. Outlook has three attribute values, and this node has three subsets.

### 3.1.2  Creating decision tree by recursive

We create three child nodes, respectively, by the Outlook nodes, and then create the decision tree recursively using the aforementioned process, until the exit signal is meeting. After the recursive invocation, IF–THEN classification rule of decision tree algorithm is output in the program console (Figure 2), and the decision tree classification algorithm is based on the rule. Figure 3 is the decision tree in Quinlan's paper. Comparing with the Classic ID3 algorithm, it is proved to be correct for the improved algorithm.

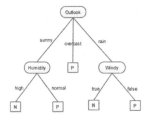

Figure 2.    Decision tree in this paper.

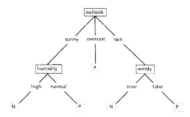

Figure 3.    Decision tree in Quinlan's paper.

## 3.2  Algorithm design and experimental analysis

To validate the efficiency of improved ID3 algorithm, this paper selected one of the most popular data Car Evaluation Data in UC Irvine Machine Learning Repository. There are 1728 data groups in data set. To make the data mining work in data preparation stage, as mentioned earlier, the data are transformed first. The conversion specifications are shown in Table 2.

In the same external environment, that is, the same computer hardware configuration, the same development environment, and the same data set, for the two algorithms before and after the improvement, results were analyzed in comparison.

Through random sampling of data numbers 460, 1152, and 1728, consisting of three data sets, in the same external conditions, each data set separately used two algorithms with 10 experiments and calculated the average value of experimental results. Decision tree leaf node number and time consumption are shown in comparison.

### 3.2.1  Comparison of the leaf nodes

In Table 3, we can see that the numbers of leaf nodes of decision tree are almost the same.

The accuracy of classification of the improved algorithm can be trusted; it can achieve the same classification effect as the original algorithm.

### 3.2.2  Comparison of time consumption

As we can see that the time consumption of the improved ID3 algorithm is smaller than that of the

Table 2. Data conversion table.

| Name of attribute | Original property value | Attribute value after conversion |
|---|---|---|
| **buying** | vhigh | 1 |
| | high | 2 |
| | med | 3 |
| | low | 4 |
| **maint** | vhigh | 1 |
| | high | 2 |
| | med | 3 |
| | low | 4 |
| **doors** | 2 | 1 |
| | 3 | 2 |
| | 4 | 3 |
| | 5more | 4 |
| | 2 | 1 |
| | 4 | 3 |
| | 5more | 3 |
| **lug_boot** | big | 1 |
| | med | 2 |
| | small | 3 |
| **safety** | high | 1 |
| | med | 2 |
| | low | 3 |

Table 3. Comparison of the leaf nodes.

| Data set | Amount of data | Original ID3 leaf node | Improved ID3 leaf node |
|---|---|---|---|
| 1 | 460 | 46 | 46 |
| 2 | 1152 | 142 | 143 |
| 3 | 1728 | 297 | 295 |

original algorithm with the same data set, this preliminary verifies the efficiency of the algorithm (Table 4).

Comparison of the time consumption of improved decision tree algorithm on decision-making is shown in Figure 4.

Figure 4 shows the time consumption to establish a decision tree before and after the improvement of ID3 algorithm; we can find from the comparison chart that, with the same training samples, the time consumption of improved algorithm is shorter than the initial algorithm. Moreover, with the increasing samples number, the time difference increases. This shows that the improved ID3 algorithm is more superior in efficiency and performance than the original ID3 algorithm with more large-scale training sample set in the decision tree construction process. The improved ID3 algorithm also shows its good scalability owing to the increase in the amount of data in the sample.

Table 4. Comparison of the time consumption on decision-making.

| Data set | Amount of data | Original (ms) | Improved (ms) |
|---|---|---|---|
| 1 | 460 | 30.3 | 25.4 |
| 2 | 1152 | 41 | 33.1 |
| 3 | 1728 | 56 | 41.2 |

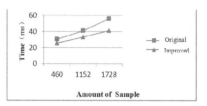

Figure 4. Comparison of the time consumption of decision-making.

## 4 CONCLUSION

This paper chooses the classic GOLF data set and Car Evaluation data set in UCI machine learning database as the object of study on decision tree construction. According to the problem of low efficiency, poor scalability, and integration with ID3 algorithm, we put forward an improved ID3 algorithm based on the efficiency of database querying and the flexibility of JAVA operating environment, and use the algorithm and database integration to avoid the time consumption of inside and outside storage exchanging into the training data, to improve the algorithm efficiency. Experiments show that, time consumption of the improved ID3 decision tree algorithm is less than that of the original ID3 algorithm in decision tree generation. In the condition of the similar attribute number and attribute value with samples, with the amount of training data increasing, the improved ID3 algorithm is more obvious and scalable.

REFERENCES

[1] Zhou Shibing, Xu Zhenyuan, Tang Xuqing. Optimal number of clustering based on affinity propagation algorithm to determine the comparative research method of [J]. computer science, 201, 38 (2): 225 228.
[2] Quinlan J R. Discovering rules by induction from large collections of examples[J]. In Michie D. Expert systems in the micro electronic age[M]. UK.
[3] Quinlan J R.c4.5:Programs for Machine Learning[M]. San Mateo Calif: Morgan-Kaufmann.
[4] Wang Xiaowei, Jiang Yuming. Analysis and improvement of decision tree with ID3 algorithm[J]. engineering and design of computer, 2011:32–9.

[5] HUANG A H,CHEN X T, An Improved ID3 Decision Trees Algorithm [J],COMPUTER ENGINEERING&SCIENCE,2009,(31):109–111.

[6] CHAI R M, WANG M. A more efficient classification scheme for ID3[J]. Computer Engineering and Technology,2010,(1):329–332.

[7] YujunTong. An Enhanced ID3 Algorithm Based on Attribute Attention[J]. IEEE 2011.

[8] Wang Yongmei, Hu Xuegang. Research on ID3 decision tree algorithm[J]. Journal of Anhui University (NATURAL SCIENCE EDITION), 2011 (5): 35–3.

[9] Duan Rujiao, Zhao Wei, Huang Songling, Chen Jianye. Fast corner detection algorithm based on Fuzzy ID3 decision tree [J]. Journal of Tsinghua University (NATURAL SCIENCE EDITION), 2011:51–12.

[10] Liu Fengjuan Li Jianlei, Study on the simplified and improved ID3 algorithm [J]. value engineering, 2013 (3).

[11] Shannon C E.A Mathematical Theory of Communication[J].ACM SIGMOBILE Mobile Computing and Communications Review,2001,5(1):3–55.

[12] Liu Qinghe, Liang Zhengyou. A method based on feature selection of information gain[J]. computer science, 2011 (4).

[13] Jiang Shengyi. Data mining theory and practice [J].: Electronics industry publishing company of Beijing, 2011:54–51.

[14] Lining, Le Qi. Lining, Solvation of the common problems of decision tree algorithm[J]. Computer and digital engineering, 2005,03:60–64.

[15] Liu Qi. Research on the improvement of decision tree of ID3 algorithm [J]. Computer software & theory, 2009.

[16] Luo Yuzi, Fu Xinghong. Decision tree classification algorithm of ID3 and improvement based on Data Mining[J]. Application of computer system, 2013:22–10.

[17] Yang Yizhan, Li Xiaoping, Duan Xiaxia. An improved decision tree algorithm based on database query, computer engineering & application [J]. (5, 2008).

*Multimedia, Communication and Computing Application – Leung (Ed.)*
© 2015 Taylor & Francis Group, London, ISBN 978-1-138-02775-6

# Visualizing seismic waves through SDP graphs matrix for discriminating of earthquake and explosion

H.M. Huang, Y. Tian & C.J. Zhao
*College of Computer Sci. and Info. Engineering, Guangxi Normal University, Guilin, China*

X.B. Liu & X.H. Shi
*College of Physics Science and Technology, Guangxi Normal University, Guilin, China*

ABSTRACT: This paper made an attempt to visually discriminate earthquake and explosion through Symmetric Dot Pattern (SDP) graphs matrices. SDP—a visualization scheme of seismic wave data—was proposed through our experiments in this paper that it can be successfully applied in the discrimination of natural earthquake and man-caused explosion. The results of experiments on 35 events of earthquakes and 27 events of explosion showed that the SDP graphs matrices of multistations seismic wave data were collectively expressive for the discrimination of earthquake events and explosion event. This implies the underlying information of a large amount seismic wave data can be efficiently expressed by some aesthetically pleasing graphics in a simple, efficient, and novel way, and deserves further investigations for more practical explosion and earthquake events.

KEYWORDS: Symmetric Dot Pattern (SDP), SDP graphs matrix, Seismic source type, Wave features, Visualization.

## 1 INTRODUCTION

Seismic signal is originated from natural earthquake events or man-made earthquake events. If seismic wave signals are appropriately processed or properly transformed, the explosion events can be effectively distinguished from earthquake events with high probability (Huang 2009; Wu, 2012).

Explosion events can only be detonated at most several tens of meters underground. The initial blast of an explosion is air-swelling, and its focal mechanism is expanding source. The release of the initial energy explosion is very rapid, causing surrounding rocks compressed in all directions. All observed initial movements in up–down components of seismic waves are up (or position direction in $z$-axis). There are exuberant P-waves in the observed seismic waves for explosion events. If P-waves travel more than 50 km distance from its epicenter, the Rayleigh wave often can be observed (Shearer, 2009). For explosion, the magnitude of P-wave is commonly larger than that of S-wave. An explosion's high-frequency components being nearly absorbed by the soft and non-uniform shallow pseudorock layer, observed seismic waves attenuate quickly, so the seismic phases are short and volatile (Bennett, Murphy, 1986).

For an earthquake, its focal mechanism is more complex with longer lasting-time source process

and deeper hypocenter, whereas that of explosion is only expanding with more nearly momentary source process and shallower hypocenter. Thus, earthquake incurs more dispersive spectrum and tardy attenuation seismic signals. The difference of seismic source mechanisms and its ensuing distinction of seismic signal are the bases for the discrimination between earthquake and explosion.

This paper makes an attempt to visually distinguish earthquake and explosion events by using symmetric dot pattern (SDP) graphs matrices. SDP is a polar coordinate graph proportionally projected from temporal seismic signal by some suitable algorithm. The algorithm aims to enhance the visual difference between earthquake and explosion when the same channel's seismic signals in all observatory stations are plotted together through a SDP graphs matrix for each event.

## 2 SYMMETRIC DOT PATTERN (SDP)

The SDP algorithm is schematically shown in Fig. 1. The aim of this SDP drawing algorithm is to strengthen the discriminatory capabilities of the seismic waves between earthquakes and explosions, and we hope to draw some obviously distinguished SDP graphs for these two types of events.

The main steps are listed as follows:

1 *Start* reading out seismic wave data. The seismic wave data are read out by the "Decision Support System (Huang, 2011) for Classification and Recognition of Earthquakes and Explosions" from .evt format files. In the .evt format file, all seismic wave data which observed in 107 observatory stations for one event are kept in one single file, and the sampling rates are 50 Hz for all observatory stations. Each station has 3 channels seismic wave data: vertical (up–down), horizontal south–northern, and horizontal east–western. In an experiment, only one channel (here the vertical channel) data are read out.

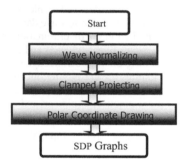

Figure 1.   Flow-chart of the SDP Algorithm.

For a channel in one station, the readout seismic wave data can be expressed as

$$S(i)|_{i=0,1,\cdots,N-1} = |h_i|, \quad i=0,1,\cdots,N-1 \tag{1}$$

where $N$ is total sample count, $h_i$ is the recorded $i$th sample value in the designated channel, and $|\bullet|$ is the absolute value operator.

2 *Wave normalizing* temporal wave normalization. Because of obvious disparate magnitude scales among different explosion or earthquake events, it is necessary to normalize each recorded seismic wave data for every channel in all stations for an event. For a readout seismic wave data $S(i)|_{i=0,1,\cdots,N-1}$, the corresponding normalized wave data is $S'(i)|_{i=0,1,\cdots,N-1}$:

$$S'(i)|_{i=0,1,\ldots,N-1} = \frac{(Max(S')-Min(S'))\times(S(i)-Min(S))}{(Max(S)-Min(S))} + Min(S') \tag{2}$$

where $Max(S)$ and $Min(S)$ are, respectively, the maximum and minimum of readout seismic

signal $S(i)|_{i=0,1,\cdots,N-1}$; $Max(S')$ and $Min(S')$ are, respectively, the maximum and minimum of normalized seismic signal $S'(i)|_{i=0,1,\cdots,N-1}$, which are often designated, respectively, as $Max(S')=1$ and $Min(S')=0$ or $Max(S')=1$ and $Min(S')=-1$; and other values are also possible when necessary.

3 *Clamped projecting*. This projecting is proposed to enhance the distinguishing capability of the SDP graphs matrices between explosion events and earthquake events. If the normalized temporal seismic signal $S'(i)|_{i=0,1,\cdots,N-1}$ is directly projected into polar coordinate graph, the overwhelming majority of signal may be concentrated at a very narrow neighbor region near the central region of the SDP graph, resulting in illegibility among obviously different temporal seismic signals.

By inspecting each temporal seismic wave graph of all 107 stations for each available event, it is not difficult to find that the maximal amplitude value of temporal seismic signal is often extremely larger than most other amplitude values, and the wave signal attenuates rapidly from its maximum. Thus, to visually express as more as possible information resided in the seismic signal, the normalized wave data $S'(i)|_{i=0,1,\cdots,N-1}$ are further processed as follows with an adjustable clamped value *Amp*, and resulting in the clamped projecting temporal seismic wave data $Y(i)|_{i=0,1,\cdots,N-1}$:

$$Amp = (Max(S')-Min(S'))\times W \ (0<W\le1) \tag{3}$$

$$Y(i) = |Amp - S'(i)| \ (i=1,\,2,\cdots,N-1) \tag{4}$$

where $S'$ or $S'(i)$ is the normalized wave data calculated in (2); $Max(S')$ and $Min(S')$ are, respectively, the maximum and minimum of $S'(i)|_{i=0,1,\cdots,N-1}$; and the $W$ is the adjustable weight value between 0 and 1, here $W=0.5$. By adjusting the value of $W$, different amplitude portion can be protruded in the final SDP graph.

4 *Polar coordinate drawing*. Polar coordinate drawing is aimed at drawing out the polar coordinate graph from the clamped temporal seismic wave data. This final polar coordinate graph is just right the proposed SDP graph. The SDP graph is a circularly integrated graph which is actually evenly distributed $K$ copies of one polar coordinate graph. The polar coordinate graph is transformed from the clamped projecting temporal seismic wave data $Y(i)|_{i=0,1,\cdots,N-1}$. The sample order number $i$ is transformed as $K$ angle values (in radian) $\theta(i,k)$ in the $K$ circularly even-distributed polar coordinate graphs, and the temporal wave amplitude is directly regarded as the polar coordinate radius $\rho$.

Thus, every sample value in temporal wave now is mapped into $K$ points in the SDP graph:

$$\theta(i,k)\big|_{i=0,1,\cdots,N-1} = \hat{\theta} \times i + \frac{2\pi}{K} \times k, \quad k = 0,1,\cdots,K \tag{5}$$

$$\rho(i)\big|_{i=0,1,\cdots,N-1} = Y(i)\big|_{i=1,2,\cdots,N-1} \tag{6}$$

where $\hat{\theta}$ is angular difference for two successive sample points: $\hat{\theta} = 2\pi / (N \cdot K)$; $K$ is the petal number, piece number, or copies number of the configuration of one SDP graph. Here, the petal number $K = 8$ and $N$ is the total sample count.

## 3 THE VISUALIZATION OF SEISMIC WAVE

### 3.1 *The SDP matrices of earthquake and explosion*

Seismic wave data are provided by China Earthquake Data Center (http://data.earthquake.cn/). The seismic data selected were that of 35 earthquake events occurred during 2003–2007, with their epicenters at neighbors of metropolitan Beijing (at about longitude 116°E, latitude 40°N). The 27 explosion events occurred from 2002 to 2008, with their epicenters at neighbors of metropolitan Beijing (at about longitude 115°E, latitude 40°N). The fundamental information of some these events—3 earthquakes and 3 explosions—is exemplified in Table 1.

On discriminating earthquake and explosion by directly inspecting SDP graphs matrices, when petal number (*n*) is 8 and weight value (*W*) is 0.5, the distinction of earthquake and explosion is intuitively quite obvious. For each of the 6 events listed in Table.1, 10 stations seismic wave data are randomly selected and their SDP graphs matrices are shown in Fig. 2. With these SDP graphs matrices, it is very clear that earthquake events show much more diversity than explosion events in SDP graphs matrices.

Table 1. Fundamental Information of Some Earthquake and Explosion Events.

| Date | Lati | Longi | Magni |
|------|------|-------|-------|
| 2003-06-19 | 40°06.95′ | 116°18.83′ | ML1.6 |
| 205-03-01 | 40°02.48′ | 115°57.93′ | ML2.3 |
| 2007-05-05 | 40°32.54′ | 115°78.53′ | ML2.4 |
| 2003-06-28 | 40°25.84′ | 115°28.89′ | ML1.1 |
| 2005-03-12 | 40°31.45′ | 115°29.61′ | ML1.2 |
| 2007-06-11 | 40°30.05′ | 115°25.83′ | ML1.3 |

T: event Type; Q: earthQuake; P: exPlosion.

Despite the distinction is obviously clear for the 10 stations SDP graphs matrices as a whole for every

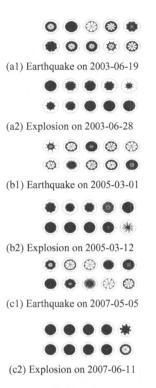

(a1) Earthquake on 2003-06-19

(a2) Explosion on 2003-06-28

(b1) Earthquake on 2005-03-01

(b2) Explosion on 2005-03-12

(c1) Earthquake on 2007-05-05

(c2) Explosion on 2007-06-11

Figure 2. SDP Graphs Matrices for 10 Stations of 3 Earthquake Events and 3 Explosion Events.

single event, it is natural to conclude that if more stations seismic signals were used, the diversity of the SDP graphs matrices for earthquake events may be severely weakened. Actually, we can deny this doubt by constructing SDP graphs matrices by more seismic wave signals randomly selected from more stations. Fig. 3 is an exemplification of SDP graphs matrices, whose seismic wave signals are randomly selected from 35 stations for the two type events, earthquake and explosion.

It is not difficult to find that when more stations data are selected, there are more different SDP graphs, in spite of earthquake or explosion. However, the diversity property is still kept: the SDP graphs matrix of earthquake is more diverse and complex than that of explosion.

### 3.2 *The possibility of recognizing explosion by SDP matrices*

Many experiments were conducted for the drawings and comparisons of SDP graphs, which were drawn from randomly selected seismic wave signals of different stations for each of the 62 events (35 earthquakes and 27 explosions). Owing to the space

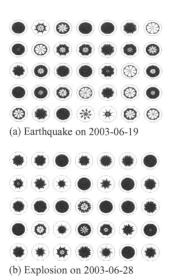

(a) Earthquake on 2003-06-19

(b) Explosion on 2003-06-28

Figure 3. SDP Graphs Matrices for 35 Stations of One Earthquake and One Explosion.

limitation of this paper, only 3-pair SDP graphs matrices examples with 10 stations are shown in Fig. 2. These SDP graphs can be roughly classified to 2 types: (1) symmetric stencil polypetals graphs [Fig. 4(a)] and (2) dull filling cake-liked graphs [Fig. 4(b)].

(a) Stencil Polypetal    (b) Filling Cake-Liked
    Graphs                    Graphs

Figure 4. Two Typical SDP Graphs.

The following results are derived from our SDP graphs drawings and comparisons experiments for 62 events:

(1) There are more diverse symmetric petal-style SDP graphs for earthquake than for explosion.
(2) There exist some transitional trends for SDP graphs of different station seismic data of earthquakes, such as the trend from nonpetals maps to stencil big polypetal maps. For different station locations of an earthquake event, SDP graphs commonly show obvious diverse styles. For explosion event, SDP graphs of different stations show less changeable styles, with unapparent transitional trends.

On discriminating the 35 earthquake events and 27 explosion events only by using SDP graphs visually and manually, 54 events can be easily recognized visually, 8 events are hard to be distinguished, that is, the correct recognition is 54/62 = 87%, overall error rate is 13%. This implies that SDP graphs provide a novel,

feasible, visual, and also rapid mean to recognized explosion event.

## 4 CONCLUSION

When seismic signals of different observatory stations for the same event are viewed as a whole in the form of the SDP graphs matrix, the distinction between earthquake and explosion is significantly obvious, revealing the underlying difference of seismic source mechanisms of these two events: dislocation source for earthquake versus expansion source for explosion. The recorded seismic signal at respective different observatory station for the same event can be regarded as a filtered signal of the same source (dislocation source or expansion source) signal passing through different paths (filter). Owing to the high diversity and extreme complexity of seismic ray traveling paths, with additive and other noises added in those paths, the discriminative information carried in the observed seismic waves is much more involved.

By the experiments in this investigation, it is easy to derive that while there are sufficient observatory stations suitably evenly surrounding the event source with adequate epicenter distances, SDP graph is an effectual mean to distinguish earthquake and explosion. When all the observed seismic waves for an event viewed as a whole, waves of earthquake are much more complex, with richer and wider frequency components than those of explosion (Hong, 2013; Huang, 2009; Wei & Li, 2003; Wu, 2012; Zeng, 2008; Zhang, 2006). This conclusion can also be confirmed by the much more diverse SDP graphs matrix of earthquake than the SDP graphs matrix of explosion.

Some index number of diversity or uniformity of the SDP graphs matrix of all observed seismic waves for an event shall be calculated by some suitable algorithms. When seismic wave data from more earthquake and explosion events are investigated, such an index based on the SDP graphs matrix can provide a novel, effectual, and rapid way to discriminate earthquake and explosion. We will investigate this potential in further researches.

## REFERENCES

Bennett, T. J. & Murphy J. R. 1986. Analysis of Seismic Discrimination Capabilities Using Regional Data from Western United States Event. *Bull Seism Soc Amer*, 76: 1069–1086.

Hong, T.K. 2013. Seismic discrimination of the 2009 North Korean nuclear explosion based on regional source spectra. *J. Seismol.* 17:753–769.

Huang, H.M.; Li R. & Lu, S.J. 2009. Discrimination of Earthquakes and Explosions Using Chirp-Z Transform

Spectrum Features. *2009 World Congress on Computer Science and Information Engineering:* 210–214.

Huang, H.M.; Zhou, H.J. & Bian, Y.J. 2011.The Development of Decision Support System for Identifying Explosion Events by Seismic Waves. *JDCTA: International Journal of Digital Content Technology and its Applications*, 5(4):107–112.

Liu, X.Q.,; Shen P.; Zhang, L. & Li, Y.H. 2003. Using Method of Energy Linearity in Wavelet Transform to Distinguish Explosion or Collapse from Nature Earthquake (In Chinese). *Northwestern Seismological Journal*, 25(3):204–209.

Shearer, P. M. 2009. *Introduction to Seismology (2nd edition)*. New York: Cambridge University Press.

Steele J. & Iliinsky N. (Eds) 2010. *Beautiful Visualization*. O'Reilly Media, Inc.

Wei, F.Sh. & Li, M. 2003. Cepstrum Analysis of Source Character (In Chinese). *Acta Seismologica Sinica*, 25(1):47–54.

Wu, A.X 2012. On Quantitative Identification of Explosion Earthquake Based on Cepstrum Computation of HHT and Statistical Simulation of Sub-Cluster. In *Proceedings of the 31st Chinese Control Conference*, July 25-27,2012, Hefei, China:5311–5316.

Zeng, X.W. 2008. Discrimination between Earthquake and Explosion using Wavelet Packet Transform (Master degree thesis, in Chinese). *Lanzhou Institute of Seismology*.

Zhang, F. 2006. Seismic Signal Time-frequency Analysis and its Application to Blast Distinguishing (Master degree thesis, in Chinese). *University of Science and Technology of China*.

*Multimedia, Communication and Computing Application – Leung (Ed.)*
© *2015 Taylor & Francis Group, London, ISBN 978-1-138-02775-6*

# Research on the progress of physiological signal-based method for fatigue driving risks detection

K.J. Sun, X.X. Yang, L. Yang, J.X. Cui, S. Zhang & Y. Lu

*Biological Engineering and Life Science College, Beijing University of Technology, Beijing, China*

ABSTRACT:   The objective is to prevent fatigue driving and to reduce traffic accidents. We analyzed the risks, characteristics, causes, and domestic and international status of fatigue driving detection method. We adopted risk assessment and analyzed electroencephalography, electrocardiography, pulse wave signal, and other physiological signals. According to our study, signal-based fatigue driving detection method can prevent fatigue driving and reduce traffic accidents effectively and timely.

KEYWORDS:   Fatigue detection, Disease risk assessment, Physiological signal, Fatigued driving

## 1   INTRODUCTION

On November 18, 2012, World Day for Road Traffic Victims, with its theme of "now is the time to learn from the past," was held at Beijing, sponsored by Health Development Research Center, World Health Organization and Global Road Traffic Safety Partnership. It is reported that in China there was a total of 210,812 road traffic accidents in 2011, with 62,387 deaths, 237,421 injuries, and direct economic losses of more than 1 billion. We can see that traffic accidents have replaced suicide and became the primary cause of injury and death. Moreover, the accidents caused by fatigue driving are in a large proportion and cause serious threat to people's life and property. In the light of the necessity to avoid fatigue driving and decrease road traffic accident, the research on fatigue driving detection method is of practical significance [1].

## 2   THE CHARACTERISTICS OF FATIGUE DRIVING

Fatigue refers to physical and psychological function disorders as the result of excessive manual and mental work. Fatigue driving is also a kind of disorder in physical and mental function of drivers owing to continuous long-hour driving, which is represented as the phenomenon of declining driving skills [2]. It can be divided into short-term mild driving fatigue and long-term severe driving fatigue, according to the difference of driving time and fatigue severity. The mild driving fatigue manifests itself in drivers as follows: (1) frequent blink, a little tiredness and decreasing attention to safety; (2) untimely and inaccurate shift,

scatterbrained driving; and (3) untimely speed changes varying with different road conditions. While people who drive in severe driving fatigue show features such as: (1) yawn, numb expressions, downcast head; (2) relaxed muscles, ptosis, nearly closed eyes with redness and dryness; (3) blurred narrowing vision, easy to overlook the things ahead; (4) unresponsive, slow action, inability to concentrate; and (5) stiff, slow pace, disorientation, resulting in lane departure, blind shifting gears, unstable travel speed. If one shows the aforementioned symptoms while driving, he or she is more prone to all kinds of accidents.

## 3   THE CAUSE OF FATIGUE DRIVING

### 3.1   *Driving circumstances*

Drivers need to receive relevant information and make actions and decisions accordingly while driving. Thus, drivers are more inclined to tiresome in poor internal and external vehicle environments, for example, highway hypnosis phenomenon [1], which refers to that drivers express sleepiness and weariness because of the monotonous and tedious highway driving characterized by single-road driving, stabilized speed, a rather low vibration frequency, and traffic noise. Therefore, the driver is apt to inattention, poor judgment, slow response, and operational errors, which are the reasons for major traffic accidents.

### 3.2   *The driver himself/herself*

Driver's lack of sleep is the main source of fatigue driving. Experts hold that 15% of accidents are caused by this reason, 50% of which are due to driver's sleep of less than 6 h. Consequently, whether one can drive safely or not is directly depends on his/her sleep. Sufficient sleep is considered as one of the most elemental, significant ways for drivers to eliminate fatigue, restore physical function, and adjust psychological condition [2].

*Long-Time Driving.* The fatigue of driver's neutron and sensory organs puts strains to their central nervous system (CNS) all the time [3]. Restricted by the limited space in a car, drivers have to stay in a fixed pose for a long time which results in slow blood circulation, so some of the organs endure suppression resulting in fatigue and nervous in the muscle. Drivers will feel tired mentally and physically.

*Drug Reason.* Taking pills or even drinking beverage containing hypnosis ingredient will affect driver's CNS.

*Driver's Reason.* Driver does not have good sense of endurance or has some chronic disease. Sometimes they are not skillful enough and have little sense of safety. Emotions of fear and stress will also lead to fatigue.

## 4 STATUS QUO OF PHYSIOLOGICAL SIGNAL-BASED DETECTION METHOD

Currently, physiological signal-based method can be divided into two methods: subjective detection method and objective detection method.

### 4.1 *Subjective detection method*

Subjective detection method is classified into self-assessment and assessment from others. The assessment is made by several pieces of material such as subjective questionnaire, driver's profile record form, sleeping habit questionnaire, Stanford sleep scale form, and Pearson Fatigue Scale. Subjective method is not reliable because it is based on driver's judgment, which means it can mainly be used to test correlation of the other signal-based detection method.

### 4.2 *Objective method*

Currently fatigue driving detection method can be sorted by detection based on physiological signal, driver's operating feature, car driving condition, and multifeature method.

Examine driver's electroencephalography (EEG), electrocardiography (ECG), pulse wave signal, and other physiological signals. This method uses physiological sensor to sense variation of physiological index to judge whether driving has become fatigue. MIT Massachusetts Institute of Technology invented Smart Car with diverse built-in sensors to test ECG, electromyography (EMG), breathe, and skin resistance [4].

Test operation features of driver, including eye movement, nodding, eye closing, and grip. This method uses machine optical technology and sensor technology to test driver's facial feature. For example, eye feature, diameter of pupil, and direction of sight to study fatigue issues. In 1998, Federal Highway Administration found that PERCLOS can test the fatigue level reliably and efficiently [5]. We can judge whether the driver is sleepy or not by attaching sensor to his head which tracks movement of head from time to time. ASCI Advanced Safety concepts Inc raised an opinion of relationship between head movement sensor system and nodding. By tracking head movement, we can get driver's related signals and therefore get the relation between fatigue driving and head movement [6]. Through grip test of driver's grasping steer wheel, we could know that grip power will decrease as the driver gets tired [7].

Test driving state, including speed, position of white line, and steer wheel. These kinds of methods are used to predict driver's fatigue state by testing vehicle moving track, speed, lateral acceleration, and deviation of lanes. DAS2000, a road warning system invented by Ellison Research Labs (US), warns drivers to be careful while detecting road condition.

## 5 THE DEVELOPMENT OF PHYSIOLOGICAL SIGNAL-BASED FATIGUE EVALUATION

### 5.1 *EEG signal detection*

Through the use of brain wave experiment, Lal (USYD) The University of Sydeny found that the variety of EEG is related to fatigue, and with the aggravation of tiredness, the amplitude between waves is increasing and decreasing [8, 9]. With this relationship, according to the variety of each band, the conversion of frequency domain can define the fatigue degree [10]. We can use the ratio of average power spectrum density of different waves as the index of fatigue driving [10]. By adopting the data of stimulating driving and different experiments of EEG signal's average power, we can verify the rationality of using the EEG as the index of fatigue [12, 13].

### 5.2 *ECG signal detection*

The heart wave could be detected by using ECG instrument, and according to the verifying amplitude of heart rate, we can determine the fatigue level. By analyzing the heart rate variability (HRV) power spectrum density and using the ratio of low frequency (0.04–0.15 Hz, LF) and high frequency (0.15–0.4 Hz, HF), the fatigue degree of drivers can be determined [14]. We have developed an HRV detecting [15] and eye tracking of 10 adult male drivers in driving stimulation platform and designed the detecting system according to ECG. In addition, we verified the rationality of using ECG signal as the index of fatigue detection and analyzed influences of pressure of steering wheel, temperature difference, and driver emotion's to the experimental results [16, 17].

### 5.3 EOG signal detection

Electrooculography (EOG) is in the eyes of the placement of electrodes, in vertical direction of the data. Through the analysis of EOG waveform figure, we can conclude the peak amplitude, rise time, and fall time parameters [18]. Owing to the close of their eyes, the extreme value maximum, corresponding wave shape drawings showing the peak amplitude and rise time are the durability of the eyes. According to the peak amplitude, close your eyes, open time parameter values to classify the waveform. Draw all kinds of mental state EOG waveform. According to the provisional distribution of oscillogram, the electric eye diagram belongs to certain extent of other columns pillars, which can get a clear type and changes in time with the blink of an eye, to determine the degree of vigilance and fatigue [19].

### 5.4 EMG signal detection

EMG is an electrical diagnosis therapy which records nerve and muscle electrical activities to determine the function. EMG signal is used for fatigue test, mainly local muscle fatigue test. Measurement generally adopts the method of evoked potentials, in the muscle-surface-fixed electrode surface. To pass through the electrode surface, EMG signals to EMG data recorder to analyze. Study is about a long-distance driving fatigue process which is repetitive and discontinuous. The research shows that after the fatigue, the number of electrical sheet value increases, whereas the EMG average frequency decreases. Through the surface electric measurement of the waist muscle, its absolute integral value increases as a whole [20–21].

### 5.5 Skin electrical signal detection

Skin resistance or conductance changes over skin sweat gland muscle can be known as a galvanic skin response. With the emergence of driving fatigue phenomenon, the driver's skin electrical signal is on the rise, so the extension of fatigue driving time gradually rises and finally levels off.

### 5.6 Pulse wave signal detection

Pulse wave is in abundant physiological and pathological information, and the research has found the pulse signal by the power spectrum analysis and the relationship between the human body fatigue. Tiredness and waking the pulse signals are obviously different [22]. Wavelet transform is used to find the normal radial artery and fatigue state signal characteristic value of each frequency with the identified motorists for assessing the driver fatigue [23]. After the acquisition of 188 cases of motorists' pulse condition of signal processing and analysis of statistics, the feature has good repeatability and stability, which can judge feature vector as a driver's fatigue state.

## 6 CONCLUSION

This paper has introduced characteristics and traits of fatigue driving, and analyzed its causes to achieve real-time testing of driver's fatigue level, and provided support for the causation. In addition, on the basis of the current description of the testing of driver's fatigue level, both subjectively and objectively, these tests can be improved according to driver's physical signals. Therefore, it not only can determine future testing direction—by using touch sensors—and also can determine driver's fatigue mechanism according to the test results.

## REFERENCES

[1] ZHANG Y: 17 Solutions to limit road killers [N] People's Daily, 2012-02-08.
[2] LIN S. The harm and Prevention of fatigue driving [J]. Fujian Agricultural Machinery, 2010(9):13.
[3] WANG Z.Y. Research on the prevention of fatigue driving.
[4] SUN M.C, GUO Zhaoyong, QIAN Jiasheng, Research on fatigue driving [J]. NEI JIANG KE JI, 2012(9):13.
[5] YUAN X, SUN X.M. The development of test method on fatigue driving[J]. Automobile Research & Development, 2012, 2(3):157–164.
[6] ZHAO Z. Test Method of fatigue driving and flush-bonading [D]. Dalian Maritime University, 2010(8):2.
[7] Dinges DF, Grace R.PERCLOS: a valid psychophysiological measure of alertness as assessed by psychomotor vigilance[R]. Washington DC: US Department of Transportation, Federal Highway Administration, 1998.
[8] Philip WK, Roger DJ, Jone MC. Development of driver alertness detection Advanced Safety concepts Inc, 1998.
[9] ZHENG P, SONG Z.H, ZHOU Yiming. Research on the progress and the development of fatigue driving test [J] Journal of China Agricultural University.
[10] Lal SK, Craig A. A critical review of the psychophysiology of driver fatufye[J]. Biological Psycho logy, 2001, 55(3):173–194.
[11] Lal SK, Craig A, Brood P, et al. Development of an algorithm for an EEG-based driver fatigue countermeasure [J]. Journal of Safety Research, 2003, 34(3):321–328.

[12] Gu HS, Ji Q. An automate face reader for fatigue detection[C]. Pro of the 6th IEEE International Conference on Automatic Face and Gesture Recognition, 2004:111–116.

[13] Peng JQ. Exploring the Characters of Electroencephalogram for Fatigued Drivers[J]. Transactions of Beijing Institute of Technology, 2007, 27(7); 585–589.

[14] MAO J.K., ZHAO X.H., LIU X.M., etc. The forecast of fatigue driving based on EEG analyze [J]. Chinese Journal of Ergonomics, 2009, 15(4):25–29.

[15] FANG R.X., ZHAO X.H., RONG J., etc. The research of fatigue driving based on EEG signals [J] Public transportation technology, 2009, S1(26):12–13.

[16] JIAO K., Li Z.Y., Chen M. etc. Comprehensive effect analysis on changes of fatigue drivers' heart rate and blood pressure [J] ournal of Biomedical Engineering, 2005, 22(2):127–130.

[17] DONG Zhanxun, SUN Shouqian, WU Qun etc.relativity between heart rate and fatigue driving [J] Journal of Zhejiang University, 2010, 44(1):46–50.

[18] Jeong IC, Lee DH, Park SW, et al. Automobile Drivers Stress Index Provision System that Utilizes Electrocardiogram[C]. IEEE Intelligent Vehicles Symposium, 2007:652–656.

[19] Rohado E, Glatc IJL, Batea R, et al. D river fatigue detection system[C]. Proc of IEEE International Conference on Robotics and Biomimetics, 2009:1105–1110.

[20] Ohsuga M, Kamakura Y, Inoue Y, et al. Classification of blink waveforms toward the assessment of drivers arousal levels: an EOG approach and the correlation with physiological measures[C]. Proc of the 7th International Conference on Engineering Psycho logy and Conitive Ergonomics Berlin: Springer, 2007:787–795.

[21] Nohuch IY, Nopsuwancha IR, Ohsuga M, et al. Classification of blink waveforms towards the assessment of drivers arousal level an approach for HMM based classification from blinking video sequence[C]. Pro of 7th International Conference on Engineering Psychology and Cognitive Ergonomics. Berlin: Springer, 2007:779–786.

[22] Hostens I, Ramon H. Assessment of muscle fatigue in low level monotonous task performance during car driving [J]. J Electromyography Kinesiol, 2005, 15(3):266–274.

[23] ZHANG H.Z. Uses and methods of fatigue driving based on physical signals[D]. Haerbin: Harbin Institute of Technology, 2006:4–5.

[24] XIONG Y.X., MIAO D.H., DENG S.P. A pulse wave detection system design for fatigue[C]. Guilin:MCMI2009.

[25] CHEN X.F., ZHOU K.J. Study on the judgment of driver fatigue based on pulse wave analysis[C]. Dalian: Dalian University of Technology Institute of systems engineering.

*Multimedia, Communication and Computing Application – Leung (Ed.)*
© *2015 Taylor & Francis Group, London, ISBN 978-1-138-02775-6*

# Automatic detection of IR ship on the basis of variance statistic

Y. Chen, S.H. Wang, G.P. Wang, W.L. Chen & J.L. Wu
*Science and Technology on Optical Radiation Laboratory, Beijing, China*

ABSTRACT: Aiming at the adversary features of infrared (IR) ship image with sea-sky background, such as low contrast between ship and background and fuzzy image edges, an automatic detection method of IR ship on the basis of variance statistic is proposed in this paper. First, the preprocesses such as denoising and contrast enhancement are carried out, then the Hough transform is employed to detect the sea-sky-line, and finally, the row-variance and column-variance of the area around the sea-sky-line are calculated to detect the ship region. The experimental results show that the method can effectively detect the IR ship in complex sea-sky background, and it can lay a good foundation for the extraction and recognition of subsequent geometric features.

KEYWORDS: Variance statistic, IR ship, Automatic detection, Sea-sky-line.

## 1 INTRODUCTION

Automatic detection and recognition of infrared (IR) ship is the key component of anti-ship technology, and it plays a vital role in reliable capturing, precise tracking, and accurate striking of enemy ships. Thus, they have attracted widespread attention at home and abroad in recent years. However, because of the interferential factors, such as detector noise, sea-sky background radiance, far operating distance, and so on, the signal-to-noise ratio (SNR) and contrast of the detected IR ship image are low. Especially, if the detected signal is very weak and the background is interfered by a nonstationary clutter, then the ship could be submerged in the cluttered background, and the target detection may be difficult. Therefore, according to preprocess steps, such as denoising, contrast stretching, and sea-sky-line detection, an automatic detection method of ship on the basis of variance statistical characteristic is researched in the paper.

## 2 SHIP DETECTION STRATEGY ACCORDING TO VARIANCE STATISTICAL CHARACTERISTIC

### 2.1 Procedure of ship detection

Typical IR ship detection steps include image preprocessing, sea-sky-line detection, and the target region detection (Fig. 1).[1]

Figure 1. Flow chart of ship automatic detection.

### 2.1.1 Image preprocess

Owing to the effect of the IR device noise, the background clutter, the atmospheric transport, and other factors, the IR image is always with large noise, low contrast, and blurred edge, so that the target detection difficult. To improve the clarity of the ship region, for detection, and improve the quality of the IR image, preprocessing operations are required.[2] Aiming at the low SNR situation, because the salt and pepper noise is dominant in the IR image, the classical nonlinear median filter is employed for denoising. Aiming at the low contrast and fuzzy edge, histogram equalization is employed to enhance the ship edge.

### 2.1.2 Sea-sky-line detection

According to the characteristics of long-range IR ship which always appears near the sea-sky-line, detecting the sea-sky-line position to determine the target potential area can greatly reduce the search scope, eliminate the interference of cloud and sea clutter, and then raise the probability of detection. Considering that the Hough transform is robust to noise and discontinuous line[3, 4], we adopt it for sea-sky-line detection.

### 2.1.3 Target region detection

Target region detection is to look for potential target area near the sea-sky-line. If the false alarm rate is high, the subsequent target recognition would be difficult, whereas if the missing rate is high, even with high precision of the recognition algorithm due to the no-target potential area, the recognition rate also would not high, which could not meet the engineering needs[1, 5]. Thus, the potential region extraction is a key step in target recognition. This paper focuses on the target potential region detection method on the basis of variance statistical characteristic.

### 2.2 Target region detection according to variance statistic

Assuming that the original IR image is $G$, the row coordinate of the detected sea-sky-line described in Section 2.1 is $s$. Considering that the long-range ship to be detected generally appeared above the sea-sky-line, and the probability of appearing below the sea-sky-line is small[6], the area $X = G(1:s+n,:)$ is extracted as the tested area ($n$ is the number of rows below the sea-sky-line, which can be set according to the actual situation, $n = 10$ in this paper), which can greatly reduce the impact of sea clutter on the subsequent target detection. Because of the gray value fluctuation of the area where target exists frequently, the gray value of sky background changes gently, and this variance can measure the fluctuation of a set of data, that is, the greater the variance, the lager the fluctuation. Thus, the variance is employed to identify whether the detected region is the target or background. Therefore, for the detection of the ship, the row-variance and column-variance of $X$, and the concrete steps are calculated as follows:

**Step1:** The variance of each column in region $X$ is calculated, and the column-variance vector $V_{Xj}$ is obtained. If all column-variances in interval $[X_{j1}, X_{j2}]$ are larger than $TH_X$ (threshold $TH_X$ is the mean of vector $V_{Xj}$), the left endpoint j1 and right endpoint j2 of target region can be determined. The potential target region is constricted to $Y$, where $Y = X(:,[j1:j2])$.

**Step2:** The variance of each row in region $Y$ is calculated, and the row-variance vector $V_{Yi}$ is obtained. If all row-variances in interval $[Y_{i1}, Y_{i2}]$ are larger than $TH_Y$ (threshold $TH_Y$ is the mean of vector $V_{Yi}$), the above endpoint i1 and the below endpoint i2 of target region can be determined.

**Step3:** According to the four endpoints determined by Step1 and Step2, the ultimate target region $T$ can be determined, where $T = X([i1:i2],[j1:j2])$.

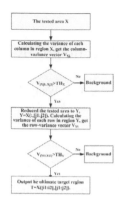

Figure 2. Flow chart of target region detection according to variance statistic.

What needs to be emphasized is that when calculating the variance, it is possible that some row-/column-variance values are larger than the threshold, whereas the other row-/column-variance values in its five neighborhood are smaller than the threshold, so this row/column is not considered to be the target region. Similarly, if some row-/column-variance values are smaller than the threshold, while the other row-/column-variance values in its five neighborhood are larger than the threshold, this row/column is also not considered to be the background.

## 3 EXPERIMENTAL RESULTS

IR ship image with typical sea-sky background is selected to verify our method. Fig. 3 is the real navigating ship IR image sized 240×320 with obvious salt–pepper noise and low contrast. Before detecting the sea-sky-line, preprocessing operations which contain median filtering and contrast stretching are executed, and the pretreated result is shown in Fig. 4.

Figure 3. Original IR ship image.

Figure 4.    Pretreated result.

Comparing with the original image, in the pretreated image, the noise is mostly removed, and the image contrast is enhanced obviously, which is beneficial to sea-sky-line detection.

After executing the edge extraction on the pretreated image, Hough transform is employed to detect the sea-sky-line of the edge extraction result, and the sea-sky-line detecting result is shown in Fig. 5.

Figure 5.    Result of sea-sky-line detection.

The entire region above the sea-sky-line and the following n rows below the sea-sky-line ($n = 10$ in this paper) are extracted as the tested area $X$, which is shown in Fig. 6.

Figure 6.    Tested area X.

According to Step1 as described in Section 2.2, by calculating the variance of each column in area $X$,

and setting the mean of all column-variance values $TH_X$ as the threshold, $TH_X$ is 0.5634. The region in which all the column-variance values are larger than $TH_X$ is considered to be the target region. However, there may be some special columns which are marked with red circles as shown in Fig. 7. Although the variance values of these two marked columns are larger than $TH_X$, the other column-variance values in their five neighborhoods are all smaller than $TH_X$. These two columns are not considered to be the target region. Thus, we can obtain the left endpoint $j1 = 129$ and the right endpoint $j2 = 160$ as shown in Fig. 7, and the potential target area is converged to $Y$, $Y = X(:,[129:160])$, which is shown in Fig. 8.

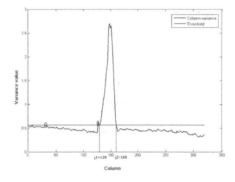

Figure 7.    Comparison of column-variance and threshold.

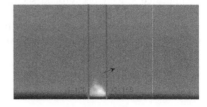

Figure 8.    Potential target area Y.

Similarly, according to Step2 as described in Section 2.2, calculating each row-variance of area $Y$, and the mean $TH_Y = 0.0406$ of all row-variance values in area $Y$ is set as the threshold. By comparing each row-variance value with $TH_Y$, we can obtain the above endpoint $i1 = 131$ and the below endpoint $i2 = 156$, and the comparing result is shown in Fig. 9.

Through two variance comparisons, the four coordinate positions can be identified, thus the final target region $T$ can be determined, $T = X([131:156],[129:160])$ and the detected target region T is marked with red rectangle as shown in Fig. 10.

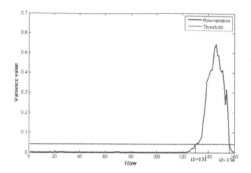

Figure 9.   Comparison of row-variance and threshold.

Figure 10.   Detected ship region.

## 4   CONCLUSIONS

Aiming at the adversary features of IR ship image under complicated sea-sky background such as low contrast between target and background and fuzzy image edges, an automatic detection method of IR ship target on the basis of variance statistic is proposed in this paper. After the preprocessing operations such as denoising and contrast enhancement, and sea-sky-line detection, column-variance and row-variance of the tested area are calculated by comparing the column-variance and row-variance with thresholds respectively, and we can determine the endpoint coordinates of the ultimate target region, and realize the automatic detection of ship. Experimental results show that the method can effectively withstand the complicated sea-sky background, and it can be used for automatic detection of the ship target.

## REFERENCES

[1] Ding-he Wang, Zhao-dong Niu, Da Tang, Zeng-ping Chen. Extracting IR ship ROI using multi-grade segmentation and fusion method[J], Laser & Infrared, 2014, 43(4): 461–465.
[2] Pei-zhi Wen, Ze-lin Shi, Hai-binYu. Automatic detection method of IR small target in complex sea background[J], Infrared and Laser Engineering, 2003, 32(6): 590–594.
[3] Song-tao Liu, Xiao-dong Zhou, Cheng-gang Wang. Robust sea-sky-line detection algorithm under complicated sea-sky background[J], Opto-Electronic Engineering, 2006, 33(8): 5–10.
[4] Rafael C.Gonzalez, Richard E.Woods, Steven L.Eddins. Digital image processing using MATLAB[M], Publishing House of Electronics Industry, 2006, 300–304.
[5] Yang Ming-yue, Yang Wei-ping. Automatic detection method of IR warship target in the complex sea-sky background[J], Infrared and Laser Engineering, 2008, 37(4): 638–641.
[6] Zou Chang-wen, Feng Li-tian, Liu Xian-zhi, Dai Jun. Infrared warship target detection based on multi-scale variance[J], Laser & Infrared, 2011, 41(6): 697–699.

*Multimedia, Communication and Computing Application – Leung (Ed.)*
© *2015 Taylor & Francis Group, London, ISBN 978-1-138-02775-6*

# A dual-window detection scheme considered motion direction based on optical flow consistency for frame deletion detection

W.T. Zhang, C.H. Feng & Z.Q. Xu
*Wuhan University, LIESMARS, Wuhan, China*

ABSTRACT: With more and more sophisticated video editing softwares appearing, the digital video forensics, especially the forensics for the videos with frame deletion which conceal some certain facts, have become an important part of information security. The video forensics based on optical flow consistency has been proposed recently, but the research still has space for improvements and the motion direction has not been considered in most of the current detection schemes. Thus in this paper a dual-window detection scheme taking motion direction into consideration, based on optical flow consistency, is proposed. The optical flow in different motion direction has different sensitivity for deletion, and dual-window makes the detection more accurately. The contrast experiment shows that our detection scheme is effective and has achieved good results.

## 1 INTRODUCTION

With the development of the information age, more and more information resources exist in the form of digital media. For example, the paper documents and books in the past have changed to electronic records widely. Millions of monitors are the loyal guardians of the city's security. More and more pictures and videos have become the key clues of solving criminal cases and the forensic evidence. All of these have shown the importance of digital media and our reliance on it. Therefore, the security of the digital media has increasingly become a widely concerned issue.

Digital video is the primary information carrier of digital media, which has become the most important record of our lives and social development, with its richer and more intuitional content compared with the image. With the advent of sophisticated video editing software, such as Photoshop, Premiere and so on, it is becoming increasingly easier to tamper videos than ever before. Some of the tampered videos departing from its original meaning may mislead the public, and even cause serious consequences. A typical video manipulation is to delete a number of frames of the sequence, so that some certain fact can be concealed. For example, in court a video-evidence recording the process of crime was deleted intentionally, which may cause the illusion of the suspect's absence and influence the trial result. Tampered videos in news reports with the distortion and fabrication of facts may exert adverse effects on the public opinion, which damages the accuracy of news. In sports competition, a tampered video with the foul's part deleted will affect the final judgment and disrupt

the game's normal order [1]. Thus it can be seen that video forensics, especially the forensics of the frame deletion, is of great significance in the information security.

In the early time, scientists detected video forgery by using active detection, such as video watermarking, digital signature and so on. To identify the integrality and authenticity of a video, some pretreatment, like a signature or watermark precisely embedded at the time of recording, must be made on the video in advance, and through specific extraction and validation algorithm it can achieve the detection effect. For example [2–3], an effective technique of embedded watermark and a video hashing based on radial projections of key frames have been proposed to detect video forgery, and they all achieves some effects. But these two methods need appropriate instrument to embed video watermark or digital signature at the end-transmission equipment, which limits these approaches to specially equip digital cameras and will increase the cost of information and delay the transmission. Different from these active detections, digital forensics detects the manipulation of visual media through the intrinsic features in the media left by acquisition devices or manipulation acts without using any pre-embedded signal, which can be seen as a passive detection. Thus video forensics has greatly increased the range of applications for detecting video forgery, and it is a promising tool in the field of authenticating digital visual media [4]. The inherent characteristics usually analyzed in the algorithm of video forensics contain the correlation between pixels, the compression effect, and certain inter-frame statistical characteristic and so on. Based on the theory, Wang and Farid [5] detected traces of tampering in de-interlaced video by using the correlations introduced by the camera or de-interlacing algorithms. Qin yunlong [6] et al. proposed a scheme which can locate the tampered point by using the statistics of motion vectors produced by inter-frame prediction.

Lately, several approaches specially for detecting frame deletion have been presented, which are usually based on some characteristics caused by the break of inter-frame content continuity. In this area Wang and Farid [7] have made the most fundamental work. They exploit the fact that when a frame deletion video is subjected to double JPEG compression, the spikes of I-frame are introduced and the larger motion estimation errors will be caused by frames moving from one GOP to another. Based on [7], Stammet al. [8] have proposed an automatic solution for detecting the residual spikes of I-frame in the mean motion, which also expanded the work in an adaptive group of picture GOP-length structure situation. Other than only detecting the existence of frame deletion, in [4] Feng et al. presented a method which can directly locate the frame deletion point by analyzing the distinguishing fluctuation feature of motion residual caused by frame deletion. In [9], Chao et al. proposed different schemes for detecting frame insertion and deletion based on the optical flow consistency, especially for the detection of frame deletion that has more improvement area because of the relatively smaller change in optical flow.

In this paper, we propose a new detection scheme based on [9]. In [9], the optical flow's law in both x and y directions is considered to be similar, and frame-to-frame optical flows and double adaptive thresholds based on optical flow consistency are applied to detect frame deletion forgery. However, this scheme has its limitation, especially when the videos contain complex and rapid motion. Firstly, the video optical flow has two aspects of the

motion information, the motion's amplitude which is expressed by the optical flow's value and the motion's direction which is expressed by the ratio between the optical flow's value in x and y directions. The scheme in [9] only analyzed the single direction of amplitude in optical flow, and ignored the property of directions. Secondly, the second threshold in [9] is used to ensure the deletion points compared with the means of all the video optical flow. However, the motion in all the video is not always the uniform motion, which makes the optical flow consistency become the partial consistency, not a global one. Despite using the double thresholds, the single window and the means of all the optical flow have led to a lot of leak detections and false detections. In view of both two points, this paper proposes a dual-window detection scheme taking motion direction into consideration based on optical flow consistency, which can select the optical flow direction automatically for detection and use dual-window to locate the deletion point. Besides, the comparative experiment has been done to prove that this video forensics scheme is more effective.

The rest of the paper is organized as follows. In Section 2, this paper's detection algorithm is presented and the reason why the algorithm needs to consider the motion direction and use dual-window will be given. The comparative experiment's results and discussions are provided in Section 3, and the conclusion is drawn in Section 4.

## 2 PROPOSED DETECTION SCHEME

In this paper, a dual-window detection scheme taking motion direction into consideration based on optical flow consistency is proposed, which contains three parts, the calculation of optical flow, the selection of optical flow based on motion direction, and the dual-window detection in Figure 1.

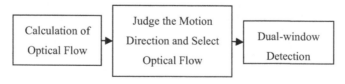

Figure 1. The flowchart of the detection scheme in this paper.

### 2.1 Optical flow consistency for video forensics

The optical flow is first proposed by Gibson in 1950, which is used to describe the velocity of pattern movement in time-varying image [10]. Because of its highly description of motion, it has been widely used in computer vision field including the target tracking, the detection and dynamic analysis of moving

objects, mosaic construction and so on. In the field of video forensics, the video optical flow consistency is first analyzed by Chao et al. in [9].

A video can be seen as an ordered sequence of images, so the changing pattern of the motion can be described as the corresponding pixel point between the adjacent two frames in the field of optical flow.

Then, for every adjacent two frames in a video, we add each absolute value of the optical flow field to simplify the data, and the video of optical flow is got. And an original video with K frames computes the video optical flow, that's to say, the kth optical flow is computed with the kth frame and the k+1th frame. We get the optical flow sequence

*OFX _ video*[*K* − 1] and *OFY _ video*[*K* − 1] in the x and y directions respectively with Equation 1 and Equation 2.

$$OFX\_sum_{(m,n)} = \sum_{i}^{width} \sum_{j}^{height} |OFX_{(m,n)}(i,j)|$$

$$OFX\_video(k) = OFX\_sum_{(k,k+1)}, k=1,2,...K-1 \quad (1)$$

$$OFY\_sum_{(m,n)} = \sum_{i}^{width} \sum_{j}^{height} |OFY_{(m,n)}(i,j)|$$

$$OFY\_video(k) = OFY\_sum_{(k,k+1)}, k=1,2,...K-1 \quad (2)$$

Where the $OFX_{(m,n)}(i,j)$ and $OFY_{(m,n)}(i,j)$ denote the values of the optical flow field between adjacent two frames in the corresponding pixel$(i,j)$ ; *width* and *height* denote the size of the frame.

The inter-frame motion in original video is continuous because of the continuity of content, thus the video has the character of optical flow consistency. However, tampering will destroy the continuity of motion, which makes the video optical flow no longer consistent. Just like Figure 2, the above (a) is the optical flow histogram of an original video, and the below (b) is the one of its tampered video. From it we can see the different consistency on optical flow.

Based on this finding, the video forensics on optical flow consistency has achieved a certain effect.

## 2.2 *Motion direction selection*

Usually for a vector $(x, y)$, the value $\sqrt{x^2 + y^2}$ is used for describing the amplitude, and the ratio $x / y$ is used for describing direction. To put it another way, the values of x and y can be seen the motion intensity in x and y directions, and if the ratio is bigger than 1 the motion direction can be seen deflect horizontally, and if the ratio is less than 1 the motion direction can be seen deflect vertically.

The same can be applied in optical flow. Because of the optical flow's description for motion, the motion direction can be described by the ratio of the optical flow values in x and y directions. If the ratio is greater than 1, the main motion direction is x direction, and on the contrary the main motion direction is y direction. That is to say, for a horizontal motion video the optical flow value in x direction is greater than the value in y direction, and for a vertical motion video the inequation is opposite. In Figure 3, the above (a) is the optical flow histogram of a horizontal motion tampered video, and the below (b) is the one of a tampered video with same deletion in the vertical motion. From them, we can see that there exists difference in the level of sensitivity to the abnormal peaks along with the difference of main motion direction. When the video is a horizontal motion one the optical flow in y direction is more sensitive with abnormal peaks, however, for a vertical motion video the optical flow in x direction is more sensitive.

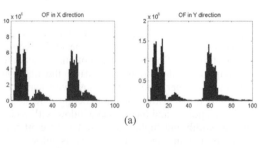

(a)

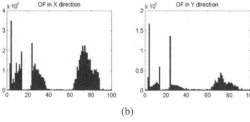

(b)

Figure 2. The optical flow histogram of an original and tampered video.

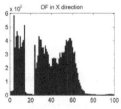 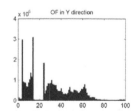

(a) The optical flow histogram of a horizontal motion tampered video.

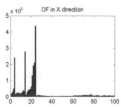 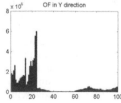

(b) The optical flow histogram of a vertical motion tampered video.

Figure 3. The optical flow histogram of different motion direction tampered video.

Take a horizontal motion video as example. Its optical flow of original video in x and y directions are $OFX\_video$ and $OFY\_video$ and in main motion direction, the greater motion intensity and more complex motion content it has, the bigger values and stronger volatility of optical flow it has. From the statistics, the mean value and variance of optical flow in x direction are all bigger than the ones in y direction. Just like the Equation 3 and Equation 4.

$$Mean(OFX\_video) > Mean(OFY\_video) \qquad (3)$$

$$Var(OFX\_video) > Var(OFY\_video) \qquad (4)$$

For the tampered video, the optical flow in x and y directions are $OFX\_video'$ and $OFY\_video'$. In the one video the abnormal peaks caused by the deletion can be seen the same. The variance reflects the level of the sample's deviation with the mean, and because of the bigger mean in x direction, the deviation of the abnormal peaks caused in x direction is smaller than in y direction. Express the deviation with Equation 5.

$$Var(OFX\_video') - Var(OFX\_video)$$
$$< Var(OFY\_video') - Var(OFY\_video) \qquad (5)$$

The inequation (5) can explain that the optical flow in y direction is more sensitive than in x direction when the main motion is horizontal, so selecting the optical flow based on the motion direction is very necessary. The selection principle is below.

If $OFX\_video'(k) > OFY\_video'(k)$, then select $OFY\_video'(k)$ to detect;

If $OFX\_video'(k) < OFY\_video'(k)$, then select $OFX\_video'(k)$ to detect.

### 2.3  Dual-window detection

Because of the break of optical flow consistency caused by deletion, the optical flow histogram will appear abnormal peaks in the deletion points. Our detection scheme's goal is to detect the abnormal peaks correctly, and for this purpose the algorithm compares the optical flow value of every frame with the optical flow means of its adjacent few frames. If some point is several times higher, this point will be predicated as the peak. The above described is called as the window-method.

In [9], to detect the deletion point it uses the scheme with single-window and double-thresholds. At first a window with first threshold can detect the lower peaks as the suspicious deletion point, and then at the suspicious points it uses the second threshold compared with the means of all the optical flow values to confirm the deletion points.

Though the above scheme from [9] uses the double-thresholds, the second threshold has not played a very good role because the optical flow consistency is only effective partly. In the common situations, the motion may happen in the halfway and is not continuous through all the video. From the low-moving motion to high-moving motion, the normal motion change may also make some peaks. In Figure 4, take a tampered video with three points deleting from frame 5 every other 10 frames as example. From it we can see that the 4th, 14th, 24th points have peaks which agree with the deletion. However, around the 60th, 140th and 155th there also exists some peaks because of high-moving motion. If the comparison value is the mean of all the optical flow, these interferential peaks will cause false detection result.

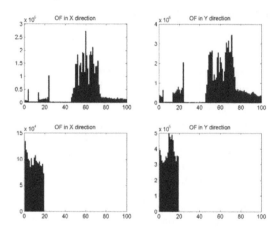

Figure 4.   The example of normal motion change making some peaks.

But from Figure 4, the interferential peaks are much smaller than the deletion point peaks. To reduce the interference of high-moving motion, we use the self-adapting dual-window to confirm the deletion point in this paper. The first window with narrower window width and lower window height detects the suspicious points, and the second window with wider window width and higher window height confirms the deletion points finally. The process is below.

1. If the previous step has selected the optical flow in x direction, for the kth optical flow value $OFX\_viedo'(k)$ we use the first window with the width win1 around it. If the value is T1 times than the mean of the first window, that is to say, it meets Equation 6, and we think this point as the suspicious point, where win1=2, T1=2.

$$OFX\_video(k) >$$
$$T1 * \sum_{i=1}^{win1} OFX\_video(k-i) + OFX\_video(k+i) \qquad (6)$$

2. In the suspicious point we use the second window with the width win2, generally win2>win1. If the optical flow value is T2 times than the mean of the second window, that is to say, it meets Equation 7, we confirm the point as the deletion point.

$$OFX\_video(k) >$$

$$T2 * \sum_{i=1}^{win2} OFX\_video(k-i) + OFX\_video(k+i) \quad (7)$$

## 3 SIMULATION EXPERIMENTS AND RESULTS

In this section, we will take a contrast experiment to evaluate the detection scheme proposed in this paper. The experiment is to detect the same video databases with the below four methods and compare the results.

1 No direction selection and single-window detection, just the scheme in [9].
2 Direction selection and single-window detection.
3 No direction selection and dual-window detection.
4 Direction selection and dual-window detection, just the scheme in this paper.

### 3.1 Evaluation standards

To evaluate the detection efficiency, the True Positive Rate (TPR) and False Alarm Rate (FAR) are used. The TPR is the percentage of correctly detected points among all the deletion frame points which is calculated with Equation 8, and it can well demonstrate the detection efficiency. The FAR is the percentage of falsely detected points among all the video frames which is calculated with Equation 9, and a low FAR can well prove the detection accuracy.

$$TPR = \frac{sum\_N_t}{sum\_N_d} * 100\% \quad (8)$$

$$FAR = \frac{sum\_N_f}{frames} * 100\% \quad (9)$$

Where $sum\_N_t$ denotes the sum number of all the correctly detected points; $sum\_N_d$ denotes the sum number of all the deletion points; $sum\_N_f$ denotes the sum number of all the falsely detected points; frames denotes the sum number of all the video frames.

### 3.2 Test video databases

Research on video forensics has just began in recent years and there are no uniform databases to test videos databases. Ensuring the effective experiment results and considering the situation of video forensics is usually done in the daily life, so here in this paper we take 100 videos in the MOV format in the busy road with the Canon LEGRIA FS406 camera. The video database is named "my videos", where the half is mainly the horizontal motion, while the left is mainly the vertical motion.

The tampered videos come from the video databases "my videos" and they are conducted as follows. The videos in the MOV format transform into videos in the YUV format, and for every video we delete frames in three points from the frame 5 every other 10frames with the deletion number 10, 20, 30 respectively. Then we get three deletion frames video databases named "cut_10", "cut_20" and "cut_30".

### 3.3 Experimental results and discussions

The four methods all use the three deletion frames video databases "cut_10", "cut_20" and "cut_30" to detect. In each method, because the first window detects the suspicious points roughly, we don't discuss it detailedly and the first window width win1=2, and the first threshold T1=2. The second window is to ensure the correct deletion points, and the window width win2=3. The second threshold we choose four groups, T2_1=3.5, T2_2=3.2, T2_3=2.9, T2_4=2.6. If there is not the second window, the second threshold is the same as the above four T2 values.

1. No direction selection and single-window detection.

This scheme is just the method in [9], and because of no direction selection, the optical flow in both x and y directions are detected respectively. The result is in Table 1.

2. Direction selection and single-window detection.

In this method, for the comparison with the scheme in [9], though using the single-window, we also

Table 1. The result of method 1.

| x direction (%) | | T2_1 = 3.5 | T2_2 = 3.2 | T2_3 = 2.9 | T2_4 = 2.6 |
|---|---|---|---|---|---|
| cut_10 | TPR | 44 | 47.3 | 51 | 54 |
| | FAR | 0.103 | 0.103 | 0.103 | 0.114 |
| cut_20 | TPR | 53 | 55 | 58.3 | 62.3 |
| | FAR | 0.089 | 0.089 | 0.096 | 0.103 |
| cut_30 | TPR | 51.7 | 54.7 | 59 | 62 |
| | FAR | 0.095 | 0.095 | 0.095 | 0.095 |
| y direction (%) | | T2_1= 3.5 | T2_2= 3.2 | T2_3= 2.9 | T2_4= 2.6 |
| cut_10 | TPR | 37.7 | 40.3 | 42.3 | 45 |
| | FAR | 0.068 | 0.068 | 0.068 | 0.074 |
| cut_20 | TPR | 42.3 | 47.7 | 51.3 | 56 |
| | FAR | 0.083 | 0.089 | 0.089 | 0.089 |
| cut_30 | TPR | 46.7 | 49.7 | 50.7 | 54.7 |
| | FAR | 0.078 | 0.078 | 0.078 | 0.078 |

detect with double-thresholds as the same with [9]. The result is in Table 2.

Table 2.    The result of method 2.

| (%) | | T2_1=<br>3.5 | T2_2=<br>3.2 | T2_3=<br>2.9 | T2_4=<br>2.6 |
|---|---|---|---|---|---|
| cut_10 | TPR | 48.7 | 50.7 | 54.7 | 57 |
| | FAR | 0.068 | 0.068 | 0.068 | 0.08 |
| cut_20 | TPR | 57.7 | 59.7 | 63.7 | 66.7 |
| | FAR | 0.062 | 0.069 | 0.076 | 0.083 |
| cut_30 | TPR | 56.3 | 60.3 | 63.3 | 65.7 |
| | FAR | 0.078 | 0.078 | 0.078 | 0.078 |

3. No direction selection and dual-window detection.
   In this method, we use the optical flow in both x and y directions respectively with dual-window. The result is in Table 3.

Table 3.    The result of method 3.

| x direction (%) | | T2_1=<br>3.5 | T2_2=<br>3.2 | T2_3=<br>2.9 | T2_4=<br>2.6 |
|---|---|---|---|---|---|
| cut_10 | TPR | 76.7 | 80.3 | 83 | 86.3 |
| | FAR | 0.011 | 0.022 | 0.04 | 0.097 |
| cut_20 | TPR | 81.7 | 85.7 | 87.7 | 88.3 |
| | FAR | 0.014 | 0.028 | 0.041 | 0.117 |
| cut_30 | TPR | 82.7 | 85 | 87 | 88.7 |
| | FAR | 0.017 | 0.035 | 0.061 | 0.104 |
| y direction (%) | | T2_1=<br>3.5 | T2_2=<br>3.2 | T2_3=<br>2.9 | T2_4=<br>2.6 |
| cut_10 | TPR | 60.7 | 67.3 | 72.3 | 78 |
| | FAR | 0 | 0.023 | 0.034 | 0.097 |
| cut_20 | TPR | 73 | 76.3 | 81 | 84.7 |
| | FAR | 0.014 | 0.048 | 0.069 | 0.124 |
| cut_30 | TPR | 73.7 | 78.3 | 83 | 86.3 |
| | FAR | 0.009 | 0.043 | 0.069 | 0.139 |

4. Direction selection and dual-window detection.
   This method is just the scheme in this paper considering the motion direction and using dual-window. The result is in Table 4.

Table 4.    The result of method 4.

| (%) | | T2_1=<br>3.5 | T2_2=<br>3.2 | T2_3=<br>2.9 | T2_4= 2.6 |
|---|---|---|---|---|---|
| cut_10 | TPR | 85 | 86.7 | 88 | 90.3 |
| | FAR | 0.011 | 0.017 | 0.029 | 0.086 |
| cut_20 | TPR | 87.7 | 90.3 | 91 | 92.7 |
| | FAR | 0.007 | 0.021 | 0.034 | 0.089 |
| cut_30 | TPR | 89 | 91 | 92.3 | 94.7 |
| | FAR | 0.009 | 0.017 | 0.043 | 0.086 |

In order to make it more clearly, we make ROC curves with the above four results in Figure 5. The

ROC curves illustrate the performance of different thresholds under different detection schemes. From it we can find that the detection scheme in this paper has achieved good results in these three deletion video databases with the TPR around 85%~95%. Along with more frames deleted, the TPR will increase. That is because more deletion frames cause bigger content difference between the frames, and the bigger difference makes higher optical flow's abnormal peak which can be detected more easily. The decrease of T2 causes the TPR and FAR both increase due to the less threshold makes the more "relatively low peaks" detected. This may increase the number of correctly detected, but also the more "interference peaks" will mistake for deletion points.

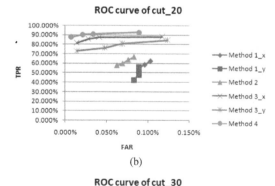

(a)

(b)

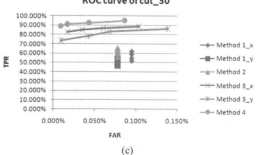

(c)

Figure 5.    The ROC curves of all the experiments.

282

For the first method, because of the different video database, in this paper we didn't make a comparison with the result in [9]. But by comparing the results with the same database in this paper, we can also draw a conclusion that the scheme in this paper highly improves the detection efficiency with the higher TPR and lower FAR.

Compared with the fourth and second method, the dual-window makes the TPR increase sharply by the same direction selection. Compared with the fourth and third method, with the same dual-window, direction selection makes the TPR increase too. Thus we can say that both direction selection and dual-window are necessary for deletion detection, but the second method has bigger space for improvements. By analyzing the reason, the direction selection can be seen the selection for preferential detection sample, and the non-main motion direction is only more sensitive to peaks. But the dual-window optimizes the detection algorithm essentially and it immediately affects the positioned pecks correctly.

## 4 CONCLUSIONS

The frame deletion video forensics is very important and necessary for our daily life. In this paper a dual-window detection scheme taking motion direction into consideration based on optical flow consistency is proposed and the contrast experiment is done to prove that this algorithm is effective. This paper notices the subtle difference between optical flow in x and y directions, and proposed a simple direction selection scheme for detection. Besides, the dual-window reduces the influence of interferential peaks and improves the detection accuracy. Experiment shows that the TPR can reach 85%~95%, and the FAR is extraordinary low, which can be considered as a significant detection scheme. Future work will focus on the more complex self-adapting direction selection and more application for different video forensics.

ACKNOLEDGEMENTS

This work is supported by the Major State Basic Research Development Program (No.2011CB302200) and the National Natural Science Foundation of China (No. 41371402).

REFERENCES

[1] Chunhui Feng, Zhengquan Xu, Xinghui Zheng, Li Jiang. The digital visual media forensics [J]. Journal on Communication, 2014(4):155–165.
[2] Lin Youru; Huang Huiyu; Hsu Wenhsing. An embedded watermark technique in video for copyright Protection, 2006.
[3] De Roover, De Vleeschouwer, Lefebvre, Macq. Robust video hashing based on radial projections of key frames [J]. IEEE Signal Processing Society, 2005(10):4020–4037.
[4] W. Wang and Hany Farid. Exposing digital forgeries in interlaced and de-interlaced video [J]. IEEE Transactions on Information Forensics and Security, 2007(3).
[5] Yunlong Qin, Guanglin Sun, Xinpeng Zhang. Exposing digital forgeries in video via motion vectors [J]. CIHW2009.
[6] W. Wang and Hany Farid. Exposing digital forgeries in video by detecting double MPEG compression. MM&Sec'06 Proceedings of the 8th workshop on Multimedia and security, 2006:37–47.
[7] M. C. Stamm, W. S. Lin, and K. J. R. Liu. Temporal Forensics and Anti-Forensics for Motion Compensated Video. IEEE Transactions on Information Forensics and Security, 7(4): 1315–1329, 2012.
[8] Chunhui Feng, Zhengquan Xu, Wenting Zhang, Yanyan Xu. Automatic location of frame deletion point for digital video forensics. The 2nd ACM Workshop on Information Hiding and Multimedia Security.
[9] J. Chao, X. Jiang, and T. Sun. A Novel Video Inter-frame Forgery Model Detection Scheme Based on Optical Flow Consistency. Digital Forensics and Watermarking, LNCS 7809: 267–281, 2013.
[10] Guoliang Yang, Zhiliang Wang, Shitang Mou, Lun Xie, Jiwei Liu. An improved optical flow algorithm[J]. Computer Engineering, 2006(8):187–188.

*Multimedia, Communication and Computing Application – Leung (Ed.)*
© 2015 Taylor & Francis Group, London, ISBN 978-1-138-02775-6

# A SVM-based spectrum sensing method

X.P. Zhai & Q.M. Liu
*Key laboratory of Specialty Fiber Optics and Optical Access Network, Shanghai University, Shanghai, China*

ABSTRACT: The detecting performance under low SNR is crucial for a spectrum sensing method. In this paper, the Support Vector Machine (SVM) had been adopted to solve the spectrum sensing problem. The model of spectrum sensing was proposed. The choice of the features and the kernel functions are discussed which affect the sensing performance greatly. The simulation results show that the proposed method owns excellent performance under low SNR conditions. Most importantly, the optimal separating hyperplane obtained by training a time can be shared between different SNRs and users.

KEYWORDS: Cognitive radio, Spectrum sensing, Support vector machine

## 1 INTODUCTION

With the rapid development of wireless communication services, the conflict between spectrum resources and customer's demand grows more and more acutely (Guo, C.L. et al. 2010). In order to improve spectrum utilization, cognitive radio (CR) was originally proposed by Dr. Joseph Mitola. Spectrum sensing is a key technology in cognitive radio.

Besides some traditional methods, such as energy detection (Haroon Rasheed et al. 2010), matched filter detection (Zhu, J.H. et al, 2010) and cyclostationary feature detection, many new methods (Wang, H.Q. et al. 2009, Perez-Neira, A.I. et al. 2009) were proposed in recent years. Zeng Y.H. & Liang Y.C. (2008) defined the ratio of Maximum-Minimum Eigenvalue (MME) as the test statistic, combing with the random matrix theory to achieve spectrum sensing. Theoretical analysis and simulation results showed that the MME method can obviously outperform the classical energy detection, and overcome the noise uncertainty problem (Liu Z.W. et al. 2010). However, it needs to make asymptotic assumption of sampling dimension. So the threshold was asymptotic threshold. It couldn't perform well in low SNR. The modified algorithm based on MME (MMME method) was proposed in (Wang R.L. et al. 2012). Compared with the MME method, it has improved sensing performance. A SVM-based spectrum sensing method in (Zhang D.D. & Zhai X.P. 2011) was proposed which

adopted the SVM to deal with the spectrum sensing problem. In (Zhang D.D. & Zhai X.P. 2011), the collected raw data were used directly as the training sets and the testing sets of SVM and no any feather was extracted and applied.

To verify the effectiveness of the SVM-based spectrum sensing method further, we have studied the effect of feather in this paper. And a SVM-based maximum-minimum Eigenvalue (SMME) method was proposed.

## 2 THE THEORY ABOUT SVM

SVM (Vapnik V.N. 2000) is a popular kernel based classification algorithm in hyperspectral image classification. It has been widely applied to face detection, web pages classification and so on (Burges J.C. 1999).

### 2.1 *Linearly separable classification*

The essential problem of SVM is to find optimal separating hyperplane. In Figure 1, two different shapes represent two groups of samples. Let training data be $D = \{(x_1, y_1), ..., (x_i, y_i)\}, x_i \in R^n, y \in \{-1, 1\}$. $x_i$ represents the training data vector and $y_i$ denotes the class of $x_i$. SVM tries to find the optimal separating hyperplane that maximizes the margin between the closest training samples and the separating hyperplane.

Then the classification is formulated as a quadratic optimization problem

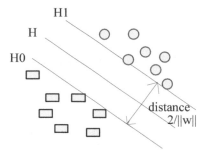

Figure 1. SVM linearly separable principle scheme.

$$\min_{w} \quad \frac{\|w\|^2}{2}$$

$$s.t. \quad y_i(wx_i + b) - 1 \geq 0, (i = 1, 2, ..., l) \quad (1)$$

Equation (1) is solved by Lagrangian function

$$L(w, b, a) = \frac{\|w\|^2}{2} - \sum_{i=1}^{l} a_i[y_i(wx_i + b) - 1], a_i > 0 \quad (2)$$

$a_i (i = 1, 2, ..., l)$ is Lagrangian factors. The dual representation of the above optimization problem is

$$\min_{w} \quad \frac{1}{2} \sum_{i=1}^{l} \sum_{j=1}^{l} y_i y_j (x_i x_j) a_i a_j - \sum_{j=1}^{l} a_j$$

$$s.t. \quad \sum_{i=1}^{l} y_i a_i = 0, a_i \geq 0 \quad (3)$$

The classifier for sample $x$ can be expressed as

$$y = \text{sgn}[\sum_{i=1}^{l} y_i a_i (x_i x) + b], \text{sgn}(a) = \begin{cases} 1 & a \geq 0 \\ -1 & a < 0 \end{cases}.$$

## 2.2 Linearly non-separable classification

Linearly separable classification can't work when two groups of data are mixed. Then the slack variables $\xi_i$ are introduced for separating input vector. So the quadratic optimization record as

$$\min_{w} \quad \frac{\|w\|^2}{2} + C \sum_{i=1}^{l} \xi_i$$

$$s.t. \quad y_i(wx_i + b) \geq 1 - \xi_i$$

$$\xi_i \geq 0, i = 1, 2, ..., l \quad (4)$$

The non-linear classifier for a sample $x$ can be expressed as

$$y = \text{sgn}[\sum_{i=1}^{l} a_i y_j (\Phi(x_i) \cdot \Phi(x_j)) + b].$$

$K(x_i, x_j) = [\Phi(x_i) \cdot \Phi(x_j)]$ is the kernel function. $x_i, x_j$ are random input vectors.

Figure 2 shows an example of a nonlinear problem in low dimensional space becoming a linear problem in high dimensional space. The left figure in Figure 2 shows the nonlinear data sets in low dimensional space. The right figure in Figure 2 shows the linear data sets in high dimensional space after the polar conversion.

## 2.3 Selection of the feature

Feature extraction is an important pre-process step. It can improve the classification performance and generalization ability.

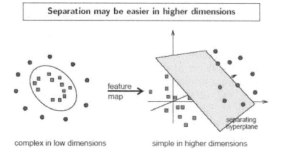

Figure 2. The polar conversion.

The main idea of SVM is to construct an optimal separating hyperplane to separate the different classes. Although all features can be put into SVM for training, inappropriate features will possess poor performance and generalization ability. In this paper, we choose the ratio of maximum-minimum Eigenvalue of covariance matrix of a signal as the valid feature.

## 2.4 Selection of the kernel function and kernel parameter

The kernel function and its parameters play an important role in SVM. Popular kernels are the Liner kernel function, the Polynomial kernel function, the Radial Basis Function (RBF) kernel, and the Sigmoid kernel function.

Lots of research has shown that the RBF $K(x_i,x) = \exp(-\|x_i - x\|^2 / 2\sigma^2)$ is preferred ($\delta$ is the kernel parameter). Firstly, whether the dimensional space is high or low, the number of samples is large or small, RBF always possesses good performance (Li P.C. & Xu S.H. 2005). KEERTHI S.S. & LIN C.J. (2003) proved that the liner kernel function and the polynomial kernel function are both the special cases of the RBF. Lin H.T & Lin C.J. (2003) stated that the performance of RBF is like the performance of the Sigmoid kernel function in some cases with different parameters. Secondly, compared with the other kernel functions, the kernel parameters are simple. So we choose RBF in this paper.

## 3   SVM-BASED SPECTRUM SENSING METHOD

The SVM-based spectrum sensing model is described as follows.

(1) According to the observations of the users' signal and the noise for a while, samples are gathered under condition the user exists and the user doesn't exist. If the primary user exists, the sampled signal is the sum of the user's signal and noise with different SNR. If the primary user doesn't exist, the sampled signal is only noise.

(2) Vectore the sampled signal and choose an appropriate feature. Compute the feature of these sampled data to get training data. These two group training data are labeled to $y=1$ and $y=-1$. Put the training set into SVM for training to construct the optimal separating hyperplane.

(3) Just like step (1), continue to gather signal samples at random under two different conditions, and compute the feature of these sampled data to get test data. we use the optimal separating hyperplane which is got in (2) to detect the test data to get the result of judgment. So we can get the conclusion whether the primary user exists or not.

To verify its feasibility, this paper chooses the ratio of maximum-minimum eigenvalue of covariance matrix of a signal as the valid feature. We call this as SVM-based maximum-minimum Eigenvalue sensing (SMME) method.

Let $y(i)$ be the received signal samples at time $i$. $n(i)$ is a white gauss noise with mean zero, variance $\delta^2$. $s(i)$ is the licensed signal to be detected. $h(i)$ is channel response between licensed user and the receiver. The detection model can be described as

$$y(i) = \begin{cases} n(i) & H_0 \\ h(i)s(i) + n(i) & H_1 \end{cases} \tag{5}$$

The receiver gets $N_t$ signal samples in sensing period.

$$Y = \begin{bmatrix} y(1) & y(2) & \cdots & y(N_t) \end{bmatrix} \tag{6}$$

Through vectorizing $Y = [y(1)y(2)\cdots y(N_t)]$, we can get matrices $Y_1$ whose dimension is $M \times N$. $N_t = M \times N$. We have defined statistical covariance matrix of licensed signals to be detected and received signals of cognitive users as $R_s = E[ss^T]$, $R = E[Y_1 Y_1^T]$, channel matrix as $H = [h_1,...,h_l]^T$ (Wang R.L. et al. 2012). Now $R$ can be expressed as

$$R = \begin{cases} \delta_n^2 I & H_0 \\ HR_s H^T + \delta_n^2 I & H_1 \end{cases} \tag{7}$$

Where $H$ denotes the channel gain matrix. Then the corresponding ratio of the maximum eigenvalue and the minimum eigenvalue of $R$ can be expressed as

$$\lambda_{max}/\lambda_{min} = \begin{cases} 1, & H_0 \\ (\rho_{max} + \delta_n^2) \Big/ (\rho_{min} + \delta_n^2) > 1, & H_1 \end{cases} \tag{8}$$

$\rho_{max}$ and $\rho_{min}$ is respectively the maximum eigenvalue and minimum eigenvalue of statistical covariance matrix $HR_s H^T$ of licensed signals over a channel. So it can judge whether licensed users exist or not by the ratio of maximum-minimum eigenvalue of $R$.

## 4   PERFORMANCE ANALYSIS

### 4.1   Simulation of kernel parameter

This paper chooses the optimal kernel parameter by Matlab simulation. The number of the training data is 400 and that of the test data is also 400. The signal of the primary user is BPSK modulated signal. The symbol rate is 1000Hz. The sampling rate is three times of the symbol rate. The noise is white Gauss noise. The value of $N$ is set to 500 and 1000. $M$ is set to 10 to make comparison with (Wang R.L. et al. 2012).

Table 1 shows the relationship between the error probability ($p_e$) and the kernel parameter $p$. SNR is respectively −20db, −17db, −15db and −10db. As is shown in the table, when $p = 0.2$, $pe$ gets the minimum value when $N$ is 500 and 1000. So we get $p = 0.2$.

Table 1. The error probability $p_{eo}$.

| $p$ | N = 500 | | | N = 1000 | | |
|---|---|---|---|---|---|---|
| | SNR = −20 | SNR = −17 | SNR = −15 | SNR = −20 | SNR = −17 | SNR = −15 |
| 0.1 | 0.42 | 0.235 | 0.17 | 0.365 | 0.145 | 0.05 |
| 0.2 | 0.36 | 0.2 | 0.095 | 0.32 | 0.135 | 0.045 |
| 0.3 | 0.365 | 0.215 | 0.105 | 0.355 | 0.145 | 0.045 |
| 0.4 | 0.4 | 0.2 | 0.095 | 0.365 | 0.145 | 0.045 |
| 0.5 | 0.415 | 0.205 | 0.09 | 0.355 | 0.145 | 0.045 |
| 0.6 | 0.415 | 0.245 | 0.009 | 0.33 | 0.145 | 0.045 |

### 4.2 Performance with different N

We choose the ratio of the maximum eigenvalue and the minimum eigenvalue of the sample covariance matrix as the feature and the RBF as the kernel function. Let $p = 0.2$. The simulation conditions are the same as above (The data of the MMME method refers to Wang R.L. et al. (2012)).

The number of signal samples is $N$ = 100, 200, 500, 1000, 2000, respectively. SNR is from −20db to 0db.The step is 1db. Figure 3 shows the detection performance of the MMME method and the proposed SMME method. As we can see, the detection performance is affected by the number of signal samples. The detection probability ($p_d$) is proportional to the number of signal samples, i.e., the detection duration. It is clear that the SMME method outperforms greatly the MMME method in low SNR.

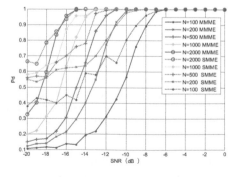

Figure 3. Detection performance comparison.

### 4.3 Performance of SMME method

Figure 4 shows the performance comparison of the SMME method and the MMME method in the same simulation condition. The false alarm probability $p_f$, the detection probability $p_d$, the missing probability $p_m$ and the error probability $p_e$ of the SMME method are shown. Only $p_d$ of the MMME method is included as we can't get other performance data from (Wang R.L. et al. 2012). Let $N = 2000$, and $M = 10$. The SNR varies from −20dB to 0dB. The step is 1db. From Figure 4 we can see that the detection performance of those two methods is similar when the SNR is greater than −14dB and the SMME method is much better than the MMME method when the SNR is lower than −14db. The $p_e$ of the SMME method is equal to 0.2 when the SNR is equal to −18dB. When the SNR is equal to −15dB, $p_d$ and $p_e$ of the SMME method reaches 1 and 0.005, respectively. So the SMME method can possess good performance in very low SNR.

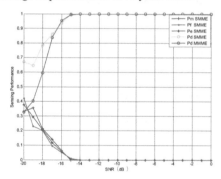

Figure 4. Performance comparison.

One key point in SVM performances lies in the separation of the learning set and the testing set. Clearly, if we use the same simulation and just split resulting dataset in two sets, it must have good performances. The following simulation illustrates that one learning set can be shared between different SNRs. Further we can derive that one learning set can be shared between different users. The simulation is as follows. The SNR varies from -20dB to 0dB. The step is 1dB. Let $N$=1000 and $M$=20. The Training data number for every SNR is 100 and the testing data number for every SNR is 400. After being trained once, the optimal separating hyperplane is obtained, and the learning machine can be used to classify signal in different SNR. Figure 5 shows the performance (the detection probability, the missing probability, the false alarm probability and the error probability) of the SMME method when one learning set is shared. We can see that the SMME method can perform very well.

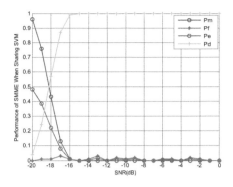

Figure 5. Performance of the SMME method when sharing the same training set.

## 5 CONCLUSION

A SVM-based sensing method is proposed. The computing complexity of the proposed method is very limited. To verify the effectiveness of SVM for spectrum sensing, we chose the ratio of the maximum-minimum Eigenvalue of covariance matrix of the received signal as the valid feature. The SMME method overcomes the defects of setting the fixed threshold in the MME method which is inaccurate and complicated. The simulations and the theoretical results showed that, with a slight increase of complexity, the proposed method greatly improves the sensing performance. The SVM-based method could be shared in different SNRs/ users after being trained one time. So it is robust to noise uncertainty. Furthermore, the ratio of the maximum-minimum Eigenvalue can be chose as the valid feature. Besides, other test statistics of spectrum sensing methods can be treated as the valid feature.

## ACKNOWLEDGMENTS

This work was supported by The National Natural Science Foundation of China (61171085), and the Shanghai Committee of Science and Technology project (Grant No. 12510500600), China.

## REFERENCES

Burges J.C. 1999. *A tutorial on Support Vector Machines for Pattern Recognition*. Kluwer Academic Publishers, Boston.
Guo,C.L. et al. 2010. *Cognitive radio network technology*. Beijing: Electronic Industry Press
Haroon Rasheed et al. 2010. Spectrum Sensing with Energy Detection under Shadow-fading Condition. *5th IEEE International Symposium on Wireless Pervasive Computing (ISWPC):* 104–109.
KEERTHI S.S. & LIN C.J. 2003. Asymptotic behaviors of support vector machines with Gaussian Kernel. *Neural Computation* 15(7):1667–1689.
Li P.C. & Xu S.H. 2005. Support vector machine and kernel function characteristic analysis in pattern recognition. *Computer Engineering and Design*, 26(2):302–304.
Lin H.T & Lin C.J. 2003. *A study on sigmoid kernels for SVM and the training of non-PSD kernels by SMO-type Methods*. Department of Computer Science and Information Engineering, National Taiwan University.
Liu Z.W. et al. 2010. Simulation and analysis of Eigenvalue-based spectrum sensing algorithms for cognitive radio networks. *Journal of system simulation* (12): 2805–2808.
Perez-Neira, A.I. et al. 2009. Correlation matching approach for spectrum sensing in open spectrum communications, *Signal Processing, IEEE Transaction on* Vol.57: 4823–4836.
Vapnik V. N. 2000. *The Nature of Statistical Learning Theory*. Springer Inc.
Wang, H.Q. et al. 2009. Spectrum sensing in cognitive radio using goodness of fit testing. *Wireless communication, IEEE Transaction on*. Vol.8: 5427–5430.
Wang R.L. et al. 2012. Modified algorithm for spectrum sensing based on MME. *Application Research of Computers* 29(7).
Zeng Y.H. & Liang Y.C. 2008. Eigenvalue based spectrum sensing algorithm for cognitive radio. *IEEE Trans on Communication*, 57(6):1784–1793.
Zhang D.D. & Zhai X.P. 2011. SVM-based spectrum sensing in cognitive radio. *IEEE 2011 7th International Conference on Wireless Communications, Networking and Mobile Computing (WiCOM)*, Sept. 2011:1–4.
Zhu, J.H. et al. 2010. *Cognitive radio based on circulation stationary feature detection*. Network Technology.

*Multimedia, Communication and Computing Application – Leung (Ed.)*
© *2015 Taylor & Francis Group, London, ISBN 978-1-138-02775-6*

# Infrared thermal wave image segmentation based on genetic algorithm and maximal between-class variance

D.D. Wang, W. Zhang, S.J. Tao, G. Tian & Z.W. Yang
*Xi'an Research Institute of High Technology, Hongqing Town, Xi'an, China*

ABSTRACT: The traditional Otsu threshold segmentation method is only applied to the images whose target and background have an even distribution; there were some limitations when applied to image segmentation of the infrared thermal wave inspection. The basic principle of the traditional Otsu thresholding method was analyzed first, and then a new Otsu threshold segmentation method based on the Genetic Algorithm (GA) was put forward and used in the damage segmentation and extract of the infrared thermal wave detection. The results of the image segmentation showed that the disadvantages in thermal wave image segmentation of the traditional Otsu method such as quantities of calculation and low velocity were solved by the new method, and the segmentation results are better to maintain the integrity of damage and laid foundation for the infrared thermal wave detection technology in structural damage assessment.

## 1 INTRODUCTION

Infrared thermal wave inspection technology as a new style and practical nondestructive testing method had been widely used in aerospace [1], electric power [2], petrochemical [3], civil construction [4], and other industries because of its fast speed; high efficiency; test results intuitive; and easily to be identified and assessed the damage of tested structures to improve the quality of materials, design of product, manufacturing, product testing, and online in-service inspection [5]. The image sequences captured by thermal camera contain abundant defective information, which is used for recognizing the defects. However, the problems of high ground and noise and low contrast are existed in the original image sequence and seriously affect the interpretation of damage from test structures in the thermal image. So the images obtained from experiments would be dealt with at first, which would enhance the contrast of the defect in image, and also the damage regions were segmented and extracted from images in order to facilitate identification and quantitative assessment of damage in the subsequent structure.

Image segmentation of infrared thermal wave images is an important research content of the thermal image processing, and many ways to segment the thermal image had been proposed at present [6-8]. However, the threshold segmentation as a basic image segmentation method, its application in the thermal wave image segmentation has not been carried out in the corresponding research. In this paper, the low-pass filter was used to enhance the contrast

of the damage region based on the thermal image obtained from the experiments, and then the threshold segmentation method based on GA-Otsu was carried out to segment the image after the enhancement. The algorithm proposed in this paper overcomes the disadvantages of traditional Otsu threshold segmentation method, such as large calculation, and easy to fall into local optimal solution. And the segmentation tasks of thermal images were well completed and laid the foundation for the quantitative identification and assessment of the damage in the structure.

## 2 OTSU THRESHOLD SEGMENTATION METHOD

Otsu method is an adaptive threshold segmentation method, which was proposed by the Japanese scholars of Otsu in 1979 [9], referred to as OTSU method. This method is derived from the least squares method, and the processing images were divided into two parts including the target and background according to the gray characteristics of the image. The between-class variance between the target and background is more larger, which indicates the difference between the two parts constituting the image is larger, and the distinction between the two parts would be diminished in the way that part of the target could be divided into the background in error or part of the background could be divided into the target in error, so that the maximum between-class variance of segmentation threshold means that the probability of misclassification in this case is minimum, and the value of the threshold is the best segmentation threshold.

Setting the gray level of the segmented images into $L$, $n_i$ is the number of the gray value $i$, so the total number of pixels of the image are:

$$N = \sum_{i=0}^{L-1} n_i \qquad (1)$$

So the probability of the each gray value is

$$p_i = \frac{n_i}{N} \qquad (2)$$

An initial threshold value $Th$ was given firstly, and the image was divided into two parts $R_1$ and $R_2$, where the probability of those two regions is

$$P_1 = \sum_{i=0}^{Th} p_i \ , \ P2 = 1 - P1 \qquad (3)$$

The gray mean $\mu_1$ and $\mu_2$ of the image after the initial threshold value segmentation of the two regions $R_1$ and $R_2$ are

$$\mu_1 = \frac{\sum_{i=0}^{Th} n_i * i}{\sum_{i=0}^{Th} n_i}, \mu_2 = \frac{\sum_{i=Th+1}^{L-1} n_i * i}{\sum_{i=Th+1}^{L-1} n_i} \qquad (4)$$

The overall mean gray of the image is

$$\mu = \sum_{i=0}^{L-1} p_i * i \qquad (5)$$

The variance $d(Th)$ between the two parts of the image was calculated as

$$d(Th) = P_1 \cdot (\mu_1 - \mu)^2 + P_2 \cdot (\mu_2 - \mu)^2 \qquad (6)$$

The formula (5) was taken into formula (6) and could be obtained as follows

$$d(Th) = P_1 \cdot P_2 \cdot (\mu_1 - \mu_2)^2 \qquad (7)$$

The value of the threshold $Th$ was changed from 0 to $L-1$ and then the optimal segmentation threshold $Th_{new}$ was obtained in accordance with the threshold value so that the image was divided into the two regions $R_1$ and $R_2$, which were filled with $d(Th_{new}) = \max(d(Th))$. The best segmentation results of the image could be obtained through segmenting the image by the threshold $Th_{new}$.

## 3 OTSU THRESHOLD SEGMENTATION BASED ON GA

The kernel of the image segmentation method was analyzed above by calculating the between-class variance of the two parts of image, but in the course of processing, the more the gray level of image, the more the number of calculating the variance, so the time of selecting thresholding value was much longer, and the efficiency of the image segmentation was much lower. To solve the problem of achieving the best optimal segmentation value in the Otsu method, the intelligent algorithm GA was introduced into the solution process of achieving the best optimal segmentation value in order to reach the goal of global optimum and to quickly solve the problem.

### 3.1 Basic principles of GA

(See artwork document) Genetic Algorithm (GA) was first proposed by Professor John H Holland in the University of Michigan in America [10], and this kind of search algorithm of "generating + testing" characteristics was formed through simulating biological genetic and evolutionary processes in the natural environment. The parameter space of the problem was replaced by the encoding space in GA, and to the fitness function for evaluation, based on the evolution of coding group, with individuals in the group and a series of genetic operation selection and genetic mechanism, and the iterative process was built. In this process, the bit string of new generation is superior to the older generation by randomly reorganizing the important genes in the bit string, and the individual groups evolved gradually close to the optimal solution and ultimately reach the purpose of solving problems [11]. Although the genetic operations of the GA is random in the entire evolutionary process, the characteristics which showed in the process are not completely a random search, and it could effectively use historical information to predict the next generation of expected performance improved set of finding advantages. So the final convergences to generations of evolution, one of the most adapted individual to the environment, calculate the optimal solution of the problem.

In the process of solving the optimization problem using GA method, in order to achieve the maximum or minimum of the objective function, the multiply

292

optimization had begun from a set of initial values (i.e., a group), the process generally include reproductive competition, hybridization, and mutation, and the basic process of optimization operation was shown in Figure 1.

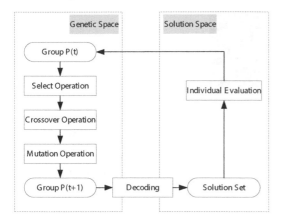

Figure 1. Diagram of the operation of genetic algorithm process.

## 3.2 Threshold segmentation method of the Otsu based on the GA

Otsu thresholding method is to seek the optimal segmentation threshold of the image in the solution space and to make its variance achieve maximum. GA, as an intelligent algorithm, is nonlinear and can quickly find the optimal solution and maximum variance of the problem, so the GA method was introduced into the traditional Otsu threshold method and uses the optimization ability of GA to solve the problem of seeking thermal wave image segmentation threshold value.

1  Numerical code of the solution space and produce the chromosome unit. The thermal images gained from the infrared thermal wave testing experiments are gray images and its gray scale range from 0 to 255, with a total of 256 gray levels, just corresponding to an 8-bit binary, which is one byte, so the chromosome can be used instead of a byte.
2  Initializing the population and producing a chromosome population with the fixed size, then each chromosome is initialized randomly, after that a number of different chromosomes could be gained. The initial value of the optimal threshold optimization process is determined in the process, and if the initial value is selected in partial, it would result in converging slowly and calculating for a longer time and other issues.
3  Encoding each chromosome. Set variance as the evaluation function (also known as the fitness function) of chromosome, and the greater the

variance of chromosome, the closer to the optimal solution.
4  Genetic operators: hybridization and mutation. The hybridization is carried out first, and its essence is that certain genes of the chromosomes exchanged, but one rate of hybridization needed to be given for controlling the number of bits. The greater the hybridization rate, the more the exchanged gene, and the faster the value, the faster the convergence rate of the solution. But the rate of hybridization is too big, and it is not conducive to get the optimal solution. Variation is carried out in accordance with the variation rate of each chromosome mutation operation and the new population of chromosomes was produced after the calculation.
5  Reproduction. Repeat the above process for the newly generated populations, with the continuous generation of populations, which results in a new value, to obtain the optimal solution with the GA method in the process of continuous reproduction.

According to the steps of the above calculation, the flowchart of the threshold segmentation method of the Otsu based on the GA was shown in Figure 2.

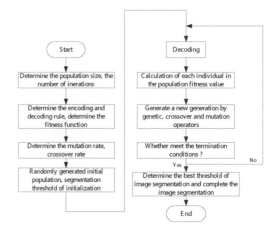

Figure 2. Flowchart of Otsu threshold segmentation based on GA.

## 4  EXPERIMENTS AND ANALYSIS OF RESULTS

### 4.1 Experiments

For the common problems of debonding and impact damage in the composite structures, the specimens of carbon fiber composite were made for the experiments, and the detected research were carried out on the infrared thermal wave nondestructive testing system by our own independent development; the original infrared thermal wave images were shown in Figure 3.

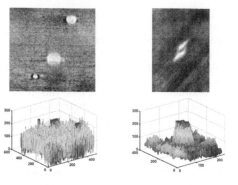

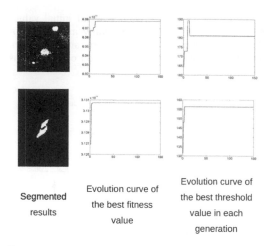

Figure 3. Original images and their three-dimensional graphs.

**Segmented results** | Evolution curve of the best fitness value | Evolution curve of the best threshold value in each generation

Figure 5. Segmented results of the infrared thermal wave images based on the method proposed in paper.

### 4.2 *Process of experimental results/deal with the experimental results*

There are some disadvantages such as high noise and low contrast in the original thermal images, so for improving the visibility and enhancing the contrast of the images and make it convenient for extracting defects in the image segmentation, the low-pass filter was used to remove noise from the image to improve image quality, and the results of image enhancement were shown in Figure 4.

In order to verify the effectiveness of the method proposed in this paper, the thermal wave images were used, and the conventional threshold segmented method was used to the same images and the segmented results were shown in Figure 6.

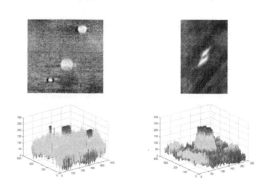

Figure 4. Images after low-pass filtering and their three-dimensional graphs.

The method proposed in this paper was used on the enhanced image for image segmentation, in which the length of chromosomes was 5, the size of the population was 10, and the probability of crossover and mutation were 0.7 and 0.4, respectively. The enhanced images were processed with the method according to the above steps and the segmented results were shown in Figure 5.

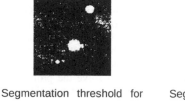

Segmentation threshold for 160 | Segmentation threshold for 100

Figure 6. Results of the defects segmentation based on threshold.

## 5 CONCLUSIONS

For the problem of the nonuniformity in the original infrared thermal wave testing images which the traditional Otsu method cannot solve, the threshold segmentation method of the Otsu based on the Genetic Algorithm was proposed in this paper. First the low-pass filter was used to eliminate the noise from the original images and improve the contrast, and then the enhanced images were segmented by the method of GA-OTSU. The segmented results showed that the new method could well extract the damage area from the image, keep the complete edge information of the defects, and lay foundation for further study on the structural damage identification and life assessment.

ACKNOWLEDGMENTS

We gratefully acknowledge the National Nature Science Foundation of China (No. 51275518, No. 51305447) and Natural Science Basic Research Plan in Shaanxi Province of China (No. 2013JM7021) for support for this project.

REFERENCES

[1] Avdelidis N. P., Almond D. P., Dobbinson A., et al, 2004. Aircraft composites assessment by means of transient thermal NDT. Progress in Aerospace Sciences, 40(3):143–162.

[2] Guoan Zhou, 2007. Application of Infrared Technology in Electrical Equipment Inspection. Infrared, 5:36–39.

[3] Yongwang Tang, 2008. Study of Infrared Detection and Diagnosis Technology in Petrochemical Hot Equipment. Kunming: Kunming University of Science & Technology.

[4] Eva Barreira, Vasco P. de Freitas, 2007. Evaluation of building materials using infrared thermography. Construction and Building Materials, 21(1): 218–224.

[5] Yanhong Li, Yuejin Zhao, Lichun Feng, et al, 2008. Pulse Phase Analysis for the Depth Measurement in Thermal Wave Nondestructive Evaluation. Transactions of Beijing Institute of Technology, 28(2):146–149.

[6] Wei Zhang, Fahai Cai, Baomin Ma, et al, 2009. Quantitative analysis of infrared thermal image defect based on mathematical morphology. Non-destructive testing, 31(8):596–599.

[7] Guofeng Jin, Wei Zhang, Zhengwei Yang, et al, 2012. Image Segmentation of Thermal Waving Inspection based on Particle Swarm Optimization Fuzzy Clustering Algorithm. Measurement Science Review, 12(6):296–301.

[8] Dongdong Wang, Wei Zhang, Guofeng Jin, et al, 2014. Application of cusp catastrophic theory in image segmentation of infrared thermal wave inspection. Infrared and Laser Engineering, 43(3):1009–1015.

[9] Otsu N, 1979. A threshold selection method from gray-level histograms. IEEE Transactions on Systems, Man, and Cyber-netics, 9(1):919–926.

[10] Holland J H, 1992. Adaptation in Nature and Artificial Systems. MA: MIT Press, 1992.

*Multimedia, Communication and Computing Application – Leung (Ed.)*
© *2015 Taylor & Francis Group, London, ISBN 978-1-138-02775-6*

# A ship emphasized bag-of-words model for ship detection in SAR images

S.H. Wan, Q. Huang, P.Q. Jin & L.H. Yue
*Key Laboratory of Electromagnetic Space Information, Chinese Academy of Sciences, School of Computer Science and Technology, University of Science and Technology of China, Hefei, Anhui, China*

ABSTRACT: Automatic detection of ships surrounded by complex sea clutters in SAR images is a challenging task. In this paper, we propose a novel ship emphasized Bag-of-Words (SEBOW) model to improve the representation of SAR images consisting of ships and sea clutters. For detecting ships, the approach includes a learning step and a detecting step. In the learning step, we train a classifier by the SEBOW-based representation of a training set. In the detecting step, we apply the classifier to each sub-image within a sliding window to check whether it contains ships, and then fuse the results to finally detect ships. The experimental results on the ERS-2 dataset testify to the effectiveness and efficiency of our method.

KEYWORDS: Ship detection, Bag-of-Words (BOW), SAR image

## 1 INTRODUCTION

Ship detection is widely used in maritime surveillance, fishery management, and vessel traffic monitoring and so on. With the significant progress of satellite-based remote sensing, vast images provided by the Synthetic Aperture Radar (SAR) are exploited for automatic ship detection, because SAR images are less influenced by illumination and weather conditions than optical images. However, ship detection in SAR images is still a challenging task in the presence of complex and diverse sea clutters.

The most common method, the Constant False Alarm Rate (CFAR), and its variant algorithms [1, 2, 3] detect ships in SAR images by searching for pixels that are brighter than their surrounding areas above a threshold. A probability density function (PDF) has to be designed in advance to model the statistic characteristics of sea clutters. However, the pre-defined distribution cannot suit all sea clutters, which vary considerably among images. Moreover, when ship targets have similar intensity with sea clutters, it is hard for the CFAR algorithm to determine a threshold to separate ships from sea clutters.

Recently, the Bag-of-Words (BOW) model, which represents an image as a histogram of quantized local image descriptors, has been introduced in the remote sensing field. For example, Xu et al. [4] proposed a BOW-based representation for object-based classification in land-use/cover mapping of aerial photographs. Lienou et al. [5] combined the BOW model with generative probabilistic model for annotation of large satellite images. For object detection, Sun et al. [6] used a sliding window method to detect airplanes

by augmenting structure information for the BOW model. However, due to the simple shapes of ship targets, the additional structure information cannot improve the performance of ship detection and much semantic information is also ignored.

In this paper, we propose a novel ship emphasized BOW (SEBOW) model to represent SAR images. Different from the conventional BOW model [7], our model labels each visual word according to its association with ships and sea clutters. Compared with the commonly used CFAR approach, our algorithm does not need to build different models for different sea clutters and thus is more robust and easy to use.

The remainder of this paper is organized as follows. In Section 2, a novel SEBOW model for ship detection is proposed. Section 3 describes the ship detection algorithm based on the SEBOW model. Finally, experimental results and conclusions are provided in Section 4 and Section 5.

## 2 SEBOW MODEL

As Figure 1 illustrates, the process in the dashed box is the BOW representation model, the gray box is our proposed SEBOW model. In our model, we need a training set $I_1, I_2, ..., I_N$ where $N$ is the number of training set. After generating the codebook $\{z_1, z_2, ..., z_K\}$, we label each visual word $z_k$ by $l_k$. Thus, we obtain the codebook with labels $\{(z_1, l_1), (z_2, l_2), ... , (z_K, l_K)\}$, where $K$ denotes the size of codebook. Finally, based on the codebook with labels, a SAR image is quantized into a vector representation $v=[v_{sea}, v_{ship1}, v_{ship2}, ... , v_{shipM}]$.

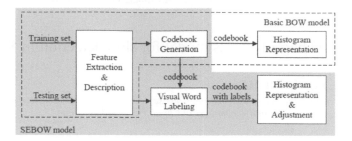

Figure 1.   Basic BOW model and SEBOW model.

## 2.1   *Feature extraction and description*

In the feature extraction and description step, we extract dense SIFT [8] features of all images in the training set. Dense SIFT is a dense version of SIFT [9], and work by Li [10] has shown that dense features work better for classification. Moreover, dense SIFT extracts SIFT descriptors at every position of the image while the regular SIFT descriptors are only extracted at some regions detected by DoG.

## 2.2   *Codebook generation and visual words labeling*

Given all descriptors extracted from the training set, we generate an initial codebook $\{z_1, z_2, \dots, z_K\}$ by K-means clustering, which consist of $K$ visual words and each visual word $z_k$ is represented by the center of each clusters in K-means.

Then, in order to augment more semantic characteristics for codebook clustered by K-means clustering, we plot all features assigned to a visual word $z_k$ in the training set to observe its distribution. As a result, there are only two kinds of distribution as is shown in Figure 2. The first kind (Fig.2 (a)) is a scatter of visual words on the entire image, while the other (Fig.2 (b)) exhibits more concentrated distribution located especially along the edges of potential ship targets. It is because dense SIFT features extracted on sea clutters are quite different with those extracted nearby ship targets that these features are usually clustered into different visual words. Therefore we label visual words which scatter sparsely as "sea clutter words", and visual words which only plot at particular areas as "ship words".

In detail, we first divide $I_i(i=1,\dots,N)$ into $m_i*n_i$ bins according to (1).

$$m_i = \left\lceil \frac{h_i}{b} \right\rceil, n_i = \left\lceil \frac{w_i}{b} \right\rceil \tag{1}$$

where $h_i$ and $w_i$ is the height and width of $I_i$, $b$ is the size of each bin.

Then we initialize the label of $z_k$, i.e. $l_k=-1$. For each $I_i(i=1,\dots,N)$, we update $l_k$ according (2).

$$l_k = \begin{cases} 1 & if & 0 < \lambda_k / (m_i \times n_i) < \omega \\ 0 & if & \lambda_k / (m_i \times n_i) \geq \omega, l_k \neq 1 \\ -1 & & othervise \end{cases} \tag{2}$$

Where $\lambda_k$ is the number of bins where the visual word $z_k$ appears.

Finally, if $l_k$ still equals $-1$, we set $l_k=1$.

## 2.3   *Histogram representation and adjustment*

It can be easily seen that the discriminated ship features and words are so few that they can be easily overlooked, if we count frequency of visual words as in the conventional BOW model. This fact leads to similar histogram representations of different SAR images with different ships. Therefore, we augment the influence of ship words in the SEBOW model. To represent a SAR image, we regard all sea clutter words $z_k$ ($l_k=0$) as one word $z_{sea}$, then count each visual word's frequency in the image. Finally we obtain the adjusted representation $v=[v_{sea}, v_{ship1}, v_{ship2}, \dots, v_{shipM}]$, where $M$ is the number of ship words in the codebook.

## 3   DETECTION APPROACH

Inspired by pedestrian detection [11], we scans SAR images by a detection window to detect ships as Figure 3 shows.

The overall detection approach consists of two steps: a learning step and a detecting step. The purpose of the learning step is to learn a classifier for the

(a) a sea clutter word          (b) a ship word

Figure 2.   Distribution of visual words.

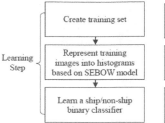

Figure 3. Ship detection scheme.

detecting step. We first create some SAR images as positive/negative training samples. The positive training images contain ship target at the center, and the negative training images do not. Figure 4 shows some representative training images. Then, we represent all training images based on the SEBOW model and learn a binary classifier using, for example, a support vector machine (SVM) with a RBF kernel.

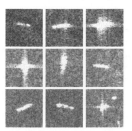 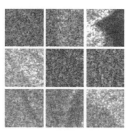

(a) Positive training images    (b) Negative training images

Figure 4. Training set.

During the detection step, each testing image is scanned by a detection window. At each state, the sub-image in the detection window is also represented as a histogram. Subsequently, the sub-image is classified for the presence/absence of ship candidates based on the learned classifier. Finally, since the detection window is sliding by an interval, it is common that some overlapping sub-images cover the same target as Figure 5, so we fuse the overlapping sub-images to achieve the final detecting results based on mean shift algorithm [12].

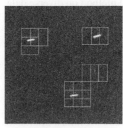 

(a) overlapped windows    (b) fusion results

Figure 5. Fusion.

## 4  EXPERIMENTS

Due to the lack of standard data sets of SAR images for ship detection, we verify the proposed method on the ERS-2 satellite data set which was acquired around southeast China in 2008 and 2009. The polarization mode is VV, and the spatial resolution is $12.5m$.

We cut 50 images from the data set to verify the effectiveness of the proposed method. In each testing image, there are several ship targets surrounded by different sea clutters. We also create 100 positive samples and 100 negative samples which are 60*60 pixels. The detection window size is also 60*60 pixels. In the fusion step, we use flat kernel and the parameter $h$ is set to be 30 in mean shift. The classifier was learned using SVM with a RBF kernel.

To quantify the detection results, we assume a ship is detected correctly if more than 75% of ship body is detected the same as [6]. We use *Recall* and *Precision* to evaluate its performance. *Recall=TP/(TP+FN)*, is the number of correctly detected ships divided by the total number of actual ships in the testing image. *Precision=TP/(TP+FP)*, is the number of correctly detected ships divided by the total number of detected targets.

### 4.4  *Comparison with CFAR algorithm*

The proposed method is compared with the classical ship detector, i.e. the CFAR detector, where Rayleigh distribution is used due to its proper description for serious sea clutters. The detection results of CFAR and proposed method are shown in Table 1. In our method, codebook size $K=300$, the interval of the sliding detection window is set to be 30. In CFAR method, the probability parameter is 0.999 and the scale parameter is computed in 128*128 pixels.

Table 1. Result of CFAR (Rayleigh) and our method.

| Detector | CFAR (Rayleigh) | Our method |
|---|---|---|
| Recall | 82.46% | 97.44% |
| Precision | 81.03% | 83.21% |

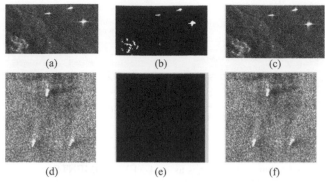

(a)    (b)    (c)

(d)    (e)    (f)

Figure 6.    Ship detection result. (a) (d) Original image. (b) (e) Result of CFAR. (c) (f) Result of our method.

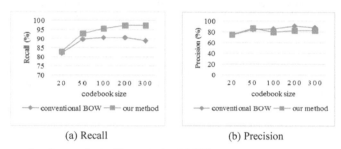

(a) Recall    (b) Precision

Figure 7.    Comparison results of our method with conventional BOW.

It can be seen that better results are obtained by our method. In particular, we use some images to exemplify the superior performance of our method in Figure 6. It is observed that the classical CFAR method cannot filter islands for Figure 6(a). In contrast, our method has detected three ship targets correctly and has taken the island as false targets. Moreover, when the target/clutter contrast is similar as is shown in the Figure 6(d), classical CFAR cannot detect any ship targets. In contrast, our method can still detect all ship targets correctly.

### 4.5    *Comparison with the conventional BOW model*

We also implemented the conventional BOW model [7] as the benchmark approach and compared our method with it. The size of codebook of two methods is set to be 20, 50, 100, 200 and 300. The interval of the sliding detection window is 30. Figure 7 illustrates the results.

It can be noticed that our method perform quite well in recall and similar results in precision. As codebook size increases, our method performs better while the conventional BOW performance stays almost the same.

## 5    CONCLUSION

In this paper, we propose a new SEBOW model to represent SAR images for ship detection. It augments more semantic characteristics for visual words and helps to separate SAR images containing ships from sea-only images.

Experimental results show that almost all the ship targets can be detected under an acceptable false alarm rate. Moreover, the comparison experiments indicate that our method outperforms the conventional CFAR algorithm and the basic BOW model.

At present, the proposed scheme is only measured under the SAR data. In the future, we will verify the effectiveness of the proposed method on diverse data, such as optical images.

## 6    ACKNOWLEDGEMENTS

This work was supported by the National Natural Science Foundation of China (Grant No. 61272317) and the General Program of Natural Science Foundation of Anhui of China (Grant No. 1208085MF90).

# REFERENCES

[1] Wang Y.H. & Liu H.W., 2012, A hierarchical ship detection scheme for high-resolution sar images, *IEEE Transactions on Geoscience and Remote Sensing*, 50:4173–4184.

[2] Ai J.Q., Qi X.Y., Yu W.D., Deng Y.K., Liu F. & Shi L., 2010,A new cfar ship detection algorithm based on 2-d joint log-normal distribution in sar images, *Geoscience and Remote Sensing Letters, IEEE*, 7( 4): 806–810.

[3] Xing X.W., Ji K.F., Zou H.X., Sun J.X. & Zhou S.L.,2011, High resolution sar imagery ship detection based on exs-c-cfar in alpha-stable clutters, *in Geoscience and Remote Sensing Symposium (IGARSS), 2011 IEEE International*. : 316–319.

[4] Xu S., Fang T., Li D.R. & Wang S.W., 2010, Object classification of aerial images with bag-of-visual words, *Geoscience and Remote Sensing Letters, IEEE*, 7(2): 366–370.

[5] Lienou M., Maitre H. & Datcu M., 2010,Semantic annotation of satellite images using latent dirichlet allocation, *Geoscience and Remote Sensing Letters, IEEE*, 7(1): 28–32.

[6] Sun H., Sun X., Wang H.Q., Li Y. & Li X.J., 2012, Automatic target detection in highresolution remote sensing images using spatial sparse coding bag-of-words model, *Geoscience and Remote Sensing Letters, IEEE*, 9(1): 109–113.

[7] Csurka G., Dance C., Fan L.X., Willamowski J. & Bray C., 2004, Visual categorization with bags of keypoints, *Workshop on statistical learning in computer vision, ECCV*, 1: 22.

[8] Vedaldi A. & Fulkerson B., 2010,Vlfeat: An open and portable library of computer vision algorithms, *Proceedings of the international conference on Multimedia. ACM*:1469–1472.

[9] Lowe, D.G., 2004, Distinctive image features from scaleinvariant keypoints, *International journal of computer vision*, 60( 2):91–110.

[10] Li F.F. & Perona P., 2005, A bayesian hierarchical model for learning natural scene categories, *Computer Vision and Pattern Recognition, CVPR. IEEE Computer Society Conference on. IEEE*, 2: 524–531.

[11] Dalal N. & Triggs B., 2005, Histograms of oriented gradients for human detection, *Computer Vision and Pattern Recognition (CVPR)*, 2005 IEEE Conference on. IEEE:886–893.

[12] Comaniciu D. & Meer P., 2002, Mean shift: A robust approach toward feature space analysis, *Pattern Analysis and Machine Intelligence, IEEE Transactions on*, 24(5): 603–619.

*Multimedia, Communication and Computing Application – Leung (Ed.)*
*© 2015 Taylor & Francis Group, London, ISBN 978-1-138-02775-6*

# The novel algorithm of dial pointer recognition based on image processing

J. Zhai, N.M. Wang & Z. Chai
*Science and Technology on Optical Radiation Laboratory, Beijing, China*

ABSTRACT: This paper proposes a novel and practical dial pointer recognition method. First, it determines the center of the label based on label positioning method using color information. As the displacement from the dial to the label is fixed, the center of the dial can be determined. Second, it gets the pointer thick end based on the circumference seeking method. Then the pointer thin end can be estimated by using the symmetric relationship. After further processing, the precise position of the pointer thin end can be got to calculate the readings of the dial. This algorithm takes advantage of the original label positioning method and circumference seeking method. It is of high stability, strong robustness and good practicability, opening up a new way of recognizing the pointer on the dial.

## 1 INSTRUCTION

Nowadays, the instruments and meters with pointer, such as: voltmeter, ammeter, oil level gauge, thermometer, etc., have been widely used in the process of industrial production. It is effective to monitor and read the operating status of the industrial production through the instruments and meters. To substitute for human to complete the heavy reading task, the meters dial recognition based on image processing emerges at the historic moment.

Many researchers are in their study (Zhou & Zhong 2003, Zhang et al. 2013, Matus et al. 2000) use the subtraction algorithm based on Hough transform. This method is of high accuracy. But it is generally not allowed to adjust the pointer to the benchmark position and Hough transform linear parameter estimation is not accurate, leading to impracticability. Paper (Sun et al. 2005) proposes the pointer angle recognition algorithm based on the searching of concentric ring. It selects the center and radius of the area and seeks the intersection of the pointer and concentric circles with certain step length. This method is simple and real-time. But in large workshops, the running large equipment will drive the dial joggled, leading to the collected images vague. At the same time, the instrument dial is often stuck with annual inspection certificate label. All of these make the algorithm powerless.

In the light of the specific applications mentioned above, this paper proposes a novel and practical method to recognize the needle of the instrument with pointer. The innovation is that this paper proposes the label positioning method and the circumference seeking method to implement the instrument pointer recognition. First, the label positioning method is used to determine the label of the circle. Then the pointer position is found by the circumference seeking method. After further processing, the precise location of the pointer can be got.

## 2 DETERMINING THE CENTER OF THE DIAL BASED ON LABEL POSITIONING METHOD

In large plants or gas supply system, the dial is always stuck with an annual check certificate label in the center and the background scene is complex, as shown in Figure 1. Restricted by the above factors, a lot of information about the dial is lost. This paper makes full use of the label which seems useless to position the dial of the circle accurately.

a        b

Figure 1. The dials in the specific circumstance.

### 2.1 *Extracting the label based on color information*

This paper uses the color extraction and image segmentation technologies to extract the label. The color extraction technology separates a certain color from other colors. Image segmentation divides the image into different characteristic parts and extracting the interesting targets. Using these two technologies lays a foundation for subsequent processing.

The label on instrument dials is generally chromatic, but the color of the dial region and background region is always gray (including white and black). When dividing the original image into R, G, B three channels, it is found that the values of the R, G, B channels of the gray region are closed but that of the chromatic region is quite different. By calculating the maximum of the difference between each two color channels, the label information can be got. See for Equation 1 below:

$$\max(|\mathbf{R} - \mathbf{G}|, |\mathbf{G} - \mathbf{B}|, |\mathbf{B} - \mathbf{R}|) \tag{1}$$

where $\mathbf{R}$, $\mathbf{G}$, $\mathbf{B}$ respectively represents the R, G, B three channels of the original image.

The label information image can be extracted, as shown in Figure 2.

a        b        c

Figure 2a.  The label information image.
Figure 2b.  The binary image of the label information image.
Figure 2c.  The label information image after morphology.

## 2.2 *Getting the position of the label*

The label information image is affected by noise and other factors, so the binarization to process the image is used. The binary image of the label information image is shown in Figure 2b. As seen in Figure 2b, some disturbance regions still exist.

As it can remove the noise and fill the holes in the image, morphology filtering is used. As seen in Figure 2c, after morphological filtering, the label information is more accurate.

The projection scanning method is used to get the position of the label accurately. Assuming that the image size is N × M, the projection scanning method refers to projecting the binary image onto the x axis and y axis and scanning the pixels respectively along the direction perpendicular to the axis. When scanning, this paper calculates the sum of pixels of which the brightness is 255, which is shown in Equation 2–4

$$n_i = \sum_{0 \le j \le M-1} f(i,j) \ , \quad 0 \le i \le N-1 \tag{2}$$

$$m_j = \sum_{0 \le i \le N-1} f(i,j) \ , \quad 0 \le j \le M-1 \tag{3}$$

$$f(i,j) = \begin{cases} 1, & I(i,j) = 255 \\ 0, & I(i,j) = 0 \end{cases} \tag{4}$$

where $n_i$ = the sum of pixel, the brightness of which is 255, when scanning the $i$ column, and $m_j$ = the sum of pixel, the brightness of which is 255 when scanning the $j$ row.

When the sum of the pixels is greater than the threshold for the first time, the label boundary is confirmed. Then the center and the radius of the label circle can be determined by the geometric method.

## 2.3 *Determining the center of the dial*

It needs the prior knowledge to determine the dial center. First, the dial center in the first frame should be got manually. Besides, the label center can be got by the method of Section 2.2. For the same dial, the relative displacement from the label center to the dial center is fixed. The relative displacement and the center position of the label to calculate the exact position of the dial center based on vector sum should be used.

The advantage is obvious. When the dial vibrates, this method can still find the dial center accurately.

## 3 DETERMINING THE POINTER POSITION BASED ON CIRCUMFERENCE SEEKING METHOD

Determining the pointer position is the key to dial pointer recognition. Traditional algorithms use Hough transform to detect pointer position regardless of the interference out of the dial, such as the pipelines in large gas supply plants. And as to get the images processed, the dial label will block part of the pointer. Besides, the images are blurred due to the vibration of the instrument. All these reasons could fail the traditional methods. To solve these problems, this paper proposes the circumference seeking method to get the pointer position.

### 3.1 *Getting the pointer thick end position*

This paper proposes the circumference seeking method based on morphology to get the pointer thick end position. First, image preprocessing technologies, such as image filtering, should be used to optimize the original image and eliminate the interference of the noise. Then the circumference of which the center is the center of the dial in Section 2.3 should be constructed. The radius of the circumference must be chosen appropriately based on the size of the label, making sure the circumference $O$ and the dial pointer intersect. As shown in Figure 3a, the gray region shows the label and the dotted circle mean the gotten circumference. And the circumference on the real image is shown in Figure 3b. The circumference includes the area of the label and intersects the pointer.

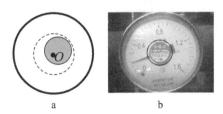

a                                    b

Figure 3.    The sketch map of the dial with the circumference.
Figure 3b.    The real image with the circumference.

After getting the circumference, there will be an intersection of the pointer and the circumference. As seen in Figure 4a, the white parts of the circumference are the intersections.

a                                    b

Figure 4a.    The intersections.
Figure 4b.    The center of the pointer thick end.

With the low number of pixels in it, the pointer thin end is easily affected by other noisy pixels or some other characters in the image. So this paper proposes determining the pointer thick end first and then getting the thin end by symmetry to reduce error remarkably.

First, the threshold should be determined to binary the pixels at the circumference. And morphology filtering should be used to remove the noise. Then the intersecting curve segments should be sought and recorded at the circumference, in which the longest one is regarded as the pointer thick end. As seen in Figure 4b, the black cross is the center of the pointer thick end.

The following is the detail introduction of circumference constructing. Circumference constructing is carried out counterclockwise respectively in accordance with four quadrants of the coordinate system.

Starting with the first pixel on the circumference in the first quadrant, the pixels in the second quadrant of the 9 neighborhood of the pixel should be sought, as

shown in Figure 5a. The distance should be calculated from the pixels to the center of the circumference, and the one of which distance is closest to the circumference radius should be regarded as the pixel on the circumference. The same goes to the rest quadrants. In the second quadrant, the pixels are sought in the third quadrant of the 9 neighborhood of the pixel, as shown in Figure 5b. In the third quadrant, the pixels in the fourth quadrant of the 9 neighborhood of the pixel are sought, as shown in Figure 5c. In the fourth quadrant, the pixels in the first quadrant of the 9 neighborhood of the pixel are sought, as shown in Figure 5d.

a            b            c            d

Figure 5a.    Circumference constructing in the first quadrant.
Figure 5b.    Circumference constructing in the second quadrant.
Figure 5c.    Circumference constructing in the third quadrant.
Figure 5d.    Circumference constructing in the fourth quadrant.

### 3.2    *Precisely calculating the center of the pointer thin end*

First, the center of the pointer thin end is estimated. Because the center of the pointer thin end is symmetric with the center of the pointer thick end with respect to the center of the dial, the center of the pointer thin end can be estimated.

Then, the center of the thin end using neighborhood information is calculated. The adjacent pixels of the center of the pointer thin end are scanned at the circumference. After getting the whole pointer thin end region, the center of the region to fine the precise location of the center is calculated. As seen in Figure 6, the cross is the center of pointer thin end.

Figure 6.    The center of the pointer thin end.

### 3.3    *Calculating the reading of the pointer*

After getting the precise position $(x_1, y_1)$ of the center of the pointer thin end, the center of the dial $O_o(x_o, y_o)$ and $(x_1, y_1)$ is corrected. This line is in the same direction with the dial line. The angle between the dial line and the x-axis is ordered in the positive direction to be $\alpha$, and the slope of the dial line to be $k$. The angle between the 0 calibration and the y axis in the negative direction is $\theta$, so the angle between the pointer and the 0 calibration is $\beta$. The variables in the coordination system described above are shown in Figure 7. Eventually, the actual reading $\gamma$ of the pointer is as shown in Equation 5.

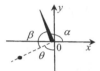

Figure 7. The variables in the coordination system.

$$\gamma = \frac{\beta}{2\pi - 2\theta} \cdot |\text{the range of the dial}|$$

(5)

## 4 EXPERIMENT AND ANALYSIS

The experiment is based on VC 6.0 experimental platform. The signal input is regarded as the reference value of the dial pointer, the experimental result is seen as the recognition value and the relative error is the ratio of the difference between them and the range of the dial. Experiments for the two different dials are done. The dials are shown in Figure 1a and 1b. The range of them is 1.6 mPa. Both of gauges are motionless and the illumination is invariant. The result of the test is shown in Table 1.

Table 1. The result of the test between two different dials.

| Number | Referenc value/mPa | Recognition value/mPa | Relative error(%) |
|---|---|---|---|
| 1^ | 0.19 | 0.1954 | 0.338 |
| 2^ | 0.82 | 0.8142 | 0.363 |
| 3^ | 1.45 | 1.4620 | 0.750 |
| 4* | 0.28 | 0.2887 | 0.544 |
| 5* | 1.27 | 1.2776 | 0.475 |
| 6* | 1.52 | 1.5318 | 0.738 |

^ The dial is shown in Figure 1a.
* The dial is shown in Figure 1b.

The experiment is also conducted in different illumination. The dial used is shown in Figure 1a, and the experiment result is shown in Table 2.

The experiment is also conducted in the different conditions of vibration. The result of the experiment is shown in Table 3.

Table 2. The result of the test under different illumination.

| Illumination/ Lx | Reference value/mPa | Recognition value/mPa | Relative error(%) |
|---|---|---|---|
| 408.2 | 0.19 | 0.1954 | 0.338 |
| 389.5 | 0.19 | 0.1953 | 0.331 |
| 353.7 | 0.63 | 0.6403 | 0.644 |
| 309.5 | 1.10 | 1.1135 | 0.844 |
| 243.9 | 1.27 | 1.2870 | 1.062 |
| 187.6 | 1.55 | 1.5311 | 1.181 |

Table 3. The result of the test when the dial vibrating.

| Vibration frequancy/Hz | Reference value/mPa | Recognition value/mPa | Relative error(%) |
|---|---|---|---|
| 0 | 0.19 | 0.1954 | 0.338 |
| 2 | 0.50 | 0.5082 | 0.513 |
| 5 | 0.68 | 0.6907 | 0.669 |
| 10 | 1.48 | 1.5000 | 1.250 |

The experiments show that this algorithm can recognize the dial pointer accurately with low error under different conditions. Even under the low illumination, the relative error is less than 1.20% within the acceptable limits. And when the joggle frequency is 10 Hz, the relative error is within 1.30%. These results mean the algorithm is robust.

## 5 CONCLUSION

Aiming to the specific applications such as large factories, gas supply systems and so on, this paper proposes a novel pointer of instrument with pointer recognition method, making use of the label positioning method and the circumference seeking method based on making the meter reading accurate. This method breaks through tradition and is quite real-time, robust and efficient.

## REFERENCES

Zhou, H. 2003. The digital process and video transmission of industry instrument dial. *Proceedings of Apparatus and Instrument* 24(4):359–370.

Zhang, J. et al. 2013. Novel automobile meter pointer detection algorithm based on computer vision. *Computer Engineering & Science* 35(3):134–139.

Matas, J. et al. 2000. Robust detection of lines using progressive probabilistic Hough transform. *Computer vison Image underst* 78(1):119–137.

Sun, F.J. et al. 2005. Studies of the recognition of pointer angle of dial based on image processing. *Proceeding of the CSEE* 25(16):73–78.

# PLL-based adaptive power line interference canceller for ECG signal

T.J. Li
*College of Electrical and Mechanical Engineering, Shandong Xiehe University, Jinan, Shandong, China*

T.H. Li
*Institute of Advanced Intelligent Machine, Hefei University of Technology, Hefei, Anhui, China*

ABSTRACT: The 50 Hz Power Line Interference (PLI) is one of the main sources of interference in the Electrocardiogram (ECG) signal process. Adaptive filtering method has become one of the most effective and popular methods for PLI cancellation. However, the main challenge is that this method is sensitive to the mismatch between the reference signal and the PLI. In this paper, we propose an effective method based on phase-locked loop (PLL). In our method the amplitude, frequency and phase of PLI are all tracked at the same time. We used MIT-BIH ECG arrhythmia database for our experimental test. Our results show that the PLL-based adaptive canceller can process frequency mismatch up to about 5Hz, and the signal-to-PLI ratio is about 30dB. The comparison shows the excellent performance of our PLL-based method.

KEYWORDS: Adaptive noise cancellation, Power line interference, Phase-locked loop, Electrocardiogram

## 1 INTRODUCTION

The ECG is a graphical representation of heart functionality and is an important tool used for diagnosis of cardiac abnormalities. For diagnostic quality ECG recordings, signal acquisition must be noise free. However, when the ECG signal is recorded, it is always corrupted by various kinds of noise, such as power line interference, baseline wandering, electrode contact noise, motion artifacts, muscle contraction, and instrumentation noise generated by electronic devices, etc. PLI is one of the most significant noises in electrocardiography. For a high quality analysis of ECG, the amplitude of the power line interference should be less than 0.5% of the peak-to-peak QRS amplitude [1]. This corresponds to a signal-to-noise ratio (SNR) about 30 dB. SNR is defined as the power ratio between the ECG signal and the power line interference. Different types of digital filters (FIR and IIR) have been used to reduce PLI [2–4]. However, since the ECG signal is non-stationary signal, it is difficult to apply these filters with fixed coefficients to reduce the PLI. Recently, adaptive filtering has become one of the effective and popular methods for the processing and analysis of the ECG signal [5]–[8]. The main drawback is that the common LMS-based adaptive filter can only estimate amplitude and phase of the reference signal, once the frequency of PLI is not stable, a frequency mismatch between PLI and reference signal leads to an inadequate reduction of the PLI.

In this paper, we proposed a PLL-based adaptive cancellation method based on PLL to remove PLI. In our method the frequency of PLI is estimated and tracked by a PLL, the PLL output is used as the reference signal of adaptive canceller, so our method can estimate all the parameters of the reference signal at the same time. As a result, the PLI can be well tracked and suppressed, and in fact, in our method we do not need an outer reference. Our experiments show the excellent performance of the new method in the removal of unstable PLI compared to current methods.

The rest of this paper is organized as follows: Section 2 describes the design of adaptive noise cancellation systems. The proposed method is described in section 3. Section 4 provides the experimental results and evaluations. And in Section 5, we come to a conclusion and discuss some future work.

## 2 THE DESIGN OF ADAPTIVE NOISE CANCELLATION SYSTEM

Adaptive filter is characterized by the fact that when the input process is unknown or variable, it can correspondingly adjust its own parameters to meet certain criteria and gain optimized filtering [9]. An adaptive noise cancellation system for power interference suppression in ECG signal is built based on adaptive filtering technology. One of the most significant methods in adaptive canceller is based on LMS algorithm, the system structure is shown in figure 1.

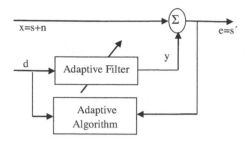

Figure 1.  Principle of adaptive noise cancellation.

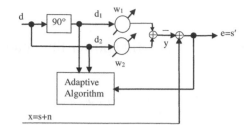

Figure 2.  The 50Hz PLI cancellation scheme.

In figure 1, the primary input consists of a useful signal which is affected by noise, and the second input is the reference signal, which is acquired from a point very close to the source of the primary signal. The adaptive noise cancellation system will produce an output signal, which is an estimate of the noise component in the primary signal, and after subtraction from primary signal, it obtains a filtered signal. What is important to note is that in this configuration, the error signal e is the filtered signal — the cleaned ECG signal.

A practical example is to eliminate the interfering signal from the power grid of 50 Hz, and the principle scheme is shown in figure 2.

The $d_1$ and $d_2$ are the 50 Hz reference sine waves which have the phase difference of 90°, the coefficient vector composed from $w_1$ and $w_2$, $w_1$ and $w_2$ are updated by the adaptive algorithm. The 50Hz reference signal y is subtracted from the noisy ECG signal x, and the result is the error signal e. According to LMS algorithm, the iterate formula is as follows:

$$w_1(n+1) = w_1(n) + \mu e(n)d_1(n) \tag{1}$$

$$w_2(n+1) = w_2(n) + \mu e(n)d_2(n) \tag{2}$$

$$y(n) = w_1(n)d_1(n) + w_2(n)d_2(n) \tag{3}$$

$$e(n) = x(n) - y(n) \tag{4}$$

Where μ is the step coefficients.

## 3  THE PROPOSED METHOD

It has been shown by Widrow and Glover [10] that once the parameter estimation is converged and fixed, an adaptive PLI canceller is approximately equivalent

to a notch filter. But in practice the frequency of PLI is not stable, which causes a mismatch between the suppression band and the PLI frequency band. In this paper, we propose a PLL-based method to solve the problem, the PLL track the input PLI frequency and the output of PLL is used as the reference signal of the adaptive noise canceller. The proposed method is shown in Figure 3.

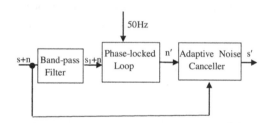

Figure 3.  The proposed method model.

In figure 3, the center frequency of the band-pass filter is 50Hz, the noisy ECG signal s+n is filtered by band-pass filter to suppress signal outside the PLI band, and the filter output s′+n is input to the PLL. According to PLL theory, when the frequency difference between the input signal frequency and the PLL center frequency is less than the pull-in range of PLL, the PLL can track and lock on the frequency of the input signal [11], and the error between the frequency estimates and the actual frequency is small enough, so the PLL gives a perfect reference signal which frequency is same as the PLI.

The most popular used PLL in practice is the second order; its closed-loop transfer function may be written as follows:

$$H(s) = \frac{2\xi\omega_n s + \omega_n^2}{s^2 + 2\xi\omega_n s + \omega_n^2} \tag{5}$$

where $\omega_n$ is the natural frequency of the PLL loop and ξ is the damping ratio. The PLL performance depends on the parameters $\omega_n$ and $\xi$, detailed discussion can be found in [11].

## 4 EXPERIMENT ANALYSIS AND RESULT

We use the benchmark MIT-BIH database ECG recordings as the reference for our work. The database consists of 48 half hour excerpts of two-channel ambulatory ECG recordings. The recordings were digitized at 360 samples per second per channel with 11-bit resolution over a 10mV range, and they have been amplified with a gain of 200 [12].

Figure 4 shows the noisy ECG signal (the sum of record 103 and PLI) and its spectrum, the record is begun after 10 second of the system starting time. It can be seen that there is a 50Hz PLI in the ECG signal, and the PLI affects the ECG wave severely.

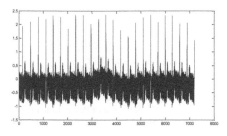

(a) The noisy ECG signal

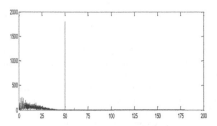

(b) Spectrum of the noisy ECG signal

Figure 4.   The noisy ECG signal.

For an accurate quantification of the cancellation performance, the SIR at the input and output of the canceller must be known. The SIR is defined as the ratio between the power of the ECG signal and the power line interference. We refer to this quantity as the input SIR ($SIR_{in}$). As a performance measure of the adaptive interference cancellation scheme, we adopt the output SIR ($SIR_{out}$), defined as the ratio between ECG signal power and residual line interference power in the error signal. Furthermore, we define $\Delta f$ as the difference between the PLL input signal frequency and PLL central frequency f, and the frequency deviation is defined as the ratio between $\Delta f$ and f.

In our simulation, we study the $SIR_{out}$ under different $SIR_{in}$ when the frequency deviation is zero, and the $SIR_{out}$ under different frequency deviation when $SIR_{in}$ is set to 0 dB. We compare the output SNR of our PLL-based adaptive canceller and the common adaptive canceller, the results are shown in Tables 1 and 2.

Table 1.   Performance for different $SIR_{in}$ (no frequency deviation).

| | $SIR_{out}$ (dB) | | |
|---|---|---|---|
| $SIR_{in}$(dB) | -20 | 0 | 20 |
| AC | 35.02 | 35.02 | 35.02 |
| PLL-based AC | 25.87 | 29.47 | 30.07 |

Table 2.   Performance for different frequency deviation ($SIR_{in}$=0dB).

| | $SIR_{out}$ (dB) | | |
|---|---|---|---|
| $\Delta f/f$ | 0.1% | 1% | 10% |
| AC | 20.11 | 2.7 | -0.92 |
| PLL-based AC | 28.16 | 28.59 | 28.42 |

When there is no frequency deviation, Table 1 gives SNR improvement for different SNR inputs in noisy situations while the common adaptive canceller and the proposed PLL-based adaptive canceller are used. It can be seen that both cancellers get good performance, the performance of PLL-based canceller is slightly worse than the common canceller because of the frequency noise introduced by the PLL. In table 2 the above two cancellers are compared under different frequency deviation when the $SIR_{in}$ is set to 0dB. It shows that the PLL-based canceller is robust to frequency deviation. The $SIR_{out}$ is about 30dB regardless of the frequency deviation. Yet, the performance of common canceller is degraded greatly when the frequency deviation becomes large.

Figure 5 shows the acquisition performance and the output spectrum of the two cancellers with certain frequency deviation and $SIR_{in}$. It can be seen that the proposed PLL-based canceller can track the frequency deviation in about 1 second, and it suppress the PLI completely after the PLL lock on the frequency of PLI, yet the common adaptive canceller cannot track and suppress the PLI.

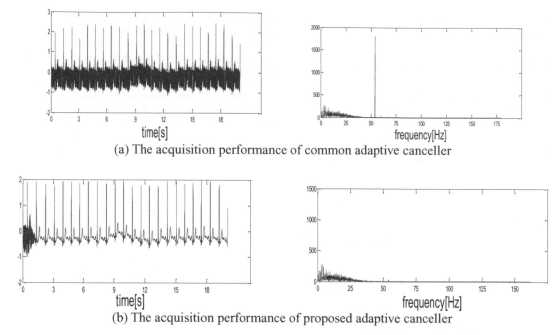

(a) The acquisition performance of common adaptive canceller

(b) The acquisition performance of proposed adaptive canceller

Figure 5.    Tracking performance comparison.
($\Delta$f/f=8%,SIR$_{in}$=0dB)

## 5    CONCLUSION

This paper proposes a PLL-based adaptive cancel-ler for the suppression of the power line interfer-ence in ECG recordings. The canceller comprises a second-order PLL, as well as the amplitude and phase. But the frequency deviation can be tracked. A significant advantage of the PLL-based adaptive canceller is that it does not need an outer reference signal. Our results show that the new method can track the frequency deviation up to 5Hz, and the lock time is about 1s. Results indicate that our method can obtain a high performance. ECG output with SIR is about 30dB. In conclusion, our proposed method is much superior to common adaptive noise canceller.

## ACKNOWLEDGMENTS

This research has been partially supported by the National High-Tech Research & Development Program of China 863 Program under Grant No. 2012AA011103.

## REFERENCES

[1] Metting van Rijn A. C. & Peper A. & Grimbergen C. A. 1990, High-quality recording of bioelectric events part 1, interference reduction, theory and practice. Med. Biol. Eng. Comput, 28:389–397.

[2] Limacher R, 1996, Removal of Power Line Interference from the ECG Signal by an Adaptive Digital Filter, in Proc. of European Tel. Conf. Garmisch-Part: 300–309.

[3] Mateo J. & Sanchez C. & Torres A. & Cervigon R. & Rieta J. J, 2008, Neural Network Based Canceller for Power Line Interference in ECG Signals. Computers in Cardiology, 35:1073–1076.

[4] Kaur M. & Singh B, 2009, Power Line Interference Reduction in ECG Using Combination of MA Method and IIR Notch Filter, Int. J. of Recent Trends in Eng., 2(6):125–129.

[5] Mitov I. P, 2004, A method for reduction of power line interference in the ECG, Medical Engineering & Physics. 26(10):879–887.

[6] Jacek M. L. & Norbert H., 2005, ECG baseline wander and powerline interference reduction using nonlinear filter bank, signal processing, 85(4):781–793.

[7] Chavdar L. & Georgy M, 2005, Removal of power-line interference from the ECG: a review of the subtraction procedure, Biomed Eng Online, 4:50.

[8] Ziarani A.K. & Konrad A., 2002, A nonlinear adaptive method of elimination of power line interference in ECG signals, IEEE Trans. Bio. & Eng. 49(6):540–547.

[9] Widrow B., 1975, Adaptive noise cancelling: principles and applications, Proc. IEEE, 63(12):1692–1716.

[10] Glover J. R., 1977, Adaptive noise canceling applied to sinusoidal interferences, IEEE Trans. Acoust. Speech. Signal Process. 25(6):484–491.

[11] Blanchard A., 1976, Phase-Locked Loops: Application to Coherent Receiver Design, New York: Wiley.

[12] Moody G. B. & Mark R. G., 2001, The impact of the MIT-BIH Arrhythmia Database, IEEE Eng. in Med. and Biol. 20(3):45–50.

*Multimedia, Communication and Computing Application – Leung (Ed.)*
*© 2015 Taylor & Francis Group, London, ISBN 978-1-138-02775-6*

# Study on fuzzy neural network controller for automatic voltage regulator

M.H. Wang & Y.Q. Yu
*Faculty of Computer Science, Guangdong University of Technology, China*

ABSTRACT: Aiming at the control problem of Automatic Voltage Regulator (AVR), the paper studied a design methodology based adaptive fuzzy neural network for development of high efficiency and reliable AVR. The AVR exciter system adopted adaptive fuzzy neural network technology to improve its flexibility. The implementation of on-line automatic adjustment of AVR parameter was done according to the fuzzy neural network control decision-making. Simulation test was made and a comparison to the conventional excitation control was performed. The simulation result showed that the fuzzy neural network control had stronger robustness and adaptability.

## 1 INTRODUCTION

Generally, marine power system applies engine generator set to produce electricity in large ship. The automatic voltage regulator (AVR) in phase compounded self-exciter system has employed switch chopper voltage regulator. The chopper is working at high switch frequency, not only the harmonics of switching frequency are generated, the higher frequency interference is also produced during the transient of switching. If the generator system is not designed properly, these high-frequency interferences can easily conduct the power mains and cause unexpected results. Control action is indispensable for maintaining power system voltage. The dynamic characteristics of a diesel-generator vary with changes in the operating conditions of the power system. AVR has been popularly used to damp out the voltage oscillations in the power system. It is difficult to design a set of excitation controller parameters which provide acceptable voltage control over the desired range of power system conditions format.

Many control strategies, applying various techniques have been proposed, such as PID regulator, self-tuning adaptive regulator, sliding mode control, robust control and model reference adaptive control [1–3]. Recently, the concepts of artificial intelligence (AI) techniques and genetic algorithms (GA) were applied, in order to obtain higher performance of robustness and adaptability [4–5].

The development of intelligent control theory provides an approach to solve the stability problem of the complicated and strong non-linear electric power systems. Especially, the fuzzy logic control is easy to be realized and has strong adaptability and high reliability. Therefore, many researchers have tried to apply fuzzy control technology to enhance the robustness of power system stabilizer. Direct fuzzy logic excitation controller for generator has also been proposed [6,7].

The performance of a fuzzy control system is determined by the performance of its fuzzy controller. Furthermore, the performance of a fuzzy controller depends on the rule sets of inference and membership functions. The design of the rules and membership functions is sophisticated and they are difficult to tune in online. In most cases, the inference methods and rule sets in the fuzzy controller are fixed after the fuzzy controller is designed. However, the fuzzy logic controller should have strong robustness and adaptability to ensure the fuzzy control system has good performance for different controlled objects or different operating conditions. So it is necessary to adjust the rules of inference and membership functions in the fuzzy controller on line.

However, in case of fuzzy control, the main problem is that the parameters associated with the membership functions and the rules depend broadly on the intuition of the engineer. To overcome this over-dependence on human intuition, adaptive fuzzy neural network is used. In adaptive fuzzy neural network, rather than choosing the parameters associated with a given membership function arbitrarily, these parameters are chosen so as to tailor the membership functions to a set of input/output data in order to account for these types of variations in the data values. This learning method is similar to that of neural networks. In this paper, zero and first order Sugeno fuzzy models with hybrid-learning algorithm are used.

This algorithm is a combination of backpropagation for the parameters associated with the input membership function and least squares estimation for the parameters associated with the output membership function. The parameters associated with the membership functions and rules change through the learning process. The computation of these parameters, or their adjustment, is facilitated by a gradient vector, which provides a measure of how well the fuzzy inference system is modelling the input/output data for a given set of parameters. Once the gradient vector is obtained, an optimization routine is applied in order to adjust the parameters so as to reduce the sum of the squared difference between actual and desired outputs.

The objective of this paper is to introduce AVR based adaptive fuzzy neural network and PSS in small-signal stability with Hybrid Learning Algorithm. In the following sections, brief introductions to Sugeno fuzzy model, ANFIS architecture and hybrid learning are given. Then the design process of ANFIS-based AVR and PSS is elaborated. Finally, in the results section, it is shown that the designed AVR and PSS can satisfactorily substitute the conventional fuzzy AVR and PSS in single-machine and multi-machine operation.

## 2  MARINE GENERATOR EXCITATION SYSTEM

Marine diesel-generator excitation system is shown in Figure 1. It is the basic scheme derived in the actual ship power station, which is a controlled phase compound excitation of the brushless excitation system. AVR consists of measuring circuit, comparator circuit, voltage error regulator, phase control, synchronization control, trigger pulse circuit and thyristor circuit.

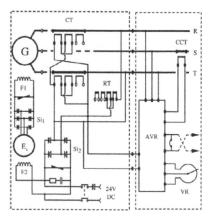

Figure 1.  Marine diesel-generator excitation system.

## 3  AUTOMATIC VOLTAGE REGULATOR BASED ADAPTIVE FUZZY NEURAL NETWORK

This design applies adaptive fuzzy neural network to AVR, the essential part of the neural fuzzy controller is structure and learning issues, which can process the

fuzzy information and the learning of the neural fuzzy controller is a kind of learning of the neural network capable of completing the fuzzy inference. This design applies Takagi-Sugeno fuzzy inference system, which consists of 25 Sugeno inference rules, for example,

If e is PB and r is PB.Then

$$\delta_i = u_i(e,r) = p_i e + q_i r + b_i, \ i=1,2,\ldots25. \qquad (1)$$

The neural network is designed with five layers, they are, input variable layer, input language layer, control rule layer, output language layer and output variable layer, as shown in figure 2.

Layer 1: fuzzify the input variables and output the membership degree of the fuzzy set. It is decided which state variables representative of system dynamic performance must be taken as the input signals to the controller. In this paper, the deviation of terminal voltage (e) and its derivative (r) are taken as input signals of the AVR based adaptive fuzzy neural network.. The language variables in these two input variables are {negative big, negative small, zero, positive small, positive big}, which is expressed with the symbols as {NB, NS, ZE, PS, PB}, every neural node represents a member function, data clustering technology is used for the clustering analysis on the sample data, so as to obtain the clustering centre of input variables. According to the clustering results, the fuzzy classification of input variables (that is, the number and initial distribution of membership functions) is made. Every mode is a self-adaptive node of the node function.

Layer 2: realize the calculation of fuzzy set in the condition part, and output the relevance weights of every rule. Every node is a fixed node marked with Π, whose output is the product of the input signals, which represents the excitation degree of a given rule.

Layer 3: Unify the relevance weights of all rules. Every node is a fixed node marked by N. The i node calculates ratio of the excitation degree of the i rule to the sum of the excitation degrees for all rules.

Layer 4: The transfer function of adaptive node is the linear function, which represents a partial linear model to calculate the output of every rule.

The parameter optimization method in the controller used in this article is Temporal Backpropagation. Widrow firstly proposed TBP, in which time is divided into sections, and the system configuration in every time section is considered as an individual node, the nodes in many time sections may be linked into a complicated time network. The error prediction is made on output status of every time section with the expected path, then this error is reversed by means of time network to adjust the network parameters, and thus reaching the learning objective, so that the output path of the controlled object can achieve the expected objective by means of the exercise of the temporal backpropagation neural network.

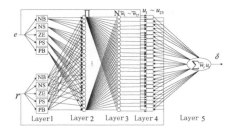

Figure 2. Structural diagram of adaptive fuzzy neural network controller for AVR.

The learning process of adaptive neural fuzzy control system based on the temporal backpropagation learning algorithm is a closed-loop dynamic simulation behaviour, in which $\Delta$ represents the time retroposition factor, $V_r(k)$ is the voltage command. Store the voltage statuses $V(k)$ at an actual motion path to form an actual motion path; the preset voltage $V_r(k)$ become an expected motion path, and their difference $\varepsilon(k)$ becomes the performance index for closed loop control based on Equation (2).

$$e_k = e(k) = V(k) - V_r(k) \tag{2}$$

$\alpha_i$ is used to represent any of pi, qi, bi in modus tollens (1) for Sugeno fuzzy rules of adaptive fuzzy neural network, and the BP algorithm gives the learning law as follows,

$$\Delta\alpha_i = -\eta\frac{\partial E}{\partial\alpha_i} = -\eta\sum_{k=1}^{n}\frac{\partial E(k)}{\partial\alpha_i} \tag{3}$$

And

$$\alpha_i^{k+1} = \alpha_i^k + \Delta\alpha_i \tag{4}$$

Equation (10) with TBP algorithm is used for optimizing the parameter $a_i$ to obtain the adaptive fuzzy neural network controller with the parameter update, and it uses this controller for the next control cycle.

The forward path calculation solves the problem of the storage and accumulation for the required data, which is the process of direct calculation, in every step of k=1, 2, … , n, and the outputs in layer 1 of adaptive fuzzy neural network are:

$$\mu_e^{NB}(e_k), \mu^{NS}(e_k), \mu^{ZE}(e_k), \mu^{PS}(e_k), \mu^{PB}(e_k)$$

And

$$\mu^{NB}(r_k), \mu^{NS}(r_k), \mu^{ZE}(r_k), \mu^{PS}(r_k), \mu^{PB}(r_k).$$

The nodes in layer 2 of adaptive fuzzy neural network give the excitation degree of every rule (the products of membership functions), for example, $\omega_1 = \mu^{PB}(e_k)\cdot\mu^{PB}(r_k), \omega_2 = \mu^{PB}(e_k)\cdot\mu^{PS}(r_k),$

$\cdots\cdots, \omega_{25} = \mu^{NB}(e_k)\cdot\mu^{NB}(r_k)$ These products are unrelated to $\alpha_i(p_i, q_i, b_i, i = 1 \sim 25)$.

The standard value for the output $\omega_i$ ($i = 1 \sim 25$) in layer 3 of ANFIS is given as follows,

$$\overline{\omega_i} = \frac{\omega_i}{\sum_{j=1}^{25}\omega_j} \tag{5}$$

The output in layer 4 of ANFIS is

$$\overline{\omega_i}u_i = \overline{\omega_i}(p_i e_k + q_i r_k + b_i) \tag{6}$$

The output in layer 5 of adaptive fuzzy neural network is the command rudder angle given by adaptive fuzzy neural network autopilot, which is influenced by the function, $p_i, q_i, b_i$, with the linear relations, that is,

$$\delta_\psi(k) = \sum_{i=1}^{25}\overline{\omega_i}(p_i e_k + q_i r_k + b_i) \tag{7}$$

The adaptive fuzzy neural network backward path calculation is realized with the TBP learning algorithm. In a sampling period, $\sum_{k=1}^{n}\partial E(k)/\partial\alpha_i$ in Equation (10) is simplified with the following symbols:

$E(k) = E_k$, $\varepsilon(k) = e_k$, $\dot\psi(k) = r_k$, $\delta_\psi(k) = \delta_{\psi k}$,

$\omega_i = \omega^i$, $\overline{\omega_i} = \overline{\omega}^i$, $\omega_i(k) = \omega_k^i$, $\overline{\omega_i}(k) = \overline{\omega}_k^i$, $f_i = f^i$,

$f_i(k) = f_k^i$.

Thus

$$\frac{\partial E}{\partial\alpha_i} = \sum_{k=1}^{n}\frac{\partial}{\partial\alpha_i}(e_k^2 t_k) = 2\sum_{k=1}^{n}\varepsilon_k t_k\frac{\partial e_k}{\partial\alpha_i} \tag{8}$$

$$\frac{\partial e_k}{\partial\alpha_i} = \frac{\partial e_k}{\partial\psi_k}\cdot\frac{\partial\psi_k}{\partial\delta_k}\cdot\frac{\partial\delta_k}{\partial\delta_{rk}}\cdot\frac{\partial\delta_{rk}}{\partial\alpha_i} = -\frac{\partial\psi_k}{\partial\delta_k}\cdot\frac{\partial\delta_k}{\partial\delta_{rk}}\cdot\frac{\partial\delta_{rk}}{\partial\alpha_i} \tag{9}$$

Since

$$\delta_{rk} = \sum_j \overline{w}_k^j f_k^j \tag{10}$$

Thus

$$\frac{\partial\delta_{\psi k}}{\partial\alpha_i} = \begin{cases} \overline{\omega}_k^i e_k, & \alpha_i = p_i \\ \overline{\omega}_k^i r_k, & \alpha_i = q_i \\ \overline{\omega}_k^i, & \alpha_i = b_i \end{cases} \tag{11}$$

313

These two quantities can be solved with approximate solutions by means of digital differential method, where it is possible to take

$$\frac{\partial \psi_k}{\partial \delta_k} \approx \frac{\psi_k - \psi_{k-1}}{\delta_k - \delta_{k-1}} \qquad (12)$$

$$\frac{\partial \delta_k}{\partial \delta_{\psi k}} \approx \frac{\delta_k - \delta_{k-1}}{\delta_{\psi k} - \delta_{\psi,k-1}} \qquad (13)$$

The combination of Equations (3)~(13) can resolve the update problem of controller parameters pi, qi, bi (i = 1~25).

Adaptive fuzzy neural network controller with the parameter optimization is used for the next control cycle, till the required control performance indexes are achieved.

What follows is the simulation test on the multimode intelligent controller in the above design in order to test the robustness of the controller as well as the stability and convergence of the control system.

## 4 SIMULATION

In order to study the performance of AVR based adaptive fuzzy neural network, an AVR system was simulated with conventional PID AVR, the conventional AVR is most efficiently tuned. Fig.3 show the step response of the AVR system with a conventional PID controller.

To study the performance of the ANFIS AVR in small signal stability, the unknown parameters are tuned in the same way as in case of AVR. Now, running the optimization routine, it is found that there is no significant change in the parameters of the Gaussian membership function initially taken, but the output constants (pi, qi and bi) of the first order Sugeno model for all the 25 rules are obtained.

The performance of the AVR based adaptive fuzzy neural network is then tested. The performances of the AVR are shown in Fig.3. It is found that the proposed adaptive fuzzy neural network based AVR is

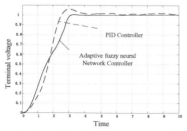

Figure 3. Terminal voltage characteristics of PID controller and adaptive fuzzy neural network controller.

performing better the conventional PID AVR in case of damping local oscillations. Experimental results show that the adaptive fuzzy neural network controller works stably and robustly. It is not necessary to identify dynamics of the plant because of the learning ability of controller.

## 5 CONCLUSION

This paper presented a practical design consideration for development of high efficiency and reliable AVR. The paper highlighted a systematic approach for designing AVR based an adaptive fuzzy neural network. AVR designed to overcome the limitation of PID AVR. The AVR designed to have more flexibility in operation with a wider range of disturbances. A good dynamic performance was obtained, which proved its superiority in signal stability of the power system.

## REFERENCES

[1] Bekkouche, H.; Charef, A., Analytical parameters tuning of the fractional PID controller for an automatic voltage regulator, International Conference on Systems and Control, 2013, pp. 253–258.
[2] Sahu, B.K.; Mohanty, P.K.; Panda, S.; Mishra, N., Robust analysis and design of PID controlled AVR system using Pattern Search algorithm, IEEE International Conference on Power Electronics, Drives and Energy Systems, 2012, pp. 1–6.
[3] Dong-Hee Lee; Jin-Woo Ahn; Tae-Won Chun, A Variable Gain Control Scheme of Digital Automatic Voltage Regulator for AC Generator, International Conference on Power Electronics and Drive Systems, 2007, pp. 1730–1734.
[4] Hasanien, H.M., Design Optimization of PID Controller in Automatic Voltage Regulator System Using Taguchi Combined Genetic Algorithm Method, IEEE Systems Journal, Vol. 7, n.4, 2013, pp. 825–831.
[5] Devaraj, D.; Selvabala, B. Real-coded genetic algorithm and fuzzy logic approach for real-time tuning of proportional-integral - derivative controller in automatic voltage regulator system, IET Generation, Transmission & Distribution, Volume: 3, Issue: 7, 2009, pp. 641–649.
[6] V. Mukherjee, S.P. Ghoshal, Comparison of intelligent fuzzy based AGC coordinated PID controlled and PSS controlled AVR system, Electrical Power and Energy Systems, vol.29 n.9, 2007, pp. 679–689.
[7] Farouk, N.; Tian Bingqi, Application of self-tuning fuzzy PID controller on the AVR system, International Conference on Mechatronics and Automation, 2012, pp. 2510 – 2514.
[8] Khezri, R. ; Bevrani, H.,Fuzzy-based coordinated control design for AVR and PSS in multi-machine power systems, Iranian Conference on Fuzzy Systems, 2013, pp. 1–5.

*Multimedia, Communication and Computing Application – Leung (Ed.)*
*© 2015 Taylor & Francis Group, London, ISBN 978-1-138-02775-6*

# Constrained-optimization-based extreme learning machine with incremental learning

R.F. Xu, Z.Y. Wang & S.J. Lee

*Department of Electrical Engineering, National Sun Yat-Sen University, Kaohsiung, Taiwan*

ABSTRACT: Extreme Learning Machine (ELM) is a single-hidden layer feed-forward network in which the input weights are randomly given and the output weights are determined analytically. Due to its simple implementation and fast learning, ELM has recently attracted a lot of attention. Huang *et al.* proposed the Constrained-optimization-based ELM (C-ELM) which can achieve similar or even better performance and learn much faster than SVM and LS-SVM. However, the number of hidden nodes in a C-ELM has to be determined by trial-and-error. In this paper, we propose an incremental learning method by which hidden nodes can be added incrementally and output weights are updated automatically. The effectiveness of our proposed method is validated by a number of experiments on classification and regression data sets.

## 1 INTRODUCTION

The main problems for a neural network with back-propagation learning mechanism are slow learning speed and local minima. Huang *et al.* have proposed extreme learning machine (ELM) (Huang et al. 2006) to deal with these problems. ELM is a single-hidden layer feed-forward network and it just needs to determine the output weights analytically. The input weights connecting the input layer to the hidden layer as well as the biases of the hidden neurons, are randomly assigned without tuning.

In 1990s, Cortes and Vapnik proposed support vector machine (SVM) (Cortes et al. 1995) and it has great ability for classification problems. Suykens and Vandewalle proposed a least square version of SVM (LS-SVM) (Suykens et al. 1999) and they adopted equality optimization constraints which only need a least square solution. Recently, Huang et al. have developed an optimization-based ELM with inequality constraints for classification (Huang et al. 2010), and later equality constrained-optimization-based ELM for classification and regression (Huang et al. 2012). For convenience, optimization-based ELM with equality constraints is abbreviated to C-ELM in this paper. C-ELM provides a unified learning framework and includes milder constraints compared with SVM and LS-SVM. The random input weights or kernel functions can also be applied to C-ELM. It has been shown that C-ELM can achieve similar or even

better generalized performance and run much faster than SVM and LS-SVM (Huang et al. 2012). Several variants of C-ELM have been proposed subsequently (Zong et al. 2013).

Although the learning speed of C-ELM is very fast, a problem must be addressed. When using C-ELM as a predicting model, the number of hidden nodes must be determined. It's very time-consuming to find an appropriate number of hidden nodes by trial-and-error. Huang et al. proposed the I-ELM (Huang et al. 2006) by which hidden nodes can be added automatically until some thresholds are achieved. Huang *et al* proposed an enhanced version of I-ELM (Huang et al. 2008) called EI-ELM. Feng et al. proposed EM-ELM (Feng et al. 2009) which can update the output weights automatically without re-computation when the number of hidden nodes changes. However, these methods can only be applied to ELM. In this paper, we develop a method for C-ELM based on EM-ELM to add hidden nodes incrementally. The output weights are automatically updated without computing from scratch after each time of addition. We demonstrate the effectiveness of our proposed method by a number of experiments on classification and regression data sets.

The rest of this paper is organized as follows. Section 2 gives a brief introduction to ELM and C-ELM. Section 3 describes our proposed incremental learning for C-ELM. Section 4 presents experimental results. Finally, concluding remarks are given in Section 5.

## 2 OPTIMIZATION-BASED ELM WITH EQUALITY CONSTRAINTS (C-ELM)

C-ELM is a single-hidden layer feed-forward network in which the input weights are randomly given and the output weights are determined analytically.

Suppose we have a set of training patterns $(\mathbf{x}_i, \mathbf{y}_i)$, $i = 1 \sim N$, $N$ is the number of training patterns, where $\mathbf{x}_i = [x_{i1}, x_{i2}, \dots, x_{ip}]^T \varepsilon \mathbf{R}^p$ and $\mathbf{y}_i = [y_{i1}, \dots, y_{im}] \varepsilon \mathbf{R}^m$ are the input vectors and the desired outputs respectively of instance $i$, $p$ is the number of features, and $m$ is the number of output neurons. Let $\mathbf{w}_j = [w_{j1}, \dots, w_{jp}]$ and $b_j$ be the random weights and bias, respectively, of hidden node $j$. For any input vector $\mathbf{x}$, the predicted output $y_k$, $k = 1 \sim m$, is $y_k = \sum_{j=1}^{J} \beta_{jk} g(\mathbf{w}_j \mathbf{x} + b_j)$

$$\mathbf{y} = [y_{i1}, \cdots, y_{im}] \tag{1}$$

where $J$ is the number of hidden nodes, $g(\cdot)$ is the activation function of the hidden nodes, $\beta_{jk}$ is the output weight, i.e., the weight vector from hidden node $j$ to output node $k$. For a regression problem, $\mathbf{y}_i = y_i$, and $m = 1$. For a classification problem, the predicted class label is the index of the maximum value in $\mathbf{y}$ shown as follows:

$$\text{label}(x) = \arg \max_{k=1,\cdots,m} y_k \tag{2}$$

The output weights $\beta_{jk}$, $j=1,\dots,J$ and $k=1,\dots,m$, can be derived by solving the following objective function

$$\text{Minimize}: L(\beta, \xi) = \frac{1}{2}\|\beta\|^2 + \frac{C}{2}\sum_{i=1}^{N}\|\xi_i\|^2$$

$$\text{subject to}: h(\mathbf{x}_i)\beta = \mathbf{y}_i - \xi_i, \quad i = 1 \sim N \tag{3}$$

where

$$\beta = [\beta_1, \cdots, \beta_m] \tag{4}$$

$$h(\mathbf{x}_i) = [g(\mathbf{w}_1 \mathbf{x}_i + b_1), \cdots, g(\mathbf{w}_J \mathbf{x}_i + b_J)] \tag{5}$$

$C$ is regularization parameter, and

$$\xi = \begin{bmatrix} \xi_1 \\ \vdots \\ \xi_N \end{bmatrix} = \begin{bmatrix} \xi_{11} & \cdots & \xi_{1m} \\ \vdots & \vdots & \vdots \\ \xi_{N1} & \cdots & \xi_{Nm} \end{bmatrix} \tag{6}$$

Note that $\beta_i = [\beta_{1i}\beta_{2i,\dots,}\beta_{Ji}]^T$ for $i = 1, \dots, m$.

The solution of Eq.(3) can be derived by constructing the following Lagrangian function:

$$L(\beta, \xi, \alpha) = L(\beta, \xi) - \sum_{i=1}^{N}\sum_{k=1}^{m} \alpha_{ik}(h(\mathbf{x}_i)\beta_k - y_{ik} + \xi_{ik}) \tag{7}$$

where $\alpha$ is a matrix of Lagrangian multipliers

$$\alpha = \begin{bmatrix} \alpha_1 \\ \vdots \\ \alpha_N \end{bmatrix} = \begin{bmatrix} \alpha_{11} & \cdots & \alpha_{1m} \\ \vdots & \vdots & \vdots \\ \alpha_{N1} & \cdots & \alpha_{Nm} \end{bmatrix} \tag{8}$$

where H is the output matrix of the hidden layer, defined as

$$\mathbf{H} = \begin{bmatrix} h(\mathbf{x}_1) \\ \vdots \\ h(\mathbf{x}_N) \end{bmatrix} =$$

$$\begin{bmatrix} g(\mathbf{w}_1\mathbf{x}_1 + b_1) & \cdots & g(\mathbf{w}_J\mathbf{x}_1 + b_J) \\ \vdots & \vdots & \vdots \\ g(\mathbf{w}_1\mathbf{x}_N + b_1) & \cdots & g(\mathbf{w}_J\mathbf{x}_N + b_J) \end{bmatrix} \tag{9}$$

$\beta$ can be obtained by

$$\beta = \left(\frac{\mathbf{I}}{C} + \mathbf{H}^T\mathbf{H}\right)^{-1}\mathbf{H}^T\mathbf{Y} \tag{10}$$

## 3 PROPOSED INCREMENTAL LEARNING

Although the learning speed of C-ELM is fast, the number of hidden nodes must be determined. It's very time-consuming to find an appropriate number of hidden nodes by trial-and-error. We develop a method to add hidden nodes incrementally. The output weights are automatically updated without computing from scratch when the number of hidden nodes changes. Therefore, time is saved for the whole learning process of C-ELM.

Assume that initially we have $J$ hidden nodes in C-ELM, and the output matrix of the hidden layer is $\mathbf{H}_1$ which is of size $N \times J$. The output weights $\beta$, according to Eq.(10), is.

$$\beta = \left(\frac{\mathbf{I}}{C} + \mathbf{H}_1^T\mathbf{H}_1\right)^{-1}\mathbf{H}_1^T\mathbf{Y} = \mathbf{H}_1^\dagger\mathbf{Y}$$

Now suppose we add $\delta J$ more hidden nodes to the C-ELM. Let $\mathbf{H}_2$ be the output matrix of the hidden layer of the resulting C-ELM. Then we have $\mathbf{H}_2 = [\mathbf{H}_1, \delta\mathbf{H}_1]$ where $\delta\mathbf{H}_1$, of size $N \times \delta J$, denotes the output matrix of the added hidden nodes, and $\mathbf{H}_2^\dagger$ can be expressed as

$$\mathbf{H}_2^\dagger = \left[\frac{\mathbf{I}}{C} + \mathbf{H}_2^T\mathbf{H}_2\right]^{-1}\mathbf{H}_2^T$$

$$= \left[\frac{\mathbf{I}}{C} + \begin{bmatrix} \mathbf{H}_1^T \\ \delta\mathbf{H}_1^T \end{bmatrix}[\mathbf{H}_1, \delta\mathbf{H}_1]\right]^{-1}\begin{bmatrix} \mathbf{H}_1^T \\ \delta\mathbf{H}_1^T \end{bmatrix} \tag{11}$$

316

where $\mathbf{I}$ is $(J+\delta J)\times(J+\delta J)$ identity matrix

$$\mathbf{I} = \begin{bmatrix} \mathbf{I}_{J\times J} & 0 \\ 0 & \mathbf{I}_{\delta J\times\delta J} \end{bmatrix} \tag{12}$$

We can rewrite Eq(11) as

$$\mathbf{H}_2^\dagger = \begin{bmatrix} \dfrac{\mathbf{I}_{J\times J}}{C} + \mathbf{H}_1^\mathsf{T}\mathbf{H}_1 & \mathbf{H}_1^\mathsf{T}\delta\mathbf{H}_1 \\ \delta\mathbf{H}_1^\mathsf{T}\mathbf{H}_1 & \dfrac{\mathbf{I}_{\delta J\times\delta J}}{C} + \delta\mathbf{H}_1^\mathsf{T}\delta\mathbf{H}_1 \end{bmatrix}$$

$$\begin{bmatrix} \mathbf{H}_1^\mathsf{T} \\ \delta\mathbf{H}_1^\mathsf{T} \end{bmatrix} = \begin{bmatrix} \mathbf{U}_1 \\ \mathbf{D}_1 \end{bmatrix}$$

Based on the block matrices inverse (Fletcher. 1981), we have

$$\mathbf{D}_1 = \left( \dfrac{\mathbf{I}_{\delta J\times\delta J}}{C} + \delta\mathbf{H}_1^\mathsf{T}\left(\mathbf{I}_N - \mathbf{H}_1\mathbf{H}_1^\dagger\right)\delta\mathbf{H}_1 \right)^{-1}$$
$$\times\delta\mathbf{H}_1^\mathsf{T}\left(\mathbf{I}_N - \mathbf{H}_1\mathbf{H}_1^\dagger\right), \tag{13}$$
$$\mathbf{U}_1 = \mathbf{H}_1^\dagger - \mathbf{H}_1^\dagger\delta\mathbf{H}_1\mathbf{D}_1$$

By doing a recursive derivation, we have the following expressions after $K$ times of adding hidden nodes, by a number of $\delta J$ neurons each time:

$$\mathbf{H}_{k+1}^\dagger = \begin{bmatrix} \mathbf{U}_k \\ \mathbf{D}_k \end{bmatrix} \tag{14}$$

$$\mathbf{D}_k = \left( \dfrac{\mathbf{I}_{\delta J\times\delta J}}{C} + \delta\mathbf{H}_k^\mathsf{T}\left(\mathbf{I}_N - \mathbf{H}_k\mathbf{H}_k^\dagger\right)\delta\mathbf{H}_k \right)^{-1}$$
$$\times\delta\mathbf{H}_k^\mathsf{T}\left(\mathbf{I}_N - \mathbf{H}_k\mathbf{H}_k^\dagger\right),$$
$$\mathbf{U}_k = \mathbf{H}_k^\dagger - \mathbf{H}_k^\dagger\delta\mathbf{H}_k\mathbf{D}_k$$

For $k = 1,\dots,K$, and the output weights associated with the resulting C-ELM after $K$ additions can be obtained by

$$\beta = \mathbf{H}_{k+1}^\dagger\mathbf{Y} = \begin{bmatrix} \mathbf{U}_k \\ \mathbf{D}_k \end{bmatrix}\mathbf{Y} \tag{15}$$

Our incremental learning can be described as follows. Initially, we have a C-ELM machine having $J$ hidden nodes. In particular, the initial machine can be an empty one, i.e., $J = 0$. Then a number of $\delta J$ hidden nodes are added and the output weights of the C-ELM after addition are computed by Eq.(15). If the machine satisfies our pre-specified criterion, e.g., classification or regression accuracy, we are done. Otherwise, another $\delta J$ hidden nodes are added

and the output weights are computed by Eq.(15). This process iterates until a satisfying C-ELM machine is obtained.

## 4 EXPERIMENTAL RESULTS

In this section, we conduct several experiments and compare the efficiency between our proposed method and C-ELM. For convenience, our proposed method is called IC-ELM. The programs are written in MATLAB R2011b and we use a computer with Intel(R) Core(TM) i5-3210M CPU, 2.5GHz, 6GB of RAM to conduct the experiments.

Firstly, we random generate a $N\times J$ matrix $\mathbf{H}$. At the beginning, we set $N = 1000$ and $J = 100$. Then we add one hidden node each time. We can see that C-ELM requires more time as the number of hidden nodes increases. From Table 1, for example, when 300 hidden nodes are added, C-ELM requires 4.541 seconds while IC-ELM only requires 1.405 seconds. IC-ELM is 3 times faster than C-ELM. When 900 hidden nodes are added, C-ELM requires 58.354 seconds while IC-ELM only requires 9.709 seconds. IC-ELM is about 6 times faster than C-ELM.

Table 1. Efficiency comparison for # of added ≤ 900.

| #of added | 100 | 300 | 500 | 700 | 900 |
|---|---|---|---|---|---|
| C-ELM | 0.63 | 4.541 | 13.965 | 30.907 | 58.354 |
| IC-ELM | 0.26 | 1.405 | 3.32 | 6.247 | 9.709 |

Next, we apply IC-ELM and C-ELM to several benchmark data sets. These data sets are downloaded from UCI (UCI & regression data set). Table 2 and Table 3 are the specifications of regression data sets and classification data sets, respectively.

The features of the data sets involved are normalized to [-1,1] and the targets of a regression data set are normalized to [0,1]. We just list the number of continuous features. The results are shown in Table 4 and Table 5. In these tables, the expression "value⇒value" associated with each data set indicates that the initial number of hidden nodes is set to be the value at the left side and the final number of hidden nodes becomes the value at the right side of. One hidden node is added each time. From Table 4, for example, for the housing data set, the training time of C-ELM is 1.815 seconds

Table 2. The specifications of regression data sets.

| Data set | Instances | Input features |
|---|---|---|
| Auto price | 159 | 14 |
| Machine cpu | 209 | 6 |
| Housing | 506 | 12 |
| Concrete | 1030 | 8 |
| Auto mpg | 398 | 4 |

317

while the training time of IC-ELM is 0.363 seconds. IC-ELM is 5 times faster than C-ELM for the housing data set. For the Concrete data set, the training time of C-ELM is 6.436 seconds while the training time of IC-ELM is 1.318 seconds. IC-ELM is almost 5 times faster than C-ELM for the Concrete data set. From Table 5, for the Vehicle data set, the training time of C-ELM is 3.436 seconds while the training time of IC-ELM is 1.236 seconds. IC-ELM is almost 3 times faster than C-ELM for the Vehicle data set.

Table 3. The specifications of classification data sets.

| Data set | Instances | Input features | classes |
|---|---|---|---|
| Glass | 214 | 9 | 6 |
| Vehicle | 846 | 18 | 4 |
| Ecoli | 336 | 7 | 8 |
| Vertebral | 310 | 6 | 2 |

Table 4. Efficiency comparison between IC-ELM and C-ELM for regression data sets.

| | IC-ELM | | C-ELM | |
|---|---|---|---|---|
| Data set | Time (s) | Accuracy (rmse) | Time (s) | Accuracy (rmse) |
| Auto price (10⇒100) | 0.061 | 0.079 | 0.181 | 0.08 |
| Machine cpu (10⇒100) | 0.068 | 0.041 | 0.207 | 0.041 |
| Housing (100⇒300) | 0.363 | 0.070 | 1.815 | 0.070 |
| Concrete (100⇒400) | 1.318 | 0.084 | 6.436 | 0.084 |
| Auto mpg (50⇒150) | 0.139 | 0.104 | 0.49 | 0.104 |

Table 5. Efficiency comparison between IC-ELM and C-ELM for classification data sets.

| | IC-ELM | | C-ELM | |
|---|---|---|---|---|
| Data set | Time (s) | Accuracy (%) | Time (s) | Accuracy (%) |
| Glass (10⇒150) | 0.161 | 67.1 | 0.372 | 67.2 |
| Vehicle (100⇒300) | 1.236 | 82.6 | 3.436 | 82.5 |
| Ecoli (50⇒200) | 0.284 | 88.0 | 0.723 | 88.0 |
| Vertebral (50⇒200) | 0.237 | 83.4 | 0.638 | 83.4 |

## 5 CONCLUSION

Extreme learning machine (ELM) is a single-hidden layer feed-forward network in which the input weights are randomly given and the output weights are determined analytically. Huang et al. proposed the constrained-optimization-based ELM (C-ELM) which can achieve similar or even better performance and learn much faster than SVM and LS-SVM. However, the number of hidden nodes in a C-ELM has to be determined by trial-and-error. We have presented an incremental learning method for C-ELM to add hidden nodes incrementally. The output weights are automatically updated without computing from scratch after each time of addition. The hidden nodes can be added one by one or group by group until some pre-specified criteria are satisfied. The effectiveness of our proposed method has been demonstrated by a number of experiments on classification and regression data sets.

REFERENCES

Boyd, S., Ghaoul, L. E., Feron, E. & Balakrishnan, V. 1981. *Linear Matrix Inequalities in system and Control Theory*. New York: Wiley.
Cortes, C. & Vapnik, V. 1995. Support vector network. *Machine Learning*, 20(3):273-297.
Feng, G., Huang, G.-B., Lin, Q. & Gay, R. 2009. Error minimized extreme learning machine with growth of hidden nodes and incremental learning. *IEEE Trans-actions on Neural Networks*, 20(8):1352–1357.
Fletcher, R. 1981. Practical Methods of Optimization. Volume 2: Constrained Optimization. New York: Wiley.
Huang, G.-B., Zhu, Q.-Y.& Siew, C.-K. 2006. Extreme learning machine: Theory and applications. *Neurocomputing*, 70:489–501. Huang, G.-B., Ding, X., & Zhous, H. 2010. Optimization method based extreme learning machine for classification. *Neurocomputing*, 74:155–163.
Huang, G.-B. Zhou, H., Ding, X. & Zhang. R..2012. Extreme learning machine for regression and multiclass classification. *IEEE Transactions on Systems, Man, and Cybernetics-Part B:Cybernetics*, 42(2):513–529.
Huang, G.-B. Chen, L. & Siew. C.-K. 2006. Universal approximation using incremental constructive feedforward networks with random hidden nodes. *IEEE Transactions on Neural Networks*, 17(4):879–892.
Huang, G.-B. & Chen, L. 2008. Enhanced random search based incremental extreme learning machine. *Neurocomputing*, 71:3460–3468.
Rong, H.-J., Huang, G.-B. Sundararajan, N. & Saratchandran, P. 2009. Online sequential fuzzy extreme learning machine for function approximation and classification problem. *IEEE Transactions on Systems, Man, and Cybernetics-Part B:Cybernetics*, 39(4):1067–1072.
Regression Datasets.
http://www.dcc.fc.up.pt/ltorgo/Regression/DataSets.html.
UCI data set. http://archive.ics.uci.edu/ml/.
Suykens, J. A. K. & Vandewalle, J. 1999. Least squares support vector machine classifiers. *Neural Processing Letters*, 9(3):293–300.
Zong,W., Huang, G.-B. & Chen, Y. 2013. Weighted extreme learning machine for imbalance learning. *Neurocomputing*, 101:229–242.

*Multimedia, Communication and Computing Application – Leung (Ed.)*
*© 2015 Taylor & Francis Group, London, ISBN 978-1-138-02775-6*

# An improved image distortion correction algorithm

S.T. Lv & Y. Ding
*Key Laboratory of Dynamics and Control of Flight Vehicle, Ministry of Education, School of Aerospace Engineering, Beijing Institute of Technology, Beijing, China*

H.L. Wei
*Department of Automatic Control & Systems Engineering, University of Sheffield, UK*

ABSTRACT: In the optical measuring apparatus based on the TV camera, the image distortion usually occurs due to the defects of optical system design and manufacture or the distortion of the image sensor. For the common barrel distortion and pincushion distortion, this paper proposes an improved distortion correction algorithm based on the equivalent spherical surface, and processes the corrected image with bilinear interpolation, scaling and edge filling algorithm. This text uses actual images with barrel distortion and pincushion distortion to conduct an experiment which verifies that the proposed algorithm can not only efficiently correct the spherical distortion in the image, but also can ensure the integrity of its information and frame.

KEYWORDS: Image processing, Distortion, Correction

## 1 INTRODUCTION

In the optical measuring apparatus based on the television camera, the object gets through the optical system and image in the target surface of television camera, which will output the TV image of an object by photoelectric conversion. The output image tends to produce significant distortion in the process of the conversion from two-dimensional to three-dimensional. These distortions may be caused by defects in the design and manufacture of the optical system or by the distortion of the image sensor, or by the changes of imaging angle.

For image-based measurement system, the spatial position of the object is obtained from the captured image; therefore the image distortion has a great impact on the measurement accuracy of the system. When the image is captured without distortion, the coordinates of the object in the object space can be calculated based on the conversion relation between the space of objects and image. When the image distortion occurs, however, the corresponding relation between object and image will become non-linear. Therefore, in the image-based measurement system, we need to correct the image distortion to reflect the truly corresponding relation between the object and image space.

Generally, the image distortion includes linear distortion and nonlinear distortion. For linear distortion, such as rotation, translation, scaling and so on, there are good correction methods. But for nonlinear distortion, we still can't figure out a good solution due to its complexity. In this paper, we focus on the correction of barrel distortion and pincushion distortion in nonlinear distortion.

The traditional method for image distortion correction usually gets the distortion correction coefficient by solving the equations. This method has advantages of high accuracy, but the process of solving these equations is complex and exact solution usually can't be concluded, especially for undetermined equations, which leads to the decline in accuracy of distortion correction. To solve this problem, a new method of geometric modelling based on the equivalent distortion surface is proposed in this paper, and the paper conducts correction for barrel distortion and pincushion distortion of an image based on this model, which achieves a good result.

## 2 AN IMPROVED DISTORTION CORRECTION MODEL BASED ON THE EQUIVALENT SPHERICAL SURFACE

Due to the fact that the optical imaging system in actual camera is not an ideal pinhole model, there is a lens distortion including three distortions such as radial distortions, eccentric distortions and thin lens, etc. The radial distortion is caused by the defects in the manufacture of the lens, and this error is characterized as like this: the farther the image point is away from the optical center, the larger the distortion is, and they appear to be symmetric along the main optical axis of the camera. The positive direction of radial distortion is called pincushion distortion, also known as barrel distortion.

Barrel distortion is a distortion phenomenon of imaging screen presenting barrel expansion caused by the lens. Pincushion distortion is a kind of scene shrinking to the middle phenomenon.

Considering the radial distortion is symmetric with the optical axis and the centre point of the image, the extent of the distortion is related to the distance from distortion centre. According to the method of using equivalent spherical surface to simulate the generation of distortion which is proposed in the literature [1], this paper presents an improved distortion correction model based on the equivalent spherical surface, and establishes the mappings relation between the points of the distorted image and points of the ideal image.

Take plane $Q$ perpendicular to the axis of camera optical lens, on which a random point $P$ is imaged as $P_1$ on the ideal imaging surface $Q_0$. Assuming that the optical system or the sensor does not produce any distortion, but the imaging plane bends, distortion generated by this curved surface is completely consistent with the distortion generated by the optical system or the sensor. Take barrel distortion correction model as an example, we denote the equivalent spherical surface by S, and the centre F represents the main point of the image side. The focal length f means the radius of the image-side, and S is tangent to the ideal imaging surface $Q_0$ in point $O$. Tile the equivalent spherical surface to the ideal imaging surface, then $P_0$ is the corresponding point of $P_1$ in ideal image. The point $O$, $P_0$, $P_1$ are in the same surface and the length of $OP_1$ is equal to the arc length as $OP_0$; besides, the corresponding distortion imaging point of $P_0$ is $P_2$. The geometric relation is given in Figure 1.

As is shown in the Figure above: $P_1(x_1,y_1)$ is the ideal imaging point of $P_0$, and $P_2(x_2,y_2)$ is the distortion imaging point of $P_0$, and $P_0 P_2 \perp OP$, in the triangle $OPP_0$. According to the law of cosines, chord length $OP_0$ could be expressed as:

$$OP_0 = \sqrt{2 \cdot f^2 \cdot (1 - \cos\alpha)} \qquad (1)$$

According to chord tangent angle theorem, we have:

$$\gamma = \frac{\alpha}{2} \qquad (2)$$

Now the distance $L_2$ between distortion point and cut-off point is given by:

$$L_2 = OP_0 \cdot \cos\gamma = \sqrt{2 \cdot f^2 \cdot (1 - \cos\alpha)} \cdot \cos(\frac{\alpha}{2}) = f \cdot \sin\alpha \qquad (3)$$

Thus, we have:

$$\alpha = \arcsin(\frac{\sqrt{x_2^2 + y_2^2}}{f}) \qquad (4)$$

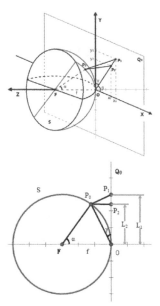

Figure 1. Barrel distortion equivalent spherical algorithm mapping relation.

In the Figure above:

$$\beta = \arctan(\frac{y_2}{x_2}) \qquad (5)$$

And

$$L_1 = OP_0 = \alpha \cdot f \qquad (6)$$

Then we have:

$$\Delta L = L_1 - L_2 = \alpha \cdot f - f \cdot \sin\alpha = f \cdot (\alpha - \sin\alpha) \qquad (7)$$

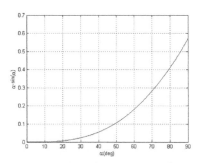

Figure 2. Relation between the distortion degree of any point and its distance away from the center of object plane.

$\Delta L$ describes the difference between the ideal imaging point and the actual imaging point. From formula (7) and Figure 2, we can conclude that for any point in the object plane $Q$, the distortion degree is related to its distance to the centre of object plane, and is symmetric to the centre point as a circularly symmetric. The closer it gets the centre point, the smaller the distortion is.

## 3 THE REALIZATION OF DISTORTION CORRECTION ALGORITHM BASED ON THE EQUIVALENT SPHERICAL SURFACE TEXT AND INDENTING

### 3.1 The coordinate computation

#### 3.1.1 The coordinate computation of barrel distortion correction point

In engineering applications we can always find barrel distortion problem, which can be easily detected when the image contains more lines. As shown in Figure 3, in order to see the image distortion in the image and the correction effect, we respectively add two red vertical baselines in the original image and the corrected image in the experiment. It shows that intersecting lines near the red baseline has undergone a significant deformation, and the deformation degree increases as the distance from the latitude and longitude lines to the centre of the image growing larger.

Figure 3.  chess-board image with barrel distortion.

Figure 4.   Image after distortion correction by the algorithm this paper proposes.

By using the improved distortion correction model based on the equivalent spherical surface, we can get the corrected position coordinates $P_1$:

$$\begin{cases} x_1 = L_1 \cdot \cos\beta = \alpha \cdot f \cdot \cos\beta \\ y_1 = L_1 \cdot \sin\beta = \alpha \cdot f \cdot \sin\beta \end{cases} \qquad (8)$$

That is,

$$\begin{cases} x_1 = \arcsin(\dfrac{\sqrt{x_2^2 + y_2^2}}{f}) \cdot f \cdot \cos(\arctan(\dfrac{y_2}{x_2})) \\ y_1 = \arcsin(\dfrac{\sqrt{x_2^2 + y_2^2}}{f}) \cdot f \cdot \sin(\arctan(\dfrac{y_2}{x_2})) \end{cases} \qquad (9)$$

Use the created barrel distortion correction model based on the spherical equivalent surface, we conduct a correct computation for the coordinate of the image distortion point by formula (9), and take the barrel distortion chess-board image acquired by a digital camera as an example. We apply Visual C++ to the correct computation in the laboratory, and the experimental results are shown in Figure 4. We can see that bending lines in the original image completely overlap with the red baselines:

#### 3.1.2 The coordinate computation of pincushion distortion correction point

Pincushion distortion is usually found in the image taken with telephoto lens, and in a similar way we can get the corrected position coordinates P1 by using the improved distortion correction model based on the equivalent spherical surface.

We denote that $P_1$ $(x_1, y_1)$ is the corresponding point of $P_0$ in ideal image, and the corresponding point of $P_0$ in distorted image is $P_2$ $(x_2, y_2)$.

$$\begin{cases} x_1 = \sin(\dfrac{\sqrt{x_2^2 + y_2^2}}{f}) \cdot f \cdot \cos(\arctan(\dfrac{y_2}{x_2})) \\ y_1 = \sin(\dfrac{\sqrt{x_2^2 + y_2^2}}{f}) \cdot f \cdot \sin(\arctan(\dfrac{y_2}{x_2})) \end{cases} \quad (10)$$

By using the formula (10) to conduct correction computation for the coordinates of the image distortion point, and by taking the coordinates with pincushion distortion (Figure 5) as an example, we apply Visual C++ to the correct computation, and the experimental results are shown in Figure 6. We can see that bending lines in the original image completely overlap with the red baselines:

### 3.2 Processing of floating point coordinates based on the bilinear difference

In the computing process of distortion correction, it is likely to find that the integer coordinates in corrected image are mapped to a non-integer coordinates in the original distorted image, and these non-integer coordinates in the mapping process are called floating point coordinates. In order to retrieve greyscale information of these floating point coordinates, we use bilinear interpolation to realize the greyscale value of floating point coordinates.

Figure 5. Coordinates with slight pincushion distortion.

Figure 6. Coordinates after correction.

Assume the $P_1 (x_1, y_1)$ represents any point in the corrected image, and after the correct computation, we can obtain its corresponding point $P_2(x_2, y_2)$ in tdistortedl distortion image. If $(x_2, y_2)$ is a floating point coordinate, we can get the largest corresponding integer point $T_0 (x, y)$ in tdistortedl distortion image based on the number rounded downwards to the nearest integer method, and we have:

$$\begin{cases} x = \lfloor x_2 \rfloor = floor(x_2) \\ y = \lfloor y_2 \rfloor = floor(y_2) \end{cases} \quad (11)$$

If $u = x_2 - x$, $v = y_2 - y$, obviously, $0 < u < 1, 0 < v < 1$.

As is shown in Figure 7, if $T_1, T_2, T_3$ are the other three po,ts near $P_2$, and their rgreyscale correspondi (g, graysca (e v, lue are (g(x, y+1), g(x+1,y+1), g(x+1,y) and the correspondi (g grayscal ( value of $T_0(x,y)$ is $g(x,y)$. By bilinear interpolation we can get the formula of approximate value of greyscale, as is shown in the formula (12).

$$g(x_2,y_2) = a \cdot g(x,y) + b \cdot g(x,y+1) + c \cdot g(x+1,y+1) + d \cdot g(x+1,y)$$

$$(12)$$

Where in, $a = (1 - v) \cdot (1 - u), b = v \cdot (1 - u), c = v \cdot u, d = (1 - v) \cdot u$.

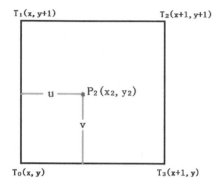

Figure 7. Sketch map of bilinear interpolation.

### 3.3 Process of scaling

From Figures 4 and 6, we know that those seriously bending lines at the edge of image improve a lot after correction, but the barrel distortion image loses its image information, as is shown in Figure 4, and around the pincushion distortion there will be a grey area with no pixel, as is shown in Figure 6. That is because the image size varies over the original one after the application of correction model, and the barrel distortion image becomes larger in size while the pincushion distortion image becomes smaller in size after the image correction. We therefore adopt the geometric scaling method to achieve the consistency of the size between corrected image and the original one. Assume that $f(x, y)$ is the grayscale value of any point in the original image, and after correction, the pixel graysc, value is $g(x, y)$, and the scaling coefficient is $K$, then we have,

$$g(x,y) = f(k \cdot x, k \cdot y) \quad (13)$$

To avoid the loss of image information after barrel distortion image correction, the $K$ we select should ensure the corner point which will change most in the correction process will just appear in the image after narrowing, and we can use the formula (14) to calculate the scaling coefficient $K$.

Denote the width and height of original image respectively by $W$ and $H$.

$$K = \frac{\frac{\arcsin(\sqrt{x_2^2 + y_2^2})}{f} \cdot f}{\sqrt{(\frac{W}{2})^2 + (\frac{H}{2})^2}} \quad (14)$$

To reduce the grey area with no pixel information around the image after pincushion distortion, we need to magnify the corrected image, and the $K$ we select should ensure the least change points which are the centre of the up and down edge of the image in the correction process will be included in the image after magnifying. We can use the formula (15) to calculate the scaling coefficient $K$.

$$K = \frac{\sin(\frac{\sqrt{x_1^2 + y_1^2}}{f}) \cdot f}{\sqrt{(\frac{W}{2})^2 + (\frac{H}{2})^2}} \qquad (15)$$

From Figure 8 we can see that there exists a pincushion area in the barrel distortion image after correction, which keeps all the information of the original image.

From Figure 9 we can see that there exists barrel area in the pincushion distortion image after correction, and the up edge of the barrel shape is just tangent to the edge of the image, which means all the information of original image is kept.

### 3.4 Edge filling

Though the image after scaling processing retains all the information of the original one, there exists an area without grayscale information at the edge of the image. To ensure the integrity of the image, we can use an edge filling algorithm. First, we expand the periphery of the image edge, then correct it and scale it, thus we can avoid the appearance of the region without grayscale information. We use the grayscale information of the image edge and expand outward. Take the left edge as an example, if $(x_0, y_0)$ is the grayscale-lacked point just near the left edge, and $(x_0+1, y_0+1)$、$(x_0+1, y_0)$、$(x_0+1, y_0-1)$ are points in the left edge, $f(x_0, y_0)$、$f(x_0+1, y_0+1)$、$f(x_0+1, y_0)$、$f(x_0+1, y_0-1)$ are their corresponding grayscale values. The value of $f(x_0, y_0)$ is given by the formula (16):

$$f(x_0, y_0) = (f(x_0+1, y_0+1) + f(x_0+1, y_0) + f(x_0+1, y_0-1)) / 3 \qquad (16)$$

Figure 8. The barrel distortion correction image after demagnified.

Figure 9. The pincushion distortion correction image after magnified.

Images in Figure 3 and Figure 5 are just shown in Figure 10 and Figure 11 after edge filling and scaling process.

Figure 10. Barrel distortion correction image after edge filling and scaling process.

Figure 11. Pincushion distortion correction image after edge filling and scaling process.

Just as is shown in above Figures, edge filling and scaling process eliminate the grayscale-lacked area at the edge of image after correction, and ensure its integrity.

### 3.5 Algorithm verification

In order to verify the correctness of the algorithm, we use the correction algorithm to correct two actual images. One is with barrel distortion and the other is with pincushion distortion. Figure 12 reflects the barrel distortion; the originally straight step edges near the red baseline deviate from the red line, and present a serious bend. After the barrel distortion correction computing, scaling and edge filling, the bending edge in Figure 12 completely overlaps with the red baselines, as is shown in Figure 13. Figure 14 reflects the pincushion distortion; the originally straight wall tile edges near the red baseline deviate from the red line, and present a serious bend. After the pincushion distortion correction computing, scaling and edge filling, the bending edge in Figure 14 completely overlaps with the red baselines, as is shown in Figure 15.

From the experiments above, it can be proved that the equivalent spherical model, this paper establishes can be not only applied to the correction of barrel distortion image, but also to the pincushion distortion image correction, therefore, this mapping model is effective to deal with any of the spherical distortion in the image.

Figure 12. The image with barrel distortion.

Figure 13. The image after distortion correction computed.

## 4 CONCLUSION

According to the idea of using equivalent spherical surface to simulate the distortion generating, this paper proposes an improved distortion correction

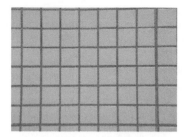

Figure 14. The image with pincushion distortion.

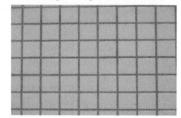

Figure 15. The image after distortion correction computed.

model based on the equivalent spherical surface, and establishes the mapping relation between coordinates in distorted images and in ideal image, and conducts experiments for the algorithm proposed in this paper. The experimental results show that the algorithm can effectively solve the correction of barrel distortion and pincushion distortion in the image, and the correction algorithm is easy to implement with simple computation, and it can be applied to the image distortion correction easily and efficiently.

## REFERENCES

[1] Guangliang Han, Jianzhong Song. An equivalent surface model of image distortion and the correction algorithm [J]. Optical Technique, 2005, 31(1):122–124.
[2] K R Castleman. Digital Image Processing [J]. Prentice Hall. 1998. 9:255–278.
[3] Y T Zhou,R Chellappa. Image restoration using a neural network [J]. IEEE Trans. Speech, Signal Processing.1988, 36(7):1141–1151.
[4] J K Palk, A K Katsaggelos. Image restoration using a modified Hopfield network. [J]. IEEE Tram. Image Processing.1992, 1(1):49–63.
[5] G R Ayers, J C Dainty. Iterative blind deconvolution method and its applications [J]. Optics Letters. 1998, 13(7):547–549.

*3. Information, electronic and mechanical engineering*

*Multimedia, Communication and Computing Application – Leung (Ed.)*
*© 2015 Taylor & Francis Group, London, ISBN 978-1-138-02775-6*

# Theoretical study on GAWBS in a highly nonlinear fiber

J. Wang
*Science and Technology on Optical Radiation Laboratory, Beijing, China*

ABSTRACT: Due to the larger birefringence effect in highly nonlinear fibers (HNLF), it is able to observe a higher scattering efficiency of GAWBS. In this paper, we theoretically analyze the polarized and depolarized GAWBS induced by transverse acoustic mode in HNLF for the first time. We deduce the displacement function and scattering efficiency of GAWBS starting from the vibration function of transverse mode. The frequency of peak scattering efficiency in HNLF is higher than single-mode fiber (SMF-28), which appears at the frequency of nearly 470 MHz. The bandwidth of scattering in HNLF is about three times than that in SMF-28. The scattering efficiency of polarized GAWBS induced by $R_{0m}$ mode is bigger than that by $TR_{2m}$ mode, which is almost vanished at higher frequency.

## 1 INTRODUCTION

Guided acoustic wave Brillouin scattering (GAWBS), also known as forward Brillouin scattering, thermally excited mechanical vibrations in glass fibers modulates the phase and polarization of the input light and can be observed in the forward direction. As a severe problem in quantum optics, firstly GAWBS in optical fibers has been studied in conventional fibers [1,2]. Some research have been done to reduce this limiting factor and to alter the fiber geometry, such as using microstructured fibers and photonic crystal fibers (PCF) [3,4]. Since then, GAWBS has been extensively explored in fibers with different core dimensions and polarization properties. The forward stimulated Brillouin scattering (FSBS) were observed recently because of the tight confinement of both light and sound [5–6], which opens new opportunities for controlling light–sound interactions. In this scattering process, the transverse acoustic mode were classified into radial ($R_{0m}$) modes and doubly degenerated torsional-radial ($TR_{2m}$) modes by the transverse displacement profile which decides the scattering characteristics by acousto-optic interaction. Recent studies show that the polarized GAWBS can enhance cascaded Stokes and anti-Stokes scattering in FSBS [7] based on the automatic phase matching.

There are polarized and depolarized GAWBS in optical fibers. The depolarized one induced only by $TR_{2m}$ modes is widely studied before, and it is easily detected by polarization techniques [8]. The polarized GAWBS is practically more important because it yields stronger and polarization-independent interactions with the fundamental optical mode. The polarization remains unaffected but only the phase of the input lights is modulated. The corresponding induced scattering is so called polarized GAWBS, which is induced by the $R_{0m}$ modes and one of the doubly degenerated $TR_{2m}$ modes. Then, a Mach-Zehnder interferometer is needed to observe this phase modulation [9]. Most researches of $TR_{2m}$ mode are studied for the depolarized GAWBS. However, few works focus on the polarized one, because the pure phase modulation induced by the $TR_{2m}$ mode does not appear to be as intense as that induced by the $R_{0m}$ modes [10]. Ref [11] reports the results that the polarized GWABS caused by the $TR_{2m}$ mode is observed in a panda fiber at lower-frequency region by employing a Sagnac loop because of the high birefringence in fiber, and then in the small core PCF [12].

In this paper, we theoretically analyze the polarized and depolarized GAWBS in HNLFs. The normalized density, displacement function, and scattering efficiency are deduced. Furthermore, we compare the results with SMF by simulation: the frequency of peak value in HNLF which appears at 470 MHz is larger than that in SMF-28; the scattering bandwidth in HNLF is about three times than that in SMF-28; the displacements of either $R_{0m}$ mode or $TR_{2m}$ mode at lower frequency are larger than that at higher frequency; the scattering efficiency of $R_{0m}$ mode is larger than that of $TR_{2m}$ mode, and it is almost vanished at higher frequency.

## 2 THEORY OF GAWBS

### 2.1 $R_{0m}$ mode and $TR_{2m}$ mode in GAWBS

GAWBS is caused by both longitudinal and shear acoustic modes propagating in the transverse direction, which is along the cross section of the fiber and perpendicular to the fiber axis. Let us recall the acoustic modes corresponding to the GAWBS in optical fibers. There are only two types of acoustic modes in optical fibers considering the electrostrictional excitation of acoustic waves by the optical fields: radial ($R_{0m}$) modes and torsional-radial ($TR_{2m}$) modes [13].

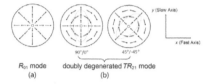

$R_{01}$ mode    doubly degenerated $TR_{21}$ mode
(a)           (b)

Figure 1. The cross section vibration patterns of the lowest order of (a) radial mode ($R_{01}$) and (b) doubly degenerated torsional-radial mode ($TR_{21}$).

Fig. 1 shows two types of the acoustic modes in the lowest mode number. Radial modes ($R_{0m}$) propagating radially are polarization insensitive longitudinal modes and the waveform of $R_{01}$ mode in fiber cross section is shown in Fig. 4 (a). They can be equally excited by circularly and linearly polarized light and do not affect the polarization of the input light. The torsional-radial modes ($TR_{2m}$) combine radial component (longitudinal) and torsional component (shear), both of which propagate radially. We get two forms of modes by combining them and the waveform of $TR_{21}$ which are shown in Fig. 4 (b). These torsional-radial modes are doubly degenerated in symmetric fibers. This motion operates as dotted ellipse in Fig. 4(b), which induces the additional birefringence to the fiber. The radial component oscillating along the fiber principal axes (axis $x$ and $y$ in Fig. 4(b)) induces the birefringence along the principal axes without changing the principal axes. The excitation efficiency of the torsional-radial modes is highly dependent on the ellipticity of the input light polarization. It is maximum for a linear polarization and zero for a circular polarization state [13–15].

One of the degenerated $TR_{2m}$ modes oscillating at 45° relative to the principal axes will rotate the principal axes of the fiber by inducing its own birefringence which does not line up with that of the fiber. This can mediate only forward stimulated interpolarization scattering (SIPS), in which processing the coherent Stokes and anti-Stokes scattering is highly suppressed [7]. However, for the other $TR_{2m}$ modes, the polarization remains unaffected by these modulations and only the phase of the input lights is modulated, which result in polarized GAWBS. So the $TR_{2m}$ modes vibrating along the 90°/0° axes can cause forward stimulated Raman-like scattering (SRIS) as $R_{0m}$ modes [16].

## 2.2 Scattering efficiency of GAWBS

We calculate the scattering efficiency based on the general scattering efficiency equation [17]. The lights with the wave vector $k_0$ are launched into a fiber of density $\rho_0$ and refractive index $n$, and the polarized and depolarized scattering efficiency induced by $R_{0m}$ modes and $TR_{2m}$ modes are

$$\eta_R = \frac{n^6 \pi k_0^2 kT (p_{11}+p_{12})^2}{2^4 \rho_0 \Omega_{Rm}^2} \cdot \frac{\left[ \int_0^a (\frac{\partial U_r^R}{\partial r} + \frac{U_r^R}{r}) E(r) r dr \right]^2}{\int_0^a \left[ (U_r^R(r) \right]^2 r dr}$$

(1)

$$\eta_T^P = \mathbf{P} \frac{n^6 \pi k_0^2 kT (p_{11}-p_{12})^2}{2^5 \rho_0 \Omega_{Tm}^2}$$

$$\cdot \frac{\left\{ \int_0^a [\frac{\partial}{\partial r}(U_r^T - U_\varphi^T) + \frac{1}{r}(U_r^T - U_\varphi^T)] E(r) r dr \right\}^2}{\int_0^a \left\{ [U_r^T(r)]^2 + [U_\varphi^T(r)]^2 \right\} r dr}$$

(2)

$$\eta_T^D = \mathbf{P} \frac{n^6 \pi k_0^2 kT (p_{11}-p_{12})^2}{2^7 \rho_0 \Omega_{Tm}^2}$$

$$\cdot \frac{\left\{ \int_0^a [\frac{\partial}{\partial r}(U_r^T - U_\varphi^T) + \frac{1}{r}(U_r^T - U_\varphi^T)] E(r) r dr \right\}^2}{\int_0^a \left\{ [U_r^T(r)]^2 + [U_\varphi^T(r)]^2 \right\} r dr}$$

(3)

Where $\mathbf{P} = |1-e^2|/(1+e^2)$, in which the ellipticity $e$ which represents the polarization ellipse is the major-to-minor-axis ratio. An ellipticity of zero or infinity corresponds to linear polarization and an ellipticity of 1 corresponds to circular polarization. The strain-optic coefficients are $p_{11} = 0.121$ and $p_{12} = 0.270$ for fused quartz; the superscripts "$P$" and "$D$" represent the polarized and depolarized one; $E(r)$ is the normalized profile of the guide acoustic wave; $U_r^{R(T)}(r)$ and $U_\varphi^{R(T)}(r)$ are the radial and azimuthal displacement components, respectively, corresponding to radius $r$ [15].

## 3 SIMULATION

### 3.1 Resonance frequency

HNLF of ofs optical fiber companycompany and SFM-28 of Corning company are used for simulation. The parameters of HNLF are 125 μm of cladding diameter, 11.5 μm² of effective fiber area, and 3% of refractive index difference; the parameters of

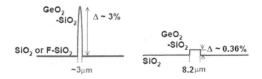

Figure 2. The configuration of two fibers.

SMF-28 are 125 μm of coating diameter, 8.2 μm of core diameter, 0.36% of refractive index difference. The configuration of two fibers is shown in Fig. 2.

The frequencies $\Omega_m = y_m V_L / a$ and $\Omega_m = y_m V_S / a$ are determined by solving the eigen equations with the boundary condition [18]. The silica rod approximation is used for the calculation corresponding to acoustic modes and both the calculated and the observed results are shown in Tables 1 and 2. The parameters for silica fibers are given by [5,6] $\gamma_e = 1.17$, $\rho_0 = 2.20 \times 10^3$ kg/m$^{-3}$, and $V_L = 5590$ m/s. Also, we use $\varepsilon_0 = 8.85 \times 10^{-12}$ F/m and $\omega_0 = 2\pi \times 193.5$ THz. It is shown that the frequency is determined by fiber dimension and the speed of acoustic mode which is approximately constant. The fiber radius is estimated by measuring the acoustic frequency in Ref [19].

## 4 GAWBS SCATTERING EFFICIENCY

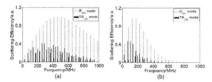

Figure 3. The scattering efficiency of polarized GAWBS induced by $R_{0m}$ and $TR_{2m}$ modes of (a) HNLF and (b) SMF.

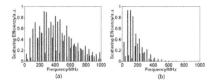

Figure 4. The scattering efficiency of depolarized GAWBS induced by $TR_{2m}$ mode of (a) HNLF and (b) SMF.

Table 1. Mode frequency of $R_{0m}$ mode.

| $M$ | 1 | 2 | 3 | 4 | 5 | 6 | 7 |
|---|---|---|---|---|---|---|---|
| $\Omega_m$(MHz) | 30.5 | 82.1 | 130.7 | 179.1 | 227.2 | 275.3 | 323.2 |
| $M$ | 8 | 9 | 10 | 11 | 12 | 13 | 14 |
| $\Omega_m$(MHz) | 371.3 | 419.4 | 467.4 | 515.3 | 563.3 | 611.4 | 659.3 |
| $M$ | 15 | 16 | 17 | 18 | 19 | 20 | 21 |
| $\Omega_m$(MHz) | 707.4 | 755.2 | 803.3 | 848.5 | 899.2 | 947.3 | 995.2 |

Table 2. Mode frequency of $TR_{2m}$ mode.

| $M$ | 1 | 2 | 3 | 4 | 5 | 6 | 7 |
|---|---|---|---|---|---|---|---|
| $\Omega_m$(MHz) | 22.3 | 39.6 | 71.0 | 81.8 | 108.5 | 127.4 | 140.0 |
| $M$ | 8 | 9 | 10 | 11 | 12 | 13 | 14 |
| $\Omega_m$(MHz) | 169.3 | 177.8 | 200.2 | 225.8 | 230.9 | 260.4 | 275.0 |
| $M$ | 15 | 16 | 17 | 18 | 19 | 20 | 21 |
| $\Omega_m$(MHz) | 290.1 | 320.2 | 324.0 | 350.6 | 372.1 | 380.7 | 410.5 |
| $M$ | 22 | 23 | 24 | 25 | 26 | 27 | 28 |
| $\Omega_m$(MHz) | 420.6 | 440.6 | 468.7 | 470.9 | 500.5 | 517.3 | 530.5 |
| $M$ | 29 | 30 | 31 | 32 | 33 | 34 | 35 |
| $\Omega_m$(MHz) | 560.3 | 565.7 | 590.3 | 613.9 | 620.3 | 650.2 | 662.3 |
| $M$ | 36 | 37 | 38 | 39 | 40 | 41 | 42 |
| $\Omega_m$(MHz) | 680.2 | 709.8 | 711.0 | 740.0 | 758.9 | 770.0 | 799.9 |
| $M$ | 43 | 44 | 45 | 46 | 47 | 48 | 49 |
| $\Omega_m$(MHz) | 807.1 | 829.8 | 855.4 | 859.8 | 889.7 | 903.7 | 919.6 |
| $M$ | 50 | 51 | 52 | 53 | 54 | | |
| $\Omega_m$(MHz) | 949.5 | 952.0 | 979.0 | 1000 | 1009 | | |

The efficiencies of the polarized GAWBS for linearly polarization in the HNLF and SMF from zero to 1 GHz are calculated and shown in Fig. 3. The experimental parameters of HNLF (cladding diameter: $125 \pm 1$ μm; effective area: $11.5$ μm$^2$) and Corning SMF-28 (core diameter: $8.2$ μm; cladding diameter: $125$ μm; refractive index different: $0.36\%$) are used for Fig. 3 (a) and (b). We notice that they are in the same cladding diameter; however, the core diameter of SMF-28 is nearly twice comparing to the HNLF in our experiment. Fig. 3 also shows that initially the efficiency increases with the frequency and then drops off after a peak value in both fibers. However, as the core radius reduces, the maximum efficiency shifts to a higher frequency. The maximum efficiencies of SMF-28 are $127$ MHz ($R_{03}$) and $105$ MHz ($TR_{25}$), which agree with the Shelby's experimental results [2]. And the peak values shift to higher frequency of $460$ MHz ($R_{0(10)}$) and $256$ MHz ($TR_{2(13)}$) in the HNLF. The smaller core fiber has a much larger overall bandwidth of the scattering and this prediction agrees with the experiment results proposed in Ref [20]. The FWHM bandwidth of $R_{0m}$ mode for HNLF is three times than that for SMF-28. It is also shown that the polarized scattering efficiency of $R_{0m}$ mode is much higher than the $TR_{2m}$ one.

When we convert the efficiency unit to "dB," the scattering efficiencies of polarized and depolarized GAWBS are shown in Fig. 5.

Figure 5. Scattering efficiency of (a) polarized GAWBS in HNLF and (b) depolarized GAWBS with unit of dBm.

The efficiency of the polarized GAWBS induced by $TR_{2m}$ modes drops off rapidly after 600 MHz in Fig. 6(a). The reason for it is that the spectrum is almost pure for $R_{0m}$ modes at higher frequencies. The 3-dB bandwidth of polarized $R_{0m}$ GAWBS is about 900 MHz for both the experiment and the calculation. The calculated efficiency shows that there is an evident envelope crossing the whole frequency range, especially for $R_{0m}$ modes, which is shown as the dotted profile in Fig. 6(a). The highest efficiency appears at around 500 MHz for $R_{0m}$ modes and 300 MHz for $TR_{2m}$ modes. All these results are in good agreement for the whole frequency range up to 1 GHz.

## 5 CONCLUSIONS

In this paper, we theoretically analyze the polarized and depolarized GAWBS induced by transverse acoustic mode in HNLF for the first time. The calculated efficiency shows that as the core radius reduces the envelope of GAWBS modes, the maximum efficiency shift to the higher range. The smaller core fiber has a much larger overall bandwidth of the scattering. The polarized scattering efficiency of $R_{0m}$ mode is higher than the $TR_{2m}$ one. The highest efficiency is around 500 MHz for $R_{0m}$ modes and 300 MHz for $TR_{2m}$ modes. The efficiency of $TR_{2m}$ modes drops off rapidly after 600 MHz. The frequency of peak scattering efficiency in HNLF is higher than single-mode fiber (SMF-28), which appears at the frequency of nearly 470 MHz. The bandwidth of scattering of HNLF is about three times than that of SMF-28. The scattering efficiency of polarized GAWBS induced by $R_{0m}$ mode is bigger than $TR_{2m}$ mode, which is almost vanished at higher frequency.

## REFERENCES

P. J. Thomas, N. L. Rowell, H. M. van Driel, and G. I. Stegeman, "Normal acoustic modes and Brillouin scattering in single-mode optical fibers," Phys. Rev. B. **15**, 4986–4998 (1979).

R. M. Shelby, M. D. Levenson, and P. W. Bayer, "Guided acoustic-wave Brillouin scattering," Phys. Rev. B **31**, 5244–5252 (1985).

J. Beugnot, T. Sylvestre, and H. Maillotte, "Guided acoustic wave Brillouin scattering in photonic crystal fibers," Opt. Lett. **32**, 17–19 (2007).

N. Shibata, A. Nakazono, N. Taguchi, and S. Tanaka, "Forward Brillouin scattering in holey fibers," IEEE Photon. Technol. Letts. **18**, 412–414 (2006).

M. S. Kang, A. Nazarkin, A. Brenn, and P. St. J. Russell, "Tightly trapped acoustic phonons in photonic crystal fibres as highly nonlinear artificial Raman oscillators," Nature Phys. **5**, 276–280 (2009).

J. Wang, Y. Zhu, R. Zhang, and D. J. Gauthier, "FSBS resonances observed in a standard highly-nonlinear fiber," Opt. Express **19**, 5339–5349 (2011). http://www.opticsinfobase.org/abstract.cfm?URI=oe-19-6-5339.

M. S. Kang, A. Brenn, and P. St.J. Russell, "All-Optical Control of Gigahertz acoustic resonances by forward stimulated interpolarization scattering in a photonic crystal fiber," Phys. Rev. Lett. **105**, 153901 (2010).

J. E. McElhenny, R. K. Pattnaik, and J. Toulouse, "Dependence of frequency shift of depolarized guided acoustic wave Brillouin scattering in photonic crystal fibers," J. of Lightwave Technol. **29**, 200–208 (2011).

K. Bergman, H. A. Haus, and M. Shirasaki, "Analysis and measurement of GAWBS spectrum in a nonlinear fiber ring," Applied Physics B. **55**, 242–249 (1992).

N. Nishizawa, S. Kume, M. Mori, T. Goto, and A. Miyauchi, "Characteristics of guided acoustic wave Brillouin scattering in polarization maintaining fibers," Opt. Rev. **3**, 29–33 (1996).

N. Nishizawa, S. Kume, M. Mori, and T. Goto, "Expriment analysis of guided acoustic wave Brillouin scattering in papda fibers," J. Opt. Soc. Am. B **12**, 1651–1655 (1995).

J. C. Beugnot, T. Sylvestre, H. Maillotte, G. Mélin, and V. Laude, "Guided acoustic wave Brillouin scattering in photonic crystal fibers," Opt. Lett. **32**, 17–19 (2007).

Y. Jaouën, and L. du Mouza, "Transverse Brillouin Effect Produced by Electrostriction in Optical Fibers and Its Impact on Soliton Transmission Systems," Optical Fiber Technology 7, 141–169 (2001).

L. du Mouza, Y. Jaouën, and C. Chabran, "Transverse Brillouin effect characterization in optical fibers and its geometrical aspects," IEEE Photon. Technol. Letts. **10**, 1455–1457 (1998).

E. A. Golovchenko, and A. N. Pilipetskii, "Acoustic effect and the polarization of adjacent bits in soliton communication lines," J. Lightwave Techonl. **12**, 1052–1056 (1994).

P. Dainese, P. St. J. Russell, G. S. Wiederhecker, N. Joly, H. L. Fragnito, V. Laude, and A. Khelif, "Raman-like light scattering from acoustic phonons in photonic crystal fiber," Opt. Express **14**, 4142–4150 (2006).

N. Nishizawa, S. Kume, M. Mori, T. Goto, and A. Miyauchi, "Characteristics of guided acoustic wave Brillouin scattering in polarization maintaining fibers," Opt. Rev. **3**, 29–33 (1996).

E. K. Sittig, and G. A. Coquin, "Visualization of plane-strain vibration modes of a long cylinder capable of producing sound radiation," J. Acoust. Soc. Am. **48**, 1150–1159 (1979).

K. Shiraki and M. Ohashi, "Sound velocity measurement based on guided acoustic-wave Brillouin scattering", 1992, IEEE Photon. Technol. Lett., Vol. 4, No. 10, pp. 1177–1180.

A. J. Poustie, "Bandwidth and mode intensities of guided acoustic-wave Brillouin scattering in optical fibers," J. Opt. Soc. Am. B 20, 691–696 (1993).

*Multimedia, Communication and Computing Application – Leung (Ed.)*
© *2015 Taylor & Francis Group, London, ISBN 978-1-138-02775-6*

# Study of the watermarking algorithm based on the mix of time domain

H.Y. Fang, X.Q. Zhao & A.H. Guo
*College of Electronics and Information Engineering, Tongji University, Shanghai, China*

ABSTRACT: This paper studies a technique of image digital watermarking—the mix of time domain. This technique can be applied to a watermarking algorithm in order to change the properties of the watermark. With the help of the Matlab, this paper carries out attack tests and a simulation experiment for the DCT–DWT–SVD digital watermarking based on the mix of time domain and compares the result with the original method. The experiment result indicates that the mix of time domain can greatly improve the imperceptibility of the watermark and enhance the sensitivity of the watermark to noise attacks, which thereby has a high application value in the field of semifragile watermark and fragile watermark.

KEYWORDS: Digital watermarking, Image watermarking, Hadamard, Information hiding, DCT

## 1 INTRODUCTION

Digital watermarking is an advanced information hiding technology, which has played an important role in copyright protection.

At the early stage of the digital watermarking, people tended to embed the watermark into the spatial domain. For example, Least Significant Bit (LSB) [1] adds the watermark information to the least significant bit of the pixel. The algorithm is simple and effective, which is also fairly perceptually invisible. However, the watermark information is vulnerable to the noise attack. As a result, LSB is suitable for making fragile or semifragile watermark. Contrast to the Time domain, the watermarking algorithm based on frequency domain not only has good invisibility but also has good robustness, which can resist several noise attacks.

A concept about the mix of time domain is proposed in this paper, which combines the time domain transformation with the frequency domain transformation when embedding the watermark. It has been proved that the application of the mix of time domain on the traditional DWT and DCT watermarking algorithm [2, 3] can improve the invisibility and robustness of the watermark. Based on the former research, this paper has made further study on the effect of the mix of time domain when used on other improved watermarking algorithm. The paper applies the mix of time domain to the DCT–DWT–SVD watermarking algorithm [4], utilizing the orthogonal matrices to transform the carrier image in time domain before executing the watermarking algorithm and then use relevant orthogonal matrix to restore. The experiment shows that the invisibility of the watermark has been

greatly improved, but watermark will be more sensitive to noise attacks.

## 2 THE APPLICATION OF ORTHOGONAL MATRICES IN IMAGE PROCESSING

Orthogonal matrices are used to make a transform process to the carrier image in the time domain. The paper executes the inverse transformation by utilizing the orthogonality of the orthogonal matrix.

### 2.1 Hadamard matrix

The Hadamard matrix is a special orthogonal matrix, which is only composed of −1 and 1. Like other orthogonal matrix, it meets the orthogonality:

$$H_n H_n^T = nI_n \qquad (1)$$

According to the recursion formula, $H_{2^{k+1}} = \begin{bmatrix} H_{2^k} & H_{2^k} \\ H_{2^k} & -H_{2^k} \end{bmatrix}$, where $k = 0,1,2\cdots$, and higher–order Hadamard Matrices could be obtained:

$$H_2 = \begin{bmatrix} H_1 & H_1 \\ H_1 & -H_1 \end{bmatrix} = \begin{bmatrix} +1 & +1 \\ +1 & -1 \end{bmatrix} \qquad (2)$$

$$H_4 = \begin{bmatrix} H_2 & H_2 \\ H_2 & -H_2 \end{bmatrix} = \begin{bmatrix} +1 & +1 & +1 & +1 \\ +1 & -1 & +1 & -1 \\ +1 & +1 & -1 & -1 \\ +1 & -1 & -1 & +1 \end{bmatrix} \qquad (3)$$

## 2.2 Orthogonal matrix pair

The orthogonal matrix pair is an extended form of orthogonal matrix, which can also be used for signal processing [5, 6]. If a two-dimensional M × M order array pairs (A, B) are orthogonal, then:

$$AB^T = cI_m \text{ or } BA^T = cI_m F \qquad (4)$$

In the equation above, c is a constant, $I_m$ is m order unit matrix, and (A, B) is called the orthogonal matrix pair. The orthogonal matrix pair (A, B) can be used to transform column vector $X$ whose length is $M$ and then generate a transform-domain signal $Y$:

$$Y = B^T X \qquad (5)$$

The original signal can be restored through the following operations:

$$X' = \frac{1}{c} AY = \frac{1}{c} AB^T = X \qquad (6)$$

If matrix $A$ is equal to matrix $B$ and the elements are only composed of $-1$ and 1, matrix $A$ and $B$ would be converted into Hadamard Matrix. As a result, a Hadamard Matrix meets the Eq. 6.

Each row of n-order orthogonal matrix pair is the cyclic shift of the n-length perfect binary sequence pair [7], so the paper gives $MA8$ of $(MA8,MB8)$ orthogonal matrix pair and the first row of $MB8$ $(MA16,MB16)$ orthogonal matrix pair.

$$MA_8 = \begin{bmatrix} 1 & 1 & 1 & -1 & -1 & -1 & -1 & -1 \\ 1 & 1 & -1 & -1 & -1 & -1 & -1 & 1 \\ & & & \vdots & & & & \\ -1 & 1 & 1 & 1 & -1 & -1 & -1 & -1 \end{bmatrix}$$

$$MB_8 = \begin{bmatrix} 1 & -1 & 1 & -1 & -1 & 1 & -1 & -1 \end{bmatrix}$$

$$MA_{16} = \begin{bmatrix} -1 & -1 & -1 & -1 & -1 & -1 & -1 & -1 & -1 & 1 & 1 & 1 & 1 & 1 & 1 \end{bmatrix}$$

$$MB_{16} = \begin{bmatrix} -1 & -1 & 1 & -1 & 1 & -1 & 1 & -1 & -1 & 1 & -1 & 1 & -1 & 1 & -1 & 1 \end{bmatrix}$$

The image matrix multiplied by one of these matrices is equivalent to operating simple additions and subtractions. The transformation thereby is very fast. In fact, besides matrices consisting of $-1$ and 1, there are many other orthogonal matrices that can also be used for the mix of time domain algorithm.

## 3 A WATERMARKING ALGORITHM BASED ON FREQUENCY DOMAIN

### 3.1 Digital watermarking algorithm combining DCT, DWT, and SVD

DCT–DWT–SVD watermarking algorithm combines DCT, DWT, SVD, the three watermarking technologies. First one-level DWT is applied to original cover image. To achieve imperceptibility, LL band is selected for second-level decomposition and HH band is selected. The experiment result shows that this watermarking algorithm has strong imperceptibility than other DCT–DWT–SVD watermarking algorithm. Meantime, the watermark is sensitive to several noise attacks. As a result, this watermark algorithm is suitable for the production of fragile watermarking and semifragile watermarking, which would be good at proving the integrity and authentication of the information. The following is the detailed process of watermark embedding and extraction.

#### 3.1.1 The embedding process of DCT–DWT–SVD watermark

1  Let $IM$ be the carrier image of size $m \times m$. Select the red channel $IMR$ and apply DWT to decompose it into four $\frac{m}{2} \times \frac{m}{2}$ sub-bands $LH$, $HL$, $HH$, and $LL$. (In this paper, Discrete Wavelet Transform uses db1 wavelets). Apply DWT to decompose the $LL$ band into four $\frac{M}{4} \times \frac{M}{4}$ sub-bands $LL\_LH$, $LL\_HL$, $LL\_HH$, and $LL\_LL$.

2  Select $LL\_HH$ band, divide it into $4 \times 4$ square blocks, and apply DCT to them. Select first DCT value of each block to get DCT coefficient matrix B. Apply SVD to B, $B = U_1 \times S_1 \times V_1^T$ and obtain $U_1$, $S_1$, and $V_1$.

3  $WR$ is the red channel of watermark $W$. The size of the watermark should be $N \times N$ ( $N = \frac{M}{16}$ ) in order to meet the size of the carrier image. Apply SVD to it, $WR = U_1 \times S_1 \times V_1^T$ and obtain $U_1$, $V_1$, and $S_1$

4  Modify $S$ with $S_1$, $S' = S + a \times S_1$. $a$ is the embedding coefficient.

5  Use $S'$, $U$, and $V$ to restore the matrix, and $B' = U \times S \times V^T$. Substitute $B'$ for the DC coefficient value in each block. Apply inverse DCT to each block.

6  Combine the block to obtain the changed $LL\_HH'$ band. Apply inverse DWT to $LL\_HH'$, $LL\_LH$, $LL\_HL$, and $LL\_HH$ to obtain matrix $LL'$. Apply inverse DWT to $LL'$, $LH$, $HL$, and $HH$ to obtain the watermarked red channel $IMR'$.

7  In the same way, obtain the blue channel $IMB'$ and red channel $IMG'$. Combine three channels to obtain the watermarked image $IM'$.

#### 3.1.2 The extraction process of DCT–DWT–SVD watermark

1  $IMR'$ is the red channel of the watermarked carrier image $IM'$. Apply DWT to decompose it in the same way as the first step in 3.1.1.

2  Select $LL\_HH'$ band, divide it into square blocks, and apply DCT to them. Select first DCT value of each block to get DCT coefficient matrix $B$.
3  Apply SVD to $B'$, $B' = U \times S' \times V^T$ and obtain $U$, $S'$, and $V$.
4  Obtain the Singular value of the watermark according to the formula, $S = (S - S')/a$.
5  Use $U_1$, $V_1$, and $S$ to obtain the red channel of the watermark according to the formula $WR = U_1 \times S \times V_1^T$. In the same way, obtain the blue channel and green channel of the watermark, $WG$ and $WB$. Combine three channels to get the extracted watermark.

## 4  THE MIX OF TIME DOMAIN ALGORITHM

### 4.1  *The principle of the algorithm*

The mix of time domain means that the image is pretransformed in the time domain before executing the watermarking algorithm. An orthogonal matrix is used to multiply the carrier image to have the pre-transformation. The process can be seen as the elements of the carrier matrix are doing addition and subtraction with each other. As a result, the energy of the carrier image distribute more evenly. The performance of the watermark would change after the watermarking algorithm with the mix of time domain. The specific process of the mix of time domain algorithm is given below.

### 4.2  *The process of the mix of time domain*

#### 4.2.1  *The watermark embedding process based on the mix of time domain*

1  Divide the red channel $PR$ of $m \times m$ color image $P$ into $n \times n$ blocks. The value of $n$ depends on the order of the orthogonal matrices. Since the paper use 4-, 8-, or 16-order orthogonal matrices, the value of n can choose 4, 8, or 16 correspondingly.
2  Multiply the orthogonal matrix $H_n$ with each block $X_i$ ($i = 1,2,3 \cdots p$, $p = \dfrac{m^2}{n^2}$), $X_i' = H_n \times X_i$. Obtain the block of mix of time domain $X_i'$.
3  Combine all the blocks of mix of time domain to obtain $PR'$.
4  According to the watermarking algorithm mentioned in section 3.1.1, embed the watermark $WR$ into the $PR'$. In the same way, embed $PG'$ and $PB'$ into $WG$ and $WB$, respectively.
5  Divide $PR'$ into blocks $Y_i$ as step 1.
6  $K_n$ is the orthogonal matrix, which is orthogonal with $H_n$, and multiply it with $Y_i$, $Y_i' = \dfrac{1}{c} K_n \times Y_i$ to obtain $Y_i'$.

7  Combine all the blocks of $Y_i'$ to obtain $TR$, the red channel of the watermarked carrier image. In the same way, $TG$ and $TB$ are acquired. Combine $TR$, $TG$, and $TB$ to get the watermarked image $T$.

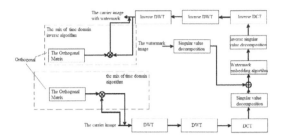

Figure 1.  Embedding schematic diagrams of watermarking technique based on the mix of time domain.

#### 4.2.2  *The watermark extraction process based on the mix of time domain*

1  Divide $TR$ to blocks $Z_i$ as step 1 in section 4.2.1.
2  Multiply $H_n$ with each block, $Z_i' = H_n \times Z_i$.
3  According to the watermark extraction algorithm in section 3.2.2, extract the red channel of the watermark from $Z_i'$. In the same way, extract the corresponding channel of the watermark from the green channel $TG$ and blue channel $TB$. Combine three channels to obtain the extracted watermark.

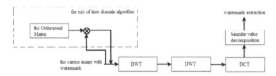

Figure 2.  Extracting schematic diagrams of watermarking technique based on the mix of time domain.

Paper A[2] and paper B[3] have proved that the traditional DCT and DWT watermarking algorithms with the mix of time domain have better imperceptibility and robustness. The following experiments test the effect of mix of time domain on watermarking based on DCT–DWT–SVD.

## 5  THE EXPERIMENTAL SIMULATION

### 5.1  *Preparation of the experiment*

The carrier images selected for the simulation experiment are four standard 512 × 512 256 color Lena, Pepper, Tiffany, and Sailboat. The watermark image use 32 × 32 256 color Baboon. Carrier images and watermark image are shown in Fig. 3.

Figure 3. The carrier images and the watermark image.

The orthogonal matrices in the simulation experiment are 4-order, 8-order, 16-order Hadamard Matrices and 8-order, 16-order orthogonal matrix pairs (*MA8,MB8*) and (*MA16,MB16*)

## 5.2 The experiment result

### 5.2.1 The comparison of the imperceptibility

The watermark realized by the orthogonal matrix and matrix pairs are shown in Fig. 5 and Fig. 6. The experiment tests the effect of the value of $a$, the embedding coefficient from 0.1 to 0.5. The result of the watermarking algorithm processed by the mix of time domain and the original algorithm is nearly the same when $a$ is close to 0.1. As a result, the value of $a$ in the experiments of imperceptibility is set to 0.5 to identify the effect of the mix of time domain algorithm more easily.

Figure 5. Watermarked image using the original algorithm.

Figure 6. Watermarked image using the algorithm based on the mix of time domain.

Figure 7. The comparison of Fig. 5 and Fig. 6 amplified to 300%.

It seems that there is little difference between the Fig. 5 and Fig. 6. However, when the images in Fig.5 and Fig. 6 are amplified to 300% in Fig.7, it is easy to notice there are some perceptible distortions near the Lena's hat in Fig. 5 as well as other watermarked images which do not apply the mix of time domain. By contrast, it is hardly to find small differences from the original carrier image in Fig. 6. PSNR and NC values are also calculated in Table.1.

Table 1. The average PSNR and NC values of different algorithms based on the mix of time domain.

| original | $H_4$ | $H_8$ | $H_{16}$ | $MA_8$ | $MA_{16}$ |
|---|---|---|---|---|---|
| 50.2 | 56.2 | 59.2 | 62.2 | 56.2 | 56.2 |
| 1.00 | 1.00 | 1.00 | 1.00 | 1.00 | 1.00 |

According to the table and the comparison in Fig. 7, the following conclusion can be attained. (a) All orthogonal matrices or orthogonal matrix pairs can improve the imperceptibility of the watermark. Among them, the effect of 16-order Hadamard matrix is the most obvious. (b) The effect of Hadamard matrices is better than that of orthogonal matrix pair when their order is the same.

### 5.2.2 The comparison of the robustness

In the robustness experiment, $a$ is set to 0.1 in the experiment of the robustness in order to easily identify the effect of noise attacks. Low-embedding coefficient is also convenient to observe the effect of the mix of time domain to the robustness.

The experiment tests the watermark extracted from the carrier image, which is under Gauss low-pass noise, Gauss noise, JPEG compression, and cropping attack. The extracted watermarks of different algorithms are compared in Table.2.

Table 2. Recovered watermarking of original algorithm and different algorithms based on the mix of time domain.

| Attack test | Original | $H_{16}$ | $MA_{16}$ |
|---|---|---|---|
| Gauss low-pass $3 \times 3$ $\delta = 0.5$ | | | |
| Gauss noise $\xi = 0$ $\delta = 5$ | | | |
| JPEG compression $Q = 60$ | | | |
| cropping attack 40% | | | |

According to Table.2, under the same noise attack, the watermark using the mix of time domain has great difference from the original watermark. Especially when attacked by the cropping attack, the extracted watermark based on the 16-order Hadamard Matrix looks much messier. The extract watermark based on the 16-order orthogonal matrix pair looks also worse. Its content is impossible to be identified.

With the conclusion of Paper A[2] and Paper B[3], it can be concluded that the mix of time domain have different effects on different watermarking algorithms. Paper A and Paper B apply the mix of time domain to the traditional watermarking algorithms. The experimental results show that the mix of time domain algorithm can improve the robustness and imperceptibility of watermark. In this paper, the mix of time domain is applied to the improved watermarking algorithm based on DCT–DWT–SVD, which combines several watermarking techniques in frequency domain. Although the imperceptibility of watermark proves to be better than the original algorithm, the ability to resist noise attack has been weakened. However, the decrease of the robustness makes it possible that the mix of the time domain can be used to produce fragile watermarks or semifragile watermarks. For example, the mix of time domain can be utilized to make authentication watermarks, which can enhance the sensitivity of the watermark to certain noise attacks. Whether the content is revised illegally by outlaws can be shown by the extracted watermark.

### 5.2.3 *The performance comparison of Hadamard matrices and orthogonal matrix pairs*

In order to compare the effect of the mix of time domain based on different orders Hadamard Matrices and Orthogonal Matrix Pairs, the robustness performance indexes of different images are calculated and then the average of the index of four images are calculated. The comparison results are shown in Table 3:

Table 3. Average PSNR and NC values of different algorithms based on the mix of time domain.

| Attack tests | Original | $H_4$ | $H_8$ | $H_{16}$ | $MA_8$ | $MA_{16}$ |
|---|---|---|---|---|---|---|
| Gauss $\xi = 0$, $\delta = 5$ | 25.5 0.99 | 21.6 0.99 | 17.7 0.97 | 15.3 0.96 | 19.6 0.98 | 19.3 0.98 |
| Lowpass $3 \times 3$, $\delta = 0.5$ | 12.3 0.96 | 12.7 0.95 | 11.1 0.86 | 9.4 0.73 | 10.1 0.83 | 9.6 0.80 |
| Rotation 180° | 286.2 1.00 | 6.3 0.39 | 6.5 0.44 | 6.8 0.48 | Inf 1.00 | Inf 1.00 |
| cropping 40% | 16.8 0.97 | 7.8 0.46 | 7.6 0.47 | 6.9 0.46 | 13.4 0.90 | 12.3 0.92 |
| JPEG Q=60 | 9.3 0.95 | 5.5 0.19 | 5.5 0.19 | 5.4 0.22 | 6.2 0.26 | 6.4 0.30 |

According to the Table 2 and Table 3, the following conclusion can be obtained. (a) Applying the mix of time to watermarking algorithm based on DCT–DWT–SVD can improve the sensitivity of watermark to noise attacks. (b) The ability of Hadamard Matrix to improve the sensitivity of watermark to noise attacks is better than orthogonal matrix pairs. (c) The higher the order of Hadamard Matrix, the higher the sensitivity of watermark to Gauss low-pass noise, JPEG compression, and cropping attack. (d) The higher the order of orthogonal matrix pair, the higher the sensitivity of watermark to Gauss low-pass noise, Gauss noise, and cropping attack. (e) The effect of 8-order and 16-order orthogonal matrix pair to the sensitivity of watermark to rotation is the same.

## 6 CONCLUSION

The paper has made further research on the effect of the mix of time domain on watermarking algorithm based on the result of paper A[2] and paper B[3]. The simulation software has been used to make the simulation experiment and attack tests. The result shows that the mix of the time domain algorithm can have significant improvement on the imperceptibility of the watermark. What is more, this algorithm can also improve the sensitivity of the watermark to certain noise attacks, which can be used to produce fragile or semifragile watermarks. This paper only focuses on the mix of time domain based on the 4-, 8-, and 16-order Hadamard Matrices and 8-, and 16-order orthogonal matrix pairs, and does not undertake relative research for other orthogonal matrices and orthogonal matrix pairs. From what has shown in this paper, there should be more research in the field of image watermarking technique based on the mix of time domain.

REFERENCES

[1] Li Xiaolong, Yang Bin, and Cheng Daofang, "A generalization of LSB matching,"IEEE Signal Processing Letters, vol. 16, 2009, pp. 69–72.
[2] Zhao Shenchen, "DWT Image Watermarking Algorithm Based on the Process in Time-domain", Beijing, Beijing University of Posts and Telecommunications.
[3] Teng Dongyang, Shi Renghui, Zhao Xiaoqun, "DCT Image Watermarking Technique Based on the Mix of Time-domain", Information Theory and Information Security (ICITIS), 2010 IEEE International Conference, 2010, pp. 826–830.
[4] Divecha, N. ; Jani, N.N., "Implementation and performance analysis of DCT-DWT-SVD based watermarking algorithms for color images", Intelligent Systems and Signal Processing (ISSP), 2013 International Conference, 2013, pp. 204–208.

[5] Liu Kai, Xu Chengqian and Li Gang, "Research on application of binary sequence pair with two-level periodic autocorrelation, "Journal of Electronics & Information Technology, vol. 31, 2009, pp. 1536–1541.

[6] Yang Yixian and Lin Xuduan, "Coding and Cryptography." Beijing: Posts and Telecom Press, 1992, pp. 125–126.

[7] Zhao Hongyang, "Research on application of low/zero correlation zone sequence pairs in spread spectrum communication system," Shanghai: Tongji University, 2009.

*Multimedia, Communication and Computing Application – Leung (Ed.)*
© *2015 Taylor & Francis Group, London, ISBN 978-1-138-02775-6*

# A method to facilitate automatic test cases generation based on inverse functions

W. Xiong, J.F. Huang, X.Z. Zhang & Y.Z. Gong
*State Key Laboratory of Networking and Switching Technology, Beijing University of Posts and Telecommunications, Beijing, China*

ABSTRACT: In the area of automatic test cases generation, it is challenging to satisfy constraints of complex expressions (such as equations with mathematical functions) in decision statements. This paper proposes a method, which based on inverse function, to conquer the challenge. The method maintains an inverse function library for different circumstances. When complex expressions encountered, this method searches the library and invokes the corresponding inverse function to make the complex expressions solved. We apply this method to some typical problems. Experiments show that this approach can achieve a high code coverage when applying it to some difficult situations, which is nearly impossible without this method.

KEYWORDS: Software testing, Inverse function, Test case generation, Unit testing

## 1 INTRODUCTION

The modern society is supported by plenty of software. Software lies everywhere from personal electric devices to something as important as aerospace engineering and national defense military system. So, maintaining the quality of software is a crucial issue. Software testing, the most common method to guarantee the quality of software, becomes more important than ever (Hirohide & Akihisa 2012).

The general aim of software testing is to find software bugs, errors, or other defects with lower amounts of time consumption and labor force (Gong et al. 2008). Statistics indicate that a well-designed strategy of software testing will save more than 40 percent of the cost used in the course of development and more than 60 percent of the cost used in maintenance the system (Huang 2004). As software development consists of several steps, such as planning, analysis, designing, and coding, several kinds of testing exist, including unit testing, integrate testing, and system testing. Unit testing is a procedure that ensures individual units of source code working properly and also serves as the foundation of other testing. Experience has shown that with unit testing it is easy to identify and correct mistakes early, and it costs relatively little.

In the path-oriented unit testing, firstly, a specific path of the program under test (PUT) is chosen. And then generation algorithm works to get a test case to satisfy the path constraints, specifically the decision statements in the path. It becomes more difficult to generate an appropriate test case as the complexity of the selected path. This paper proposes a method to facilitate the generation of test cases for paths with special mathematical functions. In our proposed method, the datum of test cases is accumulated with the help of inverse function library. This method works according to some experimental results.

## 2 THEORETICAL BACKGROUND

### 2.1 Unit testing

Unit testing is a method for modular testing of a program's functional behavior. In the test, a program is split into units. Each unit is a collection of functions (a sequence of instruction) and each unit is tested independently (Koushik et al. 2005). Unit testing requires a set of values (known as test cases) as its input. In the traditional software process, unit testing is usually practiced along with the software development by software engineers.

### 2.2 Dynamic testing and static testing

In dynamic testing, the software must be compiled and executed actually even though it is a useful way to validate the expected output and the overall performance of the software. The cost of finding and fixing defects is high and it only identifies defects in the portion of the codes which are actually executed. In contrast to dynamic test, in static testing, the code is not executed, but the structure of the code is analyzed. It s useful in discovering logical errors. And the cost of static testing is less without actually executing the code (Zheng 1992).

## 2.3 Inverse function

Inverse function is a concept in mathematics. Let f be a function whose domain is the set X, and whose range is the set Y. If f is invertible, then there exists a function g from domain Y to range X, with the property:

$$f(x) = y \iff g(y) = x \qquad (1)$$

(Ibrahim & Eddie 2004). According to the definition of inverse function, it is easy to work out the concrete value of the unknown x, if y and function f is known.

## 2.4 Constraint solving

Constraint solving problems, also known as constraint satisfaction problems, deal with a finite set of variables. Each variable is associated with a finite domain. A set of constraints restrict the values that the variables can take. The task then is to assign a value to each variable to satisfy all the constraints (Edward 1993). In our team's Code Test System (CTS, a unit testing tool developed in Java, which can generate test cases automatically), constraint solving consists of procedures, such as constraint extraction, symbolic execution, and interval arithmetic. This is the major step of test cases generation. The method proposed in this paper is also concerned with constraint solving method.

## 3 MAIN WORK

Test cases generation problems can be reduced to the constraint satisfaction problem for clarity. It should be generated to satisfy all the decision statements of a selected path. For a simple PUT shown in figure 1, if statement 4 is the target to be covered, test cases generation algorithm must find two concrete values $v1$ and $v2$ for variable $x1$ and $x2$ to satisfy two decision statements (2) and (3):

$$x_1 + x_2 == 100 \qquad (2)$$

$$x_1 - x_2 == 20 \qquad (3)$$

```
1: void test1 ( int x1, int x2 ){
2:    if ( x1 + x2 == 100 )
3:       if ( x1 – x2 == 20)
4:          printf ( "well done!" );
5: }
```

Figure 1.  Program test1.

Many algorithms have been proposed to solve this problem, such as Genetic Algorithms (Chayanika et al. 2013) and Simulated annealing Algorithms

(Nigel et al. 1998). These algorithms employ the "try and error" method to reach the target step by step and they work well at most of the time but not the circumstance of figure 2.

Ignore the precision of double value, and two decision statements of tests can be merged into "$\log10(x3) == 3$". To cover the 4th statement, $x3$ must be assigned a value of 1000. Two reasons make it difficult to generate test cases automatically for this program. Firstly, variable $x3$ has a legal range of value. In this case, $x3$ serves as the parameter of method "$\log10$," so the concrete value of it must be positive, a nonpositive value results in error. Secondly, there is only one concrete value 1000 for $x3$ that can satisfy the decision statements, and $x3$ appears only in this statements, and no useful information can be derived. The primary feature of these statements is some relatively speaking complex mathematical expressions (CEs, complex expressions) appear in it.

```
1: void test2 ( double x3 ){
2:    if ( log10(x3) – 3 > – 0.00001 ){
3:       if ( log10(x3) – 3 < 0.00001 ){
4:          printf( "Well done" );
5:    }}}
```

Figure 2.  Program test2.

How to get constraints solved is the key point of automatic test cases generation of programs with CEs. In this paper, we propose an algorithm, CSIF (constraints solving algorithm bases inverse function), which is based on the mathematical concept of "inverse function," to overcome it.

CEs are thought to be of two types in CSIF. One is expressions with ready-made inverse functions in Java API. When CSIF makes it clear that CEs belong to the first type, then it could reduce the domains of variables by invoking the inner method of Java API. Some simple examples of this kind are listed in Table 1:

Table 1.  Complex expressions with readymade inverse functions in Java API.

| Complex expression | Inverse function |
| --- | --- |
| $\sin(x)$ | $\text{asin}(x)$ |
| $\tan(x)$ | $\text{atan}(x)$ |
| $\cos(x)$ | $\text{acos}(x)$ |
| $\log(x)$ | $\exp(x)$ |

For the other type, things seem to be more complex. CSIF maintains an inverse function library for CEs. When CSIF identifies CEs as belonging to this type, CSIF searches to find the inverse functions of CEs in the library. If CSIF finds the function, invokes it, then adds it to the library and invokes it. As a result, the library is dynamically increasing.

The major steps of CSIF are listed as follows:

Algorithm 1. Constraints Solving Algorithm based on Inverse Function

Input: E, decision statement (DS)

Output: V, concrete value for variable in CE

```
begin
1: if(DS contains CE){
2:    if(CE belong to first kind)
3:       invoke inner method;
4:    else{
5:       search library;
6:       if(exist)
7:          invoke it;
8:       else
9:          add to library and invoke it;
10: }}
11: else
12:   do as normal situation;
13: return E;
end
```

The so-called inverse function in CSIF is not as strict as in mathematics. In mathematics, the input of a function is called the argument and the output is called the value. And the relationship between argument and value must be one-to-one.

But in CSIF, it is possible that more than one concrete values are available. So an extension strategy is added to make a domain convincing.

## 4 EXPERIMENTAL RESULTS

In order to observe the effects of CSIF algorithm, we carried out a number of experiments. The experiments were performed under the environment of Microsoft Windows 7 with 64-bits and run on Pentium 4 with 3.0 GHz and 4 GB memory. The algorithm was implemented in Java and run on eclipse platform.

Four programs serve as test beds as shown in table 2.

### 4.1 *Functional evaluation*

To evaluate the advantage of CSIF, we implement it in CTS and compare its effects with a widely used tool, C++ test, which has the ability of generating test cases automatically, too. In simplicity, we concentrate on the decision statements with CEs. Table 3 shows the result. In table 3, "√" indicates that the values in the generated test cases satisfy CEs, whereas "×" indicates the opposite.

After analyzing the results, we conclude that CSIF in CTS significantly outplays C++ test at certain aspect in solving CEs. C++ test generates test cases on boundary-value mechanism no matter it fits the bill or not. However, CSIF in CTS attempts to meet the constraints in CEs with the methodology of inverse function.

### 4.2 *Performance evaluation*

Before CSIF is integrated, CTS get the test datum by random method within a range, which is derived from constraints in CEs. This suffers the risk of inefficiency. On the other hand, CSIF makes a great improvement counteracting the inefficiency.

Table 4 presents results from a practical comparison of CTS with random algorithm or CSIF. It can be seen that CTS with CSIF reaches 100% successful rate for the three benchmark programs. In the great progress, CSIF acts as a constraint solver to refine the domain of variables in CEs. After executing CSIF, the constraint-satisfaction values are selected.

## 5 CONCLUSION AND FUTURE WORK

This paper presents an algorithm based on inverse function to settle constraint-solving problems of path-oriented automatic test cases generation. Experiments show that CSIF works well in solving problem with CEs. CSIF brings two innovational improvements.

First, CSIF has the ability to perceive the expressions that have been solved. That is CSIF adjusts itself according to the PUTs. CSIF knows what legal input is for PUTs and abandons illegal input. Second, CSIF is capable of refining legal input. The inverse function library is the core of CSIF and more and more elements are constantly added into the library.

Table 2. Benchmark programs used for experimental analysis.

| Program | Complex expression | Description | Source |
|---|---|---|---|
| days | year%400 == 0 | to calculate which day a specific day is in a year | refer to * |
| mathfunction | $\sin(x) < 0.00001$, $\sin(x) > -0.00001$ | test whether $\sin(x)$ equals zero | refer to ** |
| absofmath | abs(x) == 10 | to calculate absolute value of x | by authors |
| roundofmath | round(x) == 5 | returns the closest long to the argument, with ties rounding up | by authors |

* A Method of Test Cases Generation Based on Necessary Interval Set (Wang et al. 2013).
** Analysis and Design of Unit test usage automated generation (Li 2007).

Table 3. Comparison of CSIF algorithm in code test system and C++ test.

| Complex expression | C++ test | Code test system |
|---|---|---|
| $x\%400==0$ | √ | √ |
| $\sin(x) < 0.00001$, $\sin(x) > -0.00001$ | × | √ |
| $abs(x) == 5$ | × | √ |
| $round(x) == 5$ | × | √ |

Table 4. Successful rate comparison of code test system with random algorithm and CSIF.

| Complex expression | Code test system with random algorithm | Code test system with algorithm 1 |
|---|---|---|
| $x\%400==0$ | 0.21% | 100% |
| $abs(x)==5$ | 10.5% | 100% |
| $round(x)==5$ | 5.7% | 100% |

Our following research will expand CSIF to accommodate more situations. For example, we also concern with occasions that user-defined methods appeared in decision statements and methods with more than one parameter.

As a result, hopefully we will build an inverse function library, which could comprise more and more significant inverse functions.

## ACKNOWLEDGMENTS

This work was supported by the National Grand Fundamental Research 863 Program of China (No. 2012AA011201), the National Natural Science Foundation of China (No. 61202080), the Major Program of the National Natural Science Foundation of China (No. 91318301), and the Open Funding of State Key Laboratory of Computer Architecture (No. CARCH201201).

## REFERENCES

Chayanika, S. & Sangeeta, S. & Ritu, S. 2013. A Survey on Software Testing Techniques using Genetic Algorithm. *International Journal of Computer Science Issues,* 10(1):381.

Edward, T. 1993. *Foundations of constraint satisfaction.* London: Academic press.

Gong, Y.Z. 2008. *Software Testing.* Beijing: China Machine Press.

Hirohide, H. & Akihisa, S. 2012. Automatic Test Case Generation based on Genetic Algorithm and Mutation Analysis. *2012 IEEE International Conference on Control System, Computing and Engineering*: 119.

Huang, Q.Q. 2004. Introduction of Software testing and testing Methods. *Ship Electronic Engineering* 139:32.

Ibrahim, B. & Eddie, G. 2004. Understanding inverse functions: The relationship between teaching practice and student learning. *Proceedings of the 28th Conference of the International Group for the Psychology of Mathematics Education* vol2: 104.

Koushik, S. & Darko, M. & Gul, A. 2005. A Concolic Unit Testing Engine for C. *Proceedings of the 10th European software engineering conference*: 263.

Li, H. 2007. Analysis and Design of Unit test usage automated generation. *Master's thesis of Taiyuan University of Technology*: 45.

Nigel, T. & John, C. & Keith, M. 1998. Automated Program Flaw Finding using Simulated Annealing. *ACM SIGSOFT Software Engineering Notes,* 23(2):73.

Wang, Y.W. & Gong, Y.Z. & Xiao, Q. 2013. A Method of Test Case Generation Based on Necessary Interval Set. *Journal of Computer-Aided Design & Computer Graphics,* 25(4):550–556.

Zheng, R.J. 1992. *Computer software testing technology.* Beijing: Tsinghua University Press.

*Multimedia, Communication and Computing Application – Leung (Ed.)*
© *2015 Taylor & Francis Group, London, ISBN 978-1-138-02775-6*

# A method of stub generation for I/O functions in automatic test

Y.W. Yang, Y.W. Wang & Y.Z. Gong
*State Key Laboratory of Networking and Switching Technology, Beijing University of Posts and Telecommunications, Beijing, China*

Y.W. Wang
*State Key Laboratory of Computer Architecture, Institute of Computing Technology, Chinese Academy of Sciences, China*

ABSTRACT: Library functions and system calls have been major difficulties faced by automatic test. Input/output (I/O) functions are a set of common library functions. Testers have to interact with the test procedures if the test program contains I/O functions, resulting in inefficient automatic test. Unlike normal functions, the association between I/O functions and operating on a specific I/O device makes ordinary stub method disabled. Aiming at solving this problem, this paper proposes one solution by I/O device modeling and I/O function modeling. The semantics of the I/O functions are stored in the I/O function model, and the operations of I/O functions are similar to the I/O device model. The final state and properties of the I/O device are represented by two models. Finally the stub code is automatically generated by path-oriented stub generation technology. Experiments show that this method is effective.

KEYWORDS: Automatic test, Stub generation, I/O modeling.

## 1 INTRODUCTION

Software test is the primary means to find a software failure and improve software reliability, and it is also the key technology to make sure software quality. Unit test is an important stage. Testers have to finish lots of work by manual test. For example, the input for the test data, intervention during test execution, as well as the analysis of a large number of test results is very tedious but necessary. In the actual process of unit test, testers tend to realize that writing unit test code is a complex and time-consuming work. Currently, automatic test system is a better way to solve this problem.

In automatic test, coverage is an important indicator to measure the quality of the test. We could test the function unit under the statement, branch, or modified condition determination coverage criteria. The main criteria are more important than coverage criteria for white-box testing to measure the adequacy of the test. The higher coverage of a test based on a criterion is achieved, the more thorough the tests are. The main factors affecting the automatic unit test coverage is the completeness of the test cases and the accuracy of the stub functions. The main purpose of the stub functions is to substitute the original functions and isolate the test unit from other function calls.

## 2 BACKGROUND

Library functions and system calls have been major difficulties faced by automatic test. These calls do not typically provide source code if any, due to the underlying optimization or inherent implementation, it is difficult to analyze. I/O functions are a set of common library functions. In unit tests, testers always hope to test the code in automatic way. In order to make unit test run normally, testers need to do according to the test procedure (keyboard input, etc.) if the test program contains I/O functions. Thereby the efficiency of automatic test is greatly reduced. Therefore, we offer a way to transform the I/O functions. Testers will be released from the inefficient interaction with the program, but it will not change the logic of the original test program through the transformation.

When reading values from the command line, it is difficult to handle the situation of environment variables, files, network packets, and other environmental ordinary stub methods (Lin et al. 2011). Such as writing data to a file, subsequent data reading operation will be affected by this written operation; due to insufficient privileges, nonexisting file, such as failing calls, the function should return error codes, such as a NULL value or -1. Consider the following C code in Figure 1 (Sen et al. 2005), in that the program writes first 10 chars to st and then writes to the file. Relocate the file stream to the beginning, then read 20 chars in line 6. Line 7 is dependent on the written data of line 3.

At present, there are three kinds of stub methods (Galler & Aichernig 2013): (1) generating stub framework, the output value of the stub function is inputted by testers manually, known as low degree of automation. (2) The output value of the stub function is a particular value, such as it is assigned 0 if the output variable of the function is integrated. This approach is mainly used for boundary value test, resulting in low coverage. (3) Output of stub function is a random value. We can increase coverage to some extent in this way, but it is far from being inefficient. And in order to improve coverage, tests should be performed several times. General stub method analyzes one function at a time. In order to achieve the purpose of high coverage, we can change the value of the return value or argument in the stub function. But limitation exists in ordinary stub methods, due to the correlation between functions. A series of I/O functions operate on a specific I/O device and there must be links among these I/O functions. Therefore, the above stub method could not be applied directly in the I/O functions.

```
1    for (i=0; i<10; i++)
2        st[i] = ...;
3    fwrite(st,strlen(st),1,fp);
4        ...
5    //seek to the start of the file
6    ret = fgets(buf,20,fp);
7    use = buf[3];
```

Figure 1.   Program fragment.

There is also much work by other scholars on the problem of library functions and system calls. Stanford University's Christian Cadar (2008), who proposed a model environment modeling, provides a new symbolic execution tool KLEE to address the following questions: When your code calls function associates with the environment system, the function returns all of the legitimate values of its environment, rather than a single specific value, with the purpose of achieving high coverage. Qi (2012) proposed a method for optimizing environment modeling; the library functions and system calls, denoted as fc, is semiautomatically synthesized to a new function fs, and its input/output pairs are constructed as collection $E = \{(a_1, b_1), (a_2, b_2) ...\}, b_i = fs (a_i)$. Finally, they generate the result with the SMT solver. Renshaw (2011) proposed a method of expansion for KLEE file system model, increasing the semantic directory attributes and associated function.

In summary, this paper proposes the approach for I/O functions modeling and I/O device modeling,

saving the constraints in the property and state of the model by analyzing the constraints on the path; then the property and state of the model are represented in the test program by stub functions. In this way, the test program is more logical, and the stub is more accurate.

## 3   BASIC CONCEPT

There is a control flow graph $G = (N, E, n_0, n_f)$ in Definition 1 Path P, which represents the execution path sequence $<n_0, n_1, ..., n_f>$ of the control flow graph, where $n \in N$, $n_0$ and $n_f$ are graph ingress and egress nodes, and E is the set of edges. In the program, n is an assignment statement or a conditional statement, and $e \in E$ is a jump statement from $n_i$ to $n_j$.

Definition 2 is I/O device model. Physical entities interacted with the environment or external factors in the program are called I/O devices. This paper analyzes the console, file, and socket. These models are called I/O device model, which is built based on the entity I/O device.

Definition 3 Interval Set. The interval I is defined as $I = [x_i, x_j] = \{x \mid x_i <= x <= x_j\}$, the interval set is a set of disjoint sets of intervals, $IS = I_1 \cup I_2 \cup ... \cup I_n$.

Definition 4 Path is about the constraints on I/O model. Pcm = $\{<P, f, IS, M>\}$ represents the given execution path P and I/O functions f on the path, f will change the properties and the state of the I/O device model M, and the return value and argument of f must be modified to satisfy the value at the interval set IS. The interval serves as constraint of M if it fits IS, allowing the function executes through the path.

Two tables and their specification are given to describe the I/O model as shown below.

Table 1.   I/O device modeling table.

| ID | Open | State | EF | Con | <fi, IS> |
|---|---|---|---|---|---|
|  |  |  |  |  |  |

*ID* is the id of the I/O device. It is a unique identifier for an I/O device used to distinguish the different I/O devices.

*Open* indicates that the I/O device is in the open state or not and record information such as the open mode.

*State* records a series of specific states of the I/O device, such as file reading and writing position, the length of the current file, and so on of the file model.

*EF* is a collection which records all the location information of the output function where the contents have been changed. $ef_i = \{pos_i \mid pos_x <= pos_i <= pos_y\} = [pos_x, pos_y]$, $EF = ef_1 \cup ef_2 \cup ... \cup ef_n$, used to indicate the collection of all location information where output functions write data.

*Con* stores all the specific content which changes the output function <*fi, IS* > The key-value pairs of the currently named I/O function and the constraint set are the most important attributes in the model. It indicates the current function, including the function name, arguments, and return values; IS saves as a constraint set in order to make sure the implementation of the calculated execution path, each IS corresponding to *fi* is calculated based on *EF* and *Con*.

Table 2. I/O function modeling table.

| ID | Ret | Pl | Op |
|----|-----|----|----|
|    |     |    |    |

*ID* is the id of the I/O function; it is a unique identifier for I/O functions, used to distinguish the different I/O calls.

*Ret* record the return value information of the I/O function, including the type and the value.

*Pl* records argument list of the I/O function, including the type and location.

*Op* is the option based on actual I/O device semantically equivalent to the operations based on I/O device model for different I/O functions; the specific semantic equivalence is very different. I/O function simulation varies depending on the different I/O functions.

## 4 ALGORITHM

I/O model algorithm consists of two parts, one is the algorithm for the I/O device model, I/O functions and equivalent semantics, the second one record the constraints of the path and I/O devices, then the program is automatically stubbed by certain strategies.

Algorithm 1: Established the I/O model

1 Establish the device model for a specific I/O device according to the I/O device model table.
2 Establish all the corresponding I/O function models by analyzing all the I/O device-related library functions according to I/O function model tables.
3 Establish equivalent semantic information table of the I/O functions by analyzing the specific semantic I/O functions and equipment corresponding to the specific I/O device.

First, we select a target execution path based on the control flow graph, followed by analysis of the path of each node.

*isIOfunction(n)*: determine whether the current node is an I/O function call, and if so, then enter the I/O function modeling.

*Analysis (n)*: analyzes all the information of the node, including the type, number, name of the

Algorithm 2: Constraint extraction of the I/O model.

```
1   for each n in path:
2       if (isIOfunction(n)):
3           func = analysis(n)
4       if func==openfunction:
5           ID=new IOdev()
6       else if func==Inputfunction:
7           flag = isConflict(ID,EF,inputef)
8           if flag==True:
9               input=get(ID,content,inputef)
10          if flag==False:
11              input=get(ID,IS)
12      else if func==Outputfunction:
13          modify(ID,content,outputef)
14          add(ID, outputef)
```

arguments, etc. And then determine the I/O function call in the operation of opening, reading, or writing.

*IOdev()*: create a new I/O device model, including the initial state and properties of the model, the return value of the instance ID, and join the ID to a list. The list saves all active I/O devices, make inquiries with the I/O devices in subsequent operations.

*isconflict(ID, EF, inputef)*: check whether the current position of the reading content has been written before. The return value could not be set to the value at the constraint intervals, but set to the specific value corresponding to the position taken in the model *Con* offered, if the inputted *EF* have intersections.

*Get (ID, content, inputef)*: obtain the specific contents of the inputted position of the I/O device as a return value.

*Get (ID, IS)*: obtain the specific value of the constraint set IS as the return value, in order to meet the requirements to perform the path.

*Modify (ID, content)* modify the value of I/O model *Con* attributes.

*Add (ID, outputted)*: adding a location information, indicating the contents of the location is written by output function, EF = output∪EF.

Algorithm 2 describes the particular establishment and operation under I/O function calls of the I/O model for a specific program. If reading data from the position without conflict, set the value in accordance with the constraint set of the path. Otherwise set the output value in accordance with the input value. In this way, Transformation tries to sign the logical structure of the program itself as much as possible to satisfy the premise of the path constraint.

According to the algorithm, finally we get the constraint set IS of each I/O function calls. Then stub function is automatically generated with IS and guaranteed by the order of ID in accordance with corresponding constraint intervals.

## 5 CASE STUDY

The following example in figure 2 illustrates the above models and algorithms.

First, we choose a target execution path 3> 4 > 5> 6> 9> 10> 11> 12> 13> 14> 15> 18> 19> 22> 23, and the lines related to I/O functions in line 6, 12, 14, and 15. Line 6 opens a file, creates a new file model due to selection of false branch. Then ID, Open and State values are recorded. Next, line 12 explains the file model and records the corresponding values of EF. Line 14 sets the file cursor to the starting position of the file, which changes the value of State. Finally, line 15 is about the contents of the file model. According to the values calculated by $EF$ and $Con$, conflict between the value of buf[5] and "c" is impossible. The value is shown directly from the file model. According to the attribute of the file model, the stub function is defined as follows:

```
FILE* AAA_fopen("test","wr+"){
    return new FILE* a;
}
int AAB_ fwrite(st,strlen(st),1,fp){
    return strlen(st);
}
void AAC_fseek(fp,0,SEEK_SET){}
int AAD_fgets(buf,20,fp){
    fp[5]='f';
return 20;
}
```

Final execution path is 3> 4 > 5> 6> 9> 10> 11> 12> 13> 14> 15> 18> 22> 23, which does not pass through the line 19.

```
1    #include<stdio.h>
2    int f(){
3        FILE *fp;
4        char st[20],buf[20];
5        int i;
6        if((fp=fopen("test","wr+"))==NULL){
7            printf("Cannot open file!");
8            exit(1);
9        }
10       for (i=0;i<10;i++)
11           st[i]=i+'a';
12       fwrite(st,strlen(st),1,fp);
13       ... //omitted code
14       fseek(fp,0,SEEK_SET);//seek to the beginning
15       int ret = fgets(buf,20,fp);
16       /*first 10 chars are known as 'abcdefghij'
17       chars range from 10 to 20 are unknown*/
18       if (buf[5]=='c')//conflict
19           {/*do something*/}
20       else if (buf[15]>='f') // not conflict
21           {/*do something*/}
22       fclose(fp);
23  }
```

Figure 2.  C code for file operation.

Consider another path 3> 4 > 5> 6> 9> 10> 11> 12> 13> 14> 15> 18> 20> 21> 22> 23, we will go through the else branch on line 20. Analysis of the I/O model is the same as above, not repeated then. Function on line 15 reads from the file and the content of buf[15] has not been written yet calculated by $EF$ and $Con$, denoted as not conflict. So set the value of buf[15] to a random value from the interval [102,127], for the ASCII value of "f" is 102. The stub function is defined as follows:

```
int AAD_fgets(buf,20,fp){
    fp[15]='y';
    return 20;
}
```

## 6 CONCLUSION

In this paper, we adopt the method of I/O functions modeling and I/O device modeling, semantically equivalent to the operations based on I/O functions for I/O devices to I/O models. In order to record the statement of I/O device model in the symbolic execution process, recording constraint conditions of I/O device-related variables, using stub module to represent the results, and finally generating test result to execution through constraint solving.

## ACKNOWLEDGMENTS

I would like to express my sincere acknowledgment for the support of the National Grant Fundamental Research 863 Program of China (No. 2012AA011201), the National Natural Science Foundation of China (No. 61202080), the Major Program of the National Natural Science Foundation of China (No. 91318301), and the Open Funding of State Key Laboratory of Computer Architecture (No. CARCH201201).

## REFERENCE

Cadar, Cristian, Daniel Dunbar, and Dawson R. Engler. 2008. *OSDI*, KLEE: Unassisted and Automatic Generation of High-Coverage Tests for Complex Systems Programs. Vol. 8.

Galler S J, Aichernig B K. 2013: 1–25. *International Journal on Software Tools for Technology Transfer*, Survey on test data generation tools[J].

Lin, M. X., et al. 2011. *Software*, Lazy symbolic execution for test data generation, IET 5.2 (2011):132–141.

Qi, Dawei, et al. 2012. *IEEE*, Modeling software execution environment. Reverse Engineering (WCRE), 2012 19th Working Conference on.

Renshaw, David, and Soonho Kong. 2011. Symbolic Execution in Difficult Environments.

Sen K, Marinov D, Agha G. 2005. *ACM*,CUTE: a concolic unit testing engine for C[M].

*Multimedia, Communication and Computing Application – Leung (Ed.)*
*© 2015 Taylor & Francis Group, London, ISBN 978-1-138-02775-6*

# Multi-features based classification for automatic summarization

H.Y. Li, R.F. Liu & W.R. Xu
*School of Information and Communication Engineering, Beijing University of Posts and Telecommunication, Beijing, China*

R.S. Shi
*Education Ministry Key Laboratory of Trustworthy Distributed Computing and Service, BUPT*
*School of Humanities, BUPT, Beijing, China*

ABSTRACT:    That the summary of an article can be automatically generated is an important text processing technique. The technique can automatically extract the most useful information of the text and present that information, which makes it easy to navigate and judge the importance to readers. This paper is based on a variety of texts and structure features. Supporting vector-machine classification method is used to identify summary sentences, which abandoned the traditional experience weight setting to optimize. The effective system that combines keywords density and document structure with the best weights is implemented for automatic summarization. The experimental results show that our Density and Structure Based Summary system improves the automatic summarization results. It's better than the other two systems and has a big promotion on the precision, recall and F score.

## 1 INTRODUCTION

With a growing number of texts flooding into the Internet, the requirement of automatic summarization is becoming more and more important currently, which can compress information of text, and help us quickly and effectively determine whether the text is useful.

The research of automatic summarization [1] has been going on for many years. There are two main directions. Extractive summarization is concerned with what the summary content should be, based on the article structure, or the diverse features of text, to score each sentence and identify important sections or sentences. Abstractive summarization puts strong emphasis on the form, aiming to produce a grammatical summary, and produces summary in a new way based on the understanding of the article, which usually requires advanced language generation techniques. Abstraction seems fluent, but the method has low precision and high cost, which is always used for multi-document summarization [7–9].

Most of extract methods are considering structural information of the article and the features of the text separately, and score each sentence with a linear combination of these part scores. Then, the problem is defined as a classification problem, to judge which ones are summary sentences. Important features of the structure and text are words frequency, position information of sentences, keywords, cue words, etc.

How to combine these features reasonably to achieve the best results is the most urgent problem. Various methods are basically based on experience to set the weight values of every part of linear combination, and most systems assumed feature independence.

This paper presents a new approach. Although it will extract summary sentences as a binary classification problem identified, but we analyse all the features of structure and text. With these features, SVM is used for automatic classification to avoid the weights, setting of separate scores and the feature independence assumption.

## 2 RELATED WORKS

Most early works on single-document extractive summarization [1] focused on technical document or news reports. These articles have good structure and clear features that naïve-Bayes classifier is used.

Blog summarization [3] aim to extract representative sentences from a blog post that best represent the topics discussed among its comments, which are additional information for evaluating the importance of sentences. Other works implement automatic summarization based on sub topic partition [6]. These methods can improve the precision with additional information. Zhang Minghui uses the LDA topic model to generate the topic distribution [2]. By using the topic feature that LDA model generated and basic text model features such as frequency and position to complete a simple linear combination. And there is also the method which based on document topic and context [4]. Through the keywords analysis, dynamic processing with documents with abstract and concrete title, then grammar or semantic analysis NLP technology is used to link sentences smoothly. It finally adopts co-reference resolution technology. WordNet also can be used for automatic summarization [5]. In paper [3], Meishan Hu et al. keywords density is used to measure the importance and relativity of a sentence. With the calculated score and the sequence sentence in the document it selects top k sentences as the summarization.

## 3 PROBLEM DEFINITION

For a given article, it can be looked as a vector consists with sentences, $D=\{S1, S2, ... Sm\}$, m is the number of sentences in the article. Each sentence Si is a set of words, $Si=\{W1, W2, ... Wn\}$. Besides, each sentence includes other structured information, such as the position of the sentence in a paragraph and the position of the paragraph, the word frequencies. The task is to extract a subset of the sentences from D that best represents the discussion of the article.

The existing solutions have some shortages.

For vector space model based method, each sentence is represented by a high dimensional sparse binary vector $Si=(W1, W2, ... Wn)$. All the sentences have the same dimension. There are two problems concerning this method. First, for text classification problems, only considering the appearance of the word will miss the importance of the word. Second, calculating the score of high dimension sparse vectors will have high costs and low accuracy.

For the keyword density based method, the weight of keyword and the distance between keywords are used to calculate the score of the sentence. The sentences with high scores will be the summary sentences. The method takes into account the keywords effect, but neglects the article structure information. What's more, how to determine the keywords is an important factor that greatly has an impact on the results.

For the structure based method, the article title, subtitle, or the first sentences of each paragraph are selected as summary sentences. This method is suitable for the highly standardized article, for example, the news media articles, papers and other scientific literature. But the method could not get good results from the articles without clear structures.

For the method combining content and structural information, keyword frequency, cue words and sentence position are separately considered to score sentences, and then the several parts scores are combined

with different weights of experience to get the total score for each sentence. The method, although the integration, including content, structure, cues and other information, but simply linear combines the individual features could not achieve an optimal result, we need to find a way for the combination to best fit all kinds of articles.

## 4 THE AUTOMATIC SUMMARY SYSTEM

The system we proposed aims to find the most reasonable weight combination of all features to maximize the information of summary for an article. The resolution is a binary classification, the key sentence in accordance with the density score, and structured information as feature values. The flowchart shows in Fig 1.

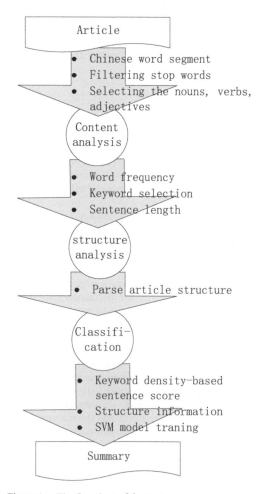

Figure 1. The flowchart of the system.

### 4.1 Keywords selection

Keywords can exhibit the topics of an article to some extent. First, we want to do pre-processing and remove stop words and punctuation to avoid the impact of our keywords selection. Secondly, we have found that the main keywords are nouns, adjectives and verbs in general, so we eliminate other categories of words. Finally TF-IDF values are used to select keywords. Where TF is the word frequency information, IDF is inverted document frequency, which is used to measure the importance of word Wi in the article D. To sort the TF-IDF values decrease, and choose the top N words as keywords.

### 4.2 Sentence scoring

Three features are considered to score sentences.

1 The sentence which contains the keyword should be the high score sentence, which is the primary criterion because the sentence should give the article topics.
2 In addition to the presence keywords, we should consider the distance of each pair of keywords.
3 The length of the sentence should be considered.

For a given set of keywords, K is the number of keywords, $score(w_i)$ is the TF-IDF value of the word $w_i$, $distance(w_i, w_{i+1})$ is the distance between the pair of keywords, $length(s_i)$ is the word number of the sentence, so the score of the sentence is defined as Equation (1).

$$score(s_i) = \begin{cases} 0 & if\ K = 0 \\ \dfrac{score(w)}{length(s_i)} & if\ K=1 \\ \dfrac{1}{K(K+1)} \sum_{i=1}^{K-1} \dfrac{score(w_i)score(w_{i+1})}{distance(w_i,w_{i+1})^2} & K > 1 \end{cases} \quad (1)$$

The score reflects the density of keywords in a sentence, so we think it can represent the importance of the sentence. To regularize the score between 0 and 1, it can be one of the features of a sentence.

### 4.3 Structure information

The position of the sentence in an article is another important feature of the sentence. In general, the title, subtitle, the first sentence of each paragraph, or the sentence in the first and last paragraph have high probability to be a summary sentence, but different kinds of articles have different structure. We think a kind of articles have similar structure. We parse the structure of articles as equation (2).

$$\{doc : \{paragraph\ 1 : \\ \{sentence\ 1, sentence\ 2, \ldots, sentence\ i, \ldots\}, \quad (2) \\ paragraph\ 2 : \{\ldots\}, \ldots, paragraph\ j : \{\ldots\}\}$$

Here we extract the information with sentence position, both counting from front and rear in its paragraph. At the same time, we extract the paragraph information as a feature, too. So, with three structural information features and one content feature we can deal the problem as a classification.

### 4.4 Multi-features based classification

Fig. 2 shows the distribution of a training dataset, X axis is the score of sentences and the Y axis is the structural value of the paragraph. Fig. 3 shows the distribution of sentences in its own paragraph

Diamonds are the sentences which can be used as summary sentences, the cross represents the sentence which cannot be used as summary sentence. We can see that it is not a linear binary classification problem, but in Fig. 2 summary sentences appear to be on the bottom of the figure. And in Fig. 3 we could find that most summary sentences exist at the front of a paragraph. On the other hand, it shows that most summary sentences are in the paragraph that only contains one or less sentences. We consider those paragraphs the sub-title of a document. So we try to use SVM radial basis function for nonlinear classification model training.

With the best hyperplane learned we can automatically extract a new document summarization which is better than the method of manually weighted.

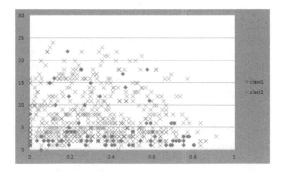

Figure 2. The distribution of sentences score and paragraph.

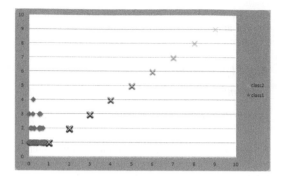

Figure 3. The distribution of sentences in its own paragraph.

## 5  EXPERIMENTAL RESULTS

### 5.1  *Experimental data*

We use the corpus supplied by Sogou Lab. There are several types of text data in the corpus. We select the well-structured ones which are type of Military. Fifty documents are used to be the training dataset. There are totally five hundred sentences, including about one hundred summary sentences. And then we measure other fifty documents that includes six hundred sentences as the test dataset. All of the summary sentences are carefully identified by a volunteer group.

### 5.2  *Evaluation of automatic summarization*

In this paper, we use precision (P), recall (R) and F-measure to evaluate our automatic summarization system. We compute P, R and F as equation (3).

$$P = |S_1 \cap S_{sum}| / |S_1|$$

$$R = |S_1 \cap S_{sum}| / |S_{sum}| \tag{3}$$

$$F = \frac{(\beta^2 + 1) \times P \times R}{\beta^2 \times P + R}$$

$S_1$ presents the sentences we classified. And $S_{sum}$ means the sentences that are summarizations in the test dataset. $\updownarrow$ is the parameter that shows the tendency of P or R. The bigger $\updownarrow$ is, the more important P is. We set the $\updownarrow$ to 1 which means we consider P and R have equal importance.

### 5.3  *Results and analysis*

To evaluate our system, we select two other algorithms to compare. One is a VSM language model that processes a document to a vector filled with 0/1 factor which presents a word's existence. Because our training data only contains fifty documents, we skip the

feature selection. We directly deal documents with word segmentation and remove the words in stop word list. And then SVM is used to classify the sentences in our training dataset. Finally, the model with the test dataset is run. The other one is using DBS(density-based summarization) to extract summaries. The experimental results are shown in Table 1.

Table 1.  The experimental results.

|      |         | P | R | F |
|------|---------|---|---|---|
| DSBS | class 1 | **0.47929** | **0.94186** | **0.635294** |
|      | class 2 | **0.989362** | **0.840868** | **0.909091** |
| DBS  | class 1 | 0.161137 | 0.303571 | 0.210526 |
|      | class 2 | 0.893151 | 0.78649 | 0.836434 |
| VSM  | class 1 | 0.33 | 0.843631 | 0.474422 |
|      | class 2 | 0.832766 | 0.803519 | 0.817881 |

In Table 1, class 1 presents the summaries, and class 2 presents the sentences that are not summaries. From the results we could see that our DSBS(Density and Structure Based Summary) system improves the automatic summarization results not only for precision but for recall ratio. It's better than the other two systems and has a big promotion on the P, R and F. So our system effectively combines the DBS algorithm with document structure information. And its best hyper plane could deal with the summary better.

## 6  CONCLUSION

Although this extraction seems less readable, it has high precision and low calculation cost. Based on the timeliness of the characteristics of high fashion structure, an effective system that combines keywords density and document structure with the best weights is implemented for automatic summarization. Here we raise the additional parts for DBS algorithm and regularize the score for better learning. With the SVM, we found the best model to associate different features including keywords and structure information. In the future, we will try to find other effective features such as similarity distribution or indicative word's existence.

## 7  ACKNOWLEDGMENT

This work was partly supported by 111 Project of China under Grant No. B08004; Chinese Universities Scientific Fund (BUPT2014RC0701).

## REFERENCES

[1] D. Das & A.F.T. Martins. A Survey on Automatic Text Summarization. Literature Survey for the Language and Statistics II course at CMU (2007).

[2] Zhang M., Wang H. & Zhou G. An automatic summarization approach based on LDA topic feature. *Computer Applications and Software*. Vol. 28 No.10, Oct. 2011:20–23. (in Chinese).

[3] M. Hu, A. Sun & Ee-Peng Lim. Comments-Oriented Blog Summarization by Sentence Extraction. *Conference on information and knowledge management (CIKM2007)*, pp. 901–904.

[4] Chen yanmin etc. Automatic Text Summarization based on Topic and Content. *Computer Engineering and Applications*. 2004:11–14. (in Chinese).

[5] Alok Ranjan Pal & Diganta Saha. An Approach to Automatic Text Summarization using WordNet.

*International Advance Computing Conference (IACC)* 2014:1169–1173.

[6] Xueming Li etc. Automatic Summarization for Chinese Text Based on Sub Topic Partition and Sentence Features. *Intelligence Information Processing and Trusted Computing (IPTC2011)*, pp. 131–134.

[7] Jade Goldstein etc. Multi-Document Summarization By Sentence Extraction. *Association for Computational Linguistics*, Morristown, NJ, 40–48.

[8] A. Celikyilmaz & D. Hakkani-Tür. A Hybrid Hierarchical Model for Multi-Document Summarization. *Proc. of Association for Computational Linguistics (ACL 2010)*, Uppsala, Sweden, pp. 1149–1154.

[9] Dingding Wang etc.Multi-Document Summarization using Sentence-based Topic Models. *Proc. of the ACL-IJCNLP 2009 Conference Short Papers*, pp. 297–300.

*Multimedia, Communication and Computing Application – Leung (Ed.)*
© *2015 Taylor & Francis Group, London, ISBN 978-1-138-02775-6*

# Construction of Uighur-Chinese parallel corpus

J.L. Song & L. Dai
*Beijing Institute of Technology, College of Computer Science, Beijing, China*

ABSTRACT: Uighur-Chinese parallel corpus is an important foundation of Uighur-Chinese cross-language information processing. As a corpus of minority language, its construction is relatively more difficult. In this paper, we discuss issues related to the construction. We firstly introduce the selection of corpus resources. Second, in order to accelerate the construction and improve the quality of the corpus, we develop an assistant construction system based on webpage content extraction and text duplication removal, etc. By using this system, we build a Uighur-Chinese parallel corpus with approximately 300,000 sentence pairs and a moderate size of dictionary of person name and place name. Finally, to evaluate the corpus, we build a demo Uighur-Chinese statistical translation system to explore the corpus. The result preliminarily verifies its effectiveness.

## 1 INTRODUCTION

Corpus linguistics is the study of language as expressed in samples (corpora) of the real world. As an important source of cross lingual information processing, bilingual parallel corpus has been widely concerned. A parallel corpus belongs to bilingual corpus, in which source text in a language is aligned to target text in another language in the sentence or paragraph level. Compared with dictionaries or monolingual corpus, parallel corpus has the unique advantage, such as bilingual content contrast and context-richness etc. So it is now widely used in conditions, like cross lingual information retrieval(Schäuble 1997), cross-language text classification(Gliozzo 2006), etc. In particular, parallel sentences have been applied more and more widely in statistical machine translation(Peter Brown,1993) and machine-aided translation researches, where it not only improves the quality of machine translation, but also improves the human-computer interaction.

Since the 1960s, many countries and regions have established a variety of monolingual corpora. But it was not until 1990s, parallel corpora are applied to text classification and statistical machine translation research(Baker 1993, Toury 1980, Malmkjar 1995). Baker (Baker 1993) puts forward special requirements and assumptions of corpus construction and corpus linguistics, from the perspective of translation studies. After that, some parallel corpora have been built, among which earlier ones included English-Norwegian and English-Italian parallel corpus (Kenny 1998). However, because the construction of parallel corpus involves the alignment of the text of two languages and requires a lot of human work, it advances slowly.

Early corpus construction researches focus on European languages. The earliest parallel corpora are mostly about western languages such as the European Parliament corpus and Hansard English-French parallel corpus (Schwenk 2008). It was not until late 1990s, Chinese related parallel corpus began to be built. In recent years, the advantages of parallel corpus have been noticed by Chinese scholars, so some universities and institutions have started to construct a parallel corpus of Chinese and other languages. Up to now, several Chinese-related parallel corpora have been built. For example, General Chinese-English corpus built by Beijing Foreign Studies University (about 30 million Chinese characters/English word), Chinese-English bilingual corpus built by the Institute of Software (150,000 pairs sentences), Babel English-Chinese corpus (544,095 words, aligned at sentence level) are all known Chinese-Foreign language parallel corpora (Feng 2002).

Although there are some progresses on Chinese-related parallel corpora construction, some problems are still left. One problem is that these corpora mainly focus on Chinese-English. Very few corpora are in other languages, especially for the urgent-needed Chinese-Minority.

China is a multiethnic country, national unity and harmony have always been important to national security and economic development. For a long time, due to geographical constraints and the complexity of ethnic languages themselves, information processing research on minority languages is inadequate, which impacts the communication among nations. As an important foundation for machine translation and cross-lingual text mining, Chinese-Minority bilingual corpus will play an important role in the promotion of communication among nations.

Uighurs mainly distribute in Xinjiang province of China, with a few scattered in Taoyuan County, Hunan Province. According to the 2011 census, the Uighur population reached 10,000,000. In the meanwhile, Xinjiang is one of China's largest provinces and has a long boundary with neighbour countries. Thus, Xinjiang and Uighur are very important for both national security and economic development in China. However, due to the narrow application scope and low level of automatic processing, it is difficult to effectively recognize and process Uighur information. Currently there is not a handy Uighur-Chinese parallel corpus that meets the demands of Uighur-Chinese translation and cross-language text mining studies.

This paper studies the related problems about Uighur-Chinese parallel corpus construction. Firstly, we introduce the selection of corpus sources. Secondly, based on the characteristics of Uighur, we develop an assistant corpus construction system, aiming to accelerate the construction and improve the quality of the corpus. We construct a middle scale Uighur-Chinese parallel corpus using this software. Thirdly, in order to verify the effectiveness of the corpus, we build a demo Uighur-Chinese statistical translation system by exploring this corpus.

## 2 CORPUS SOURCES

The selection of corpus source largely determines the quality and application scope of the corpus. A good corpus should be with proper quantity, coverage and low duplication.

For the construction of Uighur-Chinese parallel corpus, it is desperately short of handy aligned documents. The limited extant aligned documents mainly exist in bilingual newspaper offices and presses, and scattered Uighur-Chinese translation companies. The main fields of these extant documents include news reports, policies and regulations, tourism and a small amount of scientific articles.

In order to ensure the usability and coverage of the corpus, we expect the distribution of the fields of the corpus is roughly as follows: news reports (political, social, sports, travel, etc.) at 50%, proses and essays at 30%, and literatures at 20%. In addition, we hope to construct electronic dictionaries of place names and person names, covering as many place names and person names of Xinjiang and Uyghurs as possible.

Parallel corpus from translation companies is translated by human labor, thus its quality is good. But due to the consideration of the confidentiality of customers and the limitation of the fields, we abandon this choose. Our final corpus sources include: (1) Daily Press. We get bilingual articles from web or electronic news reports, extracting content automatically and aligned sentences by human labor.

(2) Xinjiang local Uighur websites. These webpages are written in Uighur without Chinese translation. We extract webpage content and then translate sentences into Chinese by human labor. (3) Paper or electronic documents, mainly literature documents in Chinese. These documents are translated by native Uighur speakers who are also good at Chinese. (4) Person names and place names. We collect Uighur person names and Xinjiang place names from paper and electronic media, and translate the items manually to form an electronic dictionary.

The final completed corpus includes about 300,000 pairs of sentences. The length of sentences distributes as Figure 1 (Chinese character length).

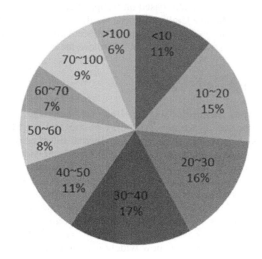

Figure 1.    Sentence length distribution of the corpus.

The person's name dictionary includes 20,000 items. Since Uighur names are relatively fixed, which are similar to English names, this person's name dictionary can cover most of Uighur names. The place name dictionary includes translated place names of Xinjiang region about 10,000 pairs, covering the village-level administrative units.

## 3 ASSISTANT CORPUS CONSTRUCTION SYSTEM

The construction of a parallel corpus is a very complicated task, and requires a lot of manpower. The limitation of the Uighur corpus source makes it demand even more manpower to construct an Uighur-Chinese parallel corpus. It is necessary to use computer tools to improve the construction efficiency and data quality. According to the actual requirements, we develop a computer-aided corpus construction system.

### 3.1 Main functions

The main functions of this assistant system are as follows.

1 *Translation and input.* Translation and input are the basic functions of the assistant system. The system provides translation and input interfaces for language engineers. Through the interfaces, Engineers translate single lingual sentences or input extant aligned sentence pairs. The system provides the context of the working sentence and subsidiary electronic dictionary to improve the efficiency and quality.
2 *Revision.* In order to ensure the quality of the translation, the system provides manual revision interface. All sentence pairs must be verified at least once before being stored into database. Verifiers can directly modify the sentences, or return to the translators for modification. The system counts each translator's error rate and rework rate, in order to improve their quality of translation.
3 *Duplication detection.* The effectiveness of duplicated sentences in a corpus is limited. When the system allocates sentences to workers, it automatically detects duplicate or similar sentences. Identical sentences are assigned only once. In the follow-up appearances, these sentences will be sent to translators as context information. For the similar sentence, their translations will be sent to translators as reference.
4 *Task management.* In order to accelerate the corpus building, the system supports multiple persons to work in parallel. The system provides task management function for translator and Revisers. The system assigns tasks to each person and records the progress and quality. It tries to push a whole document or a whole paragraph to a translator to ensure the integrity of the context information. When engineers request tasks, the system pushes several articles at a same time for selection.

In addition, the system also has functions, like data management, user management, and data extraction. Data management implements permanent data storage, retrieval, data import and export, backup and restoration. User management provides functions, such as user registration and removal, password modification. The data extraction function extracts and cleans text from web pages to get sentences.

### 3.2 System structure

The system, based on C/S architecture, consists of client part and a server part. The client part provides user interface for administrators, translators and revisers, supporting multiuser concurrent operations. The server part provides data management, data sharing, data and system maintenance and concurrency control, as well as user management and user session maintenance.

The server part modules and functions are listed as follows:

Login module. It is for verifying the user's registration and login, maintaining user sessions, and managing user's behaviour.

1 Task scheduling module. It is for allocating and reclaiming translation and revision tasks.
2 Data management module. It is for providing data access function of source text, translation text and user information, etc.
3 Duplication detection module. It is about providing the duplication detection and removal of original and translated sentences, and then extracting sentences.
4 Data extraction module. It automatically generates webpage wrappers and uses the wrappers to extract the text body from web pages.
5 Database interface. It is for providing an inner interface to access database.
6 Service interaction module. It provides server-side data communication with clients, unifies packet encapsulation formats and data transmission protocol, receives and sends data to clients.

The client modules and functions are listed as follows:

1 Login and registration module. It manages user sessions, existing user login and new user registration.
2 Task management module. It manages translation and revision tasks, uploads completed tasks to the server.
3 User management module. It provides user management interface for administrators, translators, and revisers. Among them, the administrators have the super privileges, and is responsible for system configuration and user management. Translators and revisers have permission to modify their own passwords.
4 Assistant translation module. It provides a translation interface for the user. To facilitate the translation, it also provides automatic sentence segmentation, pushes context sentences and translation references.
5 Revision module. It provides a revision interface for the user.
6 Client interaction module. It provides client-side data communication with the server, receives and sends data to the server.

### 3.3 System implementation

We use C# to implement this system in C/S architecture. For web page content extraction, we implement Valter Crescenzi's template generation algorithm

RoadRunner (Crescenzi 2001). In order to efficiently carry out duplicated and similar content detection, we implement G.S. Manku's algorithm (Manku 2007) for duplication removal in mass data storage.

## 4 PRELIMINARY VALIDATION

In order to validate the effectiveness of the Uighur-Chinese parallel corpus, we develop a demo Uighur-Chinese statistical translation system to explore it. The system is based on the Moses training platform (Koehn 2007). Moses is a phrase-based statistical machine translation system, developed by University of Edinburgh, RWTH Aachen University and other organizations. It is an open source project coded in C++, running on Linux platforms and Windows platforms.

We firstly use Moses to learn the Uighur language model and Chinese language model using N-GRAM (3-GRAM). Then we use parallel corpus to learn translation model. The system integrates a bilingual dictionary including about 200,000 words to translate words which are not covered by the corpus. When translating, sentences are passed to the translation module. The mediate results are then optimized with language model.

The system can translate documents and provide sentence level control browsing, and realize transforming traditional Uighur to Latin Uighur as well.

As it is lack of evaluation sets for Uighur-Chinese translation system, we do preliminary tests. The average readability of the results is about 60%. And it works better on news articles. This result verifies the effectiveness of the corpus.

## 5 CONCLUSION AND FUTURE WORK

In this paper, we discuss issues related to the construction of Uighur-Chinese parallel corpus. We use newspapers, publications and web pages as corpus source. Because it is short of aligned Uighur-Chinese sentences, the bilingual corpus must be built through manual translation. In order to improve the construction efficiency and corpus quality, we develop an assistant system to facilitate the construction. We construct a parallel corpus, which contains 300,000 pairs of statements to use this system. In addition, we also complete an electronic dictionary of about 20,000 person names and 10,000 place names. To evaluate the corpus, we build a demo Uighur-Chinese statistical translation system based on this corpus. Preliminary evaluation shows that the readability of the results of the demo translation system is about 60%, which verifies the corpus's effectiveness.

We are now researching on Uighur-Chinese statistical machine translation and cross-language information retrieval based on this Uighur-Chinese parallel corpus. At the same time, we continue to increase the size and improve the quality of the corpus.

## ACKNOWLEDGMENT

This work is supported by the key project of the National Natural Science Foundation (61132009), a project of the National Natural Science Foundation (60803050), and the National Key Basic Research Development Plan Project (2013CB329303).

## REFERENCES

M. Baker, G. Francis & E. Tognini-Bonelli. Corpus linguistics and translation studies: implications and applications. Text and Technology. 233–250. 1993.

V. Crescenzi, Giansalvatore Mecca, Paolo Merialdo. RoadRunner: Towards Automatic Data Extraction from Large Web Sites.Proceedings of the 27th VLDB Conference, Rome, Italy, 2001:109–118.

Z. Feng. Evolution and Present Situation of Corpus Research In China. Journal of Chinese Language and Computing 11:2, 127–136, 2002.

A. Gliozzo, C. Strapparava. Exploiting comparable corpora and bilingual dictionaries for cross-language text categorization. In Proceedings of the 21st International Conference on Computational Linguistics and the 44th annual meeting of the Association for Computational Linguistics. NJ USA: Association for Computational Linguistics, 2006:553–560.

D. Kenny. Corpora in translation studies. Bake M. Encyclopedia of Translation Studies. London: Routledge, 1998.

P. Koehn, Hoang H, Birch A, et al. Moses: Open source toolkit for statistical machine translation. Proceedings of the 45th Annual Meeting of the ACL on Interactive Poster and Demonstration Sessions. Association for Computational Linguistics, 2007:177–180.

K. Malmkjar. Linguistics and the Language of Translation. Edinburgh Textbooks in Applied Linguistics. 1995.

G.S. Manku, Arvind Jain, Anish Das Sarma. Detecting Near-Duplicates for Web Crawling. WWW, Track: Data Mining, 2007.

F. Peter Brown, Vincent J. Della Pietra, Stephen A. Della Pietra, and Robert L. Mercer. The mathematics of statistical machine translation: parameter estimation. Comput. Linguist. 19, 2. June 1993. 263–311.

P. Schäuble, Sheridan P. Cross-language information retrieval (CLIR) track overview. In Proceedings of the Sixth Text Retrieval Conference (TREC-6). 1997.

H. Schwenk, Investigations on largescale lightly-supervised training for statistical machine translation. IWSLT, 2008.

G. Toury. In Search of a Theory of Translation. Tel Aviv: The porter Institute for Poetics and Semiotics. 1980.

# Design and realization of vehicle driving simulation system based on OGRE

H.H. Li, P. Zhang & X.D. Sun
*School of Computer and Information Technology, Beijing Jiaotong University, Beijing, China*

ABSTRACT: Based on the open source 3D graphics engine Ogre, the design of the vehicle driving simulation system is explored in depth. Related technologies in Ogre are introduced in the discussion of vehicle driving simulation. An efficient method is proposed for vehicle driving simulation. The results show that the design is simple, expansible, and has realistic effects.

KEYWORDS: Driving simulation, 3D, OGRE, Terrain.

## 1 INTRODUCTION

When terrain and multi-view are taken account, vehicle driving simulation system can correctly simulate the operation of driving a car. Users could get senses of immersion and interaction from it. Currently, similar systems are still far from the ideal. Most of them are developed by commercial software and the cost is expensive. Therefore, it is necessary to design an efficient and low cost method to get realistic picture effects and provide the true feeling of driving. Based on open source 3D graphics engine Ogre, the design idea of the vehicle driving simulation system is discussed. Ogre is designed to make it easier and more intuitive for developers to produce applications utilizing hardware-accelerated 3D graphics. The class library abstracts all the details of using the underlying system libraries like Direct3D and OpenGL and provides an interface based on world objects and other intuitive classes [1].

## 2 KEY TECHNOLOGIES

### 2.1 OGRE framework

The main point of access to an Ogre application is through the Root object. It provides a convenient point of access to every subsystem in an Ogre application [2].

After creating the Root object, a typical Ogre application will load resources, such as meshes, materials, and textures. All resources are managed by a single object: Resource Group Manager. This object is responsible for locating and initializing resources.

After resources being in place, render window is needed to be created. Render window renders our scene to the screen. It looks like a canvas. All the contents should be drawn to its surface. We need at least one render window in an Ogre application. While Ogre is a 3D graphics engine, the scene to render is produced at some point from a certain viewing angle. So Ogre uses Camera to control the viewing position, the same as our eyes in the real world.

Then, the application begins rendering. In this part, a Scene Manager is used in OGRE. Scene Manager creates a series of Scene Nodes. And a Scene Node is the basic unit which is transformable and movable in the scene. All the objects in the scene need to be hooked up to a Scene Node to display. In an application, an entity of a mesh file is created by Scene Manager, and then attaches the entity to a Scene Node to display, move, and rotate the object.

Moreover, the application should be able to respond to the user's input, such as adjusting the angle of view. Frame Listener is used to deal with user's input in OGRE. A Frame Listener must correspond to a Render Window. The message will be transmitted to the Frame Listener only if it belongs to the corresponding Render Window. The method frame rendering queued of Frame Listener will be called every frame. So we override this method in response to the operation.

### 2.2 Terrain component in OGRE

OGRE provides a terrain component [2]. It makes things convenient to use terrain in an OGRE program. The terrain component of Ogre uses a level of detail (LOD) technology. LOD technology decreases the complexity of a 3D object representation according to the object's location. It helps to obtain high efficiency rendering experience. A terrain component of OGRE also supports multiple texture blending technology and real-time editing features. It is easy for us to edit and obtain the realistic effect of terrain.

Grey-scale map can be used to create terrain in OGRE. The grey value of the image is regarded as the base value of the height conversion. The height's value ranges from 0 to 255, where 0 (black) represents the lowest height and 255 (white) represents the maximum height. The basic idea to build terrain is to create a vertex grid which is the same size of the height map. The value of each pixel is deemed to be the height of each vertex grid. The vertices of the grid contain not only the location, but also information such as normal, and texture coordinates. After obtaining the coordinates of each point, we can draw the three-dimensional terrain. Then apply rocks, grass or other textures to the terrain. The construction of the terrain is completed. The terrain in this paper using ALIGN_X_Z, namely the XZ plane is the horizontal plane, Y axis is vertical axis.

### 2.3  3D model building

3D models in OGRE have particular data formats. The model can be modelled in 3ds max and converted to the OGRE mesh format through third-party plug-ins. Then you can use it directly in OGRE. Model's position and size can be adjusted in OGRE. Furthermore, a complete model in OGRE includes textures and materials in addition to mesh files.

In the development of the system, mainly used models are cars, and roadside reference, such as trees and buildings. The models of 3ds format are exported by 3ds max. And they are converted to mesh format of OGRE by a third-party format tool 3ds2mesh.

### 3  IMPLEMENTATION

### 3.1  Motion models for vehicle

To simulate driving a car is divided into two parts of this paper: car body and wheels because the movement of car body is not consistent with the movement of car tires. For example, when a car moves forward, the car body moves forward at the given speed while tires maintain forward rotation in addition to moving forward. So, they cannot be considered as an entirety.

We create scene nodes on behalf of the car body and tires respectively. And then the corresponding model entities are attached to the scene nodes, as shown in Figure 1.

When tires move forward, they produce a backward tangential force to the ground. Then the ground will produce a forward force to the wheels, so the car can run. In this article, wheel spin and tire locking are not considered. So the relationship between turns of the wheel rotation and rolling distance of the wheels can be obtained by the equation (1) shown below:

$$s = 2\pi n_w r \tag{1}$$

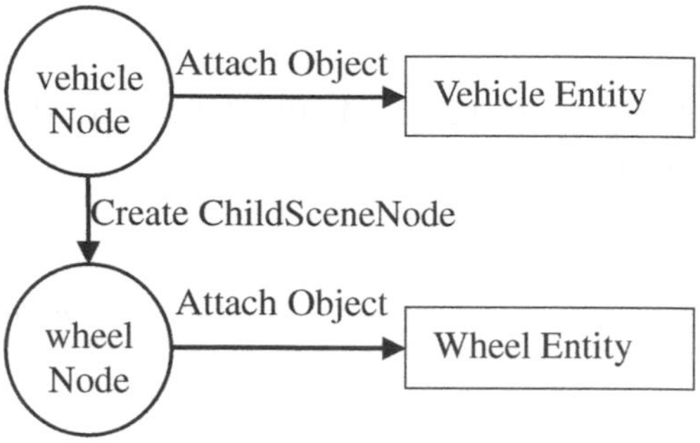

Figure 1.  Motion models for vehicle.

$n_w$ is turns of the wheels' rotation; S is the rolling distance when the turns of the wheels rotation is $n_w$ (km); r refers to the radius of wheels (km).

The relationship between driving speed and rotational speed of the tires can be obtained by the equation (1):

$$V = \frac{s}{t} = \frac{2\pi n_w r}{t} = 2\pi r * \frac{n_w}{t} \tag{2}$$

$$n = \frac{v}{2\pi r * 3600} \tag{3}$$

V is driving speed (km/h); n is the rotational speed of the tires (r/s).

In three-dimensional space of OGRE, pitch, yaw, and roll can let objects rotate around three axes, as shown in Figure 2. If we do a pitch-operation to the tire scene node point, tires will be rotated around the x-axis. The code to achieve the tire rotation is:

*Wheel Node pitch->(Ogre::Degree(360) \* evt. time Since Last Frame \* n);*

We use the time-based movement to simulate the tire rotation. evt. Time Since Last Frame records the time from the previous frame to the present. This method makes the speed of tire rotation stable, better than the method based on the frame rate. Ogre: Degree(360) indicates that one rotation is 360 degrees.

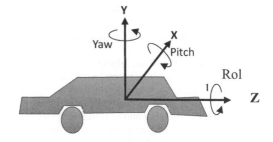

Figure 2.  Rotation in Ogre.

N is the rotational speed of tires calculated by the set speed. The rotation angle of the wheel Node for each frame can be derived by multiplying the three items.

## 3.2 Driving simulation

Operations of car driving contain pressing the accelerator, braking, shifting gears, and steering. These operations can be divided into two classes: speed control and direction control. We take two typical operations, for example, pressing the accelerator and steering. The design philosophy of the rest is similar and easy to expand. In addition, driving simulation ought to be combined with terrain. Therefore, implementation of slope-crossing are also presented.

### 3.2.1 Moving forward

Implementation of the car movement is:

1  A child node "front Node" is created by the scene node "vehicle Node" of car body.
2  Set the relative position of the front Node as the distance in front of the car body. (As is shown in Figure 3(a)).
3  Let vehicle Node get front Node's position.

As front Node is a child of vehicle Node, front-Node will get a new position. So, letting vehicleNode get front Node's position constantly is how a car moving forward works in application.

*frontNode->setPosition(vehicleNode->getPosition()*

*+Ogre::Vector3(0,0,d));*

Where d is the distance between front Node and vehicle Node. d determines the value of each time's forward movement distance, namely the speed of the car.

### 3.2.2 Steering

The direction control model is considered to be the relationship between car steering angle and steering wheel's rotation angle. And we assume that the change of the car's direction follows the change of the steering wheel angle without delay. When the steering wheel rotation angle is $d\theta$, the time is dt, the direction of the car turns $d\theta$ degrees after a period of time dt. The relationship can be described by formula (4) [4].

$$\frac{d\theta}{dt} = \frac{v_0}{I_w}\delta_w \qquad (4)$$

Where $v_0$ is the driving speed of the moment. $l_w$ is the coefficient of steering radius and steering wheel angle.

Because the system uses the XZ plane as the plane, when the automobile is steering, automobile rotates around the Y axis.

When the car body's steering, front Node steers with relative position as is shown in Figure 3(b). For car body, front Node is still right in front of the vehicle. The car can still move forward normally.

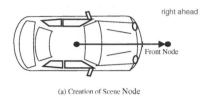

(a) Creation of Scene Node

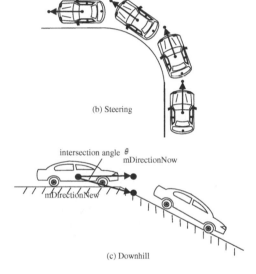

(b) Steering

(c) Downhill

Figure 3.  Fundamental of driving simulation.

### 3.2.3 Uphill/Downhill

When a car goes uphill or downhill, or travels on non-flat road, it needs to be ensured that the car is travelling on the ground. OGRE provides a terrain height query algorithm. It is achieved by ray detection: First, set the value of ray's endpoint position to be any value larger than the topography Y-axis maximum value. And then detect along the negative direction of Y-axis. The height of the coordinate points on the terrain can be determined after the surface is detected. The main code is as follows:

*ray.setOrigin(Vector3(pos.x,mTerrainPos.y+10000, pos. z));*

*ray. set Direction(Vector3::NEGATIVE_UNIT_Y);*

*Terrain Group::Ray Result ray Result = m Loader->get Terrain Group()->ray Intersects(ray);*

*Real newy = rayResult. position.y + dAboveTerrain;*

We have already obtained the height of the terrain newy where front Node locates. Then front Node's Y coordinate value is updated. So when vehicle Node replaces front Node's position, it is ensured that the car is on the ground. The car body will not fall into the terrain nor be suspended in the air.

When the car goes uphill or downhill, there will be a change in angle in addition to the coordinates in the vertical direction. Taking downhill for example, we find two vectors as is shown in Figure 3 (c). One of them is a vector that from car body's position to front Node's position. And the other is a vector from car body's position to the position where X and Z coordinate value is the same as the front Node's position(x, z) and Y coordinate value is the real height of the terrain at (x, z). There is an angle between two vectors. And this is the tilt angle when the car goes downhill.

We first obtain the dot product of the two vectors m Direction Now and m Direction New, which is the cosine value of the angle (degreeθ) between two vectors (cosθ). The degree can be obtained by the inverse cosine function. If front Node's real height is higher than the height of terrain at its position as is shown in Figure 3(c), it is downhill. The vehicle should rotate clockwise degrees. Otherwise, it is uphill. The vehicle should rotate counterclockwise θ degrees.

### 3.3 *Vision simulation*

In the car driving simulation, we need one or more viewport to observe vehicle movement. Viewport may be the car driver's perspective, or aerial perspective outside the car. Viewport may be fixed to a location in the virtual world, or follows the car moving. The camera in OGRE provides a basic method to achieve these goals.

Create Camera:
*mc = m Scene Mgr->create Camera("Car Cam");*
Set Camera position:
*mc->set Position(Vector3(x, y, z));*
Set gazing direction:
*mc->look At(Vector3(xl, yl, zl));*

If the Camera needs to follow the car, we can update Camera setting in the function of "frame Rendering Queued":
*mc->set Position(vehicle Node>get Position()+v Pos);*
*mc->look At(vehicle Node->get Position());*

v Pos is the relative position vector from Camera's position to vehicle Node's position.

### 4  SIMULATION RESULTS

The design idea of driving simulation in this paper is implemented in an application. Programming language is C++. We use the Ogre Wiki Framework [5]

which includes a number of basic functions, such as input control OIS and camera control. Rewrite the create Scene function in the framework to create and display the basic terrain which contains ramps. And in accordance with section 3.1, entities of the car body and wheels are created and attached to scene nodes. Set the coordinates of car body's scene node, which is the initial coordinates of the car in the scene. Terrain height is also needed to be detected to ensure that the car travels on the ground. After the initial static scene is created, the key Pressed function is rewritten. It is used to monitor keyboard input and listen for forward or steering commands. Rewrites frame Rendering Queued function. Moving forward, steering, going uphill or downhill and other operation is implemented in frame Rendering Queued function.

The main implementation of driving simulation in the above program can be described in a flow chart as is shown in Figure 4.

The application is written in C++. The development platform is VS2010, and it is run on an ordinary PC, the machine configuration is:

CPU: dual-core i5 2.6GHz;
Memory: 4GB;
Video card's memory: 2GB.

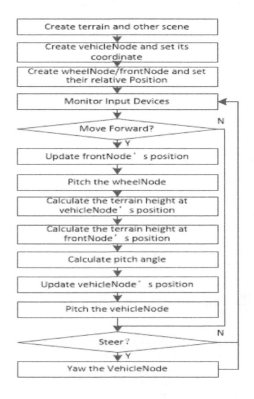

Figure 4.   Implementation process.

Figure 5 is a screenshot of the application which shows a car going uphill. The average frame per second can be 460fps or more.

In addition, based on the prototype of Santana, the motion model in this paper is validated further. In the validation process, automobile wheelbase is 14. And the actual automobile wheelbase is 3.5 meters. So the program to reality ratio is four to one. The average frame per second of verification program is from 80 to 100 fps. Update the speed for each frame:

Acceleration:

d += 0.043*evt. time Since Last Frame;

Neutral taxiing:

d −= 0.009*evt. Time Since Last Frame;

d is the distance between front Node and vehicle Node as is defined in Section 3.2.1.

Actual vehicle sample data [4] and simulation data are shown in Table 1. The two kinds of data are relatively close. It shows that the motion model and the algorithm in this paper are correct and feasible. The model can not only simplify complex situations of car driving simulation, but also be in accordance with actual requirements.

Figure 5.    Simulation results – Uphill.

Table 1.    Motion model verifications results.

| Test Items | Condition | Simulation Result | Actual Result |
|---|---|---|---|
| **Start Time/s** | *Accelerate to 40km/h* | 9.233 10.256 9.329 | 10 |
| **Time of Neutral Taxiing/s** | Initial speed is 40km/h | 51.384 53.078 52.157 | 50 |

## 5   CONCLUSION

This article describes the design ideas of driving simulation based on graphic engine OGRE. The system includes a motion model for vehicle, visual settings, and the implementation of car's moving forward and steering combined with the terrain. Experimental results show that the system can simulate the basic operation of the vehicle in the three-dimensional environment primly. Moreover, the implementation of this system is simple. Operations such as shift, reverse, brake are easy to be further improved. In addition, the perfect driving simulation [6] contains not only operation simulation and visual simulation, but also sounds simulation and environmental feedback (such as collision detection), etc. These are the further improvement of the system in future research.

ACKNOWLEDGMENT

This work has been supported by Project No. 2012AA040912 of the National High-Technology Research and Development Program of China.

REFERENCES

[1]   Gregory Junker. Pro OGRE 3D Programming [M], Apress, 2006.
[2]   Terrain, Sky, and Fog[EB/OL]. http://www.ogre3d.org/tikiwiki/tiki-index.php?page=Basic + Tutorial + 3 & structure = Tutorials.
[2]   Terrain, Sky, and Fog [EB/OL]. http://www.ogre3d.org/tiki-wki/tikiindex.php?page= Basic + Tutorial + 3 & structure = Tutorials
[3]   YU Zhi-sheng. Automobile Theory. Version 3. Beijing: China Machine Press, 2000, 1–17.
[4]   CAI Zhong-fa, ZHANG An-yuan, "Study of automobile emulated driving model and simulation," Journal of Zhejiang University (Engineering Science), Vol. 36, No. 3, pp. 327–330, May 2002.
[5]   Ogre Wiki Tutorial Framework [EB/OL].http://www.ogre3d.org/tikiwiki/tiki-index.php?page=Ogre+ Wiki + Tutorial + Framework.
[6]   CAI Zhong-fa, LIU Da-jiang, ZHANG An-yuan, "Scheme Study of Automobile Driving Training Simulator Based on Virtual Reality", Journal of System Simulation, Vol. 14, No. 6, pp. 771–774, June 2002.

Multimedia, Communication and Computing Application – Leung (Ed.)
© 2015 Taylor & Francis Group, London, ISBN 978-1-138-02775-6

# Design and simulation of parallel distributed IIR filter based on FPGA

D.X. Sun & X.F. Li

*School of Mechatronical Engineering, Beijing Institute of Technology, Beijing, China*

ABSTRACT: This paper presents a method of using FPGA to realize IIR digital filter. This method used full parallel distributed algorithm to reduce the hardware circuit scale on the premise of guaranteeing real time and to solve the problem of floating point calculation in FPGA using fixed-point binary number. Taking a second-order IIR digital filter as an example, this paper used hierarchical method to design the digital filter and its function modules and to verify the hardware circuit through simulation. The results showed that the design of IIR digital filter method has good real-time performance and the hardware circuit size is smaller. The calculation of fixed-point binary number can control the filtering precision effectively, and the lookup table structure changed conveniently, which shows the flexibility of the design.

KEYWORDS: Digital filter, FPGA, Distributed arithmetic, Fixed-point binary number, Simulation

## 1 INTRODUCTION

In the field of digital signal processing technology, the application of digital filter is very important. The key problem of using FPGA to implement IIR digital filter is the processing of filter coefficients and the realization of the multiplication module. FPGA cannot handle floating point operations, so we need to deal with filter coefficients. In addition, the IIR digital filter has more multiplication. If we use the array multiplier within hardware, it will take up too much of the hardware resources, which is not conducive to the design of large-scale integrated circuits. In a study [1], Guo-hua Wei obtained the quantization coefficients by dealing with filter coefficients six times, and the process was complex. Another study [2] mentioned the quantification of the ideal coefficients; but it may result in the zero pole position deviation of filters, which makes the differences in the frequency respond with the theoretical value. It may make filter poles out of the unit circle when serious, which lead to an unstable system [2]. Xiang-ping Li [3] proposed a serial distributed algorithm based on the read-only memory lookup table to implement the multiplication operation. Though the aim of improving the clock of circuit system through the FPGA's internal ROM has been achieved, the price to pay for it is the expense of overall data processing time [3]. A modified design of combining direct multiply accumulation and ROM lookup table has been proposed in one study [4]. Although it saves some hardware space, more time is consumed in the process [4].

This paper studies the implementation method of IIR digital filter based on FPGA. It also focuses on research of the efficient implementation of a multiplication operation, parallel distributed algorithms, and infinite precision floating-point fixed-point binary problem in hardware circuit design.

## 2 IIR DIGITAL FILTER SUMMARY

The standard transfer function of IIR digital filter can be expressed as follows:

$$H(z) = \frac{\sum_{r=0}^{M} b(r) \cdot z^{-r}}{1 - \sum_{k=1}^{N} a(k) \cdot z^{-k}} \quad (1)$$

Set the input signal for IIR digital filter as $X(z)$, and the output signal is $Y(z)$. According to the relationship between transfer function and the input-output signals, we can get

$$Y(z) = H(z)X(z) \quad (2)$$

Assume the sample period of the input signal is $T$ and do inverse transformation of $z$ transformation to (2); we get

$$y(n) = \sum_{k=0}^{M} b(k) \cdot x(n-k) - \sum_{k=1}^{N} a(k) \cdot y(n-k) \quad (3)$$

Sequences $\sum_{k=0}^{M} b(k) \cdot x(n-k)$ and $\sum_{k=1}^{N} a(k) \cdot y(n-k)$ are M-order and N-order delay structures of inputs

$x(n)$ and $y(n)$, respectively, and each order time delay is added together after being tapped; thus, it can reflect that the IIR digital filter is a feedback network.

Any high-order IIR filter can be expressed as the cascade form of a second-order system, so the second-order IIR digital filter has been taken as an example to study the FPGA implementation of IIR digital filter in this paper.

# 3 IIR DIGITAL FILTER BASED ON FPGA

The implementation of IIR digital filter based on FPGA means that the difference equation (3) is described using hardware description language. There are three important problems that need to be solved regarding FPGA hardware implementation: the representation and calculation of filter coefficients and floating-point in the input sequence, realization method of multiplications in the difference equation, and realization of the output feedback.

## 3.1 *Coefficient quantization and arithmetic based on the binary number*

During the hardware implementation of filters, all coefficients must be stored in the FPGA register in the form of a finite length binary code. While the transfer function coefficients $a(k)$ and $b(k)$ of the ideal digital filter designed by the theory are infinite precision floating-point numbers, the quantization of ideal filter coefficients needs to be done.

Using the method of fixed-point binary number to quantify filter coefficients and converting the multiplication to displacement calculation of the fixed-point binary number solved the problem that FPGA can only deal with integers [5]. It also guaranteed the precision of coefficients and improved the resource usage effectively.

According to the different precision requirements, expand data in appropriate multiples and express it by a fixed number of binary numbers. The original data can be obtained by data displacement. The bigger the expanding ratio, the smaller the truncation error. Thus, the result data is more precise [6]. The addition and subtraction operation of binary numbers is relatively simple. In multiplication, however, the result of two binary multiplications will move the decimal point position, so the operation after the multiplication cannot work normally. Aiming at this problem, the fixed-point representation of binary number is fixed, so we can cut down fixed decimal places from low order and cut down fixed integers from high order. Then, the multiplication results could be made

to maintain consistency with the fixed-point binary form.

## 3.2 *Distributed algorithm*

Equation (3) shows that the core of IIR digital filter arithmetic is multiply accumulation, and a large amount of convolution operations will take up too many hardware resources. The FPGA chip lookup table structure is very beneficial to realize the distributed algorithm, so it is widely used in multiply add. To ensure that convolution operation does not take up too much of the hardware resources, the combination of distributed algorithm and lookup table structure of FPGA chip has been proposed, which converts the convolution operation into a lookup table accumulation operation.

Due to the input sequence $x(n)$, the input sequence of equation (3) can be calculated using distributed algorithm. Set up

$$x(n) = \sum_{b=0}^{B-1} x_b(k) \cdot 2^b, x_b(k) \in [0,1] \tag{4}$$

where $x_b(k)$ shows the $b$ bit of $x(k)$. The input sequence part can be expanded as follows:

$$\sum_{k=0}^{2} b(k) \cdot \left( \sum_{b=0}^{B-1} x_b(n-k) \cdot 2^b \right) = \sum_{b=0}^{B-1} 2^b \sum_{k=0}^{N-1} b(k) \cdot x_b(n-k)$$

$$= \sum_{k=0}^{2} b(k) \cdot x(n-k) \tag{5}$$

Then, we can get the lookup table (LUT) of second-order IIR filter's input part and the whole structure of the parallel and distributed algorithm, as shown in table 1 and figure 1.

Table 1. Input LUT of second-order IIR filter.

| $x_b(n-2)$ | $x_b(n-1)$ | $x_b(n)$ | $F$ |
|------------|------------|----------|-----|
| 0 | 0 | 0 | 1 |
| 0 | 0 | 1 | b(0) |
| 0 | 1 | 0 | b(1) |
| 0 | 1 | 1 | b(0) + b(1) |
| 1 | 0 | 0 | b(2) |
| 1 | 0 | 1 | b(0) + b(2) |
| 1 | 1 | 0 | b(1) + b(2) |
| 1 | 1 | 1 | b(0) + b(1) + b(2) |

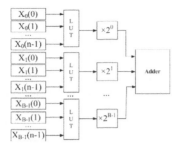

Figure 1. Structure of the full parallel distributed algorithm.

## 3.3 The realization of output feedback

The difference between IIR filter and FIR filter is that the IIR filter has a feedback loop. On the premise of realizing the same function, the introduction of the feedback loop can reduce the number of filter orders, which can improve the real-time performance of digital signal processing, but it also brings about difficulties in the FPGA design. If we design the IIR filter according to the formula, there are higher requirements in the timing cooperation of input and output feedback on FPGA implementation. We need to do parallel distributed computing to both input sequence and output feedback at the same time. Compared with the direct multiplicative algorithm, this algorithm saves hardware space, but it will produce some delay and cause bad real-time performance.

For this case, separate the difference equation (3) into input sequence and output sequence and calculate separately. Take the result of input sequence as the input of output sequence. The input sequence is calculated by parallel distributed algorithm, and the output sequence can be calculated by multiplicative method. It simplified the complexity of the design and saved hardware resources at the same time. It also improved the real-time performance of the operation.

## 4 THE FPGA DESIGN AND IMPLEMENTATION OF IIR FILTER

Because any senior IIR filter can be made up of a number of second-order filters, this paper takes the second-order IIR filter as an example and illustrates the FPGA implementation of IIR digital filter from IIR filter function module partition, module function specification, comprehensive RTL view, and simulation verification.

### 4.1 The functional modules of IIR filter

For the overall structure design of an IIR digital filter, its structure diagram is shown in figure 2: it is divided into two parts, input and output delay feedback loops,

and includes five functional modules. The main functions of each module are as follows:

1. Time delay module: using blocking assignment to achieve the time delay of the input signal, use three register variables show the result after input signal delay three times respectively, and participate in the calculation later. The structure of delay module is shown in figure 3.

2. Serial and parallel conversion module: separate each delay signal, and restrict the delay signal of each corresponding bit, which forms the input module in the lookup table, corresponding to the input of parallel distributed algorithm. The structure of series to parallel module is shown in figure 4.

3. Lookup table module: solidify table 1 into a single lookup table module, and transform the multiplication of input sequence in difference equation (3) to the combination of filter coefficients.

4. Multiplying and adding modules: let each output of the lookup table do corresponding displacement and let the outputs sum together, which can obtain the results of input sequence in the difference equation (3).

5. Output feedback module: as the input of the output feedback module changes constantly with the results of the last output, if we still use the full parallel distributed algorithm it can lead to disorder in output results. Here, we keep the form of the output sequence in difference equation, make the result of the input sequence the input for output sequence, and declare two variable registers to store the first two outputs that are used to cycle.

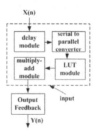

Figure 2. Overall structure.

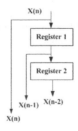

Figure 3. Structure of delay module.

365

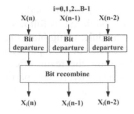

Figure 4. Structure of series to parallel module.

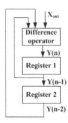

Figure 5. Structure of output feedback module.

### 4.2 Composite RTL view of IIR digital filter

According to the design index, IIR digital filter coefficients are obtained.

$$a = [1 - 0.36952737735124147 \quad 0.19581571265583314]$$

$$b = [0.206572083826 \quad 0.413144167652 \quad 0.206572083826]$$

select the input bit wide as seven which has four integer bits and three decimal bits express the filter coefficients as seven bits fixed-point binary numbers (expressed as the absolute value of coefficient):

$$a = [0001000 \quad 0000010 \quad 0000001]$$
$$b = [0000001 \quad 0000011 \quad 0000001]$$

After compiling and synthesizing in Quartus, start the waveform simulation and get the IIR digital filter RTL view, as shown in figure 6.

### 4.3 The simulation verification of IIR digital filter

Take a set of simple decimal sequences and a set of sine waves to simulate the IIR digital filter. Because the Matlab background operation is infinite precision floating-point calculation, the precision is higher. We will compare the FPGA simulation results with the Matlab simulation results.

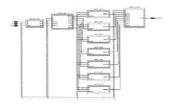

Figure 6. RTL view of IIR digital filter.

#### 4.3.1 The typical sequence simulation

Sequence $x(n) = \{1.25\ 2.25\ 3.5\ 4.5\ 5.25\ 6.75\ 7.75\ 8.5\}$ is selected to simulate the second-order IIR digital filter. The real IIR filter results in FPGA were compared with the simulation results in Matlab, and the comparison results are shown in figure 7.

The comparison results show that the results of FPGA and Matlab simulation values are almost same on the trend, but there still exists a certain deviation. In dealing with filter coefficients, certain truncation errors are produced when using a limited number of fixed-point binary numbers instead of infinite precision floating-point numbers, which lead to differences between filter frequency response and the theoretical value. So, if we want to reduce the deviation we can increase the number of bits of fixed-point binary digits to improve the computing accuracy of the filtering system.

Now, add the bit of fixed-point binary digits to 9, including 4 integer and 5 decimal places. Compare results between calculation results and the real value, as shown in figure 8.

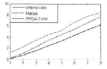

Figure 7. Typical number simulation comparison (1).

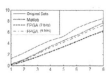

Figure 8. Typical number simulation comparison (2).

By comparing the results, it can be seen that in the case of improving the fixed number of binary digits the computing accuracy can be improved significantly.

#### 4.3.2 The typical signal simulation

For example, we use Sine-wave as an input signal, and compared the FPGA simulation results with the Matlab simulation results ; the result is shown in figure 8. According to the contrast seen, the FPGA simulation data and the data in Matlab are basically identical, but there is still a certain delay in time. Although a feedback mechanism was introduced to the IIR digital filter, resulting in a smaller delay than FIR filter while meeting the same performance requirements and technical indicators, this kind of delay cannot be avoided. Due to the blocking of the assignment statement of hardware description language causing one clock delay in each assignment, we can properly merge modules to reduce the delay time and improve the real-time performance.

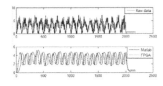

Figure 9. Typical signal simulation comparison.

## 5 CONCLUSION

This paper proposes a design method for a parallel distributed IIR digital filter based on FPGA. This method adopts full parallel distributed algorithm, which reduces the size of the hardware circuit and ensures real-time filtering. Using typical sequence and typical signal simulation, which shows that the design of IIR digital filter using the fixed point of a binary number operation can effectively control filtering precision, we can change the lookup table structure conveniently and embody the flexibility of design.

The structure of all parallel distributed algorithms is simple when the bit of fixed point of binary digits is few; but with higher filtering precision, the number of bits in fixed-point binary digits will also increase, which need more lookup table structure. This results in wastage of hardware resources. Therefore, we need to make further improvement on the filter structure and combine the parallel distributed algorithm with the serial distributed algorithm to adapt to a wider range of filtering requirements.

## REFERENCES

[1] Wei, G.H., 2003, Electronic Products, (3), 20–21.
[2] Qu, X. & Tang, N. & Yan, S., 2009, Computer Simulation, 26(8), 304–307.
[3] Li, X.P., 2005, Journal of Tianjin University of Technology and Education, 15(3), 46–49.
[4] Ni, X.D., 2009, Application of Integrated Circuit, 16, 30–33.
[5] John, G. & Proakis & Dimitris, G. & Manolakis, 2007, 3rd ed. Pearson Prentice Hall, 466–482.
[6] Zhang, D.W. & Jiang, J. & Liu, D., 2012, Marine Technology, 32(2), 24–26.

*Multimedia, Communication and Computing Application – Leung (Ed.)*
*© 2015 Taylor & Francis Group, London, ISBN 978-1-138-02775-6*

# Life parameters study on the barrel of a machine gun

Y.H. Shan
*Baicheng Ordnance Test Center, Emphases Laboratory of Killing and Wounding Technique at Terminal Point on Light Weapons of General Equipment Department, Baicheng, Jilin, China*

L.D. Tan
*Armored Force Engineering Institute, Beijing, China*

S.Q. Shan & Z.G. Nie
*Baicheng Ordnance Test Center, Baicheng, Jilin, China*

ABSTRACT: Aiming at problems of almost all firearm life tests done by whole life projectile, with long test cycles and high ammunition consumption all along, on the basis of the experimental study on machine gun barrels, the main affecting factors and their influence law to the barrel life of machine guns were determined by means of synthetic analysis. By extensive experiments and research, the correlation between barrel life of machine guns and both internal and external trajectory parameters was found. The research results recommend choosing the stress of accelerated life tests of machine guns and establishing a test scheme.

KEYWORDS: Machine gun, Barrel, Life, Mechanism, Trajectory parameters

## 1 INTRODUCTION

In the past, almost all firearm tests, especially firearm life tests, were done by whole life projectile[1], with long test cycles and high ammunition consumption. The practice shows that accelerated life test is one way to shorten test cycles and reduce consumption. Barrels, as the main components of machine guns, can be brought into harsh conditions[2] by erosion, corrosion, and high-temperature and -pressure gunpowder gases in bore and projectile abrasion. The life of barrels represents whole guns' life to a certain extent. Therefore, this paper takes the machine gun barrel as the research object and, by doing lots of experimental studies; exploring the correlation among gun barrel lives, life influence factors, and life parameter characteristics; and finding out the main controllable factors affecting gun barrel lives, discusses the correlation between the barrel lives of machine guns internal and external trajectory parameters and their changing rules. The research results can provide technical support for accelerated life testing of machine gun barrels.

## 2 EXPERIMENTAL RESEARCHES

Many factors affect the life of the barrel; broadly speaking, they include design factors, use of the environment, methods of operation, and so on. All of these will produce different effects on the life of the barrel. The style of the products, quality of its design, and processing and treatment processes determine the use of ammunition. Due to different environmental test conditions, operation methods, and other factors, the barrel life will change dramatically. A large number of theoretical and experimental experiences have shown that when barrels are shot in different specifications after the shot the barrel temperature varies greatly, resulting in a great variation in barrel life. After shooting, the higher the barrel temperature, the severer the bore erosion; further, barrel lives would be shortened [2]-[5]. Therefore, the study of the relationship between barrel lives with design specifications can approximate to the relationship between temperature after the shot and barrel shot specification.

### 2.1 *The correlation between barrel life and test stress*

Mainly, stress affects barrel lives, including environmental test temperatures, amount of cooling cycle projectiles, interval of changing bombs boxes, burst length, shooting interval, and cooling methods.

To identify the main factors of gun barrel lives, the paper carried out a large number of experimental studies about barrel firing and temperature under different conditions of stress tests in a machine gun, and the results are as follows:

(1) The impact of test temperature

When the amount of projectiles (such as 100 rounds) and the shooting mode are certain, with the rise in environmental test temperature the temperature of the barrel after shooting goes up[4]; under the conditions of same shooting methods and changing interval box, the trend of barrel temperature rising to the maximum with the increase in ambient temperature is approximately linearly upward (Fig. 1).

(2) The effect of projectiles amount

The temperature of outer barrel wall significantly increases to the maximum along with the increase in the number of shots (Fig. 2).

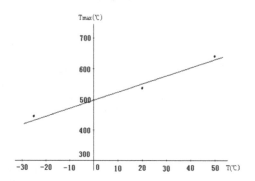

Figure 1. Maximum temperature changes of barrels with the changes in environment temperature.

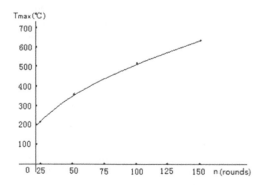

Figure 2. Maximum temperature rise on barrel with number of bullets shot.

(3) The impact of replacing caisson intervals

When the cooling cycle projectile is within 100 rounds, exchanging bomb box intervals does not cause much change in temperature of the barrel after the shot; if the cooling cycle projectiles are over 100 rounds (such as 150 rounds), the use of different bomb box intervals can make the temperature of the barrel change significantly after the shot[4]. Under certain conditions such as the same ambient temperature and the firing method, the barrel's maximum temperature goes down with the increase in exchanging bomb box intervals (Fig. 3).

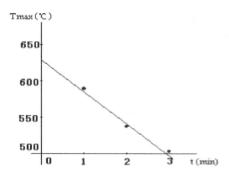

Figure 3. Maximum temperature change of barrels with the time of replacing magazine.

(4) The impact of firing length

When the ambient test temperature and the amount of cooling cycle projectiles are certain, using different barrel shots and burst lengths does not change the temperature after shot.[4].

(5) The impact of firing interval

Under the conditions of same temperature and a certain projectile, using fixed fire (2 to 3s) or radio burst mode does not change the temperature of the barrel.[4]

(6) The effects of cooling methods

In a life experiment, commonly used barrel cooling media are water and air. Because the coefficients of heat transfer of the two cooling media are very different, thermal impact that the barrel of the chromium layer gets is very different. The barrel cooling rate is faster in water than in air. When the barrel is cooled in the water after shot, breech face is susceptible to thermal impact damage. Experimental study shows that when the barrel is drowned for cooling at 350°C, the damage to the barrel from heat impact is minimal (Fig. 4); the higher the temperature is (especially above the 600°C), the more severe damage the barrel gets from thermal impact (Fig. 5). So, if water cooling is used instead of air cooling to save time the time of drowning is very important.

Based on this experimental study, when selecting the most severe test conditions (temperature is 50°C for a box interval of 0.5min) the temperature of the barrel is below 300°C when air is cooled for three minutes after the cooling cycle barrel shot (150 rounds). According to the test results, when the barrel temperature reaches 350°C or less, water cooling does not make the barrel crack due to rapid expansion, which is the basic equivalent air-cooled condition.

Test conditions: thick chromium layer; sample heating to 350°C; after heat preservation for 60 min, drown for cooling.

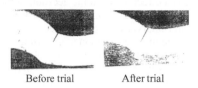

Before trial          After trial

Figure 4.   Barrel bore surface crack propagation.

Test conditions: thick chromium layer; sample heating to 350°C; after air cooled for 1min, drown for cooling.

Before trial          After trial

Figure 5.   Barrel bore surface crack propagation.

## 2.2   *Correlation between barrel life and trajectory parameters*

The results show that ambient temperature, number of cooling cycle's projectiles, and time of boxes replacement are the main stresses that determine gun barrel lives. For choosing a reasonable stress size and using scientific accelerated test methods, it is easy to determine the gun's accelerated life test program[6]–[8]. According to GJB3484-98, barrel life needs to be characterized by muzzle velocity, density, and some other ballistic parameters. In this paper, based on the thermal effects of the barrel bore, accelerated firing tests were done that use four machine guns (8 barrels), in the same amount of cooling cycle projectiles (150 rounds), under different conditions of temperature and environmental, in different combinations of shooting specification (Tab. 1), using a infrared temperature measurement device, the outer surface temperature of each barrel during firing were sustained tested. It was tested, measureed and calculated for barrel size , errosion , muzzle velocity, chamber pressure, the projectile movement time in bore, shooting intensity, inside and outside trajectory parameters etc. While shooting a certain number of bullets it was got for relevant life test parameters and statistical calculation results of each barrels in the firing progress in various stages .

Through statistical analysis of the mentioned before eight-barrel experimental data, each variation of life parameters on the conditions of accelerated life test conditions is obtained.

Table 1.   Researching test program.

| Program | Ambient temperature/°C | Barrel number | Firing mode | Interval replacing bomb box/min |
|---|---|---|---|---|
| Program I | Normal temperature (20) | 1-2 | 25×2×3 | 0.5 |
| | | 2-2 | 25×2×3 | 1 |
| | | 2-1 | 25×2×3 | 2 |
| | | 1-1 | 25×2×3 | 3 |
| Program II | High temperature (50) | 3-2 | 25×2×3 | 2 |
| | Normal temperature (20) | 2-1 | 25×2×3 | 2 |
| | Low temperature (-25) | 3-1 | 25×2×3 | 2 |
| | Low temperature (-45) | 4-2 | 25×2×3 | 2 |

Shooting mode:hand-controlling length × hand-controlling shooting times × bullets number

(1) The zone in the chamber, from casings mouth to muzzle directions, around 100 mm, is severe eroded and the erosion depth sharply decreased. Then, the erosion is flat (Fig. 6).

(2) In all combination conditions of ambient temperature and shooting methods, muzzle velocities (maximum chamber pressure) increased with the number of bombs and declined linearly (Fig. 7).

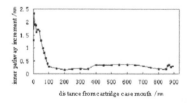

Figure 6.   Changing of inner diameter increment of the barrel of a machine gun after failing.

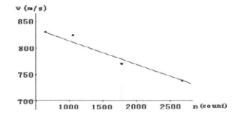

Figure 7.   Changing of beginning velocity with shooting bullets.

(3) In any combination conditions of ambient temperature and shooting methods, times of projectile bore motion drastically reduced with increase in the number of projectiles and then gradually stabilized (Fig. 8).

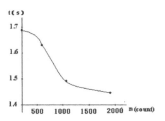

Figure 8. Changing of movement time of bullets in a barrel with shooting bullets.

(4) The intensity of R50 in 100 m distance first increased with increase in the number of projectiles, then decreased, and then gradually increased (Fig. 9).

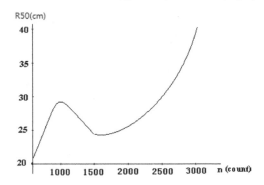

Figure 9. Changing of R50 with the number of shooting bullets.

3 CONCLUSIONS

This paper draws a conclusion as follows through the analysis of affecting factors and their effects, correlation between machine gun barrel lives, and both internal and external parameters of trajectory.

Rise in temperature after machine gun barrels shooting is a direct factor in the decision of their life; ambient temperature, cooling cycles projectiles and bombs, and interval time of replacement tank determine the size of main stress in the life of the gun barrel. The parameters such as muzzle velocity (down), chamber pressure, bore wear changes, and machine gun barrel lives have a strong regularity; time of projectile bore exercise, intensity of firing, and barrel life have a certain relevance.

These results can scientifically and reasonably determine stress of machine guns, accelerated testing, and pilot programs, as well as laying a research base.

REFERENCES

[1] Ma li etc. GJB3484-98 Firearms performance test method [s].
[2] Shan Yonghai etc. Technical analysis on relevance of Shooting specification and design life. Nanjing: Learned journal of Nanjing University of Technology and Engineering [J]. 2005, 29(4):417–424.
[3] J. Hollington. Principles of Tribology [M]. Beijing: Machinery Industry Press, 1981.
[4] Shan Yonghai etc. Integrated normal life test Research machine gun barrel. Beijing: Ordnance Journal [J]. 2013, 34(1):1–3.
[5] Shen Jinxing. Firearms Life [M]. Taiyuan: Shanxi University Associated Press, 1994.
[6] Zhang Changhua, Wen Xisen, Chen dun. Accelerated Life Test Summary [J]. Ordnance Journal. 2004, 25(4):485–488.
[7] Mao Shisong. From life test to accelerated life test [J]. Quality and Reliability. 2003, (1):8–12.
[8] Mao Shisong, Wang Lingling. Accelerated Life Test [M]. Science Press, 1997.

*Multimedia, Communication and Computing Application – Leung (Ed.)*
© *2015 Taylor & Francis Group, London, ISBN 978-1-138-02775-6*

# Load bearing analysis of the wing-fuselage connector with different structures

L. Hua, F.S. Wang, J. Liu & Z.F. Yue
*Northwestern Polytechnical University, Xian, China*

ABSTRACT: The wing-fuselage connector plays an important role in force transmitting of airplane structure. To meet the requirement of load carrying capacity, material plasticity is usually taken into consideration. A connector is analyzed to acquire the stress distribution of joint parts with a typical boundary condition. In order to improve load bearing performance, the original connector model is changed by replacing sleeve with gasket. The influence of the gap between lugs is discussed. Through the simulation, it can be found that the changed connector structure form works more effectively.

## 1 INTRODUCTION

A fixed-wing aircraft usually uses a wing-fuselage connector to transmit forces from the wing to the fuselage. Fatigue damages often occur to the connector due to overstresses. The capacity of bearing flight loads for wing-fuselage connector plays an important role in the service life of an airplane. Failure analysis of wing-fuselage connector for the agricultural aircraft was presented to estimate the first fatigue crack and total damage of connector lug [1]. Experimental data on riveted lap joints for aircraft fuselage was obtained to describe fatigue crack location and life [2].

The design of connector is limited because of the layout of integral structure. There is only partial alteration left to be allowed. The changed connector should keep the original form of transmitting loads. Usually, the whole structure accompanies the connector. Hence, friction and interaction should be considered. To predict the location of crack around the fastener holes, 3D explicit simulation was used to get edge-of-contact stresses in fretting contact condition [3]. A multiscale approach was introduced to describe rubber frictional phenomena as a process inside the bulk of the material [4].

There are several criteria for the design of wing-fuselage connector. The strengthened structure should satisfy the design requirement of planes. To take into account moving loads, there should be some safe margin in the residual strength. The improvement in the joint structure should try hard to inherit the form of the original structure, which makes sure the changes to the connector are as little as possible. The feasibility of structural processing is fully considered to satisfy the manufacturability requirements. The fatigue damage form of lugs depends on the location of the maximum concentrated stress and the direction of loads. The stretch/shear damage of lugs and bend damage of bolt are the main damage forms of the wing-fuselage connector.

Inclined lugs with wide root are used to provide the strength to resist lateral loads. Compared to straight lug and wasp waist lug, inclined lug can resist fatigue damage efficiently. Because the joint area consists of complicated structures and plenty of contacts, the wing-fuselage connector must be simulated well enough.

The wing-fuselage connector of an airplane is analyzed in this paper. The connector discussed here comprises lug-fork joints. The original model uses bushes to balance the bearing loads between different lugs and is bolted with lugs. Substitute bushes for spacers. Compare the force situation with each other. Take the engineering fabrication into consideration, and the interspace is reserved for two other kinds of models. Attention of this paper is mainly devoted to figuring out which kind of connector is more suitable for airplane under the typical load condition. What is more, the influence of reserving interspace to connector is discussed.

## 2 CONTACT SIMULATION

The interaction between two separated bodies is implemented by contact. Due to the high complexity of the connector, the numerical simulation demands a detailed model to reflect the contact characteristic [5–8].

The master/slave approach is applied to simulate the contact pair. The slave surface cannot penetrate any part of the master surface. To avoid the occurrence of the penetration phenomenon, the stiffness of the master surface must be greater than the slave surface. The grid density of the slave surface should be larger than the master surface in FEM. Coulomb's law is used here to simulate friction problems with a friction coefficient [9]. The coefficient depends on material pairing of the respective surface [4]. Small sliding is set to control the relative motion between two contact surfaces.

## 3 FEM ANALYSIS OF DIFFERENT MODELS

### 3.1 Model of sleeve/gasket

The geometry model was set up in CATIA P3 and the finite element model was simulated by ABAQUS Version 6.10, which are presented in Fig. 1. The original model consists of five parts:
1) Upper wing spar joint with two lugs;
2) Lower fuselage joint with three lugs;
3) One screw with nut;
4) One hollow pin;
5) Five sleeves.

The thickness of each lug in the upper joint is 21mm, whereas those in the lower ones are 10.5 mm, 22mm, and 10.5mm. The original bolt diameter of lug is 34mm. The combined diameter of the bolt and the hollow pin is 30mm. The thickness of the sleeve is 4mm. There is no gap between lugs and hollow pin as the whole model meshes are generated using an 8-node C3D8R solid element. The material property is shown in Table 1. Based on the model, 28 surface-to-surface contacts are set up and the displacement pattern is restricted to little slide.

Table 1. Material Property.

| Structure | Material | Young's Modulus (GPa) | Poisson Ratio |
|---|---|---|---|
| Jointhollow pin sleeve/ gasket | PH13-8Mo | 196 | 0.3 |
| Screw | 30CrMnSiA | 200 | 0.3 |

In the changed model, sleeves are replaced by gaskets. The four gaskets replaced between lugs of upper and lower joints have the same thickness as the outer loop of sleeve in the original model. Due to

the removal of the sleeves, the diameter of joint lug decreases from 34mm to 30mm. The material of gaskets is the same as that of sleeves. There are only twenty-one contacts left in the changed model.

To assemble the connector conveniently, a gap of certain size is reserved between lugs of the joint without loop of sleeve (or gasket). The thickness of the outer loop of sleeve (or gasket) is reduced correspondingly. In addition, two new connector models are established with a gap of 0.5 mm.

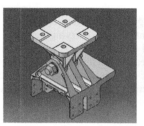 

a. The geometry model     b. The FEM model

Figure 1. Model of wing-to-fuselage connector.

### 3.2 Boundary condition

In the lower joint area, the plane of screw is simply supported. To avoid initial deflection, rotational restricts around the X direction are set in the screw plane of the upper joint. The aerodynamic load is transmitted from the wing to the upper joint. Loading condition is listed in Table 2. The horizon reference line of plane is the X axis, and the positive is the opposite of the course. The Y axis is perpendicular to the symmetry plane of fuselage, and positive direction points to the right. The vertical direction of the horizon reference line is the Z axis, and the positive direction is upward.

Table 2. Boundary Loads.

| Working condition | Fx(N) | Fy(N) | Fz(N) |
|---|---|---|---|
| Dynamic balance | 19386.75 | 27154.62 | 280492.78 |

### 3.3 Results

Four different models, (i) model with sleeve, (ii) model with gasket, (iii) gaped model with sleeve, and (iiii) gaped model with gasket, are simulated. The distributions of von Mises stress for total structure and every part are shown in Figures 2 to 5. It can be found that stresses are mainly concentrated on the joint lug and screw area.

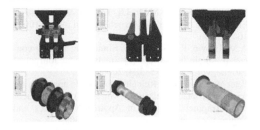

Figure 2. Distribution of von Mises stress for model with sleeve.

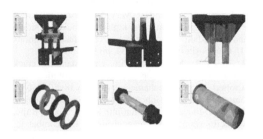

Figure 3. Distribution of von Mises stress for model with gasket.

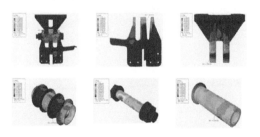

Figure 4. Distribution of von Mises stress for gaped model with sleeve.

Figure 5. Distribution of von Mises stress for gaped model with gasket.

The maximum von Mises stress of the four models is subjected to the same load case that is listed in Table 3. The maximum of the model with sleeve is located in the sleeve, and it is 1275 MPa. The location of maximum of the model with gasket is at the lower joint, and the value is 927.9 MPa. Comparing the two models, it can be found that the maximum stresses of both bolts are close. The results of the error of the upper and the lower joints are 55.25% and 23.78%, respectively, whereas the hollow pin and the gasket/sleeve are raised to 5.25 and 4.18 times. In short, the model with gasket is better in load bearing.

By comparing the model with gap to the original model, it can be found that the von Mises stress possesses the upward trend with little increments. Among all these parts, the hollow pin in the model with gasket has the most remarkable change and the difference increases to 9.13 times, which turns into the section of maximum stress. The stress of the hollow pin in the model with sleeve increases a little. According to the mechanical property of the material, the lug and sleeve of upper joint in model with sleeve are in the local plasticity phase.

Table 3. Maximum Von Mises stress of four different models (MPa).

| Model | Total | Upper Joint | Lower Joint | Hollow Pin | Screw | Sleeve Gasket |
|---|---|---|---|---|---|---|
| With sleeve | 1275 | 1027 | 749.6 | 613.8 | 354.1 | 1275 |
| With gasket | 927.9 | 661.5 | 927.9 | 98.1 | 359.3 | 245.9 |
| Gapped with sleeve | 1327 | 1123 | 789.9 | 641.4 | 374.8 | 1327 |
| Gapped with gasket | 993.6 | 706.2 | 953.1 | 993.6 | 369.3 | 211.9 |

## 4 CONCLUSION

Wing-fuselage connector is an important structure of load bearing. The nonlinear coupling effects should be taken into consideration. Based on the master/slave contact approach and material plasticity, the mechanical property of connector model subjected to a typical load case is analyzed. Two models with different parts are compared to find the better one. In addition to the practical assembly, gap is added to build new models. Conclusions can be drawn as follows:

1 The maximum of von Mises stress of the wing-fuselage connector occur in the border

region of the lug and screw. Some parts have local plasticity.

2 The model with gasket is better than the one with sleeve in carrying loads. Sleeve is the main bearing component part in the model with sleeve, whereas lug of joint bears more load than other parts in the model with gasket.

3 Gap between lugs leads to the increase of stress, which is due to the reduction of the lug. The stress distribution of model with gasket has also been changed.

## 5 ACKNOWLEDGMENT

This study was supported by the National Natural Science Foundation (No: 51210008), 111 Project (No: B07050), and Aviation Science Foundation (No: 2013ZF53068).

## REFERENCES

[1] Lucjan Witek. Failure analysis of the wing-fuselage connector of an agricultural aircraft. Engineering Failure Analysis, vol.13, pp.572–581, 2006.

[2] A. Lanciotti, F. Nigro and C. Polese. Fatigue crack propagation in the wing to fuselage connection of the new trainer aircraft M346 [J]. Fatigue & Fracture of Engineering Materials & Structures, vol.29, pp.1000–1009, 2006.

[3] A.M. Brown, P.V. Straznicky. Simulating fretting contact in single lap splices [J]. International Journal of Fatigue, vol.31, pp.375–384, 2006.

[4] Peter Wriggers, Jana Reinelt. Multi-scale approach for frictional contact of elastomers on rough rigid surfaces [J]. Computer Methods in Applied Mechanics and Engineering, vol.198, pp.1996–2008, 2009.

[5] J. Munoz, G. Jelenic and M.A. Crwasfield. Master–slave approach for the modelling of joints with dependent degrees of freedom in flexible mechanisms [J]. Communications in Numerical Methods in Engineering, vol.19, pp.689–702, 2003.

[6] B. Egan, C.T. McCarthy, M.A. McCarthy, P.J. Gray, R.M. O'Higgins. Static and high-rate loading of single and multi-bolt carbon–epoxy aircraft fuselage joints [J]. Composites: Part A, vol.53, pp.97–108, 2013.

[7] R. Jones, L. Molent, S. Pitt. Understanding crack growth in fuselage lap joints [J]. Theoretical and Applied Fracture Mechanics, vol.49, pp.38–50, 2008.

[8] Dragan Trifkovic, Slobodan Stupar, Srdjan Bosnjak, Milorad Milovancevic, Branimir Krstic, Zoran Rajic, Momcilo Dunjic. Failure analysis of the combat jet aircraft rudder shaft [J]. Engineering Failure Analysis, vol.18, pp.1998–2007, 2011.

[9] Wang Xiang-sheng, Zhao Bin, LI Yong-gang, Zhufeng YUE. Nonlinear Finite Element Analysis of Wing-Fuselage Connection Joint [J]. Computer Simulation, vol.26, pp.37–40, 2009.

*Multimedia, Communication and Computing Application – Leung (Ed.)*
*© 2015 Taylor & Francis Group, London, ISBN 978-1-138-02775-6*

# Qualitative analysis of the effect of stress coupling on fatigue life of welding zone

J.X. Kang, H.S. Gao, J. Liu, F.S. Wang & Z.F. Yue
*Department of Engineering Mechanics, Northwestern Polytechnical University, Xi'an, China*

ABSTRACT:   The fatigue tests are carried out on three different welding zones to analyze the effect of stress coupling on fatigue life of the welding zones. Three welded structures are concerned in the present study, and they are top with corners, top with trough, and top with round-hole. For each structure, the distance between the welding zone and the top is different. The results have showed that the three welded structures discussed in the present study have a strong influence on the fatigue life of welded structure.

## 1  INTRODUCTION

Nowadays, the welding technique is widely used in different industry fields. Compared with the riveted structure, the welded structure has many advantages, such as simplifying structure details, causing weight saving, etc. Meanwhile, they are also applied in energy engineering, marine engineering, aerospace engineering, petrochemical engineering, large plants, high-rise buildings, and other important structures without exception. Therefore, people pay great attention to the ways of improving the design and quality of these structures to ensure safety and reliability [1]–[2].

Fatigue failure is a typical failure mode, so it is essential to optimize design and improve fatigue performance of welded structure. To solve these problems, people have studied the effect of the welding zone on fatigue performance of welded structures in different ways. Zhang [3] has studied the influence of welding residual stress on fatigue performance; Han [4] has studied the influence of hot isostatic press on structure and properties of titanium stainless steel jointed by explosive welding. Other approaches have been used to do a few other works [5]–[7]. This paper studies the influence of stress coupling on fatigue life of welded zone by using the finite element method, and then the nominal stress method is used to analyze it qualitatively.

## 2  SPECIMEN GEOMETRIES AND EXPERIMENTAL CONDITION

### 2.1  *Specimen geometries*

The parts of the structure are shown in figure 1. The structure is divided into three parts, cylinder structure, welding circle, and top. The thickness of the cylinder was 10 mm, and the external diameter of the transversal surface was 140 mm. The three tops of the structure are shown in figure 1(a), figure 1(b), and figure 1(c). The schematic view of the welded zone is shown in figure 1(d), which is in the middle of the cylinder structure. The welding technique is downhand welding.

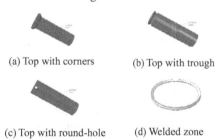

(a) Top with corners           (b) Top with trough

(c) Top with round-hole        (d) Welded zone

Figure 1.    Schematic view of the model.

### 2.2  *Mechanical parameters of the material*

The material of the cylinder is A-alloy 6082T6. Its Young's modulus is 70 GP and Poisson's ratio is 0.3; the material of welding zone is welding wire LT1. Its Young's modulus is 49 GP and Poisson's ratio is 0.3.

### 2.3  *Experimental condition and requirement*

The fatigue test was carried out on welded structure by using the FTS multichannel coordination loading system. The fatigue load spectrum was a constant-amplitude spectrum, where the stress ratio R = −1. The loading error was less than 2%, which met the loading requirement of real-time auto control coordination.

## 3 FINITE ELEMENT MODEL

### 3.1 *Load and boundary conditions*

The bottom of the cylinder was fixed, whereas the other end was free. A uniform tensile load of 150 MPa was applied to the free end.

### 3.2 *Mesh design and model classification*

The welded structure was analyzed by finite element method with the commercial software ABAQUS [8]. Hexahedron element is used, and the mesh of stress concentration parts is refined. The mesh of the model is shown in figure 2. There are 18 models, which are divided into three groups with respect to different tops. For each group, there are 6 models where the distance H of each model is varied in 20 mm. The distances from the welding zone to the top are 11.4 mm, 31.4 mm, 51.4 mm, 71.4 mm, 91.4 mm, and 111.4 mm for the six models.

Figure 2. Finite element mesh.

## 4 FINITE ELEMENT RESULT

The Von Mises stress contours of the top with corners in different distances are shown by figure 3. The Von Mises stress contours of top with trough in different distances are shown in figure 4. The Von Mises stress contours of top with round-hole in different distances are shown in figure 5.

The Von Misess contour plots of the welding zone associated with different tops are shown in figures 3, 4, and 5.

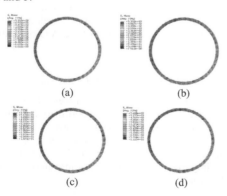

(a)　(b)

(c)　(d)

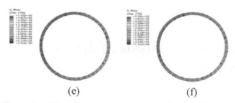

(e)　(f)

Figure 3. Von Mises stress contours of top with corner.

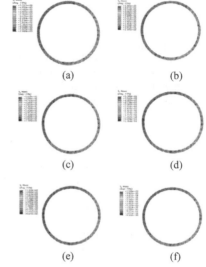

(a)　(b)

(c)　(d)

(e)　(f)

Figure 4. Von Mises stress contours of top with trough.

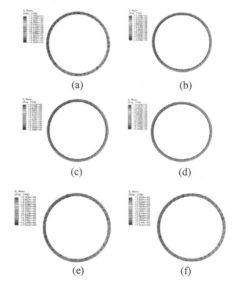

(a)　(b)

(c)　(d)

(e)　(f)

Figure 5. Von Mises stress contours of top with round-hole.

Based on the aforementioned values, the stress concentration factors are given in tables 1, 2, and 3:

Table 1. Stress concentration factors for the case of top with corner.

| Distance | Stress concentration factor, $K_T$ |
|---|---|
| H = 11.4 | 1.19 |
| H = 31.4 | 1.27 |
| H = 51.4 | 1.24 |
| H = 71.4 | 1.22 |
| H = 91.4 | 1.18 |
| H = 111.4 | 1.18 |

Table 2. Stress concentration factors for the case of top with trough.

| Distance | Stress concentration factor, $K_T$ |
|---|---|
| H = 11.4 | 1.19 |
| H = 31.4 | 1.20 |
| H = 51.4 | 1.26 |
| H = 71.4 | 1.24 |
| H = 91.4 | 1.17 |
| H = 111.4 | 1.17 |

Table 3. Stress concentration factors for the case of top with round-hole.

| Distance | Stress concentration factor, $K_T$ |
|---|---|
| H = 11.4 | 1.19 |
| H = 31.4 | 1.20 |
| H = 51.4 | 1.26 |
| H = 71.4 | 1.23 |
| H = 91.4 | 1.18 |
| H = 111.4 | 1.18 |

## 5 FATIGUE LIFE PREDICTION

It is well known that the stress concentration factor has great effect on fatigue life. Zheng [9] has given the following expression to predict fatigue crack initiation life:

$$N = C_i \left[ \Delta\sigma_{eqv}^{\frac{2}{1+n}} - \left(\Delta\sigma_{eqv}\right)_{th}^{\frac{2}{1+n}} \right]^{-2} \quad (1)$$

$$\Delta\sigma_{eqv} = \sqrt{\frac{1}{2(1-R)}} K_t \bullet \Delta S = \sqrt{(1-R)/2K_t} \bullet S_{max} \quad (2)$$

Where $C_i$ is coefficient of initial crack initiation resistance, which is a material constant dependent on tension performance; $\Delta\sigma_{eqv}$ is the equivalent stress amplitude; $\left(\Delta\sigma_{eqv}\right)_{th}$ is a threshold value of equivalent

stress amplitude, which is a material constant dependent on tension performance and fatigue limit; $n$ is strain hardening exponent; $\Delta S$ is the amplitude of nominal stress; $R$ is the stress ratio; and $K_t$ is the theoretical stress concentration factor, which can be calculated by expression (2). $S_{max}$ is the amplitude of maximal nominal stress, when $R = -1$; $S_{max}$ is the peak of maximal nominal stress.

For A-alloy, $2/(1+n) = 1.78$; take $\left(\Delta\sigma_{eqv}\right)_{th} = 20$ MPa (1/3 of yield stress of welding zone), determined by the test $C_i = 1.22 \times 10^{12}$.

Based on the aforementioned expression, the fatigue lives of structure are predicted at conditions of $S_{max} = 50$ MPa and $R = -1$. The result is shown in figures 6, 7, and 8:

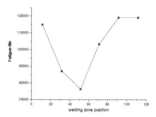

Figure 6. Effect of welding zone position on fatigue life for the case of top with trough.

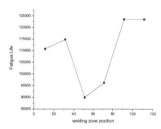

Figure 7. Effect of welding zone position on fatigue life for the case of top with corner.

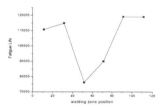

Figure 8. Effect of welding zone position on fatigue life for the case of top with round-hole.

Based on figures 6, 7, and 8, first the fatigue life starts to decrease; when the distance between the welding zone and the top is 6 times greater than the thickness of welding seam, and the fatigue life is minimum, the fatigue life starts to increase. After the distance between the welding zone and the top becomes 10 times greater than the thickness of welding seam, and the fatigue life becomes maximum, the fatigue will not change and the stress coupling effect disappears.

## 6 CONCLUSIONS

The aforementioned figures show that the fatigue life changes with the change of welding zone, and the following conclusions can be drawn from them:

1) For the three kinds of structures, with the increase of distance between the welding zone and the top, fatigue lives first become larger and then become smaller.
2) When the distance between the welding zone and the top is 10 times greater than the thickness of welding seam, the stress coupling effect disappears.
3) The minimum life expectancies of top with corners are 2 and 4 times the thickness, which are different from the other two tops whose minimum life expectancies are 4 and 6 times the thickness.

REFERENCES

[1] Radaj, Dieter. Fatigue Resistance of Welded Construction [M]. China Machine Press. 1994.
[2] CHEN Chuanyao. Fatigue and Fracture. Wu Han: Huazhong University of Science & Technology Press. 2002.
[3] ZHANG Guansheng, WANG Hongfeng, FANG Huimin. Characteristics analysis of welding deformation for larger-scale structure [J]. Group Technology & Production Modernization, 2008, 25 (4): 30–32.
[4] HAN Liqing, LIN Guobiao, Wang Guodong. Influence of hot isostatic pressing on structure and properties of titanium/stainless steel joint by explosive welding [J]. Transactions of the China Welding Institution, 2008, 29(11): 41–44.
[5] S D Korea, P Dhanesh, S V Kulkarni. Numerical modeling of electromagnetic welding [J]. International Journal of Applied Electromagnetics and Mechanics, 2010, 32: 1–19.
[6] A Ben-Artzy, A Sternb, N Frage. Wave formation mechanism in magnetic pulse welding [J]. International Journal of Impact Engineering, 2010, 37: 397–404.
[7] Ehsan Zamani, Gholam Hossien Liaghat. Explosive welding of stainless steel carbon steel coaxial pipes [J]. J Mater Sci, 2012, 47: 685–695.
[8] ZHUANG Zhuo, ZHANG Fan, CEN Song. Nonlinear finite element analysis and instances of ABAQUS [M]. Beijing: Science Press. 2005.
[9] ZHENG XiuLin. A further study on fatigue crack initiation life mechanical model for fatigue crack initiation. International Journal of Fatigue, 1986, 9(1): 17–21.

*Multimedia, Communication and Computing Application – Leung (Ed.)*
*© 2015 Taylor & Francis Group, London, ISBN 978-1-138-02775-6*

# A novel dual passband lowpass filter based on U-shaped DGS

H.Q. He, M.Q. Li, Y. Ding, H.X. Chang & F.L. Zhao
*Key Laboratory of Intelligent Computing & Signal Processing, Ministry of Education, Anhui University, Hefei, Anhui, China*

ABSTRACT:    In this paper, a novel design of the dual passband lowpass filter based on the Defect Ground Structure (DGS) is proposed. The defect ground structure (DGS) is achieved by opening a set of asymmetric U-shaped slot in the ground plane and the microstrip line using a U-strip resonator. This proposed structure is simulated by the high frequency simulation software HFSS 13.0. In comparison with the filter without U-shaped DGS, the proposed filter has low insertion loss and its frequency ranges from 8.4GHz to 12.7GHz, signal restrains from 2.7dB to 14dB. The simulated and measured results show good agreement and validate the proposed approach.

KEYWORDS:    Dual passband, Lowpass filter, U-shaped DGS.

## 1   INTRODUCTION

Defected Ground Structure (DGS)[1,2] was first proposed on the basis of discussing Photonic Band Gap (PBG) Structure by south Korean scholars J.I. Park and C.S. Kim et al in 1999. Defected Ground Structure (DGS) is a defect on the ground that can change the propagation properties of transmission line by changing the current distribution on the ground side [3]. The DGS of this filter has good band rejection and slow-wave characteristics. DGS occupying small circuit with simple structure is helpful to reduce the weight, the load and power consumption [4]. DGS structures have been widely applied in power dividers, filters, power amplifier cation circuits and other microwave circuits [5].

In this paper, we propose a novel DGS which has a set of asymmetric U-shaped slot on the ground plane to provide improved signal curbing and low insertion loss. A set of asymmetric U-shaped DGS on a dielectric substrate is used and a metal cylinder in the dielectric substrate is added. The metal cylinder not only increases the coupling ability between microstrip line and the ground efficiently, but also adjusts part of the capacitance. In addition, on the basis of this DGS, notch effect can be realized, thus dual passband characteristic is provided. The attenuation of filter is obviously increased by the simulation and adjustment, and the insertion loss of passband gets smaller. Additionally, this kind of thought can have a good reference in the design of filters in the future.

## 2   THE FILTER AND U-SHAPED DGS

The filter without U-shaped DGS is shown Figure 1. (a). To analyze the performance of the structure, a substrate material (Neltec NH9338) with permittivity of 3.38 and thickness of 1 mm is used. This U-strip line is etched on the metal sheet with thickness of 0.035 mm and the U-strip line is chosen for the characteristic impedance of 50Ω for a microstrip line. This filter is simulated by the high frequency simulation software HFSS 13.0. The S-parameters of this structure are shown in Figure 1. (b). It can be seen from Figure 1. (b) that the filter without U-shaped DGS rejects the signals only 2.7dB suppression and its insertion loss reaches 1.9dB. The proposed U-shaped DGS cell is shown in Figure 2. (a). The optimum geometry of the U-shaped DGS cell are provided: w0=0.35mm, d=4mm. The frequency characteristic of the DGS can be modeled by a parallel RLC resonance frequency. The equivalent circuit model of a U-shaped DGS cell is shown in Figure 2.(b)

The circuit parameters of the equivalent circuit are given below:

$$C = \frac{\omega_c}{2Z_0(\omega_0^2 - \omega_c^2)} \tag{1}$$

$$L = \frac{1}{\omega_0^2 C} = \frac{1}{4\pi^2 f_0^2 C} \tag{2}$$

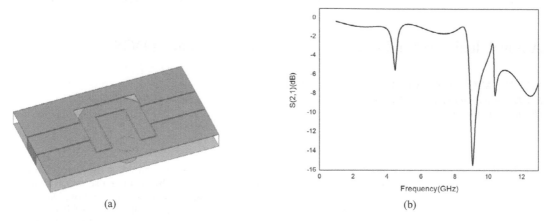

(a)                                                                 (b)

Figure 1.    (a) the filter without U-shaped DGS. (b) Simulation results of S-parameters form 1 GHz to 12 GHz.

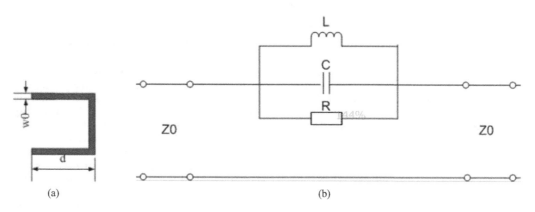

(a)                                                                 (b)

Figure 2.    (a) the U-shaped DGS cell. (b) Equivalent circuit.

$$R = \cfrac{2Z_0}{\sqrt{\cfrac{1}{S_{11}(\omega_0)^2} - \left(2Z_0\left(\omega_0 C - \cfrac{1}{\omega_0 L}\right)\right)^2} - 1}} \qquad (3)$$

where $W_0$ is the LC resonant angular frequency of the circuit in parallel, C is the equivalent capacitance and $Z_0$ is the characteristic impedance[6].

According to the equivalent circuit and formula mentioned above, the characteristics of the U-slot DGS are dependent on the structural parameters of the defect. The resonance frequency and rejection bandwidth are changed by adjusting the dimensions of slot length, width and the distance between two slots.

## 3    THE DUAL PASSBAND LOWPASS FILTER DESIGN

Figure.3 (a) shows the geometry of the top view and the bottom view of the proposed U-shaped DGS filter. The optimum dimensions of the filter with U-shaped DGS are listed as follows: sp=1.5mm, pp=5.5mm, d=0.3mm, sy=6mm, sx=6mm, D=2mm, R=0.2mm, R1=1mm, sd1=4.6mm, sd2=3mm, w2=0.1mm, w3=0.05mm and l=1.5mm. The fabricated lowpass filter is shown in Figure 3.(b) The proposed filter consists of a U-strip line and a set of U-shaped DGS on a dielectric substraten.

Figure 4 shows the simulated S-parameters of the proposed filter with the U-shaped DGS. It can be seen from Figure 4 that the S-parameters(S21) of the loss

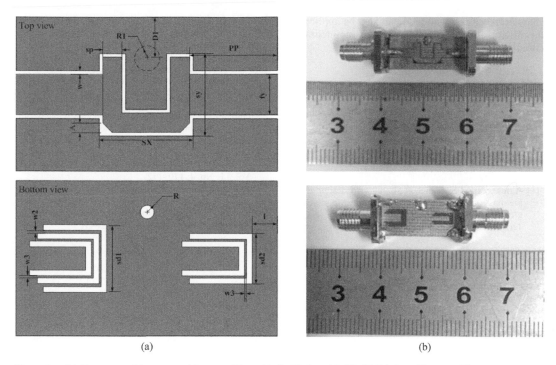

(a)                                                                    (b)

Figure 3.    (a) Geometry of the proposed lowpass filter with the U-shaped DGS. (b) Fabricated lowpass filter.

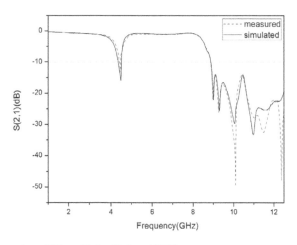

Figure 4.    Simulated S-parameters of filter with the U-shaped DGS.

of filter with the U-shaped DGS is less than -14 dB and the insertion loss of passband is limited within 1dB. Compared with the filter without U-shaped DGS, the attenuation value is increased by nearly 250%. Additionally, the dual stopband is realized with the DGS structure. It is obvious that fliters with U-DGS can implement the dual passband characteristic effectively.

4    CONCLUSION

In this paper, a novel dual passband filter with U-shaped DGS structure is presented. As can be seen from the simulated and measured results, the DGS structure of this filter can provide dual passband characteristics. This filter shows a fairly good insertion loss and has easy processing characteristics. This method can also be applied to the microwave devices and microwave integrated circuits.

## ACKNOWLEDGEMENTS

This work is financially supported by the foundation of national natural science (NO.51477001).

## REFERENCES

[1] Ahn D, Park J S, Kim J, et al, 2001 ,"A design of the low-pass filter using the novel microstrip defected ground structure," IEEE Trans Microwave Theory Tech, Vol. 49, No. 9, 83–93.

[2] Park J I, Kim C S, Kim J, et al.r, 1999, "Modeling of a photonic bandgap and its application for the low-pass filter design," Asia Pacific Microwave Conference. Singapore: The Conference Organizer.

[3] L.H. Weng, Y.C. Guo, X.W. Shi, X.Q. Chen,2008, "An Overview on Defected Ground Structure", Progress In Electromagnetics ResearchB, Vol. 7, pp: 173–189.

[4] Lim J S, Kim H S, Park J S, et al, 2008,"Vertically periodic defected ground structure for planar transmission lines," Electronics Lett, Vol. 38, No. 15, 803–804.

[5] Lin J S, Kim H S, Park J S, et al, 2001, "A power amplifier with efficiency improved using defected ground structure," IEEE Microwave Wireless Compon Lett, Vol. 11, No. 4: 170–172.

[6] Zhenfeng Yin, Zhenhai Shao,et al. 2001,"Novel U-Slot and V-Slot DGSs for Bandstop Filter With Improved Q Factor," IEEE Transactions On Microwave Theory And Techniques, Vol. 54, No. 6, 2840–2847.

*Multimedia, Communication and Computing Application – Leung (Ed.)*
© *2015 Taylor & Francis Group, London, ISBN 978-1-138-02775-6*

# The strategy of reactive power optimization considering cost of reactive power regulation in wind farms with DFIGs

Y.Q. Xu, Y. Nie, D.D. Liu & N. Wang
*North China Electric Power University, Boding, Hebei, China*

ABSTRACT: A new strategy of reactive power optimization with Doubly Fed Induction Generator (DFIG) is proposed, considering the cost of reactive power regulation and investment of reactive power compensation equipment. Two-stage optimizations, i.e., planned optimization and real-time optimization, are presented. The former is based on short-term wind power prediction information and dispatch instructions, which make substantial adjustment of reactive power and voltage in the wind farm. The latter, dynamic reactive power compensation devices, is used to make compensations for slight voltage fluctuations. The IA-PSO algorithm is adopted in the two-stage optimizations and the simulation results are carried out with the improved IEEE-30 node system, under the Power System Analysis Software Package (PSASP). The two-stage optimization strategy is proposed to meet the demand of dispatch, which has high practicability.

KEYWORDS: The IA-PSO algorithm, Doubly fed induction generators, Cost of reactive power regulation, Reactive power optimization

## 1 INTRODUCTION

Traditional reactive power optimization adjustment, which has the characteristics of discretization and slow speed of adjustment, mainly means to change the ratio of on-load voltage regulating transformer and switch capacitor banks. The wind generators have some reactive power adjustment ability. They can be used as a continuous source of reactive power, which have a fast adjustment speed [1–2]. Jianming and Fei [3] are doing research on scene decision making, based on the random changes of wind speed and probability. But this model does not consider wind farm reactive power adjustment ability. Considering the wind speed forecasting curve and the load forecasting curve, Meiyu [4] has done studies on dynamic reactive power optimization, but there are some errors that are unable to implement the real-time reactive power compensation of wind farms. Huifen and others [5] regard doubly fed wind power as a continuous controlled reactive power source that can have a large adjustment range and fast responsive speed when making reactive power optimization. Chuang and others, Jingjing and others, and Ryan and others [6–8] regard wind farms as a source of continuous reactive power, which means it participates in reactive power optimization of distribution network without considering the cost problems of wind farms in providing the reactive power.

The solution of reactive power optimization model contains the wind farm power system. This article proposes a reactive power optimization strategy that considers the cost of doubly fed generators' reactive power regulation and the investment of reactive power compensation equipment, establishing a "planned and real-time optimization." This is a two-stage optimization model using the IA-PSO algorithm to solve the objective function. Finally, do simulation analysis by using the power system analysis software package (PSASP) with improved IEEE-30 nodes. Emulational results could verify the effectiveness of the proposed strategy, which can bring online reactive power optimization into effect.

## 2 CONTAINING THE DOUBLY FED GENERATOR POWER SYSTEM REACTIVE POWER OPTIMIZATION MATHEMATICAL MODEL

### 2.1 Objective function

The goal of reactive power optimization is to maximally ensure the safety of system and improve the quality of power supply and economy of operation with minimum reactive power equipment investment and reactive power cost. Based on the aforementioned consideration, this paper established a "planned and real-time optimization" two-stage optimization model.

Planned optimization phase's target function is to meet the minimum system active loss $f_{loss}$ and the minimum investment of reactive power compensation equipment (available for capacitor and on-load voltage regulating transformer) $f_{Q\,cost}$, such as formula (1):

$$\min F = \omega_1 f_{loss} + \omega_2 f_{Q\,cost} \tag{1}$$

$$f_{loss} = \sum_{k=1}^{n} G_k (V_i^2 + V_j^2 - 2V_i V_j \cos\theta_{ij}) \tag{2}$$

$$f = \sum_{i=1}^{N} \rho_t \cdot Q_t \tag{3}$$

where $\omega_1$ and $\omega_2$ are the penalty coefficients; $n$ is the number of branches; $G_k$ is the connection of the conductance of nodes $i$ and $j$ and branch $k$; $V$ is the node voltage amplitude; $\theta_{ij}$ is the node voltage phase angle; $N_c$ is the total number of reactive power compensation nodes; $\rho_t$ is the reactive power compensation device that is used to adjust number; and $Q_t$ is the investment for reactive power compensation equipment units.

The real-time optimization phase considers the cost of wind farms with provided reactive power. Its target function is selected as the smallest active loss $f_{loss}$ and wind farm reactive power investment $f_{WF\,cost}$, which can be written as formula (4):

$$\min F = \omega_1 f_{loss} + \omega_2 f_{WFcost} \tag{4}$$

The cost of wind farms with provided reactive power $f_{WF\,cost}$ consists of three parts, namely, fixed cost, loss cost, and loss of opportunity cost (LOC)[9]. The fixed cost includes fan equipment transportation, installation, and maintenance costs, generally for 18 \$/h (equivalent to 111.6 yuan/h); the loss cost is produced in the process of wind farm energy conversion due to the demanding reactive power system side, which is usually 6 \$/h (equivalent to 37.2 yuan/h); and LOC is affected by the doubly fed wind power generator capacity curve (as shown in figure 1). When the need of reactive power is higher than what wind farms can provide, the operating point of the wind farm moves from point A to point B along the curve. As a result, wind farms have to reduce the active power output to improve the capacity of reactive power regulation; thus, the wind farms need to pay for power plants' compensation because of the economic loss. LOC is highly affected by wind farm power prediction accuracy [10]. When the prediction errors achieve 0.5 p.u., LOC would be bigger than the sum of fixed cost and loss cost, reaching 2399\$/h (equivalent to 14842.8 yuan/h).

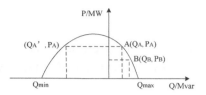

Figure 1.   Capability curve of DFIG.

Function expressions of $f_{WF\,cost}$ can be written as formula (5), and the corresponding function curve is shown in figure 2.

$$f_{WF\,cost} = \begin{cases} a_0 & 0 \leq Q \leq Q_1 \\ a_0 + b_1(Q - Q_1) & Q_1 \leq Q \leq Q_2 \\ a_0 + b_1(Q_2 - Q_1) + b_2(Q - Q_2)^2 & Q_2 \leq Q \end{cases} \tag{5}$$

where $b_1$ and $b_2$ are the weight functions and $a_0$ is the fixed fee.

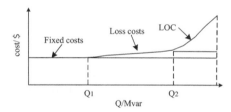

Figure 2.   Reactive power production cost function.

### 2.2   The constraint condition

(1)  Wave equation

$$\begin{cases} P_i - V_i \sum_{j=1}^{N} V_j (G_{ij} \cos\theta_{ij} + B_{ij} \sin\theta_{ij}) = 0 \\ Q_i - V_i \sum_{j=1}^{N} V_j (G_{ij} \sin\theta_{ij} - B_{ij} \cos\theta_{ij}) = 0 \end{cases} \tag{6}$$

where $N$ is the total number of nodes, $P_i$ and $Q_i$ are the injected active and reactive powers of the node i, respectively; $G_{ij}$ and $B_{ij}$ are the conductance and susceptance, respectively, between nodes $i$ and $j$.

(2)  Inequality constraints

$$T_{i,\min} \leq T_i \leq T_{i,\max} \tag{7}$$

$$Q_{ci,\min} \leq Q_c \leq Q_{ci,\max} \tag{8}$$

$$P_{WF,\min} \leq P_{WF} \leq P_{WF,\max} \tag{9}$$

$$Q_{WF,\min} \leq Q_{WF} \leq Q_{WF,\max} \tag{10}$$

$$V_{i\,\min} \leq V_i \leq V_{i\,\max} \tag{11}$$

where $T_i$ is the on-load voltage regulating transformer tap changing column vector; $Q_c$ is injected reactive column vector of the reactive power compensation nodes; $P_{WF}$ is the active power output column vector of wind farms; $Q_{WF}$ is the reactive power output column vector of wind farms; and $V_i$ is every node's voltage column vector.

(3) The limit of doubly fed wind power generator reactive power

The reactive power value range of a doubly fed wind power generator can be written as [11]

$$Q_{w\,\min} \leq Q_w \leq Q_{w\,\max} \qquad (12)$$

$$Q_{w\,\min} = -3\frac{U_s^2}{X_s} - \sqrt{\left(3\frac{X_m}{X_s}U_s I_{r\,\max}\right)^2 - \left(\frac{P_w}{1-s}\right)^2} \qquad (13)$$

$$Q_{w\,\min} = -3\frac{U_s^2}{X_s} + \sqrt{\left(3\frac{X_m}{X_s}U_s I_{r\,\max}\right)^2 - \left(\frac{P_w}{1-s}\right)^2} \qquad (14)$$

where $P_w$ is the output active power of doubly fed wind turbine; $I_{r\,\max}$ is the maximum current of rotor side converter; $U_s$ is the stator side voltage; $I_s$ is the stator winding current; $X_s$ is the stator leakage reactance; and $X_m$ is the excitation reactance.

# 3 IMMUNE PARTICLE SWARM ALGORITHM

## 3.1 Basic principle of the algorithm

The PSO (Particle Swarm Optimization) algorithm was proposed by Eberhart and Kennedyin in 1995, which is a global optimization method for random searches. Its basic idea is to find the optimal solution through collaboration and contribution of information between individuals and groups [13].

Make the search space $D$-dimensional, and let the total number of particles be $N$, the variable speed for particle $i$ be $V_i = (v_{i1}, v_{i2}, \ldots v_{iD})$, and the position of variables be $X_i = (x_{i1}, x_{i2}, \ldots x_{iD})$. The personal best position of particle $i$ is $P_{besti} = (P_{i1}, P_{i2}, \ldots, P_{iD})$, and all of the particles are to find the best location for $G_{besti} = (g_{i1}, g_{i2}, \ldots g_{iD})$. The $i$ particle's velocity i and position updating formula are as follows:

$$V_i^k = \omega V_i^{k-1} + c_1 r_1 (p_i - X_i^{k-1}) + c_2 r_2 (g_i - X_i^{k-1}) \qquad (15)$$

$$X_i^k = X_i^{k-1} + V_i^k \qquad (16)$$

The executive speed of PSO algorithm in the early stage is very fast; but later, when it is carried out, the convergence speed becomes slow for the particles are easy to fall into the local optimal solution. The efficiency of updating is also low, which affects the performance of the algorithm [14]. The immune memory and self-regulation mechanism of immune algorithm is introduced into PSO algorithm, which can overcome the precocious phenomenon during the optimization process, ensuring fast convergence of the global optimal solution.

## 3.2 The improved particle swarm optimization algorithm

(1) The generation of new particles

In each iteration, generating new particles usually follows two ways: generate $N$ particles by updating the formula of particle swarm optimization, and generate $M$ particles randomly.

(2) Dilution strategy to maintain the diversity of the particles

This paper adopts the way of choosing the elimination mechanism based on particle density. The $(N+M)$ particles are sorted by density, and then the $M$ particles are randomly eliminated at large, leaving the $N$ particles. According to the similarity of the particles, the density of particle $P_i$ can be written as

$$P_i = \sum_{j=1}^{N} \frac{1}{1 + \sum_{h=1}^{H} (x_{i,h} - x_{j,h})^2} \quad i, j = 1, 2, \ldots N \qquad (17)$$

where $N$ is the number of particles and $H$ is the dimension for particles.

Calculating the density of each particle based on (17), particles will be ordered according to the density. Leave $N$ particles by selecting whichever could make particles, which are under different fitness levels, a new generation of a particle group. The purpose of this is to maintain a certain density and ensure diversity of the population.

(3) The immune vaccine inoculation

To accelerate the convergence and guide the search direction roughly through the initial particle vaccination, this article will make gbest a vaccine. It will randomly extract R particles from the particle swarm and vaccinate some of the particles by extracting the former vaccine. And in this way, N particles could get updated.

(4) Immune selection

If the fitness value of particle after vaccinating is not as good as its parent, then it should cancel the vaccination; otherwise, keep the particles by forming a new generation of particle swarm.

## 4 POWER SYSTEM CONTAINING DOUBLY FED WIND FARM CREW TWO-STAGE OPTIMIZATION

### 4.1 Planned optimization

Planned optimization stage needs to comprehensively consider security, economy of power grid, and wind farm operation. Set the wind farm-connected node as a balanced node, and control the voltage level at 1.0 p.u.. Choose the minimum of the system active network loss $f_{loss}$ and freactive power compensation equipment investment $f_{cos\,t}$ as the objective function. According to the short-term wind power prediction information and power grid scheduling instructions, wind farm automatic voltage controlling system uses the IA-PSO algorithm to optimize calculation every 15 minutes. Getting the next 4-6 hours' action information of capacitor beforehand and transforming the tap can inhibit sharp wind farm voltage fluctuation to improve voltage quality.

### 4.2 Real-time optimization

Real-time optimization phase mainly controls dynamic reactive power compensation device, such as DFIG, SVC, SVG, and STATCOM, and adjusts the power grid dispatching command. In general, the control scheme of discrete device is determined and it no longer participates in the rest of the compensation. The lack part of capacity are compensated by dynamic reactive power compensation device Choose the minimum of the system active network loss $f_{loss}$ and reactive power cost provided by wind farms $f_{WF\,cos\,t}$ as the objective function, which are shown in formula (4). In real-time optimization, you need to set up a certain optimization and conditions of the starting control.

(1) Real-time optimization start conditions

After the regulation of capacitor and on-load voltage regulating transformer, if the system power loss and node voltage deviation exceed the set value start real-time optimization. Coordinate the dynamic reactive power compensation device and the reactive power output of DFIG, and suppose if the DFIG response speed is slow and the wind farm output in 10 s is basically unchanged the control cycle of the dynamic reactive power compensation device and the DFIG could be set as 1 s and 10 s, respectively.

(2) Real-time control threshold

IA-PSO algorithm analysis and calculation could determine whether the power systems that contain wind farms need real-time control or not. According to the results, the controlled threshold value is as follows:

$$(\Delta p_{loss} > a) \cup (\Delta \sum |\Delta V| > b) \tag{18}$$

where $\Delta p_{loss}$ is a former active network loss reduction after optimization; $\Delta \sum |\Delta V|$ is the voltage deviation before and after; and $a$ and $b$ are the set values.

When the optimization results meet formula (18), it means that the power grid mode changes or forecast errors exist in the actual operation. Now, it needs to send control instructions, making small adjustments to capacitor banks to meet the operation requirements of system voltage.

(3) Distribution of the amount of reactive power compensation and of the doubly fed wind power generation fleet.

According to the hierarchical clustering method in the study by Xu Yuqin and Wang Na [12], the doubly fed generator wind farm model is established through a clustering tree diagram. The total compensation amounts are distributed among all participating reactive power and voltage regulation DFIGs according to a certain group of wind turbines' optimization principle. The overall principle is to try to make a DFIG whose active outputs are small and have a more reactive output with the purpose of fully utilizing the capacity of the inverter of DFIG. According to the real-time power output showed by the information wind farm SCADA system, if someone wants to make a generation fleet whose active outputs are small and more reactive, it is beneficial to make better use of the doubly fed wind power generator's reactive power adjustment ability.

The process used for closed loop operation of real-time optimization is shown in figure 3.

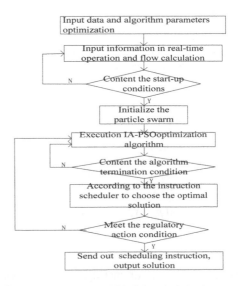

Figure 3. The flowchart of real-time optimization.

### 4.3 Immune particle swarm optimization in the application of reactive power optimization

The position of the particles in the search space corresponds to the reactive power optimization control variables. Select the capacitor compensation capacity $Q_c$ and the change of the on-load voltage regulating transformer ratio $T$ as controlled variables in the planned optimization phase; The real-time optimization phase provides the reactive power by setting doubly fed generator wind farms, $Q_{WF}$, as control variables, namely,

$$x = [Q_{c1}, ..., Q_{cnc}, T_1, ..., T_{NT}, Q_{WF1}, ... Q_{WF_{NF}}] \qquad (19)$$

where $Q_{c1}, ... Q_{cNc}$ is the capacitor reactive power compensation; $N_c$ is capacitor number; $T_1, ... T_{NT}$ is the number of on-load voltage regulating transformers; $N_T$ is the number of on-load voltage regulating transformers; $Q_{WF1}, ..., Q_{WF_{NF}}$ is the reactive power provided by doubly fed generator wind farms; and $N_F$ is the number of doubly fed generator wind farms.

Load node voltage $V$ are stated variables.

## 5 THE EXAMPLE ANALYSIS

Use PSASP to build a wind farm access IEEE-30 nodes system to verify the effectiveness of the reactive power optimization strategy. IEEE-30 nodes system data can be seen in_ieee30 file of Matpower4.0. The wind farm access system line parameter is 10.6 + 24.3 Ω. The wind farm exports transformer reactance is 20.17 Ω. A wind farm consists of sixty doubly fed wind power generators, where each one is 1.5 MW. They are divided into 6 rows whose row space is 120 m. The rated voltage of machine side is 690 V. The cut-in wind speed, key rated wind speed, and cut-out wind speed of DFIG are, respectively, 3 m/s, 11.8 m/s, and 25 m/s. The three parameters of wind speed $v_0$, $c$, and $k$ in Weibull distribution are 3.0, 10.7, and 3.97, respectively. The wind farm is connected to 28 nodes with 110-kV lines through the export transformer (ratio is 110 kV/10.5 kV). In figure 4, wind farm exports transformer's low-voltage side bus is node 33. Its connected node is the new node 32; the public join point is new node 31.

Figure 4. Diagram of wind farm integrated to IEEE-30.

The ratio range of on-load voltage regulating transformer is 0.9–1.1, up or below the gear number, ±8; the stepped size is 1.25%. $1 M_{var} \times 10$ is the maximum compensation capacitor capacity, and the capacity of SVC is $\pm 4 M_{var}$. Load expectations are shown by IEEE-30. Load variance are generated randomly between [0,0.001]p.u.. The simulation calculation uses actual wind farms for 4 hours to roll wind power forecast and the actual value in a day, which is shown in figure 5. The prediction error is 15%.

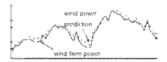

Figure 5. The prediction value and actual value of wind farm.

In planned optimization and real-time optimization, parameters of IA-PSO algorithm are set as follows: particle swarm size is $N = 50$; number of particles of vaccination is $R = 20$; learning factors are $c_1 = c_2 = 2.0$, $w_{max} = 0.9$, and $w_{min} = 0.4$; and the maximum number of iterations is $W_{max} = 0.9$.

In the optimization stage, the short-term wind power prediction information could get all the reactive power capacity of DFIG. Use IA-PSO algorithm, and solve the formula (1). In this formula, capacitor unit regulating investment is 3 yuan/time, and on-load voltage regulating transformer unit regulating investment is 6 yuan/time. Every 15 minutes, make an optimization calculation that can get the information of capacitor and transformer tap before action.

In real-time optimization phase, adjusting the dynamic reactive power compensation capacity of DFIG and SVC should be under the command of the power grid dispatch. Use the IA-PSO algorithm to solve formula (4) where the cost of reactive power provided by wind farms $f_{WF\,cost}$ is 273 yuan/hour, including the fixed cost, which is 112 yuan/hour; the loss cost is 37 yuan/hour. Because the wind farm power prediction error is 15%, the LOC is 124 yuan/hour.

The numbers of actual groups and plans for all capacitors switching in the system are shown in figure 6. According to the actual operation situation and the control instruction, the time of capacitor switching is adjusted in order to make up for the prediction error and to improve the control performance.

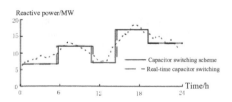

Figure 6. The plan and actual action of capacitors.

Based on the DFIG, active power data collected by SCADA system could get the reactive power regulation limits from every DFIG. Considering the capacitor switching conditions, getting the DFIG active and reactive power could control the variables. Finally, the deviation value of reactive voltage target is tracked by SVC; all the reactive power outputs of DFIG and reactive power outputs of SVC are shown in figure 7.

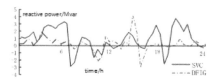

Figure 7. The reactive power output of SVC and DFIGs.

## 6 CONCLUSIONS

This paper puts forward an optimization strategy. Real-time online optimization control is available on the basis of the plan optimization stage. Making full use of the characteristics of different reactive power compensation devices could coordinately optimize economy and safety. Real-time control introduces the multiobjective function. The dispatcher can adjust the weights according to the needs of real time with the purpose of better adjusting the system requirements. Using the IA-PSO algorithm to calculate the objective function and establishing the immune information processing and self-regulation mechanism can ensure the ability of global convergence of the algorithm. Simulation results show that the IA-PSO algorithm has stable convergence ability, and the "planned and real-time optimization" two-stage optimization model improves the efficiency of the reactive power optimization.

## 7 ACKNOWLEDGMENTS

We are thankful to State Key Laboratory of Alternate Electrical Power System with Renewable Energy Sources (North China Electric Power University) support some simulation data for us.

## REFERENCES

[1] Zhu Xueling, Zhang Yang, Gao Kun, et al. Research on the compensation of reactive power for wind farms [J]. Power System Protection and Control, 2009, 37(16): 68–76.

[2] Tapia G, Tapia A, Ostolaza, J. X. Proportional-integral regulator-based approach to wind farm reactive power management for secondary voltage control [J]. IEEE Transactions on Energy Conversion, 2007, 22(2): 448–459.

[3] Yu Jianming, Liu Fei. Research on reactive power optimization of the distribution network with wind power generation base on tabu search [J]. Journal of Xi'an University of Technology, 2013, 29(1): 70–75.

[4] Deng Meiyu. Considering wing power transmission network reactive power optimization [D]. Chengdu: Southwest Jiao Tong University, 2011.

[5] Chen Huifen, Qiao Ying, MIN Yong, et al. Study on coordinated control strategy of dynamic and static reactive compensation in wind farm [J]. Power System Technology, 2013, 37(1): 248–254.

[6] Li Chuang, Chen Minyou, Fu Ang, et al. Reactive power optimization strategy in distribution network with wind farm [J]. Power System Protection and Control, 2013, 41(9): 100–105.

[7] Zhao Jingjing, Fu Yang, Li Dongdong. Reactive power optimization in distribution network considering reactive power regulation capability of DFIG wind farm [J]. Automation of Electric Power Systems, 2011, 35(11): 33–38.

[8] Ryan J, Pradip Vijayan, Venkataramana Ajjarapu. Extended reactive capability of DFIG wind parks for enhanced system performance [J]. IEEE Transactions on Power Systems, 2009, 24(3): 1346–1355.

[9] Augusto C. Rueda-Medina, Antonio-Feltrin Padilha. Distributed generators as providers of reactive power support-A market approach [J]. IEEE Transactions on Power Systems, 2009, 28(1): 490–502.

[10] Nayeem Rahmat Ullah, Kankar Bhattacharya, Torbjörn. Wind farms as reactive power ancillary service providers—Technical and economic issues [J]. IEEE Transactions on Energy Conversion, 2009, 24(3): 661–672.

[11] Yan Gangui, Wang Maochun, Mu Gang, et al. Modeling of grid-connected doubly-fed induction generator for reactive power static regulation capacity study [J]. Transactions of China Electrotechnical Society, 2008, 23(7): 98–104.

[12] Xu Yuqin, Wang Na. Study on dynamic equivalence of wind farms with DFIG based on clustering analysis [J]. Journal of North China Electric Power University (Natural Science Edition), 2013, 40(3): 1–5.

[13] Liu Shicheng, Zhang Jianhua, Liu Zongqi. Application of parallel adaptive particle swarm optimization algorithm in reactive power optimization of power system [J]. Power System Technology, 2012, 36(1): 108–112.

[14] Jiang Yuewen, Chen Chong, Wen Buying. Particle swarm research of stochastic simulation for unit commitment in wind farms integrated power system [J]. Transactions of China Electrotechnical Society, 2009, 24(6): 129–137.

*Multimedia, Communication and Computing Application – Leung (Ed.)*
© 2015 Taylor & Francis Group, London, ISBN 978-1-138-02775-6

# Research on application of multi-agent immune algorithm in reactive power optimization distribution network under the condition of distributed generation

Y.Q. Xu, Y. Nie & D.D. Liu
*North China Electric Power University, Baoding, Hebei, China*

ABSTRACT: By taking the reactive power of the Distributed Generators (DGs) as the controlled variable, the reactive power optimization problem with the condition of the distribution power system with DGs is discussed. Multi-agent immune algorithm based on the mechanism of antibody cluster and compete expansion is proposed to solve the reactive power optimization problem. Its model based on multi-agent immune algorithm is established. The antigens can be changed dynamically as quickly as it can in the process of searching for the best solution due to the efficient multi-agent system. To quickly obtain the global optimum and local optimum, this algorithm adopts the ideas of antibody cluster, compete expansion and antibody mixed mutation in the process of affinity maturation. Through those operators, the variety of antibodies are enhanced. Simulation is carried out based on a modified IEEE 33-bus system, and the results validate the efficiency and quickness of the algorithm.

KEYWORDS: Distribution system, Reactive power optimization, Multi-agent, Immune algorithm, Distributed generation

## 1 INTRODUCTION

Currently, distributed generation (DG) technology is developing rapidly in the world. On the basis of large electrical grid, introducing power supply with smaller capacity (generally less than 50 MW) in the distribution system on the side close to the users could comprehensively utilize existing resources and equipments. It can also provide reliable quality power to users. Thus, the structure of the distribution network changes to a complex network that the power supplies and the loads spread all over the network. Planning, operation and protection of the traditional distribution network face a great change.

IEEE1547 recommend that after small capacity DG is connected to the grid, it should avoid participating in voltage control. Although the standards ensure maximum security of the electric power system, to some extent the limited auxiliary services DG can be provided to the electric power system [4]. It is not good to take full advantage of DG technologies. The DG of photovoltaic power generation and fuel cells connect the grid through an inverter. Geothermal energy and ocean energy connect grid through a generator that its excitation voltage can adjust. This DG has certain reactive capacity, active power and reactive power both can transfer capacity into grid [5]. If the reactive power compensation ability could be played fully, the operational level of the distribution network will be improved. Therefore, to the network for DG, to realize the reactive power optimization by combining reactive power regulation with traditional voltage, a regulation method has been an important issue in related research fields.

Reactive power optimization is a multi-variable, multi-constraint mixed linear programming problem. Currently, there are many algorithms about reactive power optimization. Traditional optimization methods include linear programming and quadratic programming. With the development of artificial intelligence and computer technology, some new optimization algorithms such as expert systems [8], Tabu search [9], genetic algorithm [10] and immune algorithm [11] achieved some results. Distributed artificial intelligence techniques has broken the traditional centralized treatment of artificial intelligence from top to bottom, and has good robustness, scalability, flexibility, adaptability, so it attracts people's attention [12].

This paper presents a new multi-agent immune algorithm based on clustering and clone competition mechanism, to realize dynamic, changing antigen in the optimization process by using a multi - agent system, introduce the clustering competitive mechanism and hybrid mutation in the process of affinity maturation, and realize the fast convergence of the algorithm. This paper solves the problem of reactive power optimization under the condition of distributed generation systems by using new multi-agent immune algorithm. The IEEE33 node system is simulated, and the result is satisfying.

## 2 THE MATHEMATICAL MODEL OF REACTIVE POWER OPTIMIZATION OF DISTRIBUTION NETWORK WITHOUT DISTRIBUTED GENERATION CONDITION

### 2.1 *Objective function*

Reactive power capacity of $Q_{DG}$, reactive power compensation capacity of $Q_C$ and the ratio of the VR transformer of DG are selected as control variables, load node voltage U is selected as a state variable. Selecting the minimum active power loss in power system as reactive power optimization objective function, and embedding load bus voltage limits as penalty function, comprehensive objective function can be expressed in the form of:

$$F = \min\left\{ P_L + \lambda \sum_{i=1}^{n} \left( \frac{\Delta U_i}{U_{i\max} - U_{i\min}} \right)^2 \right\} \tag{1}$$

Among this equation:

$$\Delta Ui = \begin{cases} U_{i\min} - U_i & U_i \prec U_{i\min} \\ 0 & U_{i\min} \leq U_i \leq U_{i\max} \\ U_i - U_{i\max} & U_i \succ U_{i\max} \end{cases}$$

Where $P_L$ is the active power loss; $\lambda$ is the node voltage penalty co-efficient; n is load nodes; max, min are the variables, the lower limit value.

### 2.2 *Constraint conditions*

The equality constraints in the model for power flow equations.

$$P_i = V_i \sum_{j=1}^{n} V_j (G_{ij} \cos \theta_{ij} + B_{ij} \sin \theta_{ij}) = 0$$

$$Q_i = V_i \sum_{j=1}^{n} V_j (G_{ij} \sin \theta_{ij} - B_{ij} \cos \theta_{ij}) = 0 \tag{2}$$

The control variable constraints are as follows.

$$\begin{cases} Q_{DGk\min} \leq Q_{DGk} \leq Q_{DGk\max} \\ Q_{Ci\min} \leq Q_{Ci} \leq Q_{Ci\max} \\ VR_{j\min} \leq VR_j \leq VR_{j\max} \end{cases} \tag{3}$$

In these equations, $Q_{DGk\min}$ and $Q_{DGk\max}$ are minimum and maximum of reactive capacity of DG. $Q_{Ci\min}$ and $Q_{Ci\max}$ are minimum and maximum of reactive capacity of reactive-load compensation equipment. $VR_{j\min}$ and $VR_{j\max}$ are minimum and maximum of the transformer ratio of on load tap changing transformer.

The inequality constraint of state variables is as follows:

$$U_{i\min} \leq U_i \leq U_{i\max} \tag{4}$$

In this inequality, $U_{i\max}$ and $U_{i\min}$ are higher limit and lower limit of node voltage.

## 3 MULTI-AGENT IMMUNE ALGORITHM BASED ON CLONE CLUSTERING AND COMPETITION MECHANISM

### 3.1 *The living environment of multi-agent*

The shape space model of modern immunology is an important mechanism for abstract modelling and representation on immune cells and molecules. The structure multi-agent immune model [13] is shown in figure 1. Each agent has a corresponding shape space, which is equivalent to the local environment of every Agent. In figure 1 Ab is an antibody, cross line is the recognition of antigen (Ag). For an Agent, all the antibodies and antigens are located in the space of a subset Vi, each antibody can identify all antigens located in the recognition region Ve. The running of multi-agent immune model is an iterative process of co-evolution. In each iteration, the Agent interacts with each other through the exchange of information. The exchange information is a subset of each antibody set, called "antibodies represent" set, its union corresponding to Representatives set in Fig. 1. In the model, the Agent's own antibodies are regarded as anti-idiotypic antibody; representatives from other Agent are regarded as an abstraction of the current state of other Agent, which is regarded as a unique position antibody. Thus, the processing and collaborative process of Agent on information from another Agent is corresponding to the immune recognition process of its own anti-idiotypic antibody on its unique position antibody in their local environment (morphological space). After continuous iteration, many Agents collaborate on problem solving. Immune network theory will regard unique position antibodies as an antigen, the antigen presenting process of multi-agent immune model will correspondingly map the representative set of antibodies from another Agent as a set of antigens set. So, for an Agent, the antigen of its morphological space is the image of the another agent state in the agent, which is formed by antigen presenting, and is constantly changing.

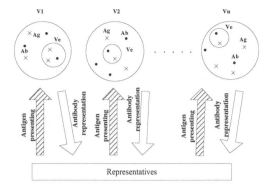

Figure 1.   Living environment of agents.

## 3.2   Multi-agent immune algorithm

For an Agent, some "antibodies represent" antibodies from another Agent antibody set form an antigen set after antigen presentation, then the single Agent implement immune recognition, including antigen presentation, affinity calculation, affinity maturation and immune complement, eventually completing the immune process.

## 4   BASED ON THE DISTRIBUTED POWER GENERATION CONDITIONS OF MULTI-AGENT IMMUNE ALGORITHM FOR DISTRIBUTION NETWORK WITH NO POWER OPTIMIZATION

### 4.1   A reactive power optimization of multi-agent immune model

For the problem of reactive power optimization of distribution network under the condition of distributed generation, there are three control variables, which are the reactive power capacity $Q_{DG}$ of each DG, all reactive power compensation equipment output $Q_C$ and the ratio of on-load tap changer VR, regarding them as an agent, the antibody of each agent is expressed as the value of each control variable in the value range. Corresponding antigen set of each agent is an antigen presenting for an optimal antibody of other Agents. Since DG has a more and more high permeability into the distribution network, the number of corresponding. Agents in the multi-agent immune model are very large. In the process of the algorithm, every Agent exchanges information with each other, updates their respective antibodies. It is the existence of multiple Agent that the advantage of distributed computation can be fully reflected, so the speed of solving the problem of reactive power optimization has been faster than the traditional artificial intelligence methods.

### 4.2   Antibody coding

As mentioned above, this algorithm uses three kinds of Agent, respectively represent reactive power capacity of DGQDG, reactive power compensation equipment output QC and the ratio of on-load tap changer VR, respectively named Agent QDGi, Agent QCj and Agent VR (i=1,2,…,m; j=1,2,…,n; m, n, representing the number of DG and in active power compensation equipment). Antibody coding in each Agent adopts real-coded. In this paper, reactive power compensation equipment takes shunt capacitor for an example, the QDG are continuous variables, QC and VR are discrete variables.

### 4.3   Initial population generation

Initial population generation of Agent $Q_{DGi}$ generates N random numbers in the range of reactive power capacity of DG; the initial population of Agent $Q_{Cj}$ and Agent VR is the corresponding parallel capacitor capacity and all possible values of on-load tap changer, and once formed, antibody in Agent will not change in the process of algorithm. For example, some parameters of on-load tap changer, initial population size of Agent VR is17, which are {0.9,0.9125,0.925,… , 1.075,1.0875,1.1}; if the parameter of shunt capacitors is 150kvr x 4, the initial population size of Agent QC is 5, which are{0,150,300,450,600}, and the antibody of Agent VR and Agent QC does not change in process of the algorithm.

### 4.4   Affinity calculation

The antibody in every agent has to do the affinity calculation, the objective function of the reactive power optimization will be adopted as an affinity calculation formula. It needs to do the power flow calculation, for the DG in this paper has some power regulating capacity, it will be a PQ node in the power flow calculation.

### 4.5   Antigen presenting

Assuming that the problem has only n Agents, which are: Agent1, Agent2, Agent3,…… , Agent$_i$,…… , Agent$_n$, for Agent$_i$, its antigen presenting process are as follows: firstly, it selects optimal antibody from another agent, then the combination of the n-1 antibodies is the obtained antigen of Agent$_i$ from this information interaction, and adopt the N-1 dimension array antigens for storage.

### 4.6   Affinity maturation

Since the initial population of Agent QCj and Agent VR will not be changed if they are formed,so after antigen presentation they will not enter the affinity maturation and immune complement steps. Affinity maturation and immune complement are concerned only for Agent $Q_{DGi}$. Affinity maturation is consisted of clustering, clone proliferation and hybrid mutation.

1  Antibody cluster and competitive expansion

In order to cluster antibodies successfully, we need to calculate distance between antibodies. This text adopts a real number encoding, on the basis of configuration space theory it adopts the Euclidean distance formula. Assuming coordinates of antibody u are given by $(au_1, au_2, \ldots, au_i)$, and coordinates of antibody v are given by $(av_1, av_2, \ldots, av_i)$. So the distance between them can be expressed by the following formula.

$$dis_{uv} = \sqrt{\sum_{i=1}^{L} (au_i - av_i)^2} \qquad (5)$$

Agent adopts the ideas of antibody cluster and competitive expansion in the process of affinity maturation [14]. These can build a good antibody cluster, prompt antibodies which have high affinity and low concentration into clonal selective augmentation quickly. At the moment of choosing antibodies which have high affinity, these ideas can restrain other similar antibodies from getting into clone augmentation. First of all, choosing antibodies which have high affinity from N antibody clusters, as the first cluster centre. And then, calculating the sum of distance from the others to the centre and choosing the farthest antibodies as the new centre until the number get to general standard number M. Finally, calculating distance from the others to M cluster centres. And according to the principle of nearest to cluster centre, we classify antibodies to the cluster, which is represented by the nearest cluster centre.

Every cluster of Agent adopts competitive expansion. Firstly, choosing an antibody which has the highest affinity in cluster and taking out a cluster centre. They can make up an excellent antibody cluster and its scale is T (T=2*M). Finally, every antibody in the excellent antibody cluster is operated by competitive expansion.

$$N_i(k) = round\left( N \times \frac{F(A_i(k))}{\sum_{j=1}^{T} F(A_j(k))} \right) \qquad (6)$$

In this equation, $N_i$ is the expansion scale of the i-th antibody in Agent, $round(\bullet)$ is rounding function

and $i = 1, 2, \cdots, T$. According to this equation, the antibody which has higher affinity has bigger scale of expansion and can clone more antibodies.

Similar antibodies belong to one cluster through clustering and cloning. Only these antibodies which have the highest affinity and stand for cluster Centre can be chosen and start cloning augmentation process. The ideas of competitive expansion provide excellent antibodies of every cluster a chance to do clone augmentation and make affinity mature. That will improve diversity of antibody cluster distribution greatly, balance immune algorithm between depth-first search and breadth-first search and improve algorithm development and exploration competence effectively.

2  Antibody mixed mutation

Mixed double mutation, Gaussian mutation and Cauchy mutation are adopted. Gaussian mutation can make individual mutate in a small neighbourhood scale in high probability. And it has strong ability of searching partially. But Cauchy mutation has a higher probability of producing big mutation than Gaussian mutation. It gets rid of local extremum points easily and has better global convergence.

The equation of Gaussian mutation:

$$u_i = u_i + \left\{ \begin{array}{l} \sigma_i \times \exp(\frac{1}{\sqrt{2}} \times N(0,1) \\ + \frac{1}{\sqrt{2}} \times N_i(0,1)) \end{array} \right\} \times N_i(0,1) \qquad (7)$$

In this equation, $u_i$ is the i-th antibody and $\sigma_i$ is a mutation step of $u_i$. $N(0,1)$ and $N_i(0,1)$ are normal distribution random variable. Their expected value is 0 and standard deviation is 1. $N_i(0,1)$ will produce new normal distribution random variable for every antibody and every antibody mutation step $\sigma_i$.

The equation of Cauchy mutation:

$$u_i = u_i + \left\{ \begin{array}{l} \sigma_i \times \exp(\frac{1}{\sqrt{2}} \times \eta(0,1) \\ + \frac{1}{\sqrt{2}} \times \eta_i(0,1)) \end{array} \right\} \times \eta_i(0,1) \qquad (8)$$

In this equation, $\eta(0,1)$ and $\eta_i(0,1)$ are Cauchy distribution random variable. Their central value is 0 and proportional coefficient is 1. $\eta_i(0,1)$ will produce new Cauchy distribution random variable for every antibody and every antibody mutation step $\sigma_i$. On the basis of existing excellent antibodies, we can find more excellent antibodies with higher probability by double mutation and process of affinity maturation.

### 4.7 *Immune recruitment*

Immune system metabolizes everyday. Useless and effete cells die gradually. And immune system will produce new cells randomly for supplement. To some extent, that maintains diversity of species [9]. In the same way, in order to maintain diversity of antibody, this text adopts a random way to produce a new individual to replace the one which has low affinity.

## 5 REALIZATION OF ALGORITHMIC

1 Parameter setting. We should confirm cluster scale N of every Agent, cluster evolving algebra $N_g$. And then we should set cluster standard number M of Agent $Q_{DG}$, Gauss mutation probability $P_{mg}$, Cauchy mutation probability $P_{mc}$ and immune recruitment number $N_r$.
2 Initialization. Producing initial cluster randomly for every Agent $Q_{DG}$, and the number of clusters is N. Enumerating all value in the domain of definition for Agent VR and Agent $Q_c$.
3 Affinity calculation. According to equation 4, we can calculate every Agent's affinity and arrange every Agent in ascending sort of affinity.
4 Antigen presentation. Every Agent needs antigen presentation. Agent VR and Agent $Q_c$ jump to step (7).
5 Affinity maturation. Operating every Agent as follow: firstly, it clusters and clones multiplication; then it forms M cluster centres and according to these centres, cluster excellent antibody cluster to do competitive expansion; finally antibodies accomplish mixed double mutation.
6 Immune recruitment. $N_r$ low affinity antibodies will be replaced by those antibodies producing randomly.
7 Stop conditional judging. If evolving algebra reaches set value $N_g$, outputting the first antibody of every Agent. These antibodies are the final results. If not, jumping to step (3).

## 6 ANALYSIS OF EXAMPLES

There are some changes about IEEE33 node distribution system, as shown in Figure 2. The system adds to load tap changing transformer, two DG and two groups of parallel compensation capacitor to maintain line parameter changeless [15]. The transformer ratio is 0.9 to 1.1. The up number of gears is 8 and down a number of gears is -8. And stepped size is 1.25 %. The active power capacity of every DG is 1000 kW, and reactive power capacity is 500 kVar. Capacities of two groups of parallel compensation

capacitor are 150 kVar*4 and kVar*7. By using multi-agent immune algorithm based on antibody cluster and competitive expansion, the number of Agent is 5. They are Agent VR, Agent $Q_{DG1}$, Agent $Q_{DG2}$, Agent $Q_{C1}$ and Agent $Q_{C2}$.

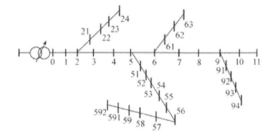

Figure 2. Diagram of the sample distribution branch.

Setting initial parameter of algorithm: initial population number of Agent $Q_{DG1}$ and Agent $Q_{DG2}$ is 50, population evolution algebra is 200, the number of cluster centre is 3, and Gauss mutation probability $P_{mg}$ is 0.8, Cauchy mutation probability $P_{mc}$ is 0.8, immune recruitment number $N_r$ is 10; initial population scale of Agent $Q_{DG1}$ is 5, antibodies are {0,150,300,450,600}; initial population scale of Agent $Q_{DG2}$ is 8, antibodies are {0, 150, 300, 450, 600, 750, 900, 1050}. Antibodies of Agent VR, Agent $Q_{DG1}$ and Agent $Q_{DG2}$ won't change through the algorithm process.

Optimization results are shown in table 1. Active loss of the entire system decreased from 99.2005 kW to 76.5714 kW. It dropped 22.81%. The voltage of every node is within the scope of the constraint. We can know that by using reactive compensation capacity of DG, we can adjust voltage within the scope of the constraint and reduce system active loss largely.

Table 1. The value of control variables before and after optimization.

|  | VR | $Q_{DG1}$ | $Q_{DG2}$ | $Q_{C1}$ | $Q_{C2}$ |
|---|---|---|---|---|---|
| Before | 1 | 0 | 0 | 600 | 600 |
| After | 1.0625 | 496.6 | 289.8 | 600 | 750 |

Optimization results in this text are the same to multi-agent immune algorithm (ICE) proposed in text [8]. Convergence properties of the two algorithms are shown in figure 3.From figure 3, it can be seen that algorithm in this text is based on multi-agent immune algorithm. It introduces cluster and competitive expansion, and uses antibody mixed mutation. Under this circumstance that system has the same rate of descent, optimization speed of the algorithm in this text is faster than ICE. In other words, this algorithm has better astringency. Algorithm in this text went through several generations' evolution optimization, system losses decrease to a value which is close to the optimal solution. So that accelerates algorithmic convergence.

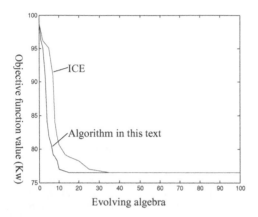

Figure 3. The curve of convergence characteristic of methods.

## 7 CONCLUSIONS

By using reactive power compensation capacity of DG, a multi-agent immune algorithm is proposed which is based on antibody cluster and competitive expansion. This algorithm solves the problem about reactive power optimization in power distribution network under condition of distributed generation. This idea combines multi-agent system and immune algorithm, and builds a model based on artificial immunity and efficient ability of solving problem. As for single agent, in the process of affinity maturation, it introduces antibody cluster, complete expansion and antibody mixed mutation to provide excellent antibodies of every cluster a chance to do clone augmentation and make affinity mature. That increases diversity of antibody cluster distribution, and reaches a balance between depth-first search and breadth-first search. Simulating calculation proves that this idea has high speedability and validity.

## ACKNOWLEDGMENTS

We are thankful for our tutor Y.Q. Xu, she gives some useful comments on our paper so that our thesis can become better than before.

## REFERENCES

[1] Zhiqun Wang, Shouzhen Zhu, Shuangxi Zhou, et al. Impacts of distributed generation on distribution system voltage profile. Automation of Electric Power Systems, 2004, 28(16): 56–60.
[2] Lin Chen, Jin Zhong, Yixin Ni, et al. Optimal reactive power planning of radial distribution system with distributed generation. Automation of Electric Power Systems, 2006, 30(14): 20–24.
[3] Wei Pei, Kun Sheng, Li Kong, etc. Impact and improvement of distributed generation on distribution network voltage quality. Proceedings of the CSEE, 2008, 28(13): 152–157(in Chinese).
[4] Vu Van Thong, Johan Driesen, Ronnic Belemans. Using distributed generation to support and provide ancillary services for the power system[C]. 2007 International Conference on Clean Electric Power, Leuven, Belgium, 2007: 159–163.
[5] M. Braun. Reactive power supply by distributed generators[C]. 2008 IEEE Power and Energy Society General Meeting, Pittsburgh, USA, 2008: 1–8.
[6] Taher Niknam. A novel approach based on ant colony optimization for daily volt/var control in distribution networks considering distributed generators[J]. Energy Conversion and Management, 2008, 49: 3417–3424.
[7] Niknam T, Ranjbar AM, Shirani AR. Impact of distributed generation on volt/var control in distribution networks[C]. IEEE Bologna Power Tech Conference Proceeding, 2003: 1–6.
[8] Wei Yan, Yujiang Sun, Chunlei Luo, etc. Ep based on specialist experience and its application to var optimization[J]. Proceedings of the CSEE, 2003, 23(7): 76–80(in Chineses).
[9] Yutian Liu, Li Ma. Reactive power optimization based on Tabu search approach. Automation of Electric Power Systems, 2000, 24(1): 61–64.
[10] Tieyuan Xiang, Qingshan Zhou, Fupeng Li, et al. Research on Niche Genetic Algorithm for reactive power optimization[J]. Proceedings of the CSEE, 2005, 25(17): 48–51(in Chinese).
[11] Hugang Xiong, Haozhong Cheng, Hongzhong Li. Multi—objective reactive power optimization based on immune algorithm [J]. Procedings of the CSEE, 2006, 26(11): 102–109(in Chinese).
[12] A. H. Bond, L. Gasser. editors. Reading in Distributed Artificial Intelligence. Morgan Kaufmann Publishers: San Mateo, 1988.
[13] Hai Qian, Jianhui Ma, Xufa Wang. Novel immune cooperative multi-agent model and its simulation analysis. Journal of System Simulation, 2008, 20(13): 3436–3444.
[14] Xuesong Xu, Jing Zhang, Qing He. Novel immune clonal selection optimization programming. Journal of System Simulation, 2008, 20(6): 1536–1540.
[15] Jian Liu, Pengxiang Bi, Haipent Dong. Complex distribution system simplified analysis and optimization. Beijing: China Electric Power Press, 2002.

*Multimedia, Communication and Computing Application – Leung (Ed.)*
© *2015 Taylor & Francis Group, London, ISBN 978-1-138-02775-6*

# Using virtual instruments and real-time simulation techniques for off-line tuning of electro-hydraulic rotational speed regulators

D.D. Ion-Guta

*INCAS - The National Institute of Aerospace Research "ELIE CARAFOLI", Bucharest, Romania*

ABSTRACT:   This paper presents a real-time simulation system developed for electro-hydraulic rotational speed regulators systems used in the energy field. The system developed allows modeling of control upon the rotational speed of a hydro unit and developing of a process control algorithm. The system is used for off-line testing/debugging of the firmware required by the equipment for automation/adjustment of these types of processes - speed regulators for Kaplan hydro unit.

KEYWORDS:   Virtual instrument, Real-time simulation, Electro-hydraulic servomechanism

## 1   INTRODUCTION

The rapid evolution of digital industrial computers and major progress achieved in the field of electro-hydraulic systems have been imposed as the natural solution in the use of electro-hydraulic equipment in development or upgrading of speed and power regulators in the energy field. Thus the special static and dynamic performances of hydraulic actuation elements are well mixed with the flexibility and reliability of digital control systems.

In Fig.1 we can see the schematic diagram of the system. It consists of two main components: the hydro unit (Kaplan turbine) and the process controller (industrial computer). The process is complex, being composed of turbine + electric generator, the actuation system of rotor blades and the actuation system of the driving device. Process controller is to ensure

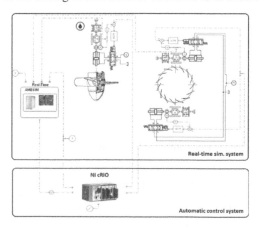

Figure 1.   Schematic diagram of the system.

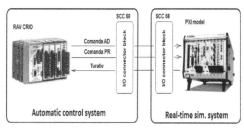

Figure 2.   Real time simulation system.

the correct sequence for on/off of the unit and control the turbine shaft speed.

The objective of this work is to develop a computer system used for tuning and automatic adjustment of electro-hydraulic systems in the field of energy. Automatic adjustment means all actions performed on a process for it need to behave in a desired manner. For practical application of automatic adjustment, it takes up a few basic steps, namely:

- modeling of processes;
- identification of processes based on experimental data and estimation of parameters;
- processing of signals by filtering, prediction, state estimation;
- design of control signals for automatic management

The system developed for tuning and automatic adjustment of electro-hydraulic systems must allow running of the afore-mentioned steps. By means of this system, it must be possible to identify the adjustment law and compliance parameters which provide for the whole system a certain unit step response, i.e. satisfying certain required transient and stationary performances. Also the system should allow the selection of a desired response in relation to the actuation of the disturbance values, preserving at the same time certain performances relative to the input value.

The system must provide capabilities for interfacing with physical processes and obviously for driving them in real time. Based on information gathered from the process, models of the process can be created to be used during the stage of off-line testing on the adjustment laws adopted. Off-line testing will involve real-time simulation of the process in parallel with execution of the adjustment law adopted.

## 2 REAL-TIME SIMULATION OF ELECTROHYDRAULIC SERVOMECHANISMS WITH POSITION REACTION

Dynamic systems modeling and simulation are techniques widely used in computer-assisted analysis of systems. it also represents an important step in the design (synthesis) assisted by computer systems.

Dynamic mathematical model of an electro-hydraulic servomechanism with position response (Fig. 3) comprises the following equations:

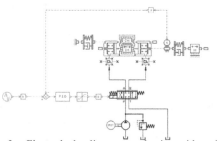

Figure 3. Electro-hydraulic servomechanism with position reaction.

- Equation of slide valve displacement
- Equation of position transducer
- Equation of electronic comparator
- Continuity equation of subsystem directional control valve-hydraulic cylinder
- Equation of current generator of proportional compensator
- Motion equation of hydraulic cylinder's piston
- Characteristic of directional control valve

The linear mathematical model:

$$H_0(s) = \frac{z(s)}{y(s)} = \frac{K_v \omega_h^2}{s^3 + 2\zeta\omega_h^2 s^2 + \omega_h^2 s + K_v \omega_h^2}$$

where:

$\omega_h$ : Hydraulic natural pulsation
$\zeta$ : Damping factor

$K_v$ : Speed amplification factor

Using a numerical system for interfacing with the physical part requires sampling the analog signal acquired from the process. The mathematical model must be transformed from the complex space s into z, determining the equivalent transfer function in z. If it is intended for the output of the continuous system to coincide with the output of the system with sampling during sampling moments, there is introduced an extrapolator for signal reconstruction.

For a continuous-time process, having transfer function H (s), the discrete equivalent of transfer function H(z) must also include the transfer function of the zero order extrapolator (ZOE).

Calculation of transfer function H(z) from H(s) according to

$$H(z) = \frac{z-1}{z} Z\left[\frac{H(s)}{s}\right] \tag{1}$$

can be performed analytically by rests or there can be used continuous-to discrete-time models functions.

It is considered:

$$b_0 = \mu \frac{K_{Qx}}{A_p}; a_3 = \frac{m}{R_h}; a_2 = m\frac{K_P}{A_p^2}; a_1 = 1; a_0 = 0$$

$$\frac{z(s)}{\varepsilon(s)} = \frac{b_0}{a_3 s^3 + a_2 s^2 + a_1 s + a_0} \tag{2}$$

The calculated coefficients of the linear model of electro-hydraulic system can be found in Table 1. In Fig. 4 it is presented that the response over time to step-type excitation signals.

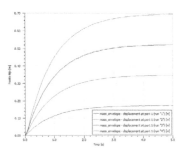

Figure 4. Response over time to step-type excitation signals.

Table 1. Coefficients of the electro-hydraulic system.

| a3 | a2 | a1 | a0 | b0 |
|---|---|---|---|---|
| 7.89e-4 | 1.23e-2 | 1 | 0 | 0.335 |

## 3 SIMULATION NETWORK AND VIRTUAL INTERFACE OF THE ADJUSTMENT MODEL

To identify the optimal system adjustment process an application for simulating, the process has been developed using AMESim and LabVIEW simulation environments. In its structure, there are found the models of the hydro unit and of the electro-hydraulic drive system and the software component for automation/ adjustment of the process. Hydraulic drive systems have been simulated in AMESim, while the adjustment algorithm has been developed using the LabVIEW software.

Real-time simulation has been performed using the mathematical model of the system (Fig. 5) consisting of:

a) model of the hydro unit (TH & GS)
b) model of the speed regulator (RAV - process computer)
c) model of the actuation system of the driving device (AD)
d) model of the actuation system of rotor blades (PR)

The dynamic mathematical model was developed using specific mathematical equations from the referenced literature [1], [2] si [3].

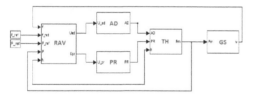

Figure 5. Schematic diagram of mathematical model.

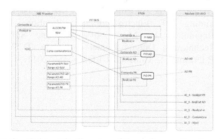

Figure 6. Schematic diagram of the software developed on RAV-cRIO.

PID loops were used for control (PI for speed controller and PID for electro-hydraulic systems). Control loops and the specific software algorithm necessary for proper functioning of the controller were implemented on cRIO process computer (x86-FPGA mixed system developed by National Instruments). The control loops (PI and PID) were implemented on FPGA and the control algorithm on x86. In Fig. 6, we can see the application software architecture developed on RAV-cRIO.

PID controller studied results from discretizing a continuous PID controller with the independent actions P, I and D.

Continuous controller [5]:

$$H_{PID}(s) = K \left[ 1 + \frac{1}{T_i s} + \frac{T_d s}{1 + \frac{T_d}{N} s} \right] \quad (3)$$

Parameters:

K – proportional amplification

Ti – integral action

Td – derivative action

Td/N – derivative action filtering

In FIG. 7 and Fig. 8 we can see the control panel of the LabVIEW application (virtual interface of the adjustment model) and application block diagram.

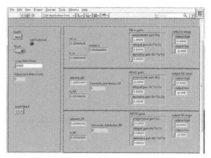

Figure 7. Control Panel.

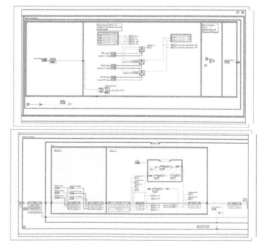

Figure 8. Application Block Diagram.

Tuning of the regulator parameters was performed taking into consideration the system stability and restrictions related to its dynamic behavior. Objectives targeted were to improve the dynamic characteristics and response to disturbances. Analysis criteria used were chosen taking into consideration the possibility to obtain information on system stability, and also on influence of the parameters which are to be optimized.

## 4  DYNAMIC CHARACTERISTIC OF THE SYSTEM

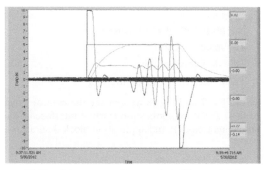

(a) Unstable system – Kp_speed=1, Ki_speed=0.2

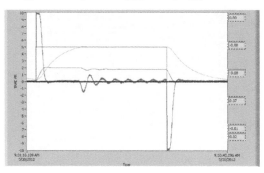

(b) Stable system - Kp_speed=10, Ki_speed= 0.00781

Figure 9.  Start with continuance at idling.
(a-unstable system; b-stable system)

In Fig. 9 we can see the evolution of the process parameters: the command signal, the speed achieved, the position of the driving device (AD), and the position of the actuation system of rotor blades (PR).

There are two cases: (a) bad choice of PID parameters -system unstable and (b) good choice of PID parameters - stable system. The first case (a) can be very dangerous for a real system; undamped oscillations of the driving device AD can cause serious damage to hydro unit.

## 5  CONCLUSION

Real time simulation of the analyzed system presents obvious advantages. One may test, with a low level of risk, different ways of adjustment and control. Although tuning on site is inevitable, the real time analysis process allows the decrease of the time necessary for testing and tuning in the real process. The decrease of time necessary for achieving these objectives leads to minimizing costs and, at the same time, decreasing of the possibility of physical damage of test equipment.

The developed system can also be used for off-line testing/debugging of the firmware required by the equipment for automation/adjustment of these types of processes.

## REFERENCES

[1] System Identification and Control Design, Landau, I.D. Prentice-Hall, 1990.
[2] Power System Stability and Control, P. Kundur McGraw-Hil, 1994.
[3] Basic concepts of real-time simulation), Dragoş ION GUŢĂ, U.P.B. Sci.Bull., Series D, Vol. 70, Iss. 4, 2008.
[4] Tuning of Electro-hydraulic Rotational Speed Regulators by Means of Real-Time Simulation Techniques, Dragos Daniel Ion Guta, T. Popescu, P. Drumea, C. Chirita, EUROSIS, ISC 2011.

*Multimedia, Communication and Computing Application – Leung (Ed.)*
© 2015 Taylor & Francis Group, London, ISBN 978-1-138-02775-6

# Autonomous vehicle: From a cognitive perspective

X.L. Zhu & S.M. Tang
*High-Tech Innovation Engineering Center, CASIA, Beijing, China*

ABSTRACT: To make an Autonomous Vehicle (AV) more cognitive, it needs the implementation of advanced cognition theories and artificial intelligence theories. In this paper, we first make a brief overview of current advanced theories of cognition in psychology and computer science. We analyze and compare particularly the architectures of the AVs winning DARPA Challenges. The layout of sensors and the design of software system are critical to the winning AVs. However, recently no paper has been published to compare the different architectures of AVs. By comparing different AVs, we find only few points are shared by them but have more differences due to the various layouts of sensors and the differences among cognition architectures, which could give some valuable directions to the researchers in both computer science and cognition fields.

KEYWORDS: Autonomous Vehicle (AV), Cognition, Perception, Sensors

## 1 INTRODUCTION

An autonomous vehicle (AV) or a self-driving car is a vehicle mounted with sensors, such as cameras, light detection and rangings (LIDARs), ultrasonic sensors, microwave sensors, global positioning system (GPS), wheel odometers, magnetic compasses, and so on, to sense their surroundings and get useful information about the road and the position of the vehicle for autonomously safe driving (Shuming Tang 2010). The promotion of AVs will be a good potential solution to the existing traffic problems, such as traffic jams, traffic accidents, energy consumption, and so on, in the whole world currently. In China, the famous event "Future Challenge" (FC), which is supported by the National Natural Science Foundation of China, has been held for five consecutive years (2009–2013). The event is partly intended to promote the development of AV in China, and we can see a great progress made in AV development over the past 5 years.

There are a lot of papers published on AVs, some of which show how to create an AV (Thrun, Sebastian, et al. 2006, Leonard, John, et al. 2008, Monternerlo, Michael, et al. 2008, Urmson, Chris, et al. 2009), and some of which mainly focus on partial but detailed technologies in an AV (Chaturvedi, Pooja, et al. 2001, Discant, Anca, et al. 2007, Enzweiler, Markus, et al. 2009, Darms, Michael, et al. 2008, Mogelmose, Andres, et al. 2012a, Mogelmose, Andres, et al. 2012b, Mathias, Markus, et al. 2013). In other words, to our best knowledge, recently there is no paper published to compare the different architectures of AVs, and only few papers are published on AV in cognition topics.

We try to solve the two aforementioned problems. In this paper, we for the first time combine AV architectures with cognition theories and compare different architectures of AVs (Fig. 1 shows the cognition architecture of AVs discussed in the following sections.) In other words, our contributions are as follows: (1) comparing different architectures of AVs; (2) combining cognition theories with AV technology toward a more cognitive AV system; and (3) giving some possible research directions to make a more intelligent AV.

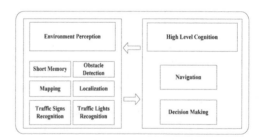

Figure 1. Cognition architecture of AV.

The following text is divided into five sections as follows: Section 2 is a brief on definitions and main research points of cognition in psychology and computer science. Sections 3–5 mainly compare the different architectures of AVs winning the DARPA Challenges under the embodied cognition theory. Finally, Section 6 concludes with some valuable research directions in both AV and cognition.

## 2 COGNITION

Cognition can be briefly defined as acquisition of knowledge (Reed, Stefen K. 2007). However, various disciplines hold different views toward intrinsic definition of cognition.

In psychology, researches on cognition mainly focus on pattern recognition, attention, memory, vision image, language, problem solving, and decision making. They consider vision, audition, tactile, olfaction, and gustation as low-level perception of cognition, and categorize language understanding, problem solving, and decision making as high-level cognition. The views on issues such as how to link low-level perception with high-level cognition and how human intelligence forms have led to different research topics in psychology. In this paper, we chronologically classify the different research topics into three categories: cognitivism, connectionism and embodied cognition (Anderson, Michael L. 2003). Cognitivism is a theoretical framework for understanding the mankind's mind, which is a response to behaviorism. While the main research points in behaviorism focus on the corresponding external relationships between perception and action, cognitivists try to disclose the internal relations between perception and action. Cognitivism believes that symbol computing is the core of intelligence. Connectionism believes that the network of numerous connected units is the basis of generating intelligence. It tries to use artificial neural network to interpret the intelligence of human. The network is formed by connecting weighted nodes hierarchically. The weights represent the mutual influences between synapses in the brain (Zhongzhi Shi 2008). Embodied cognition is developed when traditional symbol operation based on cognition theories is questioned by phenomena such as Chinese room (Barsalou, Lawrence W. 2010). In the embodied cognition theory, cognition is not about intellectual demonstration but is related to the body and its surrounding physical environment (Anderson, Michael L. 2003, Mahon, Bradford Z., et al. 2008). Therefore, intelligence can be described as the sum of the physical experience the body obtains when it interacts with the environment, that is, intelligence is related to the interaction between body and the environment. Embodied cognition gives a framework on bridging the low-level perception and high-level cognition (Mahon, Bradford Z., et al. 2008).

The research on cognition in computer science is mainly focused on artificial intelligence (AI), where the researches on cognition are to improve the intelligence of machine. According to Stuart Russell and Peter Norvig (2009), we can classify the definitions of cognition in AI into four categories: (1) thinking like a human; (2) acting like a human; (3) thinking reasonably; and (4) acting reasonably. Currently, the major researches in computer science focus on the imitation of human mind by creating intelligent software and hardware rather than understanding it. For example, a hidden Markov model obtains good results in speech recognition; however, no scientific experiment can show human also recognize speech similarly (Stuart Russell & Peter Norvig, 2009). AI also combines the methods used in symbolism and connectionism by using big data to research and improve machine intelligence.

## 3 THE ARCHITECTURE OF AV

Shuming Tang (2011) presents a view on the cognition system of an AV, where the cognition of an AV is divided into three stages: perception, cognition, and decision, which is different from the view in this paper. According to the embodied cognition theory, the cognition system of an AV can be divided into two parts: environment perception and driving decision. The two parts in the vehicle interact with each other to ensure the vehicle move to the destination safely. The two parts correspond to the low-level perception and high-level cognition separately. Environment perception via sensors belongs to the low-level cognition, and navigating and decision making belong to the high-level cognition. In the following sections, we mainly study the architecture of an AV body, which is composed of hardware and software, and see how it interacts with the environment to drive smartly.

### 3.1 Hardware

As the DARPA Challenges is the largest demonstration of the technology of AV to date (Levinson, Jesse, et al. 2011), we mainly compare the hardware and software architecture of AVs on attending the DARPA Challenges. The hardware specifications are similar among these cars. The basic requirements of an AV in hardware include: (1) enough space for the sensors and computers; (2) electronically actuated throttle, brake, shifter, and steering system (Levinson, Jesse, et al. 2011); (3) various sensors such as Velodyne, SICK, camera, and so on; (4) power unit(s) for the equipment and sensors. By surveying the teams winning the DARPA Challenges, we find that the most popular used sensors were GPS/INS and SICK, and other sensors were Velodyne, radar, and camera. Although they use the same kind of sensors, the layout of these sensors differs a lot among the teams (Montemerlo, Michael, et al. 2008, Urmson, Chris, et al. 2009, Bacha, Andrew, et al. 2008). Section 4 describes the impacts on the algorithm of different layouts when same problems are tackled.

The specification of computers in the vehicle is not a problem, because volume and price are decreasing whereas their computing power is increasing year by year.

## 3.2 Software

The low-level software configuration involves operating system selection, communication methods between different processing modules, and so on. This paper mainly focuses on the cognitive modules in the vehicle. [For a more in-depth study on the low-level software configurations, please refer to Thrun, Sebastian, et al. (2006), Leonard, John, et al. (2008), and Montemerlo, Michael, et al. (2008)].

According to the embodied cognition theory and the definition of intelligence as acting reasonably in AI, an AV, moving in an urban environment or off-road environment, should have the ability to sense the environment and then make an optimal output for the motion system. Figure 2 shows the workflow of most advanced AVs surveyed in this paper. First, the modules of the low-level environment perception output the information of environment and the vehicle's pose to high-level navigation and decision making module. Then, the high-level cognition module generates the information of basic vehicle control such as speed, direction, acceleration speed, and so on by using the information of destination, map, or routine and the outputs from low-level perception processing modules. After the execution of the commands from high-level cognition and the change of the AV-faced environment, the environment perception modules output the latest information to navigation and decision-making module, and then, a cognition loop forms.

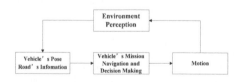

Figure 2.  Workflow of AV.

## 4 PERCEPTION MODULES

After surveying the basic architecture of an AV, now we discuss more detailed environment perception modules by comparing the vehicles that attended 2005 Grand Challenge and 2007 Urban Challenge. As pointed out previously, the perception modules sense the surroundings and generate useful information for high-level cognition. The perception modules include short-memory systems, obstacle detection, localization and mapping, and some necessary but not discussed modules in the vehicles we surveyed, such as traffic lights detection, traffic sign detection, and recognition.

### 4.1 Short-memory systems

A short-memory system of a vehicle can be analogical to the short-memory system of a human brain. It helps the vehicle to understand its surrounding environment and provides enough information for its decision-making module. The remarkable characters of a short-memory system include a wealth of information and its real-time updated information. Almost all teams have their own short-memory system although they are with different names, for example, Stanley and Boss call it Grid Map, Junior calls it 2-D Map, Odin calls it World Frame, and Talos names it Local Frame. The main information contained in the short-memory system includes position of static obstacles, trajectory, position of the moving obstacles and curbs position, and so on.

### 4.2 Obstacle detection

Detecting and avoiding collision to obstacles is a major mission when an AV is moving on the road. There are two types of obstacles in the driving environment: (1) static obstacles like trees and buildings; and (2) dynamic obstacles like moving cars and pedestrians.

According to Discant, Anca, et al. (2007), there are two types sensors for objects detection. One is active sensors like radars and the other is passive sensors like cameras. Generally, vehicles like Stanley and Odin use both types of sensors to work together to get a reliable result. Boss, Junior, and Talos use active sensors for the obstacle detection (Leonard, John, et al. 2008, Montemerlo, Michael, et al. 2008, Urmson, Chris, et al. 2009). No vehicle uses passive sensors for the obstacle detection.

We have pointed out in Section 3 that different vehicles are mounted with sensors in different positions. The problems they meet during data processing are different, and a brief overview of these is as follows: Junior uses Velodyne mounted on the roof of the vehicle as a primary scanner. It estimates whether there is any obstacle within the distance of two adjacent rings (generated by adjacent lasers) with the threshold value. However, this method may suffer from the pitching and rolling of the vehicle. By setting the distance to next laser as a function of range rather than a fixed value, it might address this problem (Montemerlo, Michael, et al. 2008).

Talos combines with SICKs, Velodyne, and radars for obstacle detection. When processing data clouds from Velodyne, it employs probabilistic method to build a nonparametric ground model. Here, we mainly discuss how it reliably differentiates between obstacles and nonflat terrain in its layout of sensors. By putting two SICK in parallel, it could measure the changes in vertical direction, so the ambiguity between obstacles and nonflat terrain is resolved (Figure 3).

For dynamic obstacle detection, the winning teams use a short-memory system surveyed to record the past detected obstacles, then it employs different methods such as point model, box model in Boss (Urmson, Chris, et al. 2009), and particle filter in Junior to detect and track dynamic obstacles (Montemerlo, Michael, et al. 2008).

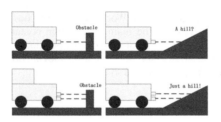

Figure 3.    Position configuration of SICKs in Talos.

### 4.3    Localization and mapping

For the vehicles competed in DARPA Challenges, routines are generally predefined. Therefore, when AVs are driving on road, the first thing they should do is to localize themselves on the routine with the help of various sensors. The most used sensors are GPS, inertial measurement unit, and wheel odometer. Junior also uses RIGEL for curb detection. Stanley, Odin, and Talos use camera to detect the road and localize themselves. To safely navigate in dynamic urban environments, AVs need higher localization accuracy rather than the one available from GPS-based inertial guidance systems (Levinson, Jesse, et al. 2011). In other words, the vehicles need to calculate the accurate positions of themselves with the help of established short-memory systems. The most used methods are to use Kalman filter to estimate the position and to use interpolation methods to smooth the routines [for detailed information please refer to Thrun, Sebastian, et al. (2006), Leonard, John, et al. (2008), Monternerlo, Michael, et al. (2008), Urmson, Chris, et al. (2009)]. Levinson, Jesse, et al. (2011) try to use SLAM technology to help the localization achieve centimeter accuracy. However, the result is got under the given prerecorded laser map. It is not suitable for AV to drive in new environment for the first time.

### 4.4    Other modules

We have reviewed the most important perception modules in the AVs attending DARPA Challenges. While the DARPA Challenges is the largest demonstration of AV technology to date, the traffic scenes are simplified in the competitions (Levinson, Jesse, et al. 2011). During the Urban Challenge, it is unavailable for pedestrians and bicyclists without traffic lights and traffic signs.

In this section, we review some state-of-the-art methods for those critical problems of driving in cities and off-road environments. For simplicity, we choose the computing speed and accuracy as our main criteria.

*Bicyclist and Pedestrian Detection:* The most used sensors in these two topics are cameras. In recent years, combining LIDARs with cameras, or by using 3D laser data, good results are obtained (Premebida, Cristiano, et al. 2009, Szarvas, Mate, et al. 2006). Bicyclist and pedestrian detection can be viewed as a different topic in computer vision community, still they share some common methods and algorithms, as Enzweiler, Markus, et al. (2009) and Cho H, et al. (2010) showed that HOG/linSVM has a clear advantage in the accuracy and speed compared with other methods, such as wavelet-based AdaBoost cascade. Benenson, Rodrigo, et al. (2012) showed that by using stixels without image resizing, their algorithm could achieve 20 times faster than previous methods while keeping the same accuracy.

*Traffic lights detection:* In traffic lights detection, the most used sensor is camera. By extracting the color information and processing with the shape information, they could achieve a good result. But the popular problem among the detection systems is that the methods greatly affected by the light condition cannot be popularized (Yu, Chunhe, et al. 2010, Diaz-Cabrera, et al. 2012).

*Traffic signs detection and recognition:* After several decades of development in both computer vision community and intelligent transportation system community, the results of traffic signs detection and recognition tested in public datasets have showed they are good enough (approximately above 95% in accuracy) (Mogelmose, Andreas, et al. 2012). However, some problems remain to be solved, such as how about the result when the systems work in adverse weather conditions.

## 5    NAVIGATION AND DECISION MAKING

We surveyed the environment perception modules in AVs, and now we step into another important modules in AVs—navigation and decision-making modules, which belong to the high-level cognition in the embodied cognition theory. The modules of navigation and decision making are to generate the basic vehicle control information to make the vehicle drive safely and reasonably. Different vehicles have different separation of navigation periods (see Table 1 for detailed information). Generally, each vehicle would first use the information provided by Road Network Definition File (RNDF) to generate a series of subgoals represented in points in the RNDF for the vehicle to achieve. Then they set a cost function for the vehicle to choose a path which is assumed to be

optimal. Vehicle like Junior implements global planning by using dynamic programming algorithm, which has achieved good results. For free-form navigation in parking lots, it is different from the on-road navigation, because the movement of the vehicle is far less constrained (Montemerlo, Michael, et al. 2008). Junior uses hybrid A* algorithm to search a near-cost optimal path to the goal and Boss uses Anytime D* algorithm to get a suboptimal path (Urmson, Chris, et al. 2009).

In addition to generate an optimal or suboptimal path for the vehicle, during the running of the vehicle, the states of vehicles may change from one to another, such as from stopping to driving. There are also many exceptional states that may occur. It's necessary for an AV to know which state it is in now, when to change to the next state and how to recover from exceptional states. Junior has a finite state machine to manage the vehicle's states and during the DARPA Challenge event, it almost never entered any of the exception states (Montemerlo, Michael, et al. 2008). However, other vehicles such as Talos are not so lucky and have an accident with the vehicle from Cornell due to sticking into exceptional states (Fletcher, Luke, et al. 2008).

## 6  CONCLUSION

We have mainly surveyed the advanced cognition theories and combined them to analyze and compare the state-of-the-art AVs winning the DARPA Challenges. From different layouts of sensors and software architectures, we have found that the advanced cognition theories such as the embodied cognition theory can also be used as the system-design guideline of an AV. Meanwhile, we have found that various AVs, though they may be different significantly in the sensors layouts and system designs, have a lot of basic knowledge in common, such as using the short-memory system for the short memory and using the probability theory and machine learning to tackle the uncertainty.

At last, from the survey, we still find some problems in the development of a more cognitive AV, which may be valuable directions for the related researches, as follows: (1) no systematic discussion of the robustness of the AV; (2) as 80% of information obtained by a human driver is from his/her vision, it is valuable for researches in computer vision field to improve reliability of computer vision methods; and (3) enough data are not yet reported regarding the evaluation of the reliability of the vehicle's cognition level. Finally, further research is required on the cognition of pedestrians, bicyclists, and vehicles' behaviors as the AV needs to interact with them properly.

Table 1. Compare of vehicles' high-level cognition architectures

| Vehicle | Mission Planning | Behavioral Planning | Error Recovery |
|---------|------------------|---------------------|----------------|
| Boss | • Generating graph to encode the dynamic environment and RNDF points<br>• Generating paths with different costs<br>• Detecting blockages | • Executing the policy generated by the mission planner<br>• Making lane-change, precedence, and safety decisions, respectively, on roads, at intersections, and at yields | Responding to and recovering from anomalous situations |
| Junior | • Planning paths from every location in the map to the next checkpoint<br>• Computing the costs of different paths | • Generating different policies according to the situation<br>• Generating smooth paths for vehicle to execute<br>• Making lane-change, crossing the intersection and merging to the traffic, etc.<br>• Using FSM to switch between different driving states and tackle the exceptions | |
| Talos | • Generating shortest route to the next MDF checkpoint<br>• Making intersection precedence, crossing, and merging<br>• Passing<br>• Blockage replanning<br>• Generating goals for the motion planner<br>• Generating fail-safe timers<br>• Turn signaling. | | Using fail-safe timers to help error recovery |

In the future, we will research on vehicle detection, pedestrian detection, and static obstacles detection to make the AV more cognitive.

## ACKNOWLEDGMENTS

This work was supported in part by the National Natural Science Foundation of China under Grants 97720008 and 61220002.

## REFERENCES

Anderson, Michael L. 2003. Embodied cognition: A field guide. Artificial intelligence149.1: 91–130.

Artificial Intelligence: A modern Approach: 3rd edition.

Bacha, Andrew, et al. 2008. Odin: Team VictorTango's entry in the DARPA Urban Challenge. Journal of Field Robotics 25.8: 467–492.

Barsalou, Lawrence W. 2010. Grounded cognition: past, present, and future. Topics in Cognitive Science 2.4: 716–724.

Benenson, Rodrigo, et al. 2012. Pedestrian detection at 100 frames per second.Computer Vision and Pattern Recognition CVPR, IEEE Conference on. IEEE, 2012.

Chaturvedi, Pooja, et al. 2001. Real-time identification of drivable areas in a semi-structured terrain for an autonomous ground vehicle. Aerospace/Defense Sensing, Simulation, and Controls. International Society for Optics and Photonics.

Cho H, Rybski P E, Zhang W. 2010. Vision-based bicyclist detection and tracking for intelligent vehicles[C] Intelligent Vehicles Symposium IV, IEEE. IEEE, 2010: 454–461.

Darms, Michael, Paul E. Rybski & Chris Urmson. 2008. An adaptive model switching approach for a multisensor tracking system used for autonomous driving in an urban environment.

Diaz-Cabrera, Moises, Pietro Cerri & Javier Sanchez-Medina. 2012. Suspended traffic lights detection and distance estimation using color features. Intelligent Transportation Systems ITSC, 15th International IEEE Conference on. IEEE, 2012.

Discant, Anca, et al. 2007. Sensors for obstacle detection-a survey. Electronics Technology, 30th International Spring Seminar on. IEEE.

Enzweiler, Markus & Dariu M. Gavrila. 2009. Monocular pedestrian detection: Survey and experiments. Pattern Analysis and Machine Intelligence, IEEE Transactions on 31.12: 2179–2195.

Fletcher, Luke, et al. 2008. The MIT–Cornell collision and why it happened. Journal of Field Robotics 25.10: 775–807.

http://baike.baidu.com/view/4572422.htm (in Chinese).

http://en.wikipedia.org/wiki/Chinese_room

http://en.wikipedia.org/wiki/Cognitivism_psychology

http://plato.stanford.edu/entries/connectionism/

Leader, Workpackage, Contractual Delivery Date, and Actual Delivery Date. 2009. State of the Art Report and Requirement Specifications.

Leonard, John, et al. 2008. A perception-driven autonomous urban vehicle. Journal of Field Robotics 25.10: 727–774.

Levinson, Jesse, et al. 2011.Towards fully autonomous driving: Systems and algorithms. Intelligent Vehicles Symposium IV, IEEE.

Mahon, Bradford Z. & Alfonso Caramazza. 2008. A critical look at the embodied cognition hypothesis and a new proposal for grounding conceptual content.Journal of Physiology-Paris 102.1: 59–70.

Mathias, Markus, et al. 2013. Traffic sign recognition— How far are we from the solution?. Neural Networks IJCNN, The 2013 International Joint Conference on. IEEE, 2013.

Møgelmose, Andreas, Mohan M. Trivedi Thomas B. Moeslund. 2012. Vision-Based Traffic Sign Detection and Analysis for Intelligent Driver Assistance Systems: Perspectives and Survey: 1–14.

Mogelmose, Andreas, Mohan M. Trivedi & Thomas B. Moeslund. 2012. Vision-based traffic sign detection and analysis for intelligent driver assistance systems: Perspectives and survey. Intelligent Transportation Systems, IEEE Transactions on 13.4: 1484–1497.

Montemerlo, Michael, et al. 2008. Junior: The Stanford entry in the urban challenge.Journal of Field Robotics 25.9: 569–597.

Premebida, Cristiano, Oswaldo Ludwig & Urbano Nunes. 2009. LIDAR and vision-based pedestrian detection system. Journal of Field Robotics 26.9: 696–711.

Reed, Stefen K. 2007. Cognition: Theory and applications. CengageBrain. com. page 8.

S.M Tang, et al. 2010. The current state and future prospect of the research on the evaluation of autonomous vehicle , 2010 Cognitive Computing of Visual and Auditory Information Symposium, NSFC(in Chinese), vol 2: pp.320–325.

S.M Tang. 2011. Research on the evaluation of autonomous vehicle's cognition and the environment design, NSFC Proposal(in Chinese).

Szarvas, Mate, Utsushi Sakai & Jun Ogata. 2006. Real-time pedestrian detection using LIDAR and convolutional neural networks. Intelligent Vehicles Symposium, IEEE. IEEE, 2006.

Thrun, Sebastian, et al. 2006. Stanley: The robot that won the DARPA Grand Challenge. Journal of field Robotics 23.9: 661–692.

Urmson, Chris, et al. 2009. Autonomous driving in traffic: Boss and the urban challenge. AI Magazine 30.2: 17.

C.H He, C Huang & Y Lang. 2010. Traffic light detection during day and night conditions by a camera. Signal Processing ICSP, 2010 IEEE 10th International Conference on.

Z. Z Shi. 2008. Cognitive Science, University of Science and Technology of China Press(in Chinese).

*Multimedia, Communication and Computing Application – Leung (Ed.)*
© 2015 Taylor & Francis Group, London, ISBN 978-1-138-02775-6

# Demand response and economic dispatch of power systems considering large-scale plug-in hybrid electric vehicles/electric vehicles

J.Y. Hu, B. Li, C. Liu, L. Xia & C.H. Li
*China Electric Power Research Institute, Beijing, China*

ABSTRACT: Increasing concerns about global environmental issues have led to the rapid development of green transportation. Our government should encourage the prosperity of the plug-in hybrid electric vehicles/ electric vehicles (PHEVs/EVs) industry in the near future. PHEVs/EVs are not only an alternative to gasoline but are also burgeoning units for power systems. The impact of large-scale PHEVs/EVs on power systems is of profound significance. This paper discusses how to use PHEVs/EVs as a useful new resource for system operation and regulation from a review of recent studies and mainly considers two mainstream methods: demand response and economic dispatch. The potential of using PHEVs/EVs to coordinate renewable energy resources is also discussed in terms of accepting more renewable resources without violating the safety and the reliability of power systems or increasing the operation cost significantly.

KEYWORD: Plug-in hybrid electric vehicles/electric vehicles, Demand response, Economic dispatch, Renewable energy, Smart grid

## 1 INTRODUCTION

Growing awareness about environment issues, especially in fragile urban areas, has inspired enthusiasm in investment and research on alternative fuel vehicles (AFVs) [1]. Avoiding serious if not catastrophic climate change and reversing the business-regular growth path over the next two decades requires vehicles that are much more environmental. The most promising solution is replacing internal combustion engine (ICE) vehicles with AFVs, such as plug-in hybrid electric vehicles (PHEVs) or electric vehicles (EVs). Governments all over the world consider the transportation electrification industry of strategic importance as national energy security [2]. Venture capitalists and financiers view this industry as a vital part of the greatest economic growth engine in the future: alternative energy. Thus, the next few decades might be the golden age for electrified vehicles. In the U.S., the Department of Energy (DOE) projected that 1 million PHEVs/EVs will be on the road and 425,000 PHEVs/EVs will be sold in 2015 alone, accounting for 2.5% of all car sales [3]. In China, according to a recent government plan, 100 billion RMB will be invested in new energy vehicles, including PHEVs, hybrid electric vehicles (HEVs), EVs and other new types of green vehicles. By 2015, there might be 1.5 million new energy vehicles on the road, making China the largest green vehicle inventory, and by 2020, the inventory might be as large as 5 million [4]. The European Union (EU) developed a roadmap that

defines the development of PHEVs/EVs in the EU by a three-stage program. The goal for the next decade is to set 5 million green vehicles on the road [5]. Other countries such as Korea and Japan have also proposed their own plans of new green vehicle development [6,7]. AFVs have also drawn the attention of some emerging economies: Iran has undertaken a practical exploration in the promotion of AFVs.

Transportation electrification is one of the indispensible trends of the future energy revolution, to reshape the traditional view of industrial power systems. On the one hand, vehicle electrification aims to change the manner in which energy is consumed, replacing fossil-based fuel with multi-source renewable electricity and thus encouraging a new boom in power grid development and construction. On the other hand, the upcoming boom of electrified vehicles will make old power systems generate new problems and new resources, such as portable small electricity storage. It is generally known that the disordered charging of a large number of PHEVs/EVs will jeopardize these power systems. The optimal charge control and dispatch of PHEVs/EVs are becoming increasingly relevant. In particular, when Kempton revealed the idea of vehicle-to-grid (V2G) and its profound significance, PHEVs/EVs were no longer regarded as only loads but also as large capacity energy carriers that can feed power back to the grid if necessary. Kempton also spoke of the fundamentals of V2G applications, including capacity calculations and net revenue. He discussed the possible implementation of V2G, from stabilizing the system to supporting large-scale renewable energy. PHEVs/EVs can provide utility services by using demand response programs or dispatch strategies. These services provide extra safety, reliability and efficiency for future smart grids with multiple new integrations, such as that of large-scale renewable energy.

## 2 IMPACT OF LARGE-SCALE PHEVS/EVS ON POWER SYSTEMS

### 2.1 *Characteristics of phevs/evs*

According to reference, a PHEV can be defined as follows: (1) a battery storage system with the capacity of 4 kWh or more; (2) capable of recharging from other power sources; and (3) capable of driving at least 10 miles without consuming gasoline. A PHEV/EV must obtain energy from the grid to remain functional. However, the methods of interaction with the grid vary. According to an estimate by the U.S. Transportation Department, home-charged PHEVs/EVs will account for 70% of all PHEVs/EVs. This type of interactive facility is widely distributed, which makes it very difficult to follow centralized control. Other PHEVs/EVs will be charged at concentrated facilities served by specific charging service providers (CSPs): (1) large parking lots, such as the parking lots of the central business district (CBD), large corporations or government facilities; (2) exclusive parking lots for taxies, buses, cargo fleets, *etc.*, which provides charge services; and (3) stations that include normal charging stations, fast charging stations, battery exchange stations, and intelligent integration stations. These facilities are easily available resources to meet system requests.

At the grid level, PHEVs/EVs are special load/power sources when they perform charging/discharging operations. For sustainable working, PHEVs/EVs harvest electrical energy like other regular electric equipment. However, PHEVs/EVs have unique characteristics that distinguish them from other loads: (1) energy storage: a certain amount of energy can be stored within PHEVs/EVs, e.g., the Nissan Leaf has a battery capacity of 24 kWh. Energy storage is the basis for the flexibility of PHEVs/EVs because charging and driving (that is, consuming) are two separate processes, unlike traditional electric equipment that obtaining and consuming energy almost at the same time, which suggests that there is possibility for adjustments; (2) bidirectional power flow: the stored energy in PHEVs/EVs is not only required for driving but can also greatly helpful for power systems. PHEVs/EVs can be both power consumers and backup power sources; (3) idle most of the time: according to recently conducted transportation research, in the U.S., private cars are idle 96% of the day, and in the U.K., the figure is 94.8%. When idle, PHEVs/EVs are free to respond to instructions from the power grid; and (4) social attributes: the behavior of PHEVs/EVs is mostly determined by the intent of the owners, which is the major reason for the stochastic nature of the PHEV/EV loads. When a large number of PHEVs/EVs plug into a system randomly, significant uncertainty will be introduced into the system. Social principles, such as those applied in sociology, psychology, and economics, can be implemented to analyze the behavior of PHEVs/EVs.

### 2.2 *Impacts caused by the charging of phevs/evs*

The large market penetration of these vehicles is likely to change the configuration of power systems. The impact of PHEVs/EVs penetration on power systems has been measured and calculated by multiple authorities in different locations using many different tools that range from analytical techniques to simulations. Most work has focused on the issue of breaking down the established balance of power supply and demand due to the charging of PHEVs/EVs. In the long run, old-fashioned power systems are not capable of handling the upcoming abundance of PHEVs/EVs, so it is necessary to determine whether the existing or planned generation capacity is able to meet the rapid development of PHEVs/EVs Therefore, it is very important to calculate the potential capacity of the PHEVs/EVs load and its distribution in both time and space. Hajimiragha conducted simulations and concluded that 6% uniform penetration of PHEVs by 2025 can be realized in Ontario, Canada. Steen estimated the charging behavior using demographic statistics data. Ikegami assumed trip patterns of 10 million EVs in Tokyo power systems and concluded that charging-time control is urgently needed for the power systems to remain effective. Oak Ridge National Laboratory recently proposed a general comprehensive report that covers 13 regions in the U.S. defined by Natural Environment Research Council (NERC) and DOE Energy Information Administration (EIA). This report addresses the impact of PHEVs/EVs with three main issues: the characteristics and market of PHEVs/EVs, regional power supply/demand and dispatch, and a regional analysis of the price of electricity, generation structure and emission level. It could be concluded that additional generation capacity and demand response programs (DRPs) designed for EVs are necessary for the large population of PHEVs/EVs.

Other scholars are concerned about changes in system feature attributes and are worried about system inefficiency or even failure when a large number of PHEVs/EVs are charging without intervention. Hu conducted a Monte Carlo simulation to calculate the power demand of PHEVs/EVs in China by 2015, 2020 and 2030; the simulation results show that a new peak hour will occur and that the difference between the power demand peaks and valleys will become even larger due to disordered PHEVs/EVs charging. This difference will shrink the abundance of power generation, raising the possibility of losing loads. Even when the generation capacity of the entire area is sufficiently large, the off-limit alarms still sound to be at some local substations, transmission lines and transformers because most of the local PHEVs/EVs will be charging at the same time, which may affect distribution transformers in particular. Assessments of the life reduction of transformers caused by the fluctuation of the power demand of PHEVs/EVs have been conducted. In addition some researchers have focused on the harmonic pollution caused by power electronic equipment in recharge facilities, especially large-capacity fast chargers. The simulation results indicate that chargers is a great threat to power quality, and additional compensation devices or harmonic filters are required. Some scholars have focused on the transient nature of power systems with PHEVs/EVs penetration, such as the stability of the voltage of the power systems.

The impact of PHEVs/EVs on power systems is of profound significance. The recent increase in research of this field has been driven by the promising potential of the upcoming increase in PHEV/EV numbers. Studies in this line of research provide a basic view of this problem: existing power systems are not sufficient due to the rising power demand of the growing population of PHEVs/EVs, especially when a large number of PHEVs/EVs charge at the same time or at the same location. Several solutions have also been proposed: increased spending on power systems construction and updating or optimizing the charge behavior of PHEVs/EVs. Despite initiative and motivation, most studies have been preliminary: due to the lack of real time data, many computer-based simulations have been based on a set of assumptions, and the methods should be further improved.

# 3 DEMAND RESPONSE (DR) OF PHEVS/EVS

One of the new philosophies of power systems states that a system will be most efficient if fluctuations in demand remain as small as possible. Because electricity consumption on the demand side is flexible and highly capital-intensive, demand response is one of the cheapest resources available for system operation. Demand response can be defined as incentive payments designed to induce lower electricity use

at times of high wholesale market prices or when system reliability is jeopardized. PHEVs/EVs are one of the most promising demand response resources for future smart grids.

## 3.1 Introduction of PHEVs/EVs in DR

Figure 1 shows several mainstream DRPs that may be suitable for PHEV/EV applications. The DRPs are divided into two major groups, incentive-based programs (IBPs) and price-based programs (PBPs). PBPs use dynamic electricity price rates to induce customers to change their energy consumption patterns. The basic type of PBP is time of use (TOU) [35], namely, the price of per unit consumption differs in different blocks of time. TOU is higher in peak blocks and lower in off-peak blocks. Real time programs (RTPs) are based on TOU but in a real-time environment with a small time interval between price adjustments [36]. Critical pricing program (CPP) adds additional penalty costs to electricity consumption in critical blocks. IBPs reward those customers who participate in the programs with a credit bill or a discount rate. Direct load control programs allow utilities to shut down contracting equipment on short notice. Programs that can be curtailed to encourage participants to reduce their loads to a predefined value; otherwise, extra penalties will be charged. Demand bidding programs require customers to bid for certain load reductions in an electricity wholesale market. When a bid is accepted, customers should curtail their load to the amount specified in the bid or face a penalty.

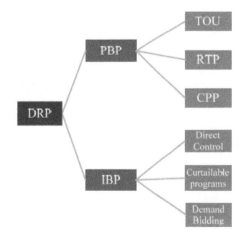

Figure 1.   Catalogue of DRPs.

PHEVs/EVs are valuable resources for demand response, and their importance cannot be overlooked in future demand side management (DSM).

In addition, a large penetration of PHEVs/EVs is impossible without optimal dispatch through demand response. DRPs will also make adjustments to cope with different market penetrations of PHEVs/EVs.

### 3.2  *Response of a Single PHEV/EV*

In the future, millions of home charge private PHEVs/EVs will have the choice either to join the programs of charging service providers (CSPs) or to plan their own charging schedule, as shown in Figure 2. CSPs are aggregator agencies who act as a liaison between system operators and terminal customers; they provide ready-made charging programs for customers to join. If customers decide to purchase energy from the day-ahead market at its transaction price instead of joining CSP programs, on the next day, options are open for simply using the energy or selling it back to grid when the price of demand bidding is tempting and the power shortfall is acceptable. Once the bid is accepted, users adjust their charging power sequence accordingly. For owners, the question is to measure whether participation is profitable, and offers will be made as long as the demand bidding price is higher than expected. Another option is to purchase energy in the real-time market and adjust their plans with fluctuations in the real-time price.

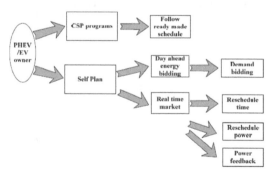

Figure 2.   Choice of Response.

## 4   CONCLUSIONS

System operation and regulation will face challenges and opportunities with the upcoming boom in PHEVs/EVs. This paper focuses on this topic and discusses the mainstream methods implemented to make use PHEVs/EVs. First, the impact of large-scale PHEVs/EVs and the potential benefits of the optimal dispatch of PHEVs/EVs were discussed. Second, DRPs were introduced to induce PHEVs/EVs to change their energy supply patterns, and the response of single home-charging PHEVs/EVs and the community of PHEVs/EVs is demonstrated in detail. Research in this field and the PHEV/EV industry is still preliminary. There are many applications of PHEVs/EVs that have yet to be discovered. Taking full advantage of PHEVs/EVs requires highly advanced compementary methods for both the electricity market and power system operation.

## REFERENCES

[1] Romm, J. The car and fuel of the future. *Energy Policy* 2006, *34*, 2609–2614.
[2] Greene, D.L. Measuring energy security: Can the United States achieve oil independence *Energy Policy* 2010, *38*, 1614.
[3] Sikes, K.; Gross, T.; Lin, Z.; Sullivan, J.; Cleary, T.; Ward, J. *Final Report: Plug-In Hybrid Electric Vehicle Market Introduction Study*; U.S. Department of Energy: Washington, DC, USA, 2010.
[4] The Central People's Government of the People's Republic of China. Avaliable online: http://www.gov.cn/zwgk/2012-07/09/content_2179032.htm (accessed on 26 Janaury 2013).
[5] Su, W.C.; Rahimi-Eichi, H.; Zeng, W.T.; Chow, M.Y. A Survey on the electrification of transportation in a smart grid environment. *IEEE Trans. Ind. Inform.* **2012**, *8*, 1–10.
[6] Orecchini, F.; Santiangeli, A. Beyond smart grids—The need of intelligent energy networks for a higher global efficiency through energy vectors integration. *Int. J. Hydrog. Energy* **2011**, *36*, 8126–8133.
[7] Kempton, W.; Letendre, S.E. Electric vehicles as a new power source for electric utilities. *Transp. Res. Part D Transp. Environ.* **1997**, *2*, 157–175.

# High-precision parameters estimation of electric power harmonic on the basis of all phase technology

L.J. Li & Y. Li

*Beijing NARI SmartChip Microelectronics Company Limited, China*

ABSTRACT: For high-precise instruments of harmonic measuring and high-precise applications of harmonic measurement, hardware and software costs are very large to achieve approximately synchronous sampling to high-precise harmonic measurement. Combined with the characteristics of the signal in the grid, this article describes the data preprocessing in all phases and the parameter estimation of phase differences caused by time delay. After discrete Fourier transform/fast Fourier transform signal processing, it would achieve high-precise parameters estimation of electric power harmonic.

KEYWORDS: Spectrum leakage, All phase technology, Hanning window, discrete Fourier transform (DFT), fast Fourier transform (FFT)

## 1 THEORY OF DATA PREPROCESSING IN ALL PHASE

Assume the data length of signal vector $x = [x(n+N-1), ..., x(n), ..., x(n-N+1)]$ as $2N-1$, the preprocessing is used to transfer it to the required signal vector $x_1 = [x_1(0), x_1(1), ..., x_1(N-1)]$. The vector $x_1$ can be seen, to make the data weight for vector $X$ by convolution window $w_c$, and then to

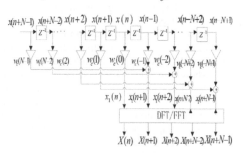

Figure 1. DFT/FFT spectrum analysis is based on data preprocessing in all phases.

To translate the data in the left side to the right of $N$ delay units, finally to add together with the data in the overlap, as shown in Figure 1.

$$w_c = \begin{cases} f(n)*b(-n) = \sum_k f(k)b(k+n)........n\in[-N+1,N-1] \\ 0........................N \le |n| \end{cases} \quad (1)$$

Convolution window $w_c$ is the convolution of front window $f$ and reversed back window $b$ (for details, see [1]).

In preprocessing, it is concerned with all segments with the length $N$, which contain the sample point $x(n)$. So this data preprocessing is called "data preprocessing in all phases."

Assume single-frequency complex exponential signal as

$$x(n) = e^{j(w_0 n + \varphi)} \quad (2)$$

where $w_0 = \beta 2\pi / N$ with $\beta$ as the real number.

So the discrete Fourier transform (DFT) formula of $x(n)$ is as follows:

$$X(k) = \frac{1}{N} \sum_{n=0}^{N-1} e^{j(w_0 n + \varphi_0)} e^{-j2\pi kn/N}$$

$$= \frac{1}{N} \cdot \frac{\sin[\pi(\beta - k)]}{\sin[\pi(\beta - k)/N]} \cdot e^{j(\varphi_0 + \frac{N-1}{N}(\beta - K)\pi)}$$

$$(k = 0, 1, ..., N-1) \quad (3)$$

After data preprocessing in all phases, the DFT formula is as follows:

$$X_{ap}(k) = \frac{1}{N} \sum_{i=0}^{N-1} X_i'(k)$$

$$= \frac{1}{N} \sum_{i=0}^{N-1} X_i(k) e^{j2\pi ik/N} \quad (4)$$

$$= \frac{e^{j\varphi_0}}{N^2} \cdot \frac{\sin^2[\pi(\beta - k)]}{\sin^2[\pi(\beta - k)/N]}$$

$$(k = 0, 1, ..., N-1)$$

The spectral amplitude of DFT after-data preprocessing in all phases, as shown (3) and (4), is the square of spectral amplitude of traditional DFT, that is, compared with the traditional DFT, spectral curve line of side-lobe leakage implements is the square of the decay. So DFT after-data preprocessing in all phases has better inhibition effect of spectrum leakage.

From (4) and (3), it can also be found that every line of the spectral curve of traditional DFT has relationship with its $\beta - k$, but the spectral phase of DFT after-data preprocessing in all phases, $\varphi_0$, is the theoretically spectral phase of the point x(n) of central sample, and has nothing to do with frequency deviation. DFT after-data preprocessing in all phases has invariant phase properties.

## 2 THEORY OF PARAMETER ESTIMATION OF PHASE DIFFERENCES CAUSED BY TIME DELAY

The phase and frequency are closely linked. According to DFT/fast Fourier transform (FFT) spectrum which can be calculated from two initial phases with time delay, $\varphi_1(k^*)$ and $\varphi_2(k^*)$. The phase differences of this signal of time delay is $\Delta\phi$ from $\omega^* = [\varphi_1(k^*) - \varphi_2(k^*)]/n_0 + 2k^*\pi/N$, and then the frequency of signal is obtained according to $\Delta\phi$ and time delay. Finally, Fourier transform function of the window segment which achieves the truncation of signal can precisely estimate the amplitude of signal. This is the basic theory of parameter estimation of phase differences caused by time delay.

In the spectral analysis by signal x(n) and $n_0$ delay signal, first, the difference from the phase of main spectral line is considered, then the frequent difference according to phase differences is obtained, finally, the frequency and amplitude are determined. As discussed in Section 6, DFT after-data preprocessing in all phases can achieve the phase of main spectral line, which is the point of the central sample, so delayed phase difference can be precisely estimated in the theory, further to make a basis for estimation of other parameters.

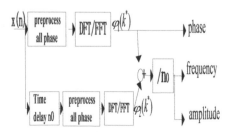

Figure 2. Parameter estimation of the phase difference caused by time delay.

Assume phase of signal x(n) is $\varphi_1(k^*)$, after $n_0$ delayed, it is $\varphi_2(k^*)$, so the difference of phase is $\Delta\phi$.

$$\Delta\phi = \phi_1(k^*) - \phi_2(k^*) + 2n_0 k^*\pi/N = \omega * n_0 \quad (5)$$

where $2n_0 k^*\pi/N$ is the phase compensation, so the estimation formula of digital angular frequency is as follows:

$$\hat{\omega}^* = [\varphi_1(k^*) - \varphi_2(k^*)]/n_0 + 2k^*\pi/N \quad (6)$$

Frequency deviation $\Delta\omega$ of main spectral line is obtained:

$$\Delta\omega = \hat{\omega}^* - 2k^*\pi/N = [\varphi_1(k^*) - \varphi_2(k^*)]/n_0. \quad (7)$$

Then, it is easy to estimate the signal amplitude, and the formula is as follows:

$$\hat{A} = \frac{|Y(k^*)|}{F_g^2(\Delta\omega)} \quad (8)$$

where $F_g(\omega)$ is the Fourier transform function of window segment which achieves the truncation of signal.

## 3 INHIBITION OF SPECTRUM LEAKAGE

Periodic signal must be truncated first, followed by DFT/FFT Digital Systems. For the signal that cannot be truncated in the integral period, spectrum leakage is inevitable. The technology of all phases is concerned with all possible situations of points of the input central sample and equivalent to smoothing the truncation, so to some extent the side-lobe leakage is inhibited. Window sequence leads to the side lobe. Figure 3 shows the Fourier transform spectrum of rectangular windows and hanning windows. It is obvious that side lobe of rectangular window is bigger than that of hanning window; consequently, the truncation of signal by using rectangular window will lead to bigger side-lobe leakage.

If the processing signal is only a single-frequency signal, the calibration (e.g., the difference of phase described earlier) can be used for a precise estimation of signal parameters; however, in addition to the fundamental signal there exists a lot of harmonic signals in the grid. This requires us to use the phase delay time parameter estimation method for harmonic, at the same time, it is necessary to choose the window function whose side lobe has small amplitude and fast decay. The purpose is to minimize interference between adjacent spectral lines to improve estimation of the accuracy of phase differences in main spectral line. This is basic theory for estimation of harmonic parameters. The performance of the common window is summarized in [3] which provides a basis for the selection of the window.

The Hanning window with good side-lobe attenuation characteriwstic, MATLAB provides two hanning windows—hann(N, 'symmetric'), hann(N, 'periodic')—and their spectrum is shown in Figures 3 and 4.

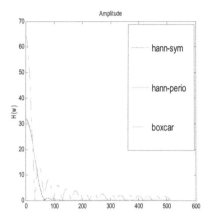

Figure 3. Amplitude spectrum comparison [hanningsym: hann(N, 'symmetric'), hanningperio: hann(N, 'priodic')].

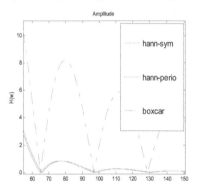

Figure 4. Partial enlarged amplitude spectrum.

From Figures 3 and 4, in hann(N, 'periodic') and hann(N, 'symmetric') the width of the main lobe is twice that of the rectangular window; hence, at the same point DFT, both sides of the main spectral line in main lobe will appear a vice line, however, with respect to the rectangular windows they have good characteristics of the side-lobe attenuation. In noninteger period, the main impact of spectrum leakage is the vice main spectral line in main lobe; however, in process of harmonic analysis with DFT/FFT, the space of adjacent main spectral lines has at least one spectral line. If concerned with the impact of 2 or 3 side lobes, among the main spectral lines there should be at least three spectral lines, which can ensure good accuracy.

Because hann(N, 'periodic') with rectangular windows has the same amplitude spectrum over the zero position (except for the first over the zero position in the rectangular window), it complies with the law of small leaks in the side lobes when the main lines has smaller deviation. So it is more suitable for combining with DFT/FFT. Approximate formula of the Hanning window (8) and substituting (7) can estimate amplitude of harmonics:

$$F_g(\omega) = \sin(\pi w)/(2\pi(1-w^2))  \qquad (9)$$

## 4 SIMULATION OF PARAMETERSESTIMATION ON THE BASIS OF MATLAB

Assume Power fundamental frequency is 50 Hz and amplitude is 220 V. 21 times less than harmonic measurement is necessary to meet the request that the accuracy is 0.1% when harmonic is 10%.

The program is as follows: because it is necessary to measure 21 harmonics, the signal bandwidth is $w_c = 50 \times 21 = 1050$ Hz. From the Nyquist sampling theorem, the sampling rate must be greater than 2.1 kHz, so sampling rate $f_s$ selected as 3.2 kHz. $N$ in DFT is 320, so DFT frequency resolution $\Delta f$ is 10Hz, that is, the main spectral line of harmonic spaces 4. It uses data preprocessing in all phases based on the hanning window and estimates electric power harmonic from Figure 2, $n_0 = N = 320$.

Assume signal in grid:

$$s(n)=220\cos(2\pi f_1 n/f_s + \pi/3) + 22\cos(2\pi f_2 n/f_s + \pi/6)$$
$$+2.2\cos(2\pi f_3 n/f_s + \pi/2) + 0.22\cos(2\pi f_5 n/f_s + \pi/4)$$

MATLAB simulations are on the basis of data in Tables 1–3:

As illustrated in Table 1, when the fundamental frequency is 0.001 Hz, the fundamental side-lobe leakage is smaller and the harmonic parameter estimation precision is above $10^{-11}$.

As illustrated in Table 2, frequency offset is 5 Hz, the most serious spectrum leakage appears at this time, the second harmonic is influenced mainly by the effects of the most fundamental side-lobe leakage, in this case, the measurement accuracy is lower than other harmonic measurement accuracy, but still can achieve $10^{-5}$ accuracy.

As illustrated in Table 3, frequency offset is 0.5 Hz, this time for the power grid generally allows deviation, in this case, it is significantly better than the most serious leaks, and error precision is improved by one order of magnitude to $10^{-6}$.

Table 1. MATLAB simulation data of small side-lobe leakage of fundamental signal (phase/frequency/amplitude units: °/Hz/V).

| Signal | Parameter | Calculated value | Real value | Tolerance |
|---|---|---|---|---|
| 2 times harmonic | Phase | 30.00000000022940 | 30.0 | 7.646689565174587e-12 |
| | Frequency | 100.0020000000000 | 100.002 | 0 |
| | Amplitude | 22.00000000013925 | 22.0 | 6.329643879739293e-12 |
| 5 times harmonic | Phase | 45.00000000003357 | 45.0 | .458047333067340e-13 |
| | Frequency | 250.0049999999999 | 250.005 | −5.684228201516771e-16 |
| | Amplitude | 0.22000000000026 | 0.22 | .182766006347823e-12 |

Table 2. MATLAB simulation data of the biggest side-lobe leakage of fundamental signal (phase/frequency/amplitude units: °/Hz/V).

| Signal | Parameter | Calculated value | Real value | Tolerance |
|---|---|---|---|---|
| 2 times harmonic | Phase | 30.00113803178707 | 30.0 | 3.793439290233812e-5 |
| | Frequency | 109.9999367738558 | 110.0 | −5.747831290198684e-7 |
| | Amplitude | 22.00033290376860 | 22.0 | 1.513198948188688e-5 |
| 5 times harmonic | Phase | 45.00005787228557 | 45.0 | 1.286050790333039e-6 |
| | Frequency | 275.0 | 275.0 | 0 |
| | Amplitude | .22000046328926 | 0.22 | 2.105860275426374e-6 |

Table 3. MATLAB simulation data of side-lobe leakage in the situation of maximum deviation in grid (phase/frequency/amplitude units: °/Hz/V).

| Signal | Parameter | Calculated value | Real value | Tolerance |
|---|---|---|---|---|
| 2 times harmonic | Phase | 30.00005910824569 | 30.0 | 1.970274856609593e-6 |
| | Frequency | 100.9999991606607 | 101.0 | −8.310290453583449e-9 |
| | Amplitude | 22.00003622995033 | 22.0 | 1.646815924193458e-6 |

In Tables 1–3, along with the increase of harmonic parameters, the estimation accuracy of side-lobe leakage is low, but the estimation accuracy is still high.

## 5 STORAGE SPACE AND COMPUTATIONAL COMPLEXITY

*Storage:* the analysis of $N$ data space is according to quasi-synchronous harmonic, the phase delay of phase differences use $3N$ - 1 data space.

*Computation analysis:* the phase shift and phase differences will be synchronous in harmonic analysis, with their capacity mainly concentrated on the two DFT/FFT calculations. Currently in the market, most of the DSP processors calculate the 1024-point FFT in the level below 1 ms.

## 6 CONCLUSION

Signals of the electric power harmonic are shown in Figure 2 of all phase preprocessing, and the estimation accuracy of the harmonic parameter can be reached at a very high level. Thus, the storage space and the measurement cost are smaller in the current processor easily. Currently, the grid allows frequency deviation within 0.5 Hz, and the accuracy of the parameter estimation goes up. Therefore, it is suitable for power system to achieve high accuracy in harmonic measurement.

## REFERENCES

[1] Wang Zhaohua, Huang Xiangdong. Digital signal of all phase spectrum analysis and [M]. Filter Technology Electronics Industry Press,. 2009. 2.
[2] Ding Kang, LuoJiangkai, Xie Ming. Discrete spectrum shift phase difference correction method [J]. Applied Mathematics and mechanics, 2002. 23. (7):729–735.
[3] Huang Xiangdong, Wang Zhaohua. The all phase spectrum analysis phase difference correction method based on [J]. Journal of electronics and information technology, 2008, 30 (2):293–297.

*Multimedia, Communication and Computing Application – Leung (Ed.)*
© 2015 Taylor & Francis Group, London, ISBN 978-1-138-02775-6

# Coordination of multi-agent systems with high-order dynamic

N. Na, Y.X. Zhang & T. Wang
*Beijing Institute of Aerospace Control Device, Beijing, China*

ABSTRACT: In this paper, we investigate the coordination of high-order multi-agent systems, where each agent can only access to the relative position information from its neighbors. Several conditions are derived to make all agents asymptotically reach consensus. Finally, simulation results are given to show the effectiveness of the obtained results.

KEYWORD: Coordination, Multi-agent systems, High-order dynamics

## 1 GENERAL INSTRUCTIONS

Coordination problems in multi-agent systems have received considerable attentions from many fields such as physics, biology, control theory, robotics and computer science. During the past decade, numerous results have been obtained for coordination problems [1]–[7]. For example, in [1], Jadbabaie et al. focus on attitude alignment for network of agents with an undirected graph in which each agent has a discrete-time integrator dynamics. In [2], Olfati-Saber et al. investigate a system atical framework of consensus problems with directed interconnection graphs or time-delays by a Lyapunov-based approach. Also, Ren et al. [5] and Lin et al. [6] propose several dynamic consensus algorithms for second-order multi-agent systems and derived sufficient conditions for state consensus of the system.

In this paper, our motivation is to extend the work of [6] to high-order multi-agent systems. A new feedback dynamic neighbor-based protocol is introduced which only uses relative information of the first states of agents. Sufficient conditions are derived as all agents reach consensus asymptotically. Different from the existing protocols in [5], our protocol does not need any information except the relative information of the first states of agent.

The following notations will be used throughout this paper. $\mathbb{R}^n$ denotes the set of all n dimensional real column vectors; $\mathbb{Z}_+$ denotes positive integer; $\mathbf{I}_n$ denotes the n dimensional identity matrix; we use $\mathbf{1}_n$ to denote 1 with dimension n; 0 denotes zero value or zero matrix with an appropriate dimension; $\varphi_n = \{1, 2, \cdots, n\}$ denotes index set; $\otimes$ denotes the kronecker product; * denotes the symmetric term of a symmetric matrix.

## 2 PROBLEM FORMULATION

Suppose that the multi-agent systems under consideration consist of n agents, each agent can be regarded as a node of directed graph G. The dynamics of the ith agent is described as follows.

$$
\begin{aligned}
\dot{x}_i^1(t) &= x_i^2(t) \\
\dot{x}_i^2(t) &= x_i^3(t) \\
&\vdots \\
\dot{x}_i^{m-1}(t) &= x_i^m(t) \\
\dot{x}_i^m(t) &= u_i(t)
\end{aligned}
\tag{1}
$$

where $x_i^j(t) \in \mathbb{R}$ ($j = \varphi_m$ )is the jth component of $\chi_i$, $\chi_i = \left[ x_i^1, x_i^2, \cdots, x_i^m \right]^T$ ($i = \varphi_n$) is the state of agent i, and $u_i(t) \in \mathbb{R}$ is the control input.

We can say the protocol $u_i$ asymptotically solves the coordination problem, i.e., the agreement of the position states and the synchronization of the speed states, if and only if the states of agents satisfy:

$$
\lim_{t \to \infty} [\chi_i(t) - \chi_j(t)] = 0, \ i, j \in I.
\tag{2}
$$

Lemma 1.[2] If the undirected graph G is connected, then its Laplacian L satisfies:

(1) Rank(L)=n-1;
(2) Zero is an eigenvalue of L, and 1n is the corresponding eigenvector;
(3) The rest n-1 eigenvalues of L all have positive and real.

Lemma 2.[8]. Let n be a positive integer and let M(s) be a stable polynomial of degree n-1:

$M(s) = m_0 + m_1 s + \cdots + m_{n-1} s^{n-1}$ with all $m_i > 0$. Then there exists an $\alpha > 0$ such that $N(s) = M(s) + m_n s^n$ is stable if and only if $m_n \in [0, \alpha)$.

Lemma 3.[6] Consider the matrix $\Psi_n = n\mathbf{I}_n - \mathbf{1}_n \mathbf{1}_n^T$. Then there exists an orthogonal matrix $U_n \in \mathbb{R}^{n \times n}$, so $U_n^T \Psi_n U_n = \text{diag}\{n\mathbf{I}_{n-1}, 0\}$ and the last column of U is $\mathbf{1}_n / \sqrt{n}$. Given a matrix $\Upsilon \in \mathbb{R}^{n \times n}$, so $\mathbf{1}_n^T \Upsilon = 0$ and $\Upsilon \mathbf{1}_n = 0$, then $U_n^T \Upsilon U_n = \text{diag}\{\bar{U}_n^T \Upsilon \bar{U}_n, 0\}$, and $\bar{U}_n$ denotes the first n-1 columns of U. For convenience of discussion, denote $U = \begin{bmatrix} U_1 & \bar{U}_1 \end{bmatrix}$, where $\bar{U}_1 = 1/\sqrt{n}$ is the last column of U and U1 is the rest part.

## 3 CONTROL PROTOCOL

To find a suitable distributed state feedback controller for each agent, we take the following control protocol:

1) Zero communication time-delay

$u_i(t) = -k_1 x_i^2(t) - k_2 x_i^3(t) - \ldots - k_{m-1} x_i^m(t) + k_m \Sigma_{s_j \in N_i}$

$\qquad f_{ij}(x_j^1(t) - x_i^1(t)) - p_i$

$\dot{p}_i(t) = -\gamma p_i - \Sigma_{s_j \in N_i} f_{ij}(x_j^1(t) - x_i^1(t))$ \hfill (3)

2) Nonzero communication time-delay

$u_i(t) = -k_1 x_i^2(t) - k_2 x_i^3(t) - \ldots - k_{m-1} x_i^m(t) + k_m \Sigma_{s_j \in N_i} f_{ij}$

$\qquad (x_j^1(t - \tau_{ij}) - x_i^1(t - \tau_{ij})) - p_i$

$\dot{p}_i(t) = -\Sigma_{s_j \in N_i} f_{ij}(x_j^1(t - \tau_{ij}) - x_i^1(t - \tau_{ij})) - \gamma p_i$

\hfill (4)

where $\gamma > 0, k_i > 0 (i = \varphi_{m-1}), k_m = 1$, $p_i > 0$ is an auxiliary variable with $p_i(0) = 0$ $(i = \varphi_n)$ $\tau_{ij}(t) \in \mathbb{Z}_+ (i \neq j)$ denotes the communication time-delay and satisfies $\tau_{ij} = \tau_{ji}$.

Denote $\xi = [\chi_1^T, p_1, \chi_2^T, p_2, \ldots, \chi_n^T, p_n]^T$,

$A = \begin{bmatrix} A_{11} & A_{12} \\ A_{21} & A_{22} \end{bmatrix}_{(m+1) \times (m+1)}$, $B = \begin{bmatrix} B_1 & B_2 \end{bmatrix}_{(m+1) \times (m+1)}$

where $A_{11} = \begin{bmatrix} 0 & \mathbf{I}_{(m-1)} \\ 0 & k \end{bmatrix} \in \mathbb{R}^{m \times m}$,

$k = \begin{bmatrix} -k_1 & -k_2 & \cdots & -k_{m-1} \end{bmatrix}$, $A_{12} = \begin{bmatrix} \mathbf{0}_{(m-1) \times 1} \\ -1 \end{bmatrix}$,

$A_{21} = \mathbf{0}_{1 \times m}$, $A_{22} = -\gamma$, $B_1 = \begin{bmatrix} 0 & \cdots & 0 & k_m & 1 \end{bmatrix}^T$,

$B_2 = \mathbf{0}_{(m+1) \times m}$.

Using protocol (3), the network dynamics is

$\dot{\xi}(t) = (\mathbf{I}_n \otimes A - L \otimes B)\xi(t)$. \hfill (5)

Similarly, using protocol (4) the network dynamics is

$$\dot{\xi}(t) = (\mathbf{I}_n \otimes A)\xi(t) - \sum_{m=1}^{M} (L_m \otimes B)\xi(t - \tau_m),$$ \hfill (6)

where $M \leq n(n-1)/2$, $m = \varphi_M$, $\tau_m \in \tau_{ij} : i, j = \varphi_n$.

It is easy to see that $L_m = L_m^T$, $L_m \mathbf{1}_n = 0$, and $\sum_{m=1}^{M} L_m = L$.

Lemma 4. Let $\beta(t) = e^{A_{11} t} \begin{bmatrix} \beta^1 & \beta^2 & \cdots & \beta^m \end{bmatrix}^T$ where $\beta^j = \frac{1}{n} \sum_{i=1}^{n} x_i^j(0), j = 1, \cdots, m$, $\delta(t) = \xi(t) - \mathbf{1}_n \otimes \begin{bmatrix} \beta^T(t) & 0 \end{bmatrix}^T$. Then $(\mathbf{1}_n^T \otimes \mathbf{I}_{m+1})\delta(t) = 0$. Moreover, the system is equivalent to

$$\dot{\delta}(t) = (\mathbf{I}_n \otimes A - L \otimes B)\delta(t)$$ \hfill (7)

And

$$\dot{\delta}(t) = (\mathbf{I}_n \otimes A)\delta(t) - \sum_{m=1}^{M} (L_m \otimes B)\delta(t - \tau_m)$$ \hfill (8)

Proof: Since G is undirected, it follows from eqn (5) that $\sum_{i=1}^{n} \dot{p}_i(t) = A_{22} \sum_{i=1}^{n} p_i(t)$. Because $p_i = 0$ for any i, we have $\sum_{i=1}^{n} \dot{p}_i(t) = 0$. Similarly, it is easy to see that

$$\dot{X}(t) = A_{11} X(t)$$ \hfill (9)

where

$$X(t) = \begin{bmatrix} \frac{1}{n} \sum_{i=1}^{n} x_i^1(t) & \frac{1}{n} \sum_{i=1}^{n} x_i^2(t) & \cdots & \frac{1}{n} \sum_{i=1}^{n} x_i^m(t) \end{bmatrix}^T.$$

416

So, $X(t) = e^{A_{11}t} \begin{bmatrix} \beta^1 & \beta^2 & \cdots & \beta^m \end{bmatrix}^T = \beta(t)$.

Evidently, $(1_n^T \otimes I_{m+1})\delta(t) = 0$. Since $\delta(t) = \xi(t) - 1_n \otimes \begin{bmatrix} \beta^T(t) & 0 \end{bmatrix}^T$ and eqn (5), we have

$$\dot{\delta}(t) = (I_n \otimes A - L \otimes B)\xi(t) - 1_n \otimes \begin{bmatrix} \dot{\beta}^T(t) & 0 \end{bmatrix}^T \quad (10)$$

By Lemma 1, we see that $L1_n = 0$. So

$$(I_n \otimes A - L \otimes B)1_n \otimes \begin{bmatrix} \beta^T(t) & 0 \end{bmatrix}^T$$
$$= 1_n \otimes \begin{bmatrix} (A_{11}\beta(t))^T & 0 \end{bmatrix}^T \quad (11)$$

We also have

$$1_n \otimes \begin{bmatrix} \dot{\beta}^T(t) & 0 \end{bmatrix}^T = 1_n \otimes \begin{bmatrix} (A_{11}\beta(t))^T & 0 \end{bmatrix}^T \quad (12)$$

So, eqn (5) is equivalent to eqn (7). Similarly, by simple computation, eqn (6) is equivalent to eqn (8). This completes the proof.

## 4 MAIN RESULTS

**Theorem 1.** Consider a network of high-order agents with a fixed topology G that is connected. Given the protocol (3), the multi-agents system (5) can reach consensus if $k_1 > 1$ and there exist $\alpha_i > 0, i = 2,3,\cdots,m$ such that $\gamma k_i + k_{i-1} \in [0, \alpha_i)$.

Proof: By Lemma 1, we can denote the eigenvalues of L as $0 = \lambda_1 < \lambda_2 \leq \cdots \leq \lambda_n$. There exists an orthogonal matrix $W \in \mathbb{R}^{n \times n}$ such that $W^T L W = diag\{0, \lambda_2, \cdots, \lambda_n\}$. It follows that

$$(W^T \otimes I_{m+1})(I_n \otimes A - L \otimes B)(W \otimes I_{m+1})$$
$$= diag\{A, A - \lambda_2 B, L, A - \lambda_n B\} \quad (13)$$

By Lemma 1 again, we see that the first column of W is $1/\sqrt{n}$. Let $\bar{W}$ denote the rest n-1 columns of W and $\bar{\delta}(t) = (\bar{W} \otimes I_{m+1})^T \delta(t)$. By Lemma 2, $\delta^T(t)(W \otimes I_{m+1}) = [0_{1 \times (m+1)} \bar{\delta}^T(t)]$. Thus,

$$(W \otimes I_{m+1})^T \dot{\delta}(t) = diag\{A, A - \lambda_2 B, \cdots, A - \lambda_n B\}$$
$$(W \otimes I_{m+1})^T \delta(t) \quad (14)$$

It is easy to see that eqn (5) is equivalent to

$$\dot{\bar{\delta}}(t) = F\bar{\delta}(t) \quad (15)$$

where $\Phi = diag\{A - \lambda_2 B, \cdots, A - \lambda_n B\}$.
The characteristic polynomials for $A - \lambda_i B$ are

$$N_i(s) = s^{m+1} + (\gamma k_m + k_{m-1})s^m + (\gamma k_{m-1} + k_{m-2})s^{m-1}$$
$$+ \cdots + (\gamma k_2 + k_1)s^2 + (\gamma k_1 + \lambda_i)s + \lambda_i + \gamma\lambda_i \quad (16)$$

When m=2, the characteristic polynomials for $A - \lambda_i B$ are

$$N_i(s) = s^3 + (\gamma + k_1)s^2 + (\gamma k_1 + \lambda_i)s + \lambda_i + \gamma\lambda_i \quad (17)$$

We can see that $\gamma + k_1 > 0, \gamma k_1 + \lambda_i > 0, \lambda_i + \gamma\lambda_i > 0$. And the Hurwitz determinant composed of the characteristic polynomials coefficients is

$$\begin{bmatrix} \gamma + k_1 & \lambda_i + \gamma\lambda_i & 0 \\ 1 & k_1 + \lambda_i & 0 \\ 0 & \gamma + k_1 & \lambda_i + \gamma\lambda_i \end{bmatrix} = (\lambda_i + \gamma\lambda_i)(\gamma^2 k_1 + k_1\gamma^2 + k_1\lambda - \lambda)$$

Then the second-order sequential principal minor is $\gamma^2 k_1 + k_1\gamma^2 + k_1\lambda - \lambda$. In terms of Hurwitz stability criterion, all agents asymptotically reach consensus if $k_1 > 1$. Then using Lemma 2, if there exist $\alpha_i > 0, i = 2,3,\cdots,m$ such that $\gamma k_i + k_{i-1} \in [0, \alpha_i)$, then $N_i(s)$ is stable. It is easy to see that $\lim \bar{\delta}(t) = 0$, i.e., $\lim \delta(t) = 0$. Therefore, all agents asymptotically reach consensus. This completes the proof.

**Theorem 2.** Consider a network of high-order agents with a fixed topology G that is connected. Given the protocol (4), the multi-agents system (6) can reach consensus if there exist symmetric matrices $0 < \bar{P} \in \mathbb{R}^{(m+1)(n-1) \times (m+1)(n-1)}$, $0 < \bar{Q}_m \in \mathbb{R}^{(m+1)(n-1) \times (m+1)(n-1)}$, $0 < \bar{R}_m \in \mathbb{R}^{(m+1)n \times (m+1)n}$ such that

$$J = \begin{bmatrix} J_{11} & J_{12} & J_{13} \\ * & J_{22} & 0 \\ * & * & J_{33} \end{bmatrix} < 0 \quad (18)$$

where

$$J_{11} = \bar{P}(I_{n-1} \otimes A - \bar{L} \otimes B) + (I_{n-1} \otimes A - \bar{L} \otimes B)^T \bar{P}$$
$$+ \sum_{m=1}^{M} \bar{Q}_m + \sum_{m=1}^{M} \tau_m (U_1 \otimes A)^T R_m (U_1 \otimes A)$$

417

$$J_{12} = -\sum_{m=1}^{M} \tau_m (U_1 \otimes A)^T R_m \Omega$$

$$J_{13} = \begin{bmatrix} \tau_1 \bar{P}(\bar{L}_1 \otimes B) & \cdots & \tau_M \bar{P}(\bar{L}_M \otimes B) \end{bmatrix}$$

$$J_{22} = -diag\{\bar{Q}_1, \bar{Q}_2, \ldots, \bar{Q}_M\} + \sum_{m=1}^{M} \tau_m \bar{\Gamma}^T \times (U_1 \otimes \mathbf{I}_3)^T$$
$$R_m (U_1 \otimes \mathbf{I}_3) \bar{\Gamma}$$

$$J_{33} = -diag\{\tau_1(U_1 \otimes \mathbf{I}_{m+1}) R_1 (U_1 \otimes \mathbf{I}_{m+1}), \ldots,$$
$$\tau_M (U_1 \otimes \mathbf{I}_{m+1}) R_M (U_1 \otimes \mathbf{I}_{m+1})\}$$

$$\Omega = \begin{bmatrix} (U_1 \bar{L}_1) \otimes B & (U_1 \bar{L}_2) \otimes B & \cdots & (U_1 \bar{L}_M) \otimes B \end{bmatrix}$$

$$\bar{\Gamma} = \begin{bmatrix} \bar{L}_1 \otimes B & \bar{L}_2 \otimes B & \cdots & \bar{L}_M \otimes B \end{bmatrix} \text{ with}$$
$$\bar{L} = U_1^T L U_1, \bar{L}_m = U_1^T L_m U_1.$$

Proof: Define a Lyapunov function for the system (6) as follows

$$V = \delta^T(t) P \delta(t) + \sum_{m=1}^{M} \int_{t-\tau_m}^{t} \delta^T(s) Q_m \delta(s) ds$$
$$+ \sum_{m=1}^{M} \int_{-\tau_m}^{0} \int_{t+\theta}^{t} \dot{\delta}^T(s) R_m \dot{\delta}(s) ds d\theta \qquad (19)$$

where $P = P^T \geq 0 \in \mathbb{R}^{(m+1)n \times (m+1)n}$, $Q_m = Q_m^T \geq 0 \in \mathbb{R}^{(m+1)n \times (m+1)n}$, $R_m = R_m^T \in \mathbb{R}^{(m+1)n \times (m+1)n}$, $(U_1 \otimes \mathbf{I}_{m+1})^T R_m (U_1 \otimes \mathbf{I}_{m+1}) > 0$ $P(\mathbf{1}_n \otimes \mathbf{I}_{m+1}) = Q_m(\mathbf{1}_n \otimes \mathbf{I}_{m+1}) = 0$ and $rankP = rankQ_m = (m+1)(n-1)$.

The time derivative of V is

$$\dot{V} = 2\delta^T(t) P(\mathbf{I}_n \otimes A) \delta(t) - 2\delta^T(t) P \sum_{m=1}^{M} (L_m \otimes B) \delta(t - \tau_m) +$$
$$\sum_{m=1}^{M} \delta^T(t) Q_m \delta(t) - \sum_{m=1}^{M} \delta^T(t - \tau_m) Q_m \delta(t - \tau_m) +$$
$$\sum_{m=1}^{M} \tau_m \dot{\delta}^T(t) R_m \dot{\delta}(t) + \sum_{m=1}^{M} \int_{t-\tau_m}^{t} \dot{\delta}^T(s) R_m \dot{\delta}(s) ds$$

(20)

Then by Newton-Leibniz formula, we have

$$-2\delta^T(t) P(L_m \otimes B) \delta(t - \tau_m) = -2\delta^T(t) P(L_m \otimes B) \delta(t)$$
$$+ \int_{t-\tau_m}^{t} 2\delta^T(s) P(L_m \otimes B) \dot{\delta}(s) ds$$

for $m = 1, 2, \ldots, M$. And since $\sum_{m=1}^{M} L_m = L$, we have

$$\dot{V} = 2\delta^T(t) P(\mathbf{I}_n \otimes A - L \otimes B) \delta(t) + \int_{t-\tau_m}^{t} 2\delta^T(s)$$
$$P(L_m \otimes B) \dot{\delta}(s) ds + \sum_{m=1}^{M} \delta^T(t) Q_m \delta(t)$$
$$- \sum_{m=1}^{M} \delta^T(t - \tau_m) Q_m \delta(t - \tau_m) + \sum_{m=1}^{M} \tau_m \dot{\delta}^T(t) R_m \dot{\delta}(t)$$
$$+ \sum_{m=1}^{M} \int_{t-\tau_m}^{t} \dot{\delta}^T(s) R_m \dot{\delta}(s) ds$$
$$= \frac{1}{\prod_{m=1}^{M} \tau_m} \int_{t-\tau_1}^{t} \cdots \int_{t-\tau_M}^{t} \eta^T(s) \Phi \eta(s) ds_M \cdots ds_1$$

(21)

where $\eta^T(s) = [\delta^T(s), \delta^T(s - \tau_1), \cdots, \delta^T(s - \tau_M),$
$\dot{\delta}^T(s_1), \cdots, \dot{\delta}^T(s_M)]$

$$\Phi = \begin{bmatrix} \Phi_{11} & \Phi_{12} & \Phi_{13} \\ * & \Phi_{22} & 0 \\ * & * & \Phi_{33} \end{bmatrix}$$

$$\Phi_{11} = P(\mathbf{I}_n \otimes A - L \otimes B) + (\mathbf{I}_n \otimes A - L \otimes B)^T P$$
$$+ \sum_{m=1}^{M} Q_m + \sum_{m=1}^{M} \tau_m (\mathbf{I}_n \otimes A)^T R_m (\mathbf{I}_n \otimes A),$$

$$\Phi_{12} = -\sum_{m=1}^{M} \tau_m (\mathbf{I}_n \otimes A)^T R_m \Gamma,$$

$$\Phi_{13} = \begin{bmatrix} \tau_1 P(L_1 \otimes B) & \cdots & \tau_M P(L_M \otimes B) \end{bmatrix},$$

$$\Phi_{22} = \sum_{m=1}^{M} \tau_m \Gamma^T R_m \Gamma - diag\{Q_1, Q_2, \cdots, Q_M\}$$

and $\Phi_{33} = -diag\{\tau_1 R_1, \tau_2 R_2, \cdots, \tau_M R_M\}$ with
$\Gamma = \begin{bmatrix} L_1 \otimes B & L_2 \otimes B & \cdots & L_M \otimes B \end{bmatrix}$.

Let $H = \mathbf{I}_{2M+1} \otimes (U_1 \otimes \mathbf{I}_{m+1}), \hat{\delta}(t) = (U_1 \otimes \mathbf{I}_{m+1}) \delta(t)$. Since $L_m = L_m^T$ $L_m \mathbf{1} = 0$ $m = 1, 2, \ldots, M$, it follows that

$$\eta^T(s) H H^T \Phi H H^T \eta(s) = \hat{\eta}^T(s) J \hat{\eta}(s) \qquad (22)$$

where $\hat{\eta}^T(t, s) = [\hat{\delta}^T(s), \hat{\delta}^T(s - \tau_1), \cdots, \hat{\delta}^T(s - \tau_M),$
$\hat{\delta}^T(s_1), \cdots, \hat{\delta}^T(s_M)]$ and J is as defined in (18) with
$\bar{P} = (U_1 \otimes \mathbf{I}_{m+1})^T P(U_1 \otimes \mathbf{I}_{m+1}) > 0$, $\bar{Q}_m = (U_1 \otimes \mathbf{I}_{m+1})^T$
$Q_m(U_1 \otimes \mathbf{I}_{m+1}) > 0$, $m = 1, 2, \ldots, M$.

It is easy to see that $\bar{P} > 0, \bar{Q} > 0$. Note that V can be rewritten as

$$V = \hat{\delta}^T(t)\bar{P}\hat{\delta}(t) + \sum_{m=1}^{M} \int_{t-\tau_m}^{t} \hat{\delta}^T(s)\bar{Q}_m\hat{\delta}(s) \tag{23}$$
$$+ \sum_{m=1}^{M} \int_{-\tau_m}^{0} \int_{t+\theta}^{t} \hat{\delta}^T(s)R_m\hat{\delta}(s)dsd\theta$$

Then, eqn (18) guarantees $\dot{V} < 0$ and by Lyapunov theory, we have $\lim_{t\to\infty}\hat{\delta}(t) = 0$. This implies $\lim_{t\to\infty}\delta(t) = 0$. That is, eqn (6) can reach consensus under eqn (18). This completes the proof.

## 5   SIMULATION EXAMPLES

We verify the correctness of the method through the simulation examples. Fig. 1 shows a graph with n=4 nodes. Suppose that the weight of each edge is 1, each agent has second-order dynamics.

Figure 1.   The network topology.

The Laplacian with respect to the graph G

$$L = \begin{bmatrix} 1 & -1 & 0 & 0 \\ -1 & 2 & -1 & 0 \\ 0 & -1 & 2 & -1 \\ 0 & 0 & -1 & 1 \end{bmatrix}.$$

There are 3 nonzero eigenvalues, which are $\lambda = \{0.5858, 2, 3.4142\}$. From Theorem 1, we choose $\gamma = 1, k = \{k_1, k_2\} = \{1.5, 1\}$. The initial condition is set as

$$x_1(0) = \begin{bmatrix} 6 \\ 2 \end{bmatrix}, x_2(0) = \begin{bmatrix} -3 \\ 5 \end{bmatrix}, x_3(0) = \begin{bmatrix} -4 \\ 3 \end{bmatrix},$$
$$x_4(0) = \begin{bmatrix} 4 \\ 5 \end{bmatrix}$$

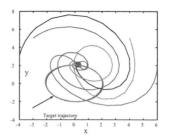

Figure 2.   Consensus of second-order multi-agent systems.

Suppose that the weight of each edge is 1 and each agent has three-order dynamics. The initial condition is set as

$$x_1(0) = \begin{bmatrix} 4 \\ 1 \\ -4 \end{bmatrix}, x_2(0) = \begin{bmatrix} -4 \\ 6 \\ 3 \end{bmatrix}, x_3(0) = \begin{bmatrix} -5 \\ 2 \\ 7 \end{bmatrix},$$
$$x_4(0) = \begin{bmatrix} 5 \\ -7 \\ 2 \end{bmatrix}$$

By Theorem 1, we choose $\gamma = 2, k = \{k_1, k_2, k_3\} = \{3, 2, 1\}$ and we can find $\alpha_2 = 8$ to satisfy the condition of Theorem 1. Clearly, all agents reach consensus, see Fig 3.

For the case of time-delay, we first assume that all time-delays are equal $\tau = 0.3$. Fig. 4-6 show the states trajectories of the network using eqn (4) with $(k_1, k_2, k_3, \gamma) = (5, 2, 1, 2)$ which satisfies eqn (24). Fig. 7-9 show the states trajectories of the network with multiple communication time-delays $(0.3, 0.5, 0.2, 0.1, 0.2, 0.4)$.

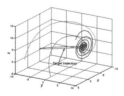

Figure 3.   Consensus of third-order multi-agent systems.

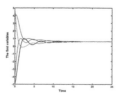

Figure 4.   The first variables with single time-delay.

419

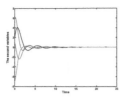

Figure 5.   The second variables with single time-delay.

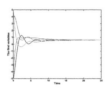

Figure 6.   The third variables with single time-delay.

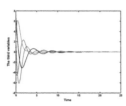

Figure 7.   The first variables with multiple time-delays.

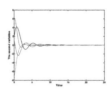

Figure 8.   The second variables with multiple time-delays.

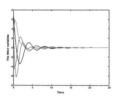

Figure 9.   The third variables with multiple time-delays.

# 6   CONCLUSIONS

In this paper, we investigate the consensus of high-order multi-agent systems. Sufficient conditions are derived under which all agents reach consensus asymptotically. Finally, simulation results are included to show the effectiveness of the obtained results.

## REFERENCES

[1] Jadbabaie, A., Lin, J. & Morse, S.A., Coordination of groups of mobile agents using nearest neighbor rules, IEEE Transaction on Automatic Control, Vol. 48, pp. 988–1001, 2003.
[2] O-Saber, R. & Murray, R.M., Consensus problems in net-works of agents with switching topology and time-delays, IEEE Transaction on Automatic Control, 49(9), pp. 1520–1533, 2004.
[3] Lawton, J.R.T., Beard, R.W. & Young, B.J., A decentralized approach to formation maneuvers, IEEE Transaction on Robotics Automat, Vol. 19, pp. 933–941, 2003.
[4] Mesbahi, M. & Hadegh, F.Y. On consensus algorithms for double-integrator dynamics, AIAA Journal of Guidance, Control, and Dynamics, 24(2), pp. 369–377, 2000.
[5] Ren, W., Distributed multi-vehicle coordinated control via local information exchange, Proceedings of the 46th IEEE Conference on Decision and Control New Orleans, pp. 2295–2300, 2007.
[6] Lin, P. & Jia, Y.M., Further results on decentralized coordination in networks of agents with second-orser dynamics, IET,Control Theory and Applications, 3(7), pp. 957–970, 2009.
[7] Ren, W., Decentralization of Coordination Variables in Multi-vehicle Systems, ICNSC '06, Proceedings of the 2006 IEEE International Conference on Networking, Sensing and Control., pp. 550–555, 2006.
[8] Wang, J.H., Cheng, D.Z. & Hu, X.M., Consensus of multi-agent linear dynamic systems, Asian Journal of Control, 10(2), pp. 144–155, 2008.

*Multimedia, Communication and Computing Application – Leung (Ed.)*
© *2015 Taylor & Francis Group, London, ISBN 978-1-138-02775-6*

# Quantitative study of oil ingress in honeycomb structure by using pulse thermography

D.P. Chen, G. Zhang & X.L. Zhang
*Science and Technology on Optical Radiation Laboratory, Beijing, China*

X.L. Li
*Beijing Key Laboratory for Terahertz Spectroscopy and Imaging, Department of Physics, Capital Normal University, Beijing, China*

ABSTRACT: Honeycomb structure has an increasing tendency of applications in aircraft manufacture, owing to its special advantages. However, because of their hollow structure, this kind of materials are susceptible to liquid ingress, such as oil leakage occurs in the honeycomb structure of the rudder, wing, and other key parts, could mean a serious mechanical failure. In this paper, according to the quantitative measurement of oil in the honeycomb structure, the relationship between surface temperature and the thickness of the second layer is studied by analyzing the theoretical model of heat conduction in a double-layer structure. An algorithm based on the theoretical $\Delta T$ equation is proposed to calculate the oil quantity in the honeycomb structure. A honeycomb sample is manufactured with only one side of skin, and the pulsed thermography is applied for the oil quantitative study. Finite element method is used for the simulation, and the simulated result is compared with the experimental results.

## 1 GENERAL INSTRUCTIONS

Composite honeycomb sandwich structure has an increasing tendency of applications in aircraft manufacture, owing to its high bending stiffness to weight ratio, superb fatigue resistance, and low manufacturing cost [1]. However, it is widely recognized that because of the hollow structure, this kind of materials are susceptible to liquid ingress, and water and oil ingress are two kinds of typical defects. Especially, if a small amount of oil leakage occurs in the honeycomb structure of the rudder, wing, and other key parts, it could mean oil spill of the hydraulic system, which plays an important role in landing gear, brake system, steering system, and engine thrust reverser. Thus, the loss of the hydraulic oil would lead to serious mechanical failure of the hydraulic components [2, 3]. Therefore, hydraulic oil ingress in the honeycomb structure could endanger the flying safety, and a practical nondestructive method to detect oil ingress quantitatively in the honeycomb structure is needed.

*The traditional methods:* Ultrasound, X-ray, and eddy current, and so on, are usually used for the qualitative detection of delaminations, cracks, and water ingress in the honeycomb structure. Each method has its advantages and disadvantages, and the liquid quantitative study is difficult by using the traditional methods. Infrared thermography has been developing since 1990s, with the characteristic of noncontact and

fast, has special advantages for composites testing, and it has been proven as an effective method to detect water or oil ingress in the honeycomb structure [4]. In Wayne State University, pulse thermography is used to detect water ingress in honeycomb panel from three sides of the sample (front, back, and vertical), and it is also used for the real-time testing of aircraft honeycomb structure [5]. Airbus uses three thermal excitation methods (heating blanket, oven, and fridge) to detect fluid ingress in the honeycomb structure of the wing [6]. Ibrra-Castanedo et al. [7] applied PPT method to detect honeycomb sandwich structure, and showed the water ingress in phase images; they also used passive infrared thermography to detect the water ingress in the honeycomb structure, and estimated the water quantity by the time of temperature rising [7]. Guo and Xu [8] studied the surface temperature changes of different water quantities in pulse thermography by using finite element simulation.

We have reported the method of liquid ingress (water/oil) recognition in the honeycomb structure [9] and then studied the method for the quantitative measurement of the thickness of the second layer in pulsed thermography by using a 20-mm-thick steel slab with four 1.1-mm-depth circular flat-bottom holes [10]. Furthermore, in this paper, the pulsed thermography is also applied, and quantitative testing of oil ingress in the honeycomb structure is studied by theory, experiments, and finite element simulation.

## 2 THEORY

To get the surface temperature expression of different quantities of oil in the pulsed thermography, theoretical model of the transient heat conduction in semi-infinite composite region should be considered first, as shown in Figure 1, a semi-infinite composite region $-\infty < x < d_0$, of which $0 < x < d_0$ is of one medium and $x < 0$ is of another. We write $k_1$, $\alpha_1$, $\rho_1$, $c_1$, and $T_1$ for conductivity, diffusivity, density, specific heat, and temperature in the region $0 < x < d_0$, and $k_2$, $\alpha_2$, $\rho_2$, $c_2$, $T_2$ for the corresponding quantities in $x < 0$. Consider a unit instantaneous plane source at $t=0$ at $x=d_0$ in the region $x>0$, then the temperature expression at $x=d_0$ can be written as [9]

$$T_{Isurf} = \frac{C}{2\sqrt{\pi\alpha_1 t}}(1 + 2\sum_{n=1}^{\infty} R^n e^{-n^2 d_0^2 / \alpha_1 t}) \qquad (1)$$

$R = (k_1\sqrt{\alpha_2} - k_2\sqrt{\alpha_1})/(k_1\sqrt{\alpha_2} + k_2\sqrt{\alpha_1})$, which can be seen as the thermal reflection coefficient at the interface between the two materials, $n$ is the number of reflection, and $C$ is a constant which has relationship with the heat power and the absorptivity of the surface.

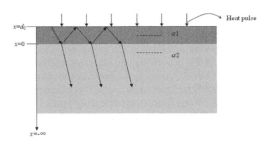

Figure 1. Theoretical model of heat conduction in double-layer structure with infinite thickness of the second layer.

Furthermore, we assume the thickness of the second layer is finite $0 < x < d_0$, then the heat conduction of theoretical model at this situation is as shown in Figure 2.

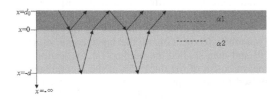

Figure 2. Theoretical model of heat conduction in double-layer structure with finite thickness of the second layer.

The heat reflection and transmission could be occurred at $x=0$, and the transmission coefficient can be expressed as $1 - R$. The heat would transfer in the second layer continuously, and be reflected at the interface of the second layer (Figure 3). We assume the thickness of the second layer is $d$, and adiabatic boundary condition at its outer interface $x=-d$, and neglect higher order ($n > 1$) reflection, then the surface temperature $x=d_0$ can be given as [10, 11]

$$T_{IIsurf} = \frac{C}{2\sqrt{\pi\alpha_1 t}}(1 + 2R e^{-d0^2/\alpha_1 t} + 2(1 - R^2)e^{-d0^2/\alpha_1 t} e^{-d^2/\alpha_2 t}). \qquad (2)$$

The temperature difference between defect and sound area is widely used to directly measure the depth of a one-layer structure or the first layer of a composite structure [12]. To obtain the depth information of the second layer for a composite structure, first a similar method needs to be considered, for which the temperature difference between $T_{Isurf}$ and $T_{IIsurf}$ is

$$\Delta T = \frac{C}{\sqrt{\pi\alpha_1 t}}(1 - R^2)e^{-d0^2/\alpha_1 t} e^{-d^2/\alpha_2 t}. \qquad (3)$$

Because of three-dimensional (3D) heat diffusion, the obtained heat signals for deep defects may not have enough signal-to-noise ratio (SNR), therefore, the pulsed thermography is normally used for surface or subsurface defects detection. Depth prediction methods for one-layer structure, such as peak of thermal contrast and peak slope of thermal contrast [13], may not be applicable for oil quantity measurement, because the obtained thermal signal is too weak to extract the peak information. Therefore, a new method should be considered on the basis of (3).

First, multiply both sides of (3) with $\sqrt{t}$, and define a new time-dependent function $f(t)$ as

$$f(t) = \Delta T(t) \cdot \sqrt{t} = \frac{C}{\sqrt{\pi\alpha_1}}(1 - R^2)e^{(-d_0^2/\alpha_1 - d^2/\alpha_2)/t}. \qquad (4)$$

The expression has relationship with the thickness $(d_0, d)$ and heat diffusion coefficient $(\alpha_1, \alpha_2)$ of the two layers. Then, define another new variable

$$w = (d_0^2/\alpha_1 + d^2/\alpha_2)/t. \qquad (5)$$

Then, (4) can be written as

$$f(w) = \frac{C}{\sqrt{\pi\alpha_1}}(1 - R^2)e^{-w}. \qquad (6)$$

It is obvious that (6) is a monotonically decreasing function, we further assume

$$f(w_0) = v_0. \qquad (7)$$

where $v_0$ is a predefined value, and $w_0$ is supposed to be the single solution of this equation. Then, we have

$$t_0 = (d_0^2 / \alpha_1 + d^2 / \alpha_2) / w_0. \qquad (8)$$

Suppose $A = d_0^2 / \alpha_1$, $B = 1 / \alpha_2$, where A and B are constants, then (8) can be simplified as

$$t_0 = A + Bd^2. \qquad (9)$$

This equation means that when it is to predict the thickness of the second medium, the obtained $t_0$ corresponding to the time that satisfies (9), is linearly related with $d^2$, and its intercept is not zero.

When adopting this method for the thickness prediction of the second medium, it is first to set a value $v_0$, and then extract the corresponding time $t_0$. If the thermal diffusivity and $w_0$ are known, $d$ can be obtained using (9). However, in practical applications, it is very difficult to calculate $w_0$ because $v_0$ and $w_0$ depend on C and $\alpha_1$. C may have very big difference for difference samples owing to its surface condition, emissivity, and so on. Therefore, if we directly use (9) to predict the thickness of the second medium, the uncertainty or the error of parameter $w_0$ could not be neglected. However, the contrast value of A and B is a certain value: $A/B = (d_0^2 \times \alpha_2) / \alpha_1$, and has no relationship with $w_0$. Therefore, the following quantitative study of oil ingress in the honeycomb structure depends on the value of A/B.

## 3 SAMPLE AND EXPERIMENTAL DESIGN

The honeycomb sample is designed with glass fiber skin of 1.2 mm thickness on only one side to inject oil, and the core material is Nomex. The height and side length of the hexagon core are 12 and 3 mm, respectively. Several separated regions in the sample are filled with oil of different quantities from 10 μl to full in each cell, as shown in Figure 3, and the detailed oil quantities in each region are also shown in Table 1.

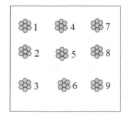

Figure 3. Experimental design of the quantitative study of oil ingress in honeycomb structure.

Table 1. Oil quantity in each cells that are shown Figure 0.

| Regions | 1 | 2 | 3 | 4 | 5 | 6 | 7 | 8 | 9 |
|---------|----|----|----|----|----|----|----|----|------|
| I (μl) | 10 | 15 | 20 | 25 | 30 | 35 | 40 | 50 | Full |
| II(μl) | 60 | – | – | – | 70 | – | – | – | Full |

## 4 EXPERIMENTAL RESULTS ANALYSIS

The glass fiber skin is placed at the bottom surface, a high-power flash of 9.6 kJ is applied as heat source, and an infrared camera is used to capture temperature variations of the surface. The sampling frequency and time is fixed at 30 Hz and 60 s, respectively. The 5th original thermal image after heat stimulation is shown in Figure 4. It can be seen that oil of different quantities have similar appearance and it is difficult to distinguish them by using thermal images, so the data should be further processed.

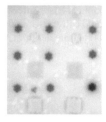

Figure 4. Fifth original thermal image.

The quantitative algorithm described above is adopted to process the original temperature dissipation data: the temperature of the core with full oil (thickness of the second layer is infinite) is used as a reference, and the temperature difference between cores of finite and full oil can be calculated. The temperature difference versus time ($\Delta T - t$) curves shown in Figure 5, which are normalized at the 40th frame and 8th order polynomial fitted. As can be seen from the $\Delta T - t$ curves, cores filled with different quantities of oil from 10 to 40 μl can be recognized, but it becomes difficult to distinguish when the quantity of oil exceeds 40 μl.

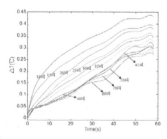

Figure 5. Experimental $\Delta T - t$ curves of different quantities of oil.

The $\Delta T - t$ data are multiplied with $\sqrt{t}$, and the $f - t$ data are obtained, whose curves are shown in Figure 6, then chose the value of $v_0=0.8$, and the corresponding $t_0$ is extracted, the experimental extracted $t_0 - d^2$ result for 10–70 μl of water is shown in Figure 7, and the linear fitted lines for 10–40 μl and 40–70 μl of oil are also shown as a comparison.

As can be seen from Figure 7, the characteristic time $t_0$ is changing with 10–40 μl of oil, and it shows good linearity with 10–40 μl of oil; however, $t_0$ does not change with $d^2$ when oil quantity exceeds 40 μl, whose depth is about 1.7 mm (>1/8 height of the core). The intercept ($A/w_0$) and slope ($B/w_0$) value of the fitted line of 10–40 μl is extracted, and the average value of four experiments is A/B=4.3885×10⁻⁷, which is basically fitted with the theoretical value of $A/B=(d_0^2 \times \alpha_2)/\alpha_1 =3.6273\times10^{-7}$. The error generated is mainly because the theory is based on an instantaneous heat pulse and 1D heat conduction; however, there is 3D heat diffusion in the experiments, and the practical heat source has a pulse width of 2 ms, which could cause error in the normalization. Another reason is that the depth of the oil cannot be calculated exactly.

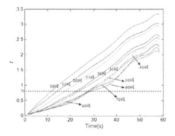

Figure 6.   Experimental $f - t$ curves.

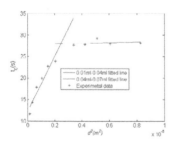

Figure 7.   Linear fitted $t_0$ versus square $d$ curves.

As can be seen from the experimental results, the characteristic time $t_0$ has linear relationship with the square thickness of the second medium ($d^2$) when the oil quantity is less than a certain value (40 μl). However,

if the water quantity exceeds the value, the value of $t_0$ does not increase. So, we set this value as the oil quantity threshold and denote it as $d_T$, and the corresponding characteristic time threshold as $t_T$. For practical applications, standard sample should be manufactured first, then the quantitative algorithm described above is applied to process the temperature data and find the corresponding $d_T$ and $t_T$. If $t_0 > t_T$, the oil quantity is considered as greater than $d_T$, otherwise, the oil quantity can be calculated by the obtained linear relation.

## 5   FINITE ELEMENT SIMULATION

ANSYS tool is used to simulate the oil quantitative experiments, choose Solid70 elements, and build the 3D model of the honeycomb structure, which is shown in Figure 8. The thickness of the glass fiber skin is set as 1.2 mm, the height and the side length of the core is also set as 12 and 3 mm, respectively. Different quantities of oil (10—70 μl-full) are filled at different areas. A 2 ms heat flux is loaded on the surface of the model to simulate the pulse heat source. The transient heat conduction for 30 s is calculated, and the simulated data of surface temperature dissipation is extracted.

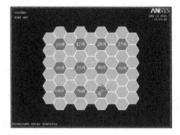

Figure 8.   3D model of the finite element simulation for quantitative study.

Then the quantitative algorithm is adopted to process the simulated temperature dissipation data; temperature of the core with full oil is also used as a reference, and the simulated temperature difference versus time ($\Delta T - t$) curves are shown in Figure 9. The $\Delta T - t$ data are multiplied with $\sqrt{t}$, and the simulated $f - t$ curves are obtained as shown in Figure 10.

As can be seen from the simulated $\Delta T - t$ curves and $f - t$ curves, the simulated results coincide with the experimental results. The oil of different quantities can be obviously recognized when the quantity is 10—40 μl, but the calculated error increasing when the quantity of oil exceeds 40 μl, especially the data of 60 and 70 μl.

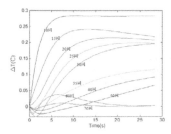

Figure 9. Simulated temperature difference versus time $(\Delta T - t)$ curves.

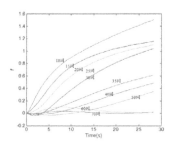

Figure 10. Simulated $f - t$ curves.

## 6 CONCLUSION

According to the quantitative study of oil ingress in the honeycomb structure by the pulsed thermography, the expression of the relationship between the surface temperature and the oil quantity is obtained by solving the 1D differential equation under an instantaneous plane source; a quantitative algorithm is proposed. Temperature of the cores with full oil (infinite oil) is used as a reference, and the thermal contrast between finite and infinite oil quantity is obtained, which is then multiplied with the square root of time to obtain a new time sequence $f$. A characteristic value $v_0$ is set for $f$ and the corresponding characteristic time $t_0$ is obtained, and a linear relationship between $t_0$ and the square thickness of oil: $t_0 = A + Bd^2$ is found. For practical application, the characteristic value $v_0$ should be determined first, and then the corresponding characteristic value of time $t_0$ should be extracted, the linear relationship between $t_0$ and $d^2$ can be used for quantitative study. The experimental results of $t_0 - d^2$ curve shows that the characteristic time $t_0$ has a good linear relationship with the square thickness of oil layer $(d^2)$ when the water quantity is less than a certain value, which also shows in the theoretical equation. However, the value of $t_0$ does not increase, if the water quantity exceeds the certain value. The intercept A and slope B of the linear fitted line is extracted, and the value of A/B is calculated and compared with the corresponding theoretical value; it is found that they are basically coincident. ANSYS is used to simulate the experiment for oil quantitative study, and the $\Delta T - t$ and $f - t$ results shows that it is coincident with the experimental results, and the reliability of the quantitative method is verified.

For the aircraft honeycomb structure, when a small amount of oil in rudder, wing, and other key parts results in a serous mechanical failure. The nondestructive method developed in this paper, which can quantitatively detect the oil ingress, with significance for promoting the reliability and extending the life of aircraft.

## REFERENCE

[1] Shen Jun, Xie Huanqin. Development of research and application of the advanced composite materials in the aerospace engineering[J]. Materials Science and Technology, 2008, V16(5):737–740.

[2] Cise D.M., Lakes R.S. Moisture Ingression in Honeycomb Core Sandwich Panels: Directional Aspects[J]. Journal of Composite Material, 1997, V31(22): 2249–2263.

[3] Shafizadeh J.E., Seferis J.C. Evaluation of the Mechanisms of Water Migration through Honeycomb Core[J]. Journal of Material Science, 2003, V38(11): 2547–2555.

[4] Thomas R.L., Han X.Y., Favro L.D., et al. Thermal Wave Imaging of Aircraft for Evaluation of Disbonding and Corrosion[A]. Presented at 7th European Conference on Non-destructive Testing[C]. Copenhagen: NDT, 1998:126–130.

[5] Han X.Y., Favro L.D., Kuo P.K., et al. Early-time Pulse-echo Thermal Wave Imaging[J]. Review of Progress in Quantitative Nondestructive Evaluation, 1996, V15(10):519–524.

[6] Bisle W. NDT Toolbox for Honeycomb Sandwich Structures - a Comprehensive Approach for Maintenance Inspections[A]. ATA NDT Forum[C]. America: ATA NDT Forum, 2010:1–8.

[7] Ibrra-Castanedo C., Brault L., Marcotte F., et al. Water Ingress Detection in Honeycomb Sandwich Panels by Passive Infrared Thermography Using a High- resolution Thermal Imaging Camera[A]. Thermosense, Thermal Infrared Applications XXXIV[C]. Orlando SPIE, 2012, V8354(05):1–8.

[8] Guo Xingwang, Xu Wenhao. Heat Transfer Analysis of Pulsed Thermography for Water Detection in Honeycomb Structures[J]. Infrared Technology, 2011, V33(5):275–280.

[9] Chen D.P., Zeng Z., Zhang C.L., Zhang Z. Liquid Ingress Recognition in Honeycomb Structure by Pulsed Thermography[J]. European Physical Journal-Applied Physics, 2013, V62(2):20401–8.

[10] Chen D.P., Zeng Z., Zhang C.L. Jin X.Y., Zhang Z. Quantitative Study of Water Ingress in Pulsed Thermography[J]. Insight, 2013, V55(5):1–7.

[11] Carslaw H.S., Jaeger J.C. Conduction of Heat in Solids[M]. New York: Oxford University Press, 1959: 182–290.

[12] Favro L.D., Han X.Y., and Kuo P.K. and Thomas R.L. Defect Depth Determination by Thermal-Wave Imaging[J]. Progress in Natural Science, 1997, V6(S): 139–141.

[13] Zeng Z., Li C.G., Tao N., et al. Depth Prediction of Non-air Interface Defect Using Pulsed Thermography[J]. NDT&E International, 2012, V48:39–45.

*Multimedia, Communication and Computing Application – Leung (Ed.)*
© *2015 Taylor & Francis Group, London, ISBN 978-1-138-02775-6*

# A kind of automatic pulse wave collection device connected to the air sac on wrist

A.R. Wang, W.W. Wang, Y.M. Yang, S. Zhang, X.W. Li & L. Yang
*College of Life Science and Bio-engineering, Beijing University of Technology, Beijing 100124, China*

ABSTRACT:   Hypertension and cardiovascular pathology are the most important factors causing cardiovascular diseases. A kind of monitoring equipment for cardiovascular function to evaluate cardiovascular pathology is urgently needed which can calculate blood pressure and report pulse wave. The piezoelectric signal collected from air sac on wrist was processed into two kinds of signals including pressure of air sac signal and pulse wave signal, which were transferred by analog-to-digital converter into microcontroller unit (MCU). The MCU can make and break operation of the pump and air valve to calculate the pressure in air sac and amplitude of the pulse wave peak. Finally, the MCU can calculate mean arterial pressure and collect radial artery pulse wave. Therefore, the device can collect pulse wave on the constant air sac pressure which is mean arterial pressure. This device can provide a way to measure pulse wave automatically while the blood pressure is measured.

KEYWORDS:   Air sac, Pulse wave, Blood pressure, Automatic collection

## 1   INTRODUCTION

The United Nations High-Level Meeting on Non-Communicable Diseases in 2011 declared that the increasing global crisis in noncommunicable disease is a barrier to the development of goals including poverty reduction, health equity, economic stability, and human security, mainly from heart disease, stroke, cancer, diabetes, and chronic respiratory disease [1]. Chronic noncommunicable disease is a worldwide public health problem, and cardiovascular disease is a priority for all. There is a rising incidence and prevalence of cardiovascular disease, with rising living standards, increasing social competition pressure, decrease of physical activity, and the change of dietary structure. Nowadays, cardiovascular disease not only happens among the elderly, but also may tend to younger people. Prospective cohort studies have indicated that there is a strong, linear, and independent relationship among blood pressure levels, the risk of cardiovascular disease incidence, and mortality among the general population [2]. Many studies have reported that hypertension has a stronger effect on cardiovascular disease risk and lowering high-normal blood pressure can reduce the risk of cardiovascular disease [3–5]. Therefore, early monitoring of hypertension and cardiovascular hemodynamic information is essential to prevent the occurrence and development of cardiovascular disease.

There are many methods of blood pressure measurement, and with the continuous development in electronic medical equipment, the indirect method of blood pressure measurement has been accepted because of its convenience and noninvasive measurements. Therefore, it is very suitable for nonmedical people and beneficial to the realization of the family medical instrument. The function of medical electronic instruments is single in the current market, that is, can only measure blood pressure and heart beat. Some researches indicated that the pulse wave contains rich cardiovascular physiological and pathological information and pulse waveform characteristics are closely bounded up with cardiovascular function [6–8]. Therefore, a device to automatically collect pulse wave for acquiring cardiovascular parameters while collecting the blood pressure is needed, which will be significant to realize the monitoring of cardiovascular function.

## 2   MEASUREMENTS OF BLOOD PRESSURE AND PULSE WAVE

This study indirectly calculated the systolic and diastolic pressures according to pressure vibration of air sac corresponding to the maximum vibration of mean arterial pressure in discharging process. The location of the air sac is not important in that the top of the brachial artery has no transmitter which is one of the advantages of this method. Another advantage is small external noise interference, which greatly reduces the measurement errors caused by technology operators.

When the pressure of air sac is much higher than systolic pressure, pulse wave will disappear and then appears with the decrease of pressure of air sac. When the pressure of air sac decreases from a higher to lesser level than that of systolic pressure, the pulse wave will gradually increases from zero. Then, the pulse wave will be reduced with the decrease in pressure. Oscillographic method of blood pressure measurement is according to the relation between amplitude of pulse wave and pressure of air sac. The maximum amplitude of pulse waveform is corresponding to the mean pressure Pm, systolic pressure Ps, and diastolic pressure Pd, respectively, determined using the corresponding pulse wave maximum amplitude [9]. We adopted the amplitude coefficient method, and clinical value are 0.48 (systolic pressure amplitude coefficient) and 0.58 (diastolic pressure amplitude coefficient). Then, we collected radial pulse wave on the constant air sac pressure which is mean arterial pressure.

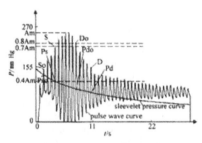

Figure 1.　Relation between amplitude of pulse wave and pressure of air sac according to the oscillographic method.

## 3　PARAMETERS CALCULATION

The change in pulse waveform characteristics will first be shown in the pulse wave form change area. To use a simple characteristic to describe these changes, Prof Zhichang Luo from the Beijing University of Technology proposed pulse waveform characteristic parameter $K$ [10]:

$$K = \frac{Pm - Pd}{Ps - Pd}.\qquad(1)$$

According to the pulse wave theory, the mean arterial pressure Pm in the radial artery is

$$Pm = \frac{1}{T}\int_0^T P(t)\,dt\qquad(2)$$

where P(t) is the curve of the pressure pulse wave, Ps is the systolic pressure, and Pd is the diastolic pressure, as shown in Figure 2.

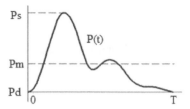

Figure 2.　Pulse wave characteristics.

The value of $K$ is influenced only by the area of pulse wave graph, which is a scalar quantity and is not related to the absolute values of Ps and Pd. It is equal to the percentage of mean pulse wave pressure pulsation component (Pm – Pd) in the maximum pulsation component (Ps – Pd). The pulse waveform and area will be different at different physiological and pathological states, and the change can be described by the value of $K$.

## 4　HARDWARE CIRCUIT STRUCTURE

Hardware circuit is composed by the baroceptor, signal-processing circuit, microcontroller unit (MCU), and voltage conversion circuit. The piezoelectric signal collected from air sac on wrist is processed into two kinds of signals: pressure of air sac signal and pulse wave signal. The radial pulse wave signal will be delivered to the MCU for signal processing after filtration and two-stage amplification; then, we monitor the pressure of air sac in real-time through the other air sac pressure signal closed to DC and transmit the signal to the MCU. The signals are transferred into MCU by analog-to-digital (A/D) converter for collecting, saving, and calculating, while both of the charging and discharging of wrist air sac are controlled by D/A. MCU connects to the PC through the serial communication.

### 4.1　Pressure sensor

The radial pulses were digitally recorded using a baroceptor (air sac) (NovaSensor Pty Ltd., NPC-1220) which has different measurements: 5, 10, 15, and 20 spi (1 mmHg = 0.019 spi). We choose the NPC 1220-005 (Figure 3) with measurement 0–250 mmHg to meet actual measurement 0–200 mmHg.

This baroceptor sensor is a kind of piezoresistive pressure sensor with high integration amplifier circuits which is easy to handle. It has good moisture-proof ability and excellent medium compatibility. It is suitable for the nonmedical people, who can operate and measure. The adaptability of the radial pulse wave can be improved only when put on the air sac.

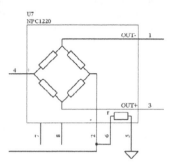

Figure 3.   Schematic diagram of NPC 1220-005.

### 4.2   *The signal-processing circuit*

The signal-processing circuit is shown in Figure 4, including the front differential amplifier circuit [Figure 4(a)] and filter and amplifier circuit of pulse wave [Figure 4(b)].

The radial pulse wave is a millivolt-level signal, and its frequency is 0.1–8 Hz. The daily breathing and motion artifacts of measurement can cause the signal interference, so this device uses the chip INA118 with high gain and common-mode rejection ratio to constitute the front differential amplifier circuit.

In the filter and amplifier circuit of pulse wave, we choose four-channel operational amplifier chip TL084. Two stages of amplification were used in the operational amplifier circuit, with the primary amplification is 20 and the second is 47. The filter range of two-stage amplifier is 33 and 15.9 Hz, then the total amplification is 940 and filter range is 0–15.9 Hz.

As the MCU has a reference voltage input value, it will not read the signal while the voltage in the input end is higher or less than this range value. The MCU may be burn out with the voltage of signal being overtop. To avoid these issues, we design a muzzle circuit after the filter and amplifier circuit. Voltage follower has the functions of isolating and buffering which can increase the input impedance and reduce the input capacitance and output impedance.

### 4.3   *MCU*

Pulse wave signal processing mainly includes the pulse wave signal and pressure signal of air sac,

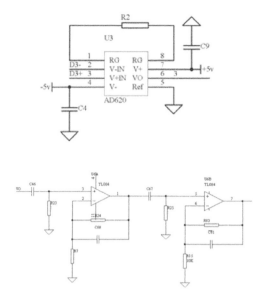

Figure 4.   Signal-processing circuit. (a) Front differential amplifier circuit (b) Filter and amplifier circuit of pulse wave

which are separated from the front differential amplifier circuit. MCU needs to have many features like low power, high memory space, high computing speed, and with minimum two 12-bit A/D converters. We use ARM chip STM32F103RBT6, which can meet the requirements of the mentioned signal processing and possesses the features of high instantaneity, reliability, and high integration density to reduce the size of device and to enable device to achieve small.

Real-time data collection and judgment is critical to confirm the pulse wave pressure collection value that of the mean arterial pressure. Using the two parallel 12-bit A/D I/O ports, the pressure signal of air sac approaching direct current and radial pulse wave signal is collected.

MCU can collect pulsating pulse wave signal when the pump pressure lets the air sac pressure value to reach the diastolic pressure. While the pressure of air sac increases, the pulse wave peak value also increases, and the peak reaches the maximum while the pressure of air sac is equal to the subjects' mean arterial pressure. MCU records the pressure of air sac corresponding to the pulse wave maximum peak value which is the subjects' mean arterial pressure. It controls the pump and the disabled air valve to keep the pressure of air sac in a state of mean arterial pressure. The sampling rate is 100 Hz, lasting all about 20 seconds to collect enough steady pulse wave form.

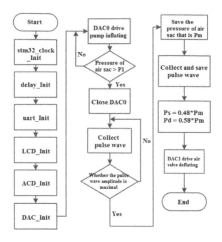

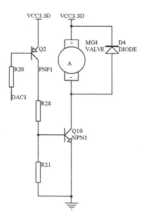

Figure 5. Relationship of function modules in the MCU.

## 4.4 Pump and air valve drive circuit

The pressure of air sac is controlled by the pump, solenoid valve, and uniform vent valve. The pump and air valve in our device are supplied 3 V power, and the on–off control is realized by the I/O port in MCU. In the meantime, the drive circuit for providing working voltage of pump and air valve is necessary to reach the working current of pump and air valve, and achieve the current protection (Figure 6).

Figure 6. Pump or air valve drive circuit.

## 5 CONCLUSION

The noninvasive wrist blood pressure determination and radial pulse wave collection will be realized at the same time according to the relation between amplitude of pulse wave and pressure of air sac by using the oscillographic method. This will lay the foundation for the comprehensive blood pressure and cardiovascular hemodynamics.

With the continuous development of electronic medical technology and further promotion and attention in home health care, realizing the home health-monitoring device with real-time dynamic monitoring technology is more conducive to public health, and offers a way for the early detection of cardiovascular disease.

## ACKNOWLEDGMENT

This work was supported by the Beijing Municipal Commission of Education Foundation under Grant PXM2013_014204_07_000069.

## REFERENCES

Beaglehole R, Bonita R, Horton R, et al. 2011. Priority actions for the non-communicable disease crisis. The Lancet, 377(9775): 1438–1447.

Gu D, Kelly T N, Wu X, et al. 2008. Blood pressure and risk of cardiovascular disease in Chinese men and women. American journal of hypertension, 21(3): 265–272.

Vasan R S, Larson M G, Leip E P, et al. 2001. Impact of high-normal blood pressure on the risk of cardiovascular disease. New England Journal of Medicine, 345(18): 1291–1297.

Selmer R, Tverdal A. 1995. Body mass index and cardiovascular mortality at different levels of blood pressure: a prospective study of Norwegian men and women. Journal of epidemiology and community health, 49(3): 265–270.

Carman W J, Barrett-Connor E, Sowers M, et al. 1994. Higher risk of cardiovascular mortality among lean hypertensive individuals in Tecumseh, Michigan. Circulation, 89(2): 703–711.

Hasegawa S, Sato S, Numano F, et al. 2011. Characteristic alteration in the second derivative of photoplethysmogram in children. Pediatr Int. 53(2): 154–158.

Foo J Y, Lim C S, Wilson S J. 2008. Photoplethysmographic assessment of hemodynamic variations using pulsatile tissue blood volume. Angiology. 59(6): 745–752.

Westerhof B E, Guelen I, Stok W J, et al. 2007. Arterial pressure transfer characteristics: effects of travel time. Am J Physiol Heart Circ Physiol. 292(2): H800–H807.

Lass J, Meigas K, Karai D, et al. 2004. Continuous blood pressure monitoring during exercise using pulse wave transit time measurement. 26th Annual International Conference of the Engineering in Medicine and Biology Society. IEEE, 1: 2239–2242.

Song J, Zhang S, Qiao Y, et al. 2004. Predicting pregnancy-induced hypertension with dynamic hemodynamics. European Journal of Obstetrics & Gynecology and Reproductive Biology, 117(2): 162–168.

*4. Management engineering and computer application*

*Multimedia, Communication and Computing Application – Leung (Ed.)*
*© 2015 Taylor & Francis Group, London, ISBN 978-1-138-02775-6*

# A robust forward-secure threshold signature schemes based on polynomial secret sharing without a trusted dealer

D.J. Lu & Y. Wang
*Department of Mathematics, Qinghai Normal University, Xining, China*

X.Q. Zhang
*Financial and Economic College, Qinghai University, Xining, China*

ABSTRACT: Based on polynomial secret sharing, we construct a robust forward-secure threshold signature schemes without a trusted dealer. This scheme has the following property: even if a lot of players are compromised, which is more than the threshold number, it is not possible to forge signatures related to the past. This property is achieved while keeping the public key fixed and updating the secret keys at regular intervals. The proposed scheme is based on the Bellare-Miner forward-secure signature scheme, which is known to be secure in the random oracle model, assuming that factoring is hard. Furthermore, the proposed scheme can tolerate the malicious adversaries.

## 1 INTRODUCTION

Exposure of a secret key for "non-cryptographic" reasons is the greatest threat to many cryptographic protocols. The most commonly proposed remedy is the distribution of the secret key across multiple servers via secret sharing. For digital signatures, the main instantiation of this idea is threshold signature schemes [1].

The distribution of the key makes it harder for an adversary to learn the secret key, but it does not remove this risk. Common mode failures-flaws may be present in the implementation of the protocol or the operating system may be run on all servers, implying that breaking into several machines may not be much harder than breaking into one.

A common principle of security engineering is that one should not rely on a single line of defence. We suggest a second line of defence for threshold signature schemes which can mitigate the damage caused by complete key exposure. The idea is to provide forward security.

Forward security for digital signature schemes was suggested by Anderson [2], and solutions were designed by Bellare and Miner [3]. The idea is that a compromise of the present secret signing key does not enable an adversary to forge signatures pertaining to the past. Bellare and Miner [3] focus on the single-signer setting and achieve this goal through the key evolution paradigm: the user produces signatures using different secret keys during different time periods while the public key remains fixed. The integrity of documents signed before the exposure remains intact. After Bellare and Miner [3], Abdalla [4], Itkis [5], Kozlov [6] proposed some improved schemes. These improvements make the above schemes more simple and practical.

Combining forward security and threshold cryptography will yield a scheme that can provide some security guarantees even if an adversary has taken control of all servers and, as a result, it has completely learned the secret. In particular, she cannot forge signatures as if they were legitimately generated before the break-in. The complete knowledge of the secret signing key is useless for her with regard to signatures from "the past".

In this paper, based on the contribution of Abdalla [4], a robust forward-secure threshold signature scheme is considered. Firstly, the proposed scheme avoids the trusted dealer, who selects and distributes the secret key. Secondly, it shows how to share the secret key (chosen by the members) so that each member of the group can verify that the share is correct. In the next section some definitions and notation are introduced. Then a robust forward security-threshold signature scheme is presented, and in Section 4, security of the proposed scheme is discussed.

## 2 DEFINITIONS AND NOTATION

In this section, we describe our communication model and the types of adversaries. We then explain what it means by a forward-secure threshold signature scheme.

## 2.1 Communication model

We assume that our computational model is composed of a set of $n$ players which are connected by a complete network of private point-to-point channels. In addition, the players have access to a dedicated broadcast channel. We assume a fully synchronous communication model and that they are simultaneously delivered to their recipients. This model is not realistic enough for many applications, but it is often assumed in the literature.

## 2.2 Types of adversaries

We categorize the types of adversaries according to the power an adversary can have over a compromised player, as outlined in [4]. Firstly, an adversary may be eavesdropping, meaning that she may learn the secret information about a player, but may not affect his behaviour in any way. A more powerful adversary is one that not only can eavesdrop, but also can stop the player from participating in the protocol. We refer to such an adversary as a halting adversary. Finally, the most powerful notion in this category is a malicious adversary, who may cause a player to deviate from the protocol in an unrestricted fashion.

## 2.3 Forward-secure threshold signature schemes

In this paper, we are concerned with forward-secure threshold signature schemes. This scheme makes use of the key evolution paradigm, and their operation is divided into time periods. Throughout the lifetime of the scheme, the public key is fixed, but the secret key changes at each time period. In a forward-secure scheme, there is a key generation protocol, a signing protocol, a verification algorithm, and an update protocol. A $(t, k_s, k_u, n)$ key-evolving threshold signature scheme can tolerate at most $t$ corrupted players and works as follows.

Firstly, there is a key generation phase. Given a security parameter $k$, the public and the secret keys are generated and distributed to the players. This can be accomplished jointly by the players. At the start of a time period, an update protocol is executed among any subset of $k_u$ non-corrupted players. After the update protocol is executed, each non-corrupted player will have a share of the new secret for that time period. To generate signatures, a subset of $k_s$ players executes the signing protocol, which generates a signature for a message $m$ using the secret key of the current time period. The players which sign can be any subset of the set of players not corrupted by the adversary. The signature is a pair consisted of the current time period and a tag.

The verifying algorithm can be executed by any individual who possesses the public key. It returns either "Accept" or "Reject" to specify whether a particular signature is valid for a given message.

Furthermore, in a forward-secure threshold signature scheme, if an adversary learns more than $t$ shares of the secret signing key for a particular time period $\gamma$, it should be computationally infeasible for her to generate a signature $\langle j, tag \rangle$ for any message $m$ such that $\text{verify}_{\text{PK}}(m, \langle j, tag \rangle) = 1$ and $j < \gamma$, where verify is the scheme's verification algorithm.

## 3 PROPOSED SCHEME

In this section, we present RPFSTS, our $(t, 2t+1, 2t+1, 3t+1)$-threshold scheme based on polynomial secret sharing, forward-secure against malicious adversaries. Its main advantage is that it does not require the presence of all players during signing or key update, and it also does not require a trusted dealer; only about two thirds of the players are needed in any of these cases. Its construction is shown in Section 3.1 and relies on several standard building blocks tailored for our purposes. These tools are described in Section 3.2. Finally, Section 4 gives details about the security of our scheme.

## 3.1 Construction

We start by running a key generate stage where each player $P_i$ commits to a $t$-degree polynomial ($t$ is the scheme's threshold) $f_i(z)$ whose constant coefficient is the random value $z_i$, contributed by $P_i$ to the jointly generated secret $x$. Each player's share of the base key $SK_0^{(\rho)}$ includes each of his shares of the $S_{i,0}$ values (there are $l$ of them), so player $\rho's$ secret key is then $\left( N, T, 0, S_{1,0}^{(\rho)}, S_{2,0}^{(\rho)}, \cdots S_{l,0}^{(\rho)} \right)$. At the beginning of each time period, the evolution of the secret key is accomplished via the key update protocol in which exactly $2t+1$ players must participate. At the start of the protocol in time period $j$, each player who participated in the previous update protocol has $SK_{j-1}^{(\rho)}$, i.e. his share of the previous time period's secret. The new secret key is computed by squaring the $l$ values in the previous secret key. The players compute this new secret key using the Robust-Mult-SS protocol (as described in Section 3.2) $l$ times. At the end of the protocol, player $\rho$ holds $SK_j^{(\rho)}$, and each player immediately deletes his share of the secret key from the previous time period.

Like the update protocol, signing does not require participation by all of the n players, only $2t+1$ active players are required. Because it is the threshold version of Bellare-Miner [3], this protocol is based on a commit-challenge-response framework, but the various steps are accomplished by the group using sub-protocols described in Section 3.2.

### 3.1.1 Protocol RPFSTS.keygen(k,t)

1. We picks random, distinct k/2-bit primes p, q, each congruent to 3 mod 4, and set $N = pq$.
2. for $i = 1, 2, \cdots, l$ do
   a. All player use Pedersen-VSS over $Z_N$ to create shares $S_{i,0}^{(1)}, S_{i,0}^{(2)}, \cdots, S_{i,0}^{(n)}$ of $S_{i,0}$. In the description of underlying Pedersen-VSS, we denote $S_{i,0}^{(\rho)}$ as $x_\rho$ and denote $S_{i,0}$ as $x$ respectively.
   b. All player computes $U_i = S_{i,0}^{2^{(T+1)}}$
3. for $\rho = 1, 2 \cdots, n$ do
   a. $P_\rho$ sets $SK_0^{(\rho)} = \left( N, T, S_{1,0}^{(\rho)}, S_{2,0}^{(\rho)}, \cdots, S_{l,0}^{(\rho)} \right)$, where $SK_0^{(\rho)}$ as $P_\rho$'s secret key.
   b. Every player set $PK = \left( N, T, U_1, U_2, \cdots, U_l \right)$, where $PK$ is public key.

### 3.1.2 Protocol RPFSTS. sign(m,j)

1. Using Joint-Pedersen-VSS, player generates random value $R \subset Z_N$ so that player $\rho$ gets share $R^{(\rho)}$ of $R$.
2. Players compute $Y = R^{2^{(T+1-j)}}$ using Robust-mult-SS and their shares of $R$.
3. Each player $\rho$ computes $c_1 c_2 \cdots c_l = H(j, Y, m)$.
4. Each player $\rho$ executes $Z^{(\rho)} = R^{(\rho)}$ so that $Z$ is initialized to $R$.
5. for $i = 1, 2, \cdots, l$ do
   if $c_i = 1$, then players compute $Z = Z \cdot S_{i,j}$ using Robust-mult-SS.
6. The signature of $m$ is set to $\langle j, (Y, Z) \rangle$, and is made public.

### 3.1.3 Algorithm RPFSTS.verify$_{PK}$ $(m, \sigma)$

1. Parse $\sigma$ as $\langle j, (Y, Z) \rangle$.
2. if $Y \equiv 0$, then return 0.
3. $c_1 c_2 \cdots c_l = H(j, Y, m)$.
4. if $Z^{2^{(T+1-j)}} \equiv Y \cdot \prod_{i=1}^{l} U_i^{c_i}$, then return 1, else return 0.

### 3.1.4 Protocol RPFSTS.update(j)

1. if $j = T$, then return the empty string. Otherwise, proceed.
2. Players compute updated secret key shares $S_{1,j}, S_{2,j}, \cdots S_{l,j}$ by squaring the previous values $S_{1,j-1}, S_{2,j-1}, \cdots S_{l,j-1}$ using Robust-mult-SS.
3. Each player $\rho$ deletes $SK_{j-1}^{(\rho)}$.

### 3.2 Building blocks

#### 3.2.1 Pedersen-VSS

Pedersen's Verifiable Secret Sharing protocol uses $p'$ and $q'$ as two large primes such that $q'$ divides $p' - 1$, $G_{q'}$ is the unique subgroup of $Z_{p'}^*$ of order $q'$, and $g$ is a generator of $G_{q'}$. It is assumed that the adversary cannot find the discrete logarithm of $h$ relative to the base $g$. We show how $n$ members of a group $(P_1, P_2, \cdots, P_n)$ can select a public of the form $(p', q', g, h)$, where $g, h \in G_{q'}$ and the corresponding secret key is $x = \log_g h$. Then the keys are selected as follows: $P_i$ chooses $x_i \in Z_{q'}$ at random (uniform distribution) and computes $h_i = g^{x_i}$. When all n members have broadcast a commitment, each opens $h_i$. The public key, $h$ is computed as $h = \prod_{i=1}^{n} h_i$. Next it shows how $x$ can be distributed among all $n$ members. Player $P_i$ distributes $x_i$ as follows ($h_1, h_2, \cdots, h_n$ are publicly known). Firstly, $P_i$ chooses at random a polynomial $f_i(z) \in Z_{q'}(z)$ of degree at most $t$ such that $f_i(0) = x_i$. Let $f_i(z) = f_{i0} + f_{i1}z + \cdots + f_{i,k-1}z^t$ where $f_{i,0} = x_i$. Secondly, $P_i$ computes $F_{ij} = g^{f_{ij}}$ for $j = 0, 1, \cdots, t$ and broadcasts $\left( F_{ij} \right)_{j=1,2,\cdots,t}$ ($F_{i,0} = h_i$ is known beforehand). Then, when everybody have sent these $t$ values, $P_i$ sends $S_{ij} = f_i(j)$ secretly to $P_j$ for $j = 1, 2, \cdots n$ (in particular $P_i$ keeps $S_{ii}$). Next $P_i$ verifies that the share received from $P_j \left( s_{ji} \right)$, is consistent with the previously published values by verifying that $g^{s_{ji}} = \prod_{l=0}^{t} F_{jl}^{i^l}$. If this fails, $P_i$ broadcasts that an error has been found, publishes $s_{ij}$ and then stops. At last, $P_i$ computes his share of $x$ as the sum of all shares received $x_i = \sum_{j=1}^{n} s_{ji}$. We notice that, in our case, we need $q' = N$ where $N = pq$. Fortunately, as in the case of the modified Pedersen-VSS protocol, this is not a problem since $Z_N$ is an excellent approximation of a field.

#### 3.2.2 Robust-mult-SS

In order to make our protocol work even in the presence of malicious adversaries, we present the robust multiplication protocol given in [7], which we call Robust-Mult-SS, in which secrets are shared using the Pedersen-VSS protocol. This variant is very similar in spirit to the multiplication protocol given in [8]. Let $\alpha$ and $\beta$ be the two shared secret values being multiplied. Let $\alpha^{(i)}$ and $\beta^{(i)}$ be the shares of $\alpha$ and $\beta$ held by player i respectively. The protocol works as follows. Firstly, each player $i$ commits to the value $\alpha^{(i)}\beta^{(i)}$ by using the Pedersen-VSS protocol [9, 10] and then proves in Zero Knowledge (ZK) that the value it has just committed to is the correct one. Examples of such proofs can be found in [7, 11]. Having done that, then each player can compute locally their shares of the new secret $\alpha\beta$ by computing a linear combination of correct shares it has received. As in the modified version of the multiplication protocol presented in the previous section, the exact values of the coefficients for the linear combination will depend on which set of qualified players is considered in the computation. Like what in previous cases, we would have to use $q' = N$. in the Pedersen-VSS protocol. Clearly, this protocol involves a lot of interaction due to the ZK proofs contained in it. To avoid that, one can resort to solutions for robust multiplication like the one presented in [12] which avoid the use of ZK proofs altogether.

### 3.2.3 Joint-pedersen-VSS

This protocol works as follows. In a first step, each player $i$ commits to a random value in $z_i \in Z_{q'}$ using Pedersen-VSS protocol. Then, in a second step, each player verifies the commitments sent by all other players and builds a set of good players containing all those players whose commitments passed the test. All good players will agree on the same set, which we call QUAL. At last step, all players $i$ in QUAL will compute their shares $R^{(i)}$ of the common shared secret $R = \sum_{i \in QUAL} z_i \mod q'$ by using the secret information they received in the first step. More specifically, if $z_i^{(j)}$ is the share of $z_i$ received by player $j$ from player $i$, then $R^{(i)} = \sum_{j \in QUAL} z_i^{(j)} \mod q'$. Again, the only modification we would need to make in our case is to have $q' = N$.

## 4 SECURITY

In this section, we give several statements about the security of our RPFSTS scheme. Firstly, Lemma 4.1 demonstrates the correctness of our construction. Then, Lemma 4.2 states the threshold-related parameters of our scheme. Finally, Theorem 4.3 relates the forward security of our construction to that of the underlying signature scheme given in [3]. It shows that, as long as we believe that the Bellare-Miner scheme is secure, any adversary working against our scheme would have only a negligible probability of success forging a signature with respect to some time period prior to that in which it gets the secret key.

**Theorem 4.1** Let $PK = (N, T, U_1, U_2, \cdots, U_l)$ and $SK_0^{(\rho)} = (N, T, S_{1,0}^{(\rho)}, S_{2,0}^{(\rho)}, \cdots, S_{l,0}^{(\rho)})$ $(\rho = 1, 2, \cdots, n)$ be the public key and player $\rho's$ secret key generated by RPFSTS.keygen, respectively. Let $\langle j, (Y, Z) \rangle$ be a signature generated by RPFSTS.sign on input $m$ when all $n$ players participated in the distributed protocol. Then RPFSTS.verify$_{PK}$ $(m, \langle j, (Y, Z) \rangle) = 1$.

**Proof:** From the description of the protocol, we know that R is a random element in $Z_N$, $Y = R^{2^{(T+1-j)}} \pmod{N}$, $c_1 c_2 \cdots c_l = H(j, Y, m)$, and $Z \equiv R \prod_{i=1}^{l} S_{i,j}^{c_i} \pmod{N}$. Hence, $Z^{2^{(T+1-j)}} \equiv \left( R \prod_{i=1}^{l} S_{i,j}^{c_i} \right)^{2^{(T+1-j)}} \equiv R^{2^{(T+1-j)}} \prod_{i=1}^{l} S_{i,j}^{c_i 2^{(T+1-j)}}$

$\equiv Y \prod_{i=1}^{l} S_{i,j}^{c_i 2^{(T+1-j)}} \equiv Y \prod_{i=1}^{l} U_i^{c_i} \pmod{N}$, and RPFSTS.verify$_{PK}$ returns 1 on input $(m, \langle j, (Y, Z) \rangle)$ as desired, that is, RPFSTS.verify$_{PK}$ $(m, \langle j, (Y, Z) \rangle) = 1$, we have verified that $(m, \langle j, (Y, Z) \rangle)$ is a valid signature.

**Theorem 4.2** Let RPFSTS be our key-evolving $(t, 2t+1, 2t+1, 3t+1)$-threshold digital signature scheme and let FS-SIG be the (single-user) digital signature scheme given in [3]. Then, RPFSTS is a forward-secure threshold digital signature scheme in the presence of malicious adversaries as long as FS-SIG is a forward-secure signature scheme in the standard (single-user) sense.

**Proof:** The proof similarly follows the description of PFST-SIG in [4].

## ACKNOWLEDGMENTS

Dianjun Lu was financially supported by the Ministry of Education Chunhui Program([2014]1310) and Qinghai Normal University Teaching Research Funds ([2013]43). Xiaoqin Zhang was financially supported by the Young and Middle-aged Scientific Research Funds (2012-QGY-3), and in part by the Ministry of Education Chunhui Program.

## REFERENCES

[1] Desmedt, Y. &Frankel, Y. 1991, Shared generation of authenticators and signatures. *Advances in Cryptology-CRYPTO'91, Santa Barbara*, 576(8):457–469.

[2] Anderson, R. 1997, Two remarks on public-key cryptology. *Relevant material presented by the author in an invited lecture at the ACM CCS'97: 4th Conference on Computer and Communications Security, Zurich, Switzerland*, 1997(4):1–4.

[3] Bellare, m. & miner, S. 1999, A forward-secure digital signature scheme. *Advances in Cryptology-CRYPTO'99, Santa Barbara*, 1666(8):431–448.

[4] Abdalla, M. & Miner, S. & Namprempre, C. 2001, Forward-secure threshold signature schemes. *Cryptology-CT-RSA'2001, Berlin Germany*.

[5] Itkis, G. & Reyzin, L. 2001. Forward-secure signatures with optimal signing and verifying. *CRYPTO 2001, Berlin Germany*, 2139:499–514.

[6] Kozlov, A. & reyzin, L. 2002. Forward-secure signatures with fast key update. *Security in communication networks, Berlin Germany*, 2576:247–262.

[7] Gennaro, R. & Rabin, M.O. & Rabin T. 1998. Simplified VSS and fast-track multiparty computations with applications to threshold cryptography. *In 17th ACM Symposium Annual on Principles of Distributed Computing, Puerto Vallarta, Mexico*, 1998:101–111.

[8] Canetti, R. & Gennaro, R. & Jarecki, S. & Krawczyk, H. & Rabin, T. 1999. Adaptive security for threshold cryptosystems. *In M.J. Wiener, editor, Advances in*

Cryptology - CRYPTO'99, volume 1666 of Lecture Notes in Computer Science, pages 98–115, Santa Barbara, CA, USA, Aug. 15–19, 1999. Springer-Verlag, Berlin, Germany.

[9] Pedersen, T.P. 1991. A threshold cryptosystem without a trusted party. In D.W. Davies, editor, Advances in Cryptology - EUROCRYPT'91, volume 547 of Lecture Notes in Computer Science, pages 522–526, Brighton, UK, Apr. 8-11, 1991. Springer-Verlag, Berlin, Germany.

[10] Gennaro, R. & Jarecki, S. & Krawczyk, H. & Rabin T. 1999. Secure distributed key generation for discrete-log based cryptosystems. In J. Stern, editor, Advances in Cryptology-EUROCRYPT' 99, volume 1592 of Lecture Notes in Computer Science, pages 295–310, Prague, Czech Republic, May 2–6, 1999. Springer-Verlag, Berlin, Germany.

[11] Cramer, R. & Damgard, I.B. 1998. Zero-knowledge proofs for finite field arithmetic, or: Can zero-knowledge be for free? In H. Krawczyk, editor, Advances in Cryptology - CRYPTO'98, volume 1462 of Lecture Notes in Computer Science, pages 424–441, Santa Barbara, CA, USA, Aug. 23–27, 1998. Springer-Verlag, Berlin, Germany.

[12] Abe, M. 1999. Robust distributed multiplication without interaction. In M.J. Wiener, editor, Advances in Cryptology -CRYPTO'99, volume 1666 of Lecture Notes in Computer Science, pages 130–147, Santa Barbara, CA, USA, Aug. 15–19, 1999. Springer-Verlag, Berlin, Germany.

*Multimedia, Communication and Computing Application – Leung (Ed.)*
© *2015 Taylor & Francis Group, London, ISBN 978-1-138-02775-6*

# Fuzzy hazmat transport network problem with random transport risk

Y. Ge, Y.B. Xu & Y.S. Xie
*Beijing Municipal Institute of Labour Protection, Beijing, China*

ABSTRACT: In this paper we consider the problem of the following hazmat transport network design. There are multiple supply points and multiple demand points. For each supply-demand point pair, the transport risk is a random variable. Preference of each route is attached. Our model has two objectives, namely: minimize the transport risk target subject to a chance constraint and maximize the minimal preference among the used routes. Since a transport pattern optimizing two objectives simultaneously usually does not exist, we define non-domination in this setting and propose an efficient algorithm to find some non-dominated transport patterns. We then show the time complexity of the proposed algorithm. Finally, a numerical example is presented to illustrate how our algorithm works.

KEYWORDS: Random transport risk, Chance constraint, Preference of routes, Non-dominated transport pattern

## 1 INTRODUCTION

The transportation of hazardous materials is an important industrial activity, due to the volumes transported and the risks associated with such transports (Erkut 2007). For a comprehensive literature survey on this problem the reader may be addressed to Erkut et al (2007). Hazmat network design has received the attention of researchers only recently. Kara & Verter (2004) is the first work to pose the hazmat network design problem as a bilevel problem. They formulated the design problem as a bilevel integer programming model, where hazmats are grouped into categories based on risk impact, and a network is designed for each group. Erkut & Alp (2007) considered a single-level network design problem. They restricted the network to a tree, so that there is a single path between each origin and destination. This restriction, in effect, removes the second level. The carriers have no alternate paths on the tree. They formulated the tree design problem as an integer programming problem with an objective of minimizing the total transport risk. A simple heuristic for a bilevel network design problem for hazmat transportation is developed (Erkut & Gzara 2008). Verma (2009) developed a bi-objective optimization model, where cost is determined based on the characteristics of railroad industry and the determination of transport risk incorporates the dynamics of railroad accident.

The model considered in this paper is an extension of these previous models. We extend the hazmat transport network problem by considering randomness of transport risk and preference of route. Randomness means that the transport risk may change according to many factors. The preference of route reflects the degree of satisfaction with respect to the chosen route. Therefore two criteria are taken into account. One is to minimize the transport risk target subject to a chance constraint. The other is to maximize the minimal preference among the used routes. But usually a transport pattern optimizing two objectives simultaneously does not exist. So we seek some non-dominated transport patterns.

The rest of this paper is organized as follows. Our problem is formulated in Section 2, and then in Section 3 we present an efficient algorithm to find some non-dominated transport patterns. Section 4 shows how our algorithm works using a numerical example. Finally, Section 5 concludes this paper and discusses further research problems.

## 2 PROBLEM FORMULATION

The following hazmat transport network problem with random transport risk and preference of route is considered.

(C1) There exist $m$ supply points $\{S_1,...,S_m\}$ and $n$ demand points $\{T_1,...,T_n\}$. Let $(i, j)$ denote the route from supply point $S_i$ to demand point $T_j$, $i = 1,...,m$, $j = 1,...,n$.

(C2) The total upper limit provided from each supply point $S_i$ is $a_i$ and the total lower limit to each demand point $T_j$ is $b_j$. Further, we assume that these $a_i$, $b_j$ are positive integers and

$$\sum_{i=1}^{m} a_i \geq \sum_{j=1}^{n} b_j \qquad (1)$$

(C3) Preference of route is attached and it is denoted by $\mu_{ij}$ in (0, 1]. It reflects the degree of satisfaction with respect to the chosen route.

(C4) For each route $(i, j)$, the transport risk $r_{ij}$ is an independent random variable according to a logarithmic normal distribution $LN(m_{ij}, \sigma_{ij}^2)$ with mean $m_{ij}$ and variance $\sigma_{ij}^2$. We denote the transport quantity using the route $(i, j)$ by $x_{ij}$ and assume that these $x_{ij}$ are nonnegative decision variables. The following chance constraint is attached:

$$\Pr\{r_{ij} \leq F\} \geq \alpha, \ (i,j)\,|\,x_{ij} > 0 \tag{2}$$

where $\alpha > 1/2$ and $F$ is also a decision variable denoting the target of bottleneck transport risk to be minimized.

(C5) We consider two criteria: one is to maximize the minimal preference among the used routes and the other is to minimize $F$.

Under the above setting, our fuzzy hazmat transport network problem with random transport risk can be formulated as follows.

FHTNP: minimize $F$
maximize $\min_{i,j}\{\mu_{ij}\,|\,x_{ij} > 0\}$
subject to $\Pr\{r_{ij} \leq F\} \geq \alpha, \ (i,j)\,|\,x_{ij} > 0$

$$\sum_{j=1}^{n} x_{ij} \leq a_i, i = 1,\ldots,m \tag{3}$$

$$\sum_{i=1}^{m} x_{ij} \geq b_j, j = 1,\ldots,n$$

$x_{ij}$ : nonnegative, $i = 1,\ldots,m, j = 1,\ldots,n$

Since

$$\Pr\{r_{ij} \leq F\} \geq \alpha \Leftrightarrow \Pr\left\{\frac{\ln r_{ij} - m_{ij}}{\sigma_{ij}} \leq \frac{\ln F - m_{ij}}{\sigma_{ij}}\right\} \geq \alpha \tag{4}$$

and $(\ln r_{ij} - m_{ij})/\sigma_{ij}$ is a random variable according to the standard normal distribution $N(0,1)$, the chance constraint (2) is equivalent to the following deterministic constraint:

$$\frac{\ln F - m_{ij}}{\sigma_{ij}} \geq K_\alpha \Leftrightarrow F \geq e^{m_{ij} + K_\alpha \sigma_{ij}} \tag{5}$$

where $K_\alpha$ is the $\alpha$ percentile point of the cumulative distribution function of the standard normal distribution and note that $K_\alpha > 0$ since $\alpha > 1/2$.

Since $F$ should be minimized, then problem FHTNP reduces to:

P: minimize $\max_{i,j}\left\{e^{m_{ij} + K_\alpha \sigma_{ij}}\,|\,x_{ij} > 0\right\}$
maximize $\min_{i,j}\left\{\mu_{ij}\,|\,x_{ij} > 0\right\}$
subject to $\sum_{j=1}^{n} x_{ij} \leq a_i, i = 1,\ldots,m$

$$\sum_{i=1}^{m} x_{ij} \geq b_j, j = 1,\ldots,n \tag{6}$$

$x_{ij}$ : nonnegative, $i = 1,\ldots,m, j = 1,\ldots,n$

Next, we define the bi-objective vector $v(\mathbf{x}) = (v(\mathbf{x})_1, v(\mathbf{x})_2)$ of a transport pattern $\mathbf{x} = (x_{ij})$ feasible for P as

$$v(\mathbf{x})_1 = \max_{i,j}\{e^{m_{ij} + K_\alpha \sigma_{ij}}\,|\,x_{ij} > 0\}, \ v(\mathbf{x})_2 = \min_{i,j}\{\mu_{ij}\,|\,x_{ij} > 0\} \tag{7}$$

Generally, a transport pattern optimizing two objectives simultaneously does not exist. Therefore, we seek some non-dominated transport patterns, the definition of which is given as follows.

**Definition 2.1** Let $\mathbf{x}^a$, $\mathbf{x}^b$ be two transport patterns that are feasible for P. Then, we say that $\mathbf{x}^a$ dominates $\mathbf{x}^b$, if

$$v(\mathbf{x}^a)_1 \leq v(\mathbf{x}^b)_1, v(\mathbf{x}^a)_2 \geq v(\mathbf{x}^b)_2 \tag{8}$$

and

$$(v(\mathbf{x}^a)_1, v(\mathbf{x}^a)_2) \neq (v(\mathbf{x}^b)_1, v(\mathbf{x}^b)_2) \tag{9}$$

If there is no transport pattern dominating $\mathbf{x}$, then $\mathbf{x}$ is called a non-dominated transport pattern.

## 3 SOLUTION PROCEDURE

Sorting $\mu_{ij}$, $i=1,\ldots,m$, $j=1,\ldots,n$, and let the result be

$$0 < \mu^1 < \cdots < \mu^g \leq 1 \tag{10}$$

where $g$ is the number of different values of them.

Compute $m_{ij} + K_\alpha \sigma_{ij}$, $i = 1,\ldots,m$, $j = 1,\ldots,n$, and arrange these values in ascending order as follows:

$$c^1 < \cdots < c^l \tag{11}$$

where $l$ is the number of different values of them. Let $\mathbf{U} = (\mu_{ij})_{m \times n}$, $\mathbf{C} = (m_{ij} + K_\alpha \sigma_{ij})_{m \times n}$.

For $u = 1,\ldots,g$, $k=1,\ldots,l$, set

$$c_{ij}^{u,k} = \begin{cases} 0 & \text{if } \mu_{ij} \geq \mu^u, m_{ij} + K_\alpha \sigma_{ij} \leq c^k \\ M & \text{otherwise} \end{cases} \tag{12}$$

where M is a sufficiently large value.

For $u = 1,\ldots,g$, $k = 1,\ldots,l$, denote the cost minimizing transportation problem with the above defined cost values as $P_u^k$:

$P_u^k$: minimize $\sum_{i=1}^{m} \sum_{j=1}^{n} c_{ij}^{u,k} x_{ij}$
subject to $\sum_{j=1}^{n} x_{ij} \leq a_i, i = 1,\ldots,m$

$$\sum_{i=1}^{m} x_{ij} \geq b_j, j = 1,\ldots,n \tag{13}$$

$x_{ij}$ : nonnegative, $i = 1,\ldots,m, j = 1,\ldots,n$

For fixed $u$ and $k$, note that $P_u^k$ is a restricted transportation problem, therefore we don't always have a feasible solution with optimal value 0. If there exists a feasible solution, then it is feasible only using the route $(i, j)$ with

$$\mu_{ij} \geq \mu^u, \ m_{ij} + K_\alpha \sigma_{ij} \leq c^k \tag{14}$$

Denote

$$
\begin{aligned}
&S(u) = \{(i,j) \mid \mu_{ij} = \mu^u, i = 1,\ldots,m, \ j = 1,\ldots,n\} \\
&T(k) = \{(i,j) \mid m_{ij} + K_\alpha \sigma_{ij} = c^k, i = 1,\ldots,m, \ j-1,\ldots,n\} \\
&p = \max\{t \mid \sum\nolimits_{r=t}^{g} |S(r)| \geq n\} \\
&q = \min\{t \mid \sum\nolimits_{r=1}^{t} |T(r)| \geq n\}
\end{aligned} \tag{15}
$$

It is obvious that $P_u^k$ is infeasible when $u$ belongs to $\{p+1,\ldots,g\}$ or $k$ belongs to $\{1,\ldots,q-1\}$.

For each $u$ in $\{1,\ldots,p\}$, we then give the algorithm to find the smallest $k$ in $\{q,\ldots,l\}$ so that $P_u^k$ is feasible. If such $k$ exists, denote it by $k_u$. Otherwise, we can see the following Remark 3.1.

**Remark 3.1** If there exists $u_0$ in $\{1,\ldots,p\}$ so that $P_{u_0}^k$ is infeasible for any $k$ in $\{q,\ldots,l\}$, then $P_u^k$, $u=u_0+1,\ldots,p$ are also infeasible for any $k$ in $\{q,\ldots,l\}$.

For each $u$ in $\{1,\ldots,p\}$, we need to find the smallest $k_u$ so that $P_u^{k_u}$ is feasible. The main idea is to find the smallest $k_u$ so that $P_u^{k_u}$ is feasible is based on a binary method, which is given as follows in detail.

**Algorithm 3.1**

(To find the smallest $k$ so that $P_1^k$ is feasible)

**Step 1.** Set $L = q$ and check whether $P_1^L$ is feasible or not. If feasible, terminate after setting $k_1 = L$. Otherwise, set $U = l$ and check whether $P_1^U$ is feasible or not. If feasible, go to Step 2. Otherwise, terminate due to infeasibility.

**Step 2.** When $U - L > 1$, set

$$K = \lfloor (L+U)/2 \rfloor \tag{16}$$

which denotes the greatest integer not greater than $(L+U)/2$. Check whether $P_1^K$ is feasible or not. If feasible, set $U = K$ and repeat Step 2. Otherwise, set $L = K$, repeat Step 2. When $U - L = 1$, go to Step 3.

**Step 3.** If $P_1^L$ is feasible, set $k_1 = L$. Otherwise, set $k_1 = U$.

For $P_u^k$, $u=2,\ldots,p$, the algorithm is very similar to that of $P_1^k$, the only difference is that we first set $L=k_{u-1}$.

Denote $A=\{(u,k_u)|k_u$ exists, $u = 1,\ldots,p\}$. If there exists $(u_1, k_{u1})$ and $(u_2, k_{u2})$ in $A$, such that $u_1 \neq u_2$ but

$k_{u1}=k_{u2}$, then delete $(\min\{u_1,u_2\},k_{\min\{u1,u2\}})$ from $A$. Let obtained set after deletion be $B$. Note that all elements in $B$ have different first components and also different second components.

For all $(u, k_u)$ in $B$, solve problems $P_u^{k_u}$'s, and let denote the optimal transport pattern by $\mathbf{x}_u^{k_u}$'s. Then we find a set of some non-dominated transport patterns and that of the corresponding bi-objective vectors of problem P, denoted by NDT and NDV respectively.

The validity of our solution procedure is shown in the following proposition.

**Proposition 3.1** The solution procedure for P is valid.

**Proof:** For each $u$, the algorithm to find the smallest $k$ so that $P_u^k$ is feasible is a binary feasibility checking method. For each $(u, k_u)$ in $A$, $P_u^{k_u}$ is feasible, conversely, for each feasible $P_u^k$, we have $(u, k)$ belonging to $A$. For each $(u, k_u)$ in $B$, an optimal transport pattern $\mathbf{x}_u^{k_u}$ of $P_u^{k_u}$ is a non-dominated transport pattern of P and $(e^{cku}, \mu^u)$ is the corresponding bi-objective vector, that is,

$$
\begin{aligned}
&\mathrm{NDT} = \{\mathbf{x}_u^{k_u} \mid (u,k_u) \in B\} \\
&\mathrm{NDV} = \{(e^{c^{k_u}}, \mu^u) \mid (u,k_u) \in B\}
\end{aligned} \tag{17}
$$

Therefore, our solution procedure is valid.

Next we show the time complexity of our solution procedure for P.

**Theorem 3.1** The time complexity of our solution procedure for P is $O(mn(m+n)^3 \log(m+n))$.

**Proof.** Note that $g = l = O(mn)$, so sorting $\mu_{ij}$ and $m_{ij} + K_\alpha \sigma_{ij}$ both takes at most $O(mn\log(mn))$ operations. For each $u$, the time complexity of the algorithm to find the smallest $k$ in $\{q,\ldots,l\}$ so that $P_u^k$ is feasible follows from the fact that the binary search over $l$ values has time complexity $O(\log l)$, and each feasibility checking takes $O(mn)$ because at most $O(mn)$ elements should be checked. So for each $u$, checking totally needs $O(mn\log(mn))$ computational times. The algorithm is executed at most $O(g)$ times to find the smallest $k$ in $\{q,\ldots,l\}$ so that $P_u^k$ is feasible. So checking totally needs $O((mn)^2\log(mn))$. Solving each feasible classical transportation problem takes at most $O((m+n)^3\log(m+n))$ (Ahuja et al. 1989) and totally at most $O(mn)$ classical transportation problems should be solved. Therefore this part takes at most $O(mn(m+n)^3\log(m+n))$ computational times. Consequently, the time complexity is

$$
\begin{aligned}
&O(\max\{(mn)^2 \log(mn), mn(m+n)^3 \log(m+n)\}) \\
&= O(mn(m+n)^3 \log(m+n)).
\end{aligned} \tag{18}
$$

## 4 NUMERICAL EXAMPLE

Consider FHTNP with 5 supply points and 5 demand points, set $\alpha = 0.9987$. The preferences of routes are given in the following matrix:

$$U = \begin{pmatrix} 0.5 & 0.8 & 0.4 & 0.6 & 0.5 \\ 0.75 & 0.6 & 0.7 & 0.8 & 1 \\ 0.85 & 1 & 0.6 & 0.8 & 0.4 \\ 0.6 & 0.4 & 0.5 & 0.75 & 0.6 \\ 0.7 & 0.5 & 0.6 & 0.5 & 1 \end{pmatrix}$$

The values of $a_i$, $b_j$ and $r_{ij} \sim LN(m_{ij}, \sigma_{ij}^2)$ are given in Table 1.

Our problem FHTNP reduces to problem P:

P: minimize $\max_{i,j}\{ m_{ij} + 3.0\sigma_{ij} \mid x_{ij} > 0 \}$

maximize $\min_{i,j}\{ \mu_{ij} \mid x_{ij} > 0 \}$

subject to $\sum_{j=1}^{5} x_{ij} \le a_i, i = 1,\ldots,5$

$\qquad\qquad \sum_{i=1}^{5} x_{ij} \ge b_j, j = 1,\ldots,5$

$\qquad\qquad x_{ij}$ : nonnegative, $i, j = 1,\ldots,5$

Sorting $\mu_{ij}$, $i, j = 1,\ldots,5$, we obtain

$0 < \mu^1 = 0.4 < \mu^2 = 0.5 < \mu^3 = 0.6 < \mu^4 = 0.7 < \mu^5$
$= 0.75 < \mu^6 = 0.8 < \mu^7 = 0.85 < \mu^8 = 1$

Compute $m_{ij} + 3.0\sigma_{ij}$, $i, j = 1,\ldots,5$, we obtain

$$C = \begin{pmatrix} 4.5 & 8.2 & 7.6 & 8.2 & 5.9 \\ 8.4 & 5.9 & 3.1 & 5.9 & 7.6 \\ 7.9 & 5.8 & 11 & 4.5 & 11 \\ 7.9 & 5.9 & 8.4 & 5.8 & 7.9 \\ 4.5 & 8.4 & 8.2 & 7.6 & 11 \end{pmatrix}$$

Arrange these values in ascending order, that is,

$c^1 = 3.1 < c^2 = 4.5 < c^3 = 5.8 < c^4 = 5.9 < c^5 = 7.6$
$< c^6 = 7.9 < c^7 = 8.2 < c^8 = 8.4 < c^9 = 11$

For $u = 1,\ldots,8$, $k = 1,\ldots,9$, set

$$c_{ij}^{u,k} = \begin{cases} 0 & \text{if } \mu_{ij} \ge \mu^u, m_{ij} + 3.0\sigma_{ij} \le c^k \\ M & \text{otherwise} \end{cases}, i, j = 1,\ldots,5$$

where M is a sufficiently large value.

It is obvious that $p = 6$, $q = 3$.

For $u = 1,\ldots,6$, $k = 3,\ldots,9$, $P_u^k$ has the following form:

$P_u^k$: minimize $\sum_{i=1}^{5} \sum_{j=1}^{5} c_{ij}^{u,k} x_{ij}$

subject to $\sum_{j=1}^{5} x_{ij} \le a_i, i = 1,\ldots,5$

$\qquad\qquad \sum_{i=1}^{5} x_{ij} \ge b_j, j = 1,\ldots,5$

$\qquad\qquad x_{ij}$ : nonnegative, $i, j = 1,\ldots,5$

Next we give the solution procedure for P.

Find the smallest $k$ in $\{3,\ldots,9\}$ so that $P_1^k$ is feasible.

Step 1. Set $L = 3$ and $P_1^3$ is infeasible. Set $U = 9$, $P_1^9$ is feasible. Go to Step 2.

Step 2. $U - L = 6 \ne 1$. Set $K = 6$ and $P_1^6$ is feasible. Set $U = 6$, repeat Step 2.

Step 2. $U - L = 3 \ne 1$. Set $K = 4$ and $P_1^4$ is infeasible. Set $L = 4$, repeat Step 2.

Step 2. $U - L = 2 \ne 1$. Set $K = 5$ and $P_1^5$ is feasible. Set $U = 5$, repeat Step 2.

Step 2. $U - L = 1$, so go to Step 3.

Step 3. $P_1^4$ is infeasible, so set $k_1 = 5$.

Find the smallest $k$ in $\{5,\ldots,9\}$ so that $P_2^k$ is feasible.

Step 1. Set $L = 5$ and $P_2^5$ is feasible. Set $k_2 = 5$.

Find the smallest $k$ in $\{5,\ldots,9\}$ so that $P_3^k$ is feasible.

Step 1. Set $L = 5$ and $P_3^5$ is infeasible. Set $U = 9$, $P_3^9$ is feasible. Go to Step 2.

Step 2. $U - L = 4 \ne 1$. Set $K = 7$ and $P_3^7$ is feasible. Set $U = 7$, repeat Step 2.

Step 2. $U - L = 2 \ne 1$. Set $K = 6$ and $P_3^6$ is infeasible. Set $L = 6$, repeat Step 2.

Step 2. $U - L = 1$, so go to Step 3.

Step 3. $P_3^6$ is infeasible, so set $k_3 = 7$.

Table 1. The values of $a_i$, $b_j$ and the distribution of $r_{ij}$.

| $i\backslash j$ | 1 | 2 | 3 | 4 | 5 | $a_i$ |
|---|---|---|---|---|---|---|
| 1 | $LN(3,0.5^2)$ | $LN(7,0.4^2)$ | $LN(4,1.2^2)$ | $LN(1,2.4^2)$ | $LN(5,0.3^2)$ | 50 |
| 2 | $LN(6,0.8^2)$ | $LN(5,0.3^2)$ | $LN(1,0.7^2)$ | $LN(2,1.3^2)$ | $LN(7,0.2^2)$ | 85 |
| 3 | $LN(7,0.3^2)$ | $LN(4,0.6^2)$ | $LN(8,1.0^2)$ | $LN(3,0.5^2)$ | $LN(2,3.0^2)$ | 30 |
| 4 | $LN(4,1.3^2)$ | $LN(2,1.3^2)$ | $LN(3,1.8^2)$ | $LN(4,0.6^2)$ | $LN(1,2.3^2)$ | 60 |
| 5 | $LN(3,0.5^2)$ | $LN(6,0.8^2)$ | $LN(4,1.4^2)$ | $LN(1,2.2^2)$ | $LN(5,2.0^2)$ | 35 |
| $bj$ | 60 | 35 | 55 | 50 | 40 | – |

Find the smallest $k$ in $\{7,8,9\}$ so that $P_4^k$ is feasible. Step 1. Set $L = 7$ and $P_4^7$ is infeasible. Set $U = 9$, $P_4^9$ is infeasible. Therefore, there exists no $k$ in $\{7,8,9\}$ so that $P_4^k$ is feasible. From Remark 3.1, such a case also holds for $P_5^k$, $P_6^k$.

Therefore B=$\{(2,5), (3,7)\}$, solve $P_2^5$ 和$P_3^7$, we obtain the optimal transport patterns $\mathbf{x}_2^5$ 和$\mathbf{x}_3^7$:

$$\mathbf{x}_2^5: \quad x_{11} = 25, x_{15} = 15, x_{22} = 5, x_{23} = 55, x_{25} = 25, x_{32} = 20,$$
$$x_{44} = 50, x_{51} = 35, \text{ other } x_{ij} = 0$$

$$\mathbf{x}_3^7: \quad x_{14} = 50, x_{22} = 5, x_{23} = 55, x_{25} = 25, x_{32} = 30, x_{41} = 25,$$
$$x_{45} = 15, x_{51} = 35, \text{ other } x_{ij} = 0$$

which are the non-dominated transport patterns of P, and the corresponding bi-objective vectors are $(e^{7.6}, 0.5)$ and $(e^{8.2}, 0.6)$, respectively. That is,

$$\text{NDT} = \left\{\mathbf{x}_2^5, \mathbf{x}_3^7\right\}, \quad \text{NDV} = \left\{\left(e^{7.6}, 0.5\right), \left(e^{8.2}, 0.6\right)\right\}.$$

## 5 CONCLUSIONS

In this paper, we have considered a bi-criteria fuzzy hazmat transport network problem with random transport risk and developed an algorithm to find some non-dominated transport patterns. Further, we have shown the validity and time complexity of the algorithm. Besides, our algorithm is illustrated by using a numerical example. As a further research problem, we should consider the flexibility of supply and demand quantity, which is the case that the total quantity from supplies is less than that to demand customers. This case makes the problem three criteria one. Additionally, there remain many other variants of hazmat transport network problem to be considered and solved.

## ACKNOWLEDGEMENTS

This research has been supported by Beijing Postdoctoral Work Foundation (No.B181) and Beijing Finance Project "Creation of Fine Management Mode for Safety Production in Industrial Park".

## REFERENCES

Ahuja, R.K., Orlin, J.B., Tarjan, R.E. 1989. Improved time bounds for the maximum flow problem. *SIAM Journal on Computing* 18:939–954.

Erkut, E. 2007. Introduction to the special issue. *Computers & Operations Research* 34(5):1241–1242.

Erkut, E. & Alp, O. 2007. Designing a road network for hazardous materials shipments. *Computers & Operations Research* 34(5):1389–1405.

Erkut, E. & Gzara, F. 2008. Solving the hazmat transport network design problem. *Computers & Operations Research* 35:2234–2247.

Erkut, E., Tjandra, S., Verter, V. 2007. Hazardous materials transportation. In C. Barnhart & G. Laporte (eds), *Handbooks in Operations Research and Management Science: Transportation* 14:539–621. Amsterdam: Elsevier.

Kara, B.Y. & Verter, V. 2004. Designing a road network for hazardous materials transportation. *Transportation Science* 38(2):188–196.

Verma, M. 2009. A cost and expected consequence approach to planning and managing railroad transportation of hazardous materials. *Transportation Research Part D* 14: 300–308.

*Multimedia, Communication and Computing Application – Leung (Ed.)*
© *2015 Taylor & Francis Group, London, ISBN 978-1-138-02775-6*

# Multi-objective hazmat transport network problem with random and fuzzy factors

Y. Ge, Y.S. Xie & B.X. Song
*Beijing Municipal Institute of Labour Protection, Beijing, China*

ABSTRACT: This paper considers a multi-objective hazmat transport network problem from both random and fuzzy factors. There exists $m$ supply points with flexible supply quantity and $n$ demand points with flexible demand quantity. For each supply-demand point pair, the transport risk is a random, variable and existent possibility that denotes the preference choosing this route is attached. Satisfaction degree about the supply and the demand quantity are attached to each supply and each demand point. Respectively, they are denoted by membership functions of corresponding fuzzy sets. Our model has three criteria, namely, minimize the transport risk of target subject due to chance constraint, maximize the minimal preference among the used routes in transportation and maximize the minimal satisfaction degree among all supply and demand points. Since a transport pattern optimizing three objectives simultaneously does not exist, we define non-domination in this setting and propose an efficient algorithm to find some non-dominated transport patterns. We then show the time complexity of the algorithm. Finally, a numerical example is presented to illustrate how our algorithm runs.

KEYWORDS: Random transport risk, Preference of routes, Fuzzy supply, Fuzzy demand, Non-dominated transport pattern

## 1 INTRODUCTION

The transportation of hazardous materials is an important industrial activity, both due to the volumes transported and the risks associated with such transports (Erkut 2007). As a comprehensive literature survey on this problem the reader may be addressed to Erkut et al (2007). Hazmat network design has received the attention of researchers only recently. Kara & Verter (2004) is the first work to pose the hazmat network design problem as a bilevel problem. They formulated the design problem as a bilevel integer programming model, where hazmats are grouped into categories based on risk impact, and a network is designed for each group. Erkut & Alp (2007) considered a single-level network design problem. They restricted the network to a tree, so that there is a single path between each origin and destination. This restriction, in effect, removes the second level; the carriers have no alternate paths on the tree. They formulated the tree design problem as an integer programming problem with an objective of minimizing the total transport risk. A simple heuristic for a bilevel network design problem for hazmat transportation is developed (Erkut & Gzara 2008). Verma (2009) developed a bi-objective optimization model, where the determination of cost is based on the characteristics of railroad industry and the determination of transport risk incorporates the dynamics of railroad accident.

The model considered in this paper is an extension of these previous models. We propose a hazmat transport network problem from randomness of transport risk, preference of route and also flexibility of supply and demand quantity. Randomness means that the transport risk may change according to many factors. The preference of route reflects on the satisfaction degree by using the route. The flexibility reflects on the actual situation that total quantity from suppliers is less than the demand of customers. So three objectives are considered, namely, minimize the transport risk of target subject due to chance constraint, maximize the minimal preference among the used routes and maximize the minimal satisfaction degree among all supply and demand points. But usually a transport pattern optimizing three objectives simultaneously does not exist. So we seek some non-dominated transport patterns.

The rest of this paper is organized as follows. Our problem is firstly formulated in Section 2, and then in Section 3, we present an algorithm to find some non-dominated transport patterns, demonstrate its validity and study its time complexity. Section 4 shows how our algorithm runs by using an example. Finally, Section 5 concludes this paper and discusses further research problems.

## 2 PROBLEM FORMULATION

In this paper, we consider the following multi-objective hazmat transport network problem from fuzzy and random factors:

(C1) There exists $m$ supply points $\{S_1,\ldots,S_m\}$ and $n$ demand points $\{T_1,\ldots,T_n\}$. Let $(i,j)$ denote the route from supply point $S_i$ to demand point $T_j$, $i=1,\ldots,m$, $j=1,\ldots,n$.

(C2) Let $s_i$, $t_j$ be the total flow value sent from $S_i$ and to $T_j$, respectively. Upper limits of supply quantity for each supply point and lower limits of demand quantity for each demand point are flexible. They are expressed by the following two kinds of membership functions $\mu_{S_i}(s_i)$ and $\mu_{T_j}(t_j)$ for fuzzy supply quantity from $S_i$ and fuzzy demand quantity to $T_j$, respectively. They characterize the satisfaction degrees of supply and demand points.

$$\mu_{S_i}(s_i) = \begin{cases} 1 & (s_i \le a_i) \\ \dfrac{b_i - s_i}{b_i - a_i} & (a_i < s_i < b_i) \\ 0 & (s_i \ge b_i) \end{cases}$$

$$\mu_{T_j}(t_j) = \begin{cases} 0 & (t_j \le d_j) \\ \dfrac{t_j - d_j}{e_j - d_j} & (d_j < t_j < e_j) \\ 1 & (t_j \ge e_j) \end{cases} \tag{1}$$

where $a_i < b_i$, $d_j < e_j$ and $a_i$, $b_i$, $d_j$, $e_j$ are positive integers.

We assume that

$$\sum_{i=1}^{m} a_i < \sum_{j=1}^{n} e_j, \ \sum_{i=1}^{m} b_i > \sum_{j=1}^{n} d_j. \tag{2}$$

Note that

$$s_i = \sum_{j=1}^{n} x_{ij}, i=1,\ldots,m, \ t_j = \sum_{i=1}^{m} x_{ij}, j=1,\ldots,n. \tag{3}$$

(C3) Preference of route is attached and it is denoted by $\mu_{ij}$ in $(0, 1]$. It reflects the degree of satisfaction relates to the chosen route.

(C4) For each route $(i,j)$, the transport risk $r_{ij}$ is an independent random variability according to a logarithmic normal distribution $LN(m_{ij}, \sigma_{ij}^2)$ with mean $m_{ij}$ and variance $\sigma_{ij}^2$. We denote the transport quantity using the route $(i,j)$ by $x_{ij}$ and assume that these $x_{ij}$ are nonnegative integer decision variables. The following chance constraint is attached:

$$\Pr\{r_{ij} \le F\} \ge \alpha, \ (i,j)\,|\,x_{ij} > 0 \tag{4}$$

where $\alpha > 1/2$ and $F$ is also a decision variable to be minimized.

(C5) Our model has three criteria, namely: minimize $F$, maximize the minimal preference among the used routes and maximize the minimal satisfaction degree with to the flexibility of supply and demand quantity.

Under the above setting, our multi-objective hazmat transport network problem can be formulated as follows.

MHTNP: minimize $F$

maximize $\min_{i,j}\{\mu_{ij} \,|\, x_{ij} > 0\}$

maximize $\min_{i,j}\{\mu_{S_i}(s_i), \mu_{T_j}(t_j)\}$

subject to $\Pr\{r_{ij} \le F\} \ge \alpha, \ (i,j)\,|\,x_{ij} > 0$ (5)

$$\sum_{j=1}^{n} x_{ij} = s_i, i=1,\ldots,m$$

$$\sum_{i=1}^{m} x_{ij} = t_j, j=1,\ldots,n$$

$$x_{ij} : \text{nonnegative}, i=1,\ldots,m,$$
$$j=1,\ldots,n$$

Since

$$\Pr\{r_{ij} \le F\} \ge \alpha \Leftrightarrow \Pr\left\{\frac{\ln r_{ij} - m_{ij}}{\sigma_{ij}} \le \frac{\ln F - m_{ij}}{\sigma_{ij}}\right\} \ge \alpha \tag{6}$$

and $(\ln r_{ij} - m_{ij})/\sigma_{ij}$ is a random variable according to the standard normal distribution $N(0,1)$, the chance constraint (4) is equivalent to the following deterministic constraint:

$$\frac{\ln F - m_{ij}}{\sigma_{ij}} \ge K_\alpha \Leftrightarrow F \ge e^{m_{ij} + K_\alpha \sigma_{ij}} \tag{7}$$

where $K_\alpha$ is the $\alpha$ percentile point of the cumulative distribution function of the standard normal distribution and note that $K_\alpha > 0$ since $\alpha > 1/2$.

Since $F$ should be minimized, then MHTNP reduces to:

P: minimize $\max_{i,j}\{e^{m_{ij} + K_\alpha \sigma_{ij}} \,|\, x_{ij} > 0\}$

maximize $\min_{i,j}\{\mu_{ij} \,|\, x_{ij} > 0\}$

maximize $\min_{i,j}\{\mu_{S_i}(s_i), \mu_{T_j}(t_j)\}$ (8)

subject to $\sum_{j=1}^{n} x_{ij} = s_i, i=1,\ldots,m$

$$\sum_{i=1}^{m} x_{ij} = t_j, j=1,\ldots,n$$

$$x_{ij} : \text{nonnegative}, i=1,\ldots,m, j=1,\ldots,n$$

Next, we define the three-objective vector $v(\mathbf{x})$ of a transport pattern $\mathbf{x} = (x_{ij})$ feasible for P as

$$v(\mathbf{x}) = \begin{pmatrix} v(\mathbf{x})_1 \\ v(\mathbf{x})_2 \\ v(\mathbf{x})_3 \end{pmatrix}^{\mathrm{T}} = \begin{pmatrix} \max_{i,j}\{e^{m_{ij} + K_\alpha \sigma_{ij}} \,|\, x_{ij} > 0\} \\ \min_{i,j}\{\mu_{ij} \,|\, x_{ij} > 0\} \\ \min_{i,j}\{\mu_{S_i}(s_i), \mu_{T_j}(t_j)\} \end{pmatrix}^{\mathrm{T}} \tag{9}$$

Generally, a transport pattern optimizing three objectives simultaneously does not exist. Therefore, we seek some non-dominated transport patterns, the definition of which is given as follows.

**Definition 2.1** Let $\mathbf{x}^a$, $\mathbf{x}^b$ be two transport patterns that are feasible for P. Then, we say that $\mathbf{x}^a$ dominates $\mathbf{x}^b$, if

$$v(\mathbf{x}^a)_1 \leq v(\mathbf{x}^b)_1, v(\mathbf{x}^a)_2 \geq v(\mathbf{x}^b)_2, v(\mathbf{x}^a)_3 \geq v(\mathbf{x}^b)_3 \quad (10)$$

and at least one inequality holds as a strict inequality. If there exists no transport pattern dominating $\mathbf{x}$, $\mathbf{x}$ is called a non-dominated transport pattern.

## 3 SOLUTION PROCEDURE

Note that $s_i$, $t_j$ are integers, then we can denote ranges of $\mu_{Si}(s_i)$ and $\mu_{Tj}(t_j)$ with $\{\mu_{Si,1},...,\mu_{Si,ki}\}$ and $\{\mu_{Tj,1},...,\mu_{Tj,lj}\}$, respectively, $i = 1,...,m$, $j = 1,...,n$. Now sorting them, let the result be

$$0 < \mu^1 < \cdots < \mu^g \leq 1 \quad (11)$$

where $g$ is the number of different values of them.

For fixed $u$ in $\{1,...,g\}$, we only consider

$$\mu_{S_i}(s_i) \geq \mu^u, \mu_{T_j}(t_j) \geq \mu^u \quad (12)$$

that is,

$$s_i \leq b_i - \mu^u(b_i - a_i), i = 1,...,m$$
$$t_j \geq d_j + \mu^u(e_j - d_j), j = 1,...,n \quad (13)$$

As it is easily seen, the total supply quantity should be not less than the total demand quantity. Otherwise, the problem becomes infeasible. So we assume

$$\sum_{i=1}^{m}\{b_i - \mu^u(b_i - a_i)\} \geq \sum_{j=1}^{n}\{d_j + \mu^u(e_j - d_j)\} \quad (14)$$

that is,

$$\mu^u \leq \frac{\sum_{i=1}^{m}b_i - \sum_{j=1}^{n}d_j}{\sum_{i=1}^{m}(b_i - a_i) + \sum_{j=1}^{n}(e_j - d_j)} \triangleq \mu^*. \quad (15)$$

Denote

$$u_0 = \arg\{u \in \{1,...,g\} \mid \mu^u \leq \mu^* < \mu^{u+1}\}$$
$$h = \max\{u \mid \sum_{i=1}^{m}s_i(u) \geq \sum_{j=1}^{n}t_j(u), u = 1,...,u_0\} \quad (16)$$

where

$$s_i(u) = \lfloor b_i - \mu^u(b_i - a_i) \rfloor, t_j(u) = \lceil d_j + \mu^u(e_j - d_j) \rceil \quad (17)$$

Here the symbols denote the greatest integer not greater than $b_i - \mu^u(b_i - a_i)$ and the smallest integer not smaller than $d_j + \mu^u(e_j - d_j)$, respectively.

Sorting $\mu_{ij}$, $i = 1,...,m$, $j = 1,...,n$, and let the result be.

$$0 < \beta^1 < \cdots < \beta^k \leq 1 \quad (18)$$

where $k$ is the number of different values of them.

Compute $m_{ij} + K_\alpha\sigma_{ij}$, $i = 1,...,m$, $j = 1,...,n$, and arrange these values in ascending order as follows:

$$c^1 < \cdots < c^l \quad (19)$$

where $l$ is the number of different values of them. Let $\mathbf{U} = (\mu_{ij})_{m \times n}$, $\mathbf{C} = (m_{ij} + K_\alpha\sigma_{ij})_{m \times n}$.

For $p = 1,...,k$, $q = 1,...,l$, set

$$c_{ij}^{p,q} = \begin{cases} 0 & \text{if } \mu_{ij} \geq \beta^p, m_{ij} + K_\alpha\sigma_{ij} \leq c^q \\ \infty & \text{otherwise} \end{cases} \quad (20)$$

For $u = 1,...,h$, $p = 1,...,k$, $q = 1,...,l$, denote the cost minimizing transportation problem with the above defined cost values as $P_u^{p,q}$:

$$P_u^{p,q}: \text{ minimize } \sum_{i=1}^{m}\sum_{j=1}^{n}c_{ij}^{p,q}x_{ij}$$

$$\text{subject to } \sum_{j=1}^{n}x_{ij} \leq s_i(u), i = 1,...,m$$
$$\sum_{i=1}^{m}x_{ij} \geq t_j(u), j = 1,...,n \quad (21)$$
$$x_{ij} : \text{nonnegative}, i = 1,...,m,$$
$$j = 1,...,n$$

Denote

$$S(p) = \{(i,j) \mid \mu_{ij} = \beta^p, i = 1,...,m, j = 1,...,n\}$$
$$T(q) = \{(i,j) \mid m_{ij} + K_\alpha\sigma_{ij} = c^q, i = 1,...,m, j = 1,...,n\}$$
$$k^* = \max\{t \mid \sum_{r=1}^{k}\mid S(r)\mid \geq n\}$$
$$l^* = \min\{t \mid \sum_{r=1}^{t}\mid T(r)\mid \geq n\} \quad (22)$$

For $u = 1,...,h$, $p = 1,...,k^*$, we propose an algorithm to find the smallest $q$ in $\{l^*,...,l\}$ such that $P_u^{p,q}$ is feasible. If such $q$ exists, denote it by $q(u, p)$. Otherwise, see the following Remark 3.1.

**Remark 3.1** For any fixed $u$, if exists $p_0$ in $\{1,...,k^*\}$, such that $P_u^{p0,q}$ is infeasible for any $q$ in $\{l^*,...,l\}$, then $P_u^{p,q}, p = p_0 + 1,...,k^*$ are also infeasible for any $q$ in $\{l^*,...,l\}$.

**Algorithm 3.1**
(To find the smallest $q$ such that $P_u^{1,q}$ is feasible)

**Step 1.** Set $L = l^*$ and check whether $P_u^{1,L}$ is feasible or not. If feasible, terminate after setting $q(u, 1) = L$. Otherwise, set $U = l$ and check whether $P_u^{1,U}$ is feasible or not. If feasible, go to Step 2. Otherwise, terminate as infeasible.

**Step 2.** When $U - L > 1$, set

$$K = \lfloor (L+U)/2 \rfloor \quad (23)$$

447

and check whether $P_u^{1,K}$ is feasible or not. If feasible, set $U = K$ and repeat Step 2. Otherwise, set $L = K$, repeat Step 2. When $U - L = 1$, go to Step 3.

**Step 3.** If $P_u^{1,L}$ is feasible, set $q(u,1) = L$. Otherwise, set $q(u,1) = U$.

For $P_u^{p,q}$, $p = 2,\ldots,k^*$, the algorithm is very similar to that of $P_u^{1,q}$, the only difference is we first set $L = q(u, p - 1)$.

**Remark 3.2** $P_u^{p,q(u,p)}$ is feasible means that there exists a feasible transport pattern, which is obviously optimal for $P_u^{p,q(u,p)}$, and the optimal value is 0.

Denote

$$A = \{(u, p, q(u, p)) \mid q(u, p) \text{ exists}, u = 1,\cdots,h,$$
$$p = 1,\cdots,k^*\}$$
$$B = \{(u, p, q(u, p)) \in A \mid \text{there exists an optimal} \quad (24)$$
$$\text{transport}$$

pattern $\mathbf{x}_u^{p,q(u,p)}$ of $P_u^{p,q(u,p)}$ such that $v(\mathbf{x}_u^{p,q(u,p)})_3 = \mu^u\}$.

For $(u, p, q(u, p))$ in $B$, let $\mathbf{x}_u^{p,q(u,p)}$ be an optimal transport pattern of $P_u^{p,q(u,p)}$ such that $v(\mathbf{x}_u^{p,q(u,p)})_3 = \mu^u$, and the corresponding three-objective vector is ($\exp (c^{q(u,p)})$, $\beta^p, \mu^u$). Then using the definition of non-domination, we find a set of some non-dominated transport patterns and that of the corresponding three-objective vectors of P, denoted by NDT and NDV respectively.

The validity of our solution procedure for P is shown in the following proposition.

**Proposition 3.1** The solution procedure for P is valid.

**Proof:** For each pair $(u, p)$, the algorithm for finding the smallest $q$ such that $P_u^{p,q}$ is feasible is a binary method. For each $(u, p, q(u, p))$ in $B$, $P_u^{p,q(u,p)}$ is feasible and there exists an optimal transport pattern $\mathbf{x}_u^{p,q(u,p)}$ of $P_u^{p,q(u,p)}$ such that $v(\mathbf{x}_u^{p,q(u,p)})_3 = \mu^u$. From the definition of non-domination, we can obtain some non-dominated transport pattern together with the corresponding three-objective vectors for P. Therefore, our solution procedure is valid.

Next we show the time complexity of our solution procedure for P.

**Theorem 3.1** The time complexity of our solution procedure for P is

$$O(Mmn(m + n)^3 \log(m + n)) \quad (25)$$

where

$$M = \sum_{i=1}^{m}(b_i - a_i) + \sum_{j=1}^{n}(e_j - d_j).$$

**Proof.** Note that $h$ is not less than M and so sorting $\mu^1,\ldots,\mu^h$ takes at most $O(M\log M)$ operations. Further, $k = l = O(mn)$ and sorting both $\mu_{ij}$ and $m_{ij} + K_\alpha\sigma_{ij}$ takes at most $O(mn\log(mn))$ operations. For each pair $(u, p)$, the time complexity of the algorithm to find the smallest $q$ in $\{l^*,\ldots,l\}$ such that $P_u^{p,q}$ is

feasible follows from the fact that the binary search over $l$ values has time complexity $O(\log l)$, and each feasibility checking takes $O(mn)$. Thus for each $(u, p)$, checking requires $O(mn\log(mn))$ computational times totally. The algorithm to find the smallest $q$ in $\{l^*,\ldots,l\}$ such that $P_u^{p,q}$ is feasible executes at most $O(hk)$ times. Therefore checking totally needs $O(M(mn)^2\log(mn))$. Solving each feasible classical transportation problem takes at most $O((m+n)^3 \log(m+n))$ (Ahuja et al. 1989) and totally at most $O(Mmn)$ classical transportation problems should be solved, therefore this part takes at most $O(Mmn (m+n)^3\log(m+n))$ computational times. And the non-domination checking needs at most $O(Mmn)$. Consequently, the time complexity of our solution procedure for P is

$$O(\max\{M(mn)^2 \log(mn), Mmn(m + n)^3 \log(m + n)\})$$
$$= O(Mmn(m + n)^3 \log(m + n)). \quad (26)$$

## 4 NUMERICAL EXAMPLE

Consider MHTNP with 5 supply points and 5 demand points, set $\alpha = 0.9987$. The preferences of routes are given in the following matrix:

$$U = \begin{pmatrix} 0.5 & 0.8 & 0.4 & 0.6 & 0.5 \\ 0.75 & 0.6 & 0.7 & 0.8 & 1 \\ 0.85 & 1 & 0.6 & 0.8 & 0.4 \\ 0.6 & 0.4 & 0.5 & 0.75 & 0.6 \\ 0.7 & 0.5 & 0.6 & 0.5 & 1 \end{pmatrix}$$

The values of $a_i$, $b_i$, $d_j$, $e_j$ and $r_{ij} \sim LN(m_{ij},\sigma_{ij}^2)$ are given in Tables 1-2.

Table 1. The values of $a_i$, $b_i$, $d_j$, $e_j$.

|       | 1  | 2  | 3  | 4  | 5  |
|-------|----|----|----|----|----|
| $a_i$ | 10 | 12 | 5  | 7  | 6  |
| $b_i$ | 14 | 18 | 8  | 10 | 12 |
| $d_j$ | 12 | 6  | 10 | 7  | 4  |
| $e_j$ | 15 | 8  | 13 | 9  | 7  |

Our problem MHTNP reduces to problem P:

P:  minimize $\max_{i,j}\{m_{ij} + 3.0\sigma_{ij} \mid x_{ij} > 0\}$

maximize $\min_{i,j}\{\mu_{ij} \mid x_{ij} > 0\}$

maximize $\min_{i,j}\{\mu_{S_i}(s_i),\mu_{T_j}(t_j)\}$

subject to $\sum_{j=1}^{5} x_{ij} = s_i, i = 1,\ldots,5$

$\sum_{i=1}^{5} x_{ij} = t_j, j = 1,\ldots,5$

$x_{ij}$ : nonnegative, $i, j = 1,\ldots,5$

**Table 2.** The distribution of $r_{ij}$.

| $i\backslash j$ | 1 | 2 | 3 | 4 | 5 |
|---|---|---|---|---|---|
| 1 | $LN(3,0.5^2)$ | $LN(7,0.4^2)$ | $LN(4,1.2^2)$ | $LN(1,2.4^2)$ | $LN(5,0.3^2)$ |
| 2 | $LN(6,0.8^2)$ | $LN(5,0.3^2)$ | $LN(1,0.7^2)$ | $LN(2,1.3^2)$ | $LN(7,0.2^2)$ |
| 3 | $LN(7,0.3^2)$ | $LN(4,0.6^2)$ | $LN(8,1.0^2)$ | $LN(3,0.5^2)$ | $LN(2,3.0^2)$ |
| 4 | $LN(4,1.3^2)$ | $LN(2,1.3^2)$ | $LN(3,1.8^2)$ | $LN(4,0.6^2)$ | $LN(1,2.3^2)$ |
| 5 | $LN(3,0.5^2)$ | $LN(6,0.8^2)$ | $LN(4,1.4^2)$ | $LN(1,2.2^2)$ | $LN(5,2.0^2)$ |

Since $s_i$, $t_j$ are integers, sorting the values of $\mu_{S_i}(s_i)$ and $\mu_{T_j}(t_j)$, we obtain

$$0 < \mu^1 = 1/6 < \mu^2 = 1/4 < \mu^3 = 1/3 < \mu^4 = 1/2$$
$$< \mu^5 = 2/3 < \mu^6 = 3/4 < \mu^7 = 5/6 < \mu^8 = 1$$

Note that, $\mu^* = 23/35$, so $\mu^4 < \mu^* < \mu^5$ and $\mu_0 = 4$.
Further, we get $h = 4$.
Sorting $\mu_{ij}$, $i, j = 1,\ldots,5$, we obtain

$$0 < \beta^1 = 0.4 < \beta^2 = 0.5 < \beta^3 = 0.6 < \beta^4 = 0.7$$
$$< \beta^5 = 0.75 < \beta^6 = 0.8 < \beta^7 = 0.85 < \beta^8 = 1$$

Compute $m_{ij} + 3.0\sigma_{ij}$, $i, j = 1,\ldots,5$, we obtain

$$C = \begin{pmatrix} 4.5 & 8.2 & 7.6 & 8.2 & 5.9 \\ 8.4 & 5.9 & 3.1 & 5.9 & 7.6 \\ 7.9 & 5.8 & 11 & 4.5 & 11 \\ 7.9 & 5.9 & 8.4 & 5.8 & 7.9 \\ 4.5 & 8.4 & 8.2 & 7.6 & 11 \end{pmatrix}$$

Arrange these values in ascending order, that is,

$$c^1 = 3.1 < c^2 = 4.5 < c^3 = 5.8 < c^4 = 5.9 < c^5 = 7.6$$
$$< c^6 = 7.9 < c^7 = 8.2 < c^8 = 8.4 < c^9 = 11$$

For $p = 1,\ldots,8$, $q = 1,\ldots,9$, set

$$c_{ij}^{p,q} = \begin{cases} 0 & \text{if } \mu_{ij} \geq \beta^p, m_{ij} + 3.0\sigma_{ij} \leq c^q \\ \infty & \text{otherwise} \end{cases},$$
$$i, j = 1,\ldots,5.$$

It is obvious that $k^* = 6$, $l^* = 3$.
For $u = 1,\ldots,4$, $p = 1,\ldots,6$, $q = 3,\ldots,9$, $P_u^{p,q}$ has the following form:

$$P_u^{p,q}: \text{ minimize } \sum_{i=1}^{5}\sum_{j=1}^{5} c_{ij}^{p,q} x_{ij}$$
$$\text{subject to } \sum_{j=1}^{5} x_{ij} \leq s_i(u), i = 1,\ldots,5$$
$$\sum_{i=1}^{5} x_{ij} \geq t_j(u), j = 1,\ldots,5$$
$$x_{ij} : \text{nonnegative}, i, j = 1,\ldots,5$$

For $u = 1,\ldots,4$, $p = 1,\ldots,6$, we find the smallest $q$ in $\{3,\ldots,9\}$ such that $P_u^{p,q}$ is feasible and the results are shown in Table 3.

**Table 3.** The smallest $q$ such that $P_u^{p,q}$ is feasible.

| $u\backslash p$ | 1 | 2 | 3 | 4 | 5 | 6 |
|---|---|---|---|---|---|---|
| 1 | 4 | 4 | 6 | 7 | | – |
| 2 | 4 | 4 | 7 | 7 | – | – |
| 3 | 4 | 4 | 7 | 7 | – | – |
| 4 | 4 | 4 | 7 | 7 | – | – |

Form Table 3, we obtain

$$A = \{(1,1,4),(1,2,4),(1,3,6),(1,4,7),(2,1,4),(2,2,4),$$
$$(2,3,7),(2,4,7),(3,1,4),(3,2,4),(3,3,7),(3,4,7),$$
$$(4,1,4),(4,2,4),(4,3,7)\}$$

For each $(u,p,q(u,p))$ in $A$, solve problem $P_u^{p,q(u,p)}$ and check whether there exists an optimal transport pattern $\mathbf{x}_u^{p,q(u,p)}$ such that $v(\mathbf{x}_u^{p,q(u,p)})_3 = \mu^u$ or not, the results are shown in Table 4.

**Table 4.** An optimal transport pattern $\mathbf{x}_u^{p,q(u,p)}$ of $P_u^{p,q(u,p)}$ such that $v(\mathbf{x}_u^{p,q(u,p)})_3 = \mu^u$.

| $(u,p,q(u,p))$ | $\mathbf{x}_u^{p,q(u,p)}$ | $(c^{q(u,p)}), \beta^p \mu^u)$ |
|---|---|---|
| (1,1,4) | $\mathbf{x}_1^{1,4}$: $x_{11} = 2$, $x_{15} = 5$, $x_{23} = 11$, $x_{32} = 7$, $x_{44} = 8$, $x_{51} = 11$ | (5.9,0.4,1/6) |
| (1,2,4) | $\mathbf{x}_1^{2,4}$: $x_{11} = 2$, $x_{15} = 5$, $x_{23} = 11$, $x_{32} = 7$, $x_{44} = 8$, $x_{51} = 11$ | (5.9,0.5,1/6) |
| (1,3,6) | $\mathbf{x}_1^{3,6}$: $x_{23} = 11, x_{24} = 1, x_{25} = 5, x_{32} = 7, x_{41} = 2, x_{44} = 7, x_{51} = 11$ | (7.9,0.6,1/6) |
| (1,4,7) | $\mathbf{x}_1^{4,7}$: $x_{12} = 7$, $x_{23} = 11$, $x_{25} = 5$, $x_{31} = 2$, $x_{44} = 8$, $x_{51} = 11$ | (8.2,0.7,1/6) |
| (2,1,4) | $\mathbf{x}_2^{1,4}$: $x_{11} = 8$, $x_{15} = 5$, $x_{23} = 11$, $x_{32} = 7$, $x_{44} = 8$, $x_{51} = 5$ | (5.9,0.4,1/4) |
| (2,2,4) | $\mathbf{x}_2^{2,4}$: $x_{11} = 8$, $x_{15} = 5$, $x_{23} = 11$, $x_{32} = 7$, $x_{44} = 8$, $x_{51} = 5$ | (5.9,0.5,1/4) |
| (2,3,7) | $\mathbf{x}_2^{3,7}$: $x_{12} = 5, x_{14} = 8, x_{23} = 11, x_{25} = 5, x_{32} = 2, x_{41} = 3, x_{51} = 10$ | (8.2,0.6,1/4) |
| (2,4,7) (3,1,4) | – $\mathbf{x}_3^{1,4}$: $x_{11} = 3$, $x_{15} = 5$, $x_{23} = 11$, $x_{32} = 7$, $x_{44} = 8$, $x_{51} = 10$ | – (5.9,0.4,1/3) |

*(Continued)*

Table 4. (continued)

| $(u,p,q(u,p))$ | $x_u^{p,q(u,p)}$ | $(c^{q(u,p)}),\beta^p\mu^\mu)$ |
|---|---|---|
| (3,2,4) | $x_3^{2,4}: x_{11} = 3, x_{15} = 5, x_{23}$ $= 11, x_{32} = 7, x_{44} = 8, x_{51}$ $= 10$ | (5.9,0.5,1/3) |
| (3,3,7) | $x_3^{3,7}: x_{14} = 8, x_{23} = 11,$ $x_{25} = 5, x_{32} = 7, x_{41} = 3,$ $x_{51} = 10$ | (8.2,0.6,1/3) |
| (3,4,7) | $x_3^{4,7}: x_{12} = 7, x_{23} = 11,$ $x_{25} = 5, x_{31} = 3, x_{44} = 8,$ $x_{51} = 10$ | (8.2,0.7,1/3) |
| (4,1,4) | – | – |
| (4,2,4) | – | – |
| (4,3,7) | – | – |

From Table 4, we obtain

$$B = \{(1,1,4),(1,2,4),(1,3,6),(1,4,7),(2,1,4),(2,2,4),$$
$$(2,3,7),(3,1,4),(3,2,4),(3,3,7),(3,4,7)\}$$

Then using the definition of non-domination, we obtain a set of some non-dominated transport patterns and that of the corresponding three-objective vectors of P, denoted by NDT and NDV respectively, which are given as follows:

$$\text{NDT} = \{x_1^{3,6}, x_3^{2,4}, x_3^{4,7}\},$$
$$\text{NDV} = \{(e^{7.9}, 0.6, 1/6), (e^{5.9}, 0.5, 1/3), (e^{8.2}, 0.7, 1/3)\}.$$

## 5 CONCLUSIONS

In this paper, we have considered the hazmat transport network problem with three criteria from both random and fuzzy factors and developed an algorithm to find some non-dominated transport patterns. Furthermore, we have shown the validity and time complexity of the algorithm. Also it was illustrated by using a numerical example. However, there remains many other network problems from both random and fuzzy factors to be investigated.

## ACKNOWLEDGEMENTS

This research has been supported by Beijing Postdoctoral Work Foundation (No.B181) and National Key Technology R&D Program (2012BAK13B05).

## REFERENCES

Ahuja, R.K., Orlin, J.B., Tarjan, R.E. 1989. Improved time bounds for the maximum flow problem. *SIAM Journal on Computing* 18:939–954.

Erkut, E. 2007. Introduction to the special issue. *Computers & Operations Research* 34(5):1241–1242.

Erkut, E. & Alp, O. 2007. Designing a road network for hazardous materials shipments. *Computers & Operations Research* 34(5):1389–1405.

Erkut, E. & Gzara, F. 2008. Solving the hazmat transport network design problem. *Computers & Operations Research* 35:2234–2247.

Erkut, E., Tjandra, S., Verter, V. 2007. Hazardous materials transportation. In C. Barnhart & G. Laporte (eds), *Handbooks in Operations Research and Management Science: Transportation* 14:539–621. Amsterdam: Elsevier.

Kara, B.Y. & Verter, V. 2004. Designing a road network for hazardous materials transportation. *Transportation Science* 38(2):188–196.

Verma, M. 2009. A cost and expected consequence approach to planning and managing railroad transportation of hazardous materials. *Transportation Research Part D* 14:300–308.

# Research and implementation of early-warning of remote environmental monitoring based on IOT

Y.F. Li & L.Q. Tian
*Computer School, North China Institute of Science and Technology, Beijing, China*

ABSTRACT:   Early-warning of environmental monitoring was implemented on the basis of Bayesian network theory. The early-warning model of environmental monitoring was constructed based on Bayesian network, and early-warning algorithm was designed. Then an early-warning system of remote environmental monitoring based on IOT was implemented. Intelligent data processing of IOT on the remote environmental monitoring platform was achieved. Real-time early-warning of remote environmental monitoring was realized by the early-warning system.

## 1   INTRODUCTION

Compared with traditional methods of environmental monitoring, it is more accurate, more reliable, better real-time, and covers a wider range to do environmental monitoring based on IOT (Internet of Things) technology (Ma, 2011). And it has become an important field of application of IOT. It was an important part for more thorough perception of the physical world to achieve intelligent processing of massive monitoring data generated by the IOT monitoring platform (Liu, 2012). So it was a research direction of practical significance and application value to achieve real-time early-warning based on the monitoring data of IOT.

There were several methods for early-warning. Vu (2007) used threshold-based method to evaluate the monitoring data for early-warning. Liang (2005) dynamically calculated the threshold value to improve the adaptability of early-warning. Also, pattern recognition techniques were introduced to early-warning, like a decision tree for fire detection (Bahrepour, 2012), to support vector machine (Li, 2012) and Naïve Bayesian classifier (Fan, 2012) for water quality early-warning. Since there were many uncertain factors in early-warning of environment monitoring, Bayesian network was used in this paper. Bayesian network was one of the most effective theoretical models in the areas of the representation and reasoning uncertain knowledge. Bayesian network was a direct graphic description based on the network structure, which combined artificial intelligence, probability theory, graph theory, and decision theory.

It provided a convenient structure to represent causal relationships so that uncertainty reasoning became more clear and easier to understand logically.

An early-warning algorithm for environmental monitoring was provided. On the basis of the reliability of monitoring data, real-time early-warning based on Bayesian networks was implemented. Combined with the actual requirements of remote environmental monitoring platform on IOT which was in construction, an early-warning system was developed to process the monitoring data for early-warning.

## 2   EARLY-WARNING MODEL OF ENVIRONMENTAL MONITORING BASED ON BAYESIAN NETWORK

### 2.1   *Bayesian network*

Bayesian network had become one of the most important methods of dealing with uncertain issues in artificial intelligence, pattern recognition, machine learning and data mining, etc. In Bayesian network, the relationship and the degree of influence between the various information elements were expressed by direct graph with network structure. The information element was expressed by node variable, the relationship between these elements was expressed by direct edge, and the degree of influence was expressed by a conditional probability table. It combined with prior knowledge and sample information and could be used as an inference engine in knowledge-based system (Bai, 2005).

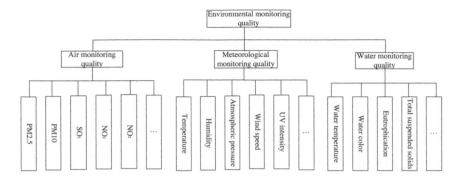

Figure 1. Early-warning index system.

There were four stages to build Bayesian network.

1 To define the domain variable: in a particular field in which variables would be used to describe the various parts of the field, and what was the exact meaning of each variable.
2 To determine the network structure: what was the network structure of the field based on the dependencies between the variables.
3 To determine the conditional probability distribution: how to quantify the dependencies between variables through the network structure.
4 Reasoning for application: to use the Bayesian network in actual system and optimize it based on the data generated.

## 2.2 Construction of early-warning model of environmental monitoring

### 2.2.1 Early-warning index system of environmental monitoring

The selection of early-warning indexes was very important in constructing the early-warning model of environmental monitoring. Accordance with the principles of science, systematic, operability, and target, an early-warning index system was established referring to the relevant national standards (National Environmental Protection Agency of China. 1996 & 2002) and studies (Qin, 2012 & Zhang 2013). Fig.1 showed the early-warning index system.

### 2.2.2 Bayesian network model of environmental monitoring early-warning

The Bayesian network model of environmental monitoring early-warning was a direct acyclic graph, composed of nodes which represented variables and directed edges connecting these nodes. The variable nodes included early-warning object E, which was the quality of environmental monitoring, and early-warning sub-object, such as air monitoring quality A, meteorological monitoring quality M and

water monitoring quality W and so on. Direct edges between nodes represented the dependencies, from parent nodes to offspring nodes. Fig.2 showed the Bayesian network model of early-warning of environmental monitoring.

## 3 EARLY-WARNING ALGORITHM OF ENVIRONMENTAL MONITORING

### 3.1 Level division of monitoring quality

In order to use Bayesian network to achieve early-warning, the monitoring quality of the nodes E, A, M, W etc. were divided into L levels.

These levels were sequentially numbered to integer i, $i \in [1, L]$. Being supposed to average divide, the range of monitoring quality was represented by these levels from low to high were

$$\left[0, \frac{E_{max}}{L}\right], \left[\frac{E_{max}}{L}, \frac{2E_{max}}{L}\right], ..., \left[\frac{(L-1)E_{max}}{L}, 1\right]$$

Emax was the optimal value of monitoring quality. Once an early-warning was achieved, the total number of early-warning, n, would increase by 1. Meanwhile, the number of certain ranges would also increase by 1 when the value of monitoring quality of each node was in the corresponding range. Also, the number of two or more certain range of different nodes when they were in the corresponding ranges simultaneously should be saved to achieve early-warning with multiple conditions. The number could be stored in two-dimensional arrays or more dimensional arrays.

The name of the array represented different nodes, and the array subscript represented different range of quality monitoring value. For example, $E_i$ and $A_i$ respectively represented the i-th range of environmental monitoring quality and air monitoring quality, while $|E_i|$ and $|A_i|$ were the number of the monitoring value in the i-th range for environmental monitoring and air quality monitoring, in which values were stored in the array named Eai and Aai. P (Ei) and P (Ai) were their probability.

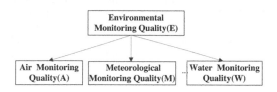

Figure 2. Bayesian network model of environmental monitoring early-warning.

## 3.2 Calculation of conditional probability

Before early-warning with Bayesian network, the priori probability and conditional probability of each node must be calculated. Bayesian formula was used:

$$P(h \mid e) = \frac{P(e \mid h)P(h)}{P(e)} \qquad (1)$$

P(h) represented priori probability of occurrence of hypothesis h. P(e) representing the priori probability of evidence e. P(h|e) represented the conditional probability of occurrence of hypothesis h in the condition of the occurrence of evidence e. P(e|h) represented the conditional probability of occurrence of e in the condition of the occurrence of hypothesis h.

The priori probability of environmental monitoring quality could be calculated by the following formula:

$$P(E_i) = \frac{|E_i|}{n}(1 \leq i \leq L) \qquad (2)$$

Where n represented the total number of early-warning, $|E_i|$ represented the number of the environmental monitoring value. The calculating method of priori probability of other nodes was the same.

Then, the conditional probability table of each node could be calculated. Supposed to divide three levels for monitoring quality, the conditional probability table of air monitoring quality node was shown in table 1.

The conditional probability of node could be calculated by the following formula:

$$P(e \mid h) = \frac{P(h,e)}{P(h)} \qquad (3)$$

For example, P(Ai|Ej) represented the conditional probability of the air monitoring quality in Ai in the condition of the environmental monitoring quality in Ej. The calculation formula was:

$$P(A_i \mid E_j) = \frac{P(E_j, A_i)}{P(E_j)} = \frac{|E_j \cap A_i|}{|E_j|} \qquad (4)$$

Table 1. Conditional probability table of air monitoring quality node.

| A \ E | $E_1$ | $E_2$ | $E_3$ |
|---|---|---|---|
| $A_1$ | $P(A_1|E_1)$ | $P(A_1|E_2)$ | $P(A_1|E_3)$ |
| $A_2$ | $P(A_2|E_1)$ | $P(A_2|E_2)$ | $P(A_2|E_3)$ |
| $A_3$ | $P(A_3|E_1)$ | $P(A_3|E_2)$ | $P(A_3|E_3)$ |

## 3.3 Early-warning algorithm

With the priori probability of environmental monitoring quality and the priori probability and conditional probability of its child nodes, the probability of certain level of environmental monitoring quality in a certain monitoring condition could be calculated to achieve real-time early-warning of environmental monitoring quality.

The early-warning algorithm of environmental monitoring was shown in Fig.3.

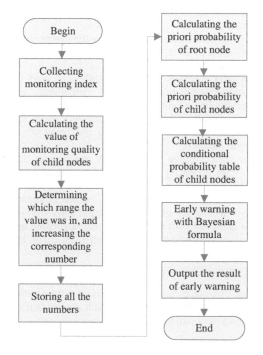

Figure 3. Early-warning algorithm of environmental monitoring.

## 4 DEVELOPMENT OF EARLY-WARNING SYSTEM OF REMOTE ENVIRONMENTAL MONITORING BASED ON IOT

### 4.1 Design of architecture of early-warning system

An early warning system of remote environmental monitoring based on IOT was developed with the

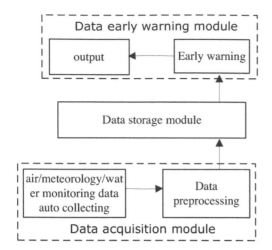

Figure 4. Architecture of early-warning system.

early-warning algorithm mentioned above. Fig.4 showed the architecture of the system.

There were three components in the system: data acquisition module, data storage module and data early-warning module.

In data acquisition module, WSN was used for automatic acquisition of the air, meteorology and water quality monitoring indexes. The data were uploaded to the local server via relaying nodes and aggregating nodes, preprocessing and being stored into the database.

In data storage module, the monitoring data and other necessary information were stored in the database.

In data early-warning module, regular automatic early warning and special event trigger early warning were achieved with the early-warning algorithm and the result would be released through SMS, email, etc.

## 4.2 *Effect of early-warning*

A test was made for the early-warning system with a set of real monitoring data. The result showed that the accuracy was satisfied.

## 5 CONCLUSION

Based on the Bayesian network theory, an early-warning algorithm was proposed. And an early-warning system of remote environmental monitoring based on IOT was developed. It met the requirement of intelligent data processing on the remote environmental monitoring platform on IOT in construction.

ACKNOWLEDGMENTS

This work was supported by the National Natural Science Foundation of China (No.61163050, No.61472137), Program for New Century Excellent Talents in University of China (NCET-10-0101), Qing Hai Province Program (No. 2012-N-525) and the Fundamental Research Funds for the Central Universities(No. 3142013074, No. 3142013098 and No.3142014125).

REFERENCES

[1] Bahrepour, M. & Neratnia, N. & Poel, M. etc. 2012. Use of wireless sensor networks for distributed event detection in disaster management applications. Int. J. Space-based and Situated Computing. 2(1):58:68.

[2] Bai, C.G. 2005. Bayesian Network based software reliability prediction with an operational profile. Journal of Systems and Software 77(2):103–112.

[3] Fan, M. & Shi, W.R. 2012. Construction and application of hierarchical naive Bayesian classifier. Chinese Journal of Scientific Instrument. 31(4):776–781 (in Chinese).

[4] Li, D.L. 2012. Introduction to Internet of Things in agriculture. Beijing: Science Press, pp. 206–217 (in Chinese).

[5] Liang, Q. & Wang, L. 2005. Event detection in sensor networks using fuzzy logic system. IEEE intl. Conference on Computational Intelligence for Homeland Security and Personal Safety.

[6] Liu, Y.H. 2012. Connecting inspiring everything. Communications of the CCF 8(3):8–10 (in Chinese)

[7] Ma, J. 2011 Introduction of Internet of Things Technology, Beijing: Mechanical Industry Press, pp. 77–83 (in Chinese).

[8] National Environmental Protection Agency. 1996. Ambient air quality standard GB 3095-1996 (in Chinese).

[9] National Environmental Protection Agency. 2002. Environmental quality standard for surface water GB3838-2002 (in Chinese).

[10] Qin, Z. 2012. Research and implementation of wireless meteorology monitoring system based on Internet of things. Dissertation of Huazhong University of Science and Technology (in Chinese).

[11] Vu, C.T. & Beyah, R.A. & Li, Y. 2007 composite event detection in wireless sensor networks. Proc. of the IEEE International Performance, Computing, and Communications Conference.

[12] Yu, C.X. & Xu, L.Y. & Xing B. etc. 2009. Design and Implementation of Intensive Agriculture water quality early warning system. Computer Engineering 35(17):268–270 (in Chinese).

[13] Zhang, Z.T. & Cao, Q. & Xie, T. 2013. Design of water quality monitoring and warning system of drinking water source. Environmental Protection Science 39(1):61–64 (in Chinese).

*Multimedia, Communication and Computing Application – Leung (Ed.)*
*© 2015 Taylor & Francis Group, London, ISBN 978-1-138-02775-6*

# Theoretical analysis on high-end equipment manufacturing growing: From the view of 3D-spiral technology collaborative innovation

K. Li

*The CPC Heilongjiang Province Committee, Harbin, Heilongjiang Province, China School of Management, Harbin Institute of Technology, Harbin, Heilongjiang Province, China*

B. Yu

*School of Management, Harbin Institute of Technology, Harbin, Heilongjiang Province, China*

Q.J. Li

*Research Office of Heilongjiang Provincial Government, Harbin, Heilongjiang Province, China*

ABSTRACT: The growing of High-end Equipment Manufacturing (HEM) industry is one of the important research topics in technology innovation-driven development strategy of China. The paper gives game analysis on the problems occurred in the technology innovation-driven collaboration among HEM companies, universities and government, from the view of 3D-Spiral-Based Technology Collaborative Innovation. The research shows that because of the adjustment of the rule from the zero-sum game with positive sum game, 3D spiral technology collaborative innovation becomes a dynamic game equilibrium among the performance improvement of high-end equipment manufacture companies, university knowledge production and government macro-control. This discovery may help to reveal the mechanical structure of growing mechanism design and path choice of HEM industry.

KEYWORDS: High-end equipment manufacturing industry, Stakeholders' game, 3D spiral technology collaborative innovation

## 1 INTRODUCTION

The growing of high-end equipment manufacturing (HEM) industry is one of the important research topics in technology innovation-driven development strategy of China. HEM industry is one of the strategic emerging industries led by high technology, and consists 5 industries, including aviation equipment, satellite and its application, rail transportation equipment, marine engineering equipment, and intelligent manufacturing equipment. The growing of these industries is critical to the quality and speed of the transforming from "Made in China" to "Created in China", as well as the sustainable development national economic society. Since the world financial crisis, the HEM industry is growing as the developed countries and rich regions are developing strategies such as manufacturing renaissance, re-industrialization, low carbon technology, next generation energy, and smart Earth. The growing of the industry is helpful in improving the nation's international competitiveness[1][2]. It aims to form apical dominance for self-innovation in international trade competition. According to the experience of global development, literature review discovered that in order to gain the sustainable development of the industry, the growing of HEM industry is composed of three kinds of manufacturing, which are manufacturing led by nation, by industry and by companies themselves. Therefore, the driven mechanism of HEM's technology collaboration innovation is mainly about the cooperation and innovation gaming among technology innovation alliances of stakeholders which are government, client, and third party innovation body [3]. From the 3D technology collaboration innovation experience of product innovation, process innovation and market innovation in electronics and telecommunication industry (intelligent manufacturing equipment industry), the empirical analysis of Yulin Zhao[4] showed that the innovation collaboration and industry growing of these three factors are highly convergent. Aiqi Wu[5] took the growing experience of the advanced manufacturing industry in Zhejiang Province as an example to prove the existing of the game relationship among the stakeholders of collaborative innovation in secondary network, such as companies, government, universities and research institutes. Because of the technology collaborative innovation system of HEM is open and non-linear[6], the construction of communication mechanism is to provide an unblocked information pathway for the efficiency improvement of difference innovation bodies[7], and to enhance the performance of every innovation system in the 3D innovation space made of market, science and technology, and administration to form a 3D spriral uptrend[8]. Nevertheless, the growing of HEM is still achieved by the 3D spiral technology collaborative innovation of companies, universities and government. There are still gaps in this research area, as the technology collaborative innovation of HEM is happened on an interest-based platform by different stakeholders, which causes the experience different in organization learning. From the present literature review, it is hard to find the systematic research on how to achieve technology collaborative innovation among different stakeholders, especially on the game analysis on interest development. This paper makes a game equilibrium analysis on the stakeholders in 3D spiral technology collaborative innovation, tries to internalize the game rule of collaborative innovation, observes the interaction result after the adjustment of game rule from zero-sum game to positive-sum game, simulates the stakeholders' game equilibrium caused by collaborative innovation with improvement of company performance, university knowledge production and government macro-control of in HEM growing, and provides evidence for mechanism design and path choice of HEM industry growing.

## 2 RESEARCH DESIGN

The practice of development showed that the industry growing of HEM is upon to whether there is a stable, collaborative game mechanism in the 3D spiral technology collaborative innovation formed by companies, universities and government. The paper will build a 3D spiral technology collaborative innovation function to explain the mathematical relationship of the critical factors in the HEM collaborative innovation system. However, in the present theory of 3D spiral technology collaborative innovation, the research of its innovation mechanism is not adequate because of the incomplete classification of the roles in collaborative innovation, the unclear of the division of innovation product, and the inadequate of the analysis of interaction mechanism, especially the mathematical relationship of the interaction mechanism in the collaborative innovation has not been built. This is the core task of the paper. The paper supposes the growing of HEM industry is based on the collaborative innovation of company, university and government, and abstracts a collaborative innovation production function with upward inclination screw under the 3 dimensions' motivation and restriction. As a new

production function, the variable and invariables of the technology collaborative innovation can indicate the 3D interactive mathematical relationship among the interest game of company, university and government, as well as the spiral growth of the collaborative innovation. Bo Zou & Bo Yu's research (2010) proved the motivation and restriction of the operation mechanism of 3D spiral collaborative innovation. They believed that the creating of the mechanism might be caused by the motivation system of company, university and government. It is supposed in this paper that the 3D spiral collaborative innovation is the interaction relationship of 3 dimensions in the collaborative innovation system, the nature of which is decided by the expectation of the contract system, other than "prisoner's dilemma". From this point of view, the paper makes the following classification to the 3D spiral collaborative innovation bodies.

1 Company dimension: commercial body of technology innovation. Because of the profit priority and the pressure of survival of the HEM companies, their direction in collaborative innovation is to achieve the interaction of technology innovation and organization innovation through the endogenization of technology innovation by institutional innovation. With government, they have the economic interest of tax, while also have the game relationship of interest. The aim is to seek bigger policy interest. With the university, they have the demand of technology, while gaming on the share of knowledge profit, which seeks the low cost for knowledge innovation.

2 Government dimension: policy body of technology innovation. Technology innovation has the common nature of the social economy resource and the externality of technology innovation resource distribution. Government tries to handle as much public resource and power resource as possible to promote new industrialization, informatization, urbanization and agricultural modernization, and create a huge policy market of government procurement for high-end equipment. The growth of HEM industry will provide a policy order form and create competitiveness to enter the international market, through the strategic positioning, planning, social promoting, industry regulation and preferential policies from the government, to achieve the institutional agreement of company's private innovation product and university's public innovation product.

3 University dimension: knowledge drive body of technology innovation. The university needs to form an alliance with a company for the transferring from knowledge innovation to technology innovation, to create organization integration of industry, study, research and application. On the

other hand, it is the university's social public responsibility to public knowledge produce, public technology breakthrough, teaching and research on basic theoretical innovation, talent training and high-level academic training authorized by government on behalf of the society.

4  Interaction behaviour locus of 3D spiral collaborative innovation bodies. The growth of HEM needs three bodies: company, university, and government. Under the motivation and the restriction on collaborative innovation target, the collaborative innovation appears to ascend like spiral because of the three bodies' nature, as shown in Fig. 1.

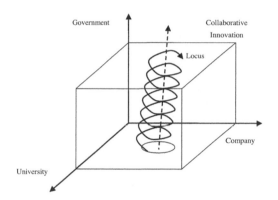

Figure 1. Demonstration of spiral technology collaborative innovation in the growth of HEM industry.

## 3  THEORETICAL DERIVATION

1  The unlock of path dependence during the interaction of technology innovation and institutional innovation. When there is path dependence between technology innovation and institutional innovation, technology innovation will appear to be increasing return, which means that the technology innovation will cause the decrease of unit production cost because of the scale economy and the first mover advantage. Once the HEM industry becomes a dominant industry of a region, especially when forming the technology collaborative innovation, it will strengthen itself, and may come across the problem of technology lock-in. During the institutional change, the self-strengthening mechanism will be obtained by HEM company who has a dominant position in the industry. The company will try to improve the present situation and prevent further technology innovation. At the same time, there will be disagreement between the company and university during the collaborative innovation: the university's income from knowledge production and innovation will continuously

decrease because the company wants to have monopoly profit. Therefore, the university will use disruptive innovation to change the situation and seek more weight in the 3D spiral collaborative innovation. From the government side, for the sake of tax and employment, it would like to be the balancer and re-balancer between the company and university. When there's an opportunity of "unlocked" path in the "path dependence", it is the government's responsibility to re-distribute the resource to reduce the uncertainty of technology collaborative innovation.

2  Further explanation of theory hypothesis. In order to achieve the institutional "unlock" of "path dependence", the following two hypotheses are needed.

Hypothesis 1: If the government has no policy or financial support to the growth of HEM industry, the companies will lack the policy motivation to collaborate with universities.

Hypothesis 2: If the government has no policy or financial support for the growth of HEM industry, the universities will have no policy support to collaborate with companies.

3  The adjustment of game rules. The research supposes if there is no government policy supporting the industry update, the collaborative R&D in a technology innovation project between company and university is a dual-game.

In this game model, there is a possibility of achievement in technology collaborative innovation between university and company. For the two game players, the income of the collaborative R&D is $q_1$ and $v_1$. When the company has no financial investment for university's industry updating research, the income of the university is $q_3$ as there is no market support. If the university does not research on the industry update, it will have incomes of $q_2$ from other researches. $q_1 > q_2 > q_3$. For the company, the income of R&D investment without university's support is $v_3$, while the income of the present is $v_2$. $v_1 > v_2 > v_3$.

| Company | | Investment | No investment |
|---|---|---|---|
| University | R&D | q1, v1 | q3,v2 |
| | No R&D | q2,v3 | q2,v2 |

Figure 2. Collaborative game equilibrium between company and university.

From the equilibrium result of the game model, there are two Nash Equilibriums (R&D, investment) and (no R&D, no investment). If there is no external regulation guarantee, the equilibrium is not clear and depends on the income of equilibrium. In this game, the government coordinates the collaborative innovation between the company and the university via regulation, to ensure university's R&D income and company's investment income. Therefore, to increase the frequency of the solution of (R&D, investment) is a good way. The game under this kind of regulation is called Coordination Game. In fact, this is the adjustment to the game rules. Because of the common interest, the stakeholders' game in 3D spiral technology collaborative innovation turns from the irrational zero-sum game into rational positive-sum game.

Compared to economic or social organizations like companies or universities, government has many differences in this game. The Government has the power and authority to distribute the social public resource. From the view of model building, government's function to economy development focuses on the decision to game rule. It will influence the collaborative innovation in HEM industry growing. The controls of game players' expected income and innovation performance by government are two interacting factors. In the game rule making, government is the decision maker, as well as a participant of the game through the way of subsidy and tax. The collaborative game among company, government and university now begins. The Government will benefit from making distribution rules, as well as obtain the payoff from the company and university. Although, the government in the game appears like "winner takes all" and "multi-win", it greatly stimulates the collaboration game between the company and the university.

4 Mathematical derivation of collaborative game model. The reason why 3D spiral collaborative innovation among companies, government and university can achieve spiral ascend, is that the strategic relationship among the three parties improves the efficient integrity and the distribution of the innovation resource. It maximizes the profit of all the three parties. In this collaborative innovation structure, if the participants only consider the profit of themselves, it is not possible to achieve the maximum for every participant, they, themselves, will come across the restriction of performance development. Therefore, it is necessary to have a complete information exchange and capital cooperation among the three parties. When such alliance sets up, the game model will turn to collaborative game model, during which, participants will take the utility maximization as the key decision-making basis.

Set model variables. In the structure of ascending 3D interactive spiral technology collaborative innovation formed by companies, government and university, the variables are defined as follows.

i. The investment of HEM company is $I_e$. The investment is continuous. The limit of the investment is $I_{eh}$. $I_e \leq I_{eh}$;

ii. The investment cost of HEM company is $C(I_e)$. It is an increasing function, which means the marginal cost is increasing. $C_e' > 0$, $C_e'' > 0$;

iii. The investment of government is $I_g$. The investment is continuous. The limit of the investment is $I_{gh}$. $I_G \leq I_{Gh}$;

iv. The investment cost of government is $C(I_g)$. It is an increasing function, which means the marginal cost is increasing. $C_g' > 0$, $C_g'' > 0$;

v. The R&D investment of university to the industry innovation is $R$, the maximum is $R_h$;

vi. Suppose the university's income from R&D investment is in the form of equity. The university's ratio in this innovation project is $p$ $(0 \leq p \leq 1)$. The rest of the equity belongs to the company.

vii. The success probability of the alliance technology innovation is $m$;

viii. The expected value of industry innovation is $V$. $V(R, I_e, I_g, m)$ is the probability function of government's investment, university's R&D investment, the company's investment and innovation success probability;

ix. The success of technology innovation depends on the success probability of the three game players $j(R)$, $k(I_e)$, and $l(I_g)$. All the first derivatives of the three functions are all positive and all the second derivatives are negative;

x. The tax of the company is $t$, $0 \leq t \leq 1$.

5 The achievement of the optimum solution under the strategic equilibrium of collaborative game. The research has the following further hypotheses. Company, government and university signed an agreement to form the alliance; university signed venture capital agreement, stating the university's R&D investment is $R$ and its equity is $p$; the three game players have a common belief in the value of collaborative innovation, and the investment will end until the innovation success; the probability of success will increase with the investment of company and government, but it is restricted by the diminishing returns to scale of capital.

Based on the above hypotheses, a collaborative innovation game model of a company, government and university is set up. The game players are university $U_r$, company $U_e$ and government $U_g$. The strategies for university are $R$ and $p$. The strategy for the company is $I_e$. The strategy for government is $I_g$. The utility functions of the game players are as follows.

University's utility function: $f_{Ur} = j(R) \ k(I_e) \ l(I_g)$ $V(R, I_e, I_g, m)(1-t)p-R$;

Company's utility function: $f_{Ue} = j(R)\ k(I_e)\ l(I_g)$
$V(R, I_e, I_g, m)(1-t)(1-p)-C(I_e)$;

Government's utility function: $f_{Ug} = j(R)\ k(I_e)\ l(I_g)$
$V(R, I_e, I_g, m)t-C(I_g)$;

The entire utility function of technology collaborative innovation:

$f_T = f_{Ur} + f_{Ue} + f_{Ug} = j(R)\ k(I_e)\ l(I_g)\ V(R, I_e, I_g, m)-$
$R-C(I_e)-C(I_g)$;

Solving the model:

$\max f_N = \max[\ f_{Ur} + f_{Ue} + f_{Ug}] = \max[\ j(R)\ k(I_e)\ l(I_g)$
$V(R, I_e, I_g, m)-R-C(I_e)-C(I_g)]$;

When meet the entire utility maximization, that is

;

;

.

Now, the optimal solution of the 3D spiral technology collaborative innovation game of company, government and university is $(I_{emax}, I_{gmax}, R_{max})$. It means when the utility is maximized, for HEM company and government, every increase on the unit collaborative innovation investment will equal to the marginal cost of increase the investment; while for university, to invest the collaborative innovation will give up same rate of return, which equals to the marginal income from other alliance members' investment.

## 4 DYNAMIC EQUILIBRIUM, CONCLUSIONS AND POLICY IMPLICATION

1 Dynamic equilibrium of collaborative game. The optimal solution under strategy equilibrium of collaborative game only explains the fact of collaborative innovation game of company, government and university. It is a static equilibrium, but it cannot explain the ascending mechanism of the 3D spiral collaborative innovation. Because in the motivation structure of this collaborative game, the reason why the static equilibrium is broken is that through the adjustment of game strategy, the game players discover that if they increase the investment to the alliance, the entire income of the alliance will be increased, as well as that of their own. Based on this, the three parties further adjust the game rule: increase collaborative investment to gain more alliance income. Therefore, the adjustment will focus on the increasing of alliance income in different places, to achieve the continuous increase of the entire performance in the form of 3D interactive spiral shape.

Suppose the university's investment increases while the investment of company and government (y axis and z axis) stay the same, which means

$R_{stage} > R_{max}$ (max is the equilibrium status), but smaller differences. Thus the university's utility is:

$f_{Urstage} = j(R_{stage})\ k(I_{emax})\ l(I_{gmax})\ V_{stage}(1-t)p-R_{stage}$
$f_{Urmax} = j(R_{max})\ k(I_{emax})\ l(I_{gmax})\ V_{max}(1-t)p-R_{max}$

In order to simplify the analysis, suppose $V_{stage}$ equals to $V_{max}$. Considering the effect of increasing returns to scale of R&D investment or the combined effect which the HEM industry growth brings to the researches. Suppose $j''(R) > 0$, $R_{stage} > R_{max}$, and $j'(R_{stage}) > j'(R_{max})$ so () > (). That means when other games players keep the same strategy, R&D department changes the strategy and increases the income of its own.

Therefore, the theoretical explanation of optimal solution under the dynamic equilibrium of collaborative innovation is obtained: one game player can change the strategy to get more income for itself while the other two game players stay the same strategy. On the whole, it is a spiral-ascending locus in the 3D collaborative innovation, which means the innovation develops from one level to another higher level.

2 Conclusion and policy implication. From the above analysis, the following conclusions can be drawn.

i. The growth of HEM industry, mainly depends on the collaborative game of 3D spiral technology collaborative innovation among companies, government and university.

ii. The mechanism of this situation is the stakeholders adjust the game rule because of the positive expectation from the zero-sum game to positive-sum game.

As an important policy variable of the government control, technology collaborative innovation is an important game boggy in this process. The government is eager to achieve the game equilibrium in the static game. Usually, government will ensure the collaborative game equilibrium of dual-game between the university and company through the cooperative game. However, in order to achieve the dynamic equilibrium, the first task of government is to build a system environment of collaborative game for three game players, motivate the positive investment to the collaborative game via maintaining the stability, continuity and sustainability of the policy to ensure the Pareto improvement of other game players.

## REFERENCES

[1] Porter, M E. 1990.The Competitive Advantage of Nations. The Free Press.
[2] Hausmann, R., Hwang, J., & Rodrik, D. What Your Export Mattrs Journal of Economic Growth, 2007, 12(1).

[3] Yuan Y. J. & Chen Y.Y. The Building and Operation of Technology Alliance: Research based on China HEM Industry, Beijing: The Science Press, 2012:8.

[4] Zhao Y.L.. Theory on Development of Dominant Hi-Tech Industry. Beijing: The Science Press, 2012:116.

[5] Wu A.Q.. TheoreticalAnalysis on Network Growth of ClusterCompanies and the Experience of Zhejiang. Beijing: China Social Science Press. 2007:69–70, 148–149.

[6] Zheng G.. Total Collaborative Innovation: Road to Innovative Enterprises. Beijing: The Science Press. 2006:22.

[7] Qingrui Xu, Zhu Lin & Wang Fangrui.Transformation from the Integration between R&D and Marketing to Synergy between Technological Innovation and Market Innovation.Science Research Management. 2006 Vol 27, No.5:22–30.

[8] Rao Y.D., Wang S., Xiong X.F. & Wang X.J.. Collaborative Innovation and Company's Sustainable Development. Beijing: The Science Press, 2012:45–49.

[9] Zou B. & Yu B. Triple Helix Innovation Model. Heilongjiang Social Sciences. 2010(5):36–37.

*Multimedia, Communication and Computing Application – Leung (Ed.)*
© *2015 Taylor & Francis Group, London, ISBN 978-1-138-02775-6*

# Research on digital publishing logistics standardization system and assessment model

X.H. Wang, Y.P. Du, Y.B. Zhang & W.M. Zhang
*School of Electromechanical Engineering, Beijing Institute of Graphic Communication, Beijing, China*

ABSTRACT: Digital publishing logistics is developing due to the digital technology and the Internet. Digital publishing industry in China has entered a rapid development path. However, it was still difficult to exchange and share the information, because there was no universal standard among digital providers and users. This paper, based on the current results of digital publishing logistics standard, presents the digital publishing logistics standard and builds an index system of the standard. Meanwhile, analytic hierarchy-process evaluation method is used, which is based on entropy to evaluate relevant indicators. The method determines indicators focusing on the development of standards and the data obtained will be the theoretical basis of the field study in the future.

## 1 INTRODUCTION

Digital publishing is an emerging publishing industry with digital content production, digital management process, digital product form, channels of communication networks and so on. During the years of 2002–2013, digital publishing industry's total revenue was over 260 billion *yuan* in China, and it is expected to reach 350 billion *yuan* in 2014. The digital publishing industry is the outcome of the rapid development of the channel, which is in urgent need of publishing Logistics Standardization. The integration process is an important way to improve the efficiency of logistics. Literature (Study on the Construction and Development of Digital Publishing Standard) pointed out that "standardized" is the rule of activities that is developed and reused to obtain the best order. Lv Lian pointed out that "publication Logistics Standardization system" refers to developing, publishing and implementing logistics standards of publications for repetitive logistics activities and concepts whose aim is to get the best order and social benefits in the process of logistics. It is the digital and network technologies that digital publication logistics shows the characteristics of the information transmission network, delivery virtualization and data integration, etc. which make it a trend to research and develop the publishing logistics.

Innovative development models and strategies of modern digital publishing industry were proposed in the book named "Research and development of Digital publishing industry" and exerted important academic value. Literature (Present Situation and Prospect of digital publications) pointed out the trends and patterns of the development of digital

publications. *Digital Communication and Publishing Transformation* argued that the digital technology has brought changes to the publishing industry. Literature (Digital Communication and Publishing Transformation.) pointed out the current situation and problems faced by China's digital publishing. And Literature (Research on the Construction Scheme Of Publications Logistics Standard System in China) mentioned the factors that influenced the publication of logistics standardization. The above papers show that, on one hand, digital technology has been integrated into the publishing stream. But it can't be integrated interoperability and share the information about logistics between different providers due to the lack of standardization which forms "islands of information". On the other hand, although the article mentioned factors that affect the standardization of publishing industry logistics, however, factors that needed to focus on are not given.

This article summarizes the relevant factors, affecting the logistics of digital publishing standards build a standardized evaluation system. AHP, based on entropy weight, is proposed to evaluate these factors. Finally, an example will be given.

## 2 THE ESTABLISHMENT OF DIGITAL PUBLISHING STANDARD LOGISTICS SYSTEM

### 2.1 *The relationship figure on digital publishing logistics*

Digital modelling is mainly responsible for publishing digital model of logistics resources, which is the basis for the entire logistics standardization of digital publishing. The underlying data model contains publication of the data, process of the data, resource of the data, planning of the data, standards, codes, etc. Digital publishing logistics system is actually a combination of forms of modern digital and network technologies and it will transform the information, images, sound, etc. into the data stored in the Internet network in the publishing logistics. These data are used to transfer the information of publications between publishers and readers, which is shown in Figure1:

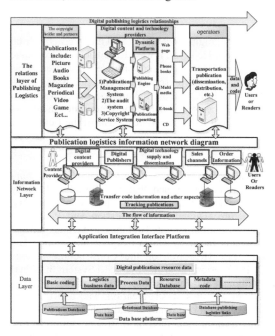

Figure 1.   Relationship map of digital publishing logistics information.

On the digital technology platform, people can get the latest and fastest information with the help of using transference of data at any time. The establishment of standard plays an extremely important role in the course of the composition of publications, not only to make the publication of Reading Terminal continuous integrated and innovated, but also tends to spread the carrier unified and virtualized data format. The standardization should be built to better share and exchange resources on digital platforms.

### 2.2   Factors affecting standardization

Figure 1 shows that information on the logistics chain is the form of the data, metadata and code. Digital publishing is the standard of logistics standardization of all publications' chain transport, storage, packaging, handling, transportation, distribution and other information whose purpose is to achieve effective transference of all kinds of information by analog - digital – analog. Factors, affected by the digital publishing logistics standards, can be summarized as shown in Figure 2:

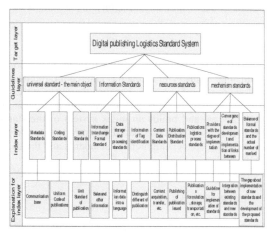

Figure 2. Factors on digital publishing logistics standardization.

### 2.3   Evaluation of the standard system of digital publications

The method used in this paper is AHP, based on the entropy weight method, to evaluate the system. Since the AHP method, in determining the use of indicators, is the result of expert advice which results in the loss of some information. So, entropy method is introduced to rectify the coefficient of right. The aim of doing this is to avoid low-level weights determined by multiple factors subjectivity, make the amount of information, increase the weight, improve reliability and to make the results be more in line with the actual situation.

## 3   EXAMPLES ON DIGITAL PUBLISHING LOGISTICS STANDARDIZATION

1   Establish a standardized index system Shown in figure 2.
2   Application of AHP to calculate the weights of the index system. According to fuzzy AHP, each order index weighting factor judgement matrix can be gotten. Weights and consistency test results are as follows.

Judgement matrix:

$$A = \begin{pmatrix} 1 & 1/2 & 3 & 4 \\ 2 & 1 & 7 & 5 \\ 1/3 & 1/7 & 1 & 1/2 \\ 1/3 & 1/5 & 2 & 1 \end{pmatrix}$$

$$A_1 = \begin{pmatrix} 1 & 1/3 & 5 \\ 3 & 1 & 7 \\ 1/5 & 1/7 & 1 \end{pmatrix} A_2 = \begin{pmatrix} 1 & 1 & 7 \\ 1 & 1 & 7 \\ 1/7 & 1/7 & 1 \end{pmatrix}$$

$$A_3 = \begin{pmatrix} 1 & 7 & 9 \\ 1/7 & 1 & 5 \\ 1/9 & 1/5 & 1 \end{pmatrix} A_2 = \begin{pmatrix} 1 & 1/4 & 1/2 \\ 4 & 1 & 3 \\ 2 & 1/3 & 1 \end{pmatrix}$$

Where: A-General objective layer;
$A_1$ – universal standard;
$A_2$ – information standards
$A_3$ – resources standards;
$A_4$ – mechanism standards
Each attribute of the largest eigenvalues and corresponding consistency index CI is gotten and be shown in Table 1:

The final results of the AHP and AHP based on the entropy weight method are compared in Table 4

Table 2. Results of weight coefficient modification.

| index level | Index Code | Entropy | Degree of deviation | Information weights | Index weights corrected |
|---|---|---|---|---|---|
| $A - A_i$ | $A_1$ | 0.809 | 0.191 | 0.252 | 0.267 |
| | $A_2$ | 0.811 | 0.189 | 0.248 | 0.473 |
| | $A_3$ | 0.835 | 0.165 | 0.218 | 0.057 |
| | $A_4$ | 0.786 | 0.214 | 0.282 | 0.203 |
| $A_1 - A_{1j}$ | $A_{11}$ | 0.662 | 0.338 | 0.440 | 0.352 |
| | $A_{12}$ | 0.752 | 0.248 | 0.323 | 0.600 |
| | $A_{13}$ | 0.818 | 0.182 | 0.237 | 0.048 |
| $A_2 - A_{2j}$ | $A_{21}$ | 0.812 | 0.188 | 0.333 | 0.465 |
| | $A_{22}$ | 0.812 | 0.188 | 0.333 | 0465 |
| | $A_{23}$ | 0.812 | 0.188 | 0.333 | 0.069 |
| $A_3 - A_{3j}$ | $A_{31}$ | 0.585 | 0.415 | 0.346 | 0.750 |
| | $A_{32}$ | 0.439 | 0.561 | 0.468 | 0.224 |
| | $A_{33}$ | 0.777 | 0.223 | 0.186 | 0.026 |
| $A_4 - A_{4j}$ | $A_{41}$ | 0.870 | 0.130 | 0.246 | 0.074 |
| | $A_{42}$ | 0.828 | 0.172 | 0.325 | 0.616 |
| | $A_{43}$ | 0.773 | 0.227 | 0.430 | 0.310 |

Table 1. Results of the AHP.

| Feature | Target | General | Information | Resources | mechanism |
|---|---|---|---|---|---|
| Value | layer | standards | standards | standards | Standards |
| $\lambda_{max}$ | 4.13 | 3.05 | 3.00 | 3.02 | 3.02 |
| CI | 0.043 | 0.025 | 0 | 0.01 | 0.01 |
| CR | 0.048 | 0.022 | 0 | 0.009 | 0.009 |

Know it from top to meet the consistency test CR<< 0.1.
The weights of level 1 is

$$w_A = (0.268 \quad 0.483 \quad 0.066 \quad 0.183)$$

The weights of level 2 is

$$W_A = \begin{pmatrix} 0.28 & 0.47 & 0.77 & 0.10 \\ 0.65 & 0.47 & 0.17 & 0.63 \\ 0.07 & 0.07 & 0.05 & 0.24 \end{pmatrix} (i=1,2,3,4)$$

3 Entropy techniques to correct the number of weights
The results of using entropy techniques for correcting weights are as follows in Table2
4 The evaluation results
The final results based on the membership functions are shown in Table 3.
5 Compare the results of different ways

Table 3. Final evaluation results.

| evaluation results | Index evaluation | Effective indicator (Max) |
|---|---|---|
| $B_1$ | $B_1 = (0.530 \quad 0.295 \quad 0.170 \quad 0.005)$ | 0.530 |
| $B_2$ | $B_2 = (0.719 \quad 0.200 \quad 0.074 \quad 0.007)$ | 0.719 |
| $B_3$ | $B_3 = (0.740 \quad 0.247 \quad 0.008 \quad 0.005)$ | 0.740 |
| $B_4$ | $B_4 = (0.468 \quad 0.215 \quad 0.255 \quad 0.062)$ | 0.468 |
| $B$ | $B = W_A \times R = [0.615 \quad 0.231 \quad 0.133 \quad 0.017]$ | 0.615 |

*R represents the membership degree evaluation matrix

Table 4. Results of different way.

| Comparison | Index layer $B_i (i = 1,2,3,4)$ | | | | Guidelines layer $B$ |
|---|---|---|---|---|---|
| AHP | 0.521 | 0.726 | 0.728 | 0.476 | 0.625 |
| AHP based on entropy | 0.530 | 0.719 | 0.740 | 0.468 | 0.615 |

Based on the above knowledge, the four-factor formula which is set to achieve the target layer plays an important role, including information standards. And the resource standard is one of the most important aspects, followed by the common standard which is the basis and mechanisms of the whole person standard. And that is the standard system to achieve a guaranteed effect. This rational development of the four factors is important in achieving objectives.

## 4 CONCLUSION

In this paper, the most fundamental factors that affect the logistics standardization are studied. On one hand, the relationship between the logistics layer, the network layer and the data layer is established on the network platform. On the other hand, the combination of the entropy and AHP is used to analyse and evaluate the related factors of standardization. This method avoids the subjectivity of experts who fails to grasp the scale of the problem. Basis data for the construction of standardized is obtained to provide a theoretical basis for the future construction standardization.

## REFERENCES

Hua Xia. 2014. Study on the Construction and Development Hua Xia. 2014. Study on the Construction and Development of Digital Publishing Standard Beijing Institute of Graphic Communication.

Xiaozhang, Huang Zhiling Zhang, Chen Dan. 2012. Research development of Digital publishing industry.The second version. Intellectual Property Press:75–111.

Haili Li. 2005. Present Situation and Prospect of digital publications. Gao School of Library and Information Forum, 4(3):30–32.

Meijuan Zhang, Guojun He, Shan Cai. 2012. Development of China's publishing practice and theory of logistics and supply chain research review published - logistics combined with the meaning of "production". Publishing Journal:72–75.

Weihua Zhou. 2011. Digital Communication and Publishing Transformation. Peking University Press:1–32.

Hongwei Liu. 2007. Research on the Construction Scheme Of Publications Logistics Standard System in China. Beijing Jiaotong University.

Hu Xye, Sheng Liu. The application of the improved fuzzy analytic hierarchy process (AHP) in the supply chain performance evaluation. Economic Management:38–39.

Gao Yang. 2011. Study on Curriculum System Structure Evaluation Based on Fuzz-AHP—Taking Profession of Electronic Information Engineering in Nan Chang for Example. Nan Chang University.

Shang Li. 2010. Research on Standardization System and Assessment Model of Public Service. China university of mining(BeiJing):156–161.

Ping Wang, Bangzhu Zhu. 2011. Entropy TOPSIS based enterprise innovation investment program evaluations. Productivity Research:173–175.

Lili Tong, Zhuhui 2010Wu. Application of Entropy Technology and AHP to Comprehensive Evaluationand Selection of Urban Afforestation Trees in Nanjing City. Journal of northeast forestry university, 38(9):58–61.

*Multimedia, Communication and Computing Application – Leung (Ed.)*
© *2015 Taylor & Francis Group, London, ISBN 978-1-138-02775-6*

# Research on the relationship between the regional information industry development and the quality of economic growth in China

Y. Liu & Y. Peng

*School of Economics and Management, Chongqing University of Posts and Telecommunications, Chongqing, China*

ABSTRACT: With the advent of information economy era, information industry has developed rapidly in China, but the extensive economic development is unsustainable, leading to high-involvement, high resource consumption, and high pollution. Under the globalization of information economy, the research on the regional economic growth in China can improve the quality of the same. With relevant theories and research methods, this article establishes evaluation index systems of the information industry and economic growth quality in China, by collecting the data from 2005 to 2010 in the 30 provincial areas. Then it measures the level of the information industry development and of the economic growth quality in the regions with EViews software from the provincial level. Finally, it quantitatively shows the relationship between information industry and economic growth quality. Thus, the development in the regional information industry in China has an improving effect on economic growth quality.

## 1 INTRODCTION

Since reform and opening up, the national economy has been developing rapidly. With the advent of the informational and economic age, informatization in countries goes very deep, and the proportion of the information industry in the national economy is increasing. At the same time, information technology has widely penetrated in various IT industries, so people are more concerned about the effect of information technology on social development and economic growth. The development level of information industry has become an important indicator to measure the regional even the national economy and social development.

However, to pursue the rapid development of information industry, people tend to ignore the quality of the economic growth. Liu Xuhong(2008)[1] believes that manufacturing industry in China has developed rapidly, but it also brings about serious damages to the environment, and thus the central government has established some legal systems to control environmental pollution. However, the local government considers that environmental protection would drag the local economy, so they often put the environmental supervision and regulation aside. Through analyzing economic growth model and strategies of China and India, Louis Kuijs (2012)[2] thinks that an increase in investment and the development of heavy industry of China has provided favorable conditions for the steady growth in economy in recent decades, but it also led to imbalance of industrial structure of

China. Therefore, Shen Lisheng(2009)[3] reported that China has been giving priority to industrial development rather than the development of service industry since 2002, which has been reducing the quality of Chinese economic growth. Nowadays, the issue of how to enhance the quality of economic growth has become one of the most important topics in the economic development of countries, and it is an outstanding issue which needs to be further addressed in China for now as well as the future. In this paper, we discuss the relationship between the development of regional information industry and the quality of economic growth in China, for which simultaneously we strive to seek countermeasures to improve.

## 2 RESEARCH METHOD

### 2.1 Measurement of information industry development level

The so-called information industry refers to a comprehensive industry group that regards information as the main production factor in computer technology, network technology, electronic communication technology, and other modern technologies. It includes research and development, collection, processing, storage, distribution and dissemination of information and information equipment for information services industry, and producing information and processing equipment for manufacturing industry, which is considered as the final product. Currently, there are a variety of measurement methods for information industry and information technology level in the academic industry, such as the Machlup's Information Economy Measurement (Wang Gang, 2008)[4], Information Technology Index (Liu Yue, Hui Meiling, 2012)[5-6], Input-output Method (Wang Gang & Wang Xin, 2007)[7], and so on. On the basis of the existing measurement methods, Jing Jipeng et al (1993)[8] as Chinese scholars put forward a set of methods of information industry comprehensive calculation—the comprehensive information industry dynamics method—from a holistic and comprehensive perspective.

Starting from evaluating the root of industry development level, we refer to the related indicators of the dynamics method of comprehensive information industry according to characteristics of regional information industry development in China. Considering related data availability, we have established an index system of information industry development as a measure model and indicators, including 2 secondary indicators and 9 tertiary indicators, as shown in Table 1.

In the multi-index model estimation on comprehensive evaluation, because the dimensions and order of magnitudes of the collected raw data for each indicator are different, these data cannot be directly summed and compared. Guo Yajun & Yi Pingtao(2008)[9], Zhang Lijun & Yuan Nengwen(2010)[10] showed that the dimensionless method of maximum difference is very effective in the linear comprehensive evaluation model, so we use the maximum difference method to standardize the obtained data. The indicators of measurement index system of information industry development level are all positive, and we should make sure that the index can be compared in a multiyear and cross-section, so we use (1) to manipulate the data by using the dimensionless method as follows:

$$Y_{ij} = \frac{X_{ij} - X_{i\min}}{X_{i\max} - X_{i\min}} \quad (1)$$

where $Y_i$ is the dimensionless data of $i$th indicator in $j$ year in some area, $X_{ij}$ is the raw data of $i$th indicator in $j$ year in the area, $X_{i\min}$ is the minimum of the raw data of $i$th indicator in sample data, and $X_{i\max}$ is the maximum of the raw data of $i$th indicator in sample data.

## 2.2 Measurement of economic growth quality

Studies on economic growth[11-15] are consistent with volume terms, but have different views in quality terms. Through summarizing views of the above scholars, we have measured the economic growth quality on the basis of the efficiency and stability of economic growth, ecological environment, living, income distribution, and industry structure terms

along with the economic growth. Currently, Chinese scholars [16-19] mostly measure the economic growth quality through establishing a comprehensive evaluation index system. When we use this system to measure the quality of economic growth, it involves two important steps: (1) on the basis of the definition of the connotation of the growth quality, we need to select and determine each dimensionality of index system and (2) synthesis of various basic indicators in the index system.

On the establishment of regional economic growth quality evaluation index system, it is based on the evaluation index system suggested by Prof Ren Baoping who proposed in his book *View the Growth by Quality: Evaluation and Reflection on Quality of China's Economic Growth*[15]. According to characteristics of regional economic development in China, with the related data available, we have established an index system of economic growth quality level measurement model, including 5 secondary indicators and 21 tertiary indicators, as shown in Table 2.

Because the indicators of economic growth quality level contain inverse indicators with binary contrast index, we use (2) to standardize the inverse indicators as follows:

$$Y_{ij} = \frac{A_{ij\max} - A_{ij}}{A_{ij\max} - A_{ij\min}}. \quad (2)$$

## 2.3 Principal component analysis

The entropy method, the relative coefficient method, and principal component analysis are the most common among the synthesis methods of indicators. The importance of the principal component analysis is determined by the characteristics of the data, avoiding the influence of subjective judgments, and the weight structure also reflects the dimensions of the contribution for various basic indicators to the total index. Therefore, we adopt the principal component analysis to evaluate the information industry and the quality of economic growth.

According to the previous established index systems of information industry development level and economic growth quality, first, we have calculated each individual indicator, and then individual indicators belonging to the same dimensionality by weight are calculated into the dimensionality index. Finally, we have converted the dimensionality index into combined index by weight. Specifically, the measurement model of information industry development level is as follows:

$$IDI = \sum_{i=1}^{m} \left( \sum_{j=1}^{n} Y_{ij} w_{ij} \right) * W_i \quad (3)$$

where IDI is the information industry development level index, $Y_{ij}$ is the normalized value of subindex $j$ in dimensionality index $i$, $w_{ij}$ is the weight of $Y_{ij}$, $W_i$ is the weight of dimensionality index $i$, $i = 1, 2, j = 1, 2, 3, \ldots, 0 \leq w_{ij}, W_i \leq 1$, $n$ is the number of subindex in dimensionality index $i$, and $m$ is the number of dimensionality index.

We use the same model to calculate the economic growth quality index.

## 3 DATA ANALYSIS AND CALCULATION

To reflect both differences of regions and the dynamic changes of each region over time, we use panel data to study the relationship between regional information industry development and the quality of economic growth in China. According to data availability, the time span of this study is limited to 2005–2010, and the objective includes 30 provinces in China (the Tibet Autonomous Region is excluded because large amount of statistics data are unavailable), with a total of 180 samples, and the amount of data are 360.

### 3.1 Calculation of information industry development level index

With the aforementioned calculation method, we input the dimensionless data into SPSS17.0, weights calculated for each individual index and dimensionality index are shown in Table 1. The weight of some indicator refers to relative important degree in the overall evaluation.

### 3.2 Calculation of regional economic growth quality index

We use the same way to measure the indicators of economic growth quality level. Weights calculated for each individual index and dimensionality index are shown in Table 2:

Table 1. Index system and weights in information industry development level measurement model.

| Index System | Weight of Dimensionality Index | Weight of Individual Index |
|---|---|---|
| **Information Industry Scale Level (X₁)** | 0.596 | |
| Proportion of Information Industry Output Value in GDP (X₁₁) | | 0.263 |
| Proportion of Information Industry Output Value Growth in GDP Growth (X₁₂) | | 0.239 |
| Number of Fixed Assets in Information Industry (X₁₃) | | 0.242 |
| Proportion of Information Industry Employment Quantity in Total Employment Quantity (X₁₄) | | 0.257 |
| **Information Industry Development Potential Level (X₂)** | 0.404 | |
| Number of College Students Per Ten Thousand People (X₂₁) | | 0.203 |
| Proportion of Technical Market Turnover in GDP (X₂₂) | | 0.093 |
| Proportion of R&D Funds in GDP (X₂₃) | | 0.209 |
| Education Expenditure Per Capita (X₂₄) | | 0.211 |
| Number of Patent Licensed Per Thousand People (X₂₅) | | 0.284 |

Table 2. Index system and weights in economic growth quality measurement mode.

| Index System | Weight of Dimensionality Index | Weight of Individual Index |
|---|---|---|
| **Economic Growth Structure (Y₁)** | 0.34 | |
| GDP Per Capita (Y₁₁) | | 0.16 |
| Comparative Labor Productivity in the First Industry (Y₁₂) | | 0.06 |
| Comparative Labor Productivity in the Second Industry (Y₁₃) | | 0.13 |
| Comparative Labor Productivity in the Third Industry (Y₁₄) | | 0.08 |
| Dual Contrastive Coefficient (Y₁₅) | | 0.14 |
| Binary Contrastive Index (Y₁₆) | | 0.21 |
| Degree of Dependence upon Foreign Trade (Y₁₇) | | 0.22 |
| **Stability of Economic Growth (Y₂)** | 0.09 | |
| Rate of Inflation (Y₂₁) | | 0.32 |
| Unemployment Rate (Y₂₂) | | 0.68 |

(*Continued*)

Table 2. (*Continued*)

| Index System | Weight of Dimensionality Index | Weight of Individual Index |
|---|---|---|
| **Welfare Change and Achievement Distribution of Economic Growth (Y$_3$)** | 0.21 | |
| Human Mortality (Y$_{31}$) | | 0.27 |
| Living Space Per Capita (Y$_{32}$) | | 0.14 |
| Deposits Per Capita of Urban and Rural Residents (Y$_{33}$) | | 0.23 |
| Engel Coefficient (Y$_{34}$) | | 0.10 |
| Income Ratio of Urban and Rural (Y$_{35}$) | | 0.26 |
| **Resource Utilization (Y$_4$)** | 0.20 | |
| Labor Productivity (Y$_{41}$) | | 0.23 |
| Capital Productivity (Y$_{42}$) | | 0.40 |
| Energy Consumption Ratio Per Unit Output (Y$_{43}$) | | 0.37 |
| **Ecological Environmental Cost (Y$_5$)** | 0.16 | |
| The Number of Exhaust Emissions of Per Unit Output (Y$_{51}$) | | 0.22 |
| The Number of Sewage Emissions of Per Unit Output (Y$_{52}$) | | 0.15 |
| The Number of Solid Waste Emissions of Per Unit Output (Y$_{53}$) | | 0.52 |
| Hazard-free Treatment Rate of Household Garbage (Y$_{54}$) | | 0.10 |

According to the above index weights by calculating the information industry development level and economic growth quality, we have measured the combined index on the basis of respective measurement model, and obtained the combined index of information industry development level and economic growth quality in 30 provinces of China among the 2005-2010, as shown in Table 3.

Table 3. Combined index of information industry development level and economic growth quality in 30 provinces of China from 2005 to 2010.

| Regions | 2005 | | 2006 | | 2007 | | 2008 | | 2009 | | 2010 | | Mean | |
|---|---|---|---|---|---|---|---|---|---|---|---|---|---|---|
| | IDI | EQI | IDI | EQI | IDI | EQI | IDI | EQI | IDI | EQI | IDI | EQI | IDI | EQI |
| China | 0.16 | 0.94 | 0.18 | 0.95 | 0.19 | 0.95 | 0.2 | 0.95 | 0.22 | 0.94 | 0.24 | 0.97 | 0.2 | 0.95 |
| Beijing | 0.45 | 1.37 | 0.5 | 1.41 | 0.46 | 1.42 | 0.44 | 1.45 | 0.44 | 1.46 | 0.49 | 1.48 | 0.46 | 1.43 |
| Tianjin | 0.36 | 1.24 | 0.43 | 1.26 | 0.39 | 1.25 | 0.37 | 1.2 | 0.35 | 1.17 | 0.4 | 1.19 | 0.38 | 1.22 |
| Hebei | 0.09 | 0.87 | 0.1 | 0.88 | 0.12 | 0.87 | 0.13 | 0.87 | 0.15 | 0.87 | 0.17 | 0.89 | 0.13 | 0.87 |
| Shanxi | 0.1 | 0.69 | 0.11 | 0.7 | 0.12 | 0.7 | 0.13 | 0.68 | 0.26 | 0.66 | 0.15 | 0.69 | 0.14 | 0.69 |
| Mongolia | 0.06 | 0.74 | 0.07 | 0.74 | 0.08 | 0.74 | 0.09 | 0.78 | 0.11 | 0.79 | 0.11 | 0.78 | 0.09 | 0.76 |
| Liaoning | 0.19 | 0.9 | 0.22 | 0.92 | 0.25 | 0.94 | 0.28 | 0.95 | 0.32 | 0.96 | 0.33 | 0.95 | 0.26 | 0.94 |
| Jilin | 0.09 | 0.91 | 0.11 | 0.9 | 0.11 | 0.92 | 0.13 | 0.93 | 0.14 | 0.93 | 0.15 | 0.95 | 0.12 | 0.92 |
| Heilongjiang | 0.13 | 1.01 | 0.11 | 0.99 | 0.15 | 0.97 | 0.14 | 0.95 | 0.15 | 0.92 | 0.14 | 0.95 | 0.14 | 0.96 |
| Shanghai | 0.49 | 1.41 | 0.53 | 1.46 | 0.57 | 1.5 | 0.5 | 1.49 | 0.53 | 1.49 | 0.61 | 1.58 | 0.55 | 1.49 |
| Jiangsu | 0.38 | 1.2 | 0.42 | 1.22 | 0.48 | 1.24 | 0.58 | 1.23 | 0.59 | 1.22 | 0.7 | 1.25 | 0.52 | 1.23 |
| Zhejiang | 0.28 | 1.21 | 0.32 | 1.27 | 0.35 | 1.3 | 0.36 | 1.31 | 0.36 | 1.31 | 0.44 | 1.35 | 0.35 | 1.29 |
| Anhui | 0.09 | 0.82 | 0.1 | 0.82 | 0.11 | 0.84 | 0.14 | 0.84 | 0.16 | 0.86 | 0.2 | 0.87 | 0.13 | 0.84 |
| Fujian | 0.2 | 1.09 | 0.2 | 1.08 | 0.2 | 1.1 | 0.21 | 1.11 | 0.21 | 1.12 | 0.256 | 1.15 | 0.21 | 1.11 |
| Jiangxi | 0.09 | 0.88 | 0.1 | 0.89 | 0.11 | 0.9 | 0.13 | 0.89 | 0.15 | 0.88 | 0.15 | 0.9 | 0.12 | 0.89 |
| Shandong | 0.19 | 1.02 | 0.21 | 1.04 | 0.24 | 1.06 | 0.26 | 1.04 | 0.31 | 1.04 | 0.32 | 1.06 | 0.25 | 1.04 |
| Henan | 0.09 | 0.89 | 0.11 | 0.9 | 0.12 | 0.91 | 0.15 | 0.93 | 0.11 | 0.88 | 0.2 | 0.91 | 0.13 | 0.9 |
| Hubei | 0.17 | 0.99 | 0.14 | 0.97 | 0.15 | 0.96 | 0.17 | 0.94 | 0.2 | 0.96 | 0.2 | 0.98 | 0.17 | 0.97 |
| Hunan | 0.1 | 0.86 | 0.11 | 0.86 | 0.12 | 0.89 | 0.15 | 0.9 | 0.18 | 0.92 | 0.2 | 0.91 | 0.14 | 0.89 |
| Guangdong | 0.4 | 1.4 | 0.43 | 1.44 | 0.45 | 1.44 | 0.49 | 1.42 | 0.48 | 1.39 | 0.6 | 1.44 | 0.48 | 1.42 |
| Guangxi | 0.06 | 0.87 | 0.07 | 0.87 | 0.08 | 0.83 | 0.1 | 0.81 | 0.11 | 0.82 | 0.12 | 0.84 | 0.09 | 0.84 |
| Hainan | 0.05 | 1.13 | 0.06 | 1.14 | 0.07 | 1.17 | 0.09 | 1.12 | 0.11 | 1.09 | 0.11 | 1.09 | 0.08 | 1.12 |
| Chongqing | 0.11 | 0.76 | 0.13 | 0.74 | 0.14 | 0.8 | 0.15 | 0.81 | 0.17 | 0.84 | 0.2 | 0.86 | 0.15 | 0.8 |
| Sichuan | 0.12 | 0.82 | 0.14 | 0.85 | 0.16 | 0.86 | 0.18 | 0.85 | 0.21 | 0.85 | 0.24 | 0.87 | 0.17 | 0.85 |
| Guizhou | 0.05 | 0.56 | 0.06 | 0.57 | 0.06 | 0.6 | 0.07 | 0.63 | 0.09 | 0.62 | 0.09 | 0.65 | 0.07 | 0.61 |
| Yunnan | 0.05 | 0.73 | 0.06 | 0.7 | 0.07 | 0.71 | 0.07 | 0.69 | 0.09 | 0.68 | 0.09 | 0.69 | 0.07 | 0.7 |

Table 3. (*Continued*)

| Regions | 2005 | | 2006 | | 2007 | | 2008 | | 2009 | | 2010 | | Mean | |
|---|---|---|---|---|---|---|---|---|---|---|---|---|---|---|
| Shaanxi | 0.16 | 0.77 | 0.17 | 0.78 | 0.17 | 0.76 | 0.18 | 0.76 | 0.2 | 0.76 | 0.22 | 0.79 | 0.18 | 0.77 |
| Gansu | 0.07 | 0.75 | 0.08 | 0.74 | 0.09 | 0.73 | 0.11 | 0.72 | 0.13 | 0.7 | 0.12 | 0.72 | 0.1 | 0.72 |
| Qinghai | 0.05 | 0.65 | 0.05 | 0.63 | 0.06 | 0.63 | 0.08 | 0.6 | 0.09 | 0.59 | 0.09 | 0.58 | 0.07 | 0.61 |
| Ningxia | 0.07 | 0.64 | 0.09 | 0.68 | 0.1 | 0.65 | 0.11 | 0.66 | 0.12 | 0.65 | 0.12 | 0.62 | 0.1 | 0.65 |
| Xinjiang | 0.05 | 0.95 | 0.06 | 0.95 | 0.07 | 0.94 | 0.08 | 0.95 | 0.13 | 0.89 | 0.1 | 0.93 | 0.08 | 0.94 |

## 4 EMPIRICAL ANALYSIS

### 4.1 *The data unit root test*

In the study of the relationship between information industry and the quality of economic growth, the spurious regression problem should be avoided and the validity of the analysis results is to be ensured. We tested the stationarity of the data by using the unit root test. We have conducted the unit root test for IDI and EQI by EViews6.0, and the test results are shown in Table 4.

Table 4. Results of the unit root test for IDI and EQI.

| Variable | IDI | | EQI | |
|---|---|---|---|---|
| Test methods | Statistic | $p^{**}$ | Statistic | $p^{**}$ |
| LLC | -12.890 | 0.0000 | -9.29599 | 0.0000 |
| ADF-Fisher | 70.182 | 0.1732 | 77.9950 | 0.0592 |
| PP-Fisher | 129.963 | 0.0000 | 125.084 | 0.0000 |

*Significant level at 10%.
**Significant level at 5%.
***Significant level at 1%

From the results shown in Table 4, not all IDI and EQI go through the unit root test, and probability value of ADF-Fisher test, namely $p^{**} > 0.05$, that is, we could not reject the null hypothesis of existing the unit root at 95% confidence level, therefore the original sequence is nonstationary. To make sequence smooth and steady, we take the first difference for IDI and EQI, then we have conducted the unit root test for the sequence ($\Delta$IDI, $\Delta$EQI) after the first difference, and the test results are shown in Table 5.

Table 5. Results of the unit root test for $\Delta$IDI and $\Delta$EQI.

| Variable | $\Delta$IDI | | $\Delta$EQI | |
|---|---|---|---|---|
| Test Methods | Statistic | $p^{**}$ | Statistic | $p^{**}$ |
| LLC | -122.288 | 0.0000 | -14.6480 | 0.0000 |
| ADF-Fisher | 127.372 | 0.0000 | 79.6468 | 0.0457 |
| PP-Fisher | 160.420 | 0.0000 | 125.488 | 0.0000 |

*Significant level at 10%.
**Significant level at 5%.
***Significant level at 1%.

From the results shown in Table 5, all $\Delta$IDI and $\Delta$EQI after the first difference have passed the test, so we can reject null hypothesis of existing the unit root at the 95% confidence level, and thus $\Delta$IDI and $\Delta$EQI are stable first-order single whole sequence.

### 4.2 *Cointegration test*

The preceding analysis shows that the indexes of the information industry development and economic growth quality are the first-order single whole sequence. To investigate the relationship between information industry and economic growth quality, it is necessary to test cointegration for data. We use Eviews6.0 software to test cointegration for data, and the results are shown in Table 6.

Table 6. Results of co-integration test.

| | Statistic | Prob. |
|---|---|---|
| ADF | -1.958452 | 0.0251 |

From the test results, we can see $p = 0.0251$ is less than 0.05 (the critical value), so the data go through the cointegration test, which indicates that there is a stable long-run equilibrium relationship between the regional information industry and the quality of economic growth in China.

### 4.3 *Granger test*

Fan Huanhuan & Zhang Lingyun (2009) [20] thought that Granger test can be used to determine whether there is a causal relationship between economic variables and the direction of its influence. For causality test between variables $X$ and $Y$, we need to establish the following regression test equations:

$$Y_t = \sum_{i=1}^{m} \alpha_i X_{t-i} + \sum_{j=1}^{m} \beta_i Y_{t-j} + u_t \quad (4)$$

$$X_t = \sum_{i=1}^{m} \lambda_i X_{t-i} + \sum_{j=1}^{m} \delta_i Y_{t-j} + v_t \quad (5)$$

469

where we assume that $\mu$ and $\nu$ are not relevant, $t$ is the length of time series, and $i$ is the lag length. Null hypothesis of Granger test is that $X(Y)$ is not the Granger cause with the change of $Y(X)$. Thus, the coefficient $\sum \alpha_i \left( \sum \delta_j \right)$ of hysteretic X(Y) estimated is not significantly different from zero statistically.

Two sets of data after settling is the first-order single whole sequence, and it meets the requirements of Granger test. Lag length of variables in the test is arbitrary, so we test multiple different lag lengths to ensure that test results are not affected by the selected lag length, and the final test results are shown in Table 7.

Table 7. Results of granger test.

| Null Hypothesis | F-Statistic | $p$ | Lag |
|---|---|---|---|
| EQI does not Granger Cause IDI | 0.85927 | 0.4366 | |
| IDI does not Granger Cause EQI | 6.45321 | 0.0060 | 2 |
| EQI does not Granger Cause IDI | 0.95265 | 0.4341 | |
| IDI does not Granger Cause EQI | 8.57016 | 0.0007 | 3 |
| EQI does not Granger Cause IDI | 1.01934 | 0.4253 | |
| IDI does not Granger Cause EQI | 4.35277 | 0.0132 | 4 |

From the test results shown in Table 7, we can conclude that the information industry development is an important contributor to improve the quality of economic growth, but it is not significant that economic growth quality promotes the development of information industry from the long-term development perspective.

## 5 CONCLUSION AND COUNTERMEASURE

### 5.1 Research conclusions

Through the aforementioned theoretical and empirical analysis, we draw the following conclusions:

First, across the whole China, the development of information industry appears a steady upward trend. From regional level in the horizontal point of view, the information industry has been developed in all regions, but the level of development has presented a lowering trend from east to west. The development of information industry presents large differences among regions, such as Yunnan, where the index of information industry development level is the minimum, and its average index is only 12.7% times than Shanghai where the index of the information industry development level is the highest in 6 years. From the pace of industrial development, the provinces which are located in central and western regions of China are fastest in the level of information industry development, and their information industry sprung up. The growth rate is relatively low in traditional provinces with strong information industry in China, which are easily affected by the financial crisis as well than the central and western provinces.

Second, the development of economic growth quality in China has changed little from the national level, and has showed a case of small amplitude fluctuations. From regional level in the horizontal point of view, the quality of economic growth in all regions also showed a fluctuating trend that economic growth quality of most regions has no obvious upward trend. Because of the impact of the financial crisis, it is lower for economic growth quality in most areas of China growth in 2008 and 2009, and economic growth quality of some areas has more drastically reduced. From the economic growth quality index level, the overall level of economic growth quality of China is lower, and more than two-third of the provinces are less than the national average in the study scope. In Guizhou economic growth, quality index was lowest is 40.9% times than highest Shanghai in 6 years, which reflects that there is a huge difference in the quality of economic growth among our regions.

Third, in terms of the impact of the information industry development on the economic growth quality, according to the results of the analysis, there are 21 provinces showing significantly boost relationship of information industry on the economic growth quality in statistically analyzed 30 provinces, whereas the remaining nine ones are not. After analysis, owing to the low level of information industry development of the nine provinces, we think that the impact capability is weak on the quality of economic growth, whereas other industries of the nine provinces are relatively developed (such as the coal industry in Shanxi, the tourism industry in Hainan, and Guangxi). Therefore, the nine provinces do not reflect the significant boost effect of information industry development on economic growth quality.

### 5.2 Countermeasures and suggestions

To make information industry of China develop better for the improvement of economy growth quality, we propose countermeasures and suggestions on the development of information industry in China, which are as follows:

#### 5.2.1 Support and guidance of strong policy
For the eastern coastal provinces of China, economic development and information industry development are in high level at average, so we can increase investment in the information industry, and in the face of

opportunities of global industry restructuring, we should strive to cultivate and develop high-tech, high value-added sectors with internet, cloud computing, integrated circuits, research and development, and trade settlement in information industry. In the central and western provinces, the bases of economic and industry development are relatively limited, but these areas generally have abundant labors and broad geographic area, and the cost of enterprises is lower in labor and land compared with the eastern provinces. In addition, central and western regions should further ease the investment policies and make more favorable preferential policies for the information industry in terms of labor, land, taxation, and financing by the Western development policy.

### 5.2.2 Strengthen the training and introduction of information talents

First, the government should proceed to develop training system of information technology personnel, and establish multilevel personnel training programs and multichannel personnel recruitment system. Second, the government should increase investment in scientific research and education, make full use of quality resources in local universities and research institutes, and create the information talents training mode that combines industry and academia with research. Third, the government should encourage the cooperation between schools and enterprises to develop vocational education of information technology. Fourth, for the purpose of development of high-level personnel, we should implement the combined mode of introducing from the exterior and training in the interior.

### 5.2.3 Accelerate the informatization construction of the society

Although the contents of the informatization and the information industry are different, they are not isolated. Currently, China's overall informatization lies in the low level especially in rural areas, which also indicates the great potential of informatization construction and broad market space for information industry products and information technology. Therefore, the country can accelerate the construction of the informatization society and promote the rapid development of information industry through interaction between informatization and information industry.

### 5.2.4 Accelerate the transformation of traditional industries

On one hand, we need to vigorously develop modern service industry and information industry. On the other hand, we must upgrade and transform traditional

industries to change the extensive mode of economic growth with high energy consumption, high pollution, and low quality. For the upgrading of traditional industries, we need advanced information technology to provide support, and need to increase the demand for information technology, thus contributing to the development of information industry. The information industry development could accelerate the transformation of traditional industries through its powerful ability to penetrate and promote the optimization and upgrade of industrial structure, and it can ultimately enhance the sustainable growth of economic and achieve the improvement of economic growth quality.

## ACKNOWLEDGMENT

This work was supported by the Ministry of Industry and Informatization of Communication Soft Science Project of China under Grant 2012-R-54.

## REFERENCES

[1] Liu Xu-hong 2008. *Industry environmental performance, economic growth and environmental regulation: evidence from China.* Boston: University of Massachusetts.

[2] Louis Kuijs 2012. Economic growth patterns and strategies in China and India: Past and future. *FUNG GLOBAL INSTITUTE Working Paper.*

[3] Shen Li-sheng 2009. The change analysis of economic growth quality and added value rate in China. *Journal Social Science of Jilin University* 49(3):126–134.

[4] Wang Gang 2008. *The research on information industry measurement and development strategy in Jilin.* Ji Lin: Northeast Dianli University.

[5] Liu Yue 2007. The construction of regional informatization index and policy analysis of western information industry development. *Journal of Information* (2):107–110.

[6] Liu Yue & Hui Mei-ling 2012. The empirical research on the relationship between regional informatization and human capital. *Areal Research and Development* 31(3):43–46.

[7] Wang Gang & Wang Xin2007. Design and research of the information industry measure index system. *Journal of Northeast Dianli Uiniversity* 27(3):16–19.

[8] Jing Ji-peng et. al. 1993. New method of the information industry measure: Comprehensive information industry dynamics method. *Information Business Research* 10(3):129–133.

[9] Guo Ya-jun & Yi Ping-tao 2008. The properties analysis of the linear dimensionless method. *Statistical Research* 25(2):93–100.

[10] Zhang Li-jun & Yuan Neng-wen 2010. Comparison and selection of index standardization methods in linear comprehensive evaluation model. *Statistics & Information Forum* 25(8):10–15.

[11] Barro R.J. 2011. Quantity and quality of economic growth. *Journal Economic Chilena* 5(2):17–36.

[12] Zhong Xue-yi 2001. *Change of the growth pattern and improvement of the growth quality*. Beijing: Economic Management Press.

[13] Li Jun-lin 2007. The connotation and evaluation of economic growth quality. *Productivity Research* 9(15):9–10.

[14] Yan Hong-mei 2008. Empirical analysis based on the factor analysis method of economic growth quality in China. *Science and Technology Management Research* 28(8):239–242.

[15] Ren Bao-ping 2010. *View the growth by quality: Evaluation and reflection on quality of new China's economic growth*. Beijing: China Economic Press.

[16] Liu Li-li & Yu Ni-sha 2007. The empirical study of the economic growth quality on cities of Shandong province. *Value Engineering* 12(6):30–33.

[17] Zhang Wei-jie & Zhang Jing 2012. The research on contribution of regional industrial transformation to economic growth quality: The experience from Beijing, Tianjin and Hebei. *Reform of Economic System* (2):44–48.

[18] Mao Qi-lin 2012. The evolution of double economy open and China's economic growth quality. *Economic Science* (2):5–20.

[19] Liu You-zhang et. al. 2011. The research based on the concept of circular economy on economic growth quality. *Statistics and Decision* (4):105–108.

[20] Fan Huan-huan & Zhang Ling-yun 2009. *Statistical analysis and application of EViews*. Beijing: Mechanical Industry Press.

# Impacts and countermeasures on the industrial security of inflow of foreign direct investment to Shandong

P.Z. Wang & J. Qian

*School of International Economics & Trade, Shandong University of Finance and Economics, China*

ABSTRACT: Shandong Province, with the third largest GDP in China, has attracted abundant foreign capital as well as the high technology inflows. This article is to empirically analyze the development of foreign capital inflows and its impacts on the industrial upgrading and its status of security in Shandong province. The article also comes up with some measures to overcome the existing problems.

KEYWORDS: Industrial security, Foreign multinational investment, Industrial co-integration

## 1 INTRODUCTION

Located in the east, Shandong province is a major economic province of China. Since the publishing of reform and opening policy in China, Shandong Province has gained great rapid economic growth in the country's rising status with the superior geographical and environmental advantages, and it has become an important economic growth area in eastern coastal areas.

As the economy continues to develop in Shandong Province, the scale of foreign investment in the industrial structure and regional distribution is increasing rapidly. A recent research shows that foreign investment in Shandong Province, is of great importance to its own security industry to explore the use of foreign capital in Shandong Province, and promote sustained and stable economic growth In this paper, we analyse empirically the impact of FDI on the restructuring of industrial settings in Shandong Province. As a result, FDI plays a positive role in the construction of Shandong's industry and other improvements.

## 2 DEVELOPMENT OF INFLOW OF FOREIGN INVESTMENT TO SHANDONG PROVINCE

In 2003, Shandong's actual utilization of foreign investment reached $11.26 billion, and it is the first time it exceeded ten billion U.S. dollars, increasing by 72.66% compared with that of 2002. Over the past decades, Shandong Province is expanding the using range of foreign capital. In 2013 newly approved foreign direct investment projects reached 1333 the contractual foreign investment was $16.56

billion, an increase of 4.9% over 2012 and finally the actual inflow of foreign investment accounted for $12.35 billion, an increase of 10.7%.

### 2.1 Geographical structure of foreign investment in Shandong

Since the implementation of reform and opening up, the number of countries and regions investing in Shandong Province has been increasing. However, the geographical structures are relatively concentrated. From Table 1 it can be seen that the top twelve countries and areas accounts for nearly 90% of the total foreign direct investment flowing into Shandong Province, while South Korea, Hong Kong and Japan reaches a total of 66%. In fact,

Shandong Province has shown a steady growth in attracting foreign investment.

### 2.2 The actual use of foreign investment in Shandong Province

Shandong's FDI lies in three features: firstly, the scale keeps a stable growth; secondly, the level of projects goes up; thirdly the third industry has attracted more FDI.

As it can be seen from the chart, Shandong Province has attracted 6.521 billion U.S. dollars, accounting for 12% of China's actual use of foreign capital. Due to the global financial crisis, it reached its floor in 2004 with 20% of the total FDI flowing into China, but a valley in 2008 and afterwards kept a steady rise till 2013 reaching $13.589 billion, an increase of 20% over the previous year.

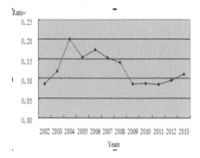

Figure 1. Actual Utilizing Ratio of FDI in Shandong.

## 2.3 The industrial structure of foreign investment in Shandong

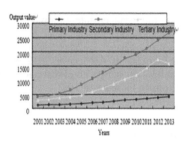

Figure 2. 2001–2013 annual output of the three industries in Shandong Province (Unit: one hundred million dollars).

It can be seen from Figure 2 that manufacturing sector still contributes to the output greatly and it rises rapidly.

However, upgrading the industrial structure is still lagging behind, although the proportion of primary industry GDP has reached a low level, but the proportion of secondary industry GDP still ranks first in the three industries with more than 50% percentage, without the development level of high value-added tertiary industry reaching a higher level. This shows that the development of the tertiary industry in Shandong Province is not sufficient, and it needs more attention and further improvement.

Shandong Province, the industrial structure adjustment in each industry among more developed countries is still lagging behind; the structure of cooperation among companies in Shandong Province is still waiting to be improved.

## 2.4 Forms of foreign investment in Shandong

From foreign investment to foreign borrowing in Shandong Province, foreign direct investment, foreign investment in the other three, this is the main form of foreign direct investment. From Shandong Foreign Direct Investment (FDI) situation, Shandong Province shows yearly changes in the following table:

As it can be seen from the table, in foreign direct investment, joint ventures and foreign companies occupy a larger proportion, while the joint-stock cooperative enterprises and foreign-invested enterprise accounts for a smaller proportion.

## 2.5 Use of the foreign capital region constitutes Shandong

Foreign investment concentrated in Shandong Province, Qingdao, Yantai, Weihai and other coastal areas, foreign direct investment has a high degree of clustering, and this feature is becoming increasingly

Table 1. Shandong Province 1990–2013 years divided by foreign investors in the form of utilization of foreign capital (unit: one hundred million U.S. dollars).

| Years | The actual utilization of foreign capital | Foreign Direct Investment | FDI accounted for the propotion of the actual use of foreign capital% | Years | The actual utilization of foregin captial | Foreign Direct Investment | FDI accounted for the proportion of the actual use of foreign capital% |
|---|---|---|---|---|---|---|---|
| 1990 | 3.15 | 1.31 | 41.59 | 2002 | 42.49 | 36.21 | 85.22 |
| 1991 | 3.11 | 1.51 | 48.55 | 2003 | 65.21 | 55.86 | 85.66 |
| 1992 | 4.68 | 1.80 | 38.46 | 2004 | 112.60 | 70.94 | 63.00 |
| 1993 | 13.77 | 9.73 | 70.66 | 2005 | 98.21 | 87.01 | 88.59 |
| 1994 | 22.61 | 18.43 | 81.51 | 2006 | 110.14 | 89.71 | 81.45 |
| 1995 | 34.01 | 25.36 | 74.57 | 2007 | 102.10 | 100.01 | 97.96 |
| 1996 | 32.67 | 26.07 | 79.80 | 2008 | 110.12 | 110.12 | 100.00 |
| 1997 | 33.94 | 25.90 | 76.31 | 2009 | 82.02 | 82.02 | 100.00 |
| 1998 | 35.84 | 25.00 | 69.75 | 2010 | 80.11 | 80.10 | 99.99 |
| 1999 | 36.10 | 22.23 | 61.58 | 2011 | 91.72 | 91.68 | 99.96 |
| 2000 | 37.45 | 24.69 | 65.93 | 2012 | 111.64 | 111.60 | 99.96 |
| 2001 | 38.12 | 29.71 | 77.94 | 2013 | 123.52 | 123.49 | 99.98 |

evident. With the opening of the three cities of the peninsula Lee and superior geographic advantages, foreign investment is widely used in developed areas, while in less developed areas, the rate is quite low.

Economic development in Shandong Province presents the eastern, central and western geographical features, while the area of foreign investment was mainly concentrated in the eastern region, in 2013, Yantai, Weihai City, the actual utilization of foreign capital accounted for 57.83 percent of the province's total actual use of foreign capital.

## 2.6 Positive impacts of foreign investment in Shandong

Foreign investment in the industry of Shandong province has lots of impact, but its distribution among industries preached extreme imbalance. Foreign investment in a small proportion of primary industry, mainly in the secondary industry can be seen from the table, in 2013 the proportion of foreign investment in the primary industry was only 2.16%, while foreign investment in the secondary industry projects accounted for 59.87% of the total number of full-year investment in the number of secondary industries contracted foreign capital of 62.71% of the total amount of money.

## 3 THE USE OF FOREIGN CAPITAL OF SHANDONG PROVINCE PROBLEMS

### 3.1 Ways of using foreign capital of Shandong Province presented a unified

East Asia is the main source of foreign investment in Shandong Province, in which Japan, South Korea, Shandong Province, has been investing more countries, nearly the size of its annual investment is relatively high, and the utilization of foreign capital in Shandong Province in these countries is mainly manufacturing transformation from high capital low technology intensity.

Europe and other developed countries are the main source of regional high-tech and advanced management philosophy. Its capital and technology are more intensive, but Shandong use of foreign investment in these countries is less than other countries, because the quality and scale of foreign investment are limited.

### 3.2 Shandong Province, irrational industrial structure of foreign capital utilization

Utilization of foreign capital in Shandong Province is mostly concentrated on the second and tertiary industries, especially the manufacturing sector accounts for a large proportion of the total foreign capital utilization, and require a lot of capital investment with a relatively long payback period agriculture, transportation, energy and other basic industries and facilities Investment areas cost shorter time, these are utilized foreign capital of Shandong Province inadequacies.

### 3.3 Shandong unbalanced regional distribution of foreign capital utilization

Economic development presents the eastern, central and western geographical features, while the area of foreign investment was mainly concentrated in the eastern region, the number of 2013, accounting for 12.63%. This imbalance exacerbated the imbalance of economic development in Shandong Province, Shandong Province, thus making the entire industrial structure irrational.

### 3.4 Shandong utilization of foreign capital contribution to GDP is not high

Shandong Province, the actual utilization of foreign capital with every percentage increase, GDP will increase Shandong 0.521 percentage points; while the country actually utilized foreign capital increased, the country's GDP will increase by 0.846 percentage points, Shandong Province, the utilization rate of foreign capital contribution to economic growth is lower than the national average.

## 4 THE SECURITY IMPLICATIONS OF FOREIGN INVESTMENT IN THE INDUSTRY IN SHANDONG PROVINCE EMPIRICAL ANALYSIS

In this paper, we use GDP data of 2000–2013 and foreign direct investment data of three industries. Foreign direct investment is used as an explanatory variable and GDP is an explained variable.

We build a model: $Log(GDP_i) = C + \beta Log(FDI_i)$, $(i = 1, 2, 3)$ which explains the first, secondary and third industries.

### 4.1 Firstly, stationary time series ADF test

Turn right sequence $Log(FDI_1)$, $Log(FDI_2)$, $Log(FDI_3)$, $Log(GDP_1)$, $Log(GDP_2)$ and $Log(GDP_3)$ ADF test, the results in the following table:

As for $Log(FDI_1)$, $Log(FDI_2)$, $Log(FDI_3)$, $Log(GDP_1)$, $Log(GDP_2)$ and $Log(GDP_3)$, the absolute value of the second-order differential sequence $t$-statistics are greater than significant level of 5 percent threshold, indicating that the 5% significance level is remarkable, so reject the null hypothesis, the second-order differential sequence is stationary,

$Log(FDI_1)$, $Log(FDI_2)$, $Log(FDI_3)$, $Log(GDP_1)$, $Log(GDP_2)$ and $Log(GDP_3)$ follow a second-order single whole ADF test conducted residuals:

Among residuals $et_1$, $et_2$ and $et_3$ ADF test results indicate as: T statistic shows three industries are different, the absolute value of $t$ is greater than the critical significance level of 1% of the value of the case, and thus denies the existence of the unit root null hypothesis, that the sequence entries $et_1$, $et_2$, $et_3$ are stable. In summary, there is a long-term stable relationship between $Log(GDP)$, $Log(FDI)$.

### 4.2 The tertiary industry Cointegration

T statistic is -5.333657 of ADF test, the critical values: 1% in significance level is -3.563915; under the 5% significance level is -2.157408; 10% significance level is -1.610463

The residuals of the ADF test sequences by comparing are stable, so there is cointegration. From the results, test value -1.610463 less than 10% threshold levels, the residual term stable, the third industry and selected economic variables are related to the continuous existence of long-run equilibrium relationship. And then follows the residual sequence.

### 4.3 The tertiary industry heteroscedasticity test: is smooth sequence, the model cointegrated

Heteroscedasticity does not exist in the 10% significance level, so the $t$ statistics and $F$ statistic can be used.

ADF unit root test results show that the time series of various economic variables exhibited non-stationary, while the first difference is that time series is stationary and it is integrated of order one I (0).

| Variable | The type of test (c, t, n) | ADF statistic | P values | Conclusion |
|---|---|---|---|---|
| Log (GDP1) | (c, 0, 1) | −0.967616 | 0.7450 | Non-stationary |
| Log (GDP2) | (c, 0, 0) | −11.91333 | 0.0000*** | Smooth |
| Log (GDP3) | (c, 0, 4) | 1.419381 | 0.9981 | Non-stationary |
| Log (FDI1) | (c, 0, 3) | −3.979170 | 0.0078*** | Smooth |
| Log (FDI2) | (c, 0, 4) | −1.658388 | 0.4374 | Non-stationary |
| Log (FDI3) | (c, 0, 0) | −5.601211 | 0.0002*** | Smooth |
| Δ Log (GDP1) | (c, 0, 0) | 0.824861 | 0.9921 | Non-stationary |
| Δ Log (GDP2) | (c, 0, 3) | −3.867178 | 0.0365** | Smooth |
| Δ Log (GDP3) | (c, 0, 0) | −2.101230 | 0.2460 | Non-stationary |
| Δ Log (FDI1) | (c, 0, 0) | −6.600100 | 0.0000*** | Smooth |
| Δ Log (FDI2) | (c, 0, 0) | −0.325617 | 0.9059 | Non-stationary |
| Δ Log (FDI3) | (c, 0, 0) | −5.357656 | 0.0003*** | Smooth |

Note: "Δ" represents the first difference, "**,***" represent 1%, 5% significance level on significant.

### 4.4 Establish a model for the quantitative analysis

According to the data processing, computing, we get under the table, and use Eviews6.0 regression Analysis.

Primary industry GDP and FDI form regression equation as:

$$Log(GDP_1) = 5.86 + 0.61 Log(FDI_1)$$
$$(0.86)(0.28)$$

$$Log(GDP_3) = 4.10 + 1.01 Log(FDI_3)$$
$$(0.51)(0.11)$$

$R^2 = 0.86$, Adjusted $R^2 = 0.89$, $F = 4.56$

Among them, the better the goodness of fit equation, these results show that the first industrial use of FDI for each 1% increase in industrial added value increased by 0.61%.

Secondary industry GDP and FDI-derived regression equation:

$$Log(GDP_2) = 3.78 + 0.92 Log(FDI_2)$$
$$(2.16)(0.36)$$

$R^2 = 0.91$, Adjusted $R^2 = 0.95$, $F = 6.43$

Among them, fit of the equation is better, these results show that the second industrial use of FDI every 1% increase in industrial added value increased by 0.92%.

We check the tertiary industry GDP and FDI regression equation:

$R^2 = 0.88$, Adjusted $R^2 = 0.87$, $F = 84.56$

Among them, goodness of fit of the equation is better, these results show that: the use of tertiary industry FDI every 1% increasing in industrial added value increased by 1.01 percent.

In summary, this description of the first industrial use of FDI to promote economic development to the most, second and tertiary industries' role in promoting the use of FDI in GDP is greater than the primary industry.

The contribution of the role of secondary and tertiary industries to GDP is greatest, but it also reveals that the FDI will help to optimize the industrial structure of Shandong Province.

## 5 CONCLUSIONS AND RECOMMENDATIONS

The current foreign investment in the province in general does not constitute the real threat to the security industry, along with the rapid entry of foreign economies, the economy in Shandong Province presents a reasonable growth and plays a positive role in promotion. Obviously, it brings the openness of foreign competition, as well as with the international under market conditions. Therefore related industries in Shandong Province should seize the opportunity to launch international industrial restructuring in order to achieve leapfrog development. It will finally result in the formation of a number of strong domestic enterprises. We figure out several measures as followings:

### 5.1 To improve the capability of the independent innovation, enhance the level of industrial technology

To increase investment in technology research and development, through the deepening of enterprise reform, Shandong Province, the formation mechanism of the enterprise innovation, through the implementation of various policies and measures to encourage innovation, and provincial government guides enterprises to increase investment in technology development.

### 5.2 To optimize the distribution of foreign origin, and expand the scope of foreign capital

The foreign investment should be actively guided from the eastern coastal areas to central and western regions, making the development of Shandong Province, east and west should implement coordinated regional economic stability and development. Firm plays an important role in eastern coastal areas, improving efficiency and quality of foreign investment, and promote coordinated economic development of the province.

### 5.3 To improve the industrial structure, optimize the investment environment

We need to reasonably guide the direction of foreign investment, foreign investment to develop appropriate land-use policies, step-by-step, there are plans to guide foreign capital, advanced technology and management experience to important industries in Shandong Province, making use of foreign capital and industrial restructuring, regional coordination structure consistent. Actively promote the "leading driven" strategy, with weak and strong, east central and western regions to promote complementary and coordinated development.

### 5.4 The contribution of foreign investment to improve and enhance the value of foreign capital utilization

Since 2003, the use of foreign capital in Shandong community is rapidly increasing compared to the proportion in other places, but the economic growth of foreign investment in Shandong Province, of the contribution is still not high. Shandong Province, a major source of foreign investment in a single, but is insufficient to absorb the high-tech and modern advanced management experience of developed countries in Europe and America.

## ACKNOWLEDGMENTS

I would like to express my acknowledgement to Shandong Natural Fund "The Marginal Estimation of Agricultural Products and Upgrading of Agricultural Product Export in Shandong (code: 2013ZRB01949)"

Wang Peizhi: Dean and Prof. of School of International Economics & Trade, Shandong University of Finance and Economics

Qian Jin: Master candidate of Shandong University of Finance and Economics

## REFERENCES

[1] King Yuqin, industrial security concept Analysis, [J] Contemporary Economic Research, 2004 (3).
[2] Lee ship, foreign investment and national economic security [J] China Industrial Economy, 1997 (8).
[3] Nain expensive, foreign investment and China Industrial Security [J] Finance, 2003 (5).

[4] Liu Yurong, Chen Xixi, foreign direct investment in Anhui industry structure and the Countermeasures [J], Heilongjiang's foreign trade; 2010 01.

[5] He Xiufeng, impact of foreign direct investment on upgrading the industrial structure of [D], Jiangsu University; 2010.

[6] Huang Yanjun, foreign direct investment (FDI) of [D] impact on China's industrial structure, China Youth Political College, 2011.

[7] Wang Guirong, Cui Bangying impact of FDI on China's economic analysis [J], Investment Research, 2010 03.

[8] Ma Ning, Countermeasures Research Manin, FDI to China's industrial structure [J], Modern Management Science, 2011 08.

[9] Chen Feng silver, measurement of FDI on Economic Growth of Shandong Province Analysis [J], School of Economics, Zhejiang Gongshang University, "Economic Research Guide" 2010 Section 32 (total 106).

*Multimedia, Communication and Computing Application – Leung (Ed.)*
© *2015 Taylor & Francis Group, London, ISBN 978-1-138-02775-6*

# The empirical research about relations between brand image, brand experience and customer value

J.B. Tu

*School of Economics and Management, North China University of Technology, China*

ABSTRACT: Research on brand experience and customer value becomes very important now. Since the relationship between brand image and customer value is still not clear now, enterprise should use brand image to enhance customer value with the help of brand experience. This paper established a framework of relationship among brand image, brand experience and customer value, and hypothesis was proposed. Empirical research of structural equation model was applied to verify the relations among brand image, brand experience and customer value. The results show that company image, user identity and product appearance have significant effect on customer value with the help of brand experience; brand personality has no significant effect on brand experience, but brand personality has significant effect on hedonic value.

KEYWORDS: Brand image, Brand experience, Customer value

## 1 INTRODUCTION

Gadena and Liway suggested the concept of brand image by 1950[1], which was an important issue of marketing research.

Researchers are studying the concept of brand image from different angles now, but the notions of research are different. Brand image was the overall perception for consumer to understand brand, and it effected buying and consuming behaviour of consumers[2]. Perceiving the company's brand image by the customers, it is used to choose the proper product and satisfy the benefit of the consumers.

In the mobile phone industry, consumers choose the brand of mobile phone according to the identity of consumers, functional demand and emotional demand. With experience marketing by the mobile phone producer, consumers know more about the function and exterior feature of mobile phone, experience pragmatic value and enjoy it. Then the consumer can choose the suitable mobile phone. For example, Nokia has the centre of experience, customers could perceive the appearance, function and quality of Nokia mobile phone there, and enjoy better service.

In the existing research at home or abroad, researchers study the effect of brand image of movement shoes on the perception of customers[3], but unluckily the relationship between brand image and customer value is still not clear. This study is able to analyse the relationship between brand image and customer value with brand experience. And it will give useful suggestion to improve the research of brand image.

## 2 LITERATURE REVIEW AND HYPOTHESIS

### 2.1 *The research of the concept and dimension of brand image*

According to the research of brand image, Kotler suggested that brand image was a group of faith to the particular brand[4]. The faith had the relationship with the consumer demand.

Additionally, Biel (1993) proposed that brand image had attributed and related association, and it was the subjective response of customers to the brand[5].

Chinese researcher Luo believed that brand image was the subjective perception of consumers of the particular brand[6]. Formed by the marketing channel and consumption experience of enterprises, brand image has the characteristics of multidimensional combination, complex diversity, relative stability, plasticity and situationality. It means that customers can perceive brand image of enterprises in the experience marketing well. The relationship of the overall perception of consumer to the brand and perception benefit is very important to promote positive behaviour to the brand.

According to the research of the dimensions of brand image, the dimensions of brand image were described by Biel (1993) as company image, user image and the product or service image. BIEL also believed that brand image was special perception that was tangible and functional, such as product appearance.

With the development of research, user image has not only the attribute of personality or living style, but also the attribute of user identity. Aaker believed the association of measuring brand image, including product association, brand personality and organization association [7]. Because there is much difference in the research on the dimensions of brand image, the attributes of brand image have not been integrated.

This paper suggests that the dimensions of brand image are brand personality, company image, product appearance and user identity.

### 2.2 The research on customer value and the composition of it

Jackson (1985)[8] believed that value was the ratio of perceived interest and cost. Hirshman and Holbrook (1982) suggested that the experience of consumption after buying made not only the rational consumption value for the customers, but also experiencing the consumption value, such as pleasure and beauty in the consumption [9]. And the research on the dimensions of customer value appeared.

### 2.3 Babin and Darden (1994) believed that the motivation of consuming by consumers could be explained by utilitarian benefit and hedonic benefit. Gursoy, Span gen berg and Ruther for (2006) Suggested that the conception value included pragmatic value and hedonic benefit[10]. Considering the angle of experience, Jensen Hansen (2007) summed up five dimensions of conceived value:excellence, harmony, emotional stimulation, identity and environmental value[11]. So this paper uses the pragmatic value and hedonic value as the constructor of customer value. Brand experience

Brand experience seemed to be subjective and intrinsic behaviour reaction owing to the driving of brand (Brakus, 2009[12], such as the design, identity, package, communication, and environment of the brand.

### 2.4 The research of relationship of brand image, brand experience and customer value

Brand image is the attitude of the customer's to the brand, including brand cognition and feeling perception. Brand cognition is the faith of customer's to brand. They could satisfy the utility demand and practical personal demand of consumers. For another, considering the perception of the feeling of the brand, brand personality was the personal characteristic related to the brand [13]. The Attribute of brand personality can call on the feeling and believing of customers to the brand. It is very important for customers to better their value.

In order to clear the relationship of brand image and customer value, Chinese researchers L.X. Jiang and T.H. Lu (2006) found that brand image had functional factor and unfunctional factor; un functional factor had the direct effect on customer value, while functional factor had no direct effect on customer value[14]. So we believe that product appearance cannot directly affect customer value, company image, brand personality and user identity may have direct effect on customer value.

Some researchers thought that brand experience made the impression and feeling of consumers to the brand when the customers contacted it, and then understood the brand better (Guo Juan, 2010[15]. Since brand image includes the designing and distinguishing of the brand, which lead to the behaviour reaction of consumers to the brand. Hypotheses are given in this paper:

H1a: company image has direct effect on brand experience,

H1b: brand personality has direct effect on brand experience,

H1c: user identity has direct effect on brand experience,

H1d: appearance of the product has direct effect on brand experience,

H1e: brand personality has direct effect on hedonic value,

H1f: brand personality has direct effect on pragmatic value,

H1g: company image has direct effect on pragmatic value,

H1h: company image has direct effect on hedonic value,

H1i: user identity has direct effect on pragmatic value,

H1j: user identity has effect on hedonic value.

Brand experience was thought as the subjective and intrinsic behaviour reaction of customer, affected by the brand. These behaviour reactions have relationship with the conception of consumer. Hypotheses are given in this paper:

H2a: brand experience has direct effect on hedonic value,

H2b: brand experience has direct effect on pragmatic value.

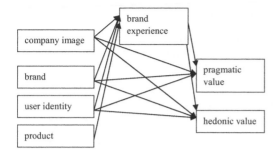

Figure 1. Research Model.

## 3 THE DESIGN OF RESEARCH

### 3.1 The design of questionnaire

This paper chooses Nokia mobile phone as the research brand. It is because that Nokia brand has the highest market share in the mobile phone industry. It is well understood by the consumers. And it gives related attribute to brand image and brand experience.

This questionnaire uses the 7scores Likert scales. The scales of brand image use the scale of brand personality of Aaker (1997) and the scale of the comprehensive evaluation model of brand image (X. C. Fan et al. 2002). The scale of brand experience uses the scale of Brakus (2009)about the brand experience. The scale of value and hedonic value are considered as the scale of customer value from W. Jia et al (2010) research [16].

### 3.2 Data collection

This study uses the method of filling in online and face to face filling out of the questionnaire. At last we get 294 questionnaires. By eliminating the useless questionnaires, 189 of them are useful. And recovery rate is 64.3%. The basic sample represents the characteristics of domestic customers who uses Nokia phone. Therefore, it meets the need of this study.

### 3.3 The method of data analysis

This paper uses SPSS17.O as the tool of exploratory factor analysis, and it also uses the method of reliability analysis. According to the variable factor of brand image, brand experience and customer value, this paper uses Lisrel 8.7 to fulfil confirmatory factor analysis. And it uses SPSS 17.0 to do regression analysis and verify the relationship of each variable.

## 4 DATA ANALYSIS AND VERIFICATION OF THE MODEL

### 4.1 Reliability and validity

This research uses Cronbach's alphas to measure reliability of scales. The Cronbach's alphas of overall scale is 0.856, which is higher than 0.7. And alphas are 0.831 for brand experience; 0.734 and 0.796 for pragmatic value and hedonic value; 0.704 and 0.857 for brand personality and company image; 0.790 and 0.773 for appearance of the product and user identity.

Cronbach's alphas of each variables are all higher than 0.7. It tells that the reliability is reasonable, and internal consistency of the scales is better.

To examine construct validity of each scale, this paper computes an exploratory factor analysis on the basis of a principal component analysis with a varimax rotation. Bartlett's test (df=276, Sig. =.000) is significant. Moreover, the Kaiser-Meyer-Olkin (KMO) value is 0.838, which presents factoring is appropriate. Confirmatory factor analysis (CFA) is used to test convergent validity and discriminant validity. We analyse the convergent validity by examining the factor loadings and squared multiple correlations. The factor loadings are greater than 0.6, so the convergent validity is significant.

### 4.2 Verify the relationship between brand image brand experience and customer value

This study uses SPSS 17.0 to conduct hypothesis testing.

And regression analysis is used to verify the relationship of the dimensions of brand image, brand experience and customer value. As a result, appearance of the product, company image and user identity have direct effect on brand experience (Coefficients of them are 0.326, 0.208, 0.354, Sig. are significant, T-value are 3.195, 3.903, 7.065, all higher than the standard of 1.96), hypothesis H1a, H1c, H1d were supported; but brand personality had no significant correlation with brand experience (Coefficients=0.191, Sig. is not significant, T-value=1.419<1.96), H1b was not supported; brand personality has direct effect on hedonic value (Coefficients=0.449, Sig. is significant., T-value= 4.926>1.96), but it has no significant effect on pragmatic value (Coefficients=0.114, Sig. is not significant, T-value=3.213), H1e was supported, while H1f was not supported; company image has direct effect on pragmatic value and hedonic value (Coefficients=0.299 and 0.323, Sig. is significant, T-value=4.919 and 5.384), user identity has direct effect on pragmatic value and hedonic value (Coefficients=0.264 and 0.341, Sig. is significant, T-value=3.878 and 4.871>1.96), H1g, H1h, H1i, H1j were supported. Brand experience has significant correlation with hedonic value and pragmatic value. (Coefficients are 0.532 and 0.602, Sig. is all significant, T-value are 10.473 and 9.993>the standard of 1.96), H2a and H2b were supported.

## 5 CONCLUSION

### 5.1 Implication of research

This paper discusses the impact machanism of brand image, brand experience and customer value. Through the analysis of previous literature, and literature review on the dimensions of brand image, we propose the hypothesis of brand experience machanism. At the same time, by referring to other literatures on measurement of brand experience, we also propose the hypothesis of the impact machanism of brand experience on the dimensions of customer value. Carrying on the empirical study through investigation and quantitative analysis to test hypothesis, our findings have the following contributions to theories.

Firstly, brand image as a marketing item, is getting more and more attention. The focus of brand image should be on brand personality, company image, user identity and appearance of product. We can improve these four items so as to improve brand image. Brand image is significant for the marketer to get the benefit of the consumer.

Secondly, some dimensions of brand image are extremely related to customer value. Like brand personality, it has no relationship with brand experience. Using brand experience, work cannot construct and improve the relationship of brand personality and customer value. Brand personality has a relationship with hedonic value, while it has no relationship with pragmatic value.

Thirdly, with the help of brand experience, some dimensions of brand image have indirect relationship with customer value, such as appearance of the product. The company can use brand experience to make an appearance of the product be known, so that customer can buy the product. Brand experience can improve consumers' understanding of the company, so as to better the customer value.

### 5.2 Managerial implications

Enterprises aim at making brand experience with customers who can obtain insights from this research, especially the enterprises in service markets. Enterprises can build different brand image to generate different customer value for customers. The conclusions of this research have certain directive meanings for enterprise marketing practice. In order to improve customer value, enterprises should provide brand experience to the customer, and then raise its value.

## ACKNOWLEDGMENT

The author would like to thank the anonymous reviewers and the editors for their constructive criticism and comment.

## REFERENCES

[1] Park C W, Jaworski B J, MacInnis D. 1986. Stragegic Brand concept-Image Management. *Journal Of Marketing.* 50(12):135–145.

[2] X.C. Fan, J. Chen. 2002. Brand image Comprehensive evaluation model and its Application. Journal of Nan Kai: The edition of Philosophy and Social Science. 2(3):65–71.

[3] H. Qin, H.L. Qiu, L.Z. Wu. 2011. The Research of the effect of brand Image of sport shoes on the relationship perception, Satisfaction and loyalty. *Chinese Journal of Management*, 63(2):123–132.

[4] Kotler. 1991. Marketing Management: Analysis, Planning, Implementation, and Control (8 Th Ed.). *Prentice-Hall,* INC: 442.

[5] Biel A L. 1993 How brand image drivers brand Equity. *Journal of Advertising Research*, 6 (11/12):6–12.

[6] Z.M. Luo. 2001. The Composition and Measurement of brand image. *Journal of Beijing Technology and Business University: The Edition of Philosophy and Social Science*, 16(4):19–22.

[7] Aaker D A. 1996. Measuring Brand Equity across Products and Markets. *California Management Review*, 38(3):102–120.

[8] Jackson. 1985. Build Customer Relationships That Last. *Harvard Business Review.* 2 (11-12):120~128.

[9] Holbrook, Hirschman. 1982. The Experiential Aspects of Consumption: Consumer Fantasies, Feelings, and Fun. *Journal of consumer research.* 9(2):132~140.

[10] Gursoy, Spangenberg, D. Rutherford. 2006. The Hedonic and Utilitarian Dimensions of Attend Attitudes toward Festivals. *Journal of Hospitality & Tourism Research.* 30(3):279~294.

[11] Jensen, Hansen. 2007. Consumer Values among Restaurant Customers. *International Journal of Hospitality Management.* 26(3):603~622.

[12] J. Josko Brakus, Bernd H. Schmitt, & Lia Zarantonello. 2009 Brand experience: What is it? How Is it measured? Does it affect loyalty? *Journal of Marketing.* 3(1):52~68.

[13] Aaker J L. 1997. Dimensions of Brand Personality. *Journal of Marketing*, 34(8):347–356.

[14] L.X. Jiang, T.H .Lu. 2006. Can image create value?. *The World of Management,*2 (4):106–129.

[15] Guo Juan, Xu Yuanyuan, Huang Kunlei. 2010. The mechanism research of effect of brand Experience on brand loyalty: The survey of Nokia Brand as an example. *Journal of Graduate student of Zhongnan University of Economics and law.*

[16] W.Jia, M.L. Zhang, D.Li. 2010. The effect of Psychological contract of customer participation on customer value creation. Journal of industrial engineering and engineering management 24(4):20–28.

*Multimedia, Communication and Computing Application – Leung (Ed.)*
© *2015 Taylor & Francis Group, London, ISBN 978-1-138-02775-6*

# Cost optimization model of exponential Weibull distribution accelerated life simulation

Y. Wang & D.J. Lu
*Department of Mathematics, Qinghai Normal University, Xining, China*

X.Q. Zhang
*Financial and Economic College, Qinghai University, Xining, China*

ABSTRACT:   Remaining life evaluation of products based on the accelerated life testing is becoming a hot spot of research. The way to improve the cost of evaluation is one of the most important issues. The product is shipped to the customer if it passes the burn-in test. The cost of per unit consists of three parts. They are the burn-in cost in the burn-in period, the loss of goodwill cost when the product is damaged during the warranty period and the operating failure cost which occurs when the product is impaired during the servicing period. Our goal is to find the appropriate testing parameters to minimize the total testing, manufacturing, quality and reliability costs. The upper and lower bounds for the optimal burn-in time are derived.

## 1 INTRODUCTION

The product is shipped to the customer if it passes the burn-in test. The cost of per unit consists of three parts. They are the burn-in cost in the burn-in period, the loss of goodwill cost when the product is damaged during the warranty period and the operating failure cost which occurs when the product is impaired during the servicing period. Our goal is to find the appropriate testing parameters to minimize the total of testing, manufacturing, quality and reliability costs. The upper and lower bounds for the optimal burn-in time are derived. In the literature, it is often assumed that the failure pattern follows Exponential Weibull distribution and the burn-in process is operated under approximately the same environment in the early operating life of the product. In this paper, we require that the distribution life of the product should have some specific properties. The burn-in process is operated under severe (stress) conditions involving in high temperature, voltage, etc. And the product's residual life depends on the burn-in stress level and the length of burn-in period. Accelerated burn-in before shipment will reject poor-quality products and improve product reliability within a warranty period. Accelerated burn-in saves time, but may sometimes cost more. Our goal is to find the appropriate testing parameters to minimize the total of testing, manufacturing, quality and reliability costs. The upper and lower bounds for the optimal burn-in time are derived.

Many products have a high failure rate in their early operating lives. Burn-in has been widely accepted as a method of screening out defects before the product is delivered to the customer. A common practice is to test the product until it reaches the change-point where the product failure rate decreases in the infant mortality stage from a constant level in the normal stage [1–3]. [4–6] study the effects of burn-in about the mean residual life of the product. [7–9] study economic designs of burn-in procedures. [10–11] study general discussions about burn-in. [12–15] study failure rate model. [16] studies the optimal designs of accelerated life tests and exponentially distributed lifetimes under progressive censoring. The search for the optimal ALT plans which minimize the asymptotic variance of the estimated means or maximizes the determinant of Fisher matrix. [17] studies the double truncation data elimination method based on model transformation. The Weibull distribution was converted into exponential distribution because the numerical control system lifetime obeys the Weibull distribution.

## 2 NOTATIONS AND ASSUMPTIONS

We assume that each finished product is subjected to an accelerated burn-in test and the product is non-repairable. There is a known relationship between stress conditions and the product life distribution. Let $\beta_0$ be the stress level of normal operation and $\beta_1$ be the stress level of burn-in test respectively. These levels may be a function of several stress parameters of operating conditions. Without loss of generality, we assume $\beta_0 \leq \beta_1$. The burn-in process may affect the residual life of the product so that the life distribution of the product is not the same as that of no burn-in under normal stress level $\beta_0$. We assume that the product's residual life is equal to that one which under a more severe stress level. And stress level $\beta$ is a function of $\beta_0$ and $\beta_1$. A The typical its function form is $\beta_r = \left(\frac{\beta_1}{\beta_0}\right)^k \beta_0$, where $0 \leq k \leq 1$.

Let $t_1$ be the burn-in time. The lifetime of the product under stress levels $\beta_0$, $\beta_1$ and $\beta_r$ are denoted by $Y_0$, $Y_1$ and $Y_r$ respectively. The product is scrapped and has a lifetime $Y_1$ if the product breaks down during the burn-in period. The product will be shipped to the customer and has a residual life time $Y_r$ if the product passes the burn-in test. Products with no burn-in have life time $Y_0$.

### 2.1 Notations

These include:

$c_s$ : burn-in set-up cost;

$c_0(\beta_1)$ : burn-in cost per unit time, which is an increasing function of the stress level;

$c_1$ : burn-in failure cost per unit;

$c_2(t)$ : loss of goodwill cost when a failure occurs at time $t$ burn-in cost per unit time, which is an increasing function of the stress level;

$c_3(t)$ : operating failure cost when a failure occurs at time $t$ with the customer;

$h_i(t)$ : hazard rate function under stress level $\beta_i$, $i$=0, 1 and $r$;

$F_i(t)$ : distribution function associated with $F_i(t)$.

### 2.2 Assumptions

The Assumptions are as follows:

1  $h_i(t) = \beta_i g(\beta_i t)$, $i$=0, 1 and $r$, where $g(t)$ is a hazard rate function satisfying;

$g(t) \to \infty$ as $t \to 0$ ;

$g(t)$ is decreasing for $t \geq 0$ ;

$g(t+s)/g(t)$ is an increasing function of $t$ for $s \geq 0$

Some often used distributions to satisfy assumption (1)which are:

Exponential Weibull distribution with density function [17]

$f(y) = \alpha\theta y^{\alpha-1} e^{-y^{\alpha}} \left(1 - e^{-y^{\alpha}}\right)^{\theta-1}$ , $\alpha, \theta > 0$, lifetime function with $F(y) = \left(1 - e^{-y^{\alpha}}\right)^{\theta}$ $\alpha, \theta > 0$, hazard rate function with $g(y) = \dfrac{\theta\alpha y^{\alpha-1} e^{-y^{\alpha}} \left(1 - e^{-y^{\alpha}}\right)^{\theta-1}}{1 - \left(1 - e^{-y^{\alpha}}\right)^{\theta}}$, $\alpha, \theta, y > 0$.

2  The loss of goodwill cost $c_2(t)$ and the operating failure cost $c_3(t)$ satisfies: Cost

optimization $c_1(t)$ is decreasing for $0 \leq t \leq T_i$ and $c_i(t) = 0$, for $t > T_i$, $i = 2$ and 3.

$c_i(t)$ is differentiable for $0 \leq t \leq T_i$ and $|\partial c_i(t)/\partial t| \leq M$ ,where $M > 0$ .

Some typical function of $c_2(t)$ are constant cost:

$$c_2 = \begin{cases} c_2, & 0 \leq t \leq T_2 \\ 0, & t > T_2 \end{cases}.$$

Linear decreasing cost is

$$c_2 = \begin{cases} c_2(1 - t/T_2), & 0 \leq t \leq T_2 \\ 0, & t > T_2 \end{cases} ;$$

Exponentially decreasing cost is $c_2(t) = c_2 e^{-\delta t^a}$ ,$t \geq 0$.

If the product passes the burn-in test, then testing time $t_1$ under stress level $\beta_1$ is equal to the using r time $t_r$ which under stress level $\beta_r$,

where $F_r(t_r) = F_1(t_1)$ (1)

Here, the residual life is affected by the length of the burn-in period as well as the stress level $\beta_1$.

## 3  COST MODEL

If the product fails during the burn-in period, the cost per unit consists of the burn-in failure cost and the testing cost, the product will be shipped to the customer if it passes the burn-in test. The cost per unit consists of three parts. They are the burn-in cost in the burn-in period, the loss of goodwill cost when the product fails during the warranty period and the operating failure cost which occurs when the product fails during the servicing period. The total cost of per unit subjects to a burn-in time $t_1$ is

$TC(t_1) = c_s + c_0(\beta_1)Y_1 I(Y_1 \leq t_1) + c_0(\beta_1)t_1 I(Y_1 > t_1) + c_1 I(Y_1 \leq t_1)$
$c_2(Y_r - t_r)I(Y_r > t_r) + c_3(Y_r - t_r)I(Y_r > t_r), t_1 > 0$ (2)

where $I(\cdot)$ is the indicator function.

Taking expectation, we have

$m(t_1) = E[TC(t_1)]$
$= c_c + c_0(\beta_1)\int_0^{t_1} s\, dF_1(s) + c_0\beta_1 t_1 (1 - F_1(t_1)) + c_1 F_1(t_1)$
$+ \int_{t_r}^{t_r+T_2} c_2(s - t_r)dF_r(s) + \int_{t_r}^{t_r+T_3} c_3(s - t_r)dF_r(s)$
$= M(t_r)$ (3)

where $t_r = \beta_1 t_1 / \beta_r$ (4)

which from assumption (1) and equation (1).

The total cost of the product without burn-in is

$$TC(0) = c_2(Y_0) + c_3(Y_0) \tag{5}$$

and
$$m(0) = \int_0^{T_2} c_2(s) dF_0(s) + \int_0^{T_3} c_3(s) dF_0(s) \tag{6}$$

## 4 OPTIMAL BURN-IN TIME

The optimal burn-in time is determined by minimizing the expected total cost $m(t_1)$ when operating an item in the burn-in period and the servicing period. Taking a derivative of the expected total cost with respect to $t_r$, we have

$$\partial M(t_r)/\partial t_r = (1 - F_r(t_r))$$

$$\{c_0(\beta_1)\beta_r/\beta_1 - (c_2(0) + c_3(0) - c_1)h_r(t_r) + A(t_r)h_r(t_r)\}, \tag{7}$$

where

$$A(t_r) = \int_0^{T_r}(-\partial c_2(s)/\partial s)(f_r(t_r + s)/f_r(t_r))ds$$

$$+\int_0^{T_3}(-\partial c_3(s)/\partial s)(f_r(t_r + s)/f_r(t_r))ds$$

$$+c_2(T_r)f_r(t_r + T_2)/f_r(t_r) + c_3(T_3)f_r(t_r + T_3)/f_r(t_r), \tag{8}$$

and $f_r(t)$ is the density function of $F_r(t)$.

### Theorem 1

a. If $c_2(0) + c_3(0) \le c_1$, the optimal burn-in time is 0:
$m(0) = \min_{t_1 \ge 0}(t_1)$.

b. If $c_2(0) + c_3(0) > c_1$, there exists $t_1^*$ such as

$m(t_1^*) = \min_{t_1 > 0} m(t_1)$, and $t_1^*$ is finite if and only if that

$c_0(\beta_1)\beta_1/\beta_1 - (c_2(0) + c_3(0) - c_1 - A_0)\beta_r d > 0$,
where $d = \lim_{t \to \infty} g(t)$

$$A_0 = \lim_{t_r \to \infty} A(t_r) = \int_0^{T_2}\left(-\frac{\partial c_2(s)}{\partial s}\right)g_r(s)ds +$$

$$\int_0^{T_3}\left(-\frac{\partial c_3(s)}{\partial s}\right)g_r(s)ds + c_2(T_2)g_r(T_2) + c_3(T_3)g_r(T_3),$$

and $g_r(s) = \lim_{t \to \infty} f_r(t + s)/f_r(t)$.
The optimal burn-in time is greater than 0 when
$m(0) - m(t_1^*) > 0$.

### Proof:
Note that $f_r(t + s)/f_r(t)$ is increasing with respect to t for $s \ge 0$. Assume (2) gives and A(t) increases, for example, $t \ge 0$ and $\lim_{t \to 0} A(t) = 0$.

Let $\partial M(t_r)/\partial t_r = f_r(t_r)R(t_r)$
where $R(t_r) = (\beta_r/\beta_1)c_0(\beta_1)/h_r(t_r) - (c_2(0) + c_3(0) - c_1) + A(t_r)$
and the sign $\partial M(t_r)/\partial t_r$ depends on $R(t_r)$.

From the properties of $A(t)$ and assumption (1), we have

$$\lim_{t_r \to 0} R(t_r) = -(c_2(0) + c_3(0) - c_1)$$

$$\lim_{t_r \to \infty}(t_r) = c_0(\beta_1)/(\beta_1 d) - (c_2(0) + c_3(0) - c_1) + A_0$$

and $R(t_r)$ is increasing.
In case (a), $R(t_r) \ge 0$ for all $t_r > 0$ gives $\partial M(t_r)/\partial t_r \ge 0$. Thus, $M(t_r)$ is an increasing function of $t_r$ and $M(t_1)$ is an increasing function of $t_1$. The optimal plan is no burn-in.
In case (b), $\lim_{t_r \to \infty} R(t_r)$ is positive if and only if

$c_0(\beta_1)/\beta_1 - (c_2(0) + c_3(0) - c_1 - A_0) - \beta_1 d > 0$. If $\lim_{t_r \to \infty} R(t_r)$ is positive, there exists a unique $t_r^*$ such as $R(t_r^*) = 0$. Also $t_r^*$ is infinite when $\lim_{t_r \to \infty} R(t_r)$ is negative.

### Theorem2
If the optimal burn-in time is finite and positive, then we have

$$\frac{\beta_r}{\beta_1} f_r^{-1}\left[\frac{\frac{\beta_r}{\beta_1}c_0(\lambda_1) + c_2(0)f_r(T_2) + c_3(0)f_r(T_3)}{c_2(0) + c_3(0) - c_1}\right] \le t_1^*$$

$$\le \frac{\lambda_r}{\lambda_1} h_r^{-1}\left(\frac{\lambda_r}{\lambda_1}\frac{c_0(\lambda_1)}{c_2(0) + c_3(0) - c_1}\right)$$

### Proof:
Suppose $A(t) \ge 0$ and $R(t_r^*) = 0$. It follows like that $(\beta_r/\beta_1)c_0(\beta_1)/h_r(t_r^*) \le c_2(0) + c_3(0) - c_1$

Hence $t_r^* \le h_r^{-1}\left[\frac{\beta_r}{\beta_1}\frac{c_0(\beta_1)}{c_2(0) + c_3(0) - c_1}\right]$

Note that $t_1^* = t_r^*\beta_r/\beta_1$. We obtain the right-hand side of the inequality. The proof of the left-hand side of the inequality follows from the fact that

$A(t) \le c_2(0)f_r(T_2)/f_r(t) + c_3(0)f_r(T_3)/f_r(t)$, and
$f_r(t) \le h_r(t)$.

The results in Theorems 1 and 2 identify the appropriateness of implementing the burn-in process before shipment. Theore, 2 gives the upper and lower bounds of the optimal burn-in time for a burn-in stress level $\beta_1$. The result is useful in estimating the optimal burn-in time numerically.

## 5 EXAMPLE

Suppose that the residual stress level is

$$\beta_r = \left(\beta_1/\beta_0\right)^k \beta_0 \,,\, 0.001 \leq \lambda, \leq 0.01 \text{ and } 0 \leq k \leq 1.$$

The burn-in cost is

$$c_0\left(\beta_1\right) = 8 + 2{,}000\beta_1 + \left(1{,}000\beta_1\right)^2.$$

The expected cost reductions are $k = 0, 0.01$ and $0.1$. They are shown in Figure 1.

There exists an optimal burn-in stress level for one case.

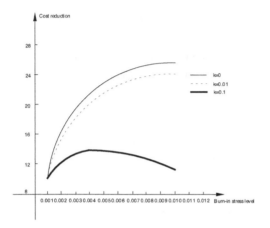

Figure 1.  The expected total cost is decreased under burn-in stress level.

## ACKNOWLEDGMENTS

Xiaoqin Zhang was financially supported by the Young and Middle-aged Scientific Research Funds (2012-QGY-3), and in part by the Ministry of Education Chunhui Program. Dianjun Lu was financially supported by the Ministry of Education Chunhui Program([2014]1310), and by Qinghai Normal University Teaching Research Funds ([2013]43).

## REFERENCES

[1] Kuo, W. &Kuo, Y. 1983. "Facing the headaches of early failures: a state-of-the-art review of burn-in decisions", *Proceedings of the IEEE*, pp.1257–66.

[2] Zacks, S. 1983. "Survey of classical and Bayesian approaches to the change-point problem", *in Rizvi, M.H, Rustagi, S. and Siegmund, S. (Eds), Recent Advances in Statistics ,Academic Press, New York, NY, pp.* 245–69.

[3] Chou, K. & Tang, K. 1992, "Burn-in time and estimation of change-point with weibull-exponential distribution". *Decision Sciences, Vol. 23, pp.* 973–90.

[4] Lawrence, M.J. 1966. "An investigation of the burn-in problem", *Technomtrics, Vol.8 No.1, pp.*61–71.

[5] Chandrasekaran, R. 1977. "Optimal policies for burn-in procedures", *OPSEARCH, Vol.4, pp.*149–60.

[6] Park, K. S. 1985. "Effect of burn-in on mean residual life", *IEEE Transactions Reliability. Vol. 34 No.5, pp.*522–3.

[7] Fox, B. 1964. "Total annual cost, a reliability criterion" *Proceedings of the National Symposium on Reliability and Quality Control, pp.* 266–73.

[8] Plesser, K.T. & Fied, T.O. 1977. "Cost-optimizied burn-in duration for repairable electronic systems", *IEEE Transactions on Reliability Vol.26 No.3, pp.*195–7.

[9] Nguyen, D. G. &Murthy, D. N. P. 1982. "Optimal burn-in time to minimize cost for products sold under warranty", *IEEE Transactions, Vol.14 No.3, pp.*167–74.

[10] Jensen, F. & Petersen, N.E. 1982. Burn-in: An Engineering Approach to the Design and Analysis of Burn-in Processes, *John Wilcy & Sons, New York, NY 1982.*

[11] Tusin, W. 1990. "Shake and bake the bugs out", *Quality Progress, september pp.*61–4.

[12] Wang, R.H. & Fei, H.L. 2004. Conditions for the coincidence of the TFR, TRV and CE models, *Statistical papers. 45:393–412.*

[13] Xu, X.L. & Fei, H.L. 2003, Approximate maximum likelihood estimate and inverse moment estimates of the parameters of the tampered failure rate model for the Weibull distribution in a step –stress accelerate life test, *Mathematics Research. 36(4):351–367.*

[14] Xu, X.L. & Fei, H.L. 2004, parameter estimation of the Weibull distribution tampered failure rate model under a normal stress, *Chinese Journal of Applied Probability and statistics. 20(2) 126–132.*

[15] Wang, R.H. & Xu, X.L. & Shi, H.W. 2009, The Statistical Analysis of Gompertz distribution based on tampered failure rate modle under Multi-step stress accelerated life testing, *Chinese Journal of Applied probability and statistics. 25(1):47–59.*

[16] Xu, X.L. & Cheng, Y.M. 2010. Optimal design of Step-tress accelerated life testing with progressive censoring, *Chinese Journal of Applied Probability and Statistics. 26(1):35–46.*

[17] Cheng, C.H. & Chen J.Y. & Li, Z.H. 2010. A New Algorithm for MLE of Exponentiated Weibull Distrbution with Censoring Data, *Mathematica applicat. 23(3):638–647.*

*Multimedia, Communication and Computing Application – Leung (Ed.)*
© *2015 Taylor & Francis Group, London, ISBN 978-1-138-02775-6*

# Based on the Chinese question relevant search algorithm research within the restricted areas

Z.X. Li & H.J. Han

*School of Computer Science& Technology, Shandong University of Finance and Economics, Jinan, Shandong, China*
*Scientific Research Office, Shandong University of Finance and Economics, Jinan, Shandong, China*

ABSTRACT: This paper presents a kind of search algorithm based on the Chinese question correlation. We determine the longitudinal correlation between the words through creating a keyword tree. We establish a longitudinal correlation matrix, and get the best answer to the query by using the correlation calculation. This method firstly extracts the semantic representation of a question sentence through "question sentence semantic representation extracting model" which extends the semantic representation of the question sentence, and then we search the answer of the question sentence by using the area keywords tree. This method adopts a strategy of keywords replacement during the process of further search. Once the keywords replacement has completed, the answer of the question sentence will be searched. This method can be integrated into an intelligent question answering system easily. Experimental results suggest that this method is effective.

## 1 INTRODUCTION

With the development of computer and digital media technology, many industries begin to use computer to deal with related affairs. And the computer is becoming more and more widely applied in intelligent public consultation. And the intelligent question answering system becomes an important research subject. After the Apache software foundation putting forward Lucene full-text search and search engine, many programmers built concrete full-text search based on its application [Yiping Yang. 1997&Chuanyuan Xu. 2008], and integrated it in a variety of software systems. Such applications segment keywords of questions and answers using word breakers, then retrieve the similarities of sentenses.. Then deposit the index file, cut the user's question after being inputted [Qiang Zhou. 2001], extracting the keywords, do the similarly comparison between questions in the index files and keywords of the answers, and at last, extract the highest weight answer as a return value.

Because of the characteristic of the Chinese language, it expresses relationships by words. But it establishes semantic relations through its statement and there is no strong lexical dependent between words. Till now, some scholars have done relevant researches of the semantic computing based on the statement at home and aboard. And most of the researches are about using the grammar and semantic information of sentence to calculate statement of

the relevance [Bin Shi. 2009&Tao Cheng. 2007]. This paper proposed a new method of using statements relevant to doing question search [Qian Wu. 2009& Wenjie Li. 2010]. Firstly, the author established the domain keywords tree after looking into a particular field, and then extracted representation of the question semantic by using the "question sentence semantic representation extracting model", expanded question semantic representations and replaced keyword based on the domain keywords tree built by the author. Finally, the paper searched the answer, returned the target answer.

## 2 CHINESE QUESTION ANALYSIS

Question semantic representation (QSR) is a question of semantic information representation by removing a question's irrelevant or interference information. It is a necessity form question semantic representation. In normal conditions, the form of a question can be a variable variable. But, it is just a semantic representation. For example: "Football is invented in which country?" and "Which country invented football?" Its semantic representational forms only one version, which is "the origin of football". Questions of semantic representation are directly related to the type of questions, Such as asking the attributes of the object, $QSR = \{property = QT, object\ name = <OBN>, attribute\ name = <QTN>\}$, Such as asking role attribute,

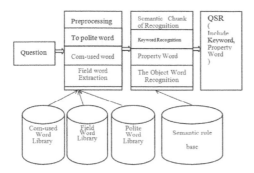

Figure 1. Question Sentence Semantic Representation Extracting Model.

QSR = {attribute = QT, Role name = <RON>, attribute name = <QTN>}.

## 2.1 Polite word filter

The difficulty to avoid in the process of the user's question is the use use of some polite words, for example: "Excuse me", "Could you please tell me...?" and so on. These kind of polite words did not help to analyse the question semantic representation. Here we just collect a polite tone and remove the polite words of the question.

## 2.2 The commonly-used words, domain words extract

Commonly-used words are defined as an industry category which are also known as general keywords. The most common major class is a noun. Filtered words contain the expression of words, the question words and so on. Commonly-used words and filed words are playing an important role in the question semantic representation. So, here just use commonly-used word libraries and special field word libraries to extract common keywords and domain words.

## 2.3 Semantic piece of identification

Semantic rule base (SRB). Here the author collects and composes a semantic rule base, according to the keywords of the problem which within the restricted areas. SRB mainly consists of the field keyword, attributes and so on. Questions of a semantic piece of identification will be passed by a Semantic rule base which is set up in advance before recognition. The semantic rules of block can be represented as $R=<t1,t2,t3,t4,t5,...,tm>$. Besides, one instance of it is $E=e1,e2,e3...em$. An analysis of the current question can be represented as $W=wi+1,wi+2,...wi+m$.

F (x) is used to identify the semantic chunks of x, d(x, y) shows the similarity between the concept of x and the concept of y. The bigger the similarity is, the more similar the meaning of x and y. The algorithm application of pseudo code is shown below:

(1)Return keywords wi equal value in R
For j=0 to m
{
If(f(wi+j))==tj;
Successful matches;
Return tj;
}
After matching is not successful;
compare d(x,y)
calculate d(x,y)

Synonym word Lin is used to calculate $R(t1,t2,t3,t4,...,tm)$ and $W(wi+1,wi+2,...wi+m)$'s d(x, y). The author uses this method to reduce the comparison between words and words.

Compare d (x, y) to obtain the biggest word as a return value

If (d(x,y)< threshold) break; When less than a certain threshold, program interrupt.

Return max(d(x,y)).

The final match result is that the word equals to keywords which exists in the R. Then calculate and compare with the keyword which exists in R, finally return keyword where d(x,y) is the largest keyword that exists in R.

## 2.4 Select Max (d (x, y)) one of the biggest words

Here, to calculate the size of the similarity between two words and exploit Synonym word Lin in extended edition, which provided by the Harbin institute of technology information retrieval laboratory, it should compare the similarity of two words. Keywords which extracted from the user's question could be compared with the word semantic rule base. If the semantic rule base contains user questions in a word, then go back to the word directly.

If the semantic rule base does not contain user question in a word, then use Synonym word Lin with computing similarity between the word, which is in the semantic rule base or the word that the semantic rule base do not contain. After the completion of the calculation the semantic rule base would select the word whose similarity is of the largest as a return value. And the usage of this analogy is used for representing question semantic representation.

## 2.5 Question semantic representation

Pass the question pre-processing is to remove the polite word. To extract general keywords and domain keywords, the method is to use semantic piece of rule base for identification. The general keywords and domain keywords which are identified as question semantic representation. Question semantic representation mainly includes the key word, attributes and so on. Question semantic representation is used in the subsequent search algorithm in computing.

## 3 CHINESE QUESTIONS RELEVANT CALCULATIONT

The paper calculated questions with semantic relevance according to the semantic representation after semantic analysis of Chinese question. It considered not only the meaning of the sentence, but also the correlation information of a sentence. Correlation responses the degree of correlation between two words, and the correlation between two words refer to the possibility of the common case of in the same context. In this paper, the longitudinal correlation and transverse correlation are real numbers between [0, 1]. For example, the correlation between "computer college" and "computer application" is particularly high.

### 3.1 The established domain keywords tree

Domain keywords tree refers to a keyword tree in a specific field. In this paper, the aim of using domain keywords of the hierarchical tree is to extract the key words.

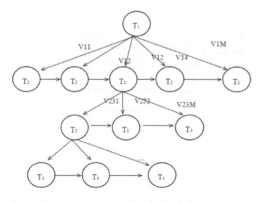

Figure 2.   Domain keywords tree.

T1,T21,T22,T2m...T4m is the tree node, and there is only one primary key in one field, level 2, level 3 and level 4 keywords can have multiples, including level 4 keywords are also known as the attribute words (used to interpret the specific meaning of primary, secondary or tertiary keywords). For example: area 1:

*** college of * * * university of * * *professional obtain employment situation. The primary keywords: * * * university; The secondary key words * * * college; the tertiary keywords * * * professional; The attribute words is employment. Area 2: * * *museum * **collection treasure * * *jade stone introduction. * * museum is primary keywords, * * * collection treasure is secondary keywords, * * * jade stone is the tertiary keywords. The attribute words are the introductions.

Definition 1: Longitudinal correlation (V) refers to the same tree areas and the correlation between the parent and child nodes.

Definition 2: Longitudinal correlation matrix M (i, j), is used to represent the correlation between a keyword i and j.

$$
\left[
\begin{array}{l}
M(T1, T2i, i \text{ from } 1\text{To } m)=1; \\
Layer\ 1\ and\ layer\ 2\ word\ relevancy \\
M(T1, T3i, i \text{ from } 1\text{To } m)=1; \\
Layer\ 1\ and\ layer\ 3\ word\ relevancy \\
M(T1, T4i, i \text{ from } 1\text{To } m)=1; \\
Layer\ 1\ and\ layer\ 4\ word\ relevancy
\end{array}
\right] \quad (1)
$$

The correlation matrix in Formula (1) is the same domain correlation value, we will make the same field of all keywords and the relevance of the roots defined as 1.

$$
\left[
\begin{array}{l}
M(T2j,j \text{ from } 1\text{To } m, T3i, i \text{ from } 1\text{To } m)=1; \\
Layer\ 2\ and\ layer\ 3\ word\ relevancy \\
M(T2j,j \text{ from } 1\text{To } m, T3i, i \text{ from } 1\text{To } m)=1; \\
Layer\ 2\ and\ layer\ 4\ word\ relevancy
\end{array}
\right] \quad (2)
$$

The correlation matrix in the formula (2) means making the second floor of a node as the root node, the child's child node or a child node of correlation with the root node is defined as 1.

$$
\left\{
\begin{array}{l}
M(T3j,j \text{from } 1\text{To } m, T4i, i \text{from } 1\text{To } m)=1; \\
Layer\ 3\ and\ layer\ 4\ word\ relevancy
\end{array}
\right\} \quad (3)
$$

The correlation matrix in the formula (3) means making the third floor of a node as the root node, its child nodes and the relevance of the root node are defined as 1. The rest did not appear in the longitudinal correlation matrix of the word and the word of the longitudinal relevance defined as 0.

Definition 3: Horizontal correlation (L) refers to the same tree areas between the brothers from the same parent node.

Definition 4: Horizontal correlation matrix H (I, j, k) refers to the horizontal correlation of the key words between I and j, and k is the i, j's parent.

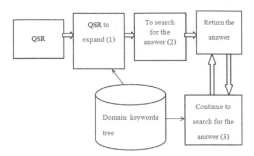

Figure 3.   Algorithm Search Model.

## 3.2   *Algorithm search model*

The expansion of question semantic representation-- the expansion of the users' question semantic representation is represented by the domain keywords tree whose aim is to expanse the users' question semantic representation via longitudinal correlation matrix and horizontal correlation matrix. The steps of expansion are followed by these: take the Q1 as an example, Q1's question semantic representation order is Q1 (W11. W12.......W1n).

(1) Firstly, judge the place where every word is in its domain keywords' tree. If calculated, there are only the first key words and attribute words can be the first, the second and attribute words and the expansion of question semantic representation are not needed. Otherwise execute (2).

(2) Use every keyword in domain keywords' tree from Q1 to judge the level that Q1 lack, and expanse it by using longitudinal correlation matrices matrices.

From the bottom level, if they lack the second level key words, there should be added some key words which are related to the third keywords and are taking taking advantage of the third words which are searched from the length ways correlation matrix. If the first level keywords lacked, there would add the first keywords directly.

2, Search answers. After expanding the question semantic representation, it's time to search answers

Calculate the relevancy of two questions--Q1 and Q2. Q1's keywords order is W11,W12,W13......W1n. Q2's keywords order is W21,W22......W2n.

That is the deepest degree of relevancy for those match words.

3, Continue searching the questions and answers

When the user finishes the first question, the second question should be continued to be asked. Then this moment the search process of the algorithm is following:

(1) Firstly, according to the keywords tree of the field, you should judge the replaced words in which layer are located.

(2) Using the keywords tree of the field, you can find keywords in this layer from the semantic representation of user questions and replace it. It should be noted that if the replaced keywords are located in the third layer, you will simultaneously replace the words which are in the second layer using longitudinal correlation matrices matrices.

(3) Execute 2 after complete replacement.

## 4   ANALYSIS OF EXPERIMENTAL RESULTS

### 4.1   *Test results of statistical*

Accuracy test of questions semantic extension. The main method is the extraction of an extended question semantic representation compared with domain keywords tree which is the accuracy of the extension. The results are shown in Table I

Table 1.   Questions semantic extension test results.

| Test Method | To test the total number | Correct number | Wrong number |
|---|---|---|---|
| Expanded Compare with the Domain keywords tree | 100 | 99 | 1 |

To test the accuracy of keywords. The main method is the statement after replacement, which is compared comprehensively with the former statement, and then domain keywords tree. To test the accuracy of the keywords replace, the results are shown in Table II:

Table 2.   Key words to replace the test results.

| Test Method | To test the total number | Correct number | Wrong number |
|---|---|---|---|
| After replacement and replace the former, the Domain keywords tree comprehensive comparison | 100 | 98 | 2 |

3. Test results of this experiment is the user's feedback. If the users are satisfied with the answer, then, the user would write 1. If users are not satisfied with the answer, the users would write 0. The results are shown in Table III:

Table 3. The experimental testing result.

| Test Method | The number of returns 1 | The number of returns 0 | Answers right |
|---|---|---|---|
| A separate test for each problem | 265 | 18 | 93% |
| On the basis of the questions to continue testing | 236 | 24 | 90% |

### 4.2 Interpretation of result

The above test results verified the accuracy and reasonableness of the comprehensive weighted algorithm based on Chinese question relevance from different angles. Summarize the advantages of the algorithm as follows:

1. The returned result of the answer is basically contained with the standard answer in people's thinking, which proved that the calculation of the algorithm was with some accuracy.

2. Semantic representation extension of the user's question showed that this algorithm can accurately made semantic representation extension for user' question by using domain keyword tree, and the extended semantic representation can still represent the meaning of the original sentence.

3. Keywords replace results showed that the algorithm can accurately satisfy users to continue the search for related issues, and the replaced question semantic representation can meet the needs of users. Typography for references.

## 5 CONCLUSIONS

First, this paper used "question sentence semantic representation extracting model" to extract semantic representation of the user's question. In the extraction process, calculate the similarity of the words using the synonyms Lin provided by the HIT information retrieval laboratory and the semantic rule base established by the author, and extract the semantic representation of the user's question. By calculating the similarity of words in the semantic rule base and the user's question. The semantic representation extraction here reduced the number of comparisons that directly compared the words in the sentence with the words in the synonyms Lin in the past. And that greatly improved the efficiency of the algorithm.

After extracting the representation, the paper expended the semantic representation and then returned the target answer through the calculation of weight. While considering question that the user continued to raise, the paper used the vertical correlation and horizontal correlation of the domain keyword tree to calculate it, and then returned the user desired target answer. Finally, the author tested a number of users, and made a conclusion that the accuracy of the algorithm can meet the users' needs and this paper would exert important significance for the study of intelligent question answering system.

## ACKNOWLEDGMENTS

This research is supported by the National Natural Science Foundation of China (No.61272431), China Postdoctoral Science Foundation (No.2013M531601), and the Natural Science Foundation of Shandong Province, China (No. ZR2012FM002), the Postdoctoral Foundation of Shandong (No.201302035).

## REFERENCES

Yiping Yang. 1997. Chinese full-text retrieval algorithms, *Computer Systems & Applications 1997, 3(9), 30–31.*

Chuanyuan Xu,Yang Zhang,Huayang Mao. 2008. Routing Algorithm in P2P Content Search Engine, *Computer Engineering, 2008, 34 (17):* 85–87.

Qiang Zhou, Weidong Zhan, Haibo Ren. 2001. Build a Large scale Chinese Function 1 Chunk Bank [A]. *BeiJing: Tsinghua University Press, 2001*:102–107.

Yi Guan, Xiaolong Wang. 2003. A Statistical Measure of Semantic Similarity Between Chinese Words[C] Miscalculated and based on the content of text processing language -- The seventh session of joint academic conference on computational linguistics. *Beijing: Tsinghua University Press, 2003*: 221–227.

Jiu-le Tian,wei Zhao. 2010. Words Similarity Algorithm Based on Tongyici Cilin in Semantic Web Adaptive Learning System, *Journal of Jilin University. 2010, 28(06)*:602–605.

Bin Shi, Jianzhuo Yan, Pu Wang. 2009. Based on the concept of ontology semantic relatedness measurements [J]. *computer engineering, 2009, 35(19)*: 83–85.

Tao Cheng, Shuicai Shi, xia Wang. 2007. Chinese text subject extraction based on synonyms Lin [J]. *Journal of Guang xi Normal University: Natural Science Edition, 2007, 25(2)*: 145–148.

Qian Wu, Yu Li, Chunyan Fang. 2009. Comparison and conversion between word processing format UOF and OOXML [J]. *Computer Application and Research, 2009(2)*:591–597.

Wenjie Li,Yan Zhao. 2010. Based on the concept of text structure between semantic similarity algorithm *[J]. Computer Engineering, 2010, 36(23),* 4–6.

Multimedia, Communication and Computing Application – Leung (Ed.)
© 2015 Taylor & Francis Group, London, ISBN 978-1-138-02775-6

# The research on attention management, optimization of Yueyahu Park in Nanjing based on the five senses

F. Wang
*School of Cultural Industry & Tourism Management, Sanjiang University, Nanjing, China*

ABSTRACT:    With the arrival of the information age, the traditional business model is being seriously questioned and influenced. In addition to the product, service, price, place, promotion and other aspects of the competition, the competition for the attention of consumers among companies have become increasingly intense. Attention operation, according to different standards, can be divided into a variety of attention business model, and the five senses of human thinking are closely related to the reference. This paper takes Yueyahu park as an empirical study, analyses the management problems based on the five senses, and finally draws the management optimization scheme.

KEYWORDS:    Attention operation, Five senses, Yueyahu Park, Optimize.

## 1   INTRODUCTION

Since Herbert Simon [1] introduced the concept of attention into management, the effect of attention on decision-making has remained one of the core problems in the researches of organizational behaviour and management [2]. In 1997, Michael H·Goldhaber, an American scholar, first proposed the concept of "the attention economy"[3]. In the same year, another American scholar named William Drake further explained and analysed the general situation of the attention economy [4]. In 1998, Michael H·Goldhaber's Attention Shopper was rendered into Chinese and published on Computer Engineering & Software. From then on, the domestic researches on the concepts of attention operation have been carried out successively. Shi Peihua summarized the fundamental theories of the attention economy[5]. Zhou Zhong et al. analysed the success and failure of attention operation of enterprises[6]. Wang Dejun also stated that attention increasingly became a kind of commodity, resource and target chased after by people in the information-alized society[7]. Thus, it can be seen that researches on attention in both domestic and foreign academic circles are more and more rational and the importance is not only attached to the discussions about the theoretical system but also to the practical applications.

## 2   INVESTIGATION INTO CURRENT SITUATION OF ATTENTION OPERATION OF YUEYAHU PARK

Yueyahu Park holds the natural beauty of lakes and mountains as well as antique ancient walls. Although Yueyahu Park is not a well-known tourist attraction, its design exhibits the essence of Chinese garden art, integrating mountains, water, city and forest as a whole one, combining traditional and modern styles and embedding cultural landscape in real mountains and waters, thus endowing lofty artistic conception in beautiful scenery.

A questionnaire consisting of 11 questions is designed from the viewpoint of visual, auditory, olfactory, gustatory and tactile senses and aims at investigating the problems in the attention operation of the Yueyahu Park from the perspective of five senses. This investigation was conducted from April 23 to April 25, 2014. The questionnaires were randomly issued to 200 tourists in the park and 200 questionnaires were taken back among which 194 of them were effective, with rate of effectiveness of 97%. The statistical result shows that 62% the of tourists are satisfied with the overall impression that Yueyahu Park leaves on them and 35% of the tourists are not quite satisfied with the current conditions of Yueyahu Park or consider the Park ordinary.

### 2.1   Current situation of visual attention operation

72% of the tourists are relatively or quite satisfied with the scenery of the Yueyahu Park at daytime, while nearly 30% of the tourists are not so satisfied with the scenery of Yueyahu Park and regard it as insipid. 61% the of tourists are relatively content with the night scenery of Yueyahu Park, but nearly 40% of the tourists complain about the night scenery the Park.

### 2.2   Current situation of auditory attention operation

71% of the tourists think Yueyahu Park is relatively quiet, but there are almost 30% of the tourists who think Yueyahu Park is buzzing and noisy. As many as 95% of the tourists have hardly heard the music played in the Park.

### 2.3 Current situation of olfactory attention operation

63% of the tourists think the water of Yueyahu is clear and odourless, while 37% of the tourists think the water of Yueyahu is smelly. 57% of the tourists think the flowers and trees in Yueyahu Park are sweet-smelling, but 43% of the tourists do not think so and they haven't noticed any scents given off by flowers and trees.

### 2.4 Current situation of gustatory attention operation

Nearly half of the tourists think the food and snacks in Yueyahu Park taste not too bad, while the other half of the tourists think the food and snacks in Yueyahu Park ordinary and featureless. Almost no one has ever tasted the food specialities in Yueyahu Park.

### 2.5 Current situation of tactile attention operation

There is no guardrail along the path around Yueyahu. 65% of the tourists feel relatively comfortable to walk on the path around Yueyahu, while 35% of the tourists think they are filled with anxiety and fear when walking on the path around Yueyahu. 61% of the tourists think it is relatively comfortable to walk in the quartzite paths, while nearly 40% of the tourists feel exhausted and uncomfortable walking on it.

### 3 PROBLEM AND INITIAL SOLUTIONS OF ATTENTION OPERATION OF YUEYAHU PARK

### 3.1 Problems and initial solutions of visual attention operation

Problems are sorted into the following aspects: (1) the water of Yueyahu becomes turbid and turns into black due to the discharge of domestic sewage of residents nearby and the blind aquaculture cause; (2) there is a large drop between the path around Yueyahu and the water level, which brings about visual fear for tourists especially when there is no guardrail; (3) the illumination facilities at night are inadequate and a match of light colors is not available; (4) though a complete variety of flowers and trees are grown with a high afforestation rate, they are short of formative design, maintenance and attraction.

The initial solution are as follows: (1) the residents nearby should be forbidden to discharge domestic

sewage into Yueyahu or carry out aquaculture with the water of the Lake, and the water should be replaced and detected regularly; (2) the guardrail should be built around Yueyahu and the water level should be raised to narrow the distance between the lake and tourists; (3) the theme night scenery should be designed in accordance with the "human-oriented and environment-foremost" principles in the initial planning and the theme night scenery can be publicized through a large-scale activity; (4) the flower, trees and lawns should be taken good care of and repaired regularly and there might be designs with special features as well.

### 3.2 Problems and initial solutions of auditory attention operation

The major problems include: (1) automobiles drive and whistle in the park and generating plenty of noise and breaking the peace and tranquility of the park; (2) there is no theme music to echo the scenery of the park and the operation of auditory attention practically remains an unworked field.

The initial solutions are as follows: (1) motor vehicles should be prohibited to enter the park in order to reduce noise and protect the quartzite paths; (2) suitable theme music and melody should be selected according to the atmosphere and artistic conception embodied in the scenery of the park and moreover the music and melody should vary along with "time (hour, solar terms and seasons)" and "situation (trend, fashion, vogue)".

### 3.3 Pproblems and initial solutions of olfactory attention operation

The major problems include: (1) a stinking smell in the air due to the water pollution of Yueya Park and this affects tourists' experience; (2) there are few flowers and trees with scents and thus the traveling flavors and emotional appeals are too limited.

The initial solutions are as follows: (1) apart from the reduction of sewage discharge mentioned in the operation of visual attention, the monitoring and guarantee of water quality of the lake should be reinforced and more aquatic plants capable of water purification should be planted; (2) more flowers and trees with scents should be cultivated in the park, like camphor tree, lotus and osmanthus tree.

### 3.4 Problems and initial solutions of gustatory attention operation

The major problems include: (1) there are many privately-owned hotels and restaurants in the park, but the food variety and flavors are similar to those in catering services outside the park and lack special features; (2) the park has no independently-operated catering points and the dishes and flavors are insipid and non-creative for they do not have any specific connotation.

The initial solutions are as follows: (1) the dish quality standards should be formulated for the privately owned restaurants and these should constantly develop innovative dishes and launch special products; (2) a "Yueyahu Gourmet Festival" should be organized to publicize the dishes with special flavors and foster a specialty-catering brand.

### 3.5 *Problems and initial solutions of tactile attention operation*

The main problems include: (1) the quartzite paths become bumpy after having been grinded by motor vehicles for a long term and thus the tourists feel uncomfortable to walk on the quartzite paths; (2) the experience of tourists in boating and playing with water is affected since there are too many suspended solids in the water in Yueyahu.

The initial solutions are as follows: (1) motor vehicles are disallowed to enter the park and the damaged stone plates should be replaced and repaired. There should be more attention to supervision and maintenance; (2) it is strictly prohibited to dump or throw sundries into the lake and the sundries in the lake should be removed, and the water should be replaced regularly.

## 4 OPTIMIZATION SCHEME FOR ATTENTION OPERATION OF YUEYAHU PARK

### 4.1 *Investigation among the tourists*

Another questionnaire with 10 questions was designed to verify the rationality of the initial solutions. This investigation was carried out from May 1 to May 3, 2014. The questionnaires were issued to 100 tourists randomly and 100 questionnaires were taken back, among which 98 questionnaires were effective, with an effective rate of 98%. Similarly, the data statistics was conducted for the questionnaires.

### 4.1.1 *Visual attention operation*

Over 80% of the tourists agree on the pavement of a path around Yueyahu and the elevation of the water level of Yueyahu. Over 60% of the tourists are in favour of the construction of large-scale night scenery, but almost 40% of the tourists object to this solution. Hence, the solution of constructing large-scale night scenery requires deliberation and it is unreasonable to put it into the design blindly in accordance with the desire of only some of the tourists.

### 4.1.2 *Auditory attention operation*

Over 90% of the tourists oppose the entry of motor vehicles into the park, which suggests this solution gets a high rate of satisfaction. 76% of the tourists think it is necessary to play theme music or melody in the park, while only 24% of the tourists object to this solution. Therefore, it can be seen that the solution has a relatively widespread foundation among the tourists.

### 4.1.3 *Olfactory attention operation*

There are 98% of the tourists who support growing flowers and trees like camphor tree, lotus and osmanthus tree in the park, so this solution is completely consistent with the desire of tourist. 100% of the tourists agree to strengthen the water quality monitoring and protection of Yueyahu.

### 4.1.4 *Gustatory attention operation*

86% of the tourists wish that restaurants in the park could develop new dishes and improve the quality of the dishes. 87% of the tourists hope that a "Yueyahu Gourmet Festival" can be organized to popularize the special snacks and this solution should be moderately considered in the final optimization scheme.

### 4.1.5 *Tactile attention operation*

Only 47% of the tourists think the quartzite paths should be repaved or replaced, so this solution is infeasible and careful consideration is required. Nearly 40% of the tourists are against adding long chairs and stone tables in the park, so this solution should be further discussed.

### 4.2 *Consultation with the staff of Yueyahu park*

Yueyahu Park administrators have put forward the following suggestions in terms of the attention operation of Yueyahu Park:

It is suitable to build guardrail around Yueyahu. Yueyahu has a relatively large water reservation capacity and can be used as a reservoir. In addition, the water level of Yueyahu varies with the rainy season and the dry season. Therefore, Yueyahu may lose its regulatory capacity if the water level is raised blindly. The large-scale night scenery requires the input of great labour, materials and financial resources and an improper design of night scenery may result in light pollution. Moreover, the tourists have no strict requirements for night scenery, so the construction of large-scale night scenery can be temporarily set aside.

Many tourists prefer self-driving tours, but there is no large parking lot in the park, so drivers enter into the park with their vehicles and park randomly. To prevent motor vehicles from entering the park, a parking lot should be built outside the park, otherwise, the attraction for tourists is too limited. Playing theme music or melody is attemptable, but attention should be paid to the selection of music style and the volume control.

A vast number of flowers and trees have been planted in the park, but there are a few flowers and trees with scents. Hence, it is appropriate to grow camphor trees, lotuses and osmanthus trees. Yueyahu Park has already begun the monitoring and protection of water quality of Yueyahu but has not yet obtained obvious effects and the maintenance and protection will be further reinforced.

Yueyahu Park has no right to force privately owned restaurants to develop new dishes or improve the dish quality because of the operating system. Instead, Yueyahu Park can only give some suggestions and opinions. As for the "Yueyahu Gourmet Festival", it is feasible and the coordination among the sponsor, co-sponsor, organizer and participants should be done well.

The quartzite paths call for repair rather than a treatment in a large scale. In addition, there are enough long chairs in the park, but these long chairs are not clean, so the tourists are unwilling to use them. Hence, efforts should be put on cleaning. Stone tables and chairs can be added where appropriate.

### 4.3   *Final optimization scheme*

The final optimization scheme for attention operation of Yueyahu Park is formulated as follows:

1 The path around the lake should be paved, but it should be better not to raise the water level of Yueyahu.
2 The construction of night scenery should be suspended before the planning, design and investment are completed.
3 A parking lot should be built outside the park and the vehicles are forbidden to enter the park to keep the tranquil environment.
4 It is appropriate to play thematic music in the park and create a melodic atmosphere. The style of background music should be selected in accordance with the environmental needs of the park and the volume should be kept within reasonable bounds.

5 The flowers and trees, like camphor tree, lotus and osmanthus tree, can be grown in the park in accordance with different seasons and the allocation of plants in the park in order to realize moderate degree and appropriate quantity.
6 The water quality monitoring system for the parks should be constantly improved and the monitoring and protection of water quality of Yueyahu should be reinforced as well.
7 The suggestions for innovation and quality promotion of dishes of privately owned catering services should be put forward.
8 Special snacks of Yueyahu can be created and activities like "Yueyahu Gourmet Festival" can be organized to popularize these special snacks.
9 The existing quartzite paths in the park should be maintained and protected.
10 The long chairs in the park should be kept clean and some stone tables can be added to satisfy the needs of tourists for having a rest. The positions of long chairs and stone tables should be determined based on the comprehensive allocation of flowers and trees so as to meet the needs of tourists for enjoying the scenery.

REFERENCES

[1] Simon, H A,1947, Administrative behavior: A study of decision-making processes in administrative organizations[M]. New York: Free Press, 23–24.
[2] Wu Jianzu, Wang Xinran, Zeng Xianju,2009,The Explorations into Current Situation and Future Prospect of Foreign Attention-based Opinions[J]. June,Vol.31, (6):58–65.
[3] Michael.H. Goldhaber,1997,Attention Shopper [J]. Hot Wired, 25:46–49.
[4] William Drake,1997,New Approaches to the Attention Economy[M]. Aspen Institute, 35–94.
[5] Shi Peihua.,2000,The Economy of Attention [M]. Beijing: Economy & Management Publishing House, 2–4.
[6] Zhou Zhong, Pu Haiyan,2000,The Competition for Eyeballs [M]. Shanxi Education Press, 5–8.
[7] Wang Dejun,2000,The Economy of Attention - New Values in the Information Age [J]. Knowledge Economy, 5:43–46.

*Multimedia, Communication and Computing Application – Leung (Ed.)*
© *2015 Taylor & Francis Group, London, ISBN 978-1-138-02775-6*

# Not every entrepreneur can be a spokesman

Z.L. Tong
*School of Economics and Management, North China University of Technology, Beijing, China*

H.L. Liu
*School of Economics and Management, Wuhan University, Wuhan, Hubei, China*

X.R. Zhang & Y.W. Yang
*School of Economics and Management, North China University of Technology, Beijing, China*

ABSTRACT:   Nowadays entrepreneurs start to become the advertising spokesperson from behind the scenes to the front in China. Although the entrepreneur endorsement will be more attractive than that of stars, it is not suitable for every entrepreneur. Generally speaking, only when the entrepreneurs are famous and have stories, they could be the right people to endorse. There are different models of entrepreneur endorsement, such as the monologue model, the persuasive endorsement model and the conversational mode.

## 1   INTRODUCTIONS

When the stars' endorsement of the firm has been difficult to attract eyeballs, entrepreneurs start to become the advertising spokesperson from behind the scenes to the front. In China the earliest entrepreneur engaged in endorsements was Wang Shi, the chairman of Vanke. As early as in 1994, Mr. Wang had already been the advertising spokesperson of MOTOROLA's mobile phone. Recently, he has started a spokesman of Jeep Grand Cherokee's advertisement.

It is undoubted that Wang Shi is a successful spokesman, because he has not only promoted the brand images he endorsed, but increased the brand awareness of himself and Vanke group. It kills three hawks with one arrow, three parties won in the end. Other entrepreneurs also act as spokesmen to advertise. Nowadays, more and more entrepreneurs start the advertising agent, such as JMEI CEO Chen Ou, Gree group chairman Dong Mingzhu, Wanda group chairman Wang Jianlin, Focus Media chairman Jiang Nanchun and so on. As more and more entrepreneurs start to be advertising spokespersons, we have to think calmly: whether every entrepreneur is fit for an advertising spokesperson? Compared with the stars, what advantages do they have as spokesmen? If the firms ask the entrepreneur endorsement, how can you improve the effect of the endorsement?

## 2   WHAT KIND OF ENTREPRENEUR CAN ACT AS A SPOKESMAN?

For example, FEIDIAO group chairman Xu Yizhong spent millions on heavily homemade enterprise advertisement in 2011, it was broadcasted in Dragon TV in January 2012, in August the same year in Heilongjiang TV, Shenzhen TV. Because the data query of open source of FEIDIAO group is limited, we use the number of weibo fans as only measurement. To July 30, 2014, the number of fans on the true-name authentication weibo of Xu Yi is 1425, the number of fans on the real-name authentication weibo of FEIDIAO group is 402. From the number of weibo fans, FEIDIAO group chairman Xu Yizhong as advertising spokesman is not successful.

### 2.1   *A necessary condition for entrepreneurs to be spokesmen: High visibility*

Look at the number of microblog fans of entrepreneurs such as Wang shi, you can find that high visibility is the basic condition which can determine whether entrepreneurs are fit for the spokesperson. The fans on the real-name authentication microblog of Wang shi account to 15.865 million, fans on the real-name authentication microblog of Chen Ou account to 3.021 million. Ironically, Dong Mingzhu microblog has no real-name certification, and there is no published articles, only the head of Dong Mingzhu and the name, the number of fans account to 23182. The contrast of Microblog followers shows that, we can at least have a snapshot of the basic conditions of an entrepreneur as a spokesman is the high profile!

### 2.2 A sufficient condition for entrepreneurs to be a spokesman: Having a story

As spokesmen, it is not enough that they just have high popularity. The key condition is that they have a story, this story can make consumers easy to think of the match between entrepreneurs and brands they have spotted, and can fully transfer the value proposition of the brands.

## 3 THE ADVANTAGE OF ENTREPRENEUR ENDORSEMENT

According to the related domestic and foreign research, compared with the stars, entrepreneurs as advertising spokespersons have a comparative advantage, which mainly reflects in credibility, persuasiveness and charm.

### 3.1 The entrepreneur endorsement has more credibility

Normally, star endorsements collect high endorsement fees, and endorsement fee is the main part of the source of stars' income. It's easy to make the consumer attribute the motivation of stars as the spokesmen of advertising to make money rather than the product itself.

But for entrepreneurs, the attribution for star endorsements is almost non-existent. Although entrepreneur endorsement will also collect lucrative endorsements, but entrepreneurs' income is mainly got by conducting an enterprise, endorsement fee is not a main source of income, the entrepreneur will not be simply for the sake of money (besides some entrepreneur endorsement without endorsement fee). Secondly, compared with the star, entrepreneurs do not regularly ads, and they won't endorse during the same period with multiple ads, so once the entrepreneur endorses a brand, consumers are likely to conjure that entrepreneurs have made a careful selection. To sum up, the entrepreneur as a spokesman, consumers are more likely to trust the product or brand they endorsed.

### 3.2 Entrepreneur endorsement has more persuasiveness

Compared with stars, entrepreneur endorsement appears more persuasive with the purchasement intention of consumers. As the first prosperous group, entrepreneurs' social influence has gone far beyond the boundary of enterprise organizations, and is on the whole society.

To sum up, entrepreneurs have influence far beyond the stars in the change of consumer lifestyles and social wealth creation. Therefore, entrepreneur endorsement appears more persuasive.

### 3.3 The entrepreneur endorsement has unique charm

Compared with stars, entrepreneurs send out a stream of unusual charm. For instance, Alibaba group chairman Ma Yun, when everyone thinks that shopping online is a fable, he created the Taobao, which made the network shopping become quite common. Again, the Giant group chairman, Shi Yuzhu borrowed 200 million to build the Giant building at the time of the rapid advances, and lost business because of capital chain rupture. People believed that Shi Yuzhu will disappear from then on, but he made a notice that he would repay the debt before bankruptcy. Eventually, he carried with brain platinum back to the market and created another miracle.

Although the stars show the charm out of the ordinary people, entrepreneurs have more powerful charm. Stars' charm no more than comes from the screen, while after all this charm is based on a fictional plot. The charm of entrepreneurs, however, is from scratch, from weak to strong, from may to possible entrepreneurial epic, the glamour is more real, and embodies the wisdom and sweat and perseverance. Therefore, when people see Wang Shi endorsement of Jeep grand Cherokee, in a low voice saying that to use experience to define yourselves, the charm is difficult to reveal the star.

## 4 THE MODE AND CHARACTERISTICS OF ENTREPRENEUR ENDORSEMENT

According to the cases of existing entrepreneur endorsement, entrepreneur endorsement can be summed up to some different types of endorsement models, and each of these modes has different characteristics.

### 4.1 The monologue model

Endorsement of a monologue type centres on entrepreneurs, by using the entrepreneurial journey as the main content of endorsements. The typical cases of the monologue endorsement type model are JMEI CEO Chen Ou.

JMEI shot a forty-second advertising into a miniature version of self-help, from be despised ridicule to succeed, CEO Chen claimed that "even if I was black and blue all over, I also want to live well. I am Chen Ou, I speak for myself." The advertisement gives a person the feeling that finding everything new and fresh, and has a strong emotional resonance for the consumers, the advertising effect is very effective.

The monologue endorsement model has the following characteristics: first, the endorsement model focuses on the recent entrepreneurship, they endorse for their company (or product) brands, the model can save the endorsement fee, also can improve the entrepreneurs' personal brand. Therefore, this model is especially suitable for stage of entrepreneurs. Second, the mode uses the internal dialogue as the main content, if the audience positioning of the advertising is accurate, it is easy to cause strong emotional resonance, which will be even greater than that of general star endorsements.

### 4.2 Persuasive endorsement model

Persuasive endorsement model centres on the entrepreneurs and the endorsed brands, entrepreneurs tell the unique value of the brand as the main content of endorsements. The typical persuasive endorsement pattern is Vanke group chairman Wang Shi.

Mr. Wang is the first Chinese entreprencurs as a brand ambassador. As early as in 2001, Mr. Wang accepted the invitation from the MOTOROLA company and endorsed for MOTOROLA cell phone A6288, the advertising appeal point is "business and sports". While Wang Shi is a well-known Chinese entrepreneur, and successfully climb some high mountains. It is suitable to convey the image of business and sports of MOTOROLA mobile phoneA6288 by Wang Shi, so the endorsement Wang Shi conveys a strong advertising persuasion.

After Mr. Wang successfully endorsed the MOTOROLA's mobile phone, he also successively served as the spokesman of the GSM, Beijing youth weekly, Lufeng car, Swiss tourism, PINGAN insurance. Besides Mr. Wang, other well-known entrepreneurs have also endorsed for brands. PanShiyi, the chairman of SOHO China, for example, for HangWang technology; Kings oft chairman Lei Jun for vancl; Innovation works, chairman Lee endorsed for tencent weibo, etc.

The prominent characteristics of persuasive endorsements model are that famous brand invites well-known entrepreneurs as a spokesperson, the model is suitable for entrepreneurs to endorse for other brands. Well-known entrepreneurs image has not only the generality of the wisdom, confidence and success, but also the character, experience, industry factors making different entrepreneurial personality. When the image of a particular entrepreneur and brand image fit, the persuasive type mode will play a strong persuasion effect. Persuasive endorsement pattern, however, when used for entrepreneurs in their company endorsement, is easy to form "claims

the melon" suspicion, and easy to form an image of prudent, and even causes antipathy.

### 4.3 The endorsement of a conversational mode

Endorsement of a conversational mode refers to the entrepreneurs and the stars, through dialogues, in the form of transferring unique value of the endorsed brand. The advertising dialogue between Gree electric chairman Dong Mingzhu and Wanda group chairman Wang Jianlin is the representative of an endorsement of a conversational dialogue model.

In this AD, Wang Jianlin asked "I heard the central air conditioning works without electricity? "Dong Mingzhu replied" Yes! Use the solar!"Mr Wang said "so I can save electricity 1billion each year". Two prominent people together quickly aroused heated debate, so that the photovoltaic Gree air conditioning spread rapidly.

Dong Mingzhu and Mr. Wang are not, in fact, the entrepreneurs of early adoption of a dialogical endorsement, as early as in 2004, Liu Xiang won the 110 m hurdles in Athens Olympic Games, Liu Xiang then signed a contract with Yili as a brand ambassador, and endorsed Yili milk with the chairman of Yili. In advertising, Mr Pan the president of Yili group looked at Liu Xiang running into the distance on a wall, the wall collapsed in front of the smiling face of Liu Xiang, more and more short Liu Xiang ran back to Mr. Pan and said: " strength and weakness are relative," Mr Pan said to Liu "I hope every Chinese is strong like you."The AD used the star attraction, also played entrepreneur's persuasiveness, thus obtaining good advertising effect.

Features of a conversational endorsement mode, on the one hand, two or even more than the couplets of celebrity endorsements can produce more attractiveness, especially the combination endorsement of entrepreneurs and stars. On the other hand, entrepreneurs talk about the brand value to avoid the image of boast and unmodest in the conversation, and product features and advantages said by entrepreneurs are more authentic and more persuasive.

## 5 THE ENTREPRENEUR ENDORSEMENT STRATEGY

Although the economic or social benefits produced by the entrepreneur endorsement are still difficult for accurate quantification, existence is reasonable in all things. Now that entrepreneur endorsement is frequent, and entrepreneur endorsement has incomparable advantages with the star endorsement, So how to improve the entrepreneur endorsement effect is the issues we had to take into account. In order to achieve high performance, entrepreneurs can make it from the following three aspects:

### 5.1 To strength personal brand symbols

As a spokesman, entrepreneurs must have a personal brand which is distinct in personality and forms a unique symbol. Reviewing the above-mentioned entrepreneurs who have a successful endorsement, you can find that they all have strong personal brands. Vanke group chairman Wang Shi loves climbing, and opens the floodgates to conquer all the peaks in the world, the mountaineering makes Mr. Wang receive unprecedented media attention, and a large number of media reports to Mr. Wang as "man", "tough guy" in the equal sign. Wang Shi "tough guy" image keeps with the brand experts Aaker mentioning "Ruggedness (strong)" brand personality, and Aaker thinks when brand personality becomes symbols, it would allow consumers not to forget in the future.

### 5.2 To adopt an appropriate way to communicate

If the entrepreneurs want to obtain superior performance on endorsement, they must adopt an appropriate way to communicate. Normally, that entrepreneurs endorse for their own enterprise is always suspected of claiming their own products and services. How to get rid of such bad associations? This will require a clever way of communication.

Entrepreneurs can use the communication means of taking dialogues to reduce the claim in the endorsement.

If the entrepreneurs use the monologue communication ways, they must not appeal of the tangible "content", and should highlight "content" of the invisibility.

### 5.3 The entrepreneur endorsement can enter or be back freely

Finally, entrepreneurs must enter or be back freely when they endorse for their own firms. Like enterprise persistent in the pursuit of longevity, it is because there is not eternal success. The same is true for entrepreneurs, entrepreneurs can be a moment of brilliant achievements, but, ebb tide that entrepreneurs cannot dominate in five-hundred years. Mostly the excellences can dominate three to five years. The frequent replacement of all sorts of rich list rank also proves this view. So, when the entrepreneurs make decisions that they will endorse for their own enterprise, they must be alert that this is just periodical endorsement strategy, after they reach the milestones, entrepreneurs have better to retreat in order not to hurt the corporation brand when they make mistakes.

### ACKNOWLEDGMENT

I would like to thank Wang Qin for commenting on an earlier draft of this paper, and Wu Dan for his useful discussion about paper writing in NCUT. Thanks also to Fang Weiyuan for her encouragement and support. They are not responsible for any errors and, of course, do not necessarily support my opinions. Sponsored by (1) Launch of the northern industrial university research and Beijing municipal education commission social science (SM201410009003; (2) Beijing municipal colleges and universities to improve science and technology innovation ability construction special fund project.

### REFERENCES

[1] Aaker, J., K. D. Vohs, et al., 2010, "Nonprofits Are Seen as Warm and For-Profits as Competent: Firm Stereotypes Matter", Journal of Consumer Research, Vol. 37(2), pp. 224–237.

[2] Anderson, John R., 1983, "A Spreading Activation Theory of Memory", Journal of Verbal Learning and Verbal Behavior, Vol. 22(3), pp. 261–295.

[3] Berens, G., C. B. M. van Riel, et al. (2005). "Corporate Associations and Consumer Product Responses: The Moderating Role of Corporate Brand Dominance." Journal of Marketing, Vol. 69(3), pp.35–18.

[4] Brown, M. W. & Cudeck, R., 1989, "Single Sample Cross-Validation Indices for Covariance Structures", Multivariate Behavioral Research, Vol. 24(4), pp. 445–455.

[5] Brown, T. J. and P. A. Dacin, 1997, "The Company and the Product: Corporate Associations and Consumer Product Responses", Journal of Marketing, Vol. 61(1), pp. 68–84.

[6] Daily, C. M. & Johnson, J. L, 1997, "Source of CEO Power and Firm Financial Performance: A Longitudinal Assessment", Journal of Management, Vol. 23(2), pp. 97–117.

[7] Drumwright, Minette E., 1996, "Company Advertising with a Social Dimension: The Role of Noneconomic criteria", Journal of Marketing, Vol. 60(4), pp. 71–87.

[8] Elena, D. & Miguel H., 2008, "Effect of Brand Associations on Consumer Reactions to Unknown On-line Brands", International Journal of Electronic Commerce, Vol. 12(3), pp. 81–113.

[9] Eichhol, Z. M., 1999, "Judging by Media Coverage? CEO Images in the Press and the Fortune 'America's Most Admired Companies' Survey", Annual Conference of the International Communication Association, San Francisco, CA.

[10] Fang Eric (er), Robert W. Palmatier, and Rajdeep Grewal, 2011, "Effects of Customer and Innovation Asset Configuration Strategies on Firm Performance", Journal of Marketing Research, Vol. XLVIII (June), pp. 587–602.

[11] Feldman, J. M. and J. G. Lynch Jr., 1988, "Self-generated Validity and Other Effects of Measurement on Belief, Attitude, Intention and Behavior", Journal of Applied Psychology, Vol. 73(3), pp. 421–435.

[12] Grandey Alicia, Lenda Fisk,Anna Mattila, Karen Jansen and Lori Sideman, 2005, "Is 'Service with a Smile' Enough? Authenticity of Positive Displays during Service Encounters". Organizational Behavior & Human Decision Processes, Vol. 96(Jan), pp. 38–55.

*Multimedia, Communication and Computing Application – Leung (Ed.)*
© *2015 Taylor & Francis Group, London, ISBN 978-1-138-02775-6*

# Research on the present situation of teachers of Gansu Normal University for Nationalities

J.Z. Gao & Q. Wang

*Gansu Normal University for Nationalities, Gansu, China*

ABSTRACT: In this paper, taking Gansu Normal University for Nationalities as the example, the plight of teaching staff construction in national regions colleges is analysed. In order to provide a new idea for the development of college faculty in ethnic regions, the theory and practice of college teachers' team construction process were discussed.

KEYWORDS: Ethnic region, College, Construction of teaching staff.

## 1 INTRODUCTION

The national universities shoulder the historic mission of providing the required high-level local personnel which help maintain local stability. The factors affecting the teaching stability are the harsh natural conditions, the backward economic and social development, the poor human environment, life stress, and so on. Gansu Normal University for Nationalities is located at an altitude of 3000 meters in the Tibetan Plateau sector. Affected by the geographical location, the faculty has poor stability. Therefore, this paper, by taking Gansu Normal University for Nationalities as as example, tries to provide a theoretical basis and practical significance for the construction of university teachers in other ethnic regions.

## 2 SUBJECTS AND METHODS

### 2.1 *Subjects*

The research object is to service teachers in the Gansu Normal University for Nationalities and the summer semesters' external experts and to analyse the teacher qualifications and teacher-student ratio titles.

### 2.2 *Methods*

The research methods include literature, questionnaires and interviews with experts.

## 3 RESULTS ANALYSIS

### 3.1 *Situation of full-time teacher qualifications*

It shows that there was a dramatic increase in the number of full-time teachers. In 2009, the number

of full-time teachers was only 384, and it reached 438 in 2013. The ratio of full-time teachers with a doctorate was 0.5%, and it reached 1.4%. In 2009, the number of full-time teachers with a master's degree was only 162, and it reached 230 in 2013. The proportion of full-time teachers with a master's degree is 52.5%. In general, Gansu Normal University for Nationalities has a higher teacher education level now.

### 3.2 *Current situation of full-time faculty professional title structure*

The study shows that there was a dramatic increase in the number of teachers' professional title. In 2009 the number of professors was only 15, and it reached 33 in 2013. The ratio increased from 4% to 7.5%. In 2009, the number of associate professors was only 100, and it reached 112 in 2013. In 2009, the number of lecturers was only 167, and it reached 209 in 2013. It showed that the structure of full-time teachers of academic title, the senior, intermediate and junior is developing proportionally.

### 3.3 *The ratio of teachers and students*

It shows that the number of full-time teachers was only 384, and the ratio of students to teachers was 20.48: 1 in 2009. The ratio of students to teachers was 18.88: 1 in 2013. It was higher than 18:1. It means that there is a severe shortage of teaching staff in the Gansu Normal University for Nationalities.

Thus, a conclusion can be easily drawn: the teacher, academic structure is relatively high the teacher title structure is relatively stable, but the shortage of teachers is serious.

## 4 INFLUENCE FACTOR ANALYSIS OF TEACHERS RESOURCE

The shortage of the faculty in minority areas is not an accidental phenomenon. The major factor analysis is summarized as follows.

### 4.1 Difficult natural conditions, poor human environment

Gansu Normal University for Nationalities, for example, is located in the Qinghai-Tibet Plateau. Some teachers, especially non-local teachers and teachers in other provinces have left because of the unbearable pain caused by the altitude sickness. In addition, the exchange of information on cultural, scientific and technological aspects is difficult because of the inconvenience of national regional transport and the occlusion of information. This also makes the academic atmosphere thin. Rich geographical lazy and other sub-culture phenomenon have affected young teachers and their interpersonal aptitudes, making them lose their interested in teaching and researching and thus abandon the cause. Coupled with the lack of talents and the imbalance of professional structure, the development of the national, regional university has less space and fewer opportunities. Training and education opportunities are also lower than the colleges in other developed areas. People with high titles, high education background cannot give full play to their talents here. So they will choose a better college to realize their career aspiration.

### 4.2 Weak economy, basic educational backwardness

Because the nation in college and more commonly known as "young and old Edge Hill poor" in the western regions, the regional economy is weak and the basic education is lagging behind. With the implementation of the western development strategy, the development of the western regions of national universities has also gained momentum. But the support to colleges in these regions is far from sufficient. Faculty allowance, bonus, housing benefits and research funding are so behind that of other universities. While taking into account the basic needs of their children's education and employment, some of the best teachers often choose to quit to teach in more developed regions, thereby forming a teacher resource drain phenomenon.

### 4.3 Internal management system needs to be improved

A small number of teachers, a large proportion of students and teachers, the harsh working environment, life stress have all increased the instability of teacher staff and the difficulty while introducing college talents to ethnic areas. With the beginning of the National Higher Education Enrollment Plan, the student-teacher ratio in Minority Areas is increasing. In order to maintain its operation, some teachers have to teach and administrate. The teachers are struggling to cope with the administrative work while teaching. In addition, because of the particularity of the work in minority areas, teachers would have to take the responsibility to maintain the campus and regional stability while teaching students, especially the counsellors and teachers who need to have a regular in-depth relationship with students in order to grasp the trend and the state of their mind.

## 5 COUNTERMEASURES OF TEACHERS

1 They need to convert their ideas, find comparative advantages and enhance the attractiveness and stability of power Construction of Teachers

Gannan is a vast region with unique natural scenery, folk customs and various tourism resources. The relatively high altitude and barrier effect increases the cultural diversity; the humane spirit of Tibetan and its culture and unique value are becoming increasingly prominent in today's word. Schools that can fully exploit these advantages and strengthen propaganda will be able to change this place with difficult living conditions in a place with rich tourism resources. As a unique highland town with a small population, the competition is not as intense as that in big cities. With the slow pace of life and work, few psychological comparisons and relatively small life pressure, it is possible to increase the overall understanding of the school. With these accomplished, Gannan will gradually transform into a school which can attract talents.

2 They should rely on geographical characteristics and tap the potential research resources.

3 They should adhere to the people-oriented policy and strengthen the institutional culture and institutional importance.

4 They should have complete and feasible policies, expand training channels, and enhance gravitational qualified personnel policies

## 6 CONCLUSIONS

Economic development and prosperity in ethnic minority areas need more talents. And university faculty in minority areas is the key to ensure the quality of personnel training. To build a solid talent poop in minority areas to help university education in Ethnic

Areas leap to a new level, they should have an accurate grasp of the law school of higher education and of the characteristics and service requirements in regional development stage. They should also follow the time tracking trends in higher education and the situation, plan ahead, and enlarge the capacity of local services.

ACKNOWLEDGMENTS

This work was financially supported by the 2013 Annual State Ethnic Affairs Commission study higher education reform project "Shared ethnic ghettos university research quality construction Teacher Resource Platform" phased research (13098) and the Gansu Normal University for Nationalities Principal Foundation (10-9).

REFERENCES

[1] G,Rajka. On the present situation of the research into turpentine allergy. Allergie und Asthma, 1970,161.

[2] Ryoichi URA. The Present Situation of Agricultural Villages and Some Problems in Improving Facilities on the Ningsho Plain in Zhejiang Sheng, China. JOURNAL OF RURAL PLANNING ASSOCIATION, 1985,34.

[3] S.L. Hu. An overview of the present situation in Chinese medical universities and colleges and academic health system. JOURNAL OF EVIDENCE-BASED MEDICINE, 2013,62.

*Multimedia, Communication and Computing Application – Leung (Ed.)*
© *2015 Taylor & Francis Group, London, ISBN 978-1-138-02775-6*

# Vertical variation model of soil water in artificial shrubland on the Loess Plateau

S.K. Liu, M.C. Guo & Z.F. Wang
*College of Science, Northwest A & F University, Yangling, Shaanxi, China*

Z.S. Guo
*Institute of Soil Water Conservation, Northwest A & F University, Yangling, Shaanxi, China*
*Institute of Soil Water Conservation, Chinese Academy of Sciences & Ministry of Water Resource, Yangling, Shaanxi, China*

ABSTRACT: In the hilly, semi-arid area of the Loess Plateau, the phenomenon of soil deterioration has appeared in most of perennial plantations. Therefore, it is necessary to study the vertical variation of soil water for the sustainable utilization of water resources. Soil water and plant growth were measured in the *Caragana korshinskii* shrubland at Shanghuang Eco-experiment Station during the period of five years from 2002 to 2006, and the data are analysed by compartment model in this paper. The results show that the change of soil water with soil depth can be expressed by the improved dynamic model and all of the goodness-of-fit is in line with the statistical requirements proves that the improved model can better describe the vertical variation of soil moisture under different situations. The new model has better application value in regulating the relationship between plant growth and soil water and in the sustainable use of soil water resources.

KEYWORDS: Loess hilly region, *Caragana korshinskii*, Soil water, Plant growth, Vertical variation model.

## 1 INTRODUCTION

Soil water resources are an important part of water resources and play an important role in the artificial vegetation restoration and plantation construction of the Loess Plateau. In the arid and semi-arid area of the Loess Plateau, China, because water is so scarce that the groundwater table is deep and there is no irrigation in most areas of this region. So precipitation becomes the only source of soil water (Yang & Shao 2000).

In recent years, the relationship between plant growth and soil water has disordered in most areas of the Loess Plateau (Guo & Li 2009). The combination of increased water use by plants and low water recharge rates has led to widespread soil deterioration occurring in the form of excessive soil drying in the perennial Caragana (*Caragana korshinskii*) forests. Such soil deterioration will adversely affect the stability of forest ecosystems and the ecological, economical and societal benefits of forests and other plant communities (Guo 2014, Guo & Shao 2013). If a dried soil layer has appeared under the maximum infiltration depth, it can not be recovered. The dried soil layer is called permanent, dried soil layer, which is irreversible (Guo & Shao 2010, Guo & Shao 2007). Thus, the study of vertical variation of soil water is necessary, which can offer certain references for the study of vegetation restoration and sustainable use of soil water resources. At present, the studies about soil water in Caragana forests in the hilly, semi-arid area of the Loess Plateau, including soil water carrying capacity for vegetation (Guo 2011, Shao et al. 2009), soil water resources use limit (Ning et al. 2013, Guo 2010), and the regular pattern of soil water supply and consumption (Guo 2009) and so on. Zhao et al. (2009) have studied the vertical change of soil water in *Robinia pseudoacacia* plantations in the loess gully slope of Loess Plateau and established a mathematical model according to the two compartment soil water models. Wang et al. (2012) applied the above model to study the change of soil water with soil depth in caragana plantations in the semi-arid region of Loess Plateau. The above model can reveal the vertical variation of soil water under some situations, but it is not suited for all cases and it should be improved.

Therefore, the purposes of this paper are to further study the change of soil water with soil depth and to establish an improved model which is best suitable for all cases under all situations. The results will lay foundation for the further study of soil water carrying capacity for vegetation, soil water resources use limit in the water-limited region of Loess Plateau and provide theoretical background for the sustainable utilization of water resources.

## 2 MATERIALS AND METHODS

### 2.1 Study area

The study was carried out at the Shanghuang Eco-experimental Station (latitude 35°59′–36°03′N, longitude 106°26′–106°30′E) in the semi-arid region of the Loess Plateau, the Eastern 20 km from Guyuan County, China. The main soil type is Huangmian soil which is developed from loess and susceptible soil. Water losses are serious problems in this region. The coefficient of variation of precipitation among the years from 1983 to 2001 was 23.8% with a median rainfall amount of 434 mm. Mean solar radiation is 5342 MJ/m$^2$; Annual average temperature is 7.0°C. The plant growing period is 152 days. The groundwater level is more than 60 meters. The experimental area was located in a 16-year-old *Caragana korshinkii* stand at the center of a west facing slope with a gradient of 0° to 15° at an elevation of approximately 1650 m. Herbaceous plants grow under the shrubs including *Stipa bungeana, Heteropappus attaicus, Artemisia giraldii, Lespedeza davurica,* and *Thymus mongolicus.*

### 2.2 Observations and measurement

Daily and monthly rainfall data in the study site from 1983 to 2001 were obtained from the Shanghuang Eco-experiment station, about 50 m far from the study site. Two 4 m long, aluminous, access tubes were inserted into the soil at the center of each experimental plot with a 2 m contour distance between them. A neutron probe, CNC503A (DR), was used for monitoring the field soil water content (Guo & Shao 2013). Measurements were made every 15 days starting at the 5 cm depth to a depth of 4 m with increment of 20 cm every time. The soil water content obtained from the depth of each measuring place was taken to be representative of the soil layer includeing the measuring point ±10 cm depth, apart from that the 5 cm depth was taken to represent the upper 10 cm of soil.

### 2.3 Statistical analysis method

The observed soil water data are analyzed by applying the compartment model in order to establish an improved model and are used by the modified model to gain fitting parameters and corresponding goodness-of-fit with SPSS16.0 software.

## 3 RESULTS AND DISCUSSION

### 3.1 The compartment model and the vertical variation of soil water model

There are 13 sets of soil water data being analyzed by using compartment model, which were under

observation from 13$^{th}$ April to 29$^{th}$ November, 2002. It shows that the soil water content between the wetting front and the deep soil layer can reach balance when the precipitation infiltrate to the wetting front. So there is $\lim_{h \to h_{max}} (f - W_c) = 0 \Rightarrow \lim_{h \to h_{max}} W_c = f$, where $Wc$ = the soil water content; $f$ = the water content at deep soil layer lower the wetting front; and $h_{max}$ = the maximum infiltration depth. The two compartment soil water models and the corresponding mathematical model are shown as follows.

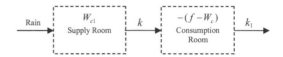

Figure 1. Schematic of the two compartment soil water models.

$$\begin{cases} \dfrac{dW_{c1}}{dh} = -kW_{c1} \\ \dfrac{d[-(f - W_c)]}{dh} = kW_{c1} - k_1[-(f - W_c)] \\ W_c|_{h=0} = W_{c0} \end{cases} \quad (1)$$

Solving equation (1), then the vertical change model of soil water can be given by:

$$W_c = ae^{-kh} - ce^{-k_1 h} + f \quad (2)$$

Where $W_{c1}$ = the soil water supply; $h$ = the soil depth; $k$ = the adsorption rate of soil particles; $k_1$ = the consumption rate of soil water; $a$ and $c$ are constant related to $k$ and $k_1$; and $W_{c0}$ = a − c + f, a > 0, c > 0, k > 0, $k_1$ > 0.

Then, using the 13 sets of observed data to fit the model (2), the results show that the model is just suited for some sets of data, and the others are unable to get correct values of fitting parameters. Thus, the model (2) should be modified in order to better illustrate the change of soil water with soil depth under various conditions in 2002.

### 3.2 The modified model of soil water

The soil ecological system is a connected body, in which the rest or excessive consumption of material and energy in the early stage will affect the soil water in the later stage. Consequently, the factors impact the change of soil water content with soil depth on that very day including precipitation, the antecedent soil water content, and the soil water consumption. The analysis of these factors found that the soil water supply and consumption can be explained by one compartment model, respectively (Fig. 2).

Figure 2.　Schematic of one compartment soil water model.

Then, following is the modified model:

$$\begin{cases} \dfrac{dW_{C1}}{dh} = -k\ln h \cdot W_{C1} \\[2mm] \dfrac{d(W_{C1} - |W_C - f|)}{dh} = -k_1 \cdot (W_{C1} - |W_C - f|) \end{cases} \quad (3)$$

Where, $W_C$ = the soil water content on that very day; $W_{CR}$ = the total effective recharge for soil water in the period; $W_{C0}$ = the antecedent soil moisture content. Denoting $W_{C1} = W_{CR} + W_{C0}$, $W_{C1}$ is the soil water supply, and $W_{C1} - W_C$ is the soil water consumption. Analysis shows that $W_{C1}$ decreases with the increasing soil depth and there is a limit, $\lim\limits_{h \to h_{max}} W_{C1} = 0$, where, $h_{max}$ is the maximum infiltration depth, and the rate of change for $W_{C1}$ is not a constant number but a variable one which is proportional to the logarithmic of soil depth $h$, denoted by $k\ln h$. The soil water consumption ($W_{C1} - W_C$) decreases with the soil depth, and the change rate can be denoted by $k_1$. In addition, with the increasing depth, the soil moisture content tends to be stable, and there is a limit $\lim\limits_{h \to h_{max}} |W_C - f| = 0$, where $f$ is the soil water content in deep soil layer bigger than $h_{max}$.

The solution of the model (3) is as follows:

$$W_c = \mp a \cdot e^{-kh(\ln h - 1)} \pm c \cdot e^{-k_1 h} + f . \quad (4)$$

Therefore, the modified vertical variation model of soil water can be given by:

$$W_c = a \cdot e^{-kh(\ln h - 1)} - c \cdot e^{-k_1 h} + f \quad a, c \in R . \quad (5)$$

### 3.3　Model analysis

If the parameters take different values, the model (5) will get different results and can show a set of curves which can be roughly divided into four categories, as illustrated in Figure 3.

Group 1: For the most part, the soil water consumption is less in the initial stage of plant growth. On the one hand, if there is no rain with greater rainfall at this time, the trend of soil water content change with soil depth is that soil water will increase first and then decrease; On the other hand, the soil water content in the upper soil layer will increase rapidly with a large amount of precipitation. So the soil water content will decrease firstly and then increase and then decrease again. In the two cases mentioned above, they are both $c < a < 0$. But the first case, is $k_1 > k > 0$, while the second case is $k > k_1 > 0$ (curve 1* and 1** in Fig. 3).

Group 2: Soil particle will have strong adsorptive power because the water content in soil is lower after a long time when there is no rain event and the plant continue to suck soil water in the growth season. Then when there is a rain event, the rainwater will be quickly adsorbed by soil, and the soil water content in the surface layer will be relatively higher. So the vertical variation features of soil water content will decrease firstly and then increase, and there is $k_1 > k > 0$ and $c > a > 0$ (curve 2 in Fig. 3).

Group 3: After heavy rain events, the soil water supply and consumption can achieve balance. If there is a big rain event again, the soil moisture content will increase quickly, especially in the upper layer, and the soil water taken up by plants can be negligible. Therefore, the change curve of soil water mainly shows a trend of monotonically decreasing, and there is c = 0, a > 0, and k > 0 (curve 3 in Fig. 3).

Group 4: After sufficient replenishment in the soil, the soil water content is relatively high. If there is no rain event for a period of time and under the case of continual soil water consumption, the soil water will not change dramatically, but in a gentle way with the increasing soil depth. Thus, the vertical variation curve of soil water has a slow ascendant trend, and there is also $k_1 > k > 0$ and $c > a > 0$. But the difference between $k_1$ and $k$ is little (curve 4 in Fig. 3).

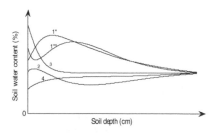

Figure 3.　Schematic of vertical variation of soil water in soil profile.

### 3.4　Model fitting

The soil water data measured in 2002 is used in model (5) with SPSS16.0 software. Then the fitting parameters and goodness-of-fit are shown in Table 1. It can be seen from Table 1 that all of the goodness-of-fit is in compliance with the statistical requirements, and indicates that the modified vertical variation model of soil water is reasonable and can better express various situations in 2002.

Through the analysis of the fitting parameters, these 13 kinds of cases in 2002 can be classified into four groups as shown below (Table 2). Table 2 shows that the features of parameters of these four groups are the same with the four categories respectively in Figure 3 which indicates that the classification is not only reasonable but also representative.

In addition, the soil water data observed on 3rd April, 2003, 15th May, 2004, 3rd July, 2005 and 2nd November, 2006 are also used in the model (5), and the goodness-of-fit is 0.798, 0.921, 0.99 and 0.935, respectively, which is in compliance with the statistics requirements which indicates that the modified model can well express the actual vertical variation of soil water with soil depth and has widely applicability.

Table 1. The fitting parameters in the modified model.

| Date | Fitting parameters | | | goodness-of-fit | | |
|---|---|---|---|---|---|---|
| | $a$ | $k$ | $c$ | $k_1$ | $f$ | $R^2$ |
| 4–13 | −50.32 | 0.01 | −56.67 | 0.023 | 7.471 | 0.859 |
| 5–30 | −33.743 | 0.008 | −36.703 | 0.021 | 7.735 | 0.742 |
| 6–30 | 28.595 | 0.003 | 30.121 | 0.009 | 9.146 | 0.835 |
| 7–15 | 29.714 | 0.002 | 35.715 | 0.007 | 10.263 | 0.943 |
| 7–31 | 13.901 | 0.002 | 16.052 | 0.008 | 9.065 | 0.754 |
| 8–15 | 10.378 | 0.013 | 0 | 0 | 7.802 | 0.945 |
| 9–01 | 25.574 | 0.001 | 32.807 | 0.004 | 12.718 | 0.883 |
| 9–15 | 3.323 | 0.059 | 0 | 0 | 7.684 | 0.722 |
| 9–30 | 20.665 | 0.001 | 27.766 | 0.004 | 12.97 | 0.889 |
| 10–15 | 10.24 | 0.001 | 15.319 | 0.004 | 11.503 | 0.749 |
| 11–01 | 2.974 | 0.02 | 0 | 0 | 7.558 | 0.64 |
| 11–15 | 8.95 | 0.003 | 10.227 | 0.007 | 9.011 | 0.665 |
| 11–29 | 15.449 | 0.003 | 17.753 | 0.01 | 8.617 | 0.805 |

Table 2. The classification of 13 sets data.

| Group | Time (Group members) | Parameters |
|---|---|---|
| Group 1 | 4–13, 5–30 | $k_1 > k > 0, c < a < 0$ |
| Group 2 | 6–30, 7–15, 7–31, 11–15, 11–29 | $k_1 > k > 0, c > a > 0$* |
| Group 3 | 8–15, 9–15, 11–01 | $k > 0, a > 0, c = 0$ |
| Group 4 | 9–01, 9–30, 10–15 | $k_1 > k > 0, c > a > 0$** |

* The difference between $k_1$ and $k$ is bigger than 0.003.

** The difference between $k_1$ and $k$ is smaller than 0.003.

# 4 CONCLUSIONS

The modified vertical variation of soil water model is $W_c = a \cdot e^{-kh(\ln h - 1)} - c \cdot e^{-k_1 h} + f$, which is a universal model, and the corresponding parameter values are not identical under different conditions. Seventeen sets of soil water data including thirteen sets of data measured in 2002 and four sets of data observed in 2003, 2004, 2005 and 2006 at Shanghuang Eco-experiment Station which respectively are used in the modified model, and all of the goodness-of-fit matches the statistical requirements. Besides, analyzing the fitting results obtained by using the data on 30th June, 2002 to fit the model (2) and (5) separately, and the goodness-of-fit is 0.799 in the model (2) and 0.835 in the model (5) which shows that the modified model can better simulate the vertical variation of soil water with soil depth under different conditions in the Caragana plantations in the water-limited region of the Loess Plateau, and it has better application value. The results will lay foundation for the deeper study of the soil water carrying capacity for vegetation, soil water resources use limit and provide theoretical ground for the water resources sustainable utilization.

## ACKNOWLEDGMENT

This study was supported by the National Science Fund of China (Project No. 41271539, 41071193), the front line of the domain of the Institute of Soil and Water Conservation, CAS. Corresponding author: Z.S. Guo Emaill-adress: zhongshenguo@sohu.com

## REFERENCES

Guo, Z.S. 2014. Theory and practice on soil water carrying capacity of vegetation. Beijing:Science Press.

Guo, Z.S. & Shao, M.A. 2013. Impact of afforestation density on soil and water conservation of the semiarid Loess Plateau, China. Journal of soil and water conservation 68(5):401–410.

Guo, Z.S. 2011. A review of soil water carrying capacity for vegetation in water-limited regions. Scientia silvae sinicae 47(5):140–144.

Guo, Z.S. 2010. Soil water resource use limit in semi-arid loess hilly area, Chinese. Journal of applied ecology 21(12):3029–3035.

Guo, Z.S. & Shao, M.A. 2010. Effect of artificial caragana korsinskii forest on soil water in the semiarid area loess hilly region. Scientia silvae sinicae 46(12):1–7.

Guo, Z.S. & Li, Y.L. 2009. Initiation stage to regulate the caragana growth and soil water in the semiarid area of Loess Hilly Region, China. Acta ecologia sinica 29(10):5721–5729.

Guo, Z.S. 2009. Using depth of soil water and water consumption by littleleaf peashrub in the semiarid area of loess hilly region. Bulletin of soil and water conservation 29(5):69–72.

Guo, Z.S. & Shao, M.A. 2007. Dynamic of soil water supply and consumption in artificial caragana shrubland. Journal of soil and water conservation 21(2):119–123.

Ning, T. et al. 2013. Soil water resources use limit in the loess plateau of China. Agricultural sciences 4:100–105.

Shao, M.A. et al. 2009. Study of soil water carrying capacity on Loess Plateau. Beijing:Science Press.

Wang, Z.F. et al. 2012. Vertical variation of soil water in shrubland of caragana microphylla in hilly and gully areas of Loess Plateau. Bulletin of soil and water conservation 32(6):71–74.

Yang, W.Z. & Shao, M.A. 2000. Study of soil water on the Loess Plateau. Beijing:Science Press.

Zhao, Z. et al. 2009. A model used to describe vertical change of soil moisture of Robinia pseudoacacia plantations growing in the loess gully slope. Scientia silvae sinicae 45(10):9–13.

*Multimedia, Communication and Computing Application – Leung (Ed.)*
*© 2015 Taylor & Francis Group, London, ISBN 978-1-138-02775-6*

# Author index

Printed and bound by CPI Group (UK) Ltd, Croydon, CR0 4YY

18/10/2024

01776219-0012